7th International Symposium

Fiber-Reinforced Polymer (FRP) Reinforcement for Concrete Structures

Volume 1

Editors
Carol K. Shield
John P. Busel
Stephanie L. Walkup
Doug D. Gremel

American Concrete Institute®
Advancing concrete knowledge

SP-230

First printing, October 2005

DISCUSSION of individual papers in this symposium may be submitted in accordance with general requirements of the ACI Publication Policy to ACI Headquarters at the address given below. Closing date for submission of discussion is March 2006. All discussion approved by the Technical Activities Committee, along with closing remarks by the authors, will be published in the July/August 2006 issue of either the *ACI Structural Journal* or *ACI Materials Journal*, depending on the subject emphasis of the individual paper.

The Institute is not responsible for the statements or opinions expressed in its publications. Institute publications are not able to, nor intended to, supplant individual training, responsibility, or judgment of the user, or the supplier, of the information presented.

The papers in this volume have been reviewed under Institute publication procedures by individuals expert in the subject areas of the papers.

Printed in the United States of America

Editorial production: Lindsay K. Kennedy

Library of Congress catalog card number: 2005932212
ISBN: 0-87031-189-1

Preface

The American Concrete Institute sponsored the First International Symposium of the Fiber-Reinforced Polymer Reinforcement for Reinforced Concrete Structures (FRPRCS) at the 1993 ACI Spring Convention in Vancouver, British Columbia, Canada. At this inaugural event, the international research community agreed to repeat these symposia every second year, rotating venues on various continents. Since 1993, the symposia have been held in Ghent, Belgium (1995); Sapporo, Japan (1997); Baltimore, Maryland, USA (1999); Cambridge, England (2001); and Singapore (2003). This volume represents the seventh in the symposium series, held in Kansas City, Missouri, USA, November 6-9, 2005.

The FRPRCS-7 Proceedings volume is divided into chapters that correspond to the sessions presented at the symposium, including a poster session representing a broad array of relevant research. The reader will find chapters devoted to materials characterization, masonry, bond, external strengthening, serviceability, design and behavior of members internally reinforced with FRP composites, FRP used for confinement, field applications, extreme events, and durability. The international flavor of research is also apparent in this volume, with papers representing work from 19 different countries on five continents.

Throughout its 12-year history, the symposium series has seen an increase in the sophistication of the application of FRP for reinforcing concrete and masonry as its commonplace use has increased on a global basis. This volume certainly continues and then expands this trend. The technical papers not only emphasize the experimental, analytical, and numerical validations of using FRP composites for externally strengthening or internally reinforcing concrete structures, but most are aimed at providing the insight needed for improving existing design guidelines. Several papers discuss the proposed design guidelines for deflections, shear strength, and reinforcing masonry, which practitioners should find especially useful. New applications are also featured, including studies and design equations for the use of near-surface-mounted FRP products, use of steel-reinforced polymer for reinforcing concrete, and the use of FRP to retrofit structures for blast mitigation. FRP composites performance is presented in chapters on durability and extreme events. Durability analysis performed on older field installations provides confidence and a basis for long-term service. It is our hope that future FRPRCS symposia will feature more papers where durability analyses are performed on aging field applications so that positive long-term performance data can lead to greater acceptance and use of FRP composites technology by practitioners.

A work of this magnitude could not be put together without the help, dedication, and cooperation of numerous people. First, we would like to thank the authors for meeting very tight deadlines for submissions, providing an opportunity for FRPRCS-7 to showcase the most current work possible at the symposium. Second, the International Scientific Steering Committee, consisting of many distinguished international researchers, including chairs of past FRPRCS symposia and key international representatives, was instrumental in attracting a large number of high-quality submissions to the symposium. This committee signifies the global expansion and advancement of FRP composites research and technology.

International Scientific Steering Committee

Baider Bakht, Canada
Chris Burgoyne, United Kingdom
Edoardo Cosenza, Italy
Charles Dolan, USA
Mamdouh El-Badry, Canada
Tim Ibell, United Kingdom
Amnon Katz, Israel
Urs Meier, Switzerland
Guilherme Melo, Brazil
Aftab Mufti, Canada
Antonio Nanni, USA
Kenneth Neale, Canada

Deric Ohlers, Australia
Kypros Pilakoutis, United Kingdom
Sami Rizkalla, USA
Carol Shield, USA
Luc Taerwe, Belgium
Bjorn Taljsten, Sweden
Kiang Hwee Tan, Singapore
Jinguang Teng, China
Thanasis Triantifillou, Greece
Tamon Ueda, Japan
Zhishen Wu, Japan

ACKNOWLEDGMENTS

We thank the many distinguished reviewers who carefully evaluated the technical papers. We especially want to recognize the fifty-six voting members of ACI Committee 440, the International Scientific Steering Committee, and the following individuals, many of whom are Committee 440 associate and consulting members, representatives of other ACI technical committees, and both research and industry colleagues who volunteered their time to review the technical papers:

Tess Ahlborn
Riadh Al-Mahaidi
Joaquim Barros
Florian Barth
Luke Bisby
Sergio Brena
Gregory Chacos
Oan Chul Choi
Mo Ehsani
Cathy French
Ted Galambos
Pawan Gupta
Neil Hawkins
Srinivasa Iyer
Howard Kliger
Venkatesh Kumar Kodur
Anthony Lamanna

Maria Lopez de Murphy
Radhouane Masmoudi
Mark Moore
Carlos Ospina
Renato Parretti
Andrea Prota
Oded Rabinovitch
David Scott
Mohsen Shahawy
John Silva
Henryk Stolarski
Jennifer Tanner
Ganesh Thiagarajan
Ahmet Tureyen
Markus Wernli
Paul Zia
Paul Ziehl

Special thanks are given to Kimberley Busel for her help in creating and maintaining the FRPRCS-7 website, http://frprcs7.ce.umn.edu, Jane Govro for her help with the paperwork for the reviews, the University of Minnesota Department of Civil Engineering for hosting the conference website, as well as Lindsay Kennedy and the other staff at ACI headquarters for their help in putting together the proceedings in time for the ACI Fall Convention in Kansas City.

The Editors

Carol Shield
Conference Chair
ACI Committee 440 - Secretary
University of Minnesota
Minneapolis, MN 55455

John Busel
Conference Co-Chair
ACI Committee 440 - Chair
American Composites Manufacturers Association
Arlington, VA 22201

Stephanie Walkup
Wiss Janey Elstner Associates, Inc.
Princeton, NJ 08550

Doug Gremel
Hughes Brothers, Inc.
Seward, NE 68434

TABLE OF CONTENTS

Chapter 3—Strengthening of Existing Masonry Structures with FRP Systems

Chapter 4— Bond of FRP Bars, Sheets, Laminates, and Anchorages to Concrete

Chapter 5—Strengthening of Existing Concrete Structures with FRP Systems, Part 1

Chapter 6—Serviceability of FRP Reinforced Concrete Structures

Chapter 7—Strengthening of Existing Concrete Structures with FRP Systems, Part II

Chapter 1

Spice Up Your Concrete with FRP Composites

An Innovative Hybrid FRP-Concrete Bridge System

by M. Elbadry, H. Abe, K. Schonknecht, and T. Yoshioka

Synopsis: An innovative corrosion-free system for short- and medium-span bridges consisting of precast prestressed concrete truss girders and cast-*in-situ* concrete deck has been developed. Advantages of the new system include reduced self-weight and enhanced durability. The girders consist of top and bottom concrete flanges connected by precast vertical and diagonal members made of fiber reinforced polymer (FRP) tubes filled with concrete. Glass FRP dowels and corrosion-resistant steel stud reinforcement are used, respectively, to connect the vertical and diagonal members to the concrete flanges. The flanges are pretentioned with carbon FRP tendons. The deck slab is reinforced with corrosion-resistant steel bars in the bottom transverse layer and with glass FRP bars in the bottom longitudinal and the top layers. The girders may be post-tensioned with external carbon FRP tendons to balance the slab weight and to provide continuity in multi-span bridges. The general details of the system and an experimental evaluation of its critical components, namely, the FRP tubes and the truss connection, are presented. Three types of FRP tube and four types of connection are investigated. The results of testing eight connection specimens under static loading are presented. The tests have shown superior performance of the connection when filament wound tubes and continuous double-headed studs are used.

Keywords: bridges; concrete-filled tubes; corrosion-free; deck slab; durability; fiber-reinforced polymers; headed studs; innovative; prestressing; truss girder

ACI member **Mamdouh Elbadry** is a professor of civil engineering at the University of Calgary, Canada. He is Fellow of CSCE and member of ASCE, PCI, IABSE and *fib*. He is member of ACI 118, Use of Computers; ACI 435, Deflection, and joint ACI-ASCE 343, Concrete Bridge Design. His research interests are in the applications of FRP in bridges and other structures. He is co-author of a book on analysis and design of concrete structures.

Hiroyuki Abe is a professional engineer with the Research Laboratory, Oriental Construction Co., Ltd., Japan. He is member of JSCE, JCI and JPCEA. His research interests are in the application of prestressed concrete in bridges and other structures.

Kyle Schonknecht is a Civil Engineering MSc student at the University of Calgary. His research is on the behavior of concrete bridges reinforced and prestressed with FRP.

Tamio Yoshioka is an engineering manager at the Overseas Department and a former manager at the Research Laboratory, Oriental Construction Co., Ltd., Japan. He is a member of JSCE, JCI and JPCEA. He received his Dr. Eng. From Kyushu Institute of Technology in 1995. His research interests include durability problems of prestressed concrete, especially under marine environments.

INTRODUCTION

Over the past decade, there has been a significant increase in the application of fiber reinforced polymers (FRPs) for strengthening and rehabilitation of aging and deteriorated infrastructure. However, the use of these innovative materials in new construction has not reached its full potential, mainly because of their high cost in comparison with other conventional construction materials such as concrete and steel. The advantages of FRPs, such as the high strength, light weight, easy handling, enhanced durability and low maintenance, can be more realized when FRPs are used as load-carrying structural elements. Reduction of construction costs can be achieved when the design offers reduced amounts of materials, simplified fabrication and construction procedures, lighter weight and shorter construction time. The use of FRP in new structures, particularly bridges, can be limited if all the structural components are made of FRP, mainly due to its relatively small rigidity. This limitation can be eliminated when the structural system combines FRPs with other conventional materials in a hybrid form.

A few researchers around the world have attempted to develop new bridge systems that employ FRP structural components as load-carrying elements. In Japan, for example, a new composite concrete-FRP prestressed box-girder bridge system has been proposed (Gossla and Yoshioka 2000; Niitani et al. 2001). In this system, the heavy concrete web of conventional box girders is replaced with pultruded glass FRP panels. In the United States, two modular bridge systems have been developed: one is made of FRP box sections supporting conventional concrete slab cast on stay-in-place FRP deck panels (Seible et al. 1998; Cheng et al. 2005), and one consists of carbon fiber shells filled with concrete and used as primary flexural members connected in the span direction along their length to a conventional reinforced concrete slab by means of steel or special FRP dowel connectors

(Zhao et al. 2001a). This latter system was used in construction of the 20 m long Kings Stormwater Channel Bridge on California State Route 86 (Zhao et al. 2001b).

The concrete-filled FRP tube system utilizes the best characteristics of both the FRP and the concrete. Under compression, the tube provides confinement to the concrete core and, hence, increases its compressive load carrying capacity. The concrete on the other hand provides local stability to the thin-walled tube and, hence, prevents premature local buckling (Fam 2000). Under tension, when properly bonded to concrete, the FRP provides corrosion-resistant tensile reinforcement. Use of concrete-filled tubes primarily as flexural members is, however, less efficient than their use as axially-loaded tension or compression members. Fam (2000) has shown that for concrete-filled FRP tubes used as flexural members, the shear strength provided by the tube is small, unless filament-wound tubes with fibers mainly oriented in the circumferential direction are used. Also, the confining effect of the tube on concrete in the compression zone of flexural members is insignificant as compared to that of similar members under pure axial compression. The FRP tubes, however, serve as stay-in-place formwork for the concrete, protect it from the environment, and enhance its durability.

This paper is concerned with the development of a system for short- and medium-span bridges. In the proposed system, the superstructure is built entirely from materials that are not vulnerable to corrosion. The system consists of precast prestressed concrete truss girders and cast-*in-situ* concrete slab (Fig. 1). Each girder has top and bottom concrete bulbs (flanges) connected by precast vertical and diagonal truss members. In addition to concrete, the materials used are FRP and stainless or any other type of corrosion-resistant steel, which are utilized as described in the following section. In addition to being immune to corrosion, the new system is light in weight and durable. The light weight reduces the load on the supports and allows for longer spans, resulting in reduction in the size of the substructure and in the number of supporting piers in multi-span bridges and, hence, reduction in the initial cost. The improved durability reduces the maintenance cost and extends the life span of the structure.

DESCRIPTION OF THE BRIDGE SYSTEM

The Truss Girders

Each truss girder consists of top and bottom concrete bulbs (flanges) connected by precast vertical and diagonal members (Fig. 2a). The concrete bulbs are pretensioned with carbon FRP tendons and provide the flexural resistance of the girder. The bulbs are provided with stainless steel stirrups and non-prestressed glass FRP longitudinal bars at the stirrup corners (Figs. 2b and c). The vertical and diagonal truss members, resisting the shear forces, are made of hollow glass FRP tubes filled with high-strength concrete. The verticals are predominantly in compression and the diagonals are mainly in tension. The diagonals are optimally set at 45 degrees. Both the verticals and the diagonals are produced prior to the bulbs. Glass FRP dowels protrude from the ends of the verticals to connect them to the bulbs (Figs. 2a and b). Stainless steel headed studs (straight bars with anchor heads at one or both ends) connect the diagonals to the bulbs (Figs. 2a and c). Alternatively, the diagonals can be pretensioned with FRP flexible tendons (e.g., Carbon Fiber Composite Cables, known as CFCC, produced by Tokyo Rope, Japan). The pretensioning will provide the diagonals with a reserve tensile capacity in case the FRP tubes are damaged, for example, by fire. The tendons protrude from the ends of the diagonals and are bent to serve as dowels connecting the diagonals to the bulbs.

Single or double-headed studs are selected for this project because of their superior anchorage properties. With an area of the head 9 to 10 times the cross-sectional area of the stem, the full yield strength of the studs can develop immediately behind the head, without the need for development length as required in conventional reinforcement (Ghali and Dilger 1998).

For ease in production, it is preferable to cast the bulbs in a rotated position, while the verticals and the diagonals lie on a horizontal surface. In case of damage by fire, the FRP tubes can be easily replaced by wrapping the concrete diagonals and verticals with FRP sheets or jackets. Stainless steel double-headed studs are used to connect the deck slab to the girders (Fig. 2b).

After casting the deck slab the precast truss girders can be post-tensioned by external carbon FRP tendons (CFCC) to counterbalance the slab weight and to provide continuity in multi-span bridges. An external tendon can be harped (held down) to the bottom bulb at one or two points within the span and held up to the top bulb at one point near the intermediate supports in continuous bridges. No deviators are required at the harping points. The horizontal parts of the tendons between the harping points pass through ducts placed inside the bottom bulb, in a single-span bridge, or inside both the top and bottom bulbs in a continuous bridge. Deviation of the tendons from the horizontal is done at the location of the truss joints. The ducts can be left ungrouted for easy replacement of the tendons, or can be grouted to achieve bond between the horizontal parts of the tendons and the concrete bulb(s). Research has shown that partially bonded external tendons enhance both the strength and ductility of externally prestressed concrete members (Hindi et al. 1996).

The Deck Slab

Particular attention is given to the bridge deck slab as it represents an important component that considerably affects the overall cost and quality of the structure. Reinforcement of the deck slab can all be glass FRP bars. Alternatively, the FRP bars can be used in both the transverse and longitudinal directions for the top reinforcement, whereas the bottom reinforcement can be composed of stainless steel bars in the transverse direction (to act as ties in the arch action) and glass FRP in the longitudinal direction (Figs. 2b and c). The glass FRP reinforcement at the bottom longitudinal and top layers is only for crack control and should be of minimum amount.

EXPERIMENTAL PROGRAM

The initial phase of this research involved computer modeling and analysis of the proposed bridge system in order to determine the optimum design of the truss girder in terms of its depth for different ranges of span and loading, dimensions of the top and bottom bulbs, amount and placement of prestressing, size of the truss elements, and number and size of the stud reinforcement required. An extensive experimental program is currently in progress at the University of Calgary to investigate performance of the various components of the bridge system and to verify the optimum design. The experimental program includes the following:

1. Axial compression tests on samples of different types of concrete-filled FRP tube.
2. Tests on bond between the FRP tube and the concrete.
3. Static and fatigue loading tests on different types of connection between the truss elements and the top and bottom bulbs.
4. Tests on large-scale truss girder specimens covered with or without a concrete slab under static and fatigue loading and long-term effects.

Details of the tests of items 1 to 3 above (except for fatigue loading) are given below.

Compression Tests on Concrete-Filled FRP Tubes

As previously mentioned, under compression, an FRP tube confines the concrete core and increases its load carrying capacity. Tests were conducted on three different types of available glass FRP tubes to determine the enhancement of the compressive resistance that could be achieved and, hence, the most suitable type for the vertical (compression) members of the proposed truss girder. One type of pultruded glass FRP tube and two types of filament-wound tubes were used. The pultruded tube contained approximately 80% of the fibers oriented in the longitudinal direction and 20% in the circumferential direction. One type of the filament-wound tube contained approximately 70% circumferential fibers and 30% longitudinal fibers. The other type contained fibers wound in the diagonal direction at approximately 45 degrees.

The pultruded tube specimens were of 80, 100, and 150 mm nominal inner diameters. All filament-wound tubes were of 85 mm nominal inner diameter. The height of all specimens was set at twice the concrete core diameter. Two groups of specimens were tested. In Group 1, the FRP tube was 10 mm shorter in height than the concrete core. This was achieved by cutting a 5 mm ring from each end of the tube. In this manner, only the concrete core would be in direct contact with the loading plate and the pure confining effect of the tube could be assessed. In Group 2, both the tube and the concrete core had the same height and the compression load was applied on both the concrete and FRP. Table 1 gives nominal dimensions of all the tested specimens. Some specimens of the filament-wound tube were also tested with a rough inner surface consisting of sand mixed with resin adhesive to obtain bond between the concrete and the tube. Plain concrete specimens, with target compressive strength between 30-35 MPa, were also tested for comparison. All specimens were instrumented with strain gauges attached in both the vertical and transverse (circumferential) directions. Figure 3 shows elevations and cross-sections of typical specimens of Groups 1 and 2.

Bond Tests on Concrete-Filled FRP Tubes

Three groups of concrete-filled pultruded tubes were tested to examine means of better bond between the concrete and the tube. In the first group (Type S), the inner surface of the tube was left without treatment for bond and was used as control specimens. In the second group (Type A), the inner surface was coated with a 50 mm wide circumferential strip of sand mixed with resin adhesive. In the third group (Type R) shear connectors were provided in the form of two 60 x 150 mm GFRP strips glued to the inner surface and fixed in position by means of rivets. Figure 4 gives dimensions of both the tubes and the concrete core. All specimens were tested under vertical load applied to the top of the concrete core.

Connection Tests

The most important component of the truss girder is the connection of the vertical and diagonal elements to the concrete bulbs. Tests were therefore carried out to examine the behavior and efficiency of different types of connection and to determine the best type to be used. Specimens of a single section of truss consisting of one vertical and one diagonal member connected to portions of the top and bottom bulbs were fabricated (Fig. 5). The specimens were sized based on a prototype design for a pedestrian bridge spanning 45 m. The bottom bulb was fixed to the strong floor and a horizontal load was applied to the top bulb to produce a compressive and tensile force in the vertical and diagonal elements, respectively.

Four different reinforcing types were used to connect the diagonal to the concrete bulbs:

1. Unbonded post-tensioned tendon passing through the centre of the diagonal and anchored outside the top and bottom bulbs (Fig. 6a).
2. Long double-headed studs with the heads embedded in the top and bottom bulbs and the stems extending through the length of the diagonal (Fig. 6b).
3. Short double-headed studs at the upper and lower ends of the diagonal, with one head embedded in the top or bottom bulb and the other head embedded in the concrete inside the FRP tube (Fig. 6c).
4. Pairs of single-headed studs, one at each end of the diagonal, with the head embedded in the top or bottom bulb and the stems spliced inside the FRP tube with deformed steel bar of the same diameter as the stem of the stud (Fig. 6d).

A total of eight specimens, two for each connection type, were fabricated and tested under static loading. The first four specimens were built using pultruded FRP tubes of 3 in. and 5 in. nominal diameters for the vertical and diagonal elements, respectively. The inner diameter and wall thickness were, respectively, 80 mm and 4.5 mm for the vertical, and 126 mm and 3.5 mm for the diagonal element. The tensile strength and modulus of elasticity of the FRP tube were, respectively, 400.5 MPa and 25.3 GPa in the longitudinal direction, and 40.8 MPa and 10.15 GPa in the circumferential direction.

In the second four specimens, filament-wound FRP tubes with 70% circumferential fibers and 30% longitudinal fibers were used for the truss elements. Tubes of 3 in. and 4 in. nominal diameters were used for the vertical and diagonal elements, respectively. The inner diameter and wall thickness were, respectively, 84 mm and 1.9 mm for the vertical, and 110 mm and 1.9 mm for the diagonal element. Nominal diameter of 5 in. was not available in this type of tube. Adequate concrete consolidation was achieved and tensile resistance was not affected with the 4-in. tube. The tensile strength and modulus of elasticity of the tube were, respectively, 240 MPa and 20.6 GPa in the longitudinal direction, and 480 MPa and 29.0 GPa in the circumferential direction.

Figure 7 shows the stud reinforcement used in this project. The double-headed studs had plain stem, whereas the single-headed studs had deformed stems. Four studs of 12.7 mm diameter were used in the diagonal of each specimen. Deformed 10M (11.3 mm diameter) were used to splice the single-headed studs. Material properties of the studs and the splice bars are given in Table 2. A single 15 mm 7-wire strand (A_{ps} = 140 mm^2) was prestressed to 0.6 F_{pu} in the specimens with post-tensioned diagonal element.

The compressive strength of concrete used in all specimens was 55 MPa, with 14 mm maximum aggregate size in the flanges, and 10 mm maximum aggregate size in the truss elements in order to obtain complete consolidation of concrete in the tubes. Figures 8 and 9 summarize the fabrication process. The studs were placed in position inside the diagonal tube, with one end of the studs extending from either end of the tube. The truss elements were then precast and placed in position in the connection specimen formwork. The flanges were then cast embedding the studs to connect the truss elements to the flanges. The bottom flange was then post-tensioned with four dywidag threadbars.

During the test, the specimen was bolted to the laboratory strong floor within the loading frame shown in Fig. 10. A horizontal load was applied through an MTS system with two 1.5 MN rams in displacement control. Displacement and rotation transverse to the load direction were prevented by means of lateral supports bearing against Teflon sheets on each side of the top flange allowing free movement only in the load direction (Fig. 10). Loading was applied until failure occurred or the stroke limit of the rams was reached. Loading the specimens with post-tensioned diagonal continued until the tendon reached 85% of its ultimate tensile strength

TEST RESULTS

Compression Tests on Concrete-Filled FRP Tubes

Table 1 summarizes all compression test results. Figure 11 depicts the failure modes of the three types of concrete-filled FRP tubes. In Table 1, the strength of confined concrete, f'_{cc}, is defined as the load carried by the concrete core divided by its cross-sectional area. For Group 2 specimens, this load is calculated from the total measured load less the contribution of the FRP tube in the axial direction. The load carried by the tube is calculated from the measured axial strain in the FRP times its modulus of elasticity in the load direction multiplied by the tube cross-sectional area. Effective confinement is defined as the ratio f'_{cc}/f'_c of the strength of confined concrete to the strength of unconfined plain concrete cylinder.

As can be seen from Table 1, the pultruded FRP tubes provide low effective confinement to the concrete core varying from 15 to 40%. This is mainly due to the small percentage of fibers oriented in the circumferential direction. The confinement is less effective in larger diameter specimens since the tube circumferential stiffness is proportional to $1/R$, where R is the tube radius (Fam 2000). However, the concrete-filled pultruded tubes can be more efficient in carrying compression load when both the concrete and the tube are in direct contact with the load (Group 2). In this case, the longitudinal fibers share the load with the concrete. Because of the small percentage of the circumferential fibers, specimens Group 1 failed by splitting of the tube over its height, whereas specimens Group 2, with both the core and the tube directly loaded, failed by buckling combined with splitting.

The two types of filament wound tubes (FWC and FWD) offer a significant increase in the strength of confined concrete as opposed to the pultruded tubes, even though the tube thickness is much smaller, raising the effective confinement ratio to approximately 3-4.5. This is due to the higher percentage of the circumferential or diagonal fibers. The results also indicate greater increase in strength in specimens with just the concrete core loaded (Group 1) than in specimens with both the concrete and the tube loaded (Group 2). This confirms the findings of Fam (2000) and Mirmiran and Shahawy (1997).

Specimens made of the filament-wound tubes, FWC and FWD, exhibited similar modes of failure (Figs. 11b and c), with fibers splitting along a diagonal line over the tube height. Failure of the specimens of FWD type, with diagonal fibers, was drastic and very brittle, with all fibers along the tube length rupturing, blowing out the crushed concrete core and allowing no residual load to be taken. On the other hand, failure of the FWC specimens, containing 30% longitudinal fibers, was less drastic (Fig. 11b) and the tubes continued to offer some confinement and load carrying capacity after the ultimate load was reached. Therefore, of the two types of filament-wound tubes, the FWC type was selected to build the connection specimens of the truss as described above.

Bond Tests on Concrete-Filled FRP Tubes

Table 3 shows results of the bond test carried out on the concrete-filled pultruded FRP tubes. As expected, the shear connectors made of FRP strips fixed to the inner surface by epoxy and rivets provided bond strength between the concrete and the tube higher than the sand/adhesive coating applied to the tube surface. Figure 12 shows failure of the two types of specimens. The increase in strength, however, was only 11%. It was therefore decided to use adhesive coating in the tubes of the truss connection specimens described above.

Connection Tests

Figure 13 compares the load-displacement behavior of the four different connections when pultruded tubes were used. Figure 14 shows the comparison for specimens built with filament-wound tubes. The displacement was measured in the direction of the applied load, at the centre of the top flange, by two transducers located on each side of the flange at the intersection of the truss elements. Failure modes of the connections are shown in Figs. 15-19.

As can be seen, the specimens built with filament-wound tubes showed better performance, particularly in terms of ductility. The maximum load that each connection type sustained was comparable for specimens of each tube type. However, the filament wound tube specimens continued to sustain load under larger horizontal displacement. This is attributed mainly to the increase in concrete strength due to better confinement.

The post-tensioned specimens performed quite similarly. The tendon in both specimens yielded at approximately the same load and displacement. The specimen with filament-wound tubes experienced more displacement after the tendon yielded, and reached a slightly higher applied load under the same tensile force in the tendon.

The connection with long double-headed studs showed the best performance in both types of specimens, with much higher ductility than the other three types of connection. The specimen with pultruded tube ultimately failed due to buckling and premature rupture of the tube of the vertical compression member (Fig. 17). The studs in the diagonal member did not reach their ultimate tensile strength. The specimen with filament-wound tube showed much better performance and continued to sustain load at displacements well beyond the pultruded tube specimen (Fig. 14). In fact, the test was stopped because the stroke limit of the rams was reached when the horizontal displacement of the specimen was approximately 85 mm. Rupture at the bottom of the vertical tube under compression was also observed.

The connection with spliced single-headed studs sustained a slightly higher load, but exhibited much less ductility and failure was caused by fracture of the studs (Fig. 18). This is possibly because the single-headed studs have deformed stems, whereas the double-headed studs have plain stems. The single-headed studs could develop bond with the concrete and share part of the load with the tube resulting in localized yielding and failure at the connection to the flange. The double-headed studs, on the other hand, developed little bond and were free to yield over a longer length (Fig. 16).

The short double-headed stud connections failed well below their yield strength. In the pultruded specimens, the FRP tube cracked longitudinally in the diagonal member, most likely due to circumferential stress induced from the transfer of force from the stud head. In the filament-wound tube specimens, the tube was able to withstand the circumferential forces and no failure in the tube took place. However, the concrete inside the tube cracked at the end of the head and the core ultimately slipped from the tube (Fig. 19). The series of peaks and drops in the load-displacement curve (Fig. 14) was due to successive failure of bond between the concrete and the tube at the locations of the sand-adhesive coating of the tube inner surface.

SUMMARY AND CONCLUSIONS

An innovative corrosion-free bridge system that combines precast prestressed hybrid FRP-concrete truss girders and a cast-*in-situ* concrete deck has been developed. The web members of the truss girders are made of concrete-filled FRP tubes connected to the prestressed concrete top and bottom flanges of the truss by means of single- or double-headed studs. Compression tests were performed on small FRP tubes filled with concrete in order to examine the effect of confinement on the truss compression members. Tests were also carried out on a single section of the truss in order to investigate the performance of different types of connections. It has been shown that filament wound GFRP tubes with fibers oriented in the circumferential direction perform extremely well in confining concrete in the truss compression members, and create a ductile connection that continues to take load well beyond yielding. Because of their excellent anchorage properties, continuous double-headed studs or spliced single-headed studs can be used efficiently to connect the truss tension members to the flanges. The studs have shown better performance than post-tensioned tendons. The studs can resist the required tensile forces and provide much more ductile connections. It is concluded that the proposed bridge system can be an efficient and more economical superstructure for short- and medium-span bridges.

ACKNOWLEDGEMENT

The authors wish to thank the following organizations for their financial support: ISIS Canada; Oriental Construction Co., Japan; Decon, Brampton, Ontario; Cement Association of Canada; City of Calgary; Ministry of Transportation of British Columbia; Alberta Transportation; Alberta Innovation and Science; Asahi Glass Matex Co. Ltd., Japan; Pipe Specialties, Calgary, Alberta; Ameron, USA; Smith Fibrecast, USA.

REFERENCES

Cheng, L., Zhao, L., Karbhari, V.M., Hegemier, G.A., and Seible, F., (2005), "Assessment of a Steel-Free Fiber Reinforced Polymer-Composite Modular Bridge System," ASCE *Journal of Structural Engineering*, V. 131, No. 3, March 2005, pp. 498-506.

Fam, A., (2000), "Concrete-Filled Fibre-Reinforced Polymer Tubes for Axial and Flexural Structural Members," Ph.D. Thesis, University of Manitoba, August 2000, 261 pp.

Ghali, A., and Dilger, W.H., "Anchoring with Double-Head Studs," *Concrete International*, V. 20, No. 11, November 1998, pp. 21-24.

Gossla, U. and Yoshioka, T., (2000), "FRP for Shear Walls of Box-Girder Composite Bridges," Proceedings of the Third International Conference on *Advanced Composite Materials in Bridges and Structures*, ACMBS-III, Ottawa, August 15-18, 2000, pp. 87-94.

Hindi, A., MacGregor, R., Kreger, M.E., and Breen, J., (1996), "Effect of Suplemental Bonding of External Tendons and Addition of Internal Tendons on the Strength and Ductility of Post-Tensioned Segmental Bridges," ACI Special Publications SP-160, on "Seismic Rehabilitation of Concrete Structures," 1996, pp. 169-189.

Mirmiran, A. and Shahawy, M., (1997), "Behavior of Concrete Columns Confined by Fiber Composites," ASCE *Journal of Structural Engineering*, V. 123, No. 5, May 1997, pp. 583-590.

Niitani, K., Yoshioka, T., Abe, H. and Gossla, U., (2001), "The Use of FRP Shear Panels in PSC Bridge Design," Proc. of the 5th International Conference on *Fibre-Reinforced Plastics for Reinforced Concrete Structures*, FRPRCS-5, Cambridge, July 16-18, 2001, pp. 1123-1131.

Seible, F., Karbhari, V.M., Burgueno, R. and Seaberg, E., (1998), "Modular Advanced Composite Bridge Systems for Short and Medium Span Bridges," Proceedings of the Fifth International Conference on *Short and Medium Span Bridges*, SMSB V, Calgary, Alberta, July 13-16, 1998, pp. 431-441.

Zhao, L., Karbhari, V.M., Seible, F., Brostorm, M., La Rovere, H., and Burgueno, R., (2001a) "Design and Evaluation of Modular Bridge Systems Using FRP Composite Materials," Proceedings of the Fifth International Conference on *Fibre-Reinforced Plastics for Reinforced Concrete Structures*, FRPRCS-5, Cambridge, July 16-18, 2001, pp. 1143-1151.

Zhao, L., Karbhari, V.M., and Seible, F., (2001b), "Development and Implementation of the Carbon Shell System for the Kings Stormwater Channel Bridge," Proceedings of the International Conference on *FRP Composites in Civil Engineering*, CICE 2001, Hong Kong, China, December 12-15, 2001, pp. 1299-1306.

Table 1 – Results of compression tests on concrete-filled FRP tubes

FRP tube type	Specimen Designation*	FRP tube size (mm)	Area of FRP (mm²)	Concrete size (mm)	Area of concrete (mm²)	P_{max} (kN)	f'_c or f'_{cc} (MPa)	$\dfrac{f'_{cc}}{f'_c}$
Pultruded (80 percent longitudinal fibers)	PC-80	–	–	75x150	4418	131.00	29.7	1.00
	PT1-80	80x150x4.5	1195	80x160	5027	210.46	41.9	1.41
	PT2-80	80x160x4.5	1195	80x160	5027	386.87	62.2	2.10
	PC-100	–	–	100x200	7854	261.95	33.4	1.00
	PT1-100	102x200x3.5	1160	102x210	8171	341.46	41.8	1.25
	PC-150	–	–	150x300	17671	583.60	33.0	1.00
	PT1-150	150x300x4.5	2184	150x310	17671	673.25	38.1	1.15
Filament wound (70 percent circumferential fibers)	PC-100	–	–	100x200	8107	278.60	34.4	1.00
	FWC1-85S	84x150x1.8	485	84x160	5542	671.54	121.2	3.53
	FWC2-85S	84x160x1.8	485	84x160	5542	667.34	111.9	3.26
	FWC1-85R	84x150x1.8	485	84x160	5542	646.47	116.7	3.39
	FWC2-85R	84x160x1.8	485	84x160	5542	576.41	96.7	2.81
Filament wound (diagonal fibers)	PC-100	–	–	100x200	8107	237.80	29.3	1.00
	FWD1-85S	85x160x2.0	547	85x170	5674	676.21	119.2	4.06
	FWD2-85S	85x170x2.0	547	85x170	5674	657.73	113.5	3.87
	FWD1-85R	85x160x2.0	547	85x170	5674	731.54	128.9	4.40
	FWD2-85R	85x170x2.0	547	85x170	5674	657.42	113.5	3.87

* The specimen designations PC, PT, FWC and FWD refer, respectively, to plain concrete, pultruded tube, filament-wound tube with 70% circumferential fibers, and filament-wound tubes with diagonal fibers. The indices 1 and 2 refer to specimens Group 1 (only concrete core loaded) and Group 2 (both tube and concrete core loaded), and S and R refer, respectively, to smooth and rough conditions of the tube inner surface.

Table 2 – Properties of stud reinforcement and splice bars

Reinforcement Type	Modulus of Elasticity (MPa)	Yield Strength (MPa)	Ultimate Strength (MPa)
Long double-headed stud	192 200	500	589
Short double-headed stud	199 000	507	602
Single-Headed Stud	184 400	504	570
10M deformed bar	201 800	445	509

Table 3 – Results of bond tests on concrete-filled FRP tubes

Specimen Designation*	Tube inner surface condition	Contact area (mm^2)	Maximum Load (kN)	Bond Stress (MPa)
S-100	Smooth	63668	4.98	0.08
A-100	Adhesive	16022	75.21	4.69
R-100	Rivets	18000	93.89	5.22

* Specimen dimensions are given in Fig. 4.

Figure 1 – Perspective view of the proposed hybrid FRP-concrete bridge system

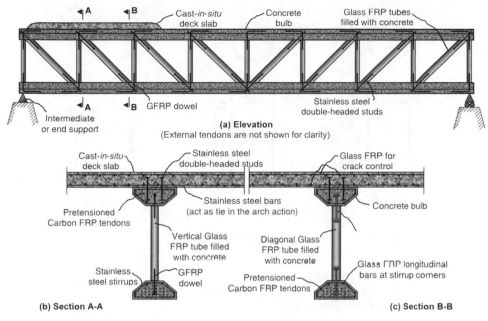

Figure 2 – Details of the proposed hybrid FRP-concrete bridge system

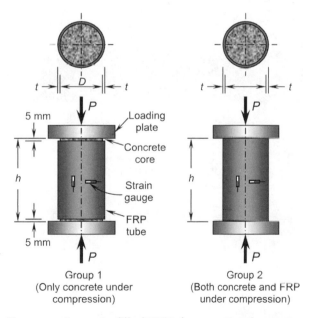

Figure 3 – Concrete-filled FRP tube compression test

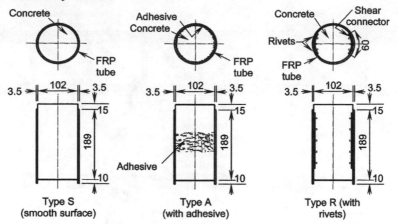

Figure 4 – Concrete-filled FRP tube bond test specimens

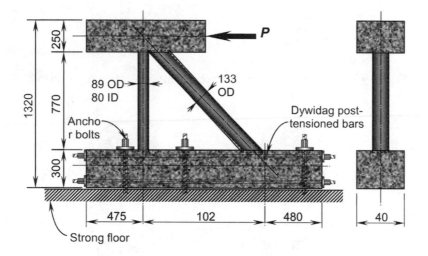

Figure 5 – Connection specimen

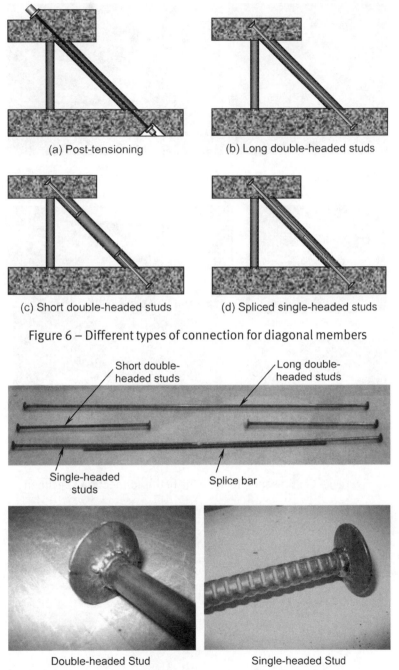

(a) Post-tensioning

(b) Long double-headed studs

(c) Short double-headed studs

(d) Spliced single-headed studs

Figure 6 – Different types of connection for diagonal members

Short double-
headed studs

Long double-
headed studs

Single-headed
studs

Splice bar

Double-headed Stud

Single-headed Stud

Figure 7 – Stud reinforcement used in different connections

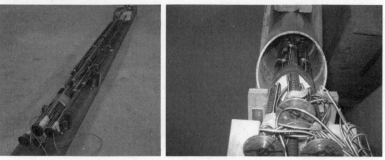

Figure 8 – Preparation of the stud reinforcement and placement inside the FRP tube

Tubes ready for casting

Tubes after casting

Flanges ready for casting

Bottom flange post-tensioned

Figure 9 – Casting and prestressing of the connection specimen

Bottom flange bolted to strong floor

Load applied to top flange

Specimen in the testing frame

Figure 10 – Connection specimen during static loading

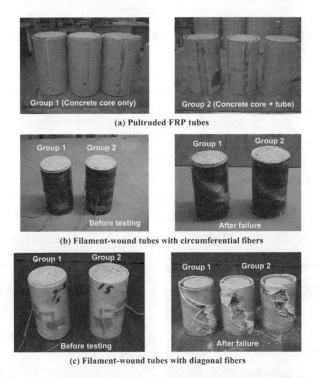

Group 1 (Concrete core only)

Group 2 (Concrete core + tube)

(a) Pultruded FRP tubes

Group 1   Group 2

Group 1   Group 2

Before testing

After failure

(b) Filament-wound tubes with circumferential fibers

Group 1   Group 2

Group 1   Group 2

Before testing

After failure

(c) Filament-wound tubes with diagonal fibers

Figure 11 – Failure of different types of concrete-filled FRP tubes in compression test

Figure 12 – Failure of concrete-filled pultruded FRP tubes in bond test

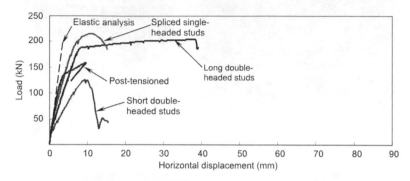

Figure 13 – Load-displacement behavior of pultruded tube connection specimens

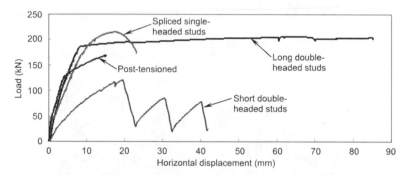

Figure 14 – Load-displacement behavior of filament-wound tube connection specimens

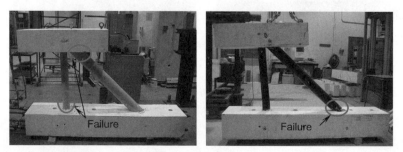

Figure 15 – Failure of connection specimens reinforced with long double-headed studs

Figure 16 – Failure of diagonals in connections reinforced with long
double-headed studs

Figure 17 – Failure of verticals in connections reinforced with long double-headed studs

Figure 18 – Fracture of spliced single-headed studs at top or bottom of diagonals

Figure 19 – Failure of connections reinforced with short double-headed studs

Proposed GFRP Connectors in Sandwich Panels

by W.A. Pong, A.F. Morgan Girgis, and M.K. Tadros

Synopsis: In this paper, a precast concrete sandwich panel system with high thermal resistance and optimum structural performance was developed. Fiber-reinforced polymer was used to provide the connector in this panel system. Each connector was prefabricated in a V-shape with both end hooks providing anchorage in the reinforced concrete wythes. The developed shear connecting system is described, along with its performance and advantages. An experimental investigation of the connecting system was conducted. The experimental program included pure-shear testing of push-off specimens with a variety of wythe connectors, flexure testing of full-scale specimens with FRP connectors, and testing of FRP material. The developed shear transfer and degree of composite were evaluated and the observed behavior of the members is presented. Design criteria are given for use by engineers and others to apply with appropriate engineering judgment.

Keywords: composite action; GFRP; horizontal shear; precast; sandwich panels

Wilast A. Pong is currently a Ph.D. student at the Civil Engineering department, University of Nebraska. He has twelve years of work experience in high rise buildings and elevated highways and three years of research experience in both analytical and experimental activities.

Dr. Amgad F. Morgan Girgis is Research Assistant Professor at the Civil Engineering Department, University of Nebraska – Lincoln, USA. He received his BS and MS from Cairo University, Egypt; and Ph.D. from University of Nebraska – Lincoln. His research interests include bridges, fiber reinforced polymers, composite structures, and self-consolidating concrete. He has twelve years of experience, including six years of research experience, three years of structural coordination, three years of design experience, and progressive research activities and publications.

Dr. Maher K. Tadros is Charles J. Vranek Distinguished Professor of Civil Engineering at the University of Nebraska-Lincoln. Dr. Tadros has been recognized on both national and international levels in numerous occasions. Most recently he received Industry Titan Precast/Prestressed Concrete Institute.

INTRODUCTION

Typical precast concrete sandwich panel (PCSP) systems[1] are composed of two concrete wythes with insulation placed at the middle, used for exterior walls. Concrete wythes are connected by metal or plastic connectors. PCSP systems can be constructed to be fully composite structural members depending on the capacity of the connectors to transfer horizontal shear generated by flexure. PCSP systems can be divided into three major categories: [2]

1. Fully composite panel
2. Non-composite panel
3. Partially composite panel

PCSP systems are considered to be structurally and thermally efficient elements; however, the effective thermal resistance can be reduced by heat transfer through metal connectors. [3] The objective of this research to study the use of glass fiber-reinforced polymer (GFRP) as a shear transfer connector. GFRP is well-known as a high tensile strength and low-conductive material.

EXPERIMENTAL PROGRAM

Testing Program

The testing program consisted of push-off and full scale flexural tests. A push-off test was designed to investigate the shear transfer capacity of the various types of shear connectors on the sandwich panel. A full-scale test was designed to investigate the composite behaviors and the degree of composite action provided by the shear connector.

The various parameters considered were concrete strength, thickness of concrete wythes, insulation, reinforcement ratio, and the effect of transverse reinforcement provided in the hook of the GFRP connector.

Push-off Specimens

The size of each push-off specimen was 2 by 4 ft comprising three concrete wythes and two insulation wythes. To compare shear transfer capacity, three material types of connectors: GFRP, welded wire and GFRP in alkali resin were used. Thirty specimens were produced, six specimens with 21 in. thickness and twenty four specimens with 13.5 in. thickness.

Full-Scale Specimens

Six specimens were produced. Each full-scale specimen was 12 in. thick, comprising 3 in. reinforcement concrete wythes and a 6 in. insulation wythe connected by a GFRP shear connector. Each concrete wythe contained a No.3 bar, Grade 60, with 4 inch spacing in both the longitudinal and transversal directions.

Three types of full-scale specimens were constructed, with two specimens of each type. Specimen serial I was horizontally constructed. Concrete wythes were connected by two rows of 2 ft. spacing with a single shear connector placed in a longitudinal direction. Specimen serial II was also horizontally constructed. Concrete wythes were connected by two rows of 2 ft. spacing with double connectors. Specimen serial III was vertically constructed with a single shear connector.

Connector Details

Three types of shear connectors were used. A GFRP-connector is a new type of fiberglass connector (see Figs. 5 through 7) that transfers both horizontal shear force and tension force between the concrete wythes. The other two commercial types of connectors are 1) GFRP embedded in an alkali resistant resin (Fig. 8), and 2) welded wire girder (Fig. 9).

PUSH-OFF TESTING

Test Setup

A test frame was installed at the structural laboratory of the University of Nebraska, Omaha. The test frame was anchored to a 500-kip tie-down accessory on the structural testing floor by two high-strength steel threaded rods. The test set up for the push-off specimen is shown in Fig. 10. The specimen was placed vertically in the test frame. Steel tubes, 4 by 4 in., were placed underneath the outer concrete wythes and another steel tube was placed on the middle concrete wythe. Rubber sheets were inserted between the steel tubes and concrete wythes.

The specimens were tested using a 400 kip hydraulic jack. The attached load cell was used to measure the force applied to the specimens.

The specimens were subjected to a concentrically vertical load applied at the middle concrete wythe, which was incrementally increased until failure occurred. The specimens failed when the applied load was reduced from its peak point.

FULL-SCALE FLEXURAL TESTING

Test Setup

The test set-up for the full-scale specimens is shown in Fig. 11. Each specimen was placed horizontally in the test frame. A deflection-measuring device was attached at the center of the bottom concrete wythe.

The specimens were tested by using a 400-kip hydraulic jack placed at the center of the panel. The attached load cell was used to measure the force applied to the specimens.

The specimens were then subjected to a vertical load. The load was incrementally increased until failure occurred. Failure of the specimens was regarded when the applied load was reduced from its peak point. The specimens were inspected and observed at specified intervals during testing.

MATERIAL PROPERTIES

Compressive strength and Modulus of Elasticity

The final compressive strength was obtained from compressive tests of concrete cylinders at the same age as their corresponding specimens. The modulus of elasticity was computed as $E_c = 57,000\sqrt{f'_c}$

TENSILE TESTING OF GFRP RODS

Tensile testing of GFRP rods was performed according to the Guide Test Methods for Fiber Reinforced Polymers (FRP) for Reinforcing or Strengthening Concrete Structures prepared by ACI Subcommittee 440. [4]

A 48-in. long GFRP rod was attached to the anchorage device to facilitate the gripping dimension, as shown in Fig. 12. An Extensometer, a strain measuring device, was mounted at the middle of the specimen, as shown in Fig. 13. Six specimens were tested under a loading rate of 6000 lbs/min. The test results are shown in Table 5. Tensile strength, ultimate strain and modulus of elasticity were calculated as follows:

Tensile strength $f_u = \dfrac{F_u}{A}$

where f_u = *tensile strength (psi)*
F_u = *load failure (lbs)*
A = *average cross section area (in^2)*

Tensile modulus of elasticity $E_L = \dfrac{F_1 - F_2}{(\varepsilon_1 - \varepsilon_2)A}$

where E_L = *axial modulus of elasticity (psi)*
A = *cross section area (in^2)*
F_1 *and* ε_1 = *load and corresponding strain, respectively, at approximately 50% of the ultimate tensile capacity or guaranteed tensile capacity*

F_2 and ε_2 = load and corresponding strain, respectively, at approximately 20% of the ultimate tensile capacity or guaranteed tensile capacity

Ultimate strain $\qquad \varepsilon_u = \dfrac{F_u}{E_L A}$

where $\qquad \varepsilon_u$ = ultimate strain of FRP bar

Ductility is the ratio of the strain at failure to the strain at yielding. Fig. 14 shows the stress-strain relationship obtained from the tensile test. Stress and strain increased in a linear proportion to failure with no yield point. The test results may be different since the tensile strength and modulus of elasticity of FRP depend on the fiber content and manufacturing process, as well as quality control.

PUSH-OFF TEST RESULTS

The failure behavior of specimens serial NU-R, NU-C, NU-CX was very similar. Failure of GFRP connector was brittle failure due to shear and flexure, as seen in Figs. 15 (a) and (b). When loading was in process, the applied load was abruptly dropped and then increased after each connector failure until ultimate load was reached.

The test results of the push-off specimens are summarized in Table 6. The relationship between ultimate load and total area of connector was computed as the ultimate load divided by area (wire or GFRP).

FULL-SCALE TEST RESULTS

The failure behavior of specimens was consistent. Visual cracks formed at the bottom fiber when the applied load was in the range of 12-15 kips. Horizontal shear failure occurred when the outermost connector failed. Crushing of concrete due to flexure failure occurred in the top concrete wythe before the second row of connectors failed. Flexure failure took place in the bottom concrete wythe when the ultimate load was reached.

Observations revealed that loading did not continuously increase. Similar to the push-off test, the applied load abruptly dropped and then increased at every single connector failure until ultimate load was reached.

The test results of the full-scale specimens in terms of ultimate load and deflection are given in Table 7.

DISCUSSION

Deflections
The load and mid-span deflection of the six specimens is shown in Fig. 16. Also shown in Fig. 16 are theoretical predictions which were calculated based on the assumption of fully composite behavior and non-composite behavior. Our predicted deflections, based on the assumption of fully composite using Equation 4 defined by ACI 318-02, [5] were underestimated. This is attributed to the fact that behavior of specimens is

partially composite. Another reason for the larger deflection in the sandwich panels with the GFRP is that the low tensile modulus of elasticity of the GFRP connectors could lead to more shear deformation that was not included in the deflection equation. To predict deflection of sandwich panels using GFRP connector, more accurate analysis procedure is required. An average modulus of elasticity is 6.548×10^6 psi obtained from GFRP tensile test results as shown in Table 5.

$$I_e = \left(\frac{M_{cr}}{M_a}\right)^3 I_g + \left[1 - \left(\frac{M_{cr}}{M_a}\right)^3\right] I_{cr} \qquad I_e \leq I_g \qquad (1)$$

$$M_{cr} = \frac{f_r I_g}{y_t} \qquad (2)$$

$$f_r = 7.5\sqrt{f_c'} \qquad (3)$$

$$\Delta_i = \frac{5 M_a L^2}{48 E_c I_e} \qquad (4)$$

where
E_c = *modulus of elasticity of concrete, psi.*
f_c' = *specified compressive strength of concrete, psi.*
f_r = *modulus of rupture of concrete, psi.*
I_{cr} = *moment of inertia of cracked section transformed to concrete, in.[4]*
I_e = *effective moment of inertia for computation of deflection, in.[4]*
I_g = *moment of inertia of gross concrete section about centroidal axis, neglecting reinforcement, in.[4]*
L = *span length of panel*
M_{cr} = *cracking moment*
M_a = *maximum moment in member at stage deflection is computed*
y_t = *distance from centroidal axis of gross section, neglecting reinforcement, to extreme fiber in tension, in.*
Δ_i = *deflection at mid span, in.*

Development of GFRP reinforcement

The test results show that no direct pull-out failure occurred. The average ultimate load of specimens serial NU-C and NU-CX was 34,526 lbs and 33,631 lbs respectively. Transverse reinforcement provided in the hook, specimen serial number NU-C, did not affect the bond characteristics of the GFRP connector. It can be concluded from the test results that 85 degree connector hooks provide sufficient strength to resist force that may cause pull-out failure in concrete wythes. However, precaution in quality control should be taken to prevent direct failure when the hook is not fully embedded in concrete.

Shear transfer

The ultimate loads obtained from the push-off and full-scale tests were much lower than those computed using Equation (5) or (6) defined by ACI 318-02. The test results showed that the GFRP connector did not fail by tension. That can be explained by the fact that in the sandwich panel there is no direct contact between the layers. Direct contact can take some of the shear stresses, as can be deduced from the first term of the ACI empirical equation 6. In addition, the contribution of steel reinforcement to shear strength is represented in the second term of the equation. The presence of the contact surface at higher level of stresses can cause separation perpendicular to the layers, which leads to yielding in steel reinforcement as in equation 5.[6] On the other hand, in the sandwich panel case, there are moments and shear forces created in the connectors in addition to the axial forces that are created from the formed truss action shown in Fig. 17. Because of that significant difference, which was confirmed by testing results, it is not recommended to use the ACI horizontal shear equation.

In the formed truss action, one link of the connector carries tensile force and the second takes the compression force as shown in Fig. 17. The compression strength of the GFRP is limited because GFRP has high tensile strength but low compressive and shear strength. Tensile strength in the range of 80-130 ksi and compressive strength in the range of 46-68 ksi have been reported for GFRP reinforcing bars. Higher compressive strengths are expected for bars with higher tensile strength.[7]
Shear friction equation:

$$V_n = A_{vf} f_y \left(\mu \sin\alpha_f + \cos\alpha_f \right) \tag{5}$$

Horizontal shear strength equation:

$$V_{nh} = \left(260 + 0.6 \rho_v f_y \right) \lambda b_v d \quad < \quad 500 b_v d \tag{6}$$

Where
A_{vf} = *area of shear-friction reinforcement, in.^{2}*
b_v = *width of cross section at contact surface being investigated for horizontal shear, in.*
d = *distance from extreme compression fiber to centroid of tension reinforcement for entire composite section, in.*
V_n = *nominal shear strength, lb*
V_{nh} = *nominal horizontal shear strength, lb*
μ = *coefficient of friction*
α_f = *angle between shear-friction reinforcement and shear plane*
ρ_v = *ratio of tie reinforcement area to area of contract surface*
λ = *correction factor related to unit weight of concrete*

Composite behavior
As shown in Table 6, the average stress of GFRP reinforcement is 31,482 psi and 38,727 psi for specimens with 6 in. and 3 in. insulation, respectively. This average stress is much lower than that of the average tensile stress of 141,212 psi, as shown in Table 5. Therefore, the GFRP connection did not fail in tension but rather in flexure-shear or compression.

Similar to the push-off test results, the average ultimate load obtained from full scale flexure tests is lower than the predicted load, calculated based on the assumption of fully composite behavior. This again can be explained by fact that the failure happened in the connectors due to moment and shear rather than tensile stresses. The current equation for predicting sandwich panel strength assumes that connectors carry tensile stresses, which results in overestimating the overall panel strength.

For example, consider the sandwich panel subjected to a vertical load, as shown in Fig 17. The applied load would create tension in the bottom fiber and compression in the top fiber, thereby causing slip in the bottom concrete,[8] insulation and top concrete wythe. The separation would create flexure and tensile stress in the connector which, in turn, would create compressive stress in another leg of the connector. The tensile stress, compressive stress and flexure occur at the same time; therefore the lowest strength governs the strength of the connector.

More experimental data are needed to verify flexure and compressive stress under the above-mentioned assumption. With lack of flexural and horizontal strength test data, designers can only use the lower bound of these test results to predict the capacity of the sandwich panel, which is constructed of the same material. Precaution should be taken in designing the thickness of the panel.

CONCLUSION AND RECOMMENDATION

The results of the presented experimental program indicate that about 75% of the full composite strength was achieved by using NU connectors every 2x2 ft. It is suggested that the number of NU connectors per square footage should be increased to achieve full composite action.

REFERENCES

1. Al-Einea, A., Salmon, D.C., Tadros, M.K., Culp, T., & Fogarasi, G. J. (1991). State-of-the-Art of Precast Concrete Sandwich Panels. *PCI JOURNAL. 36*(6), 78-98.

2. PCI Committee Report. (1997). State-of-the-Art of Precast/Prestressed Sandwich Wall Panel. *PCI JOURNAL, 42*(3), 35 pp.

3. Al-Einea, A., Salmon, D. C., Tadros, M. K., & Culp, T. (1994). A New Structurally and Thermally Efficient Precast Sandwich Panel System. *PCI JOURNAL.* Vol 39, No 4, 90-101.

4. ACI Subcommittee 440K, Draft, May 28, 2003, Guide Test Methods For Fiber Reinforced Polymers (FRP) For Reinforcing or Strengthening Concrete Structures. 21-25 and 96-103.

5. American Concrete Institute (ACI). Building Code Requirements for Structural Concrete. ACI 318-02/318R-02.

6.   Mattock, A.H., Hofbeck, J.A., & Ibrahim, I.O. (1969). Shear Transfer in Reinforced Concrete. *ACI Journal*, Vol.66, No.2, 119-128.

7.   ACI 440R 1996. State-of-the-Art Report on Fiber Reinforced Plastic (FRP) Reinforcement for Concrete Structures. American Concrete Institute Committee 440.

8.   Mast, R.F., "Auxiliary Reinforcement in Concrete Connections," Journal of the Structural Division, ASCE, V. 94, ST6, June 1968.

Table 1. Push-off specimen dimensions

	Serial No				
	NU-R	NU-C	NU-CX	DT-C	MC-C
Concrete wythe thickness (in.)	3.0	2.5	2.5	2.5	2.5
Insulation thickness (in.)	6.0	3.0	3.0	3.0	3.0
Total thickness (in.)	21.0	13.5	13.5	13.5	13.5
No. of connectors	4	4	4	4	2
Type of connectors	GFRP	GFRP	GFRP	GFRP	WW
No. of specimens	6	6	6	6	6

Table 2. Full-scale specimen dimensions

	Serial No		
Specimen size 4 by 10 ft	I	II	III
Concrete wythe thickness (in.)	3.0	3.0	3.0
Insulation thickness (in.)	6.0	6.0	6.0
Total thickness (in.)	12.0	12.0	12.0
No. of connectors	10	20	10
No. of specimens	2	2	2
Construction method	Horizontally	Horizontally	Vertically

Table 3. Final compressive strength and modulus of elasticity of push-off specimens

Specimen	NU-R		NU-C		NU-CX		MC-C		DT-C	
Concrete wythes	f'_c (psi)	E_c (ksi)	f'_c (psi)	E_c (ksi)	f'_c (psi)	E_c (ksi)	f'_c (psi)	E_c (ksi)	f'_c (psi)	E_c (ksi)
Layer 1	5,954	4,398	5,954	4,398	5,952	4,398	5,954	4,398	4,890	3,986
Layer 2	9,052	5,423	9,052	5,423	5,662	4,289	9,052	5,423	5,952	4,398
Layer 3	8,409	5,227	8,409	5,227	5,662	4,289	8,409	5,227	5,952	4,398

Table 4. Final compressive strength and modulus of elasticity of full-size specimens

Specimen	NU-R-I		NU-R-II		NU-R-III	
Concrete wythes	f'_c (psi)	E_c (ksi)	f'_c (psi)	E_c (ksi)	f'_c (psi)	E_c (ksi)
Layer 1	4,890	3,986	6,063	4,438	6,531	4,606
Layer 2	5,662	4,289	6,063	4,438	6,531	4,606

Table 5. Tensile test results

Sample No	Average Gross section area (in²)	Load Failure (lbs)	Tensile Strength (psi)	Ultimate Strain (in/in)	Modulus of Elasticity (psi)
1	0.11	15,610	141,909	0.0222	6.736×10^6
2	0.11	15,440	140,364	0.0145	6.202×10^6
3	0.11	15,500	140,909	0.0142	6.514×10^6
4	0.11	15,310	139,182	0.0284	6.628×10^6
5	0.11	15,950	145,000	0.0466	6.483×10^6
6	0.11	15,390	139,909	0.0194	6.726×10^6
Ave.		15,533	141,212	0.0242	6.548×10^6

Table 6. Push-off test results

| Specimen | | | | | Connector | | Test results | |
Serial No.	No.	Thickness Con.	Thickness Ins.	f'_c (psi)	Type	Area (in²)	Ultimate Load (lbs.)	$\dfrac{P_U}{A_S}$ (psi.)
NU-R	1	3	6	5,954	GFRP	0.88	25,417	28,883
	2	3	6	5,954	GFRP	0.88	26,233	29,810
	3	3	6	5,954	GFRP	0.88	24,910	28,307
	4	3	6	5,954	GFRP	0.88	33,923	38,549
	5	3	6	5,954	GFRP	0.88	25,130	28,557
	6	3	6	5,954	GFRP	0.88	30,613	34,788
							27,704	31,482
NU-C	1	2.5	3	5,954	GFRP	0.88	37,148	42,214
	2	2.5	3	5,954	GFRP	0.88	31,790	36,125
	3	2.5	3	5,954	GFRP	0.88	40,191	45,672
	4	2.5	3	5,954	GFRP	0.88	36,524	41,505
	5	2.5	3	5,954	GFRP	0.88	28,306	32,166
	6	2.5	3	5,954	GFRP	0.88	33,201	37,728
							34,526	39,235
NU-CX	1	2.5	3	5,662	GFRP	0.88	27,798	31,589
	2	2.5	3	5,662	GFRP	0.88	30,709	34,897
	3	2.5	3	5,662	GFRP	0.88	36,405	41,369
	4	2.5	3	5,662	GFRP	0.88	34,406	39,098
	5	2.5	3	5,662	GFRP	0.88	37,058	42,111
	6	2.5	3	5,662	GFRP	0.88	35,413	40,242
							33,631	38,218
DT-C	1	2.5	3	4,890	GFRP	1.25	25,123	20,098
	2	2.5	3	4,890	GFRP	1.25	26,196	20,957
	3	2.5	3	4,890	GFRP	1.25	28,257	22,606
	4	2.5	3	4,890	GFRP	1.25	22,727	18,182
	5	2.5	3	4,890	GFRP	1.25	23,932	19,146
	6	2.5	3	4,890	GFRP	1.25	24,079	19,263
							25,052	20,040
MC-C	1	2.5	3	5,954	WW	0.64	37,089	57,952
	2	2.5	3	5,954	WW	0.64	37,971	59,330
	3	2.5	3	5,954	WW	0.64	39,669	61,983
	4	2.5	3	5,954	WW	0.64	40,410	63,141
	5	2.5	3	5,954	WW	0.64	38,081	59,502
	6	2.5	3	5,954	WW	0.64	35,803	55,942
							38,170	59,641

Area = Total area of diagonal welded wire or GFRP
NU   Area of GFRP = (2x4x0.11) = 0.88 in²   (0.375 in Dia.)
MC   Area of Welded wire = (2x8x0.04) = 0.64 in²   (0.225 in Dia.)
DT   Area of GFRP = (4x2.5x1.25) = 1.25 in²   (0.125 in Thickness)

Table 7. Full-scale flexure test results

| Specimen | | | | | Connector | | Predicted Load (lbs) | Test results | |
Serial No.	No.	Thickness Con.	Thickness Ins.	f'_c (ksi)	Type	Area (in²)		Ultimate Load (lbs.)	Deflection (in)
I	1	3	6	4,890	GFRP	1.1	27,200	19,399	2.81
	2	3	6	4,890	GFRP	1.1	27,200	20,931	2.28
II	3	3	6	6,063	GFRP	2.2	27,300	23,891	2.55
	4	3	6	6,063	GFRP	2.2	27,300	20,681	2.95
III	5	3	6	6,531	GFRP	1.1	27,330	16,095	3.12
	6	3	6	6,531	GFRP	1.1	27,330	20,956	2.00
Ave.							27,276	20,325	2.13

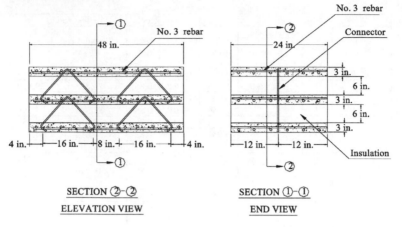

SECTION ②-②

ELEVATION VIEW

SECTION ①-①

END VIEW

Fig. 1: Details of 2 ft. by 4 ft. by 21-inch push-off specimens, FRP connectors

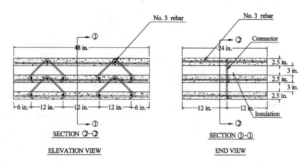

SECTION ②-②

ELEVATION VIEW

SECTION ①-①

END VIEW

Fig. 2: Details of 2 ft. by 4 ft. by 13.5-inch push-off specimens, FRP connectors

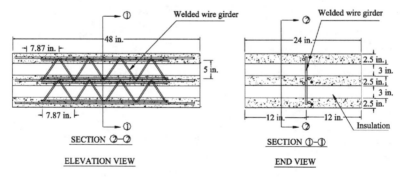

SECTION ②-②

ELEVATION VIEW

SECTION ①-①

END VIEW

Fig. 3: Details of 2 ft. by 4 ft. by 13.5-inch push-off specimens,
welded wire girder connectors

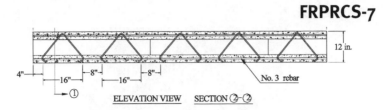

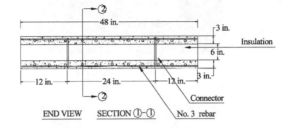

Fig. 4: Details of full-scale specimen

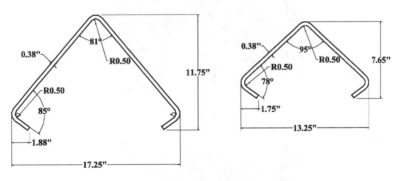

Fig. 5: Details of FRP connectors

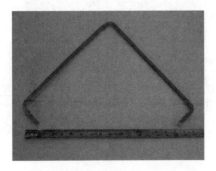

Fig. 6: FRP connector 3-6-3

Fig. 7: FRP connector 2.5-3-2.5

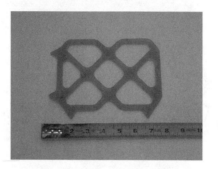

Fig. 8: FRP tie connector

Fig. 9: Welded wire girder connector

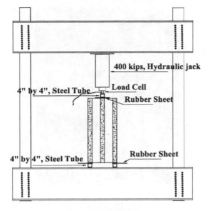

ELEVATION VIEW

Fig. 10: Push-off test set-up

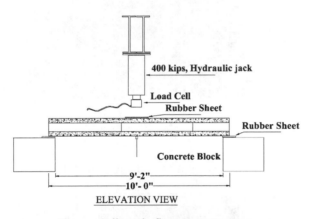

ELEVATION VIEW

Fig. 11: Full-scale flexure test set-up

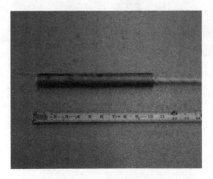

Fig. 12: Facilitate gripping

Fig. 13: Extensometer

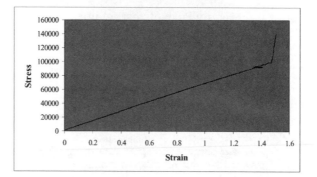

Fig. 14: FRP stress-strain relationship

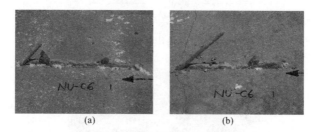

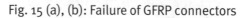

Fig. 15 (a), (b): Failure of GFRP connectors

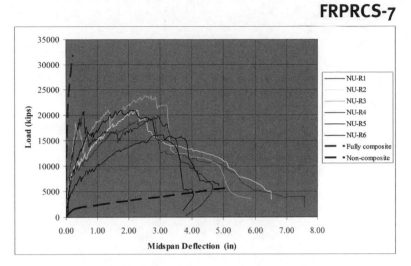

Fig. 16: Mid-span deflection

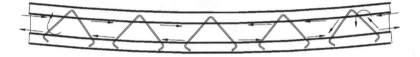

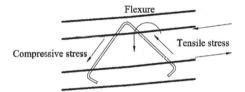

Fig. 17: Forces in the connector

Tensile Capacities of CFRP Anchors

by G. Ozdemir and U. Akyuz

Synopsis: During major earthquakes, buildings which have inadequate lateral strength may be subjected to severe damage or total collapse. To prevent such failures, buildings that are vulnerable against earthquakes must be strengthened. In reinforced concrete buildings inserting adequate amount of reinforced concrete infills is an effective strengthening technique. Another promising technique is to strengthen the existing hollow clay tile infill with diagonally placed CFRP sheets. In this technique, CFRP sheets are extended to the frame members. The connection between CFRP sheets and frame members is provided by CFRP anchors. In this strengthening technique, the effectiveness is dictated by the CFRP anchors. In this study, by means of the prepared test setup, the pull-out strength capacities of CFRP anchors are measured. The effects of concrete compressive strength, anchor embedment depth, anchor hole diameter, and number of fibers (CFRP sheet width) on the tensile strength capacity of CFRP anchors were studied.

Keywords: adhesive anchor; carbon fiber-reinforced polymer (CFRP); CFRP anchor; strengthening

Gokhan Ozdemir was born in Gaziantep, Turkey on 1981. He is graduated from Gazi University at Ankara, Turkey in 2002. He started his graduate program in the same year. Currently he is a graduate student and a research assistant in the Department of Civil Engineering at Middle East Technical University at Ankara, Turkey.

Ugurhan Akyuz was born in Ankara, Turkey on 1967. He received his Ph.D. degree in civil engineering from Middle East Technical University (METU) in 1998. He is an Associate Professor at the Department of Civil Engineering at METU. His main areas of interest are strengthening of reinforced concrete structures, seismic isolation and vibrations.

INTRODUCTION

Structures should be constructed in such a way that they have enough capacity to carry all possible loads. At the design stage, some of the loads can inadvertently be underestimated. Unintentionally created deficiencies during design and/or construction stages may cause catastrophic results when the structure is subjected to unpredicted loads. Due to these deficiencies, existing structures may require strengthening or stiffening to increase their ultimate flexural and/or shear capacity. One of the strengthening techniques is to insert enough amounts of reinforced concrete (RC) infills to meet the lateral demands of the structure. However, for the existing undamaged buildings this technique is not an occupant friendly technique.

With the improvements in technology, a promising technique was developed in the strengthening of structures aiming fewer disturbances and less time consumption. For this purpose, fiber reinforced polymers (FRP), made of high-modulus fibers bonded with a resin matrix, have been increasingly used because of its superior properties. In comparison with steel, FRP possess many advantages such as high corrosion resistance, high strength to weight ratio, electromagnetic neutrality and ease of handling. The first applications of FRP in strengthening are seen in the retrofitting of damaged columns and beams, which provide member improvements only. In recent studies, knowing that strengthening with FRP is successful for member repair, it was used in rehabilitation of undamaged structures with a new technique. In this new technique, the goal was the system improvement rather then member rehabilitation. This technique is called as seismic retrofit by carbon fiber sheets (SR-CF system).

In SR-CF system, the surface of the existing hollow clay tile infill walls are covered with diagonally glued carbon fiber sheets providing that edges of the sheet are connected to the peripheral column, beam and floor using special connections. In this method, the carbon fiber sheet behaves like a tensile bracing and increases the shear resistance of the wall. The key issue in SR-CF application is the performance of the connection between the carbon fiber reinforced polymer (CFRP) sheet and the peripheral structural members. This can be possible with proper connection details. For this purpose, special devices namely, CFRP anchors, were developed. The effectiveness of SR-CF system is highly depending on the capacity of CF anchors. When this connection is lost, load-transmitting process is failed and this causes a sudden increase in internal forces in peripheral frame members, which may cause a failure of the structure. Therefore, the behavior of these special connections, namely CFRP anchors need to be investigated for the success of this strengthening technique. In this study, only the direct tensile load capacities of CFRP anchors were determined experimentally. During the tests, the effects of concrete strength, anchor embedment depth, anchor hole diameter, and CFRP sheet width on the tensile capacity of CFRP anchors were studied.

STRENGTHENING WITH CFRP

To investigate the effect of SR-CF method on the behavior of the rehabilitated structure, an experimental study was performed in structural mechanics laboratory of Middle East Technical University (METU) (Erdem 2003). The test specimen was two story three bay 1/3 scale reinforced concrete frame. The width of the middle bay was 1000 mm while the others were 1600 mm. The first story height was 1500 mm and the second story height was 1000 mm. All the columns were 110×110 mm and all the beams were 110×150 mm. This frame had common deficiencies of the buildings in Turkey. To strengthen this frame, first the middle bay of the frame was infilled with hollow clay tiles and plastered. Then CFRP sheets were glued on both surfaces of the infill diagonally (Figure 1). The connection of the CFRP sheet with the frame members were provided by CFRP anchors (Figure 2). The specimen was subjected to quasi-static reversed cyclic lateral load at the second story level with an axial load of 9 kN applied to the columns during the test. The results of this experiment was compared with the results of the tests conducted with exactly the same frames in which the frame was a bare frame without any strengthening and a frame that was strengthened with RC infill wall inserted on the middle bay only.

In (Erdem 2003), by using the bare frame test data also, which was provided by (Canbay, Ersoy, and Ozcebe 2003), two different strengthening techniques were compared. These are namely, inserting a RC infill and inserting a hollow clay tile infill strengthened with CFRP. Both strengthening techniques provide significant increase in lateral load carrying capacity. The lateral load capacity of the frames strengthened with RC infill wall and CFRP was about 5 times greater than that of the bare frame. Test results also revealed that the increase in the initial stiffness of the frames was considerable due to application of both strengthening techniques. The significant difference observed between these two successive strengthening techniques was the failure types. As expected, strengthening with RC infill failed in flexure in a very ductile manner. However, the frame strengthened with CFRP failed due to premature failure of the CFRP anchors providing the connection at the foundation level. To give an idea about the effectiveness of the strengthening techniques the drift ratio envelopes are presented in Figures 3 and 4 (Erdem 2003) for the first and second stories, respectively. Drift ratio of a story is defined as relative displacement of that story divided by its height. As it is seen, the lateral deformations in both stories of the frame strengthened with CFRP were smaller than that of the frame with RC infill. This is because the test conducted with CFRP strengthened frame could not be continued to produce enough shear displacement due to the failure of the CFRP anchors. It is obvious that the behavior could be superior unless the CFRP anchors were ruptured. The CFRP anchors used in this study were of Type-1 anchors which will be discussed in detail in proceeding sections.

The experimental study conducted by (Erdem 2003) resulted with a very important conclusion, which emphasizes the importance of the connections, namely CFRP anchors in the effectiveness of the strengthening application with CFRP. The observed results show that the key issue in a SR-CF application is the performance of connection between the CFRP sheet and the peripheral structural members.

EXPERIMENTAL PROGRAM

Previous studies reveal that in strengthening with CFRP, connections between the frame members and CFRP sheet are the crucial points. The effectiveness of the rehabilitation depends on a proper connection. Provided that the connection has sufficient strength, then this strengthening method is satisfactory. The load transfer from hollow clay tile infill strengthened with CFRP to the RC frame members is provided by the anchors. Therefore, the capacity of the anchors needs to be known for a successful strengthening.

In the experimental study performed by Erdem, the anchors which were used in the infill were subjected to both shear and axial forces while anchors which were used on the frame were subjected to tensile force only. Due to complexity of the combined shear and axial force problem, as an initial study it is decided to investigate the uniaxial tensile capacity of the embedded CFRP anchors. Tensile capacity of a single CFRP anchor was measured throughout the experiments under the effect of aforementioned parameters. These experiments were conducted in two series of tests depending on the preparation of the CFRP anchors.

Test specimens

In order to obtain the uniaxial tensile capacities of CFRP anchors, concrete blocks with different compressive strengths were prepared as test specimens. Dimensions of these concrete blocks were 300×400×4000 mm, and holes in the desired diameters were drilled by means of a rotator hammer on it. Five concrete beams were cast with three different compressive strengths. Desired and actual 28-day compressive strengths of these blocks are given in Table 1.

CFRP anchors

In this study, MBrace produce C1-30 unidirectional fiber sheets were used. CFRP anchors were composed of an epoxy based matrix and C1-30 carbon fibers. A two component, room temperature cure epoxy resin adhesive, namely saturant, is used as the component of the CFRP. The properties of C1-30 fiber sheets and epoxy resin are given in Tables 2 and 3, respectively.

It is known that adhesive anchors transfer the load through the adhesive layer along the bond surface. Therefore, the important points for an adhesive anchor are the quality and type of the adhesive. An epoxy adhesive is a synthetic compound consisting of an epoxy resin cross-linked with a curing agent. The epoxy resin is designated as compound A and the curing agent as compound B by the manufacturer. Epoxy adhesives are thermosetting polymers; that is, they require heat to cure. This heat is generated during the exothermic reaction between the epoxy resin and the curing agent (Cook and Klingner 1992; Cook, Doerr, and Klingner 1993; Cook 1993; Cook et al. 1998).

In the literature, there are different application techniques of adhesively bonded anchors. In this experimental study, an injection type adhesive anchor application was chosen. Before the installation of the anchor, a catalyzed resin has been injected into the hole and then the anchor was pushed into the hole.

Type 1—These anchors are exactly identical to the one used in two story three bay 1/3 scale reinforced concrete frame test. While preparing these anchors, first, CFRP sheets were cut into two layers of equal width and equal height. Then, these soft CFRP sheets (Figure 5a) were rolled to have a cylindrical form. To prevent the CFRP sheet bundle deviating from its cylindrical shape, rolled sheets were tied with CFRP fibers. CFRP sheets became ready to be used as an anchor after 1 cm portion from the bottom of the rolled CFRP sheet was embedded into the epoxy resin and waited for one day. The aim of the epoxy coated bottom part of the CFRP anchor is to have a stiff part to maintain ease in installation. During the installation procedure, CFRP anchors were pushed from the hardened end into the drilled and carefully cleaned holes by using a steel wire. All of the holes were cleaned by an air pump. After pouring enough epoxy resin into the clean, dry anchor hole, epoxy coated end of the CFRP anchor was embedded into the concrete as seen in Figure 5b. The bond free part of the CFRP anchor was also perfectly bonded to a steel rod to apply a tensile force. The steel rod is used to transfer the applied load to the CFRP anchor and it is not prolong into the hole.

In Type-1 series, three different sheet width (w) (80 mm, 120 mm, and 160 mm), three different embedment depth (h) (70 mm, 100 mm, and 150 mm), three different concrete compressive strength (f_c) (10 MPa, 16 MPa, and 20 MPa), and three different anchor hole diameter (d) (12 mm, 14 mm, and 16 mm) were chosen as the study parameters.

Type 2—In the second series of tests, the CFRP anchors were prepared completely out of the hole and then installed into the concrete. This technique was developed after the first series of tests. In this technique, first the desired width of CFRP sheet was cut and coated with epoxy resin. Then, the epoxy coated CFRP sheet is rolled over a silicon rod, which is 10 mm longer than the embedment depth, and the steel rod through which the force is applied (Figures 6a-6b). The diameters of the silicon rod and the steel rod were the same. This technique was developed to have straight anchors in which the fibers of the CFRP sheets are oriented in the same alignment. The part of the anchor rolled around the silicon rod was embedded into the drilled hole and bonded there.

The extra part of the silicon rod was standing over the concrete level to prevent a discontinuity at the critical section.

The parameters studied in this series were four different embedment depths (50 mm, 70 mm, 100 mm, and 150 mm), and two different concrete compressive strengths (10 MPa, and 16 MPa). The CFRP sheet width and anchor hole diameter was chosen as 120 mm and 20 mm, respectively.

Testing apparatus and procedure

The steel frame, which was used to apply the tensile force to the embedded anchor, is shown in Figure 7. The load was applied by means of a center hole hydraulic jack and the load was measured by means of a load cell. Applied load was transferred to the CFRP anchors through the steel rod extending along the loading equipments. Load readings were taken at a rate of five samples per second.

TENSILE CAPACITY OF CFRP SHEETS

While presenting the experimental results, the measured ultimate tensile loads were divided into the ultimate capacity of a CFRP sheet with equivalent sheet width. It is well known that the tensile capacity of carbon fiber sheets can be calculated using the following equation

$$P_{FRP} = w \times t \times f_u \quad (N) \tag{1}$$

where w is the CFRP sheet width in mm, t is the thickness of the CFRP sheet in mm, and f_u is the characteristic tensile strength of the carbon fibers in terms of MPa. t and f_u are declared by the manufacturer as 0.165 mm and 3,430 MPa, respectively. Thus, for different w, the rupture capacity of CFRP sheet can easily be determined.

TEST RESULTS

The test results are going to be presented in two parts composed of the results obtained from Type-1 and Type-2 series of tests. Complete results are provided in (Ozdemir 2005). All tensile capacities observed during the tests were normalized with the capacity of carbon fibers (determined from Equation (1)) that have the same sheet width (w) with the anchors.

Type-1 test results

In Type-1, 124 tests were conducted. Results of these numerous tests are going to be discussed under the name of each parameter. Before the discussion of the results, it should be stated clearly that all anchors failed as rupture of the CFRP sheet covered with epoxy resin (Figure 8).

Effect of anchor hole diameter (*d*)--Load transfer mechanism of adhesive anchors is different from that of mechanic or headed anchors by the way how they transmit the applied tensile load. The applied load is transferred from an adhesive anchor by means of the bonding between the surface and the adhesive. That is why the bond strength is important for an adhesive anchor. The main idea to choose the hole diameter as a parameter was to investigate the effect of the free space between the anchor and concrete surface on the tensile capacity of the anchors. The procedure throughout the experiments for investigating the effect of the anchor hole diameter was to change the hole diameter while the other parameters remain unchanged. Effect of hole diameter (*d*) on 10 MPa concrete block is discussed in Figure 9 for anchors having sheet width of 80 mm, and embedment depth of 150 mm. As it is seen the normalized tensile capacity of the CFRP anchors in different hole diameters is almost constant, and it is equal to 45~48% of the capacity of the CFRP sheet calculated by Equation (1). All anchors failed in such a way that the effect of hole diameter could not be observed. CFRP sheets have reached their ultimate tensile capacity under the applied load before any bond failure occurred. So, for this series of tests it can be said that a minimum of 2-3 mm free space, which is also the suggested necessary spacing by the manufacturers, is adequate for proper bonding.

Effect of concrete compressive strength (*f$_c$*)--In previous studies (Cook et al. 1998; Makitani, Irisawa, and Nishiura 1995; Hattori et al. 1995), compressive strength of the medium, where adhesive anchors were installed, has been chosen as a parameter influencing the capacity of the anchors. Some of those studies indicate that the compressive strength does not have any significant effect on the capacity of the anchors (Makitani, Irisawa, and Nishiura 1995; Hattori et al. 1995) while some of them designate that compressive strength is influential on the capacity of the adhesive anchors (Cook et al. 1998). Figure 10 represents test results for a group of anchor which have 160 mm sheet width, 100 mm embedment depth, and 14 mm hole diameter. In Type-1 tests, for concrete compressive strengths between 10 to 20 MPa, it is not seen any significant difference between the normalized tensile capacities of CFRP anchors. The ratio of the rupture load of CFRP anchor to the tensile capacity of CFRP sheet obtained from Equation (1) is approximately 0.31 for all *f$_c$* between 10 to 20 MPa. Similar to the case in hole diameter, CFRP anchors in this category failed before the effect of concrete compressive strength can be observed. The dominant failure type is the rupture of the CFRP sheet. The load transfer mechanism at the sections very close to hole seems to be very important. The orientation of the CFRP fibers becomes very crucial since any deviation from the straight alignment of the fibers lead to premature rupture of the CFRP sheet.

Effect of CFRP sheet width (*w*) --Tensile capacity of a CFRP anchor is related with the number of fibers in that anchor. Thus, CFRP sheet width is a crucial parameter. A wider anchor contains more carbon fibers to resist applied tensile load. The ultimate tensile capacities versus CFRP sheet width are plotted in Figure 11. All anchors presented in this graph have 100 mm embedment depth, 12 mm hole diameter. They are installed into a concrete beam, which has a compressive strength of 20 MPa. As expected, test results show that load capacity increases with an increase in sheet width. However, the behavior is not linearly proportional to the sheet width. When ultimate loads, P_u, are normalized according to the capacity of a carbon fiber sheet with the same width, there is a decrease as width becomes wider (Figure 12). This shows that the tensile capacity does not increase in the same rate with the ratio of CFRP sheet widths. The normalized tensile capacity of an anchor with 80 mm sheet width is obtained in average as 0.52 while this ratio is approximately equal to 0.35 for an anchor having 160 mm sheet width.

Effect of embedment depth (*h*)--Most of the previous studies that investigate the behavior of the anchors were mostly interested in the effect of the embedment depth of the anchors (Cook and Klingner 1992; Cook, Doerr, and Klingner 1993; Cook 1993; Cook et al. 1998). The data given in Figure 13 corresponds to the anchors with 160 mm sheet width, 14 mm hole diameter, and 20 MPa concrete compressive strength. As it is seen, the normalized tensile capacity of a CFRP anchor increases as embedment depth increases from 70 mm to 100 mm. However, there is a decrease or no change in the CFRP anchor capacity after 100 mm embedment depth. Among the embedment depths studied, maximum normalized value is obtained when the embedment depth of 100 mm.

Type-2 test results

To investigate the behavior of CFRP anchors under the effect of concrete compressive strength and embedment depth of the anchor in detail, this second series of tests were carried out for only one CFRP sheet width and one hole diameter, while the embedment depth changes from 50 mm to 150 mm, including 70 mm and 100 mm. This series was studied in two concrete blocks with 10 MPa and 16 MPa compressive strengths. In this series, total of 23 tests were conducted. The normalization of the ultimate tensile capacities was also performed by dividing them to the tensile capacity of the CFRP sheet determined by Equation (1). All of the anchors tested in this part have 120 mm sheet width and installed into a hole with 20 mm diameter.

The main difference observed during this series of tests was in the failure types. A concrete cone observed for anchors with 50 mm embedment, while a combined cone-bond failure observed for 70 mm and 100 mm embedment depths. The failure of the CFRP anchors with 150 mm embedment was due to rupture of CFRP sheet.

Effect of concrete compressive strength (*f$_c$*)--Normalized tensile capacities for all embedment depths are given in Figure 14. It is seen that the highest normalized tensile capacity of a CFRP anchor is reached for an embedment depth of 100 mm. As it is also seen, normalized data for 50 mm embedment depth are so close to each other that the difference in tensile capacities of anchors installed into 10 MPa and 16 MPa concrete beams is negligible. However, as the embedment depth increases, the significance of the concrete compressive strength increases. For the highest normalized tensile capacity, i.e. for *h*=100 mm, the effect of concrete compressive strength is more significant.

Effect of embedment depth (*h*)--Figure 15 gives the normalized test results of anchors installed into a concrete beam of 10 MPa. As expected, the normalized tensile capacity of a CFRP anchor increases as the embedment depth increases. However, there is an effective depth beyond which the capacity becomes almost constant. For the parameters studied, effective embedment depth is 100 mm. The capacities of CFRP anchors that are embedded to 150 mm are almost equal to those of 100 mm embedment depth. Figure 16 shows the change in normalized tensile capacities of anchors installed into a concrete beam of 16 MPa. Similar behavior was obtained. Tensile capacities of CFRP anchors increase up to 100 mm embedment depth. Beyond this depth, the capacity becomes constant and equal to that of 100 mm. This is due to high bonding effect resulting in high load transfer to the concrete at the top of the anchor. The bond stress is no longer uniform, and if the tensile load is sufficiently high, the failure initiates with a concrete failure in the upper portion of the concrete and then the bond failure occurs in the remaining part of the embedment depth.

CONCLUSIONS

During the last decade, fiber reinforced polymers (FRP) has been widely used to strengthen bridge girders, piers, columns of structures and masonry walls in wall bearing. Usage of carbon fiber reinforced polymers (CFRP) to strengthen the existing structures is relatively new and promising technique. While strengthening the existing structures CFRP sheet are applied on the hollow clay tile infill walls diagonally. In strengthening the hollow clay tile infill by CFRP, CFRP anchorages are used both in masonry and in reinforced concrete members to provide a sufficient bond between the CFRP sheet and the masonry and the reinforced concrete member. Better connections lead higher energy dissipation and higher ductility. Thus, the capacity increase in the existing structure mostly depends on the load transfer through the CFRP anchors, or simply the increase in capacity of the structure depends on CFRP anchor capacity. In this experimental study, direct tensile capacities of CFRP anchors were investigated for different parameters. Effect of CFRP sheet width, embedment depth of adhesive anchor, hole diameter of anchor and compressive strength of concrete on the uniaxial tensile capacity of the CFRP anchors were determined. With the help of the test results, the followings are concluded.

For the embedment depth of 50 mm, uniaxial tension tests ended with a concrete cone failure. However, the anchors with 70 mm and 100 mm embedment depths formed pullout failure with a shallow concrete cone at the top. On the other hand, CFRP rupture was observed for the anchors embedded into 150 mm.

For the studied parameters, the maximum tensile load capacities are obtained for CFRP anchors which have 100 mm embedment depth. This indicates that there is an effective bond length beyond which load capacity does not increase (Cook, Doerr, and Klingner 1993; Cook 1993; Cook et al. 1998). The increase in tensile load capacities can be assumed linear up to 100 mm embedment depth.

For the shallow embedment depths, i.e. 50 mm, the effect of concrete compressive strength, in the range of 10 MPa to 16 MPa, on the tensile capacity of CFRP anchor is not significant. However, as the embedment depth increases, the effect of concrete compressive strength becomes more significant.

ACKNOWLEDGMENT

This study has been partially supported by NATO (grant no: SfP-977231) and The Scientific and Technical Research Council of Turkey (grant no: ICTAG I575).

REFERENCES

Canbay, E., Ersoy, U., and Ozcebe, G., 2003, "Contribution of Reinforced Concrete Infills to Seismic Behavior of Structural Systems", *ACI Structural Journal*, Vol. 100, No. 5, September-October, pp. 637-643.

Cook, R. A., 1993, "Behavior of Chemically Bonded Anchors", *Journal of Structural Engineering*, Vol. 119, No. 9, September, pp. 2744-2762.

Cook, R. A., Doerr, G. T., Klingner, R. E., 1993, "Bond Stress Model for Design of Adhesive Anchors", *ACI Structural Journal*, Vol. 90, No. 5, Sep.-Oct., pp. 514-524.

Cook, R. A., Klingner, R. E., 1992, "Ductile Multiple-Anchor Steel-to-Concrete Connections", *Journal of Structural Engineering*, Vol. 118, No. 6, June, pp. 1645-1665.

Cook, R. A., Kunz, J., Fuchs, W., and Konz, R. C., 1998, "Behavior and Design of Single Adhesive Anchors under Tensile Load in Uncracked Concrete", *ACI Structural Journal*, Vol. 95, No. 1, Jan.-Feb., pp. 9-26.

Erdem, I., 2003, *Strengthening of Existing Reinforced Concrete Frames*, Master's thesis, Middle East Technical University, 123 pp.

Hattori, A., Inoue, S., Miyagawa, T., Fujii, M., 1995, "A Study on Bond Creep Behavior of FRP Rebars Embedded in Concrete", *Proceedings of 2nd International RILEM Symposium on Non-Metallic FRP Reinforcement for Concrete Structures*, pp. 172-179.

Makitani, E., Irisawa, I., Nishiura, N., 1995, "Investigation of Bond in Concrete Member with Fiber Reinforced Plastic Bars", *Proceedings of International Symposium on Fiber Reinforced Plastic Reinforcement for Concrete Structures, ACI SP-138*, pp. 315-331.

Ozdemir, G., 2005, *Mechanical Properties of CFRP Anchorages*, Master's thesis, Middle East Technical University, 74 pp.

Table 1— Desired and actual 28-day compressive strengths of these blocks

Specimen	28 day f_c (MPa)	Test day f_c (MPa)
1st beam	10.7	10.7-11.3
2nd beam	15.8	15.8-16.4
3rd beam	19.4	19.4-20.1
4th beam	10.2	10.2-10.5
5th beam	16.3	16.3-16.5

Table 2— Properties of C1-30 fiber sheets

Property	Amount	Unit
Unit Weight	0.300	kg/mm^2
Effective Thickness	0.165	mm
Characteristic Tensile Strength	3,430	MPa
Charactcristic Elasticity Modulus	230,000	MPa
Ultimate Strain	0.015	mm/mm

Table 3— Mechanical properties of epoxy resin (provided by MBrace)

Property	Amount
Compressive Strength	>80 MPa
Direct Tensile Strength	>50 MPa
Flexural Tensile Strength	>120 MPa
Elasticity Modulus	>3000 MPa
Ultimate Strain	>0.025

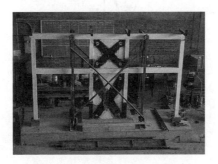

Figure 1—Test frame (taken from (Erdem 2003))

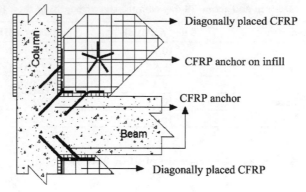

Figure 2—Connection detail (Erdem 2003)

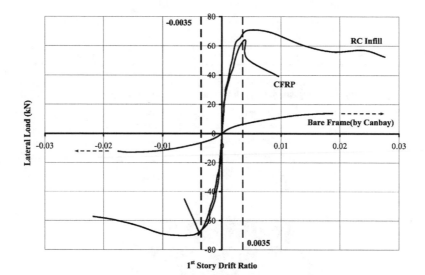

Figure 3— First story envelope curve of test frame (taken from (Erdem 2003))

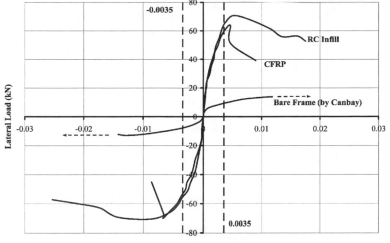
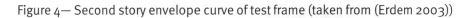

Figure 4— Second story envelope curve of test frame (taken from (Erdem 2003))

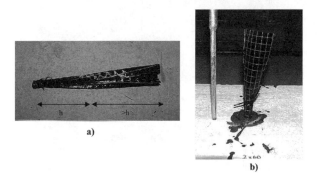

Figure 5— Type-1 series of CFRP anchors

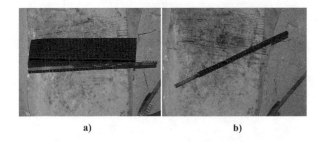

Figure 6— Type-2 series of CFRP anchors

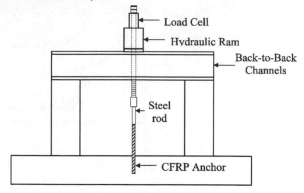

Figure 7— Schematic view of test set-up

Figure 8— CFRP rupture

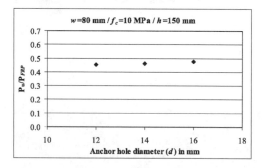

Figure 9— Anchor hole diameter versus normalized tensile capacities for w=80 mm, f_c=10 MPa, h=150 mm

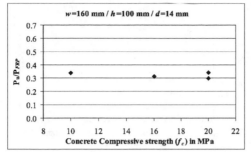

Figure 10— Concrete compressive strength versus normalized tensile capacities for *w*=160 mm, *h*=100 mm, *d*=14 mm

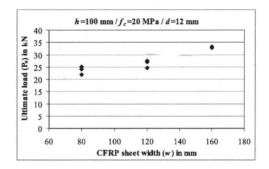

Figure 11— CFRP sheet width versus ultimate tensile capacities for *h*=100 mm, *f*_c=20 MPa, *d*=12 mm

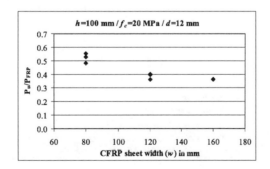

Figure 12— CFRP sheet width versus normalized tensile capacities for *h*=100 mm, *f*_c=20 MPa, *d*=12 mm

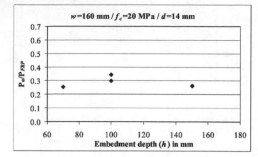

Figure 13— Embedment depth versus normalized tensile capacities for w=160 mm, f_c=20 MPa, d=14 mm

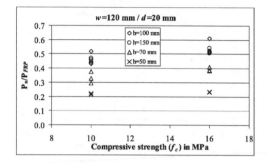

Figure 14— Concrete compressive strength versus normalized tensile capacities for w=120 mm, d=20 mm

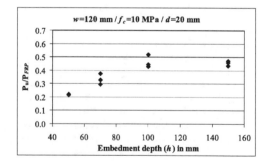

Figure 15— Embedment depth versus normalized tensile capacities for w=120 mm, f_c=10 MPa, d=20 mm

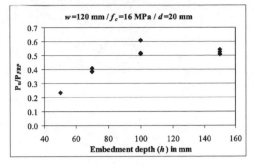

Figure 16— Embedment depth versus normalized tensile capacities for w=120 mm, f_c=16 MPa, d=20 mm

An Exploratory Study of FRP Seismic Restrainers Subjected to Dynamic Loads

by M. Saiidi, R. Johnson, and E. Maragakis

Synopsis: Glass, carbon, and hybrid (glass/carbon) fabric reinforced polymer (FRP) restrainers were developed and tested as an alternative to steel restrainers to reduce bridge hinge movement during earthquakes. The FRP bridge restrainers were dynamically tested on a representative in-span hinge, in the large-scale structures laboratory at the University of Nevada, Reno (UNR). Work included: (1) Strain rate tests on both FRP strips and FRP/concrete bond; (2) FRP restrainer development and testing; (3) Comparisons between FRP, steel, and shape memory alloy (SMA) restrainers; (4) Development and evaluation of a simple restrainer design method. Findings confirm the potential use of FRP restrainers as a viable option to steel as a restraining device for bridges. Results include: (1) FRP strength is strain-rate insensitive; (2) FRP/concrete bond strength is a function of concrete shear strength and is strain-rate sensitive; (3) FRP restrainers are easily constructed and installed; (4) A proposed restrainer design method that considers the bridge structure dynamic characteristics is demonstrated to be both simple and realistic.

Keywords: dynamic tests; fiber-reinforced polymer; FRP/concrete bond; FRP restrainers; restrainer design method; strain rate

58   Saiidi et al.

ACI member **M."Saiid" Saiidi** is Professor in the Civil Engineering Department and head of undergraduate research, University of Nevada, Reno. He received his MSc from Tehran University, Iran, and an MSc and PhD from University of Illinois in 1977 and 1979, respectively. He was the founding chair of ACI Committee 341. His research interests include earthquake engineering of bridges and buildings, experimental studies of bridges and components, and analysis and design of reinforced concrete structures.

ACI member **Rita Johnson** is employed at the office of Forbes and Dunagan Structural Engineers. She received her diploma in 2003 and her MSc in 2004 from the University of Nevada, Reno. Her research interests include the design and dynamic testing of seismic bridge restrainers made of alternative materials, such as shape memory alloy and fiber reinforced plastics.

ACI member **E. "Manos" Maragakis** is Professor and Chair in the Civil Engineering Department, University of Nevada, Reno. He received his B.S. in 1980 from the National Technical University of Athens, Greece, and his MSc and PhD from the California Institute of Technology in 1981 and 1984, respectively. His research interests include earthquake response of buildings and bridges, dynamic and earthquake response of railway bridges and connections, computer simulation, adequacy of cable restrainers for bridge performance during strong earthquakes.

INTRODUCTION

The main purpose of bridge restrainers is to prevent spans from falling off of their supports during the maximum credible earthquake. The most common type of restrainer in the United States is the steel cable restrainer. These steel cables are anchored to the diaphragms or webs of concrete bridges or to the bottom flange of steel girders (Vlassis et al 2000). Past tests were conducted at the University of Nevada, Reno (UNR) to explore the use of a shape memory alloy (SMA) restraining device to address shortcomings of traditional steel restrainers (Johnson et al 2004). The research conducted in these experiments involves the testing, manufacture, and design procedure for a bridge restrainer made with another alternative material, fiber reinforced plastics (FRP).

FRP is a material familiar in civil engineering as a means of externally increasing the strength of structures. The interest in fiber composites for construction applications is growing very rapidly. The flexibility and high strength of FRP fabric makes it an intriguing material. These qualities of high tensile strength, low weight, and flexibility, also make it a material worth investigating as a possible alternative to steel in seismic restrainers. Using FRP as a material for seismic bridge restrainers is a new and innovative idea. Beneficial aspects of FRP restrainers compared to steel restrainers are: (a) Installation of conventional restrainers is intrusive, requiring drilling of concrete, whereas, FRP restrainers are bonded directly onto the outside of the bridge after minimal surface preparation; (b) In many cases conventional restrainers are hidden, making them hard to inspect, whereas, FRP restrainers are visible; (c) Unlike conventional restrainers, FRP restrainers do not require an attachment system; (d) Availability of the material and economical manufacturing translates into low costs. FRP restrainers offer ease of installation and inspection, in addition to having high strength.

OBJECTIVES AND SCOPE

The objective of this study was to determine the feasibility of developing and testing a bridge restrainer made of FRP. To accomplish these objectives, material tests and bond tests were conducted, followed by development and shake table testing of three restrainer types. A new restrainer design method was also developed. This four-part project included:

1. Material tests on uncoated and elastomer coated glass and carbon fabric strips;
2. FRP/concrete bond tests to determine the mode of failure and bond strength between FRP and concrete;
3. Design of an FRP restrainer;
4. Development of a tentative restrainer design procedure.

The material tests and bond tests were conducted at various strain rates, from static to dynamic, to determine if there is an apparent strain rate effect for composite lamina. Strain rate effect, or the possible effect of increasing strength with increase of strain rate, is of interest in restrainer design because of the dynamic nature of earthquake movement. The strain rate could potentially affect both the FRP strength properties and bond to concrete. The mode of failure of the bond between FRP and concrete was also of interest in this study.

Results from the material tests led to the development of an FRP restrainer. Restrainer dimensions were calculated based on both current material tests and data collected during past SMA and steel restrainer tests. Three types of FRP restrainers were manufactured. They consist of glass (GFRP), carbon (CFRP), and a hybrid (GCFRP), consisting of glass and carbon composite laminate.

An identical test set-up was used for the steel, SMA and FRP restrainer tests to allow comparison among these three restrainer types. Previous steel and SMA restrainer tests, conducted at the UNR large-scale structures lab, provided the test protocol used in the FRP restrainer shake table tests. The primary focus was on the performance of FRP restrainers under longitudinal motions, but limited tests were conducted using transverse motions. Several current design procedures were investigated to determine their applicability to FRP restrainer design. A new simple force-based method was developed for restrainer design and was illustrated through an example.

MATERIAL STUDIES

Material tests at various strain rates were performed on the FRP fabrics used in the restrainer experiments. Tensile tests were carried out on fabric strips to obtain the ultimate fabric strength. Following tensile tests on uncoated (dry) fabrics, the necessity of a coating became apparent to ensure composite action among fibers. An elastomer material was chosen to coat the fabrics and ensure better strength performance. The tensile test results were analyzed to determine the effect of strain rate on FRP. The last set of material tests were conducted to determine the strain rate effect on the interfacial bond strength between externally bonded FRP fabrics and concrete.

Strain Rate Effect

Past studies have concluded that stress-strain properties are a function of loading rate. A significant increase in strength has been generally observed at higher strain rates. Past research has shown that at high strain rates, there is an effect on steel and concrete properties. At 50000 µε/sec, concrete compressive strength increases by approximately 25% (Kulkarni and Shah 1998). In steel, loading at a high rate results in f_y increasing by 10 to 20 percent (Paulay and Priestley 1992). The reason for this increase of strength with higher strain rates is believed to be from the extremely localized yielding due to enhanced bond among the molecules at high rates. Internal friction or slippage does not have time to occur under high strain rate. Data collected from previous steel (Camargo-Sanchez et al) and SMA restrainer tests revealed strain rates in excess of 100000 µε/sec and 200000 µε/sec, respectively. A test protocol was developed to assess 25.4 mm wide, 305 mm long glass and carbon fabric strips in an MTS load frame tensile testing machine at various strain rates. The maximum strain rate that could be applied was 100000 µε/sec. All three sets of material tests were conducted at a static rate and five constant dynamic rates; 1000, 5000, 10000, 50000, and 100000 µε/sec, 3 specimens per rate.

Plain Fabric Tensile Tests

Ordinary glass and carbon fabric, normally used as a reinforcing fabric in a fiber wrap, was chosen for the tensile tests, with the aim of producing a restrainer with common materials. The glass and carbon material were both unidirectional, with additional cross fibers. The carbon fabric had a glass veil backing for added fabric stability during installation. An example of the first series of material tests, seen in Fig. 1, consisted of strip of uncoated glass and carbon fabric pulled to determine failure mode, strength, elongation at break, and strain rate effect. Prior to testing, the ends of the fabric strips were dipped in resin to create the composite ends necessary to produce a successful gripping action in the load frame.

Glass fabric – The failure of the uncoated glass strips was sudden and explosive, particularly at the higher strain rates. The average measured value for tensile strength of the uncoated glass strips was approximately 266 MPa, a 42% reduction in strength from the design composite tensile strength of 460 MPa, seen in Table 1. This strength reduction appeared to be the result of the absence of load sharing among the fibers. There also appeared to be a pattern of failure occurring at the edges of the strips. Upon examination of the stress-strain results at the various strain rates, there did not appear to be a clear increase of strength with strain rate.

Carbon fabric – The test protocol used for the glass strip tests was repeated for the carbon fabric tensile tests. Unlike the glass strips, the carbon did not fail in an explosive manner. The average measured tensile strength for the carbon strips tested at a static rate was approximately 454 MPa. This was 40% lower than the expected composite strength of 745 MPa. The improvement of strength, from 42 to 40 percent reductions, from the glass and carbon tests, respectively, appeared to be from the system holding the longitudinal threads together. In the glass material, the longitudinal fibers were held together by transverse cross fibers, whereas, in the carbon material, an interlocking system of threads unified the system. It appeared that the geometrical structure of the fabric had significant influence on its tensile strength and failure mode. As in the glass strip tests, a clear trend in strength change with increase of strain rate (strain rate effect) was not demonstrated in the stress-strain graphs.

Flexible Composites

The poor strength performance seen in the uncoated fabric tensile tests confirmed the necessity for a composite action among fibers to fully mobilize the entire width. Stiff composites are commonly seen in civil engineering as a replacement for traditional materials or as an element in strengthening columns. For application as a bridge restraining device, a flexible composite action was required. It was necessary for the portion of the restrainer in the hinge area to provide high tensile strength while being sufficiently flexible to buckle under compression without building up significant stresses. A flexible composite was produced with textile fabrics and SYLGARD 184 silicone elastomer, a product of Dow Chemical. "FRP composites, such as rubber-coated fabrics, can undergo relatively large deformation. They have very good formability while retaining high strength" (Luo and Mitra 1999). These characteristics of flexible composites make them ideal as material for bridge restrainers.

The second series of tensile tests for the flexible composite strips had an identical test set-up and protocol as the first series of tests. The expectation of the flexible composite fabric strip tests was to achieve a higher strength, closer to the manufacturers design strength for glass and carbon composite. A strong mechanical interaction between elastomer and fabric occur in the initial stage of tensile testing. The elastic modulus of the elastomer is lower than the fabric and the composite is "soft" at this stage. Then, the geometrical arrangement of the fabric becomes stable and the fibers carry most of the load.

Elastomer coated glass strips – No significant strength improvement or change in failure mode was observed in the flexible glass strip tests vs. the uncoated glass strip tests. This appeared to be because of the strong influence of the geometric arrangement of the glass in these 25.4 mm strips. No apparent increase of strength with strain rate was observed in this set of tests.

Elastomer coated carbon strips -- There was a clear increase in strength between the first and second set of carbon strip tensile tests. Figure 2 shows the design strength of 745 MPa that was achieved during the flexible carbon strip tests. As in the previous fabric strip tests, no strain rate effect was observed.

FRP/Concrete Interfacial Bond Tests

Debonding failures are often brittle and occur with little or no visible warning. Studies have shown that decreasing the shear stiffness of adhesives, increasing FRP stiffness, and increasing concrete strength all play a role in bond behavior. When properly prepared, the bond strength is greater than the shearing strength of the concrete to which the FRP is attached (Teng et al 2001). Single lap tests were performed (Fig. 3) to verify that the minimum and often controlling factor of bond strength is related to the compressive strength of the concrete, to examine modes of failure, and determine if strain rate had an effect on bond strength. These tests consisted of both flexible glass and carbon strips bonded to concrete blocks, assessed at static to dynamic strain rates. The bond test set-up, seen in Fig. 4, consisted twenty concrete beams with externally bonded GRFP plates, followed by twenty concrete beams with externally bonded CFRP plates. The middle 203 mm portion of the FRP strip that was not attached to either the grips or bonded to the beams was coated with elastomer. A 279 mm x 102 mm x 12.7 mm steel plate was anchored to the concrete beams to attach the beams to the grips in the testing machine and create a rigid connection. As in the earlier fabric strip tests, a minimum of three specimens per strain rate (static, 1000, 5000, 10,000, 50,000, 100,000 $\mu\varepsilon$/sec) for each of the glass and carbon FRP/concrete bond were tested until failure.

An effective bond length of 69 mm and 76 mm was used for the GFRP and CFRP strips, respectively. Dr. Teng's formulation for bond strength and bond length (Teng et al 2001) verified that this nominal bond length would ensure bond failure in the GFRP and CFRP tests. Bond failure was desired to study both mode of failure and measured bond strength. The bond of FRP to the concrete substrate is of critical importance. There must be proper surface preparation of the beams. The method, described below, used to bond the FRP laminate to the concrete beams consisted of in-place bonding and curing of the GFRP and CFRP strips directly onto the concrete.

1.   Grind surface of concrete to an amplitude of 1.6 mm;
2.   Vacuum concrete surface to remove concrete dust;
3.   Air blast concrete surface to remove remaining dust;
4.   Fill in holes larger than 1.6 mm amplitude;
5.   Repeat steps 1 through 4 until concrete surface is smooth and dust free

Bond Test Results

Table 2 shows results of all three sets of carbon tensile tests (plain fabric, elastomer coated fabric, and interfacial bond). The tensile strength of the concrete beams was approximately 3.7 MPa. The measured bond strength of the samples was 3.7 MPa for the GFRP specimen and 4.1 MPa for the CFRP specimen when measured at a static rate. The fact that these strengths are so close confirms that concrete strength does control FRP/concrete interfacial bond strength. It is also observed in Table 2 that strain rate has an effect on the bond performance. When compared to the strength ratio vs. strain rate seen in the elastomer coated fabric strips, the effect of strain rate on strength for bond becomes evident. This is because bond is controlled by f'_c and f'_c is known to increase with increase of strain rate. Because the bond strength is controlled by concrete properties as opposed to fiber type, the data for all the bond tests were combined and an equation for the best fit line was determined (Fig 5). The ratio of dynamic strength/static strength in FRP/concrete bond was found to increase with 0.052Ln(strain rate) + 0.736.

Figures 6 and 7 show the typical modes of failure seen in the bond tests. The debonding at the interface between the concrete and the FRP plate in Fig. 6 is typical of the individual bond tests. It should be noticed that a thin layer of concrete adheres to the FRP. This can be seen in the close-up of the GFRP/concrete bond failure seen in Fig. 7. Most specimens failed by shear failure initiated in the concrete near the edge of the concrete beam. This failure zone spread quickly resulting in a sudden brittle debonding.

FRP RESTRAINERS

Results from the previous glass and carbon material tests and past steel and SMA restrainer experiments were used in design of the FRP restrainers. Three restrainer types were fabricated and tested under dynamic loading. These restrainers consisted of GFRP, CFRP, and a set of hybrid restrainers made of glass and carbon fibers. A GFRP restrainer has the advantage of maximum elongation until break (2.2%). A CFRP restrainer has greater rupture strength, and a hybrid composite restrainer has the potential of producing a system that is non-linear.

Restrainer Design Method

The premise for the design of the restrainers was force based. Because FRP is a linear material, the restrainers were designed to remain elastic. Restrainer demand was determined from earlier steel and SMA restrainer tests. A factor of safety of 2 was implemented to ensure elastic performance. Past research of bridge components strengthened using CFRP have indicated that in many cases strengthened specimens failed after the composites reached 50 to 65 % of the rupture strength of the laminates (Brena et al 2003 and Xiao et al 1997), and hence a factor of safety of 2 was believed to be reasonable.

The FRP restrainers are a two-component system. The area of the restrainer in the hinge region is flexible to allow movement in compression and tension while undergoing cyclic, dynamic motions. The restrainer ends are a stiff composite plate bonded and cured onto the concrete bridge specimen. The middle section of the restrainer in the hinge area is a flexible composite of fabric coated with SYLGARD 184 elastomer. Failure in bond is undesirable because bond failures are often brittle and occur with little or no visible warning, as observed in the FRP/concrete bond tests. The FRP restrainers were designed so if failure did occur, it would be in the flexible segment and not in the bond. The FRP/concrete bond was designed to have greater capacity than the flexible composite. An FRP plate, twice the size of the effective bond area, was used to ensure that bond failure would not occur. Research has shown that the longer the bond length, the larger the number of cycles to failure (Tan 2003). Research on the effects of adverse environmental conditions on strength capacity of externally bonded CFRP shows that the most significant reduction (33%) in strength was due to long-term exposure to 100% humidity. Less severe to the delamination process was dry heat, alkalinity, freezing and thawing, and salinity (Grace 2004). This long-term strength degradation of FRP/concrete bond from harmful weather conditions was taken into account in design of the restrainers. In calculating bond length, the compressive strength of the concrete in the bridge blocks was divided by two as an added factor of safety. Dr. Teng's bond strength model was incorporated into the design of the restrainers. An example of a final restrainer design is seen in Fig. 8.

The restrainer design methodology was used to design the glass, carbon and hybrid FRP restrainers. The glass restrainer was designed and tested first, followed by the carbon, and hybrid restrainer. The initial stage of construction of the restrainers began with the fabrication of the unidirectional piece of the restrainer, or the portion of the restrainer that contained the flexible composite section. This was a two-step process. It consisted of first coating the ends of a strip of fabric with resin. The ends were cured for approximately three days. Then the middle 10 in of the strip was coated with SYLGARD 184 (Fig. 9). This resulted in a unidirectional strip of fabric with stiff FRP composite ends and a flexible composite center (Fig 10). In the initial stage of restrainer bond strength design, Teng's model was used to determine if the restrainer segment would provide sufficient bond strength. However, the width and thickness of the restrainer was found to be insufficient. Enlarging the ends of the restrainer was necessary to produce the necessary bond capacity. The plate enlargement was determined by doubling the effective bond area, as determined in Teng's model. The FRP plate enlargement consisted of two pieces of 45 and 135-degree fabric bonded onto the ends of the restrainer segment consisting of both flexible and stiff composite. The completed bond area of the FRP restrainer met the requirement for adequate bond capacity.

EXPERIMENTAL PROGRAM

The testing program used in the FRP restrainer tests had been established in previous restrainer experiments conducted at the large-scale structures laboratory at UNR. Keeping the program the same as those of previous restrainer studies enabled comparisons among the performance of steel and SMA restrainers (Vlassis et al 2000, Johnson et al 2004, Camargo et al 2004) to that of the FRP restrainers. The test parameters used in the restrainer tests consisted of: (1) A restrainer slack of 12.7 mm; (2) An earthquake motion based on the response spectra ATC32E (Caltrans 1999 Seismic Design Criteria); (3) A period ratio between blocks of 0.6 that was determined to result in large out-of-phase motion.

Test Specimen and Instrumentation

The FRP restrainer experiments were performed on one of the 50-ton capacity biaxial shake tables in the structures lab. The test blocks and support bearings representing a typical in-span hinge were based on superstructure dimensions of representative CALTRANS bridges by Vlassis et al 2000. Each block of the test specimen, seen in Fig. 11, represent one bridge frame consisting of superstructure and columns. The specimen consisted of a light, 94 kN, and heavy, 125 kN block. The difference in weight between the two blocks was accomplished with the addition of lead bricks placed inside Block B. Two sets of elastomeric pads were used. They simulate the stiffness of the bridge superstructure. The ratio of periods between the two blocks was determined by the mass of the blocks and the stiffness of the pads.

Novotechnik LWG-225 linear extensometers were used to measure the relative hinge displacement between blocks. Unimeasure PA-40, 40-in string extensometers were used to measure absolute displacement between the blocks and a fixed frame surrounding the specimen. Crossbow CXLOZLF 1 $\pm$2g accelerometers were placed on the block and table to measure the acceleration of both the specimen and table. During the first FRP restrainer tests, three transducers, three accelerometers, and nine string extensometers measured the relative displacement, acceleration, and absolute displacement of the specimen. Additional instrumentation was added for the later tests to capture movement in the transverse direction.

Test Schedule

The FRP restrainer tests were conducted in three consecutive tests. The staggering of the restrainer test schedules allowed time for data analysis between constructions of the various restrainer types. The ATC32E-compatible synthetic record was used as the input earthquake motion for the GFRP and CFRP restrainer shake table tests. Unidirectional motion was used for all tests. Additional shake table tests in the transverse direction and an earthquake motion simulating 1994 Northridge – SYLMAR were utilized for the hybrid restrainer tests. The ATC 32E motion contains high-amplitude acceleration peaks that cause impact between the concrete blocks. The additional spectrum and motion were introduced during the last restrainer tests in hope of acquiring supplementary knowledge of FRP restrainer performance under a recorded earthquake motion as opposed to a synthetic one. The peak ground motion was increased from 0.05g to a maximum of 0.3g with incremental motion of 0.05g. The maximum peak ground acceleration of 0.3g was determined by the design displacement of the elastomeric pads attached between the blocks and the shake table.

In the GFRP restrainer tests, a PGA of 0.3g was achieved because the restrainers retained their integrity throughout the tests. For the CFRP restrainer tests, one of the restrainers ruptured during motion with a PGA of 0.2g (Fig. 12) and the testing was stopped. The hybrid FRP restrainers did not rupture during the tests of longitudinal movement. A maximum peak ground acceleration of 0.3g was achieved for both the ground motion produced by ATC 32E and that of the SYLMAR ground motion. For the hybrid restrainer, two tests were performed to examine performance in the transverse direction. Testing was discontinued due to debonding.

RESTRAINER TEST RESULTS

Shake table tests were carried out on the GFRP, CFRP and CGFRP (hybrid) restrainers to determine their dynamic performance. The test set-up seen in Fig. 11 was used for all three types of FRP restrainers. Block acceleration and restrainer elongation histories were measured every 0.01 seconds during dynamic tests. This provided the data necessary to produce acceleration and displacement histories for the FRP restrainers. Termination of the dynamic tests was either a result of restrainer failure or excessive displacement of the bearings that simulated substructure stiffness. Results of the GFRP, CFRP, and hybrid restrainers were compared. To evaluate the relative merit of different restrainer types, selected measured FRP data were compared with those of steel and SMA restrainers. Because the measured data for different FRP restrainer types were similar, the results for GFRP restrainers was assumed to represent the FRP restrainer responses and were used in this part of the study.

FRP Restrainers

Engagement of the FRP bridge restrainers during dynamic motion resulted in high restrainer forces and low relative displacements between blocks. The CFRP restrainer ruptured at a peak ground acceleration of 0.2g after reaching a maximum restrainer force of 400kN and restrainer elongation of 4.3mm. The glass restrainer was still intact at a PGA of 0.3g (Fig. 13). The GFRP restrainer tests were terminated due to large elastomeric pad displacements. The maximum restrainer forces achieved in the GFRP restrainer tests was 372 kN with a maximum restrainer elongation of 9.14 mm. The hybrid restrainer was strong and ductile enough to allow three series of tests. The first involved the same synthetic earthquake motion, ATC 32E, and direction, similar to those of the GFRP and CFRP restrainer tests. The hybrid restrainer performed very well under this motion, with a maximum restrainer force of 579 kN and maximum elongation of 3.71 mm at a PGA of 0.3g. No restrainer damage was observed at the end of the first series of hybrid tests. The second set of CGFRP restrainer tests involved longitudinal motion but with a real recorded earthquake motion, SYLMAR. The SYLMAR motion did result in comparable forces and restrainer elongation to those seen in the first series of hybrid tests but it resulted in less frequent restrainer engagement and block impact, seen in Fig. 14. The minimal restrainer elongation of this restrainer type can be noted in this elongation history.

Because ATC32E proved to be a more demanding earthquake motion, it was input into the shake table for the third set of CGFRP tests to explore the effect of transverse earthquakes on restrainers that are designed for longitudinal motions. At a PGA of 0.05g, the FRP plates of the restrainers began to unbond from the concrete blocks at the inner edge of the plates. The third set of hybrid restrainer tests ended at a PGA of 0.1g due to very large displacement of the base isolators in the transverse direction. Large FRP/concrete separation (Fig. 15) was seen. Motion in the transverse direction resulted in bond failure between the FRP plate of the restrainers and the concrete of the blocks. This is an unlikely scenario in an actual bridge, because bridges typically are equipped with transverse shear keys that prevent significant transverse relative displacements at the hinges. To improve on bond and to resist transverse movement after potential failure of shear keys, FRP anchors may be used.

Comparison of Response of FRP Restrainers with Steel and SMA Restrainers

The large discrepancy between restrainer force and maximum elongation of the FRP and other restrainer types can be noted in Table 3. The largest total restrainer force of 36 kN for steel restrainers can be compared to the largest calculated total restrainer force of 390 kN for the CGFRP restrainers. The maximum elongation of the steel cable restrainer was 38 mm, compared to the maximum GFRP restrainer elongation of 3.58 mm. Limited previous tests, conducted at UNR, to determine the performance of steel and SMA restrainers under dynamic movement have concluded that SMA is superior to steel in reducing relative displacement in the hinge area of bridges (Johnson et al 2004). The maximum displacement for blocks restrained by SMA restrainers was half that of the blocks restrained by steel restrainers. Figure 16 is a graphical presentation of the maximum elongation of the different restrainer types subjected to ATC32E, and a PGA of 0.15g. The large disparity between the maximum restrainer elongation of the FRP, SMA and steel restrainers can be observed.

RESTRAINER DESIGN APPROACH

A new design procedure for hinge restrainers was developed for design of seismic restrainers. It is a force-based design procedure, although it includes a displacement check to ensure that unseating is prevented. Several simple current methods were reviewed (CALTRANS 1990; AASHTO 1996; Trochalakis 1995; DesRoches and Fenves 1997). The use of the proposed new design method for FRP restrainer design is demonstrated.

The basic assumption in the proposed method is that FRP restrainers rigidly connect the two adjacent segments of the bridge. The linkage of the two bridge frames reduces the two-degree-of-freedom system of the bridge frames to a single-degree-of-freedom system (Fig. 17). This conversion of a 2 DOF system to a 1 DOF system makes this new method a quick, simple, and conservative design procedure. Compared to the force based AASHTO method, the proposed approach is more realistic because it accounts for the dynamic characteristics of the bridge, without introducing significant complication.

With the combined mass (m) and stiffness (k), the fundamental period of the system is calculated. Then, with the period, acceleration is determined from the response spectrum. Using the acceleration and the combined mass and stiffness, the displacement of the combined system is determined. The restrainer force is calculated by considering the free-body diagram of either one of the frames (Fig. 18).

New Design Method
The following steps are used in the new method:

1.   Determine the period of vibration for the combined system;
$T_{tot} = 2\pi * [(m_1+m_2) / (k_1+k_2) / m)]^{(0.5)}$
2.   Determine the spectral acceleration, S_a, using the period of vibration of the combined system and the acceleration spectrum for the bridge;
$S_a = S_{D1} / T_{tot}$
3.   Determine the displacement of the combined system;
$\Delta = [(m_1 + m_2) * S_a] / (k_1 + k_2)$
4.   Determine restrainer force using equilibrium;
$(m_1 * S_a) + F_r = R_1$
$(m_1 + S_a) + F_r = k_1 * \Delta$
$F_r = [k_1 * [(m_1 + m_2) * S_a] / (k_1 + k_2)] - (m_1 * S_a)$
$F_r = [[k_1 * [(m_1 + m_2) / (k_1 + k_2)] - m_1] * S_a$
5.   Design restrainer per side of bridge;
Required restrainer area = F_r/F_{yFRP}
6.   Check for unseating (CFRP has a maximum of 1.2% strain).
$\Delta_{r+s} = L_r * 1.2\%$ strain + L_s < Seat Width

The new design method was found to be most applicable to the uniqueness of FRP but it is also a simple formulation for other restrainer types. Like the force-based AASHTO method, it is easily adapted for the design of FRP restrainers but incorporation of the structural response into the formulation of the new method eliminates the flaws found in the AASHTO method.

CONCLUSIONS

The primary objective of this study was to evaluate the use of fiber reinforced plastic (FRP) fabrics as restrainers in the seismic rehabilitation of highway bridges. Both glass and carbon fibers were included in the study. Multiple stages of investigation led to the development of this new type of restrainer. The main general observations and conclusions derived from this study are outlined below:

1.   All three types of FRP restrainers showed good performance during shake table tests at limiting relative hinge displacements between adjacent blocks and producing lower block accelerations than traditional restrainers. FRP restrainers show promise as a seismic restraining device for bridges;

2.   FRP strength was insensitive to strain rate (dynamic loading). This is because unlike homogeneous materials, there is internal slippage of the fibers and the slippage is insensitive to strain rate;

3.   The method used to make a flexible restrainer using an elastomeric material was effective;

4.   The method used for the bond design and attachment of the restrainers was successful in achieving design bond strength;

5.   The new restrainer design method produced realistic restrainer demand by incorporating the uniqueness of FRP into the procedure;

6.   The performance of restrainers under a simulated recorded earthquake was similar to that of artificial earthquakes;

7.   Transverse motion produced out-of-plane motion that can debond the restrainer. Transverse shear keys need to be sufficiently strong to prevent large transverse movement effect on FRP restrainers;

8.   FRP/concrete bond strength is affected by strain rate because of its dependence to the concrete shear strength.

9.   Comparisons among FRP restrainers and steel and SMA restrainers show a minimal elongation in FRP compared to that seen in the steel and SMA restrainers under an identical test set-up.

NOTATION

f'_c =compressive strength of concrete
F_r =restrainer force
F_{yFRP} =manufacturers design strength of fiber reinforced polymer
g =gravity
Ln =natural log
L_r = predetermined length of flexible FRP

L_s = restrainer slack (typically 12.7 mm)
R_1 =restoring force
S_a =spectral acceleration
S_{DI} =design spectrum acceleration for long periods
T_{tot} =period of vibration of the total system
Δ_{r+s} =restrainer elongation + restrainer slack
$\mu\varepsilon/sec$ = micro strains per second

REFERENCES

American Association of State Highway and Transportation Officials (AASHTO), (1996), "Standard Specifications for Highway Bridges, American Association of State Highway and Transportation Officials, Sixteenth Edition, 1996.

Brena, S. F., Wood, S. L. and Kreger, M. E., "Full-Scale Tests of Bridge Components Strengthened Using Carbon Fiber-Reinforced Polymer Composites", *ACI Structural Journal*, pp 775-794, Nov-Dec 2003.

Caltrans, Bridge Design Aids, Caifornia Department of Transportation, 1990.

Caltrans (California Department of Transportation), "Caltrans Seismic Design Criteria Version 1.2," Engineering Service Center, Earthquake Engineering Branch, California, December 1999.

Camargo-Sanchez, F., Maragakis, E. M, Saiidi, M. S., Elfass, S., "Seismic Performance of Bridge Restrainers at In-Span Hinges", Civil Engineering Department, Report N. *CCEER 04-04*, University of Nevada Reno, 2004.

DesRoches, R., and Fenves, G. L., "New Design and Analysis Procedures for Intermediate Hinges in Multiple-Frame Bridges", *EERC-97/12,* College of Engineering, University of California, Berkeley, December 1997.

Grace, Nabil, F., "Concrete Repair with CFRP", *Concrete International Journal,* 26, 5, pp 500-509, April 2004.

Johnson, R. M., Maragakis, E. M., Saiidi, M. S., DesRoches, R., Padgett, J., "Experimental Evaluation of Seismic Performance of SMA Bridge Restrainers", CCEER 04-2, Department of Civil Engineering, University of Nevada, Reno, NV 2004.

Kulkarni, S. M. and Shah, S. P., "Response of Reinforced Concrete Beams at High Strain Rates," *ACI Structural Journal*, 95 (6), 705-715 (1998).

Luo, S. Y., and Mitra, A., "Finite Elastic Behavior of Flexible Fabric Composite under Biaxial Loading," *Journal of Applied Mechanics*, 66, 631-638, September 1999.

70 Saiidi et al.

Paulay, T., and Priestley, M.J.N., Seismic Design of Reinforced Concrete and Masonry Buildings, John Wiley and Sons, USA, 1992.

Saiidi, M., Randall, E. Maragakis, and T. Isakovic, "Seismic Restrainer Design Methods for Simply-Supported Bridges," *Journal of Bridge Engineering*, ASCE, Vol. 6, No. 5, September/October 2001, pp. 307-315.

Tan, K. H., "Effect of Cyclic Loading on FRP-Concrete Interfacial Bond Strength," *International Symposium on Latest Achievement of Technology and Research on Retrofitting Concrete Structures*, Japan Concrete Institute, July 2003.

Teng, J. G., Chen, J. F., Smith, S. T., Lam, L., FRP Strengthened RC Structures, John Wiley & Sons, Ltd., pp 20-27, 2001.

Trochalakis, P., Eberhard, M. O., and Stanton, J. F., "Evaluation and Design of Seismic Restrainers for In-Span Hinges," Report No. WA-RD 387.1, Washington State Transportation Center, Seattle, WA, August 1995.

Vlassis, A. G., Maragakis, E. M., Saiidi, M. S., "Experimental Evaluation of Seismic Performance of Bridge Restrainers", MCEER 00-0012, Department of Civil Engineering, University of Nevada, Reno, NV 2000.

Xiao,Y., and Wu, H., "Compressive Behavior of Concrete Stub Columns Confined by Carbon Fiber Jackets", 13th U.S.-Japan Bridge Engineering Workshop, Japan, 1997.

Table 1 -- Glass and Carbon Composite Properties

	Fiberglass Composite	
	Design	Specified
Ultimate Tensile Strength in Primary Direction	460 MPa 0.58 kN/mm width	575 MPa 0.75 kN/mm width
Elongation at Break	2.2%	2.2%
Tensile Modulus	20.9 Gpa	26.1 Gpa
Laminate Thickness	1.3 mm	1.3 mm
Primary Fiber	Glass	

	Carbon Fiber Composite	
	Design	Specified
Ultimate Tensile Strength in Primary Direction	745 MPa 0.75 kN/mm width	876 MPa 0.89 kN/mm width
Elongation at Break	1.2%	1.2%
Tensile Modulus	61.5 Gpa	72.4 Gpa
Laminate Thickness	1.0 mm	1.0 mm
Primary Fiber	Carbon	

Table 2 – Comparisons of 3 Sets of CFRP Tensile Tests

Rate	25.4 mm Fabric Strips				25.4 mm Elastomer Coated Fabric Strips				Carbon Bond Tests (25.4 mm strips)			
	Strip #	Max Force	Strength	Ave Stress	Strip #	Max Force	Strength	Ave Stress	Strip #	Max Force	Max Ave Bond Stress (Strength)	Ave Bond Strength
(με/sec)		(kN)	(MPa)	(MPa)		(kN)	(MPa)	(MPa)		(kN)	(MPa)	(MPa)
Static	1	11.73	469		1	20.92	824					
	2	7.38	295		2	19.32	761		3	8.72	4.50	
	3	15.12	610		3	18.59	732		4	6.67	3.45	
	4	10.91	441	454	4	18.99	748	766	5	8.18	4.23	4.06
5000	8	14.66	586		8	22.59	889		9	9.34	4.83	
	9	13.74	607		9	20.21	796		10	10.85	5.61	
	10	12.43	496	563	10	17.84	702	796	11	7.65	3.95	4.80
10000	11	14.27	572		11	18.61	733				0.00	
	12	13.91	531		12	19.52	769		12	12.01	6.21	
	13	13.31	496		13	17.50	689		13	8.81	4.55	
				533	14	16.49	649	710	14	9.88	5.10	5.29
50000	14	11.82	448		15	18.82	741		15	8.54	4.41	
	15	14.27	510		16	17.26	679		16	7.25	3.75	
	16	11.26	496	485	17	20.24	797	739	17	13.34	6.89	5.02
100000	17	13.28	517		18	17.05	671		18	9.61	4.96	
	18	14.11	600		19	17.09	673		19	8.85	4.57	
	19	16.12	586	567	20	16.24	639	661	20	10.28	5.31	4.95

Table 3 – Total Force vs. Displacement for Different Restrainers, Shake Table Tests

Restrainer Types	Steel Cable		SMA Cable		Glass FRP		Carbon FRP		Hybrid FRP		Hybrid FRP	
							(elastomer coated portion)					
Total Stiffness	2.8 kN/mm		2.8 kN/mm		163 kN/mm		249 kN/mm		329 kN/mm		329 kN/mm	
# of Restrainers	5		two 130-wire		2		2		2		2	
Slack (in)	0.5		0.5		0.43		0.43		0.43		0.43	
Earthquake Motion	ATC32E		ATC32E		ATC32E		ATC32E		ATC32E		Sylmar	
					Max		Max		Max		Max	
		Max		Max	Calc	Max	Calc	Max	Calc	Max	Calc	Max
	Total Force	Max Elong	Total Force	Ave. Elong	Total Force	Ave. Elong	Total Force	Ave. Elong	Total Force	Max Elong	Total Force	Ave. Elong
PGA (g)	(kN)	(mm)	(kN)	(mm)	(kN)	(mm)	(kN)	(mm)	(kN)	(mm)	(kN)	(in)
0.15	30	38.0	32	19.1	298	3.581	400	3.531	313	1.905	433	2.642
0.2	36	99.1 (Failure)	34	22.9	325	4.216	307	3.404 (Failure)	390	2.362		

Figure 1 – Material Tensile Tests

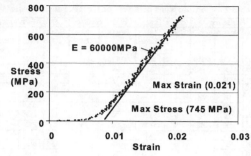

Figure 2 – Stress vs. Strain, Elastomer Coated Carbon Strip

Figure 3 — Single Lap Test

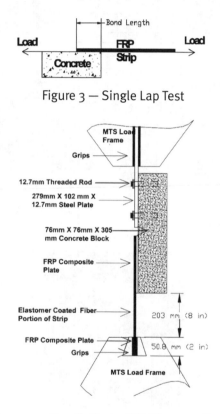

Figure 4 — Tensile Tests, FRP/Concrete Bond

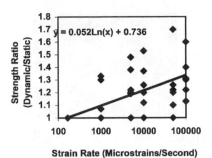

Strain Rate (Microstrains/Second)

Figure 5 – Best Line Fit of Strength Ratio Vs. Strain Rate, FRP/Concrete Bond

Figure 6 — Bond Failure

Figure 7 — Close-up of Bond Failure

Carbon Restrainer

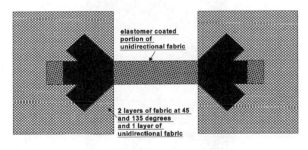

Figure 8 – Completed CFRP Restrainer Showing Effective Bond Area in Black

Figure 9 – Coating Unidirectional Segment of CGFRP Restrainer with Elastomer

Figure 10 — Illustration of Unidirectional Section of GFRP Restrainer

Figure 11 — Test Set-Up for FRP Bridge Restrainer Shake Table Experiments

Figure 12 — Rupture of CFRP Restrainers at 0.57 X ATC32E (0.2g)

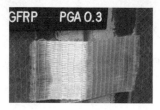

Figure 13 — GFRP Restrainers at a PGA of 0.3g

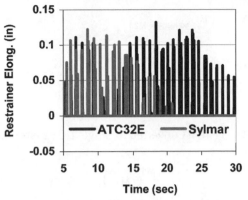

Figure 14 — CGFRP Restrainer Elongation History, PGA = 0.2g

Figure 15 — Debonding of GFRP Restrainer and Concrete Transverse Motion, 0.1g

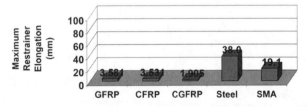

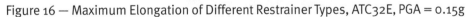

Figure 16 — Maximum Elongation of Different Restrainer Types, ATC32E, PGA = 0.15g

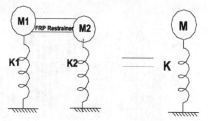

Figure 17 — Two DOF System Converted to Single DOF System with FRP Restrainer

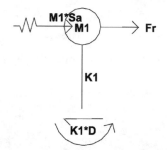

Figure 18 — Equilibrium of Forces for System with FRP Restrainers

Manufacturing, Durability and Bond Behavior of FRP Bars with Nanoclay

by P.V. Vijay, H.V.S. GangaRao, and V. Krishnaswamy

<u>Synopsis:</u> This paper focuses on the behavior of Glass Fiber Reinforced Polymer (GFRP) bars manufactured with nanoclay. Benefits of nanoclay in the resin for manufacturing FRP composites include reduced moisture ingress, better fire resistance, and enhanced resin stiffness. Several molds were developed at CFC-WVU laboratory to produce glass FRP bars with smooth surface, and crescent shaped or circular lugs. Bars were manufactured at room temperature by manual pultrusion process with varying fiber volume content and percentage of nanoclay. Vinyl ester resin exfoliated with nanoclay was used to manufacture GFRP bars with a fiber volume fraction of 44.7% and tested for tensile, bond and shear strength.

Bars manufactured in this research and tested in tension had typical ultimate stress of 108.76 ksi (749.87 MPa) 133.43 ksi (919.96 MPa) and stiffness of 4.18 msi (28.8 GPa) - 4.67 msi (32.2 GPa). Pullout tests conducted on a bar with crescent-shaped lugs and circular lugs embedded in concrete provided a maximum bond stress value of 2385 psi (16.44 MPa) and 1762.77 psi (12.15 MPa), respectively. Bars with nanoclay showed 8.3% less shear strength than those without nanoclay (24.10 ksi (166.16 MPa) vs. 26.27 ksi (181.13 MPa)). Bars with nanoclay absorbed more moisture initially than bars without nanoclay, and the moisture absorption decreased with time.

<u>Keywords:</u> bond; durability; FRP; moisture; nanoclay; pullout; pultrusion; shear; tension

Dr. P.V.Vijay is a Research Assistant Professor, Constructed Facilities Center (CFC), Department of Civil and Environmental Engineering, WVU, Morgantown, WV 26506.

Dr. Hota V.S. GangaRao is the Director and Professor, CFC, Dept. of Civil and Environmental Engineering, WVU, Morgantown, WV 26506.

V. Krishnaswamy is a graduate Research Assistant, Department of Civil and Environmental engineering, College of Engineering and Mineral Resources, West Virginia University, Morgantown, WV 26506.

INTRODUCTION

Fiber Reinforced Polymer (FRP) bars are used as reinforcement for concrete structures in which corrosion protection, magnetic transparency and electrical non-conductivity are of primary concern. Glass FRP materials are corrosion resistant, electrically and magnetically non-conductive, and exhibit several other advantages that make them suitable as reinforcement for concrete structures (ACI 440.1R-03, Mallick 1993). However, moisture ingress in concrete with alkaline environment is detrimental to mechanical properties of FRP reinforcement due to chemical reaction between silica in glass and alkaline ions. Moisture with alkalinity produces matrix softening and fiber embrittlement and damage the fiber-resin interface due to chemical attack and growth of hydration leading to loss of tensile and inter-laminar shear strength properties (Benmokrane et al. 2002). Currently, nanocomposites are being researched to improve their durability (Shah, 2002).

Nanocomposites are prepared by dispersing nanoclay into a polymer, generally at less than 5 wt% levels. Well-dispersed and exfoliated clay platelets with dimensions of the individual platelets being in nanoscale in one direction, are expected to improve the properties of composites. Exfoliation is facilitated by surface compatibilization chemistry, where individual platelets are separated from another by mechanical shear or heat of polymerization (Shah, 2002).

At present, FRP bar manufacturers are not utilizing nanoclay in their mass production process. Utilization of nanoclay in FRP bar manufacturing requires an in-depth understanding of exfoliation procedures, optimum level of nanoclay, resin-clay chemistry, including their interfacial properties, and resin viscosity. Nanoclay is hydrophilic in nature with one of the three dimensions smaller than 100 nanometers (Southern Clay products Inc., Shah, 2002). Though an initial moisture pickup is noted with the nanoclay, nanoclay provides tortuous path for moisture ingress into the FRP bar and acts as a barrier against moisture movement from surface to the core. Addition of nanoclay is also known to improve flammability resistance and resin stiffness (Hay and Shaw 2000). In this research, Glass Fiber Reinforced Polymer (GFRP) bars were manufactured with nanoclay dispersed in the vinyl ester resin. Tension, bond, shear and moisture absorption tests were conducted on GFRP bars with and without nanoclay.

Manufacturing GFRP bars with nanoclay in the Constructed Facilities Center of West Virginia University (CFC-WVU) laboratory became a necessity to achieve the research objectives with several trials of manufacturing the standardized products at a limited scale as opposed to mass manufacturing different types of bars. Significant amount of planning, revisions, and refinements were made to manufacture GFRP bars with superior properties and performance using durable vinyl ester resins and nanoclay at room temperature in laboratory conditions.

OBJECTIVES

Objectives of this study are to:
1) Manufacture GFRP bars with circumferential lugs and uniform surface finish using pulley assisted manual pultrusion process.
2) Evaluate and compare strength and stiffness of GFRP bars with and without nanoclay.
3) Evaluate bond strength between concrete and GFRP bars with nanoclay.
4) Evaluate shear strength of GFRP bars with nanoclay.
5) Compare the moisture absorption characteristics of GFRP bars with and without nanoclay.

SCOPE

Over 40 trial runs were conducted to select optimum resin chemistry, proper fiber volume fraction, fiber placement, fiber wetting, shape of lugs, reduction in void content, percentage nanoclay, mold details, and bar dimensions. GFRP bars with # 4 diameter were manufactured using nanoclay and tested as described below:
1. Tension test: This test comprised of monotonic tensile loading up to failure for establishing ultimate strength and stiffness of the bar using strain gage and data acquisition system. Tests were conducted on 20 bars with various resin chemistries, fiber content and surface configuration. Results of tests on bars developed during this research are limited to specimens after the manufacturing procedure was standardized.
2. Bond test: Cylinder pullout tests on 6 in (150 mm) and 12 in (300 mm) high specimens with embedded FRP bars were performed to evaluate the bond strength between FRP bar and concrete with and without nanoclay in 7 specimens. Slip measurements at the unloading end were noted.
3. Moisture absorption: About 20 GFRP bars with and without nanoclay of 3 in (76.2 mm) length were sealed at the ends with thin resin coatings and immersed in water and monitored for weight gain for a duration of 300 days and beyond.
4. Shear test: Double shear tests were conducted on GFRP bars with and without nanoclay on 11 bars of 6 in (150 mm) length using ½ in (12.75 mm) wide cutting tool.

Details of FRP bar constituents such as resins, fibers, nanoclay including bar manufacturing methods and tests conducted on those bars are described in the following section.

FRP BAR CONSTITUENT MATERIALS

Fibers
E-glass fibers with 113 yield were used to make GFRP bars. Up to 36 glass fiber rovings were used to manufacture FRP bars.

Resin
Vinyl ester resin supplied by Ashland chemicals was used to make FRP bars. Vinyl ester resins have following advantages:
1. Corrosion resistance to a wide range of acids, bases, chlorides, solvents and oxidizers.
2. Excellent corrosion resistance.
3. Electrical and thermal insulation.

Catalyst
Methyl Ethyl Ketone Peroxide (MEKP) was used as a catalyst to promote curing at room temperature.

Nanoclay
Nanoclay (20Å) is a natural montmorillonite modified with a quaternary ammonium salt, of which about 2%-4% was added to the resin and mixed for an extended duration. Care was taken to remove air bubbles formed during mixing by vacuuming.

Nanocomposites exhibit improvement in barrier, flame resistance, structural, and thermal properties without significant loss in impact strength. Because dimensions of the individual platelets are in nanoscale in one direction, exfoliated nanomers are transparent in most polymer systems. However, when surface dimensions extend to 1 micron, the tightly bound structure in a polymer matrix is impermeable to gases and liquids, and offers superior barrier properties. Nanocomposites also demonstrate enhanced fire resistance properties and are finding increasing use in engineering plastics. With proper choice of compatibilizing chemistries, the nanometer-sized clay platelets offer high potential for enhancing durability of composites (Southern Clay Products Inc.).

EQUIPMENT AND ACCESSORIES FOR MANUFACTURING AND TESTING

Mold for manufacturing FRP bars with nanoclay
Aluminum mold was manufactured at CFC-WVU laboratory to manufacture 4 ft (1.22 m) long FRP bar. The mold was attached with a pulley at one end to facilitate pulling of fibers in a slow, yet uniform manner. Crescent shaped lugs were created in one set of molds, whereas another set of molds consisted of circular lugs. Holes were drilled on top surface of the mold to inject additional resin into the mold. For guiding glass

fibers into the mold, Teflon plates of ½ in (12.7 mm) thickness with circular drilled holes were used. Teflon plates decrease the cling between glass fiber rovings as each roving passes through a separate hole in the Teflon plate and minimizes friction between plate hole surface and glass fibers. Holes were drilled such that the Teflon plates allow fibers to enter resin bath at an angle, stay horizontal for some distance for resin wetting, and exit out at an angle and into the mold inlet (Figure 1).

Longer mold was also manufactured to make bars of 12 ft (3.65 m) length and ½ in (12.7 mm) diameter with ribs of uniform radius as compared to the non-uniform ribs of the 4 ft (1.22 m) mold. Procedure for manufacturing the GFRP bar remained same as the old mold except a cone of 1½ in (38.1 mm) outer diameter; ½ in (12.7 mm) inner diameter and 2 in (50.8 mm) length was attached to the mold to guide fibers into the mold. Use of guiding cone helped reduce abrasion at fiber entry into the mold.

Grips

Schedule 80 steel pipes were used as grips for tension test and bond test. Split pipes of 8 in (203.2 mm) length were used as grips and attached to bars using Pliogrip™. Grips with adhesives were bonded to test specimens (GFRP bars) using C-clamps and cured for 24hrs (Figures 2 and 3).

Manufacturing process

Vinyl ester polymer was mixed with about 2%-4% of nanoclay for a duration exceeding 60 minutes using a shear mixer. After mixing, vacuum was applied to remove air bubbles formation. During manufacturing Methyl Ethyl Ketone Peroxide catalyst was used to obtain initial setting.

Mold was coated with mold-releasing agents. Glass fibers were passed through teflon plates and then through the mold. After passing through the mold, the fiber rovings were tied to an attachment in the pulley drum. Teflon plate assembly with fibers was placed inside a resin tray to allow fiber wetting. Vinyl ester resin with catalyst was transferred to the tray, and glass fibers were dipped in the resin prior to pulling. Slow pulley rotation (rate of pulling was 1 ft/min (0.3 m/min)) provided adequate wetting. Pulling was continued until wet fibers were pulled out at other end of the mold. After pulling, resin was poured through vertical holes on top of the mold to obtain good rib formation and finish. The vertical top holes in the mold reached rib channel that connected all ribs and helped fill resin in the mold. Demolding was carried out after 24 hours of room temperature curing. Bar and the ribs were well formed with proper resin wetting during this manufacturing process (Figure 4).

TEST SETUP AND TEST PROCEDURES

Tensile test setup and test procedure

GFRP bars were tested for tension using a Universal Testing Machine (UTM) having maximum load capacity of 200 kips (890 kN). GFRP bar grips consisted of schedule 80 split steel pipes of 8 in (203.2 mm) length at both ends. Splits in both end grips were aligned on the same line during grip installation (Figure 3). Bars were attached with grips and cured for 24 hours. Bar was minimally sanded using a sand paper for ½ in (12.7 mm) length in the center to facilitate strain gage placement. Sanding was carried out carefully so that no glass fibers were damaged.

Strain gauge and load cell were attached to data acquisition system. From the recorded strain and load values, Young's Modulus and maximum stress were calculated. Failure was reached in about 2 minutes (ASTM D3916-02).

Bond test setup and test procedure

Bond specimens were prepared for cylinder pullout tests with cylinder size of (6 in or 150 mm x 12 in or 300 mm). In order to control the embedment length within the cylinder, the bars were wrapped two with bond breakers, which are made of soft plastic foam placed around the rod to prevent contact between the rod and concrete. Bond breakers were wrapped around bars for a distance of 4½ in (114.3 mm) each on loaded and unloaded end providing 3 in (76.2 mm) of bond length (Figure 6). A thin coat of oil was applied on the plastic foam to make sure that no bond developed between bond breakers and concrete. Specimens were cured for 28 days and bar was attached with a grip on the pulling end with schedule 80 steel pipe of 8 in (203.2 mm) length using Pliogrip™ and was cured for 24hrs.

Bond specimens were tested using UTM (Figure 7). GFRP bar in bond specimen was attached with LVDT at its bottom to measure the slip between bar and concrete. LVDT and UTM were connected to a data acquisition system to record the values. Bond stress of a bar and bond stress-slip diagram were obtained from the recorded data. Average rate of loading for pullout test was about 15 psi (103.42 kPa) of bond stress per second.

Shear test and test procedure

GFRP bars with and without nanoclay were cut to a length of 6 in (152.4 mm) each and double shear tests were conducted. Shear test was conducted using ½ in (12.7 mm) wide cutting tool (Figure 9). Bars were positioned in the shear testing setup and anchored using bolts. Specimen was mounted such that no gap was visible between the bar and the cutting tool. Rate of loading for the shear specimen was 5-10 ksi (34.47-68.94 MPa) per minute. Shear test was conducted on bars with and without nanoclay (ASTM D4476-03).

Moisture test setup and test procedure

Moisture content of GFRP bars with and without nanoclay was compared to determine the difference in moisture absorption. FRP bars were cut to 3 in (76.2 mm) length and both ends were sealed and cured with resin to prevent water entry through sides. Bar surface and sides were examined for presence of resin flakes, which may result in incorrect calculation of moisture content. Bars were numbered (Figure 10) and their dry weights were noted before immersing them in a container having potable water. Bars were left undisturbed until a reading was taken (Figure 11). Water immerse specimens were cleaned with paper towels to absorb surface moisture and weighed regularly for an accuracy of 4 decimals. Moisture uptake data with time were plotted to compare percentage moisture absorption in GFRP bars with and without nanoclay. Weights were taken once a day for first week and time interval for measurements was increased during subsequent weeks.

TEST RESULTS AND DISCUSSIONS

Tension test results
Details of GFRP bars manufactured using 4 ft (1.22 m) and 12 ft (3.66 m) long molds and tested in this research are shown in Table 1. Results of smooth bars manufactured during preliminary trials are not included.

The GFRP bars with and without nanoclay exhibited consistent linear-elastic stress-strain behavior until failure. Failure locations on all the bars were noted to be away from the grip (Figure 13). Slope of the stress-strain graph (Figure 14) was used to find modulus of elasticity of the GFRP bar. Tensile strength was calculated using simple Equation (1).

$$\sigma = \frac{F}{S} \tag{1}$$

where,
σ = Tensile stress (ksi)
F = Load at which the stress is being calculated (kip)
S = Cross sectional area of the test (in^2)

Strength and stiffness discussion
4 ft (1.22 m) long mold-- Strength and stiffness results of GFRP bars manufactured with 4ft (1.22 m) long mold are summarized here.
1. # 4 GFRP bars without nanoclay with a fiber volume fraction of 44.7% (32 glass roving) were found to have a maximum tensile stress of 129.87 ksi (895.42 MPa) and a maximum stiffness of 4.32 msi (29.8 GPa). However, average stiffness was 4.24 msi (29.2 GPa).
2. # 4 bars with 2% nanoclay using 32 glass rovings were found to have a maximum failure tensile stress values of 113.57 ksi (783.03 MPa) and 114.08 ksi (786.55 MPa) and a stiffness of 4.28 msi (29.5 GPa) and 4.24 msi (29.2 GPa), respectively.
3. # 4 bars with 2% nanoclay using 36 glass rovings were found to have a maximum failure tensile stress value of 108.76 ksi (749.87 MPa) and a stiffness of 4.75 msi (32.8 GPa). Reduction in strength with increased rovings is due to interfacial bond strength reduction between fiber and resin due to presence of nanoclay particles. Some fiber twisting was noted during pulling process including increases friction at the mold entrance. However, stiffness increase in the bar was consistent with higher number of rovings and addition of nanoclay.

12 ft (3.66 m) long mold-- Strength and stiffness results of GFRP bars manufactured with 12ft (3.66 m) long mold are summarized here.
1. # 4 bars without nanoclay using 32 glass rovings were found to have a

maximum failure tensile stress value of 122.32 ksi (843.37 MPa) and a stiffness of 4.18 msi (28.8 GPa).

2. # 4 bars with 4% nanoclay using 32 glass rovings were found to have a maximum failure tensile stress values of 106.72 ksi (735.81 MPa) and a stiffness of 4.47 msi (30.8 GPa).

3. # 4 bars without nanoclay with fiber volume fraction of 44.7% were found to have a average maximum tensile stress of 126.22 ksi (870.26 MPa) and a stiffness of 4.24 msi (29.2 GPa).

4. # 4 bars with nanoclay with fiber volume fraction of 44.7% were found to have a average maximum tensile stress of 110.78 ksi (763.80 MPa) and a stiffness of 4.33 msi (29.9 GPa) for a nanoclay range from 2 to 4%.

5. # 4 bars with nanoclay showed a reduction in average maximum tensile stress of 110.78 ksi (763.80 MPa) compared to bars without nanoclay having average maximum tensile stress of 126.22 ksi (870.26 MPa). Reduction of tensile stress was due to exfoliation of nanoclay in resin, which reduced interfacial bond strength development and an increased resin viscosity leading to reduced fiber wetting.

6. # 4 bars with nanoclay showed a higher average stiffness of 4.33 msi (29.9 GPa) compared to bars without nanoclay having an average stiffness of 4.24 msi (29.2 GPa).

7. # 4 bars with 4% nanoclay showed a higher stiffness of 4.47 msi (30.8 GPa) compared to bars with 2% nanoclay having average stiffness of 4.26 msi (29.4 GPa). It should be noted that research is being conducted.

Commercially manufactured # 4 bars have typical tensile strength of 90-100 ksi (620.52 – 689.47 MPa) and stiffness of 5.25 to 5.5 msi (36.2 to 37.9 GPa) for 60% or higher fiber volume fraction. # 4 bars manufactured at CFC-WVU with and without nanoclay have strength of 106.72-133.43 ksi (735.81 – 919.97 MPa) with stiffness of 4.18 to 4.47 msi (28.8 to 30.8 GPa) for 44.7% fiber volume fraction.

Better strength in CFC manufactured bars is attributed to room temperature processing with following advantages.

- Longer curing time of 2 hours and demolding time of 12 hours as compared to fraction of a minute curing in commercial process.
- Fiber pulling is gentle in CFC procedure with high resin wetting and reduced abrasion between fibers.
- Additional resin saturation through vertical mold ports following pulling operation.

Bond test results
Bond test results for bar with crescent-shaped lugs and circular lugs with nanoclay are shown in Table 2. All bond test samples failed due to slipping of the GFPR bar inside the concrete (Figure 8). Bond stress was calculated using simple equation (2).

$$\tau = \frac{F}{S} \tag{2}$$

where,

    τ = Bond stress (ksi)
    F = Load at which bond failure occurs (kip)
    S = Surface area of the GFRP bar in contact with the concrete (in^2)

$$S = \pi d_x l_x \tag{3}$$

    d_x = Diameter of GFRP bar (in)
    l_x = Length embedment (in)

Bond test discussions--
1. Bars with crescent shaped lugs and nanoclay were found to have a maximum bond stress value of 2385 psi (16.44 MPa) when tested along stronger bond direction. Bars with circular shaped lugs and nanoclay were found to have an average maximum bond stress value of 1762.77 psi (12.15 MPa).
2. It should be noted that other FRP bars commercially manufactured with sand coating and ribbed surface with nanoclay were found to have maximum bond stress of 1900 - 2200 psi (13.10 – 15.67 MPa) (Vijay and GangaRao 1999). Bond stress is noted to be comparable to or better than that of steel.

Shear test results

Details of GFRP bars manufactured and tested in shear are shown in Table 3. All specimens failed at the shearing edges of the cutting tool. Shear stress was calculated using simple Equation (4).

$$\tau = \frac{P}{nA} \tag{4}$$

where,

    τ = Shear stress (ksi)
    P = Shear failure load (kip)
    A = Cross-sectional area of the test specimen (in^2)
    n = 1 for single shear test and 2 for double shear test

Shear test discussion--# 4 bars having a fiber volume fiber volume fraction of 44.7% (32 glass roving) with nanoclay showed 8.3 % reduction in shear stress (24.10 ksi (166.16 MPa) vs. 26.27 ksi (181.13 MPa)) as compared to bars without nanoclay. Reduction of shear strength in bars with nanoclay is attributed to the exfoliation of nanoclay in resin, which increases resin viscosity and creates multiple interfaces between fibers, resin and glass, leading to reduction in shear strength.

Moisture absorption test and discussions

GFRP bar without nanoclay-- Nanoclay is a hydrophilic material, which absorbs moisture initially and through exfoliation acts as a barrier against moisture movement into the bar core. Bar without nanoclay was cut to obtain 10 pieces of 3 in (76.2 mm) length and ends were coated with a thin layer of vinyl ester resin. Dry weights of each sample were taken before immersing in water. After 24 hrs, samples were taken out, wiped off clean and weighed. Samples were weighed once in a day during first week and once in two days during second week and at regular intervals of 1-2 weeks thereafter.

GFRP bar with nanoclay-- Bar with vinyl ester resin and 2% nanoclay was cut to obtain 10 pieces of 3 in (76.2 mm) long specimens whose ends were coated with a thin layer of vinyl ester resin and cured. Dry weights of the samples were taken before immersing in water. After 24 hrs, samples were taken out, wiped off clean and weighed. Samples were weighed once in a day during first week and once in two days during second week and at regular weekly intervals thereafter.

Long-term moisture absorption--
1. Moisture absorption tests were conducted on 3 in (76.2 mm) long GFRP specimens with sealed ends. Initial set of bars with high void content picked up moisture content in excess of 1% within few days. Manufacturing process was modified to eliminate presence of air voids.
2. Nanoclays are hydrophilic in nature and hence bars with nanoclay showed higher moisture absorption initially for a duration of 45 days. However, after 45 days, bars with nanoclay showed lower moisture ingress.
3. GFRP bars showed moisture gain of 1.2181 % in 283 days without nanoclay and 1.0631 % in 283 days with nanoclay.
4. GFRP bars showed moisture gain of 0.9440 % in 93 days without nanoclay and 0.9143 % in 93 days with nanoclay. This trend needs to be further researched for a long-term duration under sustained stress to identify equilibrium levels of moisture content.
5. It is evident from Figure 16 that bar with nanoclay absorbed more moisture initially and the moisture gain decreased with time. However, further research will be conducted on the moisture uptake with new set of bars having higher nanoclay content.

Manufacturing issues
Fiber volume fraction-- Fibers are the main element that provides high strength and stiffness to a composite. The fiber volume of a composite material can be determined by resin burn-off method (ASTM D2584-94) in which the matrix is burned off and fibers are weighed to find the fiber volume fraction. FRP bar specimen was weighed and placed in a weighed crucible and burned in a 600°C muffle furnace in air until only the fiber remained. The crucible was cooled and weighed. By knowing the weight and density of fiber, fiber volume fraction in the bar was calculated.

Portion of a GFRP bar measuring 2 in (50.8 mm) long was cut and its volume was measured by water displacement method and weighed to a precision of 4 decimals (Figure 17). GFRP bar was kept in furnace for 5hr at a temperature of 600 °F. Burned sample was weighed and fiber volume fraction was determined (Figure 18) as explained below:

Weight of crucible (w_1)	= 42.6628 g
Weight of crucible + GFRP bar (w_2)	= 53.6380 g
Weight of GFRP bar (w_3)	= $w_2 - w_1$

	$= 10.9752$ g
Volume of GFRP bar (v_1)	$= 5000$ mm^3
Weight of crucible + glass fibers (w_4)	$= 48.3657$
Density of glass fiber (d_g)	$= 2.549 \times 10^{-3}$ g/mm^3

$$\text{Volume of glass fiber } (v_2) = \frac{w_4 - w_1}{d_g} \tag{5}$$

$$= 2237.3 \text{cm}^3$$

$$\text{Fiber volume fraction} = \frac{v_2}{v_1} \tag{6}$$

$$= 44.7\%$$

$$\text{Fiber weight fraction} = \frac{w_4 - w_1}{w_3} \tag{7}$$

$$= 51.96\%$$

Void content-- Void content is an indication of the amount of voids in reinforced composites. High void contents can significantly reduce the composites strength and long-term mechanical properties. Monitoring on a batch-by-batch basis, void content can also act as a measure of the consistency of the composites manufacturing process.

The density of the material was determined by dry/wet weight method as per ASTM D2734-94. To calculate void content, densities of both resin and reinforcing material should be known. The individual densities were obtained from the supplier of the resin and reinforcing material.

After actual density of the material was determined, the weighed sample was placed into a weighed crucible and burned in a 600°C muffle furnace in air until only the reinforcing material remained. The crucible was cooled and weighed. The resin content (ignition loss) was calculated as a weight percent from the available data. By knowing the reinforcement and resin weight percentage, bar volume and density, void content can be calculated as shown below (ASTM D2734-94).

Weight of bar (w_1)	$= 21.5062$ gm
Weight of crucible + glass fiber (w_2)	$= 54.6894$ gm
Weight of empty crucible (w_3)	$= 42.6585$ gm
Weight of glass fiber (w_4)	$= w_2 - w_3$
	$= 12.0309$ gm

$$\text{Weight percentage of glass fiber } (g) = \frac{w_4 \times 100}{w_1} \tag{8}$$

$$= 55.9\ \%$$

$$\text{Weight of resin } (w_5) = w_1 - w_4$$

$$= 9.4753 \text{ gm}$$

$$\text{Weight percentage of resin } (r) = \frac{w_5 \times 100}{w_1} \tag{9}$$

$$= 44.1\ \%$$

To find volume of bar,

Length of the bar	= 78.74 mm
Diameter of bar (without ribs)	= 12.7 mm
Diameter of bar + rib	= 14.6 mm
Diameter of bar + projection	= 17.1 mm
Width of rib	= 2.1 mm
Depth of rib	$= \dfrac{14.6-12.7}{2} = 0.95$ mm
Width of projection	= 2.9 mm
Depth of projection	$= \dfrac{17.1-12.7}{2} = 2.2$ mm
Volume of bar	$= \dfrac{\pi(12.7)^2}{4}\,86.36 = 10939.8$ mm^3
Volume of projection	= 2.9 x 2.2 x 86.36x 2 = 1101.9 mm^3
Circumference of bar with rib	= (π x 12.7) - 2 x 2.9 = 34.098 mm
Volume of rib	= 2.1 x 0.95 x 34.098 x 9 x 0.6 = 367.3 mm^3
Total volume of bar	= 10939.8 + 1101.9 + 367.3 = 12409.0 mm^3
Density of composite (M_d)	$= \dfrac{21.5062}{12409.0} = 1.7331 \times 10^{-3}$ gm/mm^3

$$\text{Void content} = 100 - M_d \left(\frac{r}{d_r} + \frac{g}{d_g} \right) \tag{10}$$

$$= 100 - (1.73 \times 10^{-3}) \left(\frac{44.1}{1.24 \times 10^{-3}} + \frac{55.9}{2.54 \times 10^{-3}} \right)$$

$$= 0.39\ \%$$

Void content of 0.39 % indicates successful laboratory manufacturing of GFRP bars from long-term durability perspective. Other criteria of strength and stiffness were also satisfied as described in previous sections.

Resin chemistry with nanoclay-- Based on literature review and our research, nanoclay was found to be suitable for manufacturing composites with glass fibers and vinyl ester resins with up to 5% nanoclay. Based on the trials of bars manufactured with 2%, 3½%, 4% and 5% nanoclay, it was found that resin with 5% exfoliated resin had a higher viscosity, which made the resin to be less workable and resulted in more voids. SEM images were taken on 2%, 3½%, and 5% nanoclay resin to monitor exfoliation procedure Figure 19 shows the 2% nanoclay exfoliation and Figure 20 shows the 5 % nanoclay exfoliation in the resin. SEM images indicated improved exfoliation with longer duration of mixing.

Manufacturing details-- GFRP bars with and without nanoclay were successfully manufactured in the laboratory with following details.
1. Low-cost and low-volume pulley assisted manual pultrusion process was successfully developed to manufacture GFRP bars using E-glass rovings and thermoset vinyl ester resins with and without nanoclay. This process was utilized to manufacture bars at room temperature with suitable catalysts using

vinyl ester resins and nanoclay. Other fibers, thermoset resins, and additives can be substituted as necessary.

2. Three sets of molds were developed to successfully manufacture bars with smooth surface, bars with crescent shaped lugs, and bars with circular lugs in a laboratory setting.

3. Bars with circular lugs were selected for this research and molds were developed at CFC laboratories to manufacture 12 ft (3.66 m) long GFRP bars with up to 50% or higher fiber volume fraction using nanoclay.

Improving surface texture and void content--Initial GFRP bars with smooth surface texture were found to have high void content through visual observation and moisture absorption tests. After refining the mold and manufacturing process, GFRP bars were found to have maximum void content of 0.39%. In addition, uniform surface finish was obtained in the bars manufactured through pulley assisted manual pultrusion process.

CONCLUSIONS

In this research FRP bars of 4 ft (0.3 m) and 12 ft (3.66 m) length were manufactured successfully in CFC-WVU laboratories at room temperature using E-glass rovings and thermoset vinyl ester resins with and without nanoclay. Conclusions of this research are:

- # 4 bars with nanoclay showed a reduction in average maximum tensile stress of 110.78 ksi (763.80 MPa) compared to bars without nanoclay having average maximum tensile stress of 126.22 ksi (870.26 MPa). Reduction of tensile stress is attributed due to exfoliation of nanoclay in resin, which reduces interfacial bond strength development and an increased resin viscosity leading to reduced fiber wetting. Reduction in tensile strength due to addition of nanoclay is observed in the research on FRP laminates conducted by the CFC-WVU.

- # 4 bars with nanoclay showed a higher average stiffness of 4.33 msi (29.8 GPa) compared to bars without nanoclay having average stiffness of 4.24 msi (29.2 GPa).

- Bars with crescent shaped lugs and nanoclay were found to have a maximum bond stress value of 2385 psi (16.44 MPa) when tested in their stronger bond direction. Bars with circular shaped lugs and nanoclay were found to have an average maximum bond stress value of 1762.77 psi (12.15 MPa).

- # 4 bars having a fiber volume fraction of 44.7% (32 glass roving) with nanoclay showed 8.3 % reduction in shear stress as compared to bars without nanoclay (24.10 ksi (166.16 MPa) vs. 26.27 ksi (181.13 MPa)). Reduction of shear strength in bars with nanoclay is attributed to the exfoliation of nanoclay in resin, which increases resin viscosity and creates multiple interfaces between fibers, resin and glass fibers, leading to reduction in shear strength.

- GFRP bars showed moisture gain of 1.2181 % without nanoclay and 1.0631 % with nanoclay in 283 days.

Commercial production of GFRP bars with nanoclay can be carried out without altering any manufacturing setup. However, resin with exfoliated nanoclay with proper viscosities need to be used in place of the resin without nanoclay. Further experimental evaluation of beams reinforced with nanoclay GFRP bars is being carried out at the CFC-WVU to assess long-term advantages and disadvantages of nanoclay based rebars.

REFERENCES

ACI 440.1R-03: *Guide for the Design and Construction of Concrete Reinforced with FRP Bars.*

ASTM test method D2584-94, *Standard Test Method for Ignition Loss of Cured Reinforced Resins.*

ASTM test method D2734-03, *Standard Test Methods for Void Content of Reinforced Plastics.*

ASTM test method D3916-02, *Standard Test Method for Tensile Properties of Pultruded Glass Fiber Reinforced Plastic Rod.*

ASTM test method D4476-03, *Standard Test Method for Flexural Properties of Fiber Reinforced Pultruded Plastic Rods.*

Benmokrane, B., Wang, P., That, T.M., Rahman, H., and Robert, J.F., *Durability of glass fiber reinforced polymer reinforcing bares in concrete environment,* Journal of composites for construction, Vol 6, Aug 2002.

Hay J.N., Shaw S.J., *A Review Of Nanocomposites (2000),* The Institute of Nanomaterials, http://www.nano.org.uk.

Mallick. P.K., *Fiber Reinforced Composites Materials, Manufacturing and Design,* Publisher-Marcel Dekker; 2nd edition, May, 1993.

Shah A.P., Gupta R.K., Gangarao H.V.S. and Powell C.E., *Moisture Diffusion Through Vinyl Ester Nanocomposites Made with Montmorillonite Clay,* Polymer Engineering and Science, 2002.

Southern clay products Inc., Manufacturer's Data Sheet.

Vijay P.V. and GangaRao, H.V.S., *Accelerated and Natural Weathering of Glass Fiber Reinforced Plastic Bars,* Fourth Interantional Symposium, Fiber Reinforced Polymer Reinforcement for Reinforced Concrete Structures, ACI- Fall conference, SP-188, pp. 605-614, Oct-Nov., 1999.

Table 1-- Tensile test results of GFRP bars

Bar Type	Nano clay	Mold type ft (m)	No. of rovings	Fiber volume fraction	Ultimate stress ksi (MPa)	Young's modulus msi (GPa)
Bar 1		4 (1.22)			119.34 (822.87)	4.19 (28.9)
Bar 2	No		32	44.7%	122.23 (842.75)	4.18 (28.8)
Bar 3		12 (3.66)			129.87 (895.42)	4.32 (29.7)
Bar 4					133.43 (920.00)	4.29 (29.6)
					126.22 (870.26)	**4.24 (29.2)**
					6.54 (45.12)	**0.07 (0.48)**
					5.18	**1.66**
Bar 5	2%		36	50.3%	108.76 (749.88)	4.75[†] (32.8)
Bar 6		4 (1.22)			113.57 (783.05)	4.28 (29.5)
Bar 7			32	44.7%	114.08 (786.56)	4.24 (29.2)
Bar 8		12 (3.66)			112.04 (772.52)	-
Bar 9	4%				106.72 (735.83)	4.47 (30.8)
					110.78 (763.83)	**4.33 (29.9)**
					3.61 (24.93)	**0.12 (0.83)**
					3.26	**2.83**

[†] Not included for standard deviation calculation

Table 2-- Bond test results

Specimen	Bar shape	Bond strength psi (MPa)	Failure
Bar 1	Crescent shaped	2385.12* (16.44)	
Bar 2	Circular shaped	1730.28 (11.92)	Bar slip
Bar 3		1682.39 (11.59)	
Bar 4		1875.65 (12.93)	
Average		**1762.77 (12.15)**	
Std Dev		**100.64 (0.69)**	
% Std Dev		**5.70**	

*Not included for Standard deviation calculation and bond test conducted only with the strong direction

Table 3— Double shear test results of GFRP bars

Specimen	Nanoclay	Ultimate load lb (kN)	Double shear stress ksi (MPa)
Bar 1		8650 (38.48)	22.02 (151.82)
Bar 2		8450 (37.59)	21.52 (148.38)
Bar 3		9700 (43.15)	24.70 (170.30)
Bar 4	Yes	10150 (45.15)	25.84 (178.16)
Bar 5		9200 (40.92)	23.43 (161.54)
Bar 6		9450 (42.03)	24.06 (165.89)
Bar 7		9300 (41.37)	23.68 (163.27)
Bar 8		10500 (46.70)	26.74 (184.37)
Bar 9		9800 (43.59)	24.95 (172.02)
Average		**9466.66 (42.11)**	**24.10 (166.16)**
Std Dev		**661.43 (2.94)**	**1.68 (11.58)**
% Std Dev		**6.98**	**6.98**
Bar 10		10600 (47.15)	26.99 (186.09)
Bar 11	No	10200 (45.37)	25.97 (179.06)
Bar 12		10150 (45.15)	25.85 (178.23)
Average		**10316.66 (45.89)**	**26.27 (181.13)**
Std Dev		**246.64 (1.10)**	**0.62 (4.27)**
% Std Dev		**2.39**	**2.39**

Note 1: Shear stress reduction in bars with nanoclay is $\dfrac{26.27 - 24.10}{26.27} = 8.3\%$

Figure 1— Teflon plates drilled with holes for pulling fibers

Figure 2— Schematic diagram showing bar attached with grips

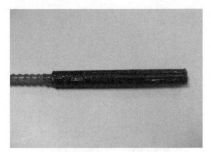

Figure 3— GFRP bar attached with steel grips

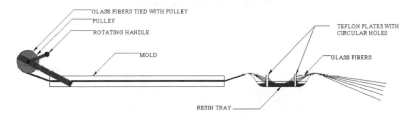

Figure 4— Schematic diagram showing pultrusion of GFRP bar

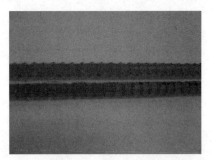

Figure 5— Crescent shaped (top) and circular shaped (bottom) lugs in bar

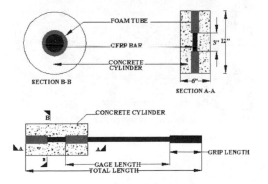

Figure 6— Bond breakers between GFRP and concrete

Figure 7— Pullout test setup

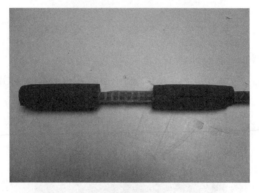

Figure 8— GFRP bar after debonding in cylinder pull-out test

Figure 9— Base fixture and cutting tool for shear test

Figure 10— Numbering of GFRP bar

Figure 11— GFRP bars immersed in water

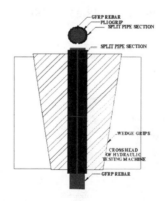

Figure 12— Diagram showing GFRP bar in wedge grips of UTM

Figure 13— GFRP bar after testing

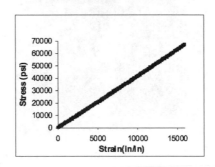

Figure 14— Stress-strain data for GFRP from tension test

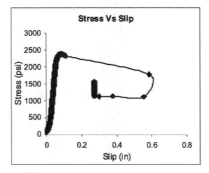

Figure 15— Bond Stress - slip (unloading end) bar graph for GFRP bar from pullout test
(<u>Note</u>-no correction for initial grip slip bond slip and elongation applied)

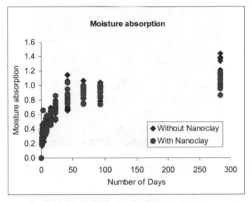

Figure 16— Comparison of moisture content in bars with and without nanoclay

Figure 17— GFRP bar before ignition

Figure 18—GFRP bar after ignition

Figure 19— SEM image of vinylester resin with partially exfoliated 2% nanoclay

Figure 20— SEM image of vinylester resin with partially exfoliated 5% nanoclay

Textile Reinforced Mortars (TRM) versus Fiber Reinforced Polymers (FRP) as Strengthening Materials of Concrete Structures

by T.C. Triantafillou and C.G. Papanicolaou

Synopsis: Fiber reinforced polymers (FRP) are investigated in this study in comparison with a new class of materials, textile reinforced mortars (TRM), for shear strengthening and/or seismic retrofitting of concrete structures. Textiles comprise fabric meshes made of long woven, knitted or even unwoven fiber rovings in at least two (typically orthogonal) directions. Mortars – serving as binders – contain polymeric additives in order to have improved strength properties. In this study, experimental investigations were carried out in order to provide a better understanding on the effectiveness of TRM versus FRP jackets as a means of increasing: (i) the axial capacity of concrete through confinement; and (ii) the load-carrying capacity of shear-critical reinforced concrete flexural members. From the results obtained it is strongly believed that the proposed TRM strengthening technique is a viable alternative to the already successful FRP strengthening technique.

Keywords: concrete; confinement; FRP; mortars; shear strengthening; textiles

ACI member **Thanasis C. Triantafillou** *is Associate Professor of Civil Engineering and Director of the Structural Materials Laboratory at the University of Patras, Greece. His main research interests include the application of advanced polymer or cement-based composites in combination with concrete, masonry and timber, with emphasis on strengthening and seismic retrofitting. Prof. Triantafillou is the convenor of the fib working party "Externally Bonded Reinforcement" of TG9.3 and member of RILEM technical committees on FRP and textile reinforcement.*

Catherine G. Papanicolaou *is Lecturer in the Department of Civil Engineering at the University of Patras, Greece. She received her diploma (1996) and PhD (2002) degrees from the University of Patras. For the period March 2003 – March 2004 she worked as Post-doctoral Fellow at the European Laboratory for Structural Assessment of the Joint Research Center, Ispra. Her current research interests are in high performance (including textile-reinforced) concrete and optimization of advanced prefabrication systems.*

INTRODUCTION AND BACKGROUND

The use of fiber reinforced polymers (FRP) in strengthening and seismic retrofitting projects has gained increasing popularity among structural engineers, due to numerous attractive features of these materials, such as: high specific strength (i.e. strength to weight ratio), corrosion resistance, ease and speed of application and minimal change of cross sections.   Despite its advantages over other methods, the FRP strengthening technique is not entirely problem-free. The organic resins used to bind and impregnate the fibers entail a number of drawbacks, namely: (a) poor behaviour at temperatures above the glass transition temperature; (b) relatively high cost of resins; (c) potential hazards for the manual worker; (d) non-applicability on wet surfaces or at low temperatures; (e) lack of vapour permeability; and (f) incompatibility of resins and substrate materials.

One possible course of action aiming at the alleviation of the afore-mentioned problems would be the replacement of organic binders with inorganic ones, e.g. cement-based mortars, leading to the substitution of FRP with fiber reinforced mortars (FRM). The problem arising from such a substitution would be the relatively poor bond conditions in the resulting cementitious composite as, due to the granularity of the mortar, penetration and impregnation of fiber sheets is very difficult to achieve. Fiber-matrix interactions could be enhanced when continuous fiber sheets are replaced by *textiles*. The latter comprise fabric meshes made of long woven, knitted or even unwoven fiber rovings in at least two (typically orthogonal) directions. The quantity and the spacing of rovings in each direction can be controlled independently, thus affecting the mechanical characteristics of the textile and the degree of penetration of the mortar matrix through the mesh openings. It is through this mechanical interlock that an effective composite action of the mortar-grid structure is achieved. For the cementitious matrix, the following requirements should be met: non-shrinkable; high workability (application should be possible using a trowel); high viscosity (application should not be problematic on vertical or overhead surfaces); low rate of workability loss (application of each mortar layer should be possile while the previous one is still in a fresh state); and sufficient shear (hence tensile) strength, in order to avoid premature debonding. In case E-glass fiber textiles are used, the cement-based matrix should be of low alkalinity.

Although research on the use of textile meshes as reinforcement of cementitious products commenced in the early 1980s [1], developments in this field progressed rather slowly until the late 1990s. But during the past five years or so, the research community has put considerable effort on the use of textiles as reinforcement of cement-based products, primarily in new constructions [2-9]. Studies on the use of textiles in the upgrading of concrete structures have been very limited and focused on flexural or shear strengthening of beams under monotonic loading and on aspects of bond between concrete and cement-based textile composites [10-11]. In the present study, the authors go a few steps further: First, textiles are combined with inorganic (cement-based) binders, named here textile reinforced mortars (TRM), to increase the strength and ductility of concrete through confinement. Next, TRM jackets are used to enhance the resistance of reinforced concrete members in shear (both monotonic and cyclic). Finally, TRM systems are compared with equivalent FRP systems, with a scope to quantify their effectiveness.

RESEARCH SIGNIFICANCE

Jacketing of reinforced concrete members in existing structures is an increasingly attractive strengthening and/or retrofit option both in non-seismic and in seismically prone areas. Among all jacketing techniques, the use of FRP has gained increasing popularity, due to the favorable properties possessed by these materials. However, certain problems associated with epoxy resins, e.g. poor behavior at high temperatures, high costs, incompatibility with substrates and inapplicability on wet surfaces, are still to be addressed. One possible solution would be the replacement of epoxies with inorganic binders, but impregnation of continuous fiber sheets with mortars is very difficult to achieve, resulting in rather poor bond between fibers and matrix. Bond conditions could be improved when textiles are used instead of fiber sheets, a concept leading to the use of textile reinforced mortar (TRM) jacketing as an alternative to FRP jacketing. It is this concept that the authors explore and study in this paper, for the confinement of concrete as well as for shear strengthening.

FRP VERSUS TRM IN CONCRETE CONFINEMENT

Test specimens and materials

The test plan included two types of specimens: (a) cylindrical specimens with diameter 150 mm and height 300 mm (Series A and B); (b) short column – type specimens with rectangular cross section 250x250 mm and height 700 mm (Series C). Each specimen series was cast using the same ready-mix concrete batch (but slightly different from series to series, in terms of water to cement ratio). All specimens were unreinforced, as the jacket – reinforcement interactions (e.g. prevention of rebar pull-out at lap splices or delay of rebar buckling) were not in the scope of the present study. The four corners of all rectangular prisms were rounded at a radius equal to 15 mm.

In the case of cylindrical specimens all confining systems were applied "as usual", that is with a single textile sheet wrapped around each cylinder until the desired number of layers was achieved. The bonding agent was either epoxy resin or inorganic mortar, applied to the concrete surface, in between all layers and on top of the last layer. Jacketing of all rectangular columns was provided using a new concept, which involved the formation of each layer through the use of a single strip. The strip was wrapped around the column in a spiral (bandage-like) configuration, starting from one end (column top) and stopping at the other (column bottom), Fig. 1a. Each successive strip was wrapped in the direction opposite to that of the previous one (Fig. 1b). The strips were attached on the concrete either through full bond (that is with resin or mortar, as in the case of cylinders), or at the ends only, using a simple method, which involved wrapping and epoxy-bonding of another strip, applied laterally in two layers at each end (top and bottom) of the column (Fig. 1c). Application of the mortars was made in approximately 2 mm thick layers with a smooth metal trowel.

Specimens in Series A are given the notation A_XN, where X denotes the type of jacket (C for the unjacketed, that is the control specimens, MI for specimens with mortar type I jackets and MII for specimens with mortar type II jackets) and N denotes the number of layers. Series B included another five different designs: the control specimens, specimens wrapped with two or three layers of textile bonded with epoxy resin and their counterparts bonded with mortar type II. Moreover, the concrete strength was a bit higher in Series B compared to Series A (due to the different water to cement ratio in the two batches). The notation of specimens in Series B is B_XN, where X and N are defined as above (R is used to denote epoxy resin and MII is used to denote mortar type II). Finally, Series C included seven different designs of short rectangular column-type specimens, as follows: The control column, columns wrapped with two or four layers of textile bonded with an epoxy resin, their counterparts wrapped with two or four layers of textile bonded with mortar type II and two more columns with two or four layers of unbonded textile, anchored at the column ends using transverse wrapping (as in Fig. 1c). The notation of columns in Series C is C_XN, where, as above, N is the number of layers and X denotes the type of jacket (C for unjacketed, R for resin-based jackets, MII for Mortar II jackets and A for jackets made of unbonded strips with end anchorage). All types of specimens used in this study are summarized in the first column of Table 1. Three and two specimens for the case of cylinders and rectangular short columns, respectively, were considered sufficient for reasonable repeatability. As a result, a total of 44 specimens were tested.

For jacketing, a commercial unwoven textile with equal quantity of high-strength carbon fiber rovings in two orthogonal directions was used (Fig. 2). The mass of carbon fibers in the textile was 168 g/m^2 and the nominal thickness of each layer (corresponding to the equivalent smeared distribution of fibers) was 0.047 mm. The guaranteed tensile strength of the carbon fibers (as well as of the textile, when the nominal thickness is used) in each direction was 3350 MPa, and the elastic modulus was 225 GPa (both values were taken from data sheets of the producer). Mortar I was a commercial low cost dry inorganic binder (suitable for plastering) containing fine cement and a low fraction of polymers. Mortar II contained cement and polymers at a ratio 10:1 by mass (higher than in Mortar I). The binder to water ratio in Mortars I and II was 3.4:1 and 3:1 by mass, respectively, resulting in plastic consistency and good workability. The 28-day compressive and tensile strength was 8.56 and 30.61 MPa, respectively, for Mortar I; and 3.28 and 4.24 MPa, respectively, for Mortar II.

Test results

The response of all specimens in uniaxial compression was obtained through monotonically applied loading at a rate of 0.01 mm/s in displacement control, using a 4000 kN compression testing machine. Loads were measured from a load cell and displacements were obtained using external linear variable differential transducers (LVDT) mounted on two opposite sides, at a gauge length of 130 mm for the cylinders and 180 mm for the rectangular columns, in the middle part of each specimen. From the applied load and average displacement measurements the stress-strain curves were obtained for each test.

Series A - cylindrical specimens (Mortar I versus Mortar II) -- Typical stress-strain plots recorded for cylinders with jackets made of textile and two different types of inorganic binders (Mortar I and Mortar II) are given in Fig. 3a, along with results for control specimens. Peak stress (confined concrete strength) values, f_{cc}, and ultimate strains, ε_{ccu}, are given in Table 1 (mean values). With one single exception, all σ-ε plots for concrete with textile confinement are characterized by an ascending branch, which nearly coincides with that for unconfined concrete, followed by a second one, close to linear, which drops rather suddenly at a point where the jacket either fractured due to hoop stresses or started debonding from the end of the lap. This notable difference in the failure mechanisms is attributed to the different mortar strengths. It is believed that the property determining which of the two failure mechanisms will be activated first is the interlaminar shear strength of the textile-mortar composite, which is proportional to the tensile (that is the flexural) strength of mortar. Note that the relatively small difference in flexural strengths between the two mortars is in agreement with the marginally higher effectiveness of jackets with Mortar II compared to those with Mortar I. The term "effectiveness" is quantified here by the ratios of confined to unconfined strength and ultimate strain. Whereas in unconfined specimens the ultimate strain is taken equal to 0.002, in confined specimens it is defined either at the point where the slope of the σ-ε curve drops suddenly or at the point where the stress drops by 20% of the maximum value.

In specimens with two layers of textile-mortar jackets the gain in compressive strength was 36% and 57% for mortar type I and II, respectively. These numbers are found by dividing the difference between confined and unconfined strength by the unconfined strength, e.g. $(20.77 - 15.24)/15.24 = 0.36 = 36\%$ for specimen A_MI2. The corresponding values in specimens with three layers were 74% and 77%. Gains in ultimate strains were much higher, with effectiveness factors (defined above as the ratio of confined to unconfined ultimate strain) around 5 or 6. Overall, it may be concluded that textile-mortar confining jackets provide substantial gain in compressive strength and deformability. This gain is higher as the number of confining layers increases and depends on the tensile (that is the shear) strength of the mortar.

Series B - cylindrical specimens (mortar versus resin) -- Typical stress-strain plots for cylinders with jackets made of textile/epoxy (FRP) (specimens B_R2, B_R3) or textile/mortar (TRM) type II (B_MII2, B_MII3) are given in Fig. 3b, along with results for control specimens; peak stresses, f_{cc}, and ultimate strains, ε_{ccu}, are given in Table 1. Specimens with resin-impregnated textiles gave a nearly bilinear response with a transition curve and failed due to tensile fracture of the jackets in the hoop direction. In these specimens the strength increased by 53% or 92% and the ultimate strain increased by a factor which exceeded 8 or 12, when the jacket was made of two or three layers, respectively.

Similarly to specimens with textile-mortar (type II) jackets in Series A, the σ-ε plots for concrete with textile confinement (B_MII2, B_MII3) are characterized by an ascending branch, which nearly coincides with that for unconfined concrete, followed by a second one, close to linear, which drops rather suddenly at a point where the jacket fractured due to hoop stresses. A point of difference is that the σ-ε curve has a first local maximum, at strain $\varepsilon_{co} = 0.002$ where unconfined concrete failed, followed by a small descending branch, which picked-up rather quickly and became ascending, until final fracture of the jacket occurred. This distinct behavior was observed only in specimens with two confining layers (and in one specimen with three confining layers), in agreement with similar observations on concrete confined with FRP jackets of low stiffness. Compared with the control specimens, in those with two-layered textile-mortar jackets the strength increased by 25% and the ultimate strain by a factor of 4.9. In specimens with three-layered textile-mortar jackets the improvement in mechanical properties was even better: the strength increased by 49% and the ultimate strain by a factor of 5.4. It should be noted that these numbers are lower than those recorded when the same jackets (two or three layers of Mortar II) were used in specimens of Series A, where concrete was of lower strength, confirming that the effectiveness of TRM jackets increases as the unconfined concrete strength decreases; the same conclusion applies to classical FRP jacketing, as suggested in numerous studies found in the literature.

A comparison of the effectiveness of mortar versus resin in textile jackets can be made by dividing the effectiveness of mortar-based jackets to that of resin-based jackets. The average value of this ratio, given in the last two columns of Table 1, is around 0.8 and 0.5 for strength and ultimate strain, respectively, and appears to decrease only marginally as the number of confining layers increases from two to three (from 0.82 to 0.77 for strength and from 0.59 to 0.42 for strain).

Another interesting observation is that, contrary to FRP jackets, TRM jackets do not fail abruptly. Their fracture in the hoop direction initiates from a limited number of fiber bundles (when the tensile stress reaches their tensile capacity) and then propagates rather slowly in the neighboring bundles, resulting in a failure mechanism which may be characterized as more ductile (compared with FRP jacketing). This fact is also reflected in the σ-ε curves, where the point of maximum stress (and the associated ultimate strain) is followed by a descending branch which keeps a nearly constant slope for a large range of strain.

Overall, it may be concluded that textile-mortar confining jackets: (a) provide substantial gain in compressive strength and deformability and (b) are characterized by reduced effectiveness, when compared with FRP jackets. The reduction in effectiveness is quite small in terms of strength and more notable in terms of ultimate strain.

Series C – rectangular columns (mortar versus resin versus end anchorage) -- Typical stress-strain plots for short column-type specimens are given in Fig. 3c, for specimens confined with two or four layers of textile impregnated with resin (C_R2, C_R4), textile impregnated with mortar (C_MII2, C_MII4) and textile strips with end anchorage (C_A2, C_A4), respectively. For the sake of convenient comparison, each figure provides also the σ-ε curves of the control (unconfined) specimens. Peak stresses, ultimate strains (defined either at the point where the slope of the σ-ε curve drops suddenly or at the point where the stress drops by 20% of the maximum value) and effectiveness ratios are given in Table 1.

Columns with resin-impregnated textile (FRP) jackets exhibited a nearly bilinear response, until tensile fracture of the jackets occurred at the corners. The strength increased by 29% or 47% and the ultimate strain increased by a factor which exceeded 6 or 10, when the jacket was made of two or four layers, respectively. The behavior of columns confined with mortar-impregnated (TRM) jackets was quite similar. The strength increased by 40% or 51% and the ultimate strain increased by a factor a little less than 6 or 9, when the jacket was made of two or four layers, respectively. Specimens with four confining layers failed in a way very similar to the ones with resin-impregnated textile jackets, whereas in those with two layers failure was gradual, starting from a few fiber bundles and propagating slowly in the neighboring fibers; as a result, the σ-ε curves of these specimens do not contain a sudden drop, which is a characteristic of excessive fiber fracture in a rather large portion of the jacket height. With regards to relative effectiveness, mortar-impregnated textile jackets were found equally good to their resin-impregnated counterparts (in fact, they were superior by 3-9%, which may be attributed to statistical error) in strength terms and marginally inferior (by 5-13%) in ultimate strain terms.

Surprisingly, spirally confined columns with unbonded strips anchored at the ends only, behaved nearly as good as those confined with fully-bonded mortar-impregnated or resin-impregnated jackets, especially in the case of four layers. The strength increased by 39% or 45% and the ultimate strain increased by a factor a little less than 4 or 9, when the jacket was made of two or four layers, respectively. Failure in these specimens developed away from the anchorages and was characterized by gradual fracture of fiber bundles, as in the case of columns with fully-bonded mortar-impregnated textile jackets. With regards to relative effectiveness, spirally applied unbonded strips with end anchorages were found equally good to their resin-impregnated counterparts in strength terms and inferior by 36%-13% (depending on the number of layers) in ultimate strain terms. When effectiveness of unbonded jacketing is compared with that of mortar-impregnated jacketing, the results are nearly identical in the case of four layers and slightly inferior in terms of ultimate strain in the case of two layers.

Overall, it may be concluded that TRM jackets are quite effective in confining columns of rectangular cross sections for strength and axial deformability. When the effectiveness is compared with that of FRP jackets, it is found nearly equal in strength terms and slightly inferior in ultimate strain terms. The same conclusion applies in the case of spirally applied unbonded strips with end anchorages, except if the number of layers is quite low, which may affect adversely the deformability.

Simple confinement model

A typical approach towards modelling confinement is to assume that the confined strength f_{cc} and ultimate strain ε_{ccu} depend on the confining stress at failure, $\sigma_{\ell u}$ as follows [12-14] :

$$\frac{f_{cc}}{f_{co}} = 1 + k_1 \left(\frac{\sigma_{\ell u}}{f_{co}}\right)^m \tag{1}$$

$$\varepsilon_{ccu} = \varepsilon_{co} + k_2 \left(\frac{\sigma_{\ell u}}{f_{co}}\right)^n \tag{2}$$

where k_1, k_2, m and n are empirical constants. The reduced effectiveness provided by jackets other than resin-impregnated ones (textile reinforced mortar jackets or unbonded strips anchored at the ends, as used in this study) may be taken into account by splitting k_1 and k_2 in two terms, as follows:

$$k_1 = \alpha k_{1,R} \tag{3}$$

$$k_2 = \beta k_{2,R} \tag{4}$$

where $k_{1,R}$ and $k_{2,R}$ are the values of k_1 and k_2, respectively, if jackets are made with resin-impregnated fibers and α, β are "effectiveness coefficients", which depend on the specific jacketing system (say α_M, β_M for mortar-based jackets and α_A, β_A for unbonded jackets anchored at the ends) and can be derived experimentally.

The literature on the precise form of confinement models for concrete is vast. Some of these models, especially the older ones, are based on the assumption that the relationship between confined strength and ultimate strain and their unconfined counterparts is linear, that is m and n are both equal to one. In other models, especially in some of the most recent ones, m and n are taken less than – but still close to – one. Whereas the main advantage of the former approach is simplicity, the disadvantage is that linear relationships between f_{cc}-$\sigma_{\ell u}$ and ε_{ccu}-$\sigma_{\ell u}$ tend to overpredict both the confined strength and the confined ultimate strain for high confining stresses. As our objective in this paper is not to elaborate on confinement models for concrete, but rather to demonstrate the procedure regarding the use of the "effectiveness coefficients" α and β for the two alternative (to epoxy-bonded) jacketing systems, we make too, for the sake of simplicity, the assumption of linearity, that is we consider m and n equal to one; but the approach presented herein is applicable without difficulty for any set of values of m and n.

The confining stress $\sigma_{\ell u}$ at failure of the jacket is, in general, non-uniform, especially near the corners of rectangular cross sections. As an average for $\sigma_{\ell u}$ in a cross section with dimensions b and h one may write:

$$\sigma_{\ell u} = \frac{\sigma_{\ell u,h} + \sigma_{\ell u,b}}{2} = \frac{1}{2} k_e \left(\frac{2t_j f_{je}}{h} + \frac{2t_j f_{je}}{b} \right) = k_e \frac{(b+h)}{bh} t_j f_{je} \tag{5}$$

where t_j is the jacket thickness, f_{je}, is the effective jacket strength in the lateral direction and k_e is an effectiveness coefficient, which, for continuous jackets with fibers in the direction perpendicular to the member axis is defined as the ratio of effectively confined area to the total cross sectional area A_g [15] :

$$k_e = \frac{1 - \left[(b - 2r_c)^2 + (b - 2r_c)^2 \right]}{3A_g} \tag{6}$$

In Eq. (6) r_c is the radius at the corners of the rectangular section. Application of Eqs. (1) - (2) to the data obtained for the specimens of Series B and C (specimens in Series A were excluded because mortar-based jackets cannot be compared with their resin counterparts) results in the plots of f_{cc}/f_{co} and ε_{ccu} versus $\sigma_{\ell u}/f_{co}$ given in Fig. 4a and Fig. 4b, respectively. The best linear fit equations to these data yield $k_{1,R} = 2.79$ ($R^2 = 0.95$), $\alpha_M = 0.68$ ($R^2 = 0.69$), $\alpha_A = 0.84$ ($R^2 = 0.84$), $k_{2,R} = 0.082$ ($R^2 = 0.99$), $\beta_M = 0.57$ ($R^2 = 0.49$) and $\beta_A = 0.82$ ($R^2 = 0.98$), which may be used along with the aforementioned confinement model. The above values state that according to this simplified model the effectiveness of TRM jackets used in this study is roughly 70% in terms of strength and 55-60% in terms of ultimate strain; the corresponding values for unbonded jackets anchored at their ends are roughly 85% for strength and 80% for ultimate strain. Of course, these values should be considered as indicative, as the test data used for calibration are relatively limited. But the method presented for obtaining these effectiveness coefficients is quite general.

FRP VERSUS TRM IN SHEAR STRENGTHENING OF RC

Test specimens and materials
The next step in this investigation was to examine the effectiveness of TRM as externally applied strengthening reinforcement of shear-critical RC members. The investigation was carried out by testing six beams deficient in shear (with a large spacing of stirrups in the shear span) in four point bending. The beams were 2.60 m long and had

a cross section of 150x300 mm.   The geometry of the beams, the details of the reinforcement and the general set-up of the test are shown in Fig. 5.

Four of the beams were tested monotonically and two of them were subjected to cyclic loading.   Three parameters were considered in the experimental investigation, namely the use of inorganic mortar versus resin-based matrix material for the textile reinforcement, the number of layers (one versus two) and the use of conventional wrapping versus "spirally applied" textiles.   By "conventional wrapping" it is implied that a single textile sheet was wrapped around the shear span until the desired number of layers was achieved (Fig. 5c).   The bonding agent was either epoxy resin or inorganic mortar, applied to the concrete surface, in between all layers and on top of the last layer. "Spirally applied" jacketing was implemented in one beam only and involved the formation of each layer through the use of a single strip, approximately 150 mm wide. The first strip was wrapped around the member in a spiral configuration, starting from one end of the shear span and stopping at the other, and the next strip was wrapped in the same configuration but in the direction opposite to that of the first one.   The two strips formed an angle of $\pm 10^{\circ}$ with respect to the transverse to the member axis.

One of the six beams was tested without strengthening, as a control specimen (C).   A second one was wrapped with two layers of mortar-based jacket in the shear span (M2). A third beam was identical to the second but with a resin-based matrix material for the textile reinforcement (R2).   In the fourth beam jacketing was provided with spirally applied strips (M2-s).   These four specimens were tested monotonically.   The next two specimens were identical to the second and third, but with one layer (instead of two) of textile in a mortar-based (M1) and a resin-based (R1) matrix, respectively.   These two specimens were subjected to cyclic loading.

Casting of the beams was made with ready-mix concrete of mean 28-day compressive strength equal to 30.5 MPa.   The steel used for longitudinal reinforcement had an average yield stress equal to 575 MPa; the corresponding value for the steel used in stirrups being 275 MPa.   Textile, mortar (type II) and resin matrices were the same materials as those in the experimental study involving confined specimens.

Test results

Specimens C, M2, R2 and M2-s were tested monotonically at a rate of 0.01 mm/s, whereas the remaining two were subjected to quasi-static cyclic loading (successive pairs of cycles progressively increasing by 1 mm of displacement amplitudes in each direction at a rate of 0.2 mm/s), all in displacement control.   The load was applied using a vertically positioned 500 kN MTS actuator and the displacements were measured at mid-span using two external linear variable differential transducers mounted on both sides of the specimens.   The load versus mid-span displacement curves for all specimens are given in Fig. 6.

The control beam (C) failed in shear, as expected, through the formation of diagonal cracks in the shear spans.   The ultimate load was 116.5 kN.   An interesting observation during this test was that no sudden drop in the load was recorded after diagonal cracking.

This is attributed to the considerable contribution to shear resistance provided by both the stirrups crossing the crack and the strong dowel action (activated by the three 160 mm diameter longitudinal rebars). The behaviour of beams R2, M2, M2-s and R1 indicated that shear failure was suppressed and that failure was controlled by flexure: cracks in the constant moment region became wide and yielding of the tension reinforcement (bottom layer in beams R2, M2 and M2-s, both layers in beams R1 and M1, depending on the sign of the force) resulted in a nearly horizontal branch of the force versus displacement curve. The maximum loads in specimens R2, M2 and M2-s were 233.4 kN, 243.8 kN and 237.7 kN, respectively, that is nearly the same. This confirms the fact that the shear strengthening scheme selected in this study did not affect the flexural resistance. But the increase in shear resistance was dramatic (more that 100%), regardless of the strengthening scheme: two layers of textile reinforcement (either in the form of continuous sheets or in the form of spirally applied strips) with the mortar binder performed equally well to the epoxy-bonded (FRP) jacket (with two layers of textile reinforcement).

Specimen R1 (one layer of textile bonded with epoxy) experienced a flexural yielding failure mode with unequal capacities in the push and pull directions (261.9 kN and 201.4 kN, respectively). This may be attributed to the (unintentionally) larger concrete cover at the top of each beam compared to the bottom (see Fig. 5b). Specimen M1 failed in shear; this was evident by diagonal cracking in the shear span as well as by the rather sudden strength and stiffness degradation. This specimen reached a peak load of 200.1 kN, corresponding to a substantial increase in shear capacity with respect to the control specimen, in the order of 70%. An interesting feature of specimen M1 was that fracture of the fibers in the mortar-based jacket was gradual, starting from a few fiber bundles and propagating slowly in the neighboring fibers. A second interesting feature was that beam cracking was clearly visible on the mortar-based jacket. This is an extremely desirable property, as it allows for immediate and easy inspection of damaged regions. Conventional FRP jackets in such regions would had been left intact after an extreme event (e.g. earthquake), thus making the assessment of damage a very difficult and rather expensive task (one that would require, for instance, non-destructive evaluation through the use of infrared thermography). When comparing these loads for specimens R1 and M1 with those of the others, it should be kept in mind that the former had, in general, slightly higher concrete strength, because they were tested a few months later. Furthermore, they were tested at a higher displacement rate.

Overall, it may be concluded that the mortar-impregnated textile jackets employed in this study were quite effective in increasing the shear resistance of reinforced concrete members. Two layers of textile reinforcement (with a nominal thickness per layer equal to only 0.047 mm in each of the principal fiber directions) were sufficient to prevent sudden shear failure, whereas one layer proved less effective compared to its resin-bonded counterpart, but still sufficient to provide a substantially increased resistance.

Modelling

Modelling of the textile-reinforced mortar jacket contribution to the shear resistance of flexural reinforced concrete members may be based on the well-known truss analogy, as proposed in the past for FRP jackets [15-18]. Assuming that the textile is made of continuous fiber rovings in two orthogonal directions (as in this study), with fibers in each direction i forming an angle β_i with the longitudinal axis of the member (Fig. 7), the TRM jacket contribution to shear resistance, V_t, can be written as

$$V_t = \sum_{i=1}^{2} \frac{A_{ti}}{s_i} \left(\varepsilon_{te,i} E_{fib} \right) 0.9d \left(\cot\theta + \cot\beta_i \right) \sin\beta_i \qquad (7)$$

where $\varepsilon_{te,i}$ = "effective strain" of the TRM in the direction i, E_{fib} = elastic modulus of fibers, d = effective depth of the cross section, A_{ti} = twice the cross section area of each fiber roving in the direction i, s_i = spacing of rovings along the member axis and θ = angle between the inclined shear crack and the member axis. Equation (7) may be extended in a straightforward way to account for textiles with more complex geometry (e.g. with fiber rovings in more than two directions). Note that if the direction i is perpendicular to the member axis, the ratio A_{ti}/s_i in the above equation equals twice the nominal thickness t_{ti} of the textile (based on the equivalent smeared distribution of fibers) in this particular direction.

The effective strain $\varepsilon_{te,i}$ in the direction i may be thought of as the average strain in the fibers crossing the diagonal crack when shear failure of the member occurs. Studies on the effective strain(ε_{fe}) for resin-based (FRP) jackets have been numerous in the past and have led to the development of semi-empirical but rather reliable formulas, which express the effective strain as a fraction of the fracture strain for the fibers. The same approach could, of course, be adopted for TRM jackets, when a substantial set of test data becomes available. Alternatively, one may treat TRM jackets exactly as their FRP counterparts (those with resin-based instead of mortar-based matrix), by multiplying the effective strain (of the FRP-equivalent) by an "effectiveness coefficient", say k.

The simple model described above is applicable to only one of the beams tested in this study, namely beam M1, as this was the only strengthened specimen that failed in shear. With $\theta = 45°$, $\beta_1 = 90°$ (fibers perpendicular to the member axis), $\beta_2 = 0°$ (fibers parallel to the member axis), d = 272 mm, E_{fib} = 225 GPa, A_{t1}/s_1 = 2x0.047 = 0.094 mm, A_{t2}/s_2 = 0 ($s_2 = \infty$) and V_t = 0.5x(200.1 kN – 116.5 kN) = 41.8 kN, the effective strain in the TRM jacket at shear failure is obtained from Eq. (7) equal to 0.8%. When the same analysis is applied to beam R1 (the resin counterpart of beam M1) with contribution of the FRP jacket to the shear resistance at least equal to 0.5x(261.9 kN – 116.5 kN) = 72.7 kN, a lower bound to the effective strain in the FRP is calculated as 1.4%. We may also note that the effective FRP strain in beam R1 has as an upper bound the fracture strain, which is about 1.5-1.6% (based on manufacturer's data). The effectiveness coefficient k of TRM versus FRP, based on the results for beams M1 and R1, can be obtained by dividing the TRM effective strain (0.8%) to the FRP effective strain (greater than 1.4% but at most equal to 1.5-1.6%); the value obtained is at least equal to 50%, with 57% being an upper bound. Hence it is concluded that the carbon fibers in the TRM jacket (with a single layer of textile reinforcement) were mobilized to a substantial degree - the average strain across the shear crack reached approximately 50% of the fracture strain of single fibers – and were a little more than 50% as efficient as their resin-impregnated counterparts. Of course, these values should be considered as indicative, until more test data become available. But the method described above for obtaining the effectiveness coefficients is quite general.

CONCLUSIONS

Based on the response of confined cylinders, it is concluded that: (a) Textile-reinforced mortar (named here TRM) confining jackets provide substantial gain in compressive strength and deformability. This gain is higher as the number of confining layers increases and depends on the tensile strength of the mortar, which determines whether failure of the jacket will occur due to fiber fracture or debonding. (b) Compared with their resin-impregnated counterparts (FRP), TRM jackets may result in reduced effectiveness, in the order of approximately 80% for strength and 50% for ultimate strain, for the specific mortar used in this study. It is believed that these numbers depend very much on the type of mortar and could be increased with proper modification of mortar composition. (c) Failure of mortar-impregnated textile jackets is less abrupt compared to that of their resin-impregnated counterparts, due to the slowly progressing fracture of individual fiber bundles.

From the response of rectangular columns it is concluded that TRM jackets are quite effective in confining columns of rectangular cross sections for strength and axial deformability. In comparison with their epoxy-based counterparts (FRP), mortar-impregnated textile jackets gave approximately the same effectiveness in strength terms and a slightly inferior one in ultimate strain terms. The same conclusion applies in the case of spirally applied unbonded strips with end anchorages, except if the number of layers is quite low, which may affect adversely the deformability. This concept of spirally applied unbonded jacketing appears to be quite interesting and certainly deserves further investigation.

From the response of RC members strengthened in shear it is concluded that closed-type TRM jackets provide substantial gain in the shear capacity. Two layers of mortar-impregnated textile reinforcement (based on carbon fibers with a nominal thickness per layer equal to 0.047 mm in each of the principal fiber directions) in the form of either conventional jackets or spirally applied strips were sufficient to increase the shear capacity of the beams tested by more than 60 kN, thus preventing sudden shear failures and allowing activation of flexural yielding (as was the case with the resin-based jacket). One layer of textile reinforcement proved less effective but still sufficient to provide a substantial shear resistance, which exceeded that of the unstrengthened beam by more than 40 kN. This corresponds to a good mobilization of the carbon fibers in the textile, at an average strain of 0.8%. However, when the performance of this jacket is compared with that of its resin-based counterpart, the TRM strengthening system is found a little more than 50% as effective as the FRP one.

Modeling of concrete confined or strengthened in shear with jackets other than resin-impregnated ones (FRP) becomes a rather straightforward procedure through the introduction of experimentally derived jacket "effectiveness coefficients", a concept developed in this study in order to compare: (a) the confining action of mortar-based jackets or spirally applied unbonded jackets to their resin-based counterparts; (b) the contribution to shear resistance of TRM and FRP jackets.

From the results obtained in this study the authors believe that TRM jacketing is a promising solution for increasing the confinement as well as the shear capacity of reinforced concrete members, of crucial importance in seismic retrofit. Naturally, further investigation is needed (part of it is already under way) towards the optimization of mortar properties, the increase of the experimental database and the understanding of jacket-steel reinforcement interactions in column-type members.

ACKNOWLEDGEMENTS

The authors wish to thank the following students who provided assistance in the experimental program: J. Boretos, D. Fryganakis, K. Karlos, T. Laourdekis, S. Magkli, S. Tambitsika and Mr. P. Zissimopoulos. The work reported herein was funded by the Greek General Secretariat for Research and Technology, through the project ARISTION, within the framework of the program "Built Environment and Management of Seismic Risk".

REFERENCES

1.   Naaman, A. E., Shah, S. P. and Throne, J. L., "Some Developments of Polypropylene Fibers for Concrete", *Fiber-Reinforced Concrete – International Symposium*, SP-81, G. C. Hoff, ed., American Concrete Institute, Farmington Hills, Michigan, 1984.

2.   Bischoff, Th., Wulfhorst, B., Franzke, G., Offermann, P., Bartl, A.-M., Fuchs, H., Hempel, R., Curbach, M., Pachow, U. and Weiser, W., "Textile Reinforced Concrete Façade Elements – An Investigation to Optimize Concrete Composite Technologies", *43rd International SAMPE Symposium*, 1998, pp. 1790-1802.

3.   Curbach, M. and Jesse, F., "High-Performance Textile-Reinforced Concrete", *Structural Engineering International*, IABSE, V. 4, 1999, pp. 289-291.

4.   Sato, Y., Fujii, S., Seto, Y. and Fujii, T., "Structural Behavior of Composite Reinforced Concrete Members Encased by Continuous Fiber-Mesh Reinforced Mortar Permanent Forms", *FRPRCS-4 Fiber Reinforced Polymer Reinforcement for Reinforced Concrete Structures*, SP-188, C. W. Dolan, S. H. Rizkalla and A. Nanni, eds., American Concrete Institute, Farmington Hills, Michigan, 1999, pp. 113-124.

5.   Brameshuber, W., Brockmann, J. and Roessler, G., "Textile Reinforced Concrete for Formwork Elements – Investigations of Structural Behaviour", *FRPRCS-5 Fiber Reinforced Plastics for Reinforced Concrete Structures*, C. J. Burgoyne, ed., Thomas Telford, London, 2001, V. 2, pp. 1019-1026.

6.   Molter, M., Littwin, R. and Hegger, J., "Cracking and Failure Modes of Textile Reinforced Concrete", *FRPRCS-5 Fiber Reinforced Plastics for Reinforced Concrete Structures*, C. J. Burgoyne, ed., Thomas Telford, London, 2001, V. 2, pp. 1009-1018.

7.   Naaman, A. E., "Progress in Ferrocement and Textile Hybrid Composites", 2nd *Colloquium on Textile Reinforced Structures*, Curbach M., ed., Dresden, 2003, pp. 325-346.

8.   Peled, A. and Bentur, A., "Quantitative Description of the Pull-Out Behavior of Crimped Yarns from Cement Matrix", *Journal of Materials in Civil Engineering*, ASCE, V. 15, No. 6, 2003, pp. 537-544.

9.   Reinhardt, H. W., Krueger, M. And Grosse, C. U., "Concrete Prestressed with Textile Fabric", *Journal of Advanced Concrete Technology*, V. 1, No. 3, 2003, pp. 231-239.

10.   Curbach, M. and Ortlepp, R., "Besonderheiten des Verbundverhaltens von Verstaerkungsschichten aus textilbewehrtem", 2nd *Colloquium on Textile Reinforced Structures*, Curbach M., ed., Dresden, 2003, pp. 361-374 (in German).

11.   Curbach, M. and Brueckner, A., "Textile Strukturen zur Querkraftverstaerkung von Stahlbetonbauteilen", 2nd *Colloquium on Textile Reinforced Structures*, Curbach M., ed., Dresden, 2003, pp. 347-360 (in German).

12.   De Lorenzis, L. and Tepfers, R., "Comparative Study of Models on Confinement of Concrete Cylinders with Fiber-reinforced Polymer Composites", *Journal of Composites for Construction*, 2003, V. 7, No. 3, 219-237.

13.   Theriault, M., Neale, K. W. and Claude, S., "Fiber-Reinforced Polymer-Confined Circular Concrete Columns: Investigation of Size and Slenderness Effects", *Journal of Composites for Construction*, 2004, V. 8, No. 4, 323-331.

14.   Lam, L. and Teng, J. G., "Strength Models for Fiber-Reinforced Plastic-Confined Concrete", *Journal of Composites for Construction*, 2004, V. 8, No. 4, 323-331.

15.   *fib* bulletin 14, *Externally Bonded FRP Reinforcement for RC Sttructures*, Technical Report prepared by the Working Party EBR of Task Group 9.3, International Federation for Structural Concrete, July 2001.

16.   Khalifa, A., Gold, W. J., Nanni, A. and Aziz, A. M. I., "Contribution of Externally Bonded FRP to Shear Capacity of RC Flexural Members", *Journal of Composites for Construction*, ASCE, V. 2, No. 4, 1998, pp. 195-202.

17.   Triantafillou, T. C. and Antonopoulos, C. P., "Design of Concrete Flexural Members Strengthened in Shear with FRP", *Journal of Composites for Construction*, ASCE, V. 4, No. 4, 2000, pp. 198-205.

18.   ACI 440.2R-02, *Guide for the Design and Construction of Externally Bonded FRP Systems for Strengthening Concrete Structures*, Reported by ACI Committee 440, American Concrete Institute, Farmington Hills, Michigan, 2002.

Table 1 – Strength and deformability of compression specimens.

Specimen notation	Compressive strength f_{cc}, MPa	Ultimate strain ε_{ccu}, (%)	$\dfrac{f_{cc}}{f_{co}}$	$\dfrac{\varepsilon_{ccu}}{\varepsilon_{co}}$	$\dfrac{f_{cc}}{f_{cc,R}}$	$\dfrac{\varepsilon_{ccu}}{\varepsilon_{ccu,R}}$
Series A						
A_C	15.24	0.20*	1.00	1.00	n.a.	n.a.
A_MI2	20.77	0.96	1.36	4.80	n.a.	n.a.
A_MII2	23.88	1.08	1.57	5.40	n.a.	n.a.
A_MI3	26.50	1.13	1.74	5.65	n.a.	n.a.
A_MII3	27.00	1.22	1.77	6.10	n.a.	n.a.
Series B						
B_C	21.81	0.20*	1.00	1.00	n.a.	n.a.
B_R2	33.47	1.67	1.53	8.35	1.00	1.00
B_MII2	27.36	0.98	1.25	4.90	0.82	0.59
B_R3	41.94	2.55	1.92	12.75	1.00	1.00
B_MII3	32.44	1.08	1.49	5.40	0.77	0.42
Series C						
C_C	14.25	0.20*	1.00	1.00	n.a.	n.a.
C_R2	18.41	1.24	1.29	6.20	1.00	1.00
C_MII2	20.00	1.18	1.40	5.90	1.09	0.95
C_A2	19.86	0.79	1.39	3.95	1.08	0.64
C_R4	20.97	2.03	1.47	10.15	1.00	1.00
C_MII4	21.56	1.76	1.51	8.80	1.03	0.87
C_A4	20.64	1.76	1.45	8.80	0.98	0.87

* The ultimate strain of control specimens is assumed equal to $\varepsilon_{co} = 0.2\%$, which agrees well with the mean value (0.22%) recorded at peak stress.

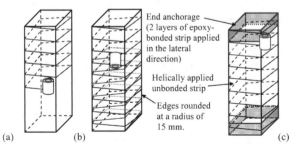

(a)   (b)   End anchorage (2 layers of epoxy-bonded strip applied in the lateral direction)

Helically applied unbonded strip

Edges rounded at a radius of 15 mm.

(c)

Figure 1 — Application of confining systems on rectangular specimens.

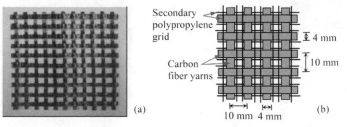

Figure 2 - (a) Photograph and (b) architecture of carbon fiber textile used in this study.

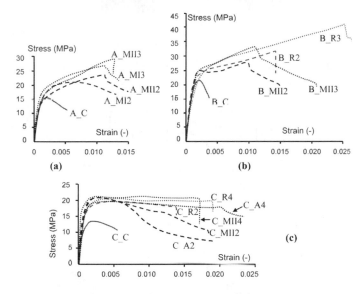

Figure 3 - Typical stress-strain curves for (a)-(b) cylinders,
(c) columns with rectangular cross sections.

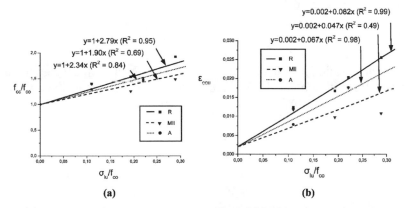

Figure 4 - (a) Normalized compressive strength and (b) ultimate compressive strain in
terms of lateral confinement.

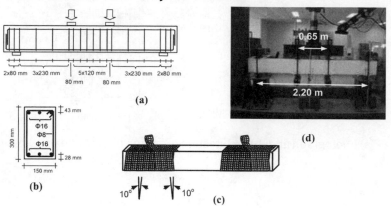

Figure 5 - (a)-(b) Geometry of the beams, (c) spiral application of strips at the shear spans and (d) general set-up of the test.

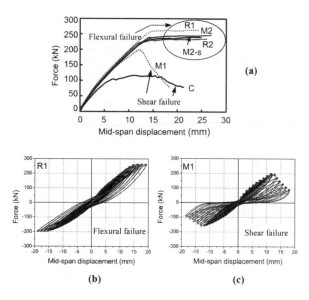

Figure 6 - Force – mid-span displacement curves: (a) for all beams tested (for beams subjected to cyclic loading the envelope curves in the push direction are given); (b) for beam R1; and (c) for beam M1.

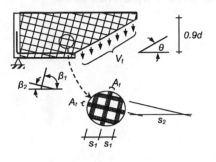

Figure 7 - Contribution of textiles with fibers in two orthogonal directions to shear resistance of RC members.

Innovative Triaxially Braided Ductile FRP Fabric for Strengthening Structures

by N.F. Grace, W.F. Ragheb, and G. Abdel-Sayed

Synopsis: This paper deals with the effectiveness of a new triaxially braided ductile Fiber Reinforced Polymer (FRP) fabric for flexural strengthening of continuous reinforced concrete beams. The tested continuous beams had two spans strengthened in flexure along their positive and negative moment regions and loaded with a concentrated load at the middle of each span. One beam was not strengthened and was tested as a control beam. The behaviors of the beams strengthened with the new fabric were investigated and compared with the behaviors of similar beams strengthened using a commercially available carbon fiber sheet. The responses of the beams were examined in terms of deflections, strains, and failure modes. The beams strengthened with the new fabric showed greater ductility than those strengthened with the carbon fiber sheet. The new fabric provided reasonable ductility due to the formation of plastic hinges that allowed for the redistribution of moment between the positive and negative moment zones of the strengthened continuous beam. Redistribution of the moment enabled the full utilization of the strength of the beam at cross sections of maximum positive and negative bending moments.

Keywords: braiding; concrete; ductility; fabric; flexural strengthening; FRP

120   Grace et al.

Nabil F. Grace is a Professor and Chairman, Department of Civil Engineering, Lawrence Technological University, Southfield, MI, USA.

Wael F. Ragheb is an Assistant Professor, Civil Engineering Department, Alexandria University, Alexandria, Egypt. He obtained his Ph.D. from the Department of Civil and Environmental Engineering, University of Windsor, Windsor, Ontario, Canada.

George Abdel-Sayed is a Professor Emeritus, Department of Civil and Environmental Engineering, University of Windsor, Windsor, Ontario, Canada.

INTRODUCTION

Ductility is an important requirement in the design of any structural element. With respect to reinforced concrete continuous beams, ductility allows the redistribution of moment between the negative and positive moment zones. The formation of plastic hinges allows the utilization of the full capacity of more cross sections of the beams. However, the beam must be able to rotate adequately at the plastic hinges in order to allow the redistribution of moment.

Fiber Reinforced Polymer (FRP) materials in forms such as pultruded plates, fabrics, and sheets have been attractive for use as strengthening materials for reinforced concrete beams. However, a large loss in beam ductility occurs when they are used for flexural strengthening of reinforced concrete beams (Saadatmanesh and Ehsani 1991; Ritchie et al. 1991; Triantafillou and Plevris 1992; Norris et al. 1997; Arduini et al. 1997; Bencardino et al. 2002). The loss in beam ductility is attributed, in part, to the mechanical characteristics of these materials. Because these materials have dissimilar behavior to that of steel, i.e. they exhibit a linear stress-strain behavior up to failure (Grace et al. 2003); they indirectly invoke brittle failures such as FRP debonding or shear-tension failure. In addition, the gain in beam yield load and stiffness after strengthening is not as significant as that of the ultimate load. Due to their high ultimate strains compared to the yield strain of steel, the FRP do not contribute with significant amounts of their strength at low strain levels such as that below the yield strain of steel.

Limited experimental investigations have been reported on the behavior of reinforced concrete beams strengthened in flexure in their negative moment regions using FRP materials. Grace et al. (1999) reported experimental investigations for reinforced concrete beams strengthened in flexure using CFRP laminates. Although increase in ultimate load was gained, large losses in ductility were experienced. The strengthened beams also showed no yield plateaus. CFRP strips have been used to strengthen the negative moment regions of reinforced concrete cantilever beams (Grace 2001). The strengthened beams experienced brittle failures as a result of strip debonding or shear-tension failure at the strip ends.

A new pseudo-ductile FRP strengthening fabric has been developed at the Structural Testing Center at Lawrence Technological University. The fabric is unique in that it exhibits a yield plateau similar to that exhibited by steel in tension. The fabric has a low yield-equivalent strain (0.35%) [the first point on the load-strain curve where the behavior becomes non-linear] that allows it to have the potential to contribute significantly to the beam load before yielding of the steel reinforcement of the strengthened beams, and a reasonable ultimate strain (around 2%), that allows the strengthened beam to exhibit adequate ductility before the fabric ruptures. This fabric was manufactured by triaxially braiding bundles of carbon and glass fibers in three different directions (+45°, 0°, -45°). These fibers were selected with different ultimate strains (0.35%, 0.8%, 2.10%) and were mixed in a way allowing them to fail successively generating a yield plateau. The fabric was designed to be used for beam strengthening for flexure and/or shear. The 0° fibers are mainly used for flexural strengthening, while the (+45°, -45°) fibers are mainly used for shear strengthening and to provide self anchoring when wrapping the beam. Fig. 1 shows details of the triaxial ductile fabric geometry and Fig. 2 shows the average tensile load-strain response of samples tested in the 0° direction, according to ASTM D 3039 specifications. Grace et al. (2003) used this fabric to strengthen reinforced concrete simple beams for flexure. The beams strengthened with the new fabric behaved in a more ductile manner than those strengthened with the carbon fiber sheets. The beams strengthened with the new fabric produced yield platcaus similar to that of the unstrengthened beam and also similar to those produced by beams strengthened with steel plates. In this paper, the effectiveness of this fabric in providing ductile behavior in reinforced concrete continuous beams strengthened in flexure is investigated.

RESEARCH SIGNIFICANCE

Ductility is a very important requirement in the design of structural elements. Ductile structures can exhibit large deformations before any potential failure and thus provide visual indicators that give the opportunity for remedial actions prior to failure. Ductility is even more important for statically indeterminate structures, such as continuous beams, as it allows for moment redistribution through the rotations of plastic hinges. Moment redistribution permits the utilization of the full capacity of more segments of the beam. A large loss in ductility is experienced when using currently available FRP materials for strengthening reinforced concrete beams for flexure. This paper investigates the capability of a new triaxial ductile FRP fabric to offer adequate ductility at the plastic hinge regions of strengthened reinforced concrete continuous beams in flexure.

EXPERIMENTAL PROGRAM

Test beams
The experimental program consisted of testing three continuous beams. All beams had identical cross sectional dimensions of 152 mm 3 254 mm (6 in. 310 in.) and lengths of 4267 mm (168 in.). The beams were symmetrically reinforced with two #5 (φ16 mm) rods at the top and the bottom. In order to avoid shear failure, the beams were over-reinforced for shear with #3 (φ 9.5 mm) closed stirrups spaced at 102 mm (4 in.). The beams were tested with two continuous spans. Fig. 3 shows the dimensions, reinforcement details, and loading set up of the test beams, respectively. The beams were prepared by sandblasting their surfaces to roughen them, cleaned with an air nozzle, and

finally wiped to remove any dust. The compressive strength of the concrete at the time the beams were tested was 41.5 MPa (6,000 psi). The steel reinforcement used had a yield stress of 490 MPa (71,000 psi).

Strengthening materials

In addition to the new triaxial ductile fabric, a commercially available carbon fiber sheet was used to strengthen similar beams in order to compare their behavior with those strengthened with the new fabric. In order to have an objective comparison, the carbon fiber sheet was selected to have a similar load-strain response to that initially exhibited by the triaxial ductile fabric (before exceeding its yield-equivalent point). The tested load-strain diagrams of the triaxial ductile fabric and the carbon fibers sheet are shown in Fig. 2 and their properties are listed in Table 1. Herein, it can be noted that the triaxial ductile fabric has a yield-equivalent load of 0.19 KN/mm (1.08 kips/in.) and an initial modulus of 50 GPa (7229 ksi), while the carbon fiber sheet has an ultimate load of 0.34 KN/mm (1.95 kips/in). Using the tensile properties of the materials, it was determined that two layers of the carbon fiber sheet would exhibit a load-strain response similar to that initially exhibited by one layer of the triaxial ductile fabric. An epoxy resin was used to impregnate the fibers and to act as an adhesive between the strengthening material and the concrete surface. This epoxy has an ultimate tensile strength of 66.2 MPa (9.62 ksi) with an ultimate strain of 4.4% and a compressive strength of 109.2 MPa (15.84 ksi).

Strengthening and set up

Test program consisted of three continuous beams. Each beam had two spans of 1981 mm (78 in.) each. The beams were loaded with a concentrated load at the middle of each span. One of these beams had no external strengthening and was tested as a control beam. The other two beams were strengthened along their negative and positive moment regions around the top/bottom face extending 152 mm (6 in.) on both sides as a U-wrap at the locations shown in Fig. 3. The first beam, beam F-CT, was strengthened using one layer of the triaxial ductile fabric that was 457 mm (18 in.) wide, U-wrapped around the tension faces and the sides, while the other beam, beam F-CTC, was strengthened using two layers of the carbon fiber sheet that were each 457 mm (18 in.) wide, with the same wrapping scheme. The deflection was measured at the middle and quarter sections of each span using string potentiometers. The FRP strain was measured at the beam tension face at the central support and at the middle of each span using electrical resistance strain gages. The reaction of the beam at the central support was measured using a load cell. Two hydraulic actuators were used to load the beam, one for each span. The load of each actuator was measured using a load cell. Table 2 summarizes the test beams.

TEST RESULTS AND DISCUSSION

Test results for the beams are shown Fig. 4 through 7, and listed in Table 3. The failed beams are shown in Fig. 8 through 11. Note that the load in Figures 4, 6, and 7 is the load at each span (P) and not the total load on the beam. The beam ductility index is calculated as the ratio between the ultimate midspan deflection and its deflection at first yield.

Control Beam B

The control beam exhibited a linear load-deflection behavior after cracking up to yielding of the tension steel at the section of the maximum negative bending moment over the central support, which occurred at a load of 92 kN (20.7 kips). After this point, a gradual decrease in the slope of the load-deflection curve was observed. The tension steel at the sections of the maximum positive bending moment yielded later, causing a significant decrease in beam stiffness as the deflection then started to increase significantly without a corresponding increase in load, as shown in Fig. 4. The beam failed by compression failure of the concrete at the midspan at a load of 127 kN (28.5 kips). A ductility index of 3.12 was observed. The beam deflection profile, shown in Fig 5, indicates that deformation of the beam at failure was very localized at the sections of maximum positive and negative moments, at the midspan and the central support, respectively.

Beam F-CT

Beam F-CT yielded at a load of 126 kN (28.3 kips) due to yielding of both tension steel and fabric over the central support. Yielding of the fabric was accompanied by the sounds of rupture of the low elongation fibers of the fabric. A gradual decrease in beam stiffness was observed, which was revealed by the decrease in the slope of the load-deflection curve, as shown in Fig. 4. A significant decrease in beam stiffness was observed after yielding of the beam at the sections of maximum positive moment, which was caused by yielding of both the tension steel and the fabric. A yield plateau similar to that exhibited by the control beam was exhibited thereafter until failure at a load of 175 kN (39.2 kips). The beam failed by tensile rupture of the fabric over the central support, followed by rupture of the fabric at midspan (see Fig. 9). A ductility index of 2.57 was exhibited, which was 18% less than that of the control beam. The load-strain diagrams of the fabric at the midspan and over the central support are shown in Fig. 6 and 7, respectively. At first failure, the fabric exhibited strain values of 1.8% and 1.47% at the sections of maximum negative and positive moments, respectively. The fact that these strain values were more than the yield-equivalent strain of the fabric indicated that full fabric strength was exploited.

Beam F-CTC

Beam F-CTC yielded at a load of 136 kN (30.6 kips), where a slight decrease in the load –deflection curve slope was exhibited caused by yielding of the tension steel at the section of the maximum negative moment over the central support. The beam exceeded the load achieved by beam F-CT and failed suddenly at a load of 185 kN (41.6 kips) by shear-tension failure at one end of the negative moment strengthening carbon fiber sheet, as shown in the photo in Fig. 10, followed by debonding of the carbon fiber sheet of the positive moment, as shown in Fig. 11. A ductility index of 1.81 was observed, which was 42% less than that of the control beam. The load-deflection curve indicates a very brittle response as shown in Fig. 4. No significant yield plateau was experienced. The load-strain curves, shown in Fig. 6 and 7, indicate that the carbon fiber sheet exhibited noticeably less strain than the triaxial ductile fabric used in beam F-CT. The maximum recorded strain values did not exceed 0.66%, which indicated that nearly half the strength of the carbon fiber sheet was not exploited.

The new triaxial ductile fabric contains bundles of fibers in the $\pm 45°$ directions. These fibers enable the fabric to have a self-anchorage along its length, when U-wrapped around the tension face and the vertical sides of the beam. As a result, anchorage failures similar to those experienced by beam F-CTC were not experienced in case of beam F-CT. On the other hand, the carbon fiber sheet used in beam F-CTC is uniaxial, and hence wrapping the beam did not enhance the anchorage. In addition, yielding of the fabric limited the increase in the tensile force developed in it. Therefore, the fabric needed less anchorage than the carbon fiber sheet, whose tensile force kept increasing until a brittle failure took place.

Using the readings of the load cell located at the central support, the actual bending moment diagram for each beam at failure was determined. Also, the bending moment diagram based on elastic analysis was determined for each beam using the value of the failure load. This is shown in Fig. 12. It is clear from the figure that unlike beam F-CTC, beam F-CT exhibited a similar level of moment redistribution to that of the control beam. The moment redistribution ratio shown in Table 3 was calculated for each beam by calculating the value of the maximum negative moment, based on the elastic analysis, and comparing it with the experimental value at beam failure. Beam F-CT had a redistribution ratio of 13.4%, which was 6% less than that of the control beam. On the other hand, beam F-CTC had a redistribution ratio of 6.5%, which was significantly less than that of beam F-CT. The ductile behavior of the new fabric resulted in a reasonable ductility in the plastic hinge regions in beam F-CT, which in turn allowed for the redistribution of moment between positive and negative moment zones.

CONCLUSIONS

The unique characteristics of the new triaxial ductile fabric helped to reduce the significant loss in beam ductility associated with the use of conventional FRP materials in flexural strengthening of reinforced concrete beams. The beams strengthened with the new fabric exhibited higher ductility index than those strengthened with the carbon fiber sheet.

The triaxial ductile fabric was successful in providing reasonable ductility at the plastic hinge regions. Therefore, the redistribution of the moment between the negative and positive moment zones of the continuous beam became possible. Redistribution of the moment allowed full utilization of the strength of the beam at the cross sections of maximum positive and negative moments.

Yielding of the triaxial ductile fabric was accompanied by various noticeably audible sounds for a long period of time that are loud enough to be considered as a warning sign.

The beams strengthened with the triaxial ductile fabric did not exhibit anchorage failures. That is attributed, in part, to its ductile behavior. The force in the fabric did not significantly increase after it yielded. Thus, it did not exceed its anchorable force limit and debonding did not take place.

The existence of bundles of fibers in the ± 45° directions enable the triaxial ductile fabric to "self anchor" itself when wrapped around the tension face and the vertical sides of the beam along its length. Therefore, it was generally less vulnerable to anchorage failures than the uniaxial carbon fiber sheet.

The strength of the triaxial ductile fabric was fully exploited as its maximum recorded strains before beam failure were much more than its yield-equivalent strain. In contrast, the maximum recorded strains of the carbon fiber sheet were noticeably less than its ultimate strain, which indicated that its strength was not fully exploited.

ACKNOWLEDGMENTS

This research has been conducted at the Structural Testing Center at Lawrence Technological University, Southfield, Michigan, USA, and was funded by the National Science Foundation under grant No. CMS-9906404, awarded to the first author. The authors wish to thank Diversified Composites Inc., Erlanger, Kentucky for manufacturing the fabric, Shelby Precast Concrete, Shelby Township, Michigan, for contributing the test beams, and Baker Concrete Technology Inc., Columbus, Ohio, for contributing the carbon fiber sheet used.

REFERENCES

Arduini, M., Tommaso, A. D., and Nanni, A., "Brittle Failure in FRP Plate and Sheet Bonded Beams," ACI Structural Journal, V. 94, No. 4, 1997, pp. 363-370.

ASTM D 3039, "Standard Test Method for Tensile Properties of Polymer Matrix Composite Materials," Annual Book of ASTM Standards, ASTM, V. 15.03, 2000, pp. 106-118.

Bencardino, F., Spadea, G., Swamy, N., "Strength and Ductility of Reinforced Concrete Beams Externally Reinforced with Carbon Fiber Fabric," ACI Structural Journal, V. 99, No. 2, 2002, pp. 163-171.

Grace, N. F., Soliman, A. K., Abdel-Sayed, G., and Saleh, K. R., "Strengthening of Continuous Beams Using Fiber Reinforced Polymer Laminates," 4th International Symposium on Fiber Reinforced Polymer Reinforcement for Reinforced Concrete Structures, SP-188, American Concrete Institute, 1999, pp. 647-657.

Grace, N. F., "Strengthening of Negative Moment Regions of Reinforced Concrete Beams Using Carbon Fiber-Reinforced Polymer Strips," ACI Structural Journal, V. 98, No. 3, 2001, pp. 347-358.

Grace, N. F., Abdel-Sayed, G., and Ragheb, W. F., "Flexural and Shear Strengthening of Concrete Beams Using New Triaxially Braided Ductile Fabric," ACI Structural Journal, V. 100, No. 6, November-December, 2003, pp. 804-814.

Norris, T., Saadatmanesh, H., and Ehsani, M. R., "Shear and Flexural Strengthening of R/C Beams with Carbon Fiber Sheets," Journal of Structural Engineering, ASCE, V. 123, No. 7, 1997, pp. 903-911.

Ritchie, P. A., Thomas, D. A., Lu, L., and Connelly, G. M., "External Reinforcement of Concrete Beams Using Fiber Reinforced Plastics," ACI Structural Journal, V. 88, No. 4, 1991, pp. 490-500.

Saadatmanesh, H., and Ehsani, M. R., "RC Beams Strengthened with GFRP Plates I: Experimental Study," Journal of Structural Engineering, ASCE, V. 117, No. 11, 1991, pp. 3417-3433.

Triantafillou, T. C., and Plevris, N., "Strengthening of RC Beams with Epoxy-Bonded-Fiber-Composite Materials," Materials and Structures, V. 25, 1992, pp. 201-211.

Table 1. Properties of the strengthening materials

Type	Yield-Equivalent Load kN/mm (kips/in.)	Yield-equivalent Strain (%)	Ultimate Load kN/mm (kips/in.)	Ultimate Strain (%)	Thickness mm (in.)
Carbon Fiber Sheet	-	-	0.34 (1.95)	1.2	0.13 (0.005)
Triaxial Ductile Fabric	0.19 (1.08)	0.35	0.33 (1.89)	2.10	1.0 (0.039)

Table 2. Summary of test beams

Beam Designation	Strengthening Scheme	Strengthening Material	Positive Moment Strengthening		Negative Moment Strengthening	
			Number of Layers	Strengthened Length	Number of Layers	Strengthened Length
Control B	N/A	N/A	None	None	None	None
F-CT	U-wrap around tension face and sides	Triaxial ductile fabric	1	1.63 m (5.33 ft)	1	1.42 m (4.67 ft)
F-CTC		Carbon fiber sheet	2		2	

Table 3. Summary of test results

Beam (1)	Yield Load kN (kips) (2)	Deflection* at Yield mm (in.) (3)	Failure Load kN (kips) (4)	Deflection* at Failure mm (in.) (5)	Middle Support Reaction at Failure kN (kips) (6)	Max. Negative Moment at Failure** kN.m (kips.in.) (7)	Max. Negative Moment (Elastic Analysis)*** kN.m (kips.in.) (8)	Moment Redistribution Ratio (%) [(8)-(7)]/(8) (9)	Ductility Index (5)/(3) (10)	Type of Final Failure (11)
Control B	92 (20.7)	9.3 (0.37)	127 (28.5)	29.1 (1.15)	168 (37.8)	40.6 (359)	47.3 (418)	14.2	3.12	Steel yield followed by concrete failure
F-CT+	126 (28.3)	9.1 (0.36)	175 (39.3)	23.4 (0.92)	232 (52.1)	56.4 (499)	65.1 (575)	13.4	2.57	Steel & fabric yield followed by fabric rupture
F-CTC++	136 (30.6)	8.9 (0.35)	185 (41.6)	16.1 (0.63)	250 (56.2)	64.4 (569)	68.9 (609)	6.5	1.81	Steel yield followed by shear-tension failure at sheet end

* Deflection at loading point(s).
** Based on the loads and the reactions in column (5) and (7).
*** Equal to 0.188 3Load 3Beam Span
+ Triaxial ductile fabric
++ Carbon fiber sheet

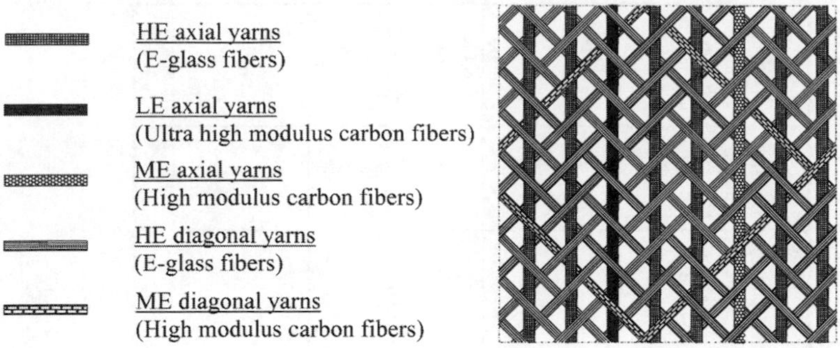

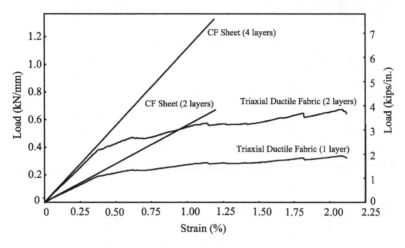

Figure 1 —Details of the triaxial ductile fabric geometry

Figure 2 —Tensile properties of materials used

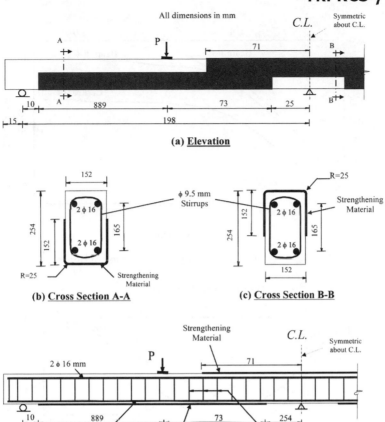

All dimensions in mm

(a) Elevation

(b) Cross Section A-A

(c) Cross Section B-B

(d) Longitudinal Section

Figure 3 —Test beam details

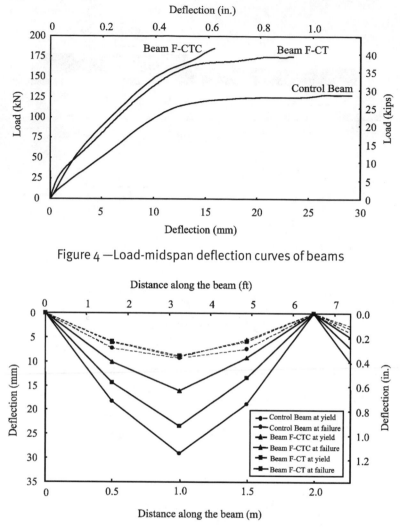

Figure 4 —Load-midspan deflection curves of beams

Figure 5 —Deflection profile of beams

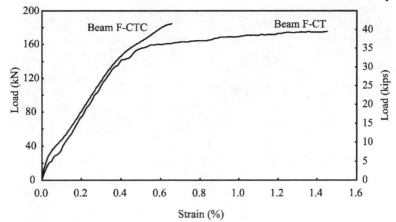

Figure 6 —FRP strain at midspan of beams

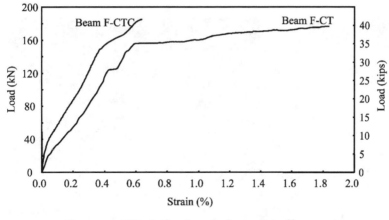

Figure 7 —FRP strain at central support of beams

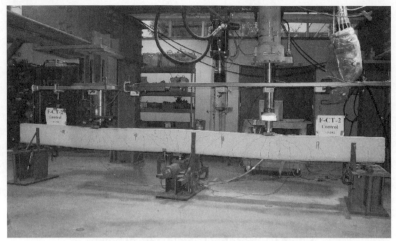

Figure 8 —Control beam B at failure

Figure 9 —Beam F-CT at failure

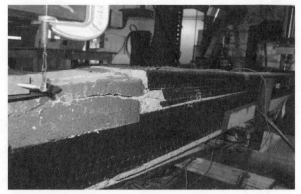

Figure 10 —Shear-tension failure at sheet end of beam F-CTC

Figure 11 —Beam F-CTC at failure

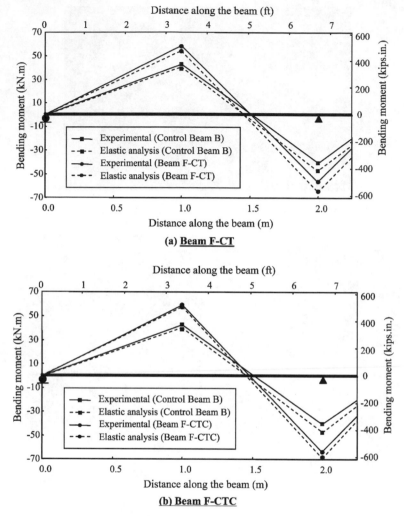

(a) Beam F-CT

(b) Beam F-CTC

Figure 12 —Elastic and experimental bending moment diagrams at failure

Chapter 2

Innovations in FRP Material Testing & Characterization

Material Characterization of FRP Pre-Cured Laminates Used in the Mechanically Fastened FRP Strengthening of RC Structures

by A. Rizzo, N. Galati, A. Nanni, and L.R. Dharani

Synopsis: The Mechanically Fastened-FRP (MF-FRP) strengthening system consists of pre-cured FRP laminates having high longitudinal bearing strength attached to the concrete surface using closely spaced steel fasteners in the form of nails or concrete wedge anchors. The connection depends on several parameters such as, the intrinsic laminates properties (tensile and bearing strength), diameter and effective embedment length of the fasteners, clamping force, presence and type of washer, presence and type of filler in the gaps between the FRP material, the fastener and accessories, and the concrete. This paper focuses on the material characterization of pultruded pre-cured laminate which was used as MF-FRP strengthening to retrofit and rehabilitate off-system bridges in Missouri, USA. Tensile tests, in the longitudinal and transverse direction of the laminate, were performed on full section and open hole coupons in order to characterize the material. Bearing tests, in the longitudinal and transverse direction of the laminate, were performed to determine the capacity of the unrestrained fastener-FRP-concrete connection.

Keywords: flexural strengthening; material characterization; mechanically fastened FRP pre-cured laminates

Andrea Rizzo is a PhD Candidate in Civil Engineering at the University of Lecce-Italy where he also received his B.Sc. in Materials Engineering. He obtained his M.Sc. degree in Engineering Mechanics at the University of Missouri-Rolla. His research interests include analysis and design of RC structures.

Nestore Galati is a research engineer at the University of Missouri-Rolla. He obtained is PhD in Civil Engineering at the University of Lecce-Italy where he also received his B.Sc. in Materials Engineering. He obtained his M.Sc. degree in Engineering Mechanics at the University of Missouri-Rolla. His research has concentrated on the area of retrofitting of masonry and upgrade and in-situ load testing of RC structures.

Antonio Nanni (FACI) is the V & M Jones Professor of Civil Engineering at the University of Missouri-Rolla. He is the founding Chair of ACI Committee 440, Fiber-Reinforced Polymer Reinforcement, and is the current Chair of ACI Committee 437, Strength Evaluation of Existing Concrete Structures. He is an active member in technical committees of ASCE (Fellow), ASTM, and TMS.

Lokeswarappa R. Dharani is professor of Engineering Mechanics and Aerospace Engineering, and Senior Investigator in Graduate Center for Materials Research, at the University of Missouri-Rolla. His research interests are micromechanics, composite materials, fatigue analysis, fracture mechanics, process modeling, wear and friction in composites, analysis and design of laminated glass units.

RESEARCH SIGNIFICANCE

The mechanical characterization of pultruded pre-cured laminate to be used in a Mechanically Fastened-FRP (MF-FRP) strengthening system is discussed. The critical parameters affecting design are investigated from a fundamental point of view, allowing for the development of FRP products usable for these types of applications.

INTRODUCTION

The Mechanically Fastened-FRP (MF-FRP) strengthening system has recently emerged as a practical alternative for the flexural strengthening of RC structures. The efficiency of this strengthening technique was demonstrated in terms of structural performances, and costs, labor and time savings by means of several laboratory (Borowicz 2002; Lamanna 2002; Arora 2003; Ekenel et al. 2005) and field applications. The first reinforced concrete bridge strengthening with the method was performed on a bridge in Edgerton, Wisconsin in 2002 (Bank et al. 2003). This was followed by applications on four rural bridges in Phelps and Pulaski Counties in Missouri (off-system project) in 2004 (Bank et al. 2004; Rizzo 2005).

Different issues were to be solved in the development of the FRP laminate to be mechanically fastened to RC members. The common existing pultruded strips intended to be adhesively bonded on the concrete surface are highly orthotropic in nature, possessing a high modulus and strength in the longitudinal direction, along the length of the laminate, and a low modulus and strength in the transverse direction, across the width of the laminate. The lack of reinforcing fibers in the transverse direction causes these strips to splinter apart when a fastener is driven through them. The pre-cured FRP laminate trademarked under the name SafStrip^{TM}, used in the off-system project (Bank et al. 2004; Rizzo 2005), was designed, produced and commercialized as the result of an investigation performed by Lamanna (2002) having as objective the development of a high bearing strength FRP plate. It consists of a glass and carbon hybrid pultruded strip embedded in a vinyl ester resin. Its thickness and width are 3.175 *mm* and 101.6 *mm* , respectively. Continuous glass fiber strand mats are used to provide transverse and bearing strength, while 16-113 yield E-glass roving and 40-48K Grafil standard modulus carbon tows are utilized to provide longitudinal strength.

Table 1 summarizes the properties of the constituent materials (Arora 2003). Tensile tests were performed on 25.4 *mm* wide full section and open hole (having 4.775 *mm* diameter hole in the middle) coupons: the average stress at failure and the modulus of elasticity resulted 844.0 *MPa* and 61.3 *GPa* (Borowicz 2002; Lamanna 2002; Arora 2003), respectively. Unrestrained bearing tests were made using 4.775 *mm* and 12.7 *mm* diameter pins (the free edge distance was equal to 50.8 *mm*): the maximum load at failure was 3.99 *kN* and 9.45 *kN* , respectively, after the material around the bolt contact area crushed resulting in an elongation of the hole (bearing failure).

A powder-actuated fastening tool was adopted by the University of Wisconsin – Madison (UW) research team for all the laboratory and field applications, in order to increase the speed of installation of the FRP laminates (Borowicz 2002; Lamanna 2002; Arora 2003). The powder-actuated system consists of pins embedded into the base material by means of a gunpowder charge. The type of fasteners was chosen depending on the compressive strength of the concrete. Concrete spalling and cratering were reduced by pre-drilling 12.7 *mm* deep holes in the concrete. A neoprene backed washer was used to prevent localized crushing and provide a clamping pressure, which is known to significantly increase bearing strength in the vicinity of the hole (Camanho et al. 1997; Cooper et al. 1995; Wu et al. 1998). Anchor bolts having a 12.7 *mm* diameter replaced pins in vicinity of the supports to accommodate for the strip delamination at the ends and to obtain extended ductility of the structural elements (Arora 2003). Shear test of single connection showed a bearing mode of failure at a maximum load of 6.08 *kN* .

The use of the powder-actuated fastening system was found efficient for low compressive strength concrete in lab (Borowicz 2002; Lamanna 2002) and field (Arora 2003) application. Nevertheless, during the installation of the FRP strengthening on the field it was found that occasionally, fasteners did not fully penetrate the concrete substrate due to the presence of obstructions (such as large aggregates), and pocket of poor consolidation and/or deteriorated concrete (factors that can be easily controlled in a lab environment) caused loosening of nails. On the other hand, in cases of compressive concrete strength higher than 17.2 *MPa* , the fastening method resulted in concrete spalling and cratering which were considered not acceptable for the full engagement of the laminate. Therefore, the fastening method developed by Bank and Lamanna was modified for the Missouri off-system bridges characterized by high compressive concrete strength and/or with large hard aggregates.

Renouncing the speed of powder-actuated system in order to have a system slower but more reliable, two different types of high strength anchor were tested: concrete wedge anchors and bolts of different size (see Fig. 1). Concrete wedge anchors are inserted into a hole drilled into concrete. The concrete wedge bolts are one piece, heavy duty anchors. The installation of wedge bolts is much easier and faster than the installation of wedge anchors since it only consists in driving it into predrilled holes.

Since the pre-cured FRP laminate SafStrip^{TM} is a relatively new material, a consistent database of its properties was not available at the onset of the project. The values of tensile and bearing properties in literature were found just for $25.4\ mm$ wide coupons, while the cross section of the laminate presents inhomogeneities that could affect the overall capacity of the strip. In addition, the documented open hole tensile strength was not completely defined since it referred to a fixed ratio between the width and the hole size of the coupon. Regarding the bearing capacity, the data available were related to the size of the particular fastener used and, therefore, they could not be adopted for different anchor diameters. Thus, it was needed to investigate the behavior of the laminate under the influence of the new connection geometric parameters chosen.

CHARACTERIZATION OF THE FRP PRE-CURED LAMINATE

Material characterization tests were conducted to determine key characteristics to the design of the strengthening system: tensile and bearing tests in the longitudinal and transverse directions, and open hole tensile tests in the longitudinal direction. For each type of test, different coupon geometries were analyzed and, if possible, at least three units for each type were tested. From pilot tests, it was observed that no tabs were necessary in order to prevent gripping damage.

Coupons were cut using common band and table saws. The bolt holes were machined using the special hole generation procedure developed at the Royal Institute of Technology in Stockholm (Persson 1997). The hole is generated with a cutting tool rotating eccentrically about a principal axis and, at the same time, about its own axis. Using this method, it is possible to machine holes with high precision without causing delamination and chip-out. Fig. 2 shows the conditions of the material around the holes corresponding to different procedures used to generate them. Delamination at the side of the holes, when a common rotary hammer drill bit was used, underlines the major problem associated with the drilling of fiber-reinforced composite materials. In addition to the reduction of the structural integrity of the material, drilling also results in poor assembly tolerance and potential deterioration of long-term performance.

Longitudinal tensile tests

Tensile tests according to ASTM D 3039 were conducted on the laminates to determine longitudinal elastic modulus and tensile strength. Since the cross section of the laminate is not homogeneous, specimens having width in the range $6.6 \div 77.3\ mm$ were tested in order to analyze size effects on the mechanical behavior of the laminate. Both ends of each sample were enclosed in the machine grips for a 76.2 *mm* length. An Instron 25.4 *mm* extensometer was clamped to the middle of each specimen.

Fig. 3 shows the representative failure modes observed in longitudinal tensile tests. The failure was explosive (see Fig. 3a) and occurred always in the gage length close to the middle of the specimens. The rupture started with single carbon fiber failures randomly distributed across the coupon width with subsequent delamination of the two external mat layers. The delamination was complete right before reaching the ultimate load. For coupons wider than 50.8 *mm*, some strips of carbon tows remained unbroken (see Fig. 3b): the number and the width of this strips increased with the width of the specimens.

The behavior of the material was linear elastic up to failure. The modulus of elasticity E was calculated according to ASTM D 3039 as the tensile cord modulus of elasticity within the strain range $1000 \div 3000$ $\mu\varepsilon$. The strain at failure ε_u was defined as the ratio between the stress at failure σ_u and the modulus E. It is possible to distinguish three different groups of coupons in which the stress at failure σ_u was about constant (see Table 2). The first and the second groups have about the same stress at failure but the former has a higher standard deviation. The third group consists of the wider coupons and it is characterized by the lowest value of stress at failure and relative standard deviation. The following observations can be made based on the tests results:

- the ultimate stress σ_u decreases by increasing of the width w;
- the scattering of the ultimate stress σ_u decreases for higher values of the coupon widths w;
- the modulus of elasticity E slightly decreases with the increasing of the width w;
- there is no correlation between the strain at failure ε_u and the width w of the coupons;
- the results obtained in the longitudinal tensile tests are consistent with the theoretical ones calculated according to the mixture rule. The theoretical stress and strain at failure, and the elastic modulus are 1094 *MPa*, 17648 $\mu\varepsilon$ and 62 *GPa*, respectively. The difference between the experimental and the theoretical results can be attributed to the presence of defects and premature failure modes that the theory does not take into account.

Transverse tensile tests

Transverse tensile tests according to ASTM D 3039 were conducted on the laminates to determine their elastic modulus and tensile strength in the transverse direction. Specimens having width in the range $14.1 \div 49.0$ *mm* were used in order to analyze the size effects on the mechanical behavior of the laminate. Both ends of each sample were enclosed in the machine grips for a 12.7 *mm* length. An Instron 25.4 *mm* extensometer was clamped to the middle of some specimens.

Fig. 4 shows the representative type of failure for the transverse tensile tests. The failure was brittle and net across the entire width of the samples: it occurred always in the gage length of the specimens (see Fig. 4a). The rupture started with the appearance of one or more splitting cracks through the carbon fibers core along the load direction, cracks that propagate until the ultimate load was reached (see Fig. 4b): to be noted that, by visual inspection, no splitting cracks were detected running through the glass roving, showing that the glass fibers-vinyl ester mechanical interface is more efficient than the carbon fibers-vinyl ester one.

Fig. 5 depicts the average stress-strain ($\sigma - \varepsilon$) curve up to failure. It is possible to state that the behavior of the material in the transverse direction can be modeled as a parabola with satisfactory approximation. The following observations can be made based on the tests results (see Table 3):

- no correlation was found between the ultimate stress σ_u and the coupons width w;
- the modulus of elasticity at origin E_o, calculated as the modulus at origin of the parabola that best fit the experimental stress-strain curve, increases with the increasing of the width w;
- no correlation was found between the strain at failure ε_u and the width w of the coupons;
- the high scattering in the stress and strain at failure can be attributed to the particular lay-up of the laminate. The non regularity in the cross section could have caused stress concentration points from where cracks initiated and then propagated along the load direction (see Fig. 4b).

Longitudinal open hole tensile tests

Longitudinal open hole tensile tests according to ASTM D 5766 were conducted on the laminates to determine the properties in the longitudinal direction based on the coupons net section. Different specimens were tested to analyze the size effects on the mechanical behavior of the laminate. In particular, the ratio $\rho = w/\phi$, defined as the ratio between the width of the specimens w and the diameter of the holes ϕ, was varied in the range 1.27 to 12.0. Both ends of each sample were enclosed in the machine grips for a length equal to $63.5\ mm$. An Instron $25.4\ mm$ extensometer was clamped to the middle of the samples.

Fig. 6 shows the representative type of failure for open hole longitudinal tensile tests (see Table 4). The failure was brittle and explosive across the entire width of the samples: it occurred always in the gage length of the specimens with extensive splitting and delamination. The failure was always anticipated by single carbon fiber failures randomly distributed in the coupon width. Fig. 6a shows the pattern of the cracks around the hole in the outer layers for different coupons. In general, it was observed that:

- for ratios $w/\phi \cong 1.0 \div 1.5$, not appreciable cracking could be seen up to failure but, reached the ultimate load, the outer layers failed across the drilled section;
- for ratios $w/\phi \cong 2.0 \div 4.0$, a crack formed at the center of each quadrant of the hole and propagated at 45° reaching, in some cases, the side of the coupon deviating from the original toward the horizontal direction;

• for ratios $w/\phi > 4.0$, the cracks, formed at low load level, tended to propagate in the longitudinal direction marking the splitting of the carbon core starting from the hole edges (see also Fig. 6a);

• a visual inspection of the specimens after failure showed the splitting of a carbon strip wide as the hole size and running across all the half length of samples (see Fig. 6b). Fig. 7 and Fig. 8 plot the average stress at failure on the net area $\sigma_{u,n}$ as a function of the width of the coupons w and of the ratio w/ϕ, respectively.

Bearing tests in the longitudinal direction of the laminate

Double-shear tensile loading tests according to ASTM D 5961 Procedure A were conducted on the laminates to determine the bearing response in the longitudinal direction. For this group of tests, designated as series A, no clamping pressure was applied on the contact area pin-FRP material (unconstrained specimens). Specimens having ratios $w/\phi \in 2.7 \div 10.5$ and $e/\phi \in 0.8 \div 7.7$ were used to analyze the effect of the geometry on the mechanical behavior of the laminate. w and ϕ are the width of the coupon, and the hole size, respectively. e is the distance, parallel to load, from center of the hole to end of specimen. Fig. 9 shows the test setup used for the bearing tests. The particular fixtures allowed using pins of different size maintaining a clear sight of the hole deformation during the loading of the specimens. Because of the particular test setup and test machine available, it was possible to record just the data related to the load and the cross-head displacement.

Fig. 10 shows the representative type of failure for the series A longitudinal bearing tests. Mainly, two different modes of failure occurred depending on the ratios w/ϕ and e/w. Maintaining constant w/ϕ, for low values of e/w coupons failed along shear-out planes parallel to the load direction (see Fig. 10a), while for high values of e/w coupons failed for bearing in the material immediately adjacent to the contact area of pin and laminate (see Fig. 10b). The shear-out failure started with the appearance of two longitudinal cracks on the hole edges at both sides of the coupon; then, the cracks ran quickly up to the free edge of the specimen. No relevant damage was found at the outer layers in contact with the pin, while the inner carbon fibers layer split along the middle plane of the laminate (see Fig. 10c). Such mode of failure was less extensive for specimens having higher distance between hole and edge. The failure due to bearing has been shown to occur through buckling and brushlike failure with consequent delamination of the outer layers in the material far away from the contact area (see Fig. 10b): commonly, carbon fibers were push out in the final part of the test (see Fig. 10d).

According to ASTM D 953 and ASTM D 5961, the experimental results can be summarized in terms of maximum bearing stress $\sigma_{br,max}$, cross-head displacement at failure Δ_u for each group of specimens, bearing strength $\sigma_{br,u}$ and the bearing chord stiffness E_{br}. To be noted that generally, the bearing failure starts at the contact points in the 0° – direction and then propagates through the rest of the contact area. This means that the stress distribution on the contact area is not constant and that the effective maximum bearing stress is higher than the average value suggested by ASTM D 953 and calculated using the conventional bearing area. The hole opening δ was estimated subtracting the elastic deformation of the test apparatus from the cross-head displacement (Rizzo 2005). The following observations can be made based on the tests results (see Fig. 11 and Fig. 12 related to $\phi = 9.525$ mm and reported as example):

- the maximum bearing stress $\sigma_{br,\max}$ strongly depends on the distance from the edge e. It is mildly influenced by the hole size ϕ, and it almost does not depend on the width w;
- the maximum bearing stress $\sigma_{br,\max}$ does not depend on the width of the coupons;
- by maintaining a constant diameter ϕ, there is a minimum value of the edge distance $e_{\min}$ beyond which the maximum bearing stress $\sigma_{br,\max}$ assumes a constant value independently from the edge distance e and below which there is a linear correlation between $\sigma_{br,\max}$ and e. Fixing an edge distance $e \leq e_{\min}$, the maximum bearing stress $\sigma_{br,\max}$ is slightly higher for smaller diameter size;
- no correlation was found between the edge distance $e_{\min}$ and the width of the coupons;
- the bearing chord modulus E_{br} is not affected by the width of the specimen and it tends to increase with the edge distance.

Double bearing tests in the longitudinal direction of the laminate

Double-shear tensile loading tests according to ASTM D 5961 Procedure A were conducted on the laminates to investigate the bearing response in the longitudinal direction in the presence of two holes at the same distance from the free edge. For this group of tests, designated as series B, no clamping pressure was applied on the contact area pin-FRP material (unconstrained specimens). The investigation was limited to specimens with full width of the laminate, and using 9.525 mm diameter holes spaced at 50.8 mm symmetrically with respect to the center of each specimen. Different values of edge distance were tested in order to analyze the effect of this geometrical parameter on the mechanical behavior of the laminate.

The mode of failure was bearing and occurred through buckling and brushlike failure with consequent delamination of the outer layers. The failure was not contemporaneous for both holes. The following observations can be made based on the tests results:

- the bearing chord modulus E_{br} tends to increase with the edge distance since more material reacts to the pressure of each pin;
- the maximum bearing stress $\sigma_{br,\max}$ is independent from the edge distance e: $\sigma_{br,\max} = 213.5 \pm 15.6$ MPa. This value is slightly smaller (about 9%) than the maximum bearing stress found for the series A specimens;
- the curves $(\sigma_{br} - \varepsilon_{br})$ are generally linear up to $60 \div 80\%$ of the maximum bearing stress with a subsequent deviation to the linearity depending on the failure mode (the non

linearity is higher for bearing failure).

Bearing tests in the transverse direction of the laminate

Double-shear tensile loading tests according to ASTM D 5961 Procedure A were conducted on laminates to investigate the bearing response in the transverse direction. For this group of tests, designated as series C, no clamping pressure was applied on the contact area pin-FRP material (unconstrained specimens). Specimens having ratios $w/\phi \in 2.5 \div 10.0$ and $e/\phi \in 1.3 \div 3.8$ were tested to analyze the effect of the geometry on the mechanical behavior of the laminate.

Fig. 13 shows the representative type of failure for series C transverse bearing tests. The ratio e/w has no effect on the type of failure since the presence of the longitudinal carbon fiber prevents the premature shear-out failure. Therefore, mainly two different modes of failure occurred depending on the ratio w/ϕ. For low values of w/ϕ, coupons failed for net tension (see Fig. 13a): for the smaller width w, traces of bearing failure damages were localized in the material immediately adjacent to the contact area of pin and laminate. For high values of w/ϕ, coupons failed for bearing in the material immediately adjacent to the contact area of pin and laminate (see Fig. 13b): for the smaller width w, net tension failure occurred right after the bearing failure. The net tension failure started with the appearance of two transverse cracks on the hole boundary at both sides of the coupon; then, the cracks ran quickly up to the free edges of the specimen. By visual inspection, no relevant damages were found in the outer layers of material in contact with the pin, except for the specimens with the highest value of w for which the initial stage of the bearing failure process could developed due to the higher net tension capacity of the coupons. The failure due to bearing has been shown to occur through brushlike failure with consequent delamination of the outer layers. It is important to note that the damaged zone is less extensive than what found in the series A longitudinal bearing tests. In fact, in the transverse bearing tests, there is not buckling but the carbon fibers are just pushed in a more stable configuration creating a natural ring in the contact area against the pin, a fact that increased the post-bearing capacity until the final secondary net tension failure. The following observations can be made based on the tests results:

- the maximum bearing stress $\sigma_{br,\max}$ depends on the geometry of the coupon: it strongly depends from the coupon width w (see Fig. 14), slightly from the edge distance e and mildly from the hole size ϕ;
- the maximum bearing stress $\sigma_{br,\max}$ depends on the width of the coupons (Fig. 14b);
- with a fixed diameter ϕ, there is a minimum value of the edge distance $e_{\min}$ beyond which the maximum bearing stress $\sigma_{br,\max}$ assumes a constant value independently from the edge distance e and below which there is a linear correlation between $\sigma_{br,\max}$ and e;
- the edge distance $e_{\min}$ slightly depends on the hole size ϕ and tends to increase with the diameter;

• the curves $\left(\sigma_{br} - \varepsilon_{br}\right)$ are generally linear up to $75 \div 90\%$ of the maximum bearing stress, while subsequent deviation to the linearity become more evident depending on the failure mode;

• the bearing capacity depends just on the contact area that the pin can provide. In fact, it was proved that the maximum bearing stress is constant for specimens of a certain hole size tested with pins of different size (Rizzo 2005).

CONCLUSIONS

Conclusions based on the material characterization tests of the SafStrip^{TM} pre-cured FRP laminate can be summarized as follows:

• the SafStrip^{TM} laminate behavior is linear elastic up to failure for tensile load in the longitudinal direction. For design purposes, the stress at failure and the elastic modulus found for full size specimens can be conservatively assumed to be $836 \, MPa$ and $62 \, GPa$, respectively. The resultant strain at failure was found to be $13809 \, \mu\varepsilon$;

• the SafStrip^{TM} laminate behavior can be modeled as a parabola for tensile load in the transverse direction: $\sigma_{av} \left[MPa \right] = -1.884 \cdot 10^{-7} \mu\varepsilon^2 + 7.151 \cdot 10^{-3} \mu\varepsilon$ with modulus at origin equal to $E_o = 7151 \, MPa$. In addition, no size effects on the mechanical behavior of the laminate for tensile load in the transverse direction have been observed. For design purpose, the average stress and the strain at failure can be conservatively assumed to be $65.6 \, MPa$ and $13498 \, \mu\varepsilon$, respectively;

• the full size SafStrip^{TM} laminate containing a hole in the middle of the width behaves as the same laminate without hole. Therefore, for design purposes, the stress on the net area, the elastic modulus and the strain at failure for full size specimens with hole in the middle of the width can be assumed to be equal to the corresponding values found for full size specimens without hole, that is $836 \, MPa$, $62 \, GPa$ and $13809 \, \mu\varepsilon$, respectively. In addition, it is reasonable to state that the same conclusions are valid for full size ($101.6 \, mm$ wide) laminates with a hole in a generic position along the width provided the ratio $w/\phi \geq 3$;

• the maximum bearing stress in the longitudinal direction for unconstrained material around the hole is equal to $235.2 \pm 17.2 \, MPa$ and it does not depend on the size of the hole. This value of bearing stress can be only reached by choosing an adequate value of the edge distance: according with the experimental results, the minimum edge distance that satisfied the previous requirement is about $25 \, mm$;

• the maximum bearing stress in the transverse direction for unconstrained material around the hole is equal to $160.1 \pm 23.2 \, MPa$ and it can be considered independent from the size of the hole. This value of bearing stress can be reached only choosing an adequate value of the edge distance ($15 \, mm$ for the tested laminate).

ACKNOWLEDGMENTS

The project was made possible with the financial support received from the UMR - University Transportation Center on Advanced Materials, Center for Infrastructure Engineering Studies at the University of Missouri-Rolla and Meramec Regional Planning Commission (MRPC). Strongwell provided the FRP materials.

REFERENCES

Arora, D., 2003, *Rapid Strengthening of Reinforced Concrete Bridge with Mechanically Fastened Fiber-Reinforced Polymer Strips*, Master's thesis, University of Wisconsin – Madison, WI.

ASTM D 953-02, 2002, *Standard Test Method for Bearing Strength of Plastics*, American Society for Testing and Materials, West Conshohocken, PA.

ASTM D 3039/D 3039M, 2000, *Standard Test Method for Tensile Properties of Polymer Matrix Composite Materials*, American Society for Testing and Materials, West Conshohocken, PA.

ASTM D 5766/D 5766M, 2002, *Standard Test Method for Open Hole Tensile Strength of Polymer Matrix Composite Laminates*, American Society for Testing and Materials, West Conshohocken, PA.

ASTM D 5961/D 5961M, 2001, *Standard Test Method for Bearing Response of Polymer Matrix Composite Laminates*, American Society for Testing and Materials, West Conshohocken, PA.

Bank, L.C.; Arora, D.; Borowicz, D.T.; and Oliva, M., 2003, "Rapid Strengthening of Reinforced Concrete Bridges*," Wisconsin Highway Research Program*, Report Number 03-06, 166 pp.

Bank, L.; Nanni, A.; Rizzo, A.; Arora, D.; and Borowicz, D., 2004, "Concrete Bridges Gain Strength with Mechanically Fastened, Pultruded FRP Strips," *Composites Fabrication Magazine*, September, pp. 32-40.

Borowicz, D. T., 2002, *Rapid Strengthening of Concrete Beams with Powder-Actuated Fastening Systems and Fiber-Reinforced Polymer (FRP) Composite Materials*, Master's thesis, University of Wisconsin – Madison, WI.

Camanho, P. P. and Matthews, F. L., 1997, "Stress Analysis and Strength Prediction of Mechanically Fastened Joints in FRP: a Review," *Composites Part A*, Vol. 28A, pp. 529-547.

Cooper, C. and Turvey, G. J., 1995, "Effects of Joint Geometry and Bolt Torque on the Structural Performance of Single Bolt Tension Joints in Pultruded GRP Sheet Material,"

146 Rizzo et al.

Composite Structures, Elsevier, Vol. 32, pp. 217-226.

Ekenel, M.; Rizzo, A.; Myers, J. J.; and Nanni,, A., 2005, "Effect of Fatigue Loading on Flexural Performance of Reinforced Concrete Beams Strengthened with FRP Fabric and Pre-Cured Laminate Systems," *Conference of Composites in Construction*, Lyon, France.

Lamanna, A.J., 2002, *Flexural Strengthening of Reinforced Concrete Beams with Mechanically Fastened Fiber Reinforced Polymer Strips*, PhD thesis, University of Wisconsin – Madison, WI.

Persson, E.; Eriksson, I.; and Zackrisson, L., 1997, "Effects of Hole Machining Defects on Strength Fatigue Life of Composite Laminates," *Composites Part A*, 28A, pp. 141-151.

Rizzo, A., 2005, *Application in Off-System Bridges of Mechanically Fastened FRP (MF-FRP) Pre-Cured Laminates*, Master's thesis, University of Missouri-Rolla, MO.

Wu, T. J. and Hahn, H. T., 1998, "The Bearing Strength of E-Glass/Vinyl-Ester Composites Fabricated by VARTM," *Composites Science and Technology*, Vol. 58, pp. 1519-1529.

Table 1 -- Components Properties of the Pre-cured FRP Strip (Arora 2003)

Component	Tensile Elastic Modulus[*] $[GPa]$	Cross Sectional Area or Thickness[*]	Tensile Ultimate Strength[*] $[MPa]$
16-113 Yield E-glass Roving	72.4	1.730 mm^2	3448
40-48K Standard Modulus Carbon Tows	234.5	1.778 mm^2	4137
42.5 *gr* Continuous Strand Mat	6.90	0.737 mm[**]	345
Vinyl Ester Resin	3.38	NA	81.4

[*] Obtained from manufacturers data.
[**] Mat thickness based on 552 *kPa* .

Table 2 -- Experimental Longitudinal Properties of the SafStrip[™] Laminate

Coupons	Failure Stress σ_u $[MPa]$	Modulus of Elasticity E $[GPa]$	Strain at Failure $\varepsilon_u = \dfrac{\sigma_u}{E}$ $[\mu\varepsilon]$
6.6 *mm* $\le w <$ 35 *mm*	950±120	68.0±32.4	14101±1639
35 *mm* $\le w <$ 60 *mm*	1002±66	62.5±4.2	16261±1920
60 *mm* $\le w \le$ 77.3 *mm*	836±30	62.7±7.6	13809±1599
6.6 *mm* $\le w \le$ 77.3 *mm* (Average Value)	937±109	65.6±5.9	14705±1938
Specimen without Hole for Longitudinal Tensile Test			

Grip Length $s = 76.2$ *mm*
Thickness $t = 3.175$ *mm*
Total Length $L = 381 \pm 5$ *mm*
Width $w \in 6.6 \div 77.3$ *mm*

Table 3 -- Experimental Transverse Properties of the SafStrip™ Laminate

Coupons	Failure Stress σ_u [MPa]	Elastic Modulus at Origin E_o [GPa]	Strain at Failure ε_u [$\mu\varepsilon$]
14.1 mm ≤ w < 49.0 mm	65.6±14.5	6.765±1.017	13498±3940
Specimen without Hole for Transverse Tensile Test			

Grip Length $s = 12.7$ mm
Thickness $t = 3.175$ mm
Total Length $L = 101.6 \pm 0.5$ mm
Width $w \in 14.1 \div 49.0$ mm

Table 4 -- Experimental Longitudinal Open Hole Properties of the SafStrip™ Laminate

Coupons	Average Failure Stress At Failure $\sigma_{u,n} = \dfrac{P}{(w-\phi)t}$ [MPa]	Elastic Modulus E_{oh} [GPa]	Strain at Failure ε_u [μc]
$w/\phi = 6$; $w \leq 67.2$ mm	$-12.681\, w\,[mm] + 991$	NA	NA
$w/\phi = 6$; $w > 67.2$ mm	787±17	70.8±3.0	11755±1467
$\phi = 6.35$ mm; $w \geq 25.4$ mm	870±91	66.4±2.8	13503±1090
$\phi = 9.525$ mm; $w \geq 25.4$ mm	882±62	71.5±3.3	12613±226
Average Values for $w/\phi \geq 3$	898±114	70.2±3.9	12594±1137
Specimen with Hole for Longitudinal Open Hole Tensile Test			

Grip Length $s = 63.5$ mm
Thickness $t = 3.175$ mm
Total Length $L = 381 \pm 5.8$ mm
Width $w \in 12.2 \div 101.6$ mm
Diameter $\phi \in 1.43 \div 12.81$ mm

a) Concrete Wedge Anchor

b) Concrete Wedge Bolt

Fig. 1 — Concrete Anchors

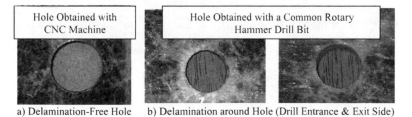

Hole Obtained with CNC Machine

Hole Obtained with a Common Rotary Hammer Drill Bit

a) Delamination-Free Hole   b) Delamination around Hole (Drill Entrance & Exit Side)

Fig. 2 — FRP Material Conditions around Holes

a) Explosive Failure   b) Failure Details for Coupons Wider than 50.8 *mm*

Fig. 3 — Failure Modes in Longitudinal Tensile Tests of SafStrip™ Coupons

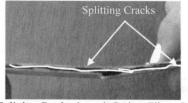

a) Net Failure across the Width   b) Splitting Cracks through Carbon Fibers Core

Fig. 4 — Type of Failure for Transverse Tensile Tests of SafStrip™ Coupons

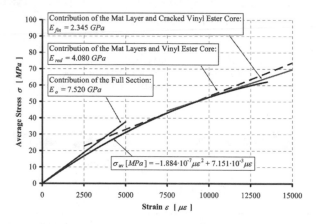

Fig. 5 — Average Stress-Strain ($\sigma - \varepsilon$) Curve for Transverse Tensile Tests

a) Cracks and Longitudinal Delamination   b) Split Strip Starting from Hole Edges

Fig. 6 — Type of Failure for Longitudinal Open Hole Tensile Tests of SafStrip™ Coupons

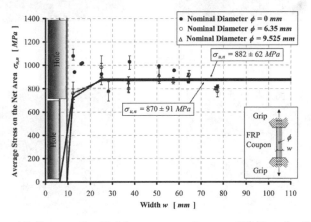

Fig. 7 — Stress at Failure on the Net Area $\sigma_{u,n}$ vs. Coupon Width w for Constant Hole Size for Longitudinal Open Hole Tensile Tests

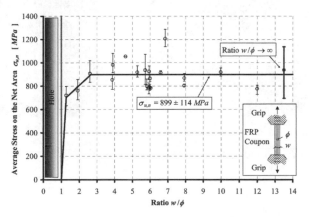

Fig. 8 — Stress at Failure on the Net Area $\sigma_{u,n}$ vs. Ratio w/ϕ for Longitudinal Open Hole Tensile Tests

a) Samples with Width $w \le 76.2$ *mm*

b) Samples with Width $w > 76.2$ *mm*

Fig. 9 — Bearing Tests Setup

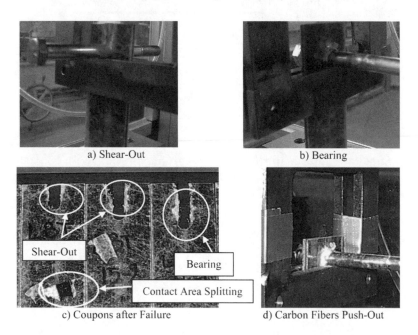

a) Shear-Out

b) Bearing

Shear-Out

Bearing

Contact Area Splitting

c) Coupons after Failure

d) Carbon Fibers Push-Out

Fig. 10 — Type of Failure for Series A Longitudinal Bearing Tests of SafStrip™ Coupons

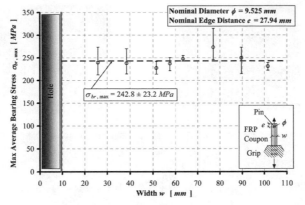

Fig. 11 — Series A Longitudinal Bearing Tests Results for w Variable,
ϕ = 9.525 mm and e = 27.94 mm

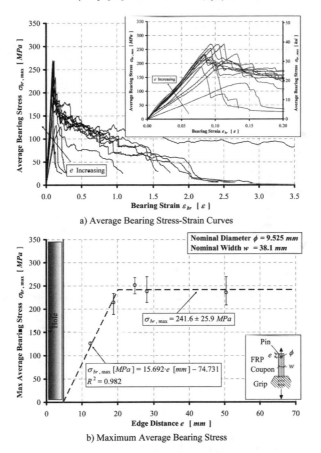

a) Average Bearing Stress-Strain Curves

b) Maximum Average Bearing Stress

Fig. 12 — Series A Longitudinal Bearing Tests Results for e Variable,
ϕ = 9.525 mm and w = 38.1 mm

a) Net Tension Failure without Bearing Traces   b) Net Tension Failure after Bearing

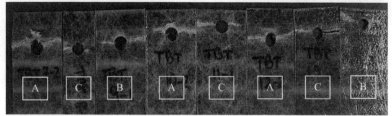

c) Coupons after Failure: Net Tension Failure with Severe (A), Minor (B) and
   Negligible (C) Bearing Damages in the Contact Area

Fig. 13 — Type of Failure for Series C Transverse Bearing Tests of SafStrip™ Coupons

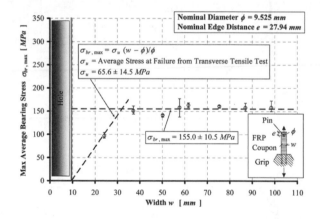

Fig. 14 — Series C Transverse Bearing Tests Results for w Variable,
$\phi = 9.525$ mm and $e = 27.94$ mm

Method for Screening Durability and Constituent Materials in FRP Bars

by D. Gremel, N. Galati, and J. Stull

Synopsis: A method for screening the durability of FRP bars under bending stress and immersion in high pH solution at elevated temperature is described. Discussion of the need for such a test, process variables affecting durability, determination of the appropriate bending radius and a description of the test method are shown. Test results from a series of eight production runs varying only one of the processes related variables, glass fiber supplier, are shown. Fiber sizing chemistry for the fiber/resin/production system is key to better durability of GFRP rebar. The bending stress durability test method helps reveal FRP bar system performance for different constituent materials and offers a more practical method for evaluating alkaline durability of GFRP bars. The method is intended as an indicator of durability performance and not a definitive evaluation.

Keywords: alkaline exposure of FRP bars; bending stress; constituent materials; durability; GFRP bar

Doug Gremel is Director of Non-metallic Reinforcing for Hughes Brothers, Inc of Seward Nebraska and Aslan Pacific Ltd of Hong Kong. is member in the technical committees of ACI and serves as an active member on ACI Committee 440, Fiber Reinforced Polymer Reinforcement.

Nestore Galati is a post doctorate research assistant at the University of Missouri Rolla. He obtained is PhD in Civil Engineering at the University of Lecce- Italy where he also received his B.Sc. in Materials Engineering. He obtained his M.Sc. degree in Engineering Mechanics at the University of Missouri-Rolla. His research has concentrated on the area of retrofitting of masonry and upgrade and in-situ load testing of RC structures.

Jim Stull is Quality Control Manager of Hughes Brothers, Inc. Seward Nebraska.

A SIMPLE METHOD FOR SCREENING FRP BARS FOR DURABILITY

The recently published ACI 440.3R-04, "Guide Test Methods for Fiber-Reinforced Polymers for Reinforcing or Strengthening Concrete Structures"[1] includes the description of an accelerated test method for measuring the alkali resistance of FRP bars. The method, test B.6, includes procedures that immerse the FRP bars in an alkaline solution at elevated temperature with and without a sustained tensile load applied. A third procedure subjects the FRP bars to a sustained tensile load while encased in moist concrete at elevated temperature. While performing these durability tests is critical and gives needed information about the long term durability prospects of a given FRP bar, the tests themselves are difficult to conduct as they require elaborate jigs, fixtures and a large physical area for maintaining the FRP bars in the prescribed conditions.

The need exists for a test that is easier to perform and gives a result that is indicative of success in the more elaborate ACI440.3R-04 test method B.6. Such a test is used at Hughes Brothers, Inc of Seward Nebraska to screen material constituents, changes in production process variables and to qualify raw material suppliers. The test method gives insight into production related variables that affect durability and has the benefit of helping improve the end product. Owners of structures that will use FRP bars, such as Departments of Transportation around the world, also have the need for a simple test to screen FRP bar suppliers and pre-qualify manufacturers for standards approval and specific projects.

It is hoped that this simple test method will be used to help improve the understanding of production process variables and interaction of the components of an FRP bar "system". Ultimately, there is a need to validate ACI 440 C_e, environmental reduction factors[2] used in the various 440 documents and for a "black-box" test that can be used for service life prediction models, which are correlated to more elaborate test methodologies.

PROCESS RELATED DURABILITY VARIABLES

If two separate GFRP rebars, each unique in their composition, were placed on the table before you, they may appear identical to the naked eye. Short-term material properties may also be within a range that would cause one to believe they are similar in their constituent make up and suitability for a given use. However, there can be a very wide range in the "durability performance" of the two bars when subjected to accelerated aging tests in a high pH environment. The situation is similar to a pair of steel bars with differing chemical composition or different carbon content, except that these differences in steel properties are now well documented and understood. The chemical composition and processing of FRP materials are not as well known to the civil engineering community as are those of steel materials. This situation can be problematic for potential owners and users of FRP bars when it is necessary to qualify differing vendors for a given project. Simply specifying constituent materials in vague terms such as "e-glass/vinyl ester resin" GFRP bars falls short of truly quantifying durability performance of the GFRP rebar "system".

Some of the production process variables that can affect the durability of the FRP bar include the following:

- Fiber Type and Supplier
- Resin Type and Supplier
- Fiber Sizing
- Processing Line Speed
- Rate of Cure
- Rebar Surface treatment

- Curing Temperature
- Thoroughness of Cure
- Resin Shrinkage
- Resin Additives and fillers
- Catalysts
- Fiber Volume

Fortunately, the pultrusion process is a fairly stable one and individual process control variables can be closely controlled. The FRP rebar supplier is free to make a number of choices among these variables in the manufacture of their product. Some choices are dictated to the supplier by their production method and the need to add processing additives to eliminate phenomenon such as resin shrinkage cracking. The use of catalysts to effect curing of the rebar and fillers to improve the process ability of the resin balanced with processing speeds and production efficiencies are all factors that weigh into the methods used. There are a number of ASTM tests that can measure process related parameters such as degree of cure. For example, ASTM D4475[3] is a simple short beam shear test that can be applied to production line specimens directly off the line and compared to specimens that are post-cured in an oven. The relative degree of cure can be measured in a straightforward comparison of the values before and after post curing the sample rebar. Parameters such as fiber volume fraction are directly related to the short-term physical mechanical properties such as ultimate tensile strength, which should also be measured for each production run by the manufacturer. The choice of fillers is also an important one for the manufacturer. When used in the proper amount, the use of fillers can improve the physical mechanical properties of the FRP Rebar and act to reduce costs by decreasing the amount of neat resin used. Fillers also have an important role to play in enhancing the durability performance of the FRP Rebar. As bar diameters increase, ensuring thoroughness of cure while avoiding the occurrence of thermal cracks, which are strength reducing defects and avenues for direct alkaline penetration of the bar, necessitate the use of resin shrinkage additives and slower line processing speeds.

With all these variables factoring into the durability performance of the FRP bar, and the physical appearance being indistinguishable from bar to bar, how is the end user to know they are getting a suitable product? It's not sufficient for the specifier to simply require the FRP bar supplier use "e-glass/vinyl ester resin" and be certain they are getting a desirable end product. A better approach is to approve the FRP bar "system" by pre-qualifying sample bars that have been subjected to a "durability by four point bending stress test" in addition to requiring material test certification and validation of a plant certified quality assurance program. The durability by bending test method[4] is intended to offer a reasonably quick and straightforward means to gage the relative durability performance of the FRP bar supplier's "system".

DETERMINATION OF STRESS BY BENDING FORCES

The problem of the evaluation of the stresses induced by the bending of an initially straight bar is fairly complicated. An exact solution can be found by using the Muskhelishvili stress functions[5]. However, if the radius of the bar is small compared to the radius of the bend, the problem can be solved with a good accuracy, by invoking the curved beam theory[6,7]. For a curved beam, the stresses are not distributed in a linear fashion across the section, although the plane section can be assumed to be plane during the deformation. Therefore the neutral axis, the location at which the stresses vanish, cannot coincide with the centroidal axes and can be determinate using the following equation:

$$r_n = \frac{\rho I_2}{I_2 + A\rho^2} \quad (1)$$

where ρ is the radius of curvature of the bar, A is the area of the bar and I_2 is given by:

$$I_2 = \iint_{Area} \frac{y^2}{g} dA = \iint_{Area} \frac{y^2}{1+ky} dA \quad (2)$$

where k is the curvature and y is the local ordinate measured from the centroid of the bar as in the case of a straight beam.

Equation (1) becomes very simple in the case of circular section with constant curvature $k = \frac{1}{R}$ and no torsion:

$$r_n = \frac{\left(\dfrac{\phi}{2}\right)^2}{2\left[R - \sqrt{R^2 - \left(\dfrac{\phi}{2}\right)^2}\right]} \quad (3)$$

where ϕ is the diameter of the bar.

Since r_n does not change during the bending of the bar, in order to hoop a bar with a final radius R, it is necessary to use a straight bar with a length of $2\pi r_n$. Consequently, the strain in the FRP can be determined by simple geometrical considerations:

$$\varepsilon_{t,max} = \frac{R + \dfrac{\phi}{2} - r_n}{r_n} 100 \quad (4)$$

Equations (3) and (4) allow determining the maximum tensile strain in the FRP given the diameters of the hoop and of the bar. In fact, assuming the diameter of the hoop and of the bar, Eq. 3 allows determining the neutral axis position, r_n, which substituted in Eq. 4 gives the maximum tensile strain $\varepsilon_{t,max}$. On the contrary, given the diameter of the bar and the maximum strain in the bar, Eq. 3 and 4 can iteratively be solved to determine the required diameter of the bend.

DESCRIPTION OF DURABILITY TEST BY BENDING STRESS

Using the concept of inducing stress in the bar by bending or hooping, the following procedure describes a method for evaluating the long-term durability (alkaline resistance) of GFRP rebar by immersion in an aqueous solution at an elevated temperature for a specified period of time. Stress in the bar is induced by bending stresses equivalent to 25% of the ultimate tensile strength of the bar for these experiments. However, changing the bending radius of the bar varies the amount of stress in the bar. The test reveals the residual tensile strength, tensile modulus and ultimate strain retention after the specified exposure to the test solution. Depending on climatic conditions under which the bar will see service, the test can be used to approximate the residual properties of the bar after approximately 50 or 75 years of service depending on the duration of this test.[8,9]. Based on the work of Porter and Barnes, the expression for simulated age is a function of time in the aqueous solution and its temperature.

$$Age\left(\frac{Days - Real\ time}{Days\ Accelerated\ Aging}\right) = \begin{cases} 0.200 \times e^{0.052(T)} & \text{if T in degrees Fahrenheit} \\ 0.200 \times e^{0.052\left(32 + \frac{5}{9}T\right)} & \text{if T in degrees Celsius} \end{cases} \quad (5)$$

This expression is based on real time exposure of Glass Reinforced Concrete or GRC (Litherland et al, 1981)[10] and is used to transform time for the GFRP bars in alkali solution. However it is not obvious that the time-temperature relationship that applies to GRC is valid for GFRP[8]. Another approach, proposed by Dejke involves using time shift factors obtained from strength reduction curves for GFRP bars exposed in alkaline solutions at differing temperatures. If these curves are of the same shape, then an Arrhenius relationship can be considered and time shift factors established for a given FRP bar. The relationship between strength retention curves on a logarithmic time scale at different exposure temperatures gives the Time Shift Factor described by Dejke as.

$$TSF = \begin{cases} e^{\frac{B}{T_1+273.15} - \frac{B}{T_2+273.15}} & \text{if } T_1 \text{ and } T_2 \text{ in degrees Celsius} \\[2ex] e^{\frac{B}{(T_1-32)\frac{5}{9}+273.15} - \frac{B}{(T_2-32)\frac{5}{9}+273.15}} & \text{if } T_1 \text{ and } T_2 \text{ in degrees Fahrenheit} \end{cases}$$

(6)

where, TSF is the Time Shift Factor and;
B is a constant determined using a logarithmic graph of strength retention curves and;
T_1, T_2 are the temperatures between with the TSF is calculated.

The bend stress test can be used to determine the time shift factors between any two-exposure temperatures. The merits of these different approaches are not discussed here.

A description of Hughes Brothers' test methodology is as follows:
1. Durability testing will be performed periodically on random production lot samples of rebar. Additional testing will be performed when new materials are used, e.g. roving, resin, fillers, for qualification purposes.
2. Test Samples:
    2.1. A minimum of twelve samples shall be selected at random from any production lot. All samples will be pre-conditioned for 12 hours at 130^0 F (54^0 C). All samples should be weighed after conditioning. (This conditioning is done to accelerate cross-linking of the polymer chains in the thermoset resin, which, similar to concrete curing can continue for up to a month after the initial chemical reaction.) Number the end of each sample with a permanent marker. Six samples will be set aside and potted with appropriate anchorage for tensile testing per ACI440.3R-04 test method B.1. The remaining six samples will be used for durability testing.
    2.2. Each sample will be cut to 28 inches (711 mm) in length using a diamond blade saw.
3. Apparatus: A 55-gallon steel drum, cleaned inside with all foreign material removed. See Figure 1.
    3.1. Electric Drum Heater with thermostat control from 60^0 F (15^0 C) to 250^0 F (121^0 C) working temperature.
    3.2. Bend Testing jig for stressing the samples to 25 % of ultimate rating. Per the previous section, the bend radius for #3 (10 mm) diameter rebar is 54 inches (137 cm). See Figure 2.
    3.3. Digital thermometer reading to $\pm$ 0.1 accuracy with a range to 250^0 F (121^0 C).
    3.4. Digital pH tester with accuracy to $\pm$ 0.2
4. Alkaline Aqueous Solution:
    4.1. Water (H_2O)-potable tap water may be used.

5. Sodium Hydroxide (NaOH) or other alkaline chemical able to achieve a minimum pH of 12.5.
6. Mixing procedure for aqueous solution:
    6.1. The steel drum should be filled to approximately 2/3 full with a measured quantity of water (gallons).
    6.2. The sodium hydroxide (NaOH) will be added at a rate of 8 to 12 ounces per gallon of water. This should result in a pH of 12.5.
    6.3. The sodium hydroxide should be added in small quantities and mixed so that it is dissolved completely.
7. Assembly of samples in bend test fixture for stressing:
    7.1. Six samples of each lot shall be mounted in the bend test jig. See Figure 3. The test jig will hold a total of 18 samples.
    7.2. Samples should not be placed in bend fixture until the temperature and pH of the aqueous solution has been stabilized.
8. Immersion Test Procedure:
    8.1. Pre-heat the test solution to 140^0 F (60^0 C). This usually will take at least 48 hours for the solution to reach temperature and stabilize. The solution should be stirred periodically to assure the sodium hydroxide remains in suspension.
    8.2. Once temperature is achieved measure the temperature and pH of the solution and record.
    8.3. Place the bend fixture with rebar into the solution and cover the steel drum with a vented lid. A regular drum lid with the small bung removed works well. The small bung also allows for subsequent temperature and pH readings to be taken.
    8.4. The samples are to remain in the test solution for a minimum of 30 days. Temperature and pH readings should be taken no less than once a week.
    8.5. At the end of the specified test period, remove the bend jig and samples from the solution. Rinse thoroughly with clean water and remove the samples from the jig taking care not to over bend or stress the samples.
    8.6. All samples shall be post conditioned at 120^0 F (48^0 C) for 12 hours.
9. Tensile Testing for Durability (retention):
    9.1. The samples taken from the solution (after post conditioning) shall be potted for tensile testing per ACI440.3R-04.
    9.2. Test all samples to determine residual properties.

SAMPLE TEST RESULTS

In November of 2004, Hughes Brothers purchased eight different brands of glass fibers from seven manufacturers of glass fiber roving. Five of the glass fiber types were considered to be traditional "e-glass" roving, two of the fiber types were considered ECR glass and one variation was an AR glass. Eight separate production runs of GFRP rebar were manufactured under identical conditions using each of these differing glass fibers. Great care was taken to ensure that the only processing variable that differed from rebar to rebar was the glass fiber itself. The resin formulation, line speed, roving end count, fiber volume fraction, curing temperature, surface treatment, fillers, catalyst and additives were all identical from bar to bar. For this experiment a #3 (10 mm) diameter GFRP bar was produced. Each of the glass fibers used was a 113 yield (113 lineal yards of glass equate one pound of roving) 4400 tex fiber with the exception of the AR-glass fiber which was a 250 Yield (2400 tex) fiber. A summary of the fibers used is given in Table 1.

During manufacturing, each separate variation of GFRP bar was given a unique color-coded polypropylene tracer element that was integrated directly into the surface of the bar to help in tracking and identifying each bar variation. After manufacturing, 12 sample bars were selected at random from each of the eight GFRP bar variations. Six bars from each variation were set aside and considered "virgin specimens" and six samples were subjected to the durability test regiment. In addition, standard production quality assurance tests were performed. Tests included Barcol hardness, fiber volume fraction, inter-laminar shear per ASTM D4475, dye-wick penetrant tests to check for thermal cracks and continuous hollow fibers. A summary of the standard quality assurance test results is shown in Table 2.

The Barcol Hardness readings were taken using a Coleman Impressor. Internal company requirements specify a minimum value of 50. The Dye Penetrant test is performed by slicing a 6-inch (150 mm) length of each bar into 1-inch (25 mm) long segments and placing them on a bed of glass beads in a fluid of Magnaflux Zyglo. After periods of 5 minutes, 10 minutes, 15 minutes and 60 minutes, a black light is shown on the specimens. Any hollow fibers or voids will be clearly evident under the black light as the phosphorescent Zyglo will appear on the top surface of the bar due to capillary action. Failure is determined when a continuous void or number of hollow fibers are present along the length of the bar as evidence in all six specimens.

After all the bars were subjected to a 30 day exposure while in the bending stress fixture in the 12.5pH solution at a temperature of 140 degrees Fahrenheit (60^0C), the bars were removed and post cured in an oven at 120^0 F (48^0 C) for 12 hours. The virgin samples and the conditioned samples each had tensile test anchorages applied and were tested for tensile strength, tensile modulus and ultimate strain properties. The average test results from each GFRP bar variation are reported in Table 3.

Due to observation from previous experiments, there is little change in the tensile modulus of elasticity and thus residual changes are not reported here.

As can be seen, using the same processing methodologies but simply varying the glass fiber results in a wide variation in initial properties of the GFRP rebar "system". This variation is compounded when the bars are subjected to a bending stress while in the aqueous solution at elevated temperature. Bars that had very good results in the virgin samples did not necessarily perform as well when subjected to the durability regiment. In addition, the GFRP bars made from Alkaline Resistant, or AR-Glass retained a much lower percentage of residual tensile strength than their e-glass counterparts. Therein lies a key clue as to why some GFRP bars perform better as a system than other bars. The AR-glass fibers are generally used bare in concrete mixes as plastic shrinkage reinforcing. Much of the existing body of research into the durability of glass fiber reinforced polymer bars borrows heavily from the long term durability exposure data published by the Pilkington Brothers Corporation based on exposure and degradation rates of bare e-glass and AR-glass fibers[10]. However, AR-glass fibers are generally not suitable for use in thermoset resin systems due to incompatibilities of the existing fiber "sizing" chemistries to polyester or vinyl ester thermoset resins. In addition, the use of glass fibers in a thermoset polymer resin matrix will likely result in a second order degradation mechanism making application of the degradation rates of bare glass fibers in concrete inappropriate for FRP bars. Since each glass fiber manufacture uses their own closely guarded fiber sizing chemistry, different fiber, resin, and processing characteristics result in a fairly wide range of performance properties. While one fiber may be perfectly well suited to one FRP application, say insulator core bars for example, its performance in a GFRP rebar "system" may not be appropriate at all. This doesn't mean that fiber isn't a good one, or should not be used for its intended pultruded profile, just that it shouldn't be used in the application a GFRP Rebar. As mentioned previously, these subtle differences are imperceptible to the naked eye, casual observer and even as a result of measuring initial properties. Since the basic chemistry make up of the glass fibers themselves are not remarkably different, the only conclusion that can be made is that fiber sizing itself is a key aspect of GFRP bar performance. A test method such as the one described herein is a way to reveal the suitability of the GFRP Rebar "system" as one that might achieve desired long term performance as a concrete reinforcing bar.

CONCLUSIONS

As evidence by the production trials using eight separate glass fiber suppliers and maintaining all other production variables the same from bar to bar, there can be a wide range in short and long term performance of an e-glass/vinyl ester GFRP bar system. A test method using bending stresses induced in an FRP bar while subjecting it to an elevated temperature and aqueous pH environment is a method for screening FRP rebar and their constituent materials for successful use in the field. The method can be used by manufacturers and users to qualify their FRP rebar systems and approve suitable vendors. A key barometer of success is the fiber sizing chemistry used by the fiber supplier and its appropriateness in the FRP bar manufacturers process.

REFERENCES

[1] ACI 440.3R-04 "Guide Test Methods for Fiber-Reinforced Polymers for Reinforcing or Strengthening Concrete Structures".

[2] ACI 440.1R-03 "Guide for the Design and Construction of Concrete Reinforced with FRP Bars".

[3] ASTM D4475-02 "Standard Test Method for Apparent Horizontal Shear Strength of Pultruded Reinforced Plastic Rods by the Short Beam Method.", ASTM International, 100 Barr Harbor Dr, PO Box C700 West Conshohocken, PA 19428

[4] C.S. Helbling, V.M. Karbhari, "Environmental Durability of e-Glass/Vinyl ester Composites under the Combined Effect of Moisture, Temperature and Stress",

Proceedings of the Second International Conference on Durability of FRP Composites for Construction, CDCC 02, Montreal Canada, May 29-31, 2002, pp 247-258.

[5] N.I. Muskhelishvili, Some Basic Problems of the Mathematical Theory of Elasticity, (translated from the Russian by J.R.M. Radok), Groningen, P. Noordhooff, 1953.

[6] Robert D. Cook, Warren C. Young, Advanced Mechanics of Material, Prentice Hall Upper Saddle River, New Jersey 1998.

[7] William B Bickford, Advanced Mechanics of Material, Addison Wesley Longman, 1998.

[8] Porter, M. L., and Barnes, B. A., "Accelerated Aging Degradation of Glass Fiber Composites," Second International Conference on Composites in Infrastructure, ICCI'98, Tucson, AZ, January 1998, pp. 446-459.

[9] Dejke V, Tepfer R, "Durability and Service Life Prediction of GFRP for Concrete Reinforcement", proceedings of the fifth international conference on fibre-reinforced plastics for reinforced concrete structures, Cambridge, UK 16-18 July 2001, pp 505-514.

[10] Litherland K L , Oakley D R, Proctor B A, "The use of Accelerated Ageing Procedures to Predict the Long Term Strength of GRC Composites", Cement and Concrete Research – An International Journal – Vol 11, No 3, New York, Pergamon Press, May, 1981, pp. 455-466.

Table 1: Glass Fibers tested

Company	Yield	Tex	Glass Type
A	113	4400	e-glass
B	113	4400	e-glass
C	113	4400	Ecr-glass
D	113	4400	e-glass
E	113	4400	e-glass
F	250	2400	AR-glass
G	113	4400	e-glass
H	113	4400	Ecr-glass

Table 2: QA processing parameters

Company	Barcol Hardness	Dye Penetrant	ASTM D4475 shear strength (psi)	Fiber Volume by weight (%)
A	55-63	No voids	7281	75.1
B	55-63	No voids	7039	73.5
C	52-60	No voids	6363	75.0
D	50-58	No voids	7316	75.7
E	54-60	No voids	7329	74.2
F	48-53	No voids	6761	73.6
G	53-60	No voids	7872	74.0
H	59-64	No voids	7993	74.0

Table 3: Tensile test results; Virgin and Aged Specimens

Company	Original Tensile Strength (psi)	Tensile Modulus of Elasticity (psi x 10^6)	Residual Tensile Strength After conditioning (psi)	%Residual Tensile Strength (psi)
A	148,000	6.90	121,500	81.87
B	145,100	6.57	98,800	68.09
C	135,800	6.73	96,900	71.35
D	125,500	6.49	89,800	71.55
E	122,600	6.63	79,800	65.09
F	104,600	5.69	69,300	66.25
G	145,000	6.64	115,600	79.70
H	139,900	6.52	82,900	59.30

Figure 1: Drum and band heater

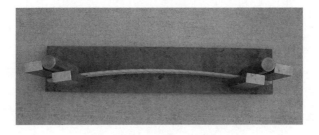

Figure 2: Steel bending fixture

Figure 3: GFRP bars in bending fixture

A New Set-Up for FRP-Concrete Stable Delamination Test

by C. Mazzotti, M. Savoia, and B. Ferracuti

Synopsis: Results of an experimental campaign on FRP – concrete delamination are presented. Two specimens have been tested by using a particular experimental set-up where a CFRP plate has been bonded to concrete and its back side fixed to an external restraining system. The adopted set-up allows a stable delamination process and transition between two limit states (perfect bonding and fully delaminated plate) to be observed. Both strain gages along the FRP plate and LVDT transducers have been used. Starting from experimental data, shear stress – slips data have been computed. A non linear interface law has been calibrated and compared with analogous results obtained by a more conventional experimental set-up. A numerical bond – slip model has been used, adopting the above mentioned law for the FRP – concrete interface to simulate experimental tests. Numerical results are found to be in good agreement with experimental results.

Keywords: concrete; delamination; experimental study; FRP; interface

Claudio Mazzotti is Assistant Professor of Structural Engineering at the University of Bologna, Italy. His main research interests include strengthening of r.c. structures, creep of concrete, damage mechanics, seismic engineering, in situ and laboratory testing of concrete and concrete structures.

Marco Savoia is Professor of Structural Engineering at the University of Bologna, Italy. He is has been an expert member of the (italian) CNR Committee for Standards in RC and PRC Civil Structures. His research interests include creep and damage of concrete structures, seismic engineering, structural reliability methods, composite materials and structures, stability.

Barbara Ferracuti is a Ph. D. Student of Structural Engineering at the University of Bologna, Italy. Her main research interests include strengthening of RC and wood structures with FRP, creep of concrete, fuzzy sets theory, structural reliability.

INTRODUCTION

When using FRP (Fibre Reinforced Polymer) – plates or sheets to strengthen reinforced concrete beams, FRP – concrete bonding is very important. Since delamination is a very brittle failure mechanism, it must be avoided in practical applications. Bonding depends on mechanical and physical properties of concrete, composites and adhesive. Definition of a correct interface law is then important to predict ultimate failure load due to delamination. It is required to estimate the effectiveness of FRP-strengthening of RC (reinforced concrete) elements also under service loadings, due to significant stress concentrations close to transverse cracks in concrete [1].

Very few experimental studies can be found in literature which can be useful to calibrate a FRP – concrete interface law; both global data (applied force) and local data (e.g. strain measurements along the plate) are necessary [2, 3, 4, 5]. Furthermore, delamination is an instable process due to sudden elastic energy release and its experimental observation is very difficult in tests under displacement control due to the occurrence of snap-back behaviors[6]).

In the present paper, an experimental set-up is presented allowing for a stable delamination process. The specimen back-side is fixed to an external retaining system, i.e. concrete and CFRP (Carbon Fiber Reinforced Polymer) plate in that section has null displacement. The delamination process was developed with constant value of applied force, which was used to define fracture energy of interface law. Two specimens have been prepared with a number of closely spaced strain gages to measure strains along the FRP plate. Moreover, LVDTs (Linear Voltage Differential Transducers) have been used to measure plate elongation. Starting from experimental data, average shear stresses between two subsequent strain gages and corresponding shear slips have been computed. These data have been used to calibrate a non linear interface law, according to a procedure described in Mazzotti et al. (2003)[7]. Comparison between present interface law and the same law calibrated with adopting conventional delamination set-ups is presented.

Finally, a bond – slip model, originally presented in Savoia et al. (2003)[6] has been used. Concrete and plates are considered elastic materials and the proposed non linear interface law is adopted between two materials. Numerical results are found to be in good agreement with experimental results.

GEOMETRY AND MECHANICAL PROPERTIES OF SPECIMENS

Plate – concrete bonding has been investigated by testing two specimens (SP1 and SP2) with a CFRP plate bonded to a concrete block. Load is applied at one end of the plate, whereas the opposite extremity of both plate and concrete specimen is clamped to an external retaining system. For both specimens, the plate is 80 mm wide and a 500 mm long , representing the bonded length (Figure 1a).

Concrete block dimensions were 150×200×600 mm. They were fabricated using normal strength concrete. Concrete was poured into wooden forms and externally vibrated. The top was steel-troweled. Five 15 cm-diameter by 30 cm-high standard cylinders were also poured and used to evaluate mechanical properties of concrete (according to Italian standards). Specimens were demoulded after 24 hours and covered with saturated clothes for 28 days; after that, they were stored at room temperature with variable humidity inside the laboratory until tests occur.

Mean compressive strength f_{cm} = 52.7 MPa from compression tests and mean tensile strength f_{ctm} = 3.81 MPa from Brasilian tests have been obtained on cylinders at an age of 20 months. Mean value of elastic modulus has been found E_{cm}=30700 MPa and Poisson ratio v=0.227.

For the composites plates, CFRP Sika CarboDur S plates, 80 mm wide (b_p) and 1.2 mm thick (h_p) have been used (Figure 1a). According to technical data provided by the manufacturer, the plates have a fiber volumetric content equal to 70 percent, and the epoxy matrix, minimum tensile strength of 2200 MPa and a mean elastic modulus E_p = 165,000 MPa.

Top surfaces of concrete blocks were grinded with a stone wheel to remove the top layer of mortar, just until the aggregate was visible (approximately 1 mm). The plates were bonded to the top surface of blocks using a 1.5 mm thickness of two – component Sikadur-30 epoxy adhesive, which does not require primer before bonding. The adhesive has a mean compressive strength of 95 MPa and a mean elastic modulus E_a = 12,800 MPa, according to the manufacturer datasheet. Curing period of all specimens was at least 1 day prior to testing.

In the present experimental investigation, the plate bonded length starts 100 mm from the front side of specimen and has a total bonded length of 500 mm (i.e. plate covers the whole concrete block (Figure 1b). Previous experimental investigations [8, 9] showed that adoption of this particular bonding system provides for a bond stress-slip behaviour and is less affected by boundary conditions and more representative of the material behaviour from cracked sections (i.e. as in the case of plate end debonding).

THE EXPERIMENTAL SETUP

Each concrete block was positioned on a rigid frame with a front side steel reaction element 60 mm high, to prevent global horizontal translation. Moreover a steel apparatus was clamped to the back side of the specimen in order to prevent, in that section, the displacements of both concrete and plate (Figure 2). Bonded length L_{tot} from initial to clamped section was 350 and 360 mm, respectively for SP1 and SP2. The opposite side of the plate was mechanically clamped with a two steel plates system that was free to rotate around the vertical axis. Traction force was then applied to the steel plates system by using a mechanical actuator (Figure 1b). Tests were performed under displacement control of the plate free end.

Instrumentation

A load cell has been used to evaluate the applied traction force. Along the CFRP plate, a series of thirteen strain gauges were placed on the centerline. In Figure 1c, spacing between strain gauges is reported, starting from the traction side of bonded part of CFRP plate. Two LVDTs were placed at the opposite sides of free bonded length in order to measure CFRP plate elongation and to verify the effectiveness of the clamping system (Figure 1b).

RESULTS OF DELAMINATION TESTS

The experimental results

Tests were carried out by performing a first load cycle up to 10 kN of traction force, followed by a monotonic loading at a rate of about 0.2 kN/s. During delamination a plate free end displacement rate of about 50 μm/s was adopted and tests conducted under displacement control.

In Figure 3, force-plate elongation curves of both specimens are reported; the curves are very similar and three main parts can be identified: the first branch is almost linear up to 70-80 % of maximum transmissible force. Beyond that value, stiffness degradation can be observed up to the onset of delamination, when the shear strength is attained at the beginning of bonding length. Subsequent delamination occurred at an almost constant value of applied force (F≈34 kN for both specimens). The delamination process lasted for about 15 seconds. Finally, when complete delamination occurred, the only resisting element was the CFRP plate, properly fixed at the extremity, and its behavior was linear elastic. Specimens were unloaded and reloaded again up to 50 kN. The unloading and reloading branches, as expected, are perfectly linear elastic and characterized by a tangent stiffness of about 42 kN/mm, and very close to the theoretical value of:

$$K_{tg} = \frac{E_p \cdot b_p \cdot h_p}{L_{tot}} = \frac{165 \cdot 80 \cdot 1.2}{355} = 44.6 \, \text{kN/mm} .\qquad(1)$$

Some analogies with the tension-stiffening effect can be drawn. Prior to delamination, a specimen can be considered in an uncraked state (usually defined as state I, where both concrete and reinforcement contribute to specimen stiffness). After complete delamination, only the CFRP plate carries the applied load (*state II*). The delamination process links these two limit states.

Effectiveness of the clamping system was verified by measurement of absolute displacement of the plate close to restrained section. The maximum displacement recorded by LVDT 2 was about 0.15 mm.

Longitudinal strains along the plate at different loading levels are reported in Figures 4a-b, both for SP1 and SP2 specimens. Corresponding values of applied force are reported in Figure 4c. Strains at $x = 0$ are obtained from values of applied force as $\varepsilon_0 = F / E_p A_p$.

For low-to-medium load levels of applied force, the effect of clamping is not significant and the classical distribution of strains along the FRP plate, already found in other experimental tests [2, 7, 9], can be observed. FRP strains are very regular showing an exponential decay starting from the loaded section ($x = 0$). This strain profile corresponds to a linear behavior of the interface. For high force levels, strains tend to be almost constant along the CFRP plate close to loaded end, due to onset of delamination phenomenon along the bonded length. Close to clamped section where delamination did not yet occur, an exponential decay behavior can be observed again.

With classical set-up [2, 7], complete delamination occurs during a snap-back branch [6] and it is not possible to conduct stable measures during the delamination process because it becomes a dynamic event, due to instantaneous releasing of elastic energy of the plate. On the contrary, the set-up adopted in this study provides a stable delamination process which can be fully observed. Experimental data measured during delamination are indicated by square marker in Figures 4a-b, whereas black dots are used for pre- and post-delamination data. The same distinction will be used in the next figures. After complete delamination, strains along the plate are constant and proportional to the applied force.

A further advantage of the proposed set-up is the possibility of obtaining a value of maximum force at delamination which can be considered equal to the asymptotic value of transmissible force by an anchorage of infinite length. Making use of the relation[10] between maximum transmissible force F_{max} and fracture energy G_f of a general non linear interface law:

$$F_{max} = b_p \sqrt{2 E_p h_p G_f} \,, \tag{2}$$

where E_p, h_p, b_p are elastic modulus, thickness and width of the plate, respectively, the value $G_{f,80} = 0.46$ N/mm was obtained. Subscript "80" indicates plate width.

Eqn (2) was previously derived by Wu et al. (2002)[11] for a bilinear interface law and by Brosens (2001)[12] in the case of a power law.

POST-PROCESSING OF EXPERIMENTAL DATA

Strains along the FRP plate at different loading levels were used to calculate shear stress – slip data. The origin of the x-axis is taken at the origin of the bonded plate. Considering an elastic behavior for the composite, the average value of shear stress between two subsequent strain gages can be written as:

$$\bar{\tau}_{i+1/2} = \frac{E_p A_p \left(\varepsilon_{i+1} - \varepsilon_i \right)}{b \left(x_{i+1} - x_i \right)}, \tag{3}$$

with A_p, E_p being cross section and elastic modulus of the composite.

Moreover, assuming that perfect bonding (no slip) occurs at the end of bonded plate and concrete strain is negligible with respect to FRP counterpart, integration of the strain profile gives the following expression for the slip at x, with $x_i \le x \le x_{i+1}$:

$$s(x) = s(x_i) + \int_{x_i}^{x} \varepsilon(x) dx = s(x_i) + \frac{\left(\varepsilon_{i+1} - \varepsilon_i \right)}{\left(x_{i+1} - x_i \right)} \frac{x^2}{2} + \varepsilon_i x, \tag{4}$$

where the initial condition $s(0) = 0$ is assumed.

Average value $\bar{s}_{i+1/2}$ of slip between x_i and x_{i+1} is then computed. The obtained shear stress-slip couples ($\bar{\tau}_{i+1/2}$, $\bar{s}_{i+1/2}$) are reported in Figures 5a, b for SP1 and SP2 specimens, respectively. Curves related to delamination phase are reported with square markers. Adoption of this set-up allows complete delamination to be observed and also the softening branch, usually identified with a limited number of very scattered points, is properly described. Note that some small negative values of shear stress can be found in the softening branch; these values are obtained from the initial portion of the plate where complete delamination has already occurred and strains should be perfectly constant according to theory. In experimental tests, small strain variations have been recorded. According to eqn. (3), these small changes can affect the shear stress sign, providing negative values when theoretically they should be null values. For this reason all experimental data were considered in the calibration procedure described in the next paragraph.

CALIBRATION OF A NON LINEAR INTERFACE LAW

According to the procedure described in Mazzotti et al.[7], shear stress – slip data are used to calibrate a non linear FRP – concrete interface law. All the data related to both experiments are grouped together (Figure 6) and an interface law recently proposed by the authors in terms of fractional law:

$$\tau = \tau_{max} \frac{s}{\bar{s}} \frac{n}{(n-1) + (s/\bar{s})^n} \tag{5}$$

is adopted, where τ_{max} is the peak shear stress, $\bar{s}$ the corresponding slip, and n is a parameter mainly governing the softening branch. Values of $n > 2$ are required in order to obtain positive and finite values of fracture energy.

Both shear stress – strain data ($\bar{\tau}_{i+1/2}$, $\bar{s}_{i+1/2}$) and fracture energy G_f are used to evaluate the three unknown parameters of interface law in eqn (5), i.e., τ_{max}, $\bar{s}$, n. A least square minimization between theoretical and experimental shear stress – strain data is performed, adopting as a constraint in the minimization procedure the value of fracture energy obtained from eqn (2). Further details on numerical procedure can be found in Mazzotti et al. [7].

The values $\tau_{max} = 6.43$ MPa, $\bar{s} = 0.044$ mm, $n = 4.437$ have been obtained. In Figure 6 the proposed interface law is reported, together with all shear stress – slip experimental data. It is worth noting that the proposed law is in good agreement with experimental data both for slips smaller than $\bar{s}$ and in the softening branch where experimental results are more scattered.

Comparison with other tests

The results presented here can be compared with results reported in Mazzotti et al. [7, 9], adopting the same concrete and specimen dimensions and the same CFRP plates. In this study, classical delamination set-up was adopted, with free back side of the plate. Different bonded lengths were considered and further details can be found in Mazzotti et al. [7, 9]. From these tests, the maximum transmissible force (considered as an asymptotic value) was evaluated by a numerical interpolation of values of applied forces at failure for different bonded lengths and a value $F_{max}=36.2$ kN was obtained. The actual maximum force value, corresponding to plateau of Figure 3, is about 34 kN with a 6% difference. According to eqn. (2), fracture energy is slightly smaller than in previous tests, where $G_{f,80} = 0.517$ N/mm was obtained.

In Figure 7, a comparison between interface laws calibrated by using present and previous experimental data is reported. The curves are very similar and only in the softening branch the present curve is more sharply descending due to better experimental description of this region. Hence, methodology of interface law identification proves to be very reliable because both variations in test method do not remarkably affect the final results.

NUMERICAL SIMULATIONS OF TESTS

Experimental tests have been numerically simulated, in order to verify the accuracy of the proposed plate – concrete interface law. A bond-slip kinematic model was adopted and originally presented in Savoia et al. (2003)[10]. The model is based on the assumption of pure extension for two different materials, concrete and FRP plate (no bending). Linear elasticity is adopted for concrete and plate, whereas the non linear law (5) is used for the interface between two materials. Then, a finite difference discretization is used for the unknown variables (axial displacements and stress resultants of concrete and plate).

In the numerical model, both concrete and FRP plate are restrained at the end section and bond length is $L = 355$ mm. Comparison between experimental and numerical results are reported in Figures 3, 8 and 9.

In Figure 3, plate elongation-force curve is compared with experimental curves. As previously described, the behaviour of the specimen is a transition between *State I* condition for low level loads (both stiffness contributions of concrete and FRP plates) and *State II* condition after FRP delamination (FRP plates only contribute to specimen stiffness). Hence, the post-delamination branch is now stable and has followed up to complete delamination in the experimental tests. The results confirm that the proposed interface law provides for a good prediction of delamination load.

Strain distributions in the FRP plate along the bonded length are reported in Figure 8. Experimental data from the SP1 specimen are considered. Numerical results are generally in very good agreement with the experimental data (considering the unavoidable scattering of experimental results at higher load level). The behavior for low load levels is well predicted. There is very good agreement between numerical and experimental results for very high loads, i.e., during plate delamination (square markers). The numerical model was able to correctly follow progressive plate delamination, as it occurred in experimental tests. Comparison between shear stress distribution along the plate, obtained by postprocessing experimental data and by numerical simulation is also reported, both for low force levels (Figure 9a) and during the delamination phase (Figure 9b). Good agreement is found with results obtained by experimental data, although the last ones are sometimes scattered; in any case, maximum value of shear stress and also its position and the gradient of shear stress distribution along the bonded length are well predicted.

CONCLUSIONS

Results from a set of experimental delamination tests have been presented. A particular set-up was designed, providing for stable delamination. Applied force, displacements and strains along FRP plate were measured. Complete progressive delamination was observed and the corresponding value of applied force was used to estimate the fracture energy of interface law.

Force-elongation curve can be divided into three main branches: the first one is almost linear up to the onset of delamination; the second one represents delamination process and has a constant value of applied force; in the third one, complete delamination occurred and the load is then directly transferred to the clamped end of CFRP plate.

An interface shear stress – slip law has then been calibrated starting from experimental data, and adopting the value of fracture energy as a constraint in the minimization procedure between experimental and predicted values. Finally, numerical simulations were performed and results are found to be in good agreement with experimental results.

Experimental tests on FRP – concrete delamination are fundamental to obtain data to establish the maximum value of strain in FRP reinforcement prior to end or intermediate debonding from concrete substrate, see for instance Refs. [13], [14], [15]. In the last two references, maximum admissible strain in FRP reinforcement has been defined as a function of mechanical/geometrical properties of reinforcement and of fracture energy of FRP – concrete interface, where the latter has been estimated from a statistical analysis of results of delamination tests. Moreover, values of maximum shear stress before delamination have been adopted in Ref. [14] as the basis for verification of FRP – strengthened structures under service loadings.

ACKNOWLEDGMENTS

The authors would like to thank the Sika Italia S.p.A. for providing CFRP plates and adhesives for the specimens. The financial supports of (italian) MIUR (PRIN 2003 Grant, FIRB 2001 Grant) and C.N.R., PAAS Grant 2001, are gratefully acknowledged.

REFERENCES

1.   FIB TG 9.3., 2001, *Externally bonded FRP reinforcement for RC structures*, FIB, technical report, Bullettin n° 14.

2.   Chajes, M.J., Finch, W.W. jr, Januska, T.F. and Thomson T.A. jr., 1996, "Bond and force transfer of composite material plates bonded to concrete," *ACI Structural J.*, Vol. 93, pp. 208-217.

3.   Miller, B., Nanni, A. and De Lorenzis, L., 2001, "Bond of FRP laminates to concrete," *ACI Material J.*, Vol. 98(3), pp. 246-254.

4.   Brosens, K. and Van Gemert, D., 1998, "Plate end shear design for external CFRP laminates," *Proceedings of FRAMCOS-3, Germany, 1998*, Eds H. Mihashi, K. Rokugo., Aedificatio Publs, Freiburg, pp. 1793-1804.

5.   Dai, J., Ueda, T., Sato, Y., 2005, "Development of the nonlinear bond stress-slip model of fiber reinforced plastics sheet-concrete interface with a simple method," *J. of Comp. For Constr., ASCE*, Vol. 9(1), pp. 52-62.

6.   Savoia, M., Ferracuti, B. and Mazzotti, C., 2003, "Delamination of FRP plate/sheets used for strengthening of r/c elements," *II International Structural Engineering and Construction: Proceedings of ISEC-02, Rome, Sept. 2003*, V.2, pp. 1375-1381.

7.   Mazzotti, C., Ferracuti, B. and Savoia, M., 2004, "An experimental study on FRP – concrete delamination," *Fracture Mechanics of Concrete and Concrete Structures: Proceedings of FraMCoS–5 Conference, Colorado, U.S.A., 2004*, V. 2, pp. 795-802.

8.   Mazzotti, C., Ferracuti, B. and Savoia, M., 2005, "FRP – concrete delamination results adopting different experimental pure shear setups", *International Conference on Fracture: Proceedings of ICF XI, Turin, March, 2005*.

174 Mazzotti et al.

9. Mazzotti, C., Savoia, M. and Ferracuti, B., 2005, "An experimental investigation on delamination process of CFRP plate bonded to concrete," in preparation.

10. Savoia, M., Ferracuti, B. and Mazzotti, C., 2003, "Non linear bond- slip law for FRP-concrete interface," *Proceedings* of *FRPRCS-6 Conference, Singapore, 2003*, Ed. K.H. Tan, pp. 1-10.

11. Wu, Z., Yuan, H. and Niu, H., 2002, "Stress transfer and fracture propagation in different kinds of adhesive joints," *J. Eng. Mech. ASCE*, Vol. 128(5), pp. 562-573.

12. Brosens, K., 2001, *Anchorage of externally bonded steel plates and CFRP laminates for the strengthening of concrete elements*, doctoral thesis, University of Leuven, Belgium, 150 pp.

13. ACI Committee 440F, 2002, *Guide for design and construction of externally bonded FRP systems for strengthening concrete structures*.

14. CNR Italian Committee, 2004, *Guide Lines for design, constructions and control of rehabilitation interventions with FRP composites* (in Italian).

15. Teng, J.G., Smith, S.T., Yao, J. and Chen, J.F., 2003, "Intermediate crack-induced debonding in RC beams and slabs". *Construction and Building Materials*, Vol. 17(6-7), pp. 447-62.

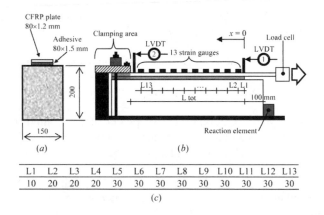

L1	L2	L3	L4	L5	L6	L7	L8	L9	L10	L11	L12	L13
10	20	20	20	30	30	30	30	30	30	30	30	30

(c)

Figure 1—Experimental set-up: (*a*) Specimen transverse section,
(*b*) side view with instrument positions and CFRP plate clamping system,
(*c*) spacing between strain gauges (mm) along the CFRP plate.

Figure 2—Experimental set-up: view of the clamping system for concrete and plate at the opposite sides of the specimen.

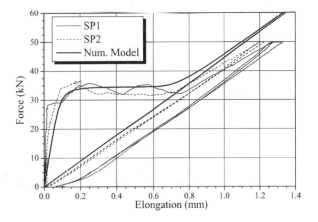

Figure 3—Force-plate elongation curves from experimental tests (specimens SP1, SP2) and numerical simulations.

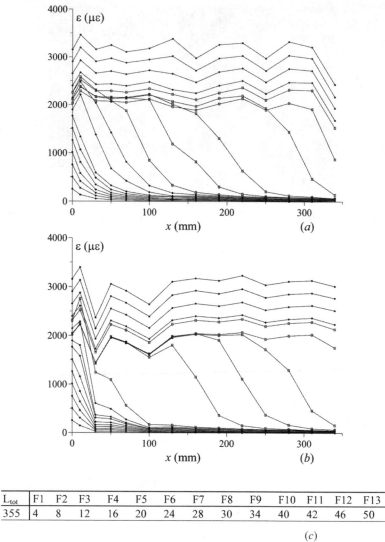

L_{tot}	F1	F2	F3	F4	F5	F6	F7	F8	F9	F10	F11	F12	F13
355	4	8	12	16	20	24	28	30	34	40	42	46	50

(c)

Figure 4—Profiles of experiemntal strains in CFRP plates along the bonded lengths of (a) SP1 and (b) SP2 speciments. (c) Loading levels; —▫— measures taken during the delamination phase (load level equal to 34 kN).

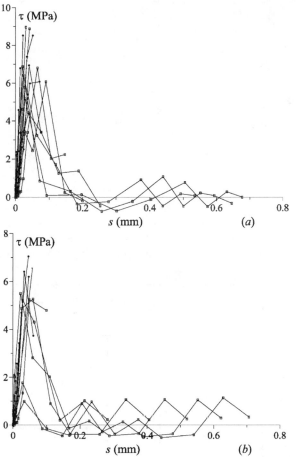

Figure 5—Shear stress—slip data obtained from post-processing experimental results from (*a*) SP1 and (*b*) SP2 specimens; —⊟— measures taken during delamination phase.

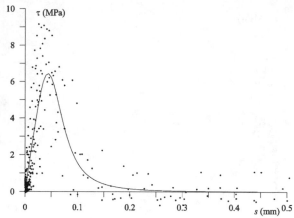

Figure 6—Shear stress-slip data obtained by post-processing experimental results from SP1 and SP2 specimens and proposed interface law.

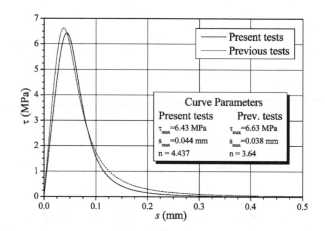

Figure 7—Comparision between interface laws calibrated with present experimental results and previous results by Mazzotti et al. [9]

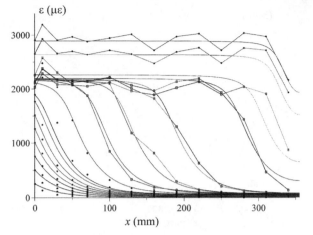

Figure 8—Profiles of strains along FRP plate: comparison between numerical and experimental results from specimen SP1; ⊟ measures taken during delamination phase.

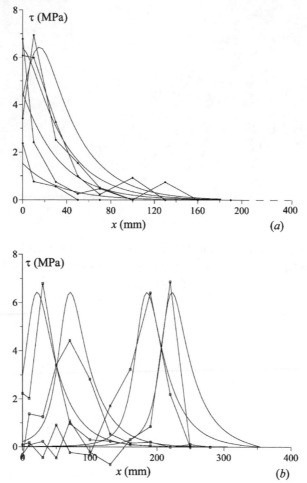

Figure 9—Profiles of shear stressed along FRP plate for specimen SP1. Comparison between numerical and experimental results for: (a) low and (b) high loading levels; ⊟ measures taken during delamination phase.

Stepped Isothermal Method for Creep Rupture Studies of Aramid Fibres

by K.G.N.C. Alwis and C.J. Burgoyne

Synopsis: Aramid fibres have been used in rope construction and for prestressing tendons, but when subjected to a constant static load the fibres creep with time and may rupture, leading to a catastrophic failure of the rope. To understand this behaviour many life-time models have been suggested but they suffer from the lack of long term creep rupture data to make firm conclusions on rupture times and load levels. Such data is expensive to obtain using conventional creep testing as it takes a long time before failure of a specimen. To overcome this problem, and to obtain the creep-rupture data at low stress levels within a reasonably short time scale (hours), accelerated testing methods, the Stepped Isothermal Method (SIM) and Time Temperature Superposition (TTSP), have been investigated. In SIM testing a single yarn specimen is tested at a specific stress level under a series of increasing temperature steps from which a single response curve, known as the master curve, is obtained which predicts the long-term behaviour. Some manipulation of the data is required, but the technique has many advantages over the TTSP and conventional creep testing and it can be automated to obtain the long-term creep-rupture data points relatively easily.

Keywords: accelerated testing; creep-rupture; master curve; stepped isothermal method (SIM)

182 Alwis and Burgoyne

K. G. N. C. (Nadun) Alwis: Obtained his BSc. (Eng) degree from University of Moratuwa Sri Lanka in 1998 and successfully completed the PhD on "Long-term stress-rupture behaviour of aramid fibres" from University of Cambridge under the supervision of Dr. C. J. Burgoyne in 2003. He is currently working at Kellogg Brown and Root, in the U.K as an Associate Structural Professional.

Chris Burgoyne is Reader in Concrete Structures at the Dept. of Engineering, University of Cambridge. He has been undertaking research into the behaviour of concrete structures reinforced or prestressed with FRP since 1982.

INTRODUCTION

If high-strength fibres, such as aramids, are to find practical application in structural engineering, it is most likely to be as non-corrodable external (or unbonded) prestressing tendons in concrete. In such applications, where the applied stress varies very little, the governing factor is not going to be the short-term strength, or modulus, but the long-term creep-rupture strength.

It takes a long time however, using conventional creep tests, to obtain creep-rupture data for aramid fibres at the low stress levels likely to be used in practical applications. As an alternative, two accelerated testing methods have been suggested to predict the creep-rupture behaviour at low stress levels: the time temperature superposition principle (TTSP) and the stepped isothermal method (SIM). These methods offer many advantages when compared to conventional creep tests as testing requires shorter time scales to obtain long-term data.

In TTSP, it is assumed that raising the temperature will increase the creep rate but not alter the mechanism. Several individual creep tests are performed at different temperature levels, to obtain strain versus logarithmic time curves. These curves can then be time-shifted, parallel to the logarithmic time axis, by an amount $log\ (a_t)$ to give a single reference curve, on which all the separate test results are superposed. This master curve applies for a certain temperature and a fixed stress level. This technique is not described in the paper but a detailed description can be found in elsewhere[1]. A comprehensive literature review on early development of the time-temperature superposition principle can also be found elsewhere[2] and there have been many applications[3,4]. In this paper, however, the creep rupture data obtained from TTSP method is used to compare with the results obtained from SIM.

Thornton et al.[5] first applied the SIM to predict the long-term creep behaviour of geogrids in soil reinforcement applications; for this application there is virtually no conventional creep data and test data derived from SIM has been accepted as the basis of design rules. The principle of the SIM is that a single element (in this case a yarn) is placed in a testing machine and loaded by a chosen force. The temperature is then raised, typically by a few °C, and kept constant for a fixed period of time, typically a few hours. The sequence is then repeated at a slightly higher temperature, on the same sample. Some manipulation of the data is required in order to compensate for the temperature steps. The SIM can be considered as a special case of the TTSP, a detailed description of which is given elsewhere[1,6]. In SIM tests, a single specimen is tested at a sequence of temperature levels under a constant load, whereas in TTSP testing different specimens are tested at each temperature level. SIM is very promising when compared to TTSP and conventional creep tests since a yarn can be tested until it fails in a much shorter time; this depends on the temperature and time steps adopted.

Three different adjustments are needed for each SIM test to produce a single master creep curve; the creep-rupture prediction comes from the end of the master curve when the specimen fails under a specific load and temperature (Figure 1). The ***vertical shift*** allows for the strains caused by the change in temperature, taking account of the creep that occurs while the temperature change is taking place. ***Rescaling*** is needed to allow for the previous history of the specimen: when the temperature changes some allowance must be made for the fact that some creep has already taken place under the previous time steps, unlike TTSP when each test is separate. This adjustment takes the form of a shift in the time direction when plotted against a *linear* time scale. The ***horizontal shift*** takes the form of a shift on a creep strain vs. *log* (time) plot and is similar to the technique used in TTSP to allow comparison of tests at different temperatures. Each of these adjustments will be described in more detail below.

RESEARCH SIGNIFICANCE

The paper presents a method that can be used to obtain creep-rupture test data for fibres in a short time-scale from which predictions can be made for the behaviour of the materials over very long time-scales in practical applications. This paper does not, of itself, provide answers to the many questions which remain about the behaviour of these fibres, but it does give a technique which can be used to address them.

MATERIALS AND EXPERIMENTAL SET-UP

In the sample tests described here, Kevlar-49 yarns were used. The average breaking load (ABL) of the yarns was 445 N, obtained from 12 short-term tests. All test results described below will be reported relative to the ABL, since it is known that size effects can be taken into account by relating all stresses to the short-term breaking load[7]. The cross sectional area of the yarn was 0.1685×10^{-6} m^2.

The tensile tests were carried out in a conventional testing machine, using round bar clamps that have also been used for long-term dead-weight testing of yarns. The load was applied by moving the cross-head of the machine at a specific rate; the cross-head movement and the load level were recorded. The testing set-up is shown in Figure 2; the oven is set-up within the test machine, with the two clamps mounted on extension pieces so that the complete test specimen lies inside the oven.

One of the difficult tasks is to determine the absolute zero of the stress-strain curve, due to initial slack and slippage of the yarn around the jaws. It is essential to know accurately the strain of the specimen just after the initial loading in order to compare the creep curves at different temperatures. A small error of this value would result in displacing the creep curves on the creep strain axis which then makes it impossible to obtain valid, smooth master curves only by making time shifts.

An extensive study was thus first carried out, using spring-steel hoops fitted with high-temperature strain gauges, to determine the jaw effect. This was carried out with yarns of different length, and with the oven set at different temperatures. This procedure allowed the SIM tests to be carried out using machine extension alone, since the clamping action on the spring-steel gauges might affect the stress-rupture lifetime of the yarns.

By separating the jaw effect from the yarn extension, it is possible to determine accurate stress-strain curves for the yarns, at different temperatures, as shown in Figure 3. These graphs were used to determine the initial strains for a given stress level at different temperatures. For example, points at which the line AB crosses the stress-strain curves are the initial strain values at 70% ABL. This process is described in detail elsewhere[8].

The initial loading rate was 5 mm/min and the specimen length was 350 mm (centre to centre distance of the jaws). In each test, load was applied only after the temperature had reached the desired value; by adjusting the initial strains for each test as described above, only time and vertical shifts were needed to obtain the master curve.

Testing procedure
A series of SIM tests were carried out at 70% ABL on Kevlar-49 at different steps of temperature over different time steps. All tests started at 25 °C as it was easy to control this temperature by heating only. The testing machine was kept in a temperature-controlled room where the temperature was maintained at 21 °C. It was not possible to carry out any tests below this value since the oven had no cooling facility.

Load was applied only after the temperature had reached 25 °C, so no initial correction for temperature was needed. Table 1 shows the temperature sequences used for the tests reported here; different sequences were used since, if the method is to be valid, similar master curves must be obtained no matter what temperature steps are used.

Each yarn was tested to failure; the failure point could be observed from the load reading of the testing machine and it was not necessary to open the oven for investigation. Two tests were carried out at each test number; to distinguish them the following identification was used:

- SIM70-01-01
- SIM70-01-02

'70' denotes the load level, the succeeding number '01' denotes the test number and the last number denotes the repetition of the test. A similar testing procedure was used to test the yarns at 50% ABL but at different time and temperature steps; space does not allow that data to be included here.

ADUSTMENT OF STRAIN FOR CHANGE IN TEMPERATURE-VERTICAL SHIFT

Figure 4 shows a schematic picture of a temperature step. The temperature is raised from T_1 to T_2 over the time, t_c. Point B represents the creep strain just after the temperature step; B' is the creep strain that would have been observed due to thermal contraction, noting that aramid fibres have a negative coefficient of thermal expansion. However, the final creep strain, B is observed due to continuing creep over time, t_c (BB'). The adjusted strain just after the temperature step ($\overline{B}$) can be found:-

(a) by adding the thermal contraction, so $\overline{B} = B + B'B''$, or

(b) by adding the creep over t_c, so $\overline{B} = B'' + B'B$

To calculate the distance B'B'' , an accurate value of the coefficient of thermal expansion is needed, but in the literature different values are stated, so Method (a) is not reliable.

In contrast, Method (b) can be performed using measured values. Changes of the creep rate over time t_c can be found by conducting separate creep tests from temperature T_1 to a variety of different temperatures. This allows the variation of creep rate with temperature to be measured; the creep over time, t_c (BB') can then be found by integration. A similar procedure has to be applied for each temperature step. This means that many subsidiary tests have to be performed, but avoids reliance on uncertain published data.

RESCALING PROCEDURE

One of the main differences with the SIM approach is that the history of the specimen at different temperatures is not the same as in TTSP. In TTSP a specimen is subjected to a certain temperature level starting from room temperature whereas in SIM the specimen already has a strain history caused by extensions that took place at previous temperature steps.

Figure 5 shows the strain response for two temperature steps. The curve OABC is the measured response of the SIM specimen through the first two temperature steps. $OA\overline{BC}$ is the response after making the vertical strain adjustment. PQ is the response of a TTSP test carried out at the higher temperature T_2. It is now necessary to determine the time t' that represents the notional starting time for a TTSP specimen that would have the same response as the SIM specimen at the higher temperature. The value $t'' - t'$ is assumed to be the time needed for a specimen which had been at T_2 to arrive at the creep state at time t''. It should be equal to t^* from the TTSP curve. The selection of t' for each temperature step has a great influence when obtaining smooth master curves. A graphical method is used to obtain an initial estimate of the time t' by extending the $\overline{CB}$ curve as smoothly as possible on to the horizontal line that passes through P, which is then refined numerically.

THE HORIZONTAL SHIFT

This step is similar to the shifting procedure as used for TTSP[1]. Once the vertical and rescaling shifts have been carried out the SIM data represent a set of creep curves, as would have been obtained using the TTSP method. The adjustment therefore takes the form of a horizontal shift on a creep strain vs. *log* (time) plot. In the SIM approach it is necessary to perform the rescaling and horizontal shifts together using a numerical procedure. Once the possible ranges of the rescaling and horizontal shifts have been identified using a graphical method, an automated numerical procedure is used by fitting a polynomial through the overlap region and adjusting the shifting and scaling parameters to minimise the lack-of-fit of the two overlapping curves. The same technique is then carried out at each temperature step which results in a single, smooth creep curve (the master curve) for a known load at a specified temperature. This master curve, examples of which are shown in Figures 6 and 7, represents the best estimate of the extension against time at the specified temperature, under the given load. If the specimen was allowed to creep until failure, the end point of the master curve gives a data point for creep-rupture.

RESULTS AND DISCUSSION

A series of conventional creep tests have been performed to check the validity of this method. These tests have been carried out in a controlled temperature (25 °C) and humidity (65% RH) environment. For comparison, SIM70-01-01 data is plotted together with the TTSP data and conventional creep data at 70% ABL (Figure 6). All curves match reasonably closely and SIM seems to be promising since the curves match both in form and position. However, even if the SIM test picks up the basic form of the results, a question remains about its repeatability. All SIM curves at 70% ABL are plotted in Figure 7; it is apparent that all curves follow the same shape which indicates its repeatability, even though different temperature steps were used for each test.

The initial part of Figure 6 shows that the conventional curve clearly follows the master curve. There is, however, speculation about the reverse curvature of the master curves between 100 to 10,000 hours. The same behaviour was observed for the master curves generated from TTSP and also for SIM tests carried out at 50% ABL[8]. The behaviour may be attributed to re-arrangement of the internal fibres and is independent of the type of the accelerating method. This reverse curvature of the creep response has not been described in the literature and this may be the first time it has been observed. It is not possible at this stage fully to understand this response since only a limited amount of testing has been carried out. Further investigation should be carried out with a variety of tests at different stress levels, different time steps and different temperature steps to come to a firm conclusion.

It is also significant to note that the horizontal shift factors needed to produce the master curves turn out to vary inversely with the absolute temperature[8]. This indicates that creep can be regarded as an Arrhenius process and is consistent with observations elsewhere[6].

The availability of SIM means that it is now possible to investigate creep-rupture behaviour in much more detail. Each of the master curves on Figure 7 ends with failure of a yarn. For comparison these failure times are plotted in Figure 8 along with the best statistical life prediction based on Kevlar rope data[8]. It is apparent that the failure times of some of the SIM data at 70% ABL lie within the confidence limits of the model, but there is more spread of the rupture times than predicted by the statistical model; the rupture times predicted by SIM are considerably longer. More testing is needed at low stress levels before firm conclusions can be reached. The results presented here do not, of themselves, answer such questions as the effect of varying loads, varying temperature or problems associated with cumulative damage. But these results do show that SIM testing provides a tool which can be used to obtain some of the necessary data, and also to provide a prediction of future behaviour against which other theories can be tested.

CONCLUSION

SIM can be readily applied to generate long-term creep-rupture data of aramid yarns and can be used to mimic the behaviour of TTSP tests. The SIM technique has many advantages over conventional TTSP. Both the test procedures and the data reduction can be automated, and a single specimen can be tested at each stress level for the entire thermal history within a reasonably short time scale; the effects due to the variability of yarns can thus be minimised.

SIM results show repeatability but there was some variation of the rupture times which may be attributed to the variability of the yarns. The technique seems to be promising and can be recommended as a basis to generate more rupture data at different stress levels.

REFERENCES

[1] Alwis, K.G.N.C and Burgoyne, C.J, 2003, "Accelerated testing to predict the stress-rupture behaviour of aramid fibres", *Fibre reinforced plastics for reinforced concrete structures (FRPRCS-6)*, Edited by Kiang Hwee TAN, Singapore, 2003, pp. 111-120.

[2] Ferry, J.D., 1970, "Viscoelastic properties of polymers", John Wiley and Sons, Inc.

[3] Povolo, F. and Hermida, E.B., 1991, "Analysis of the master curve for the viscoelastic behaviour of polymers", *Mechanics of Materials*, No. 12, pp. 35-46.

[4] Brinson, L.C. and Gates, T.S., 1995, "Effects of physical aging on long term creep of polymers and polymer matrix composites", *Int. J. Solids and Structures*, Vol. 32, No. 6/7, pp. 827-846.

[5] Thornton, J.S., Allen, S.R., Thomas, R.W. and Sandri, D., 1988, "The stepped isothermal method for TTS and its application to creep data on polyester yarn", *Sixth International Conference on Geosynthetics*, Atlanta, USA.

[6] Tamuzs, V., Maksimovs, R. and Modniks, J., 2001, July 8-10, "Long-term creep of hybrid FRP bars", 5th *International Symposium on FRP Reinforced Concrete Structures (FRPRCS-5)*, Cambridge, Vol. 1, pp. 527–535.

[7] Amaniampong, G. "Variability and viscoelasticity of parallel-lay ropes", Thesis submitted to the University of Cambridge, 1992.

[8] Alwis, K.G.N.C, "Accelerated testing for long-term stress-rupture behaviour of aramid fibres" Thesis submitted to the University of Cambridge, 2003.

Table 1 – SIM tests at different temperature steps (^{0}C) at 70% ABL

Test No.	No. of tests	Time duration for each temperature step				
		5hrs	5hrs	5hrs	5hrs	5hrs
SIM70-01-01/02	2	25	40	60	80	100*
SIM70-02-01/02	2	25	40	80	100*	-
SIM70-03-01/02	2	25	40	60	100*	-
SIM70-04-01/02	2	25	40	60	80	120*
SIM70-05-01/02	2	25	40	80	120*	-
SIM70-06-01/02	2	25	60	80	100*	-
SIM70-07-01/02	2	25	60	100*	-	-
SIM70-08-01/02	2	25	60	80	120*	-

* Final step extended until failure

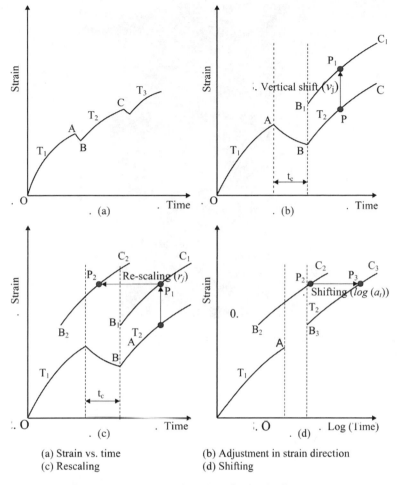

(a) Strain vs. time
(c) Rescaling

(b) Adjustment in strain direction
(d) Shifting

Figure 1 – SIM procedure in schematic diagrams

Figure 2 – Experimental set-up for tensile, TTSP and SIM test

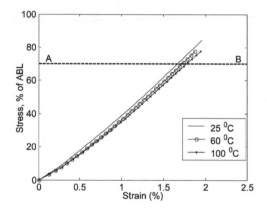

Figure 3 – Stress vs. strain curves at different temperature

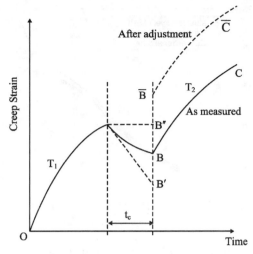

Figure 4 – Change of creep behaviour at a temperature step

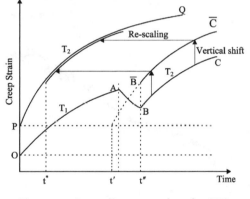

Figure 5 – Rescaling procedure for SIM

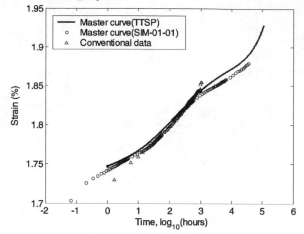

Figure 6 – Master curves with conventional creep data at 70% ABL

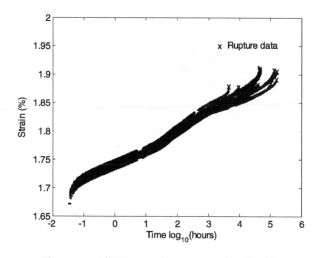

Figure 7 – All SIM master curves at 70% ABL

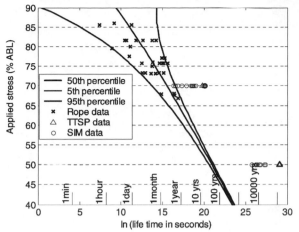

Figure 8 – Comparison of stress rupture data at 50 and 70% ABL

Fiber Optics Technique for Quality Control and Monitoring of FRP Installations

by V. Antonucci, M. Giordano, and A. Prota

<u>Synopsis:</u> Advantages such as light weight, high mechanical properties and resistance to aggressive and chemical agents, non invasive installation, low tooling and machinery costs, possibility to repair the structure without interrupting its use have determined great expectation on Fiber Reinforced Polymer (FRP) composites for the strengthening of civil structures. However, their diffusion is often limited by the lack of procedures and methods to assess the quality of installations. The main focus of this paper is to present a non destructive methodology based on a fiber optic refractometer for the quality control of FRP wet lay-up systems. After summarizing the basic parameters necessary for monitoring the cure reaction and introducing the principles of the proposed technique, some preliminary experimental tests are discussed. The analysis of laboratory outcomes allows identifying the aspects that need further research before the proposed technique can be implemented as a field tool for quality assessment of FRP installations.

<u>Keywords:</u> curing; fiber optic sensors; quality control; refractometer; wet lay-up systems

Vincenza Antonucci is permanent researcher at Institute for Composites and Biomedical Materials of Italian National Research Council. Her research interests are focused on polymer composites mainly on processes modeling, development of new technologies and manufacturing monitoring systems. Recently, the research activities have devoted also to the study of shape memory alloys for the design of adaptive structures.

Michele Giordano is permanent researcher at Institute for Composites and Biomedical Materials of Italian National Research Council. He is member of the Scientific Board Center of Excellence on Structural Composites for Innovative Construction (SCIC) based at the University of Naples. His research interests are in the area of polymer based composites, smart systems and structural health monitoring technologies.

Andrea Prota is Assistant Professor of Structural Engineering at University of Naples Federico II, Italy. He is member of *fib* WG 9.3 and Associate Member of ACI 440 Committee. His research interests include seismic behavior of RC and masonry structures, use of advanced materials for new construction and for retrofitting of existing structures, and use of innovative techniques for structural health monitoring.

INTRODUCTION

Over the last years it has been recognized that FRP materials could represent a very effective solution for the upgrade of existing structures and their installation has been done in the majority of applications by manual wet lay-up. This installation procedure is characterized by the mixture of a two-component resin, typically epoxy, that is then used to impregnate a dry FRP sheet. Once impregnated with resin, the FRP system goes through a chemical process over time that provides the cured laminate with desirable properties. The success of such chemical process is dependent upon many factors such as: use of proper mixed amounts of each resin component; storage environmental conditions of resin components; surface temperature, moisture and exposure; atmospheric conditions during application; curing of the resin (Nanni et al. 2001). This means that the structural performance of the externally bonded reinforcement are strongly dependent on the proper and full curing of the resins used to impregnate the fibers. Such aspect assumes even more importance in the case of some emerging strengthening systems where the resin is obtained by automatic mixture of the two components. The advantage of such techniques is that the resin can then be sprayed on the support with great saving of time and manpower; however, there is a high risk related to a potential improper automatic mixture that is hard to detect.

The importance of developing protocols for quality control of FRP installations is also recognized by European (*fib* bulletin 14 2001) and North American (ACI 440.2R.02 2002) design guidelines for FRP systems externally bonded to reinforced concrete structures. Even though both documents underline the need for quality control checks and suggest some standard tests, there is a lack of procedures and methods that could be used in practice to assess the quality of an FRP installation. The development of such a methodology could be key towards the definition of inspection protocols for FRP systems and it could make it possible to quantitatively assess the quality of its installation. By increasing the trust of practitioners and reducing the uncertainties and the risks related with the adoption of innovative materials, this achievement could represent a crucial step towards a wide diffusion of FRP in construction with high potential for new jobs and business opportunities in the construction industry.

The possibility of checking the quality of an FRP installation depends on the availability of a non-destructive technique (NDT) that allows monitoring some properties of FRP through the use of a certain type of sensor. During the last years, different systems have been applied to civil infrastructures. Electric sensors (Chen and Liu, 2003), Acoustic Emissions (Chang and Liu, 2003), Ultrasonic tests (Mariin et al., 2003), piezoelectric sensors (Gowripalan, 2001), termography (Starnes et al., 2003) and microwaves (Feng et al., 2002) have been used to measure the strains in the FRP composite or to detect defects at its interface with the support. Lately, the possibility of using fiber-optic sensors to measure displacements, strains and temperature has been also demonstrated (Inaudi, 2003; Katsuki, 2003; Imai et al., 2003).

A large project is under development at the Center of Excellence on Structural Composites for Innovative Construction (SCIC) based at the University of Naples. Its overall objective is the implementation of a field protocol for the quality control of wet lay-up installations by means of an innovative NDT technique based on the use of fiber-optic sensors embedded through the reinforcing fibers of the FRP laminates. The sensors are connected to a refractometer that, using the Fresnel reflection laws, makes it possible to measure the variation of the refraction index associated to the reaction of reticulation and then monitor the advancement of the cure reaction of the resin. The present paper discusses the findings of a first set of research activities that have been focused on two main lines, namely: the kinetic characterization of a typical thermoset resin used to impregnate the fibers in civil applications, and the assessment of the potential of the fiber optic refractometer to provide reliable information about the advancement of the resin cure under typical conditions of temperature and humidity.

CURE REACTION MONITORING

Refractive index as a process state parameter

The Lorenz-Lorentz law (eq.1), applied to polymeric material (clarifies the relationship between the refractive index n, the density ρ and the polarizability β (Ku and Liepins, 1987):

$$\frac{n^2 - 1}{n^2 + 2} = \frac{N}{3M\varepsilon}\rho\beta \tag{1}$$

where N is the Avogadro number, M is the molecular weight of polymer repeat unit and ε is the free space permittivity. In other terms, the refractive index reflects the variation of the polymer density that is an useful process parameter in the case of thermoset manufacturing. In fact, the cure of low molecular weight prepolymer involves the transformation of a fluid resin into a rubber then into a solid glass; this is the result of the exothermic chemical reactions of the reactive groups present in the system, which develop a progressively denser polymeric network.

The growth and branching of the polymeric chains are due to intramolecular reactions that initially occur in the liquid state until a critical degree of branching is reached and an infinite network and an insoluble material is formed (gelation or gel point). After the gelation, successive crosslinking reactions increase the crosslink density and stiffness of the polymer is steadily increased, leading, at the end of the process, to the glassy structure of the fully cured thermoset (Nicolais and Kenny, 1991). Since the curing process strongly influences the density of the material, direct relation between the extent of cure and refractive index is expected. This finding leads to the possibility to develop a compact, robust, cost effective and high resolution fiber optic sensing system for non destructive evaluation of the polymeric matrix characteristic features during the manufacturing process.

Fiber optic refractometer

The use of fiber optic sensor offers a very powerful tool to perform remote, on-line, in-situ monitoring of composite manufacturing processes (Afromowitz et al., 1995; Cusano et al., 2000a); Lew, et al., 1984; Culshaw et al., 1997; Crosby et al., 1997). As well known, in fact, the fiber optic is free from electromagnetic interference, and is characterized by high chemical and high temperature resistance. Moreover, due to the capability of the fiber optical sensors to be multiplexed in a large number of independent channels, due to the fact that fibers are readily embedded into the composite and due to their small size that make them minimally intrusive in the host structures, this approach provides useful tools for implementing integrated sensing networks within the material itself.

The proposed sensor is based on the principle Fresnel reflection, here the transducer is simply the fiber-optic/host material   interface. This leads to a more simple, less intrusive, and lower cost sensing system. The response of the sensor is a function of the mismatch in refractive index between the fiber optic and the host interface.

Figure 1 shows the optical part of the proposed sensor and Figure 2 is a picture of the optical refractometer and of the acquisition system during the consolidation of an FRP on a concrete specimen. A laser beam lights a mono mode optical fiber, which is embedded into the under test resin. At the interface between the fiber and the resin, due to the mismatch of the two refractive indexes (of the fiber core and the resin at the laser wavelength) part of the light beam is transmitted and part is reflected. The reflected signal is collected on a photodetector by means of a fibre Y-couple.

According to Fresnel's equations, the field amplitude reflection coefficients at the fiber-end/resin interface for the perpendicular and parallel polarisation, respectively r_n and r_p, can be expressed as (Cusano et al., 2000b):

$$r_n = \frac{n_f \cos \theta_1 - n_m \cos \theta_2}{n_f \cos \theta_1 + n_m \cos \theta_2}$$

(2)

$$r_p = \frac{n_m \cos \theta_1 - n_f \cos \theta_2}{n_m \cos \theta_1 + n_f \cos \theta_2}$$

where θ_1 is the angle of incident light and θ_2 is the angle of the transmitted light, n_f and n_m are respectively the effective refractive index of the fiber and the sample refractive index. If step index optical fibre is used, a number of propagating modes are guided. If a mono mode fiber is used, it can be assumed that the fundamental guided mode travels along a paraxial path in the fiber, in this hypothesis, ($\theta_1 \cong \theta_2 \cong 0°$) and $|r_p| = |r_n| = r$. As a consequence, the intensity reflection coefficient R can be expressed by (Ku and Liepins, 1987):

$$R = \left| r_p \right|^2 = \left| r_n \right|^2 = \left| \frac{n_f - n_m}{n_f + n_m} \right|^2$$

(3)

Thus, by monitoring the light intensity reflected from the fiber end/host interface detailed information about the sample refractive index are available. In fact, the signal at the photodetector V in hypothesis of monochromatic light source can be expressed as:

$$V = h \; \beta \; \alpha \; P_{INC} R$$

(4)

where h is the gain of the photodetector, α the coupling coefficient introduced by the 1x2 coupler, β accounts for signal losses along the optical chain and P_{INC} represents the optical power impinging the sample/fiber end interface.

EXPERIMENTAL RESULTS

The studied resin system was a mixture of an epoxy resin MAPEWRAP31 and its hardener (amine) supplied by Mapei (2003). The amine to epoxy volumetric ratio was 4:1 as suggested by the supplier. The complete consolidation of this resin at environmental temperature occurs in 7 days. Therefore, at first, the polymerization of the MAPEWRAP31 resin has been analysed at 70 °C, that is a sufficient level of temperature to attain a complete conversion. Its kinetic behaviour has been investigated both by measuring the refractive index with the developed fiber optic refractomer and performing a Differential Scanning Calorimetry (DSC) experimental test. The DSC is a conventional technique, suggested also by the European and American guidelines for the cure monitoring of FRP. It allows to investigate the material phase transitions and the chemical kinetic by measuring the absorbed or released heat flow during a dynamic or an isothermal test. Since the resin polymerization is an exothermal chemical reaction, the measurement of the heat released by the analysed sample gives indications on the chemical conversion.

Figure 3 compares readings of the degree of cure (DoC) over time obtained from both DSC test and refractive index measurements; at any time t, DoC is computed according to the following relationship:

$$DoC = \frac{f(t) - f(0)}{f(\infty) - f(0)} \tag{5}$$

where f(t) is the heat emission in the DSC test and the refractive index in refractometer test, measured between 0 (beginning of test) and ∞ (time when the asymptotic value is reached). A good agreement can be observed between the two sets of data confirming the capability and reliability of the fiber optic sensor apparatus. For more details about the data analysis both by DSC and fiber optic refractometer see Cusano et al., 2000.

This preliminary test was useful to know the required time for the resin to attain a specified conversion level at 70 °C. Then, various experimental tests have been performed at selected degrees of cure and at different temperature values between 10-35 °C that is the effective operative range. In particular, the resin refractive index has been measured in the temperature range 10-35 °C at conversion levels of 0, 0.3, 0.7, 1. To attain an incomplete cure reaction (i.e., DoC of 0.3 and 0.7), the analysed sample has been first subjected to an isothermal heating at 70 °C for 5 and 15 min, respectively; according to the DSC curve shown in Figure 3, these time values correspond to DoC of 0.3 and 0.7, respectively. As an example, Figure 4 reports the refractive index as function of time during the whole experimental test for the conversion level of 0.7. As expected, since the refractive index is related to the density (see eq. 1), one should observe that, at the beginning, the refractive index decreases as the temperature increases due to the resin density reduction.

Then, as the polymerization reaction takes place, the optical properties changes result dominated by the formation of cross-links between the reactive groups of the polymer resin that develop a progressively denser network. This behaviour can be observed also in the diagram of Figure 5, where the refractive index of the partially cured resin is shown as function of temperature during the cooling and heating in the temperature range 10-35 °C. Figures 6 and 7 summarize the results of whole experimental tests reporting the resin refractive index as function of temperature and of the cure degree. The availability of these diagrams could be useful during the in-situ application of this resin system allowing to know if the resin has completely reacted by measuring the environmental temperature and the refractive index.

CONCLUSIONS AND FUTURE WORK

The preliminary assessment of the potential of using the proposed system to check the degree of curing of the resin has demonstrated that it could be possible to embed fiber-optic sensors into the laminates in order to monitor the advancement of the curing reaction of the resin over time in conjunction with temperature measurements. The experimental data herein discussed represent a first set that will need to be enriched in order to calibrate a direct relationship between the degree of cure and the refractive index. Once the behaviour of the resin has been studied and a reliable tool is available to follow the evolution of the resin reaction, the next effort will be devoted to defining reference values that, for different combinations of environmental temperature and humidity, can be used to judge whether the quality of an FRP installation is structurally acceptable and eventually plan the appropriate actions necessary to overcome curing problems in order to make the installation satisfying the minimum quality requirements.

The availability of such instructions collected into a manual could be of crucial importance for contractors, practitioners and inspectors because it will contain responses to present needs, namely: how the curing reaction of the resin can be controlled after the execution of a manual wet lay-up process and which actions should be taken in order to solve issues related to an imperfect curing of the resin. However, the information obtained at material level are not enough to achieve this goal. It is necessary to move the attention from the material level to the structural level and assess which is the minimum amount of chemical imperfection that could result in a structural imperfection; this is done with the objective of avoiding negative evaluations of installations that are not perfect from a chemical standpoint but whose chemical imperfections do not affect the structural effectiveness.

In order to define these tolerance thresholds for imperfect resin curing in terms of structural performance, it is planned to perform pull-off tests on FRP laminates bonded on concrete or masonry supports. The outcomes of these tests will highlight the influence of different levels of imperfect curing on the bond of the FRP laminates and provide a database to determine at which extent a deviation of the key parameters (environmental conditions, support conditions, mixing proportions) from the standard conditions indicated by the manufacturers could be still structurally accepted. It is also planned to perform tests on structural elements in order to check the feasibility of the overall quality control protocol. FRP laminates will be installed on concrete and masonry specimens and fiber-optic sensors will be embedded through the dry fibers. Many samples will be prepared by changing environmental and concrete surface conditions. Readings of the sensor will be used to test if the technique is able to highlight the curing problems theoretically expected for each given installation condition created in the laboratory. The results will be important to check the sensitivity of the sensors to the rolling of fibers typically performed during the impregnation by manual wet lay-up.

ACKNOWLEDGMENTS

The authors would like to thank the Center of Excellence on Structural Composites for Innovative Construction at University of Naples Federico II and the Regional Government of Campania for supporting the research presented in this paper.

REFERENCES

ACI 440.2R-02, *"Guide for the Design and Construction of Externally Bonded FRP Systems for Strengthening Concrete Structures,"* Reported by ACI Committee 440, American Concrete Institute, 2002.

Afromowitz M.A., Lam K.Y., "Fiber-optic epoxy composite cure sensor. I. Dependence of refractive index of an autocatalytic reaction epoxy system at 850 nm on temperature and extent of cure", Applied Optics, 1995, 34 (25), 5635–5638.

Chang P. C. and Liu S. "Recent Research in Nondestructive Evaluation of Civil Infrastructures", ASCE Journal of Materials in Civil Engineering May/June 2003, pp. 298-304.

Chen B. and Liu J. "Damage in carbon fiber-reinforced concrete, monitored by both electrical resistance measurement and acustic emission analysis" Proceedings of the International Conference Composites in Constructions, Cosenza, Italy, 2003, pp. 253-257.

Crosby P.A., Powell, G.R. Fernando G.F., Waters D.N., France C.M., Spooncer R.C., "A comparative study of optical fiber cure monitoring methods", in: C.O. Richard and R.O. Proceeding SPIE, Smart Structures and Materials: Smart Sensing, Proceeding and Instrumentation, 1997, vol. 3042.

Culshaw B., Dakin J., "Optical Fiber Sensors: Applications, Analysis and Future Trends", Artech House Inc., Norwood, 1997.

Cusano A., Breglio G., Giordano M., Calabrò A., Cutolo A., Nicolais L., "An optoelectronic sensor for cure monitoring in thermoset based composites", Sensors and Actuators A, 2000a, 84 (3), 270–275.

Cusano A., Breglio G., Calabrò A., Giordano M., Cutolo A., Nicolais L., "A fiber optic thermoset cure monitoring sensor", Polymer Composites, 2000b, 21 523-530.

Feng M. Q., De Flaviis F., and Kim Y. J., "Use of Microwaves for Damage Detection of Fiber Reinforced Polymer-Wrapped Concrete Structures", Journal of Engineering Mechanics February 2002, pp. 172-183.

fib bulletin 14, *"Externally Bonded FRP Reinforcement for RC Structures,"* Technical Report prepared by Task Group 9.3 FRP reinforcement for concrete structures, International Federation for Structural Concrete, 2001.

Gowripalan N., "Fiber optic and piezoelectric sensors for structures with FRP: state of the art", Proceedings of the International Conference on FRP Composites in Civil Engineering, Hong Kong, China, 2001, Vol. II, pp. 1635-1641.

Imai M., Sako Y., Miura S., Yamamoto Y., Ong S.S.L., Hotate K., "Dynamic health monitoring for a building model using a BOCDA based fiber optic distributed sensor", Proceedings of the First International Conference on Structural Health Monitoring and Intelligent Infrastructure, Tokyo, Japan, 2003, pp. 241-246.

Inaudi D., "State of the art in fiber optic sensing technology and EU structural health monitoring projects", Proceedings of the First International Conference on Structural Health Monitoring and Intelligent Infrastructure, Tokyo, Japan, 2003, pp. 191-198;

Lew, et al., "Single mode fiber evanescent wave spectroscopy", Conference Proceedings of the Second International Conference on Optical-fiber Sensors, 1984.

Katsuki K., "The experimental research on the crack monitoring of the concrete structures using optical fiber sensor", Proceedings of the First International Conference on Structural Health Monitoring and Intelligent Infrastructure, Tokyo, Japan, 2003, pp. 277-284;

Ku C.C.and Liepins R., "Electrical properties of polymers", Chemical principles, Hanser publ., 1987, Munich.

Mapei, "World-wide Leader in Products for the Building Indutry", http://www.mapei.it, Milan, Italy, 2003.

Mariìn C.G., Santiago M.O., Fernàndez J.R. and Garcìa R.R., "Ultrasonic tests for the evaluation of polymer mortars", Proceedings of the International Conference Composites in Constructions, Cosenza, Italy, 2003, pp. 265-272.

Nanni A., Cosenza E., Manfredi G. and Prota A., "Composites in Construction: Present Situation and Priorities for Future Research," *Proceedings of the International Workshop on "Composites in Construction: A Reality"*, July 2001, American Society of Civil Engineers, pp. 269-277.

Nicolais L., Kenny J.M., "Processing of high performance composite materials", Science and Technology of Polymer composite, 1991, 472-525

Starnes M.A., Carino N. J. and Kausel E. A., "Preliminary Thermography Studies for Quality Control of Concrete Structures Strengthened with Fiber-Reinforced Polymer Composites", ASCE Journal of Materials in Civil Engineering May/June 2003, pp. 266-273.

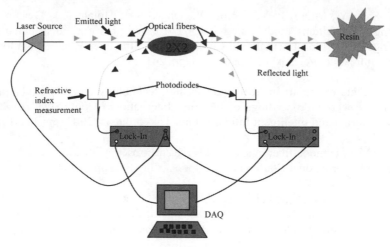

Figure 1 — Schematic of the fiber optic refractometer.

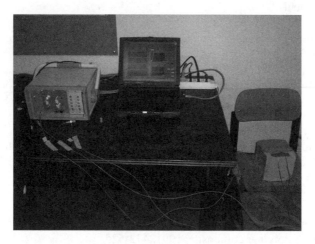

Figure 2 — Fiber optic refractometer and its in-situ application.

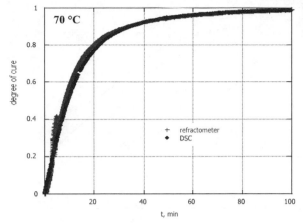

Figure 3 — Conversion evaluated by DSC test and refractive index measurements.

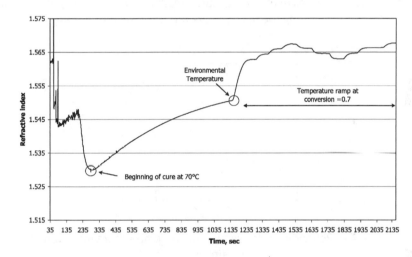

Figure 4 — Refractive index as function of time during the incomplete polymerization at 70°C and the temperature ramps.

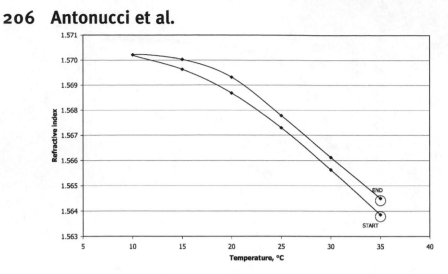

Figure 5 — Refractive index of the partial cured resin (conversion=0.7) as function of temperature during the temperature ramps in the range 10-35 °C.

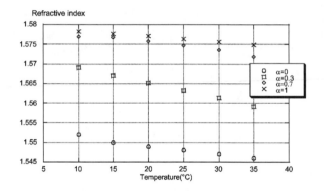

Figure 6 — Refractive index as function of temperature for different level of resin conversion.

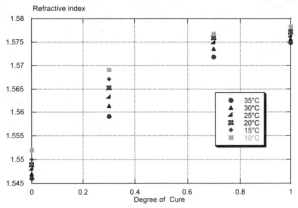

Figure 7 — Refractive index as function of cure for different temperature levels.

Gripping Behavior of CFRP Prestressing Rods for Novel Anchor Design

by A. Al-Mayah, K. Soudki, and A. Plumtree

<u>Synopsis:</u> In prestressed concrete applications, the wedge anchor system of the CFRP rod utilizes interfacial shear as the main gripping component between the rod and contacting metal. To simulate this condition, the interfacial mechanics of CFRP-metal couples under different contact pressures was studied. A range of CFRP rods with different strengths, sizes and surface profiles was investigated. The composite rods were in contact with either as-received or annealed aluminum alloy or copper sleeves. In all cases, the interfacial shear stress was found to increase linearly with contact pressure. High interfacial shear stresses were recorded when high strength rods were used. The highest shear stress was obtained by removing the surface perturbations of the rod resulting in the increase of the real contact area. Higher load oscillations, resulting from larger amounts of wear debris, were observed when lower strength and smooth rod surfaces were tested.

<u>Keywords:</u> aluminum alloy; CFRP; contact pressure; copper; hardness; interfacial stress

Adil Al-Mayah is a research assistant professor at the University of Waterloo, Canada. He received his PhD from the University of Waterloo in 2004. His research interests include fiber reinforced polymer composite applications in repair and strengthening of concrete structures, mechanics of composite materials, and numerical and mathematical modeling of contact surfaces.

ACI member Khaled Soudki, is the Canada Research Chair Professor in Innovative Structural Rehabilitation at the University of Waterloo, Canada. He is a member of ACI Committees 440, Fiber Reinforced Polymer Reinforcement; 222, Corrosion of Metals; 546, Repair of Concrete; 550, Precast Concrete Structures. His research interests include corrosion, durability, rehabilitation, and strengthening of concrete structures using fiber reinforced polymer composites.

Alan Plumtree is Distinguished Professor Emeritus, University of Waterloo. He is a Fellow of ASM International, Fellow of the Institute of Mining, Materials and Manufacturing (UK) and Honorary Fellow of the German Society for Materials Testing. His main interests include deformation and fracture of advanced materials.

INTRODUCTION

Fiber reinforced polymer (FRP) rods have a high specific stiffness and strength in addition to high corrosion resistance. These are attractive properties for use as prestressing material in concrete for new design, repair and/or strengthening of existing structures. However, successful implementation of unidirectional FRP rods depends upon their anchoring systems[1].Finding the optimum anchor system is a challenging problem due to the weakness of the material in the transverse direction[2].

Different anchors systems have been developed, including the wedge anchor, to address the gripping mechanism problem. Generally, the wedge anchor consists of a metal sleeve to encase the FRP rod. The interfacial shear stress at the FRP rod-sleeve surface has a major role on the anchor's performance. In addition to the amount and distribution of contact pressure, the shear stress depends upon other factors such as the type of contacting materials, their surface condition, and sliding direction.

To simulate the anchor mechanism of a carbon fiber reinforced polymer (CFRP) rod in the wedge anchor system, a comprehensive interfacial mechanical study has been conducted in preparation for designing the wedge anchor system.[3] An analytical investigation was carried out to determine the contact pressure distribution on the CFRP rod-metal surface.[4] The highest contact pressure occurred at the top of the rod and at the entrance to the clamping system.

Both the FRP and contacting material affect the gripping mechanism. The effect of the sleeve material has been investigated by using copper and aluminum alloys to encase a CFRP rod[5]. A lower shear stress was recorded and more slip took place with copper than with the aluminum alloy. The initial softness of the aluminum alloy allowed it to deform and fill the gaps between asperities, hence providing a larger real contact area. A similar effect was seen when the CFRP rod, encased in a series of metallic sleeves, was used in testing wedge anchor designs. When copper and aluminum alloy sleeves were used, the softer aluminum alloy allowed less rod slippage in the anchor.[6]

Giltrow and Lancaster[7] investigated the effect of carbon fiber type reinforcement in thermosetting resins contacting different metals on the coefficient of friction and wear rate. It was found that the hardness of the metal (steel, aluminum, copper and gold) had little effect on the coefficient of friction (0.15-0.3) for Type I carbon fiber (1.74 GPa ultimate tensile strength, 460 GPa modulus of elasticity) reinforced polymer. However, the hardness had a greater effect on the coefficient of friction (0.13-0.2 for hard materials and 0.4-0.7 for softer materials) in the case of Type II carbon fiber (2.88 GPa ultimate tensile strength, 262 GPa modulus of elasticity) composite. This was related to the high strength of Type II carbon fiber being in less degrading and more abrasive fiber than Type I carbon fiber.[7]

In general, unidirectional CFRP composite-steel couples displayed a low coefficient of friction and low wear rate.[8] Both of these characteristics are attributed to a high Young's modulus and interlaminar shear strength. Although, the interfacial properties of these materials are orientation dependent; the type of carbon fiber and the nature of the counterface material have a greater effect on the tribological properties of composites than fiber orientation, especially when the fiber volume fraction is increased.[7] It was observed that when the fibers were parallel to the contact surface and sliding took place in the direction of the fiber axes, the wear resistance was reduced by decreasing the hardness of the metal counterpart.[9] This experimental procedure applies to the present work.

Sliding and wear behavior has been found to change with surface roughness. The asperities of the rough metal counterface penetrate the soft polymer matrix and wear occurs by shear or micro-cutting.[10] However, when the counterface is smooth, sliding wear resulted due to localized fatigue failure. Oveart[11] reported an extensive separation between the fiber and matrix in PEEK-IM6 (graphite) composite as roughness of the counterface increased.

The sliding of composites against a metal counterpart produces a film of composite debris on the surface. Debris plays an important role in the tribological process of highly conforming contacting surfaces. The larger contacting area produces larger amounts of debris[12,13], as stated by Bowden and Tabor.[14] This effect is magnified by using a soft sleeve[9], specifically when debris piles up in front of the sliding metal.[15]

Debris creates an obstacle to sliding[16] and acts as a stress raiser.[13] When the polymer matrix material is rubbery and not brittle, the frictional force increases continuously as the slider moves over the debris.[17] By incorporating grooves in the counterface, debris removal is facilitated, thereby reducing the wear rate and sliding force.[12]

RESEARCH SIGNIFICANCE

This paper is part of a study on the interfacial contact behavior of CFRP rod- metal counterface used for prestressed concrete anchors. It presents the effect of CFRP rod type, rod topography on the sliding and, conversely, the gripping behavior of the CFRP-metal couple. These results form the basis for the design of a novel wedge anchor system.

EXPERIMENTAL PROGRAM

CFRP Rods Tested

Two types of CFRP rods were tested in this research program. Table 1 gives the material properties of the CFRP rods used in the pull-out tests.

Hughes No3 (H3): This was a single spiral indented CFRP rod of nominal diameter 9.4 mm (3/8 in). The longitudinal carbon fiber/epoxy composite was covered with a thin peel-ply. The texture of the rod surface is shown in Fig. 1. The guaranteed longitudinal strength of the rod was 1.38 GPa (200 ksi).[18] The longitudinal and transverse elastic modulii were 124 GPa (measured) and 7.4 GPa (calculated), respectively. The volume fraction of carbon fiber was 60% with a vinylester epoxy resin matrix.

Leadline (L8): Double spiral indented CFRP Leadline[TM] rod of 7.9 mm diameter. Fig. 2 shows the surface texture of the Leadline rod. A rod of a guaranteed unidirectional breaking strength of 2.12 GPa was tested. Its longitudinal and transverse modulii were E_{long} = 147 GPa and E_{tran} = 10.3 GPa, respectively.[19] The carbon fiber volume fraction is 65% in epoxy matrix.

Both rods were sectioned parallel to the fiber direction and examined under an optical microscope to determine the fiber distribution near the surface. It was found that the rods had been pressed in order to form spiral indentations. The outer layers consisted of epoxy matrix that had flowed to the outside of the rod during fabrication.

In one series of tests, the spiral indentations were removed. This resulted in a smooth surface with an outer diameter of 7.9 and 6.4 mm for Hughes No3, and Leadline rods, respectively. This process involved machining the surface of the rod to approximately the specified diameter, then finally polishing the surface in the longitudinal direction, starting with rougher and then progressively finer emery clothes (D120, C240 and 600A).

Test Systems

Fig. 3 is a schematic of the pull-out test apparatus which consisted of two hydraulic systems, one for clamping the CFRP rod-metal couple and the other for pulling the rod. A pressure dial was connected to the clamping jack to measure the applied normal load. For pulling, a load cell and linear variable differential transducer ($LVDT_1$) were connected to the sliding part. A second $LVDT_2$ was used at the free end of the rod to determine whether there was any difference in the recordings of the two LVDTs. Due to the importance of alignment in friction testing[20], the system was made self-aligning. The clamping anchor (A) was free to move in the direction normal to the sliding direction.

A 360 mm long CFRP rod was gripped using two clamping systems. One was a reusable anchor (Clamping Anchor A) and the other was the unit under test (Clamping Plates C). Anchor A was made of two grooved steel plates each 75 mm in length, 75 mm in width, and 25 mm thick. The anchor encased the CFRP rod with an aluminum sleeve of an inner diameter (ID) similar to the diameter of the rod, i.e. 7.9 mm and 9.4 mm, and outer diameter of (OD) of 9.53 mm and 12.7 for the Leadline and Hughes No.3 rods, respectively.

At the test end, Clamping Plates C consisted of two hardened steel plates 90 mm long, 45 mm wide, and 25 thick. A semicircular groove, 12.65 mm long, was machined in each plate to accommodate metal sleeves of given outer diameter OD1. The sleeve was machined with a 10 mm long collar of outer diameter (OD > OD1) which acted as a stop against the restraining plate, preventing any slippage once the pulling load was applied. The inner diameter (ID) of the sleeve was equal to the CFRP rod diameter. The sleeves used with Clamping Plates C for different rods are shown in Fig. 4. Further details of the clamping anchor, the clamping plates and their sleeves can be found in Al-Mayah et al.[21]

Test Parameters

The test matrix is given in Table 2. The parameters considered were contact pressure, type of sleeve material, sleeve hardness, and rod surface profile. The contact pressure was defined as the normal applied load divided by the projected area of the rod. For as-received rods, generally, six levels of contact pressure were investigated ranging form 50 to 350 MPa. Only three tests with 250 MPa pressure were conducted due to high grip that exceeds the machine capacity. However, for the machined rods, three contact pressures were examined with each of the four sleeves. The pull-out load and amount of rod displacement were monitored for each contact pressure.

Experimental Procedure

The CFRP rod was inserted into an aluminum sleeve and secured in the clamping anchor (A). This assembly was mounted in the test machine by passing the rod under the clamping jack (B) and the loading head was moved to its operating location near the clamping jack (B). After cleaning the plates, sleeve and rod with acetone, the test sleeve was placed over the rod under the center of the bearing plate attached to the clamping jack. The CFRP-metal couple was aligned in the vertical and horizontal directions. A known contact pressure was applied by the jack to the heat-treated steel plates (C). The rod was effectively gripped by the clamping anchor at one end and the controlled pressure at the other. The test then commenced by pulling the rod through the clamping plates (C) at a horizontal velocity of about 40 mm/sec, chosen to minimize the viscoelastic effect of the rod. The pulling load and corresponding displacement of the rod were recorded for a maximum sliding distance of 148 mm.

TEST RESULTS AND DISCUSSION

As-received rods

The effect of contact pressure on the pull-out load-displacement relationship using an annealed copper sleeve with as-received Hughes No3 rod is seen in Fig. 5. There was a rapid increase in pulling load to a maximum *(F1)* that increased with contact pressure which was then followed by a slight decrease in load with continuous displacement.

For contact pressures of 50 and 70 MPa, there was continued sliding with no obvious pull-out load oscillations. As the contact pressure increased above 70 MPa, the amplitude of the oscillations (*h*) increased. The pull-out load reached its highest value at a contact pressure of 200 MPa when the load exceeded the machine capacity. Similar behavior was observed using the Leadline rod with an annealed aluminum alloy sleeve, as shown in Fig. 6. The differences between the two rods were that the Leadline had a lower amplitude for the stick-slip cycles, an unclear peak load *(F1)*, and more broken carbon fibers. The post sliding surfaces of Hughes No3 and Leadline using annealed aluminum with contact pressure of 150 MPa are shown in Fig. 7. For Hughes No3, the rod surface was completely covered with epoxy. No carbon fibers were seen (Fig. 7-a). In contrast, an obvious exposure of fibers was observed on the surface of the Leadline rods (Fig. 7-b).

The maximum load *(F1)* from Figs 5 and 6 can be expressed in terms of shear stress (*τ*) by dividing the maximum force by the nominal surface area of the rod. Fig. 8 shows the relationship between shear stress and contact pressure for the as-received and annealed copper sleeves with Hughes No3 and Leadline rods. In general, the shear stress increased with increase of contact pressure. The Leadline rod developed higher shear stress than Hughes No. 3 rod and the annealed sleeves required a higher shear stress for the same contact pressure. Considering the as-received and annealed sleeves, it is interesting to note that the Leadline rod displayed the largest variation in shear stress. It is suggested that this effect is related to the high strength, and larger volume fraction of the carbon fibers and surface profile nature of the Leadline rod. This behavior was observed by Giltrow and Lancaster[7]. This has a bearing on future anchor design. Hence, the increased grip required will be found by using a softer sleeve. Considering the Hughes rod, work by Al-Mayah[3] showed that varying the rod diameter from 6.4 to 9.4 mm had a little effect on the shear stress and the sliding behavior.

The amplitude of the load fluctuations also increased with the higher contact pressure. This was significant at contact pressures between 100 and 200 MPa (Fig. 9) where the sleeves flowed into the rods asperities causing more ploughed debris on the surface that required higher pull-out force to be overcome. In the case of Leadline rod, little change was observed in amplitude due to higher fiber carbon content in the debris (Fig. 7-b).

Smooth Rod Surface

At a contact pressure of 50 MPa, little deformation in the form of surface wear was observed at the beginning of sliding (Fig. 10-a). After 32 mm of sliding, the fibers were more exposed to the surface due to the removal of the epoxy, as shown Fig. 10-b. In all cases and unlike the as-received CFRP rod, broken carbon fibers were observed on the surface. At the higher contact pressure of 200 MPa, agglomeration of fibers and epoxy debris were formed once sliding started (See Fig. 11-a). The extent of these debris increased as the rod continued to slide (See Fig. 11-b). In comparison to the as-received (non-machined) rod (See Fig. 7-a), broken fibers in the debris of the smooth (machined) rod were more noticeable.

For the annealed copper sleeve and for both smooth and as received H3 and L8 rods, the shear stress (based on maximum load F1, Fig. 5) increased significantly by machining the rod surface, as shown in Fig. 12. The smooth surface increased the real area of contact due to the removal of the spiral configuration on the rod surface causing the shear stress to increase.

The load oscillations following the peak load increased with the smooth rod surfaces especially at higher contact pressure, as shown in Fig. 13. This was due to increase of trapped debris between the rod and sleeve surfaces in comparison to that of the as-received rod where debris was deposited in the rod indentations.

The schematic Fig. 14 illustrates the sleeve deformation for the as-received and smooth rod surfaces. As the contact pressure was applied, the smooth rod surface was in complete contact with the sleeve. This was in contrast to the as-received rod where the spiral indentations were not completely contacting the sleeve (Fig. 14-b) and as the pull-out load was applied, the resulting debris was deposited in the spaces of the spiral. In addition, the debris began building up in front of the plate on the smooth surface (Fig. 14-c) causing a greater impediment to slippage. Also, the three dimensional arrangement of the broken fiber fragments between the contact surfaces would impede the sliding associated with the longitudinal carbon fiber lubricity.[9, 22]

ROLE OF THE PRESENT WORK IN WEDGE ANCHOR DESIGN

This work shows that by using soft metal sleeves, wedge anchor systems can be applied to successfully grip a range of CFRP rods of varying strengths, sizes and surface profiles, thereby allowing them to attain their design strength. With this general information it becomes possible to design future wedge anchor systems with greater confidence.

Each of the CFRP rod/metal couples presented in this paper simulates a segment of the wedge anchor system under different contact pressure levels. For the best anchor design for prestressing CFRP rod with no premature failure, low contact pressure is required at the location where the highest tensile stress exists in the rod to prevent any stress concentration on the rod. In the mean time, in order to have a full grip of the rod, higher contact pressure may be provided at the free end of the rod where the lower tensile stress is. Therefore, in order to simulate the anchor system, different contact pressures along the rod have to be investigated. The relationship between the contact pressure and the pull-out force can be directly applied to the anchor design. Also, since different metals may be used as a sleeve, annealed copper should be regarded as the most appropriate metal and condition due to its high interfacial shear stress and its lower galvanic corrosion potential. The findings of this study are being used in the design of a novel wedge anchor system able to carry the ultimate tensile strength of rod and pass the cyclic load.[3]

CONCLUSIONS

A wedge anchor system for a CFRP rod-metal couple was simulated by studying the interfacial behavior using two CFRP rods with two surface profiles (smooth machined and as-received) under different contact pressures. In general, the shear stress increased with contact pressure for rods and metal sleeves under consideration. The higher strength rod (Leadline) showed higher interfacial shear stress as a result of the surface profile, fiber content and fiber strength.

A smooth rod surface showed a higher shear stress than that of the as-received rod surface with perturbations due to the increased real area of contact. During slippage, higher load oscillations were observed with the smooth rods especially at high contact pressures. Scanning electron micrographs showed that the surface of the smooth rod was covered with patches of broken fibers and epoxy debris.

Higher load oscillations were observed with the lower strength rod (Hughes No3) in both as-received and machined conditions. This is related to the lower strength fiber and lower volume fraction of the fibers which leads to more debris. These results can be used in the design of a future wedge anchor system.

ACKNOWLEDGMENTS

The authors wish to thank the Natural Sciences and Engineering Research Council of Canada for financial support and Hughes Brothers Inc. NE, US for providing the CFRP rods.

REFERENCES

1. Karbhari, V.M., "Use of Composite Materials in Civil Infrastructure in Japan. International Technology Research Institute," Word Technology Division, 1998.

2. ACI Committee 440, "State-of-the-Art Report on Fiber Reinforced Plastic (FRP) Reinforcement for Concrete Structures," ACI Manual of Concrete Practice, 2002, Part 5.

3. Adil Al-Mayah, " Interfacial Behaviour of CFRP-Metal Couples for Wedge Anchor Systems," PhD Thesis, University of Waterloo, Waterloo, Canada, 2004, pp. 199.

4. A. Al-Mayah, K. Soudki, and A. Plumtree (2004a). Finite element and mathematical models of the interfacial contact behaviour of CFRP-metal couples." Composite Structures, Elsevier Applied Science. (Accepted).

5. Al-Mayah, A., Soudki, K., and Plumtree, A. (2004b). "Effect of Sleeve Material on Interfacial Contact Behavior of CFRP-Metal Couples." Journal of Materials in Civil Engineering, ASCE (Submitted).

6. Al-Mayah, A., Soudki, K., and Plumtree, A. (2001). "Mechanical Behaviour of CFRP Rod Anchors under Tensile Loading." Journal of Composites for Construction, ASCE, 5(2), 128-135.

7. Giltrow, J.P., and Lancaster, J.K. (1970). "The Role of the Counterface in the Friction and Wear of Carbon Fiber Reinforced Thermosetting Resins." Wear, 16(5), 359-374.

8. Tsukizoe, T., and Ohmae, N. (1986). "Friction and Wear Performance of Unidirectionally Oriented Glass, Carbon, Aramid and Stainless Steel Finer-Reinforced Plastic." Friction and wear of polymer composites, 205-231.

9. Jacobs, O., Friedrich, K., and Schulte, K. (1992). "Fretting Wear of Continuous Fiber-Reinforced Polymer Composites." Wear Testing of Advanced Materials, ASTM STP 1167, Philadelphia, 81-96.

10. Lancaster, J.K. (1972). "Polymer-Based Bearing Materials: The Role of Fillers and Fibre Reinforcement." Tribology, 5(6), 249-255.

11. Ovaert, T. C. (1995). "On the Wear Behavior of Longitudinally (parallel) Oriented Unidirectional Fiber Reinforced Polymer Composites." Tribology Transactions, 38 (1), 27-34.

12. Anderson, J.C., and Robbins, E.J. (1981). "The Role of Wear Debris in the Wear of Some Polymer Composites at High Loads." Wear of Materials, The International Conference on Wear of Materials, S.K. Rhee, A.W. Ruff, and K.C. Ludema, Eds., San Francisco, California, 539-541.

13. Lancaster, J.K., Play, D., Godet, M., Verrall, A.P., and Waghorne, R. (1980). "Third Body Formation and the Wear of PTFE Fibre-Based Dry Bearings." Journal of Lubrication Technology, Transactions of the ASME; 102(2), 236-246.

14. Bowden, F.P. and Tabor, D. (1974). "Friction: an introduction to tribology." Heinmann, London.

15. Landheer, D., and Zaat, J.H. (1974). "The Mechanism of Metal Transfer in Sliding Friction." Wear, 27 (1), 129-145.

16. Suh, N.P. (1985). "Tribophysics." Prentice-Hall Inc., New Jersey.

17. Suh, N.P., and Sin, H.C. (1981). "Genesis of Friction." Wear, 69 (1), 91-114.

18. Hughes Brothers Inc., (2002).

19. Sayed-Ahmed, E.Y., and Shrive, N.G. (1998). "A New Steel Anchorage System for Post-tensioned Applications Using Carbon Fibre Reinforced Plastic Tendons." Canadian Journal of Civil Engineering, 25 (1), 113-127.

20. Schön, J. (2000). "Coefficient of Friction of Composite Delamination Surfaces." Wear, 237(1), 77-89.

21. Al-Mayah, A., Soudki, K., and Plumtree, A. (2004c). "Effect of Sandblasting on Interfacial Contact Behavior of CFRP-Metal Couples." Journal of Composites for Construction, ASCE, (Accepted).

22. Schön, J. (2004). "Coefficient of friction for aluminum in contact with carbon fiber epoxy composite." Tribology International, 37(5), 395-404.

Table 1— CFRP Rod Material Properties

Properties	Hughes No.3 (H3)	Leadline (L8)
Carbon Fiber	Grafil	N/A
Epoxy Matrix	Vinylester	N/A
Volume Fraction (%)	60	65
Guaranteed Longitudinal Tensile Strength – (GPa)	1.38	2.12
Longitudinal Modulus (GPa)	124	142-150
Transverse Modulus (GPa)	7.4	10.3

Table 2 — Test Matrix

Contact Pressure (MPa)	Sleeve Material			
	As-received Aluminum	Annealed Aluminum	As-received Copper	Annealed Copper
a- Leadline Rod (L8)				
1) As-received rod				
50	2	2	2	2
75	2	2	2	2
100	2	2	2	2
150	2	2	2	2
250	2	2	2	2
350	2	2	2	2
2) Smooth Rod Surface (SM)				
50	2	2	2	2
100	2	2	2	2
250	2	2	2	2
b- Hughes Rod No3(H3)				
1) As-received rod				
50	2	2	2	2
75	2	2	2	2
100	2	2	2	2
150	2	2	2	2
200	2	2	2	2
250	2	1		
2) Smooth rod surface (SM)				
50	2	2	2	2
100	2	2	2	2
200	2	2	2	2

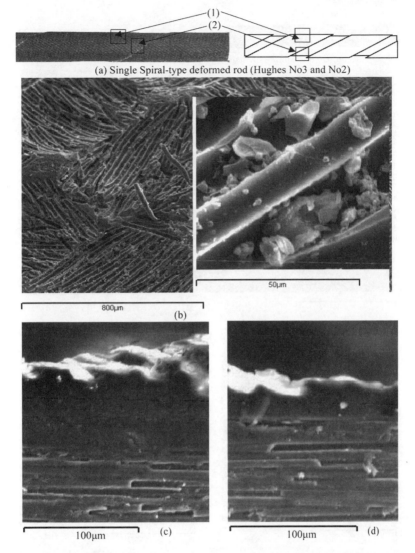

(a) Single Spiral-type deformed rod (Hughes No3 and No2)

Figure 1—Hughes Rod No3 a) Surface Profile, b) Surface Texture c) Surface Cross Section outside Spiral Deformation (1) and d) inside the Spiral Deformation (2)

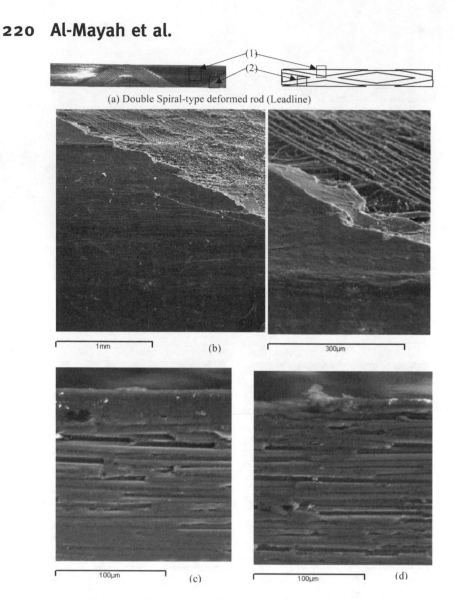

(a) Double Spiral-type deformed rod (Leadline)

1mm (b) 300µm

100µm (c) 100µm (d)

Figure 2—Leadline Rod a) Surface Profile, b) Surface Texture c) Surface Cross Section outside Spiral Deformation (1) and d) inside the Spiral Deformation (2)

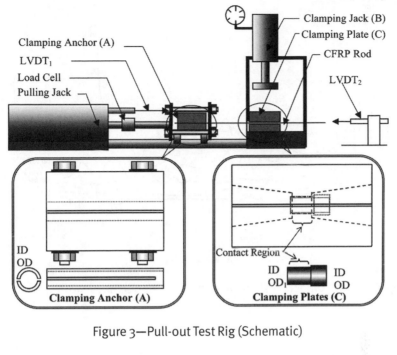

Figure 3—Pull-out Test Rig (Schematic)

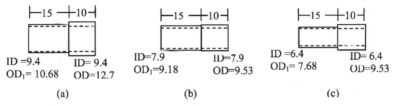

Figure 4—Sleeve Used with Clamping Plates C for a) Hughes No3, b) Smooth Hughes No3 and Leadline, and c) Smooth Leadline Rods (Dimensions in mm)

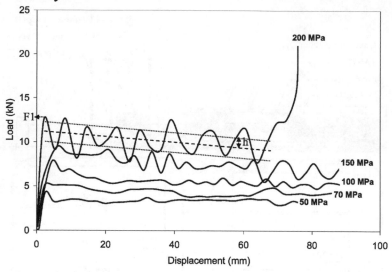

Figure 5—Effect of Contact Pressure on Annealed Copper Sleeve with Hughes No3 Rod

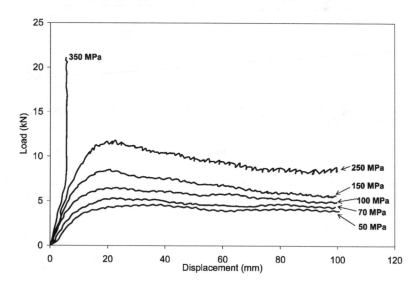

Figure 6—Effect of Contact Pressure on Annealed Aluminum Sleeve with Leadline Rod

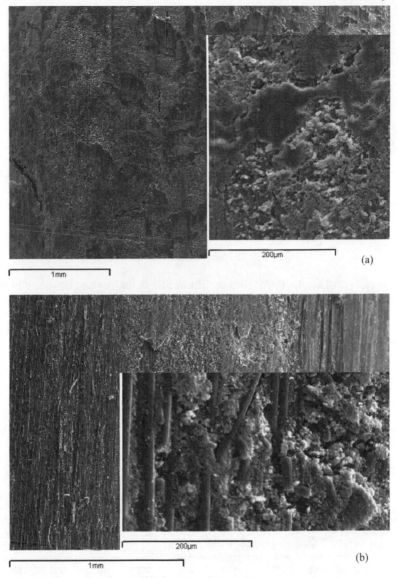

Figure 7—a) Hughes No3 and b) Leadline Rods Post-Sliding Views under 150 MPa
Contact Pressure with Annealed Aluminum Sleeve after 32 mm of Sliding

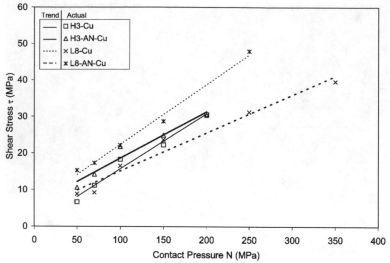

Figure 8—Effect of Contact Pressure on Shear Stress for Hughes No 3 and Leadline Rods with As-received and Annealed Copper Sleeves

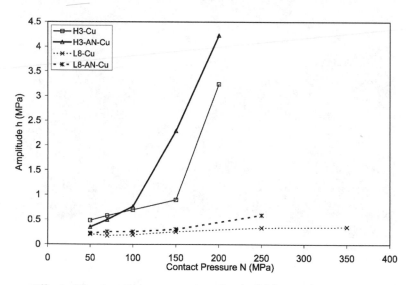

Figure 9—Effect of Contact Pressure on Amplitude (*h*) for Hughes No 3 and Leadline Rods with Annealed Copper Sleeves

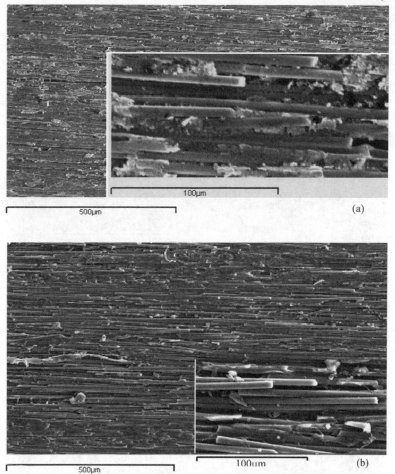

Figure 10—Smooth Rod Surface under 50 MPa Contact Pressure of Annealed Copper
Sleeve with Hughes No 3 Rod a) at the Beginning; and b) after 32 mm of Sliding

Figure 11—Smooth Rod Surface under 200 MPa Contact Pressure of Annealed Copper Sleeve with Hughes No 3 Rod a) at the Beginning; and b) after 32 mm of Sliding

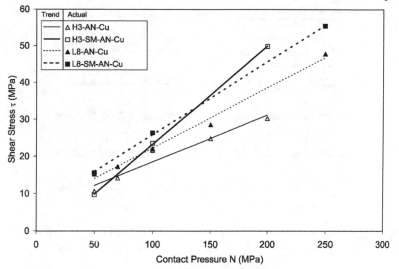

Figure 12—Effect of Rod Surface on Shear Stress for Annealed Copper Sleeve with Hughes No 3 and Leadline Rods

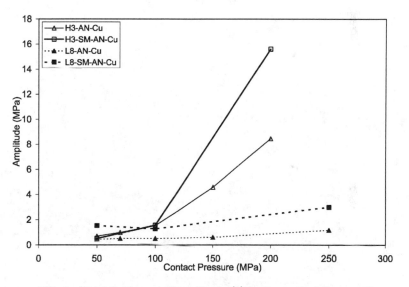

Figure 13—Effect of Rod Surface on Amplitude (*h*) for Annealed Copper Sleeve with Hughes No 3 and Leadline Rods

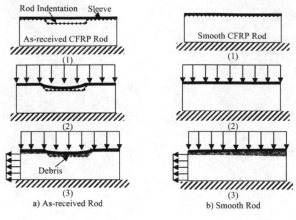

Figure 14—Sleeve Deformation of (a) As-Received and (b) Smooth Rods for 1) Unloading; 2) under Contact Pressure; 3) Under Contact Pressure and Pull-out Loading

Chapter 3

Strengthening of Existing Masonry Structures with FRP Systems

Out-of-Plane Static and Blast Resistance of Unreinforced Masonry Wall Connections Strengthened with FRP

by P. Carney and J.J. Myers

Synopsis: Recent world events have illustrated that the sustainability of buildings to blast loads is an ever increasing issue. Many older buildings contain unreinforced masonry (URM) infill walls. Due to their low flexural capacity and their brittle mode of failure, these walls have a low resistance to out-of-plane loads, which includes blast loads. As a result, an effort has been undertaken to examine retrofit methods that are feasible to enhance their out-of-plane resistance. The use of externally bonded and near surface mounted (NSM) Fiber Reinforced Polymer (FRP) laminates and rods have been proven to increase the out-of-plane load capacity.

This paper investigates the out-of-plane behavior of URM walls strengthened with FRP subjected to static and blast loading and the capability of developing continuity between the FRP strengthening material and the surrounding reinforced concrete (RC) frame system. There were two phases to this research study. Phase I evaluated strengthened URM walls' out-of-plane performance using static tests. Two strengthening methods were utilized, including the application of glass FRP (GFRP) laminates to the wall's surface and the installation of near surface mounted (NSM) GFRP rods. In both methods, the strengthening material was anchored to boundary members above and below the wall on some of the specimens in the research program. The effects of bond pattern, and the effects of FRP laminate strip width were also investigated in this phase. Phase II involved the field blast testing of two walls to dynamically study the continuity detail for laminates and verify the results obtained in Phase I. The development of continuity between the FRP materials and the surrounding framing system is one approach to improving the blast resistance of URM infill walls.

Keywords: blast resistance; FRP strengthening; masonry wall connections; masonry wall retrofits

ACI member **John J. Myers** is an Associate Professor at the University of Missouri-Rolla. He received his BAE from The Pennsylvania State University; MS and Ph.D. from University of Texas-Austin. His research interests include high performance concrete and use of fiber-reinforced polymers in structural repair and strengthening applications. He is a member of ACI Committees 201, 342, 363, 440, E801, E802, and E803. He is the current sub-committee co-chair of 440L (durability of FRP concrete structures) and chair of E801.

Preston Carney is a structural design engineer with Wallace Engineering Inc. located in Tulsa, Oklahoma. He has a BS and MS from the University of Missouri-Rolla in Civil Engineering.

INTRODUCTION

Recent events throughout the world have drawn attention to the vulnerability and sustainability of buildings and infrastructure to acts of terrorism. Our infrastructure is vital to this nation's economy and way of life. Any damage to it would and has had drastic effects on our culture. Attacks may cause a variety of results ranging from minor building damage to complete structural failure and considerable loss of life. Some examples within the United States include the bombing of the Murrah Federal Building in Oklahoma City (1995) and the bombing and attacks on the World Trade Center in New York City (1993, 2001). Abroad, numerous attacks have been directed toward embassies, and suicide car bombers have been used to target populated areas. In the cases where complete structural failure is not an issue, the dangers of flying debris have resulted in loss of life or injury to numerous civilians. Of particular concern are unreinforced masonry (URM) infill walls. Structural systems composed of a reinforced concrete (RC) framing system with URM infill walls make up a significant portion of the building inventory in the United States and around the world. Since there is no reinforcement within these walls, they have little resistance to out-of-plane loads such as a blast load. As a result, an effort has been undertaken to examine retrofit methods that are feasible to enhance their out-of-plane resistance. One method of strengthening URM walls is the application of fiber reinforced polymers (FRP) to the surface of the wall to improve their performance.

Since the effects of a blast cause a pressure to be exerted on the surface of a wall, the flexural behavior of the wall can be observed. This makes it appropriate to strengthen the walls to improve their flexural capacity. The application of externally bonded FRP materials have been shown to improve the flexural capacity of walls with and without arching action (El-Domiaty et al., 2002), but the development of continuity between the wall system and surrounding boundary members needs to be investigated.

Strengthening of walls is not the only step involved in the process of reducing a building's vulnerability to blast loadings. Proper risk assessment must also be performed to determine the level of vulnerability of a structure. One must also determine the level of damage that is acceptable for the structure to sustain. The characteristics of an explosion are key in assessing this vulnerability. The pressures that are developed as a result of an explosion are a function of the weight of the charge and the distance from the explosion, commonly called the standoff distance. The charge weight is expressed in terms of equivalent weight of trinitrotoluene (TNT). As the charge weight increases, the pressures that are developed are also increased. Similarly, as the standoff decreases, the pressures on a surface increase. For a given charge weight, the effects may be drastically different if the standoff distance is changed. For a very small standoff distance, strengthening the wall per say may have little effect; rather the addition of significant mass in the form of thick walls is often the approach. However, it may be more appropriate to try to increase the standoff distance to a facility by implementing barriers or restricting vehicular access to a structure. Wall strengthening would then allow for a compromise, that is the standoff distance would only have to be increased to the point which the strengthened wall could withstand the pressure from the design blast. With the proper assessment and an understanding of the key parameters, the strengthening of URM infill walls with FRP to improve their blast resistance has great potential.

EXPERIMENTAL PROGRAM

This research program made use of two FRP composite strengthening systems. The systems were laminate manual lay-up and near surface mounted (NSM) rods. Both systems use E-glass based fibers.

Out-of-plane testing

The testing of URM walls in the out-of-plane direction can be accomplished in a variety of ways. Several different methods of loading have been utilized in previous research to effectively apply a static load. Some research programs have applied a point load either at center span or used a device to apply two point, or line, loads on either side of the midpoint of the wall. A complex, but effective method of applying a uniform load is the use of a pressurized water chamber. A wall can be constructed between two tanks and one of them pressurized to apply a uniform load to the wall. A simpler method is the use of an airbag which is what was used in Phase I of this study. An airbag can be used to apply a load by placing the bag in contact with the test wall and a reaction structure. The airbag used in this research program had deflated dimensions of 36 in (914.4 mm) wide by 48 in (1219.2 mm) tall. They were six ply paper dunnage bags commercially produced by International Paper's Ride Rite Division. They were capable of withstanding pressures of over 20 psi (138 kPa) based on testing by the manufacturer.

Test matrix

The development of this test program extended previous research performed at UMR (El-Domiaty et al., 2002). The previous work illustrated that strengthening masonry walls with FRP materials does in fact improve their out-of-plane performance. This research was conducted to further investigate the effectiveness of strengthening URM walls with several variables.

This study was completed in two phases. Phase I was the evaluation of the retrofit techniques under static loading conditions using an airbag to incrementally load the walls to failure. Phase I was divided into two series. Series I consisted of six test walls. Series II was composed of an additional six walls based on the results obtained from Series I. Phase II was the field evaluation of two walls under actual blast loading. The walls in both phases were constructed of 4 in x 8 in x 12 in (101.6mm x 203.2 mm x 304.8 mm) CMU. The overall dimensions of the walls were 48 in (1219.2 mm) tall by 36 in (914.4 mm) wide. These dimensions result in a slenderness ratio of 12. The slenderness ratio of a 10 ft (3.05 m) tall wall constructed out of 8 in (203.2 mm) thick blocks is approximately 15. Therefore, the slenderness ratio used in this research program is comparable to what would be expected in an existing building. The 36 in (914.4 mm) wide dimension allowed for the wall to be three blocks wide giving two vertical, or head, joints in each wall. The FRP was applied along or within each of the head joints. It may be noted that 8 in (203.2 mm) thick walls were not tested due to limitations in the capacity of the air bags to meet the required failure pressure requirements of thicker wall units.

After Series I walls were tested, the test program for Series II was developed. Walls #1 and #2 served as unreinforced control walls in Phase I. Walls #7 and #8 in Series II served as strengthened FRP control walls without anchorage details. These were strengthened, but do not make use of the anchorage techniques. This allowed for a direct measure of the increase in capacity associated with the use of anchorage. The walls in Series I were all constructed using a stacked bond pattern. Since many facilities are constructed using a running, or staggered, bond, it was necessary to study the effects of bond pattern. This was done by constructing two of the walls (Walls #9 and #10) using a running bond. Both FRP retrofit techniques with anchorage were tested using this bond pattern. Series II concluded by studying the effects of the reinforcement ratio, or the width of the laminate strip. The final two walls made use of laminate strips that were 4.5 in (114.3 mm) and 6.5 in (165.1 mm) wide to investigate how the capacity of the walls changes as the amount of reinforcement on the walls increases. This change is only possible when using the laminates. The amount of reinforcement in the case of NSM bars cannot be increased without cutting additional grooves in the blocks due to the size limitations of the mortar joints.

The research program concluded with Phase II. This was the field blast testing of two walls. Under static loading, the laminates performed better than the NSM rods, so they were selected for use in this phase to evaluate their performance under short dynamic (blast) loading. One wall made use of 2.5 in (63.5 mm) laminates unanchored, while the other wall had the same reinforcement, but the FRP was anchored to the boundary members and the shear retrofit was included.

This experimental program investigated the development of continuity between the FRP and the boundary members. Several other variables were also examined, including the effects of bond pattern, and the effects of the width of the laminate strips. The test program is summarized in Table 1. Two FRP strengthening methods were utilized in this research along with anchorage techniques for both methods. The first strengthening technique is the use of externally bonded glass FRP laminates as shown in Figure 1. The laminates (fabrics) are applied vertically to the surface of the wall centered on the two head joints at the two third points. This system includes a primer, putty, saturant, and a glass fiber sheet to form the composite material. Glass fiber sheets were selected by the researchers in lieu of carbon FRP sheets based previous studies of retrofitting masonry systems by the research team. Glass fibers are more economical and provide a more compatible strength than the carbon fibers. The second method was the application of near surface mounted (NSM) glass FRP rods, illustrated in Figure 2.

Material and application material properties

The properties of the FRP rods and FRP fabric used in this study are detailed in Tables 2 and 3, respectively. Table 4 details the properties of the application materials. All material properties were evaluated using standard ASTM test methods. The mortar strength was 2000 psi, 1150 psi, and 1250 psi for Phase I-Series I, Phase I-Series II, and Phase II, respectively at test age of the walls. The compressive strength of the RC boundary elements was 4000 psi at test age. The concrete boundaries were one foot square beams, reinforced with three longitudinal #3 steel rebar top and bottom, allowing the beams to have the same strength in both directions. Shear reinforcement consisting of #3 stirrups spaced at 14 in (355.6 mm) on center were used in the boundary elements.

Phase I test set-up

Phase I tests were performed in the high-bay structural engineering laboratory at UMR. A strong wall was used as a reaction surface to load the URM walls. Concrete block was used to fill the void between the strong wall and the test location. RC beams, 12 in. (305 mm) square, were used as boundary elements on the top and bottom of the walls. The boundary members were post-tensioned to the strong floor as illustrated in Figure 3. The top member was also laterally restrained to the strong wall to limit translation at the top boundary. An air bag was placed between the test specimen and the concrete block fill to act as the loading mechanism. The bag was inflated incrementally and the pressure measured and recorded. As the bag inflated to a nominal level to bear against the wall, a strip along the edges of the wall was left unloaded. Dial gauges were used to measure the out-of-plane deflection (see Figure 3) and strain gauges were placed on the FRP to monitor the strain at each load increment where cracks were expected to form in the wall under out-of-plane load.

Phase II test set-up

The blast testing of the walls in Phase II took place at the United States Army Base at Fort Leonard Wood (FLW) near St. Roberts, Missouri. The tests were conducted on a certified military explosives range. The infill walls had boundary members (concrete beam / footing) on the top and bottom, respectively of the wall. A structural steel frame was designed to withstand the blast loading and support the boundary members and was anchored to the footings. The structural steel frame composed of 6 in × 6 in × 3/8 in (152 mm × 152 mm × 9.5 mm) tube sections and miscellaneous steel plates and angles is shown in Figure 4.

LABORATORY TEST RESULTS AND DISCUSSION

Of the twelve walls tested in this phase of the research program, none exhibited a shear problem or failure near the supports. When arching action is present, two possible failures can occur. The first is crushing of the masonry block and the second is the snapping through of the two rotating panels before crushing occurs. None of the walls showed signs of CMU crushing. The out-of-plane performance, development of arching action, and a description of the failure mode are provided for each wall.

Wall #1 -- Wall #1 was the first of the two unreinforced control walls. As testing began, an initial crack formed above the fourth course at 0.6 psi (4.1 kPa). Rotation, or the development of arching action, could be observed at the bottom of the wall. At 3.0

psi (20.7 kPa), a crack at the midspan joint occurred. After the crack occurred at midspan, a distinct development of arching action was observed just prior to failure (see Figure 5a, similar). Wall #1 failed at a pressure of 5.3 psi (36.5 kPa) with a deflection at failure of 1.3 in (33 mm).

Wall #2 -- Wall #2 was an unreinforced control with a shear retrofit. This wall performed much the same as Wall #1. Initial cracking was at the fourth course and occurred at 3.1 psi (21.4 kPa). Midspan cracking occurred at 4.0 psi (27.6 kPa) and development of arching action occurred (see Figure 5a, similar). Loading continued until failure with an ultimate load of 6.6 psi (45.5 kPa) and a deflection at failure of 0.74 in (18.8 mm).

Wall #3 -- Wall #3 was reinforced with 2.5 in (63.5 mm) wide sheets along the head joints. The sheets were anchored to the boundary members in this case. Initial cracking occurred at midspan at 3.1 psi (21.4 kPa), immediately followed by a crack above the fourth course at 3.2 psi (22.1 kPa). At 4.1 psi (28.3 kPa), a crack formed at the bottom course. Propagating cracks began to form at the intersection of the midspan crack and the GFRP sheet (similar Figure 5b). This was followed by the cracking of the block and additional propagation of cracks at midspan as well as above the fourth and fifth course. Distinct arching began to occur and a form of delamination was observed. The wall failed at 12 psi (82.7 kPa) with a deflection at failure of 1.5 in (38.1 mm). As the wall failed, pullout of the top anchorage occurred initially followed by the shearing of the sheets at the connection to the bottom boundary member. The rebar anchoring the sheets to the top member was broken near the location where the sheet was wrapped around it. The anchorage bar also pulled out of the top groove within the boundary element very clean, indicating that the bond between the rod and the paste may not have been good. The failure of the FRP can be classified as a delamination failure. However, there was no separation of the laminate from the wall. Concrete remained attached to the sheet after failure. The tensile strength of the concrete is reached before the bond breaks.

Wall #4 -- Wall #4 was also reinforced with 2.5 in. (63.5 mm) GFRP sheets. This wall had an ultimate capacity of 11.4 psi (78.6 kPa) and a maximum deflection of 1.1 in (27.9 mm). Midspan cracks formed at 3.4 psi (23.4 kPa) and a crack formed above the forth course at 3.8 psi (26.2 kPa). This wall also displayed the propagation of cracks as did the previous wall. Delamination was also present just prior to failure. Initially, the FRP ruptured at midspan, followed by shearing of the sheets at the bottom. There was a partial pullout at the top. One of the sheets pulled off the embedded rebar, while part of the other one stayed in tact. The top bar in this case also appeared to be pulling out fairly clean.

Wall #5 -- This wall was reinforced with two #2 GFRP bars along the head joints. The bars were anchored approximately three inches into the concrete boundary members. The failure occurred with the shearing of the FRP at three of the four connections. The fourth remained epoxied into the boundary member. Cracking began in this wall at 3.8 psi (26.2 kPa) with a midspan crack. Additional cracking continued with a crack above the fourth course at 4.4 psi (30.3 kPa). Several cracks formed through the blocks as shown in Figure 5b. Failure occurred at 10.2 psi (70.3 kPa) and a deflection of 1.2 in (30.5 mm).

Wall #6 -- The cracking of Wall #6 began at 2.6 psi (17.9 kPa) with a crack above the fourth course. This was followed by a midspan crack at 3.4 psi (23.4 kPa). During testing, the chain restraining translation of the top boundary broke. As a result, the pressure was decreased to a safe working level and the chain was replaced. Loading was then continued until failure was reached at a deflection of 1.1 in (27.9 mm) and a load of 11 psi (75.8 kPa). Failure occurred at midspan with the rupture of one of the FRP bars. The other bar pulled out of the bottom boundary, but remained attached at the top.

Wall #7 -- This was the first wall tested as part of Series II. This wall was reinforced with 2.5 in (63.5 mm) GFRP sheets that were not anchored to the concrete boundaries. An initial midspan crack occurred at 2.0 psi (13.8 kPa). At 2.4 psi (16.5 kPa), a crack formed above the fourth course. As was the case with the previous wall reinforced with sheets, propagating cracks began to form at midspan. Cracks illustrating the arching action can be seen in Figure 5a. Additional cracks formed through the blocks, as well as diagonal crack through the blocks visible from the side of the wall (Figure 5b). Delamination of the FRP accurred just prior to failure. A pressure of 9.6 psi (66.2 kPa) was achieved at a displacement of 1.8 in (45.7 mm). This wall test demonstrated the importance of anchoring the bonded laminates.

Wall #8 -- Wall #8 was reinforced with two unanchored #2 GFRP rebar. Initial cracks formed at midspan and above the fourth course. Propagation of cracks continued until failure at 4.8 psi (33.1 kPa), a load similar to the unreinforced condition. The deflection at failure was 0.96 in (24.4 kPa).

Wall #9 -- This wall was reinforced with NSM rods and made use of a running bond pattern, so the rods actually pass through some of the blocks. Cracking began above the fourth course at 2.2 psi (15.2 kPa) and at midspan at 3.0 psi (20.7 kPa). Extensive crack propagation was present, as were cracks through the blocks. Arching action was also clearly defined. 9.4 psi (64.8 kPa) was the failure load that occurred at 1.3 in (33.0 mm) of lateral displacement. The GFRP rods sheared off at the top, and one pulled out of the boundary at the bottom.

Wall #10 -- Wall #10 was reinforced with 2.5 in (63.5 mm) wide laminate strips anchored to the boundary. This wall was constructed using the running bond pattern. Initial cracking occurred above the fourth and fifth courses and at midspan at 1.8 psi (12.4 kPa). Propagating cracks began to occur at midspan and above the fourth course. Distinct arching action could be observed. The propagation of cracks continued with additional cracking through the blocks. Failure occurred at a pressure of 11.0 psi (75.8 kPa) and a displacement of 1.7 in (43.2 mm). Delamination was observed as were diagonal cracks through the 4 in (101.6 mm) dimension of the blocks. Pullout from the top beam occurred, but a portion of the rod remained in the beam. The FRP was sheared at the bottom boundary.

Wall #11 -- Wall #11 was reinforced with 4.5 in (111.8 mm) anchored FRP Sheets. This wall failed at a deflection of 1.9 in (48.3 mm) and a load of 12.6 psi (86.9 kPa). Cracking began at 2.0 psi (13.8 kPa) above the fourth course, followed by a crack at midspan at 2.8 psi (19.3 kPa). At 3.6 psi (24.8 kPa) there was a crack above the fifth course. Arching action and propagation of cracks were observed. There was also a shear crack through the block. Additional propagating cracks and cracks through the blocks were observed. Extensive cracking occurred prior to the delamination failure at 12.6 psi (86.9 kPa). At the failure deflection of 1.9 in (48.3 mm), the FRP sheets pulled off of the bar anchoring them to the top boundary and sheared off at the bottom. The integrity of the wall system was generally intact after failure.

Wall #12 -- Wall #12 had the highest reinforcement ratio of all of the walls tested in Phase I. This wall was reinforced with 6.5 in (165.1 mm) wide sheets anchored to the boundaries. At 2.6 psi (17.9 kPa), cracks formed at midspan and above the fourth course. This was followed by additional cracking at 4.6 and 4.8 psi (31.7 and 33.1 kPa) above the fifth and second course, respectively. Initial failure was the pullout of the FRP sheets from the bottom beam followed by pullout from the top. This failure occurred at 15.2 psi (104.8 kPa) and a deflection of 1.9 in (48.3 mm). Arching and propagation cracks and cracks through the blocks were present. The laminates held the system intact after failure (Figure 5c).

ANALYSIS OF LABORATORY DATA AND FURTHER DISCUSSION

As the results indicate, strengthening the walls with FRP materials does in fact increase the wall's resistance to out-of-plane loads. Furthermore, the anchorage details allow for the development of continuity between the FRP and the concrete boundary elements. This can be seen by comparing the results of Walls #7 and #8 to Walls #3 through #6 as illustrated in Figure 6. Walls #3 through #6 investigate the condition in which the reinforcement is anchored to the boundary members. In Walls #7 and #8, the same reinforcement is used without the anchorage. For the case of the GFRP sheets, the unanchored condition provides a capacity between the unreinforced case and the anchored case. Some benefit can be obtained for walls with arching action just by applying the laminates to the walls. When anchorage of the sheets is provided, this research suggests additional capacity is gained. This is not true in the case of the NSM rods. When the NSM rods are installed without anchoring them to the boundary, they behave in much the same way as an unreinforced wall. When anchorage is provided, continuity is developed and additional capacity is obtained.

Walls #9 and #10 examined the effects of a URM wall's bond pattern on the strength increase provided by the FRP. These two walls were constructed using a running bond, so the FRP does not follow a continuous mortar joint. The FRP was anchored to the boundary element for this case. The walls performed similarly to those using the stacked bound used in the rest of the test program. Though not examined in this research, bond pattern may have an effect on the case where unanchored NSM rods are used. In this case, the rods would run through the face of the blocks in every other row. This may provide an increase in strength over the case of a stacked bond where the rod is placed in a continuous joint.

Walls #11 and #12 evaluated the influence of the width of the GFRP laminates on the out-of-plane strength. It is evident that as the width increases, the failure load also increases. As a rule, an increase in the strain energy, or area under the load deflection curve, usually provides a more desirable mode of failure. The load versus deflection curves for each wall are provided in Figure 7. As shown in this figure, two distinctly different initial stiffness values can be observed. The Series I walls have an increased stiffness over the walls from Series II. This is due to a variation in the compressive strength of the mortar as reported previously. These two series of walls were constructed at different times, and as a result had different mortar compressive strengths even though their material compositions were similar. To allow for a more accurate comparison, a correction was performed. This was done by adjusting the deflections of the elastic portion of Series II walls. According to the Masonry Standards Joint Committee (2002), the mortar's modulus of elasticity is directly proportional to its compressive strength as shown in Equation 1. The measured deflections from Series II were corrected by multi-

$$E_g = 500 \times f_m'$$ (1)

plying them by the ratio of their mortar strength to that of the walls in Series I. Essentially Series II walls were normalized based on the modulus of Series I walls. The deflections beyond the linear range (elastic wall response) were simply shifted the amount of the correction at the end of the linear range since the mortar had cracked and no longer contributed in a significant fashion to the stiffness of the wall system. The modified load versus deflection plot provides a more representative comparison for the two series of walls and is illustrated in Figure 8.

Using the plot of pressure versus modified deflection illustrated in Figure 8, the strain energy of each wall can be calculated by estimating the area under this load versus deformation curve. For this research, Wall #1 was used as a control or benchmark on which to base a strain energy ratio. The normalized strain energy as shown in Figure 9 is the ratio of the strain energy of a given wall to that of the control wall, Wall #1. From this figure, it is clear that the laminates provide the system with the ability to absorb more energy prior to failure. This was observed visually during the out-of-plane tests as well. The walls strengthened with NSM rods failed in a sudden brittle manner. When the laminates were used, more of the wall was intact, and impending failure was apparent by the large number and wide distribution of the crack patterns. Table 5 summarizes the initial failure modes for each wall and categorizes the failure as brittle or ductile based on their normalized strain energy ratio.

Two different methods were used to compare the ductility of the reinforcing techniques. The first method is the deflection ductility. This is calculated by dividing the wall's ultimate deflection (u_f) by its deflection at the apparent yield point (u_y). The second method is the energy ductility. This is determined by dividing the total area under the pressure versus corrected deflection plot by the area under the linear portion of the plot. Often, energy ductility is used to characterize and discuss the ductility of composite systems. Figure 10 shows both the normalized deflection and energy ductility based on the control wall, as well as for the walls strengthened with 2.5 in (63.5 mm) GFRP sheets and #2 NSM GFRP rods. For comparison, the ductility has been normalized with respect to the control wall. As illustrated in the figure based on two different ductility terms, strengthening the walls with both sheets and rods provides the wall system with additional ductility. Figure 11 illustrates the relationship between increasing laminate strip width and ductility. In this figure, the ductility ratio has been normalized with respect to the wall strengthened with 2.5 in (63.5 mm) wide sheets (the lowest reinforcement ratio). Again, the ductility increases as additional strengthening is provided. In the case of the sheets, as the ductility increased, the ability of the GFRP laminates to hold the wall together also increased. As shown in the results for Walls #11 and #12 (Figure 5c), the walls were largely held in tact as the amount of reinforcement and ductility increased.

The ability of the FRP to hold the wall together upon failure is important under blast loading. People can often survive a blast, but when hit by flying objects and debris, loss of life may occur. The pressures that would cause loss of life to a human are significantly higher than those that cause catastrophic damage to a building. If the integrity of walls in a building can be maintained, there is a reduced amount of flying debris that could potentially injure the occupants of the building. Increasing the amount of GFRP laminates on the wall was shown to improve the integrity of the wall system.

In several of the cases where the GFRP laminates were anchored to the boundaries, the GFRP rod used in the anchorage detail pulled out of the groove. Upon observation of the rod after failure, it was noted that little or no epoxy paste was still attached to the rod. This indicates that integrity of the system at higher reinforcement ratios is limited by the bond of the epoxy to the rod, and should be closely studied in future research.

To allow for the correlation of these results to those predicted theoretically, an equivalent uniform load was calculated. As the air bag inflates an area around the edges of the wall is left unloaded due to the size and shape of the air bag. To develop an expression for the equivalent uniform pressure, the load area and distribution was analyzed. The moment caused by loading the reduced area was first determined. The required pressure to cause this same moment given a uniform load over the entire wall was then calculated. It was determined that the equivalent uniform pressure was 66.3% of the pressure recorded during testing. Table 6 shows the equivalent uniform pressures for each wall. The relationship shown in Equation 2 developed by Shapiro et al. (1994) was then used to predict the capacity of an URM wall. This theory makes use of three coefficients, R_1, R2, and λ. R_1 is taken as 1.0 because there is no previous cracking. R_2 is taken as the minimum value of 0.5 because there is no framing along the sides of the wall. λ was taken as 0.0496, based on the wall's slenderness ratio. This model predicted an out-of-plane capacity of 5.58 psi (38.5 kPa) as shown below.

$$w = \frac{2f'_m}{h/t} R_1 R_2 \lambda \qquad (2)$$

$h/t = 48/4 = 12$ $f'_m = 1350$ psi $R_1 = 1$ no previous cracking
$\lambda = 0.0496$ based on h/t $R_2 = 0.5$ no framing on sides

$$w = \frac{2*1350}{12}*1*0.5*0.0496 = 5.58\,psi\,(38.5kPa)$$

This value was slightly higher than the value causing failure of the test specimens in this research program. Based on the theoretical approach presented by Galati et al. (2003), the strength of the strengthened walls was predicted. The experimental and theoretical results are summarized in Table 7. Also listed in the table are the corresponding reinforcement indexes for each wall. These were determined as shown below using Equation 3.

$$\omega_f = \frac{\rho_f E_f}{f'_m (h/t_m)} \qquad (3)$$

For the 2.5 in laminates: $E_f = 10500$ ksi   Thickness = 0.0139 in

Therefore:

$$\rho_f = \frac{A_f}{b_m t_m} = \frac{2.5 * 2 * 0.0139}{36 * 4} = 0.00024$$

$$w_f = \frac{0.00024 * 10500}{1.350 * 48 / 4} = 0.1556$$

Figure 12 plots the ratio of the experimental pressure to the theoretical pressure versus the reinforcement index. This plot indicates that the theoretical approach yields reasonable results in predicting the capacity of the walls. In the worst case, the experimental result was approximately 80% of the theoretical value. At times, conservative values result. The theoretical approach suggests that the crushing of the masonry is the primary mode of failure for most of the specimens. During testing, none of the walls failed due to masonry crushing; rather most of the failures were initiated by delamination of the strengthening technique. This may be due to the slight translation of the top boundary member during testing. This translation could have prevented the crushing of the masonry and allowed the wall to resist an increased load. Had rotation been fully restrained, concrete crushing may very well have controlled.

FIELD BLAST TEST RESULTS AND DISCUSSION

To verify the performance of the strengthening systems tested in the lab under blast loads, field blast tests were conducted on two walls. One of the walls was strengthened with GFRP laminates unanchored to the boundary elements, while the other made use of the same reinforcement but included the anchorage detail. Four damage levels have been established to categorize the damage caused by a blast load to test walls [Myers et al. (2002)]. Table 8 summarizes the blast loadings undertaken by the walls in this phase, as well as the level of damage the wall sustained under each loading.

Wall #1 -- Wall #1 was strengthened with 2.5 in (63.5 mm) GFRP laminates. No anchorage detail was provided for this wall to examine the differences in behavior of un-anchored and anchored walls. The wall survived the first blast event of 2 lb (0.9 kg) with minimal cracking sustaining light damage. The second blast event made use of 4 lb (1.8 kg) of pentolite explosive and caused a failure. Extensive cracking occurred in all of the mortar joints with a sliding failure of the mortar joint between the top course of blocks and the top boundary element where the wall was not anchored. Cracks formed through the blocks and a diagonal crack formed on the side of a mid-height block.

Wall #2 -- Wall #2 was strengthened with 2.5 in (63.5 mm) GFRP laminates that were anchored to the boundary members. This wall survived the first blast of 3 lb (1.4 kg) with light damage consisting of minimal cracking in some of the mortar joints. The following blast event induced heavy damage but did not result in failure of the wall system. Failure occurred when subjected to a 5 lb (2.3 kg) charge. Lateral rotation occurred with the loss of one of block form the wall, however, the anchorage details remained intact. Propagating cracks near the midspan mortar joint indicate that the wall system was approaching the onset of delamination.

The end stack of blocks comprising the wall began to rotate. It may be noted that in a continuous full scale wall, this rotation would not have occurred due to the fact that the column of blocks would have either been supported by a vertical boundary element or bonded to the next column of blocks. There would not have been an end free to rotate as did the wall in this test program. Despite the rotation, the anchorage detail remained in tack, suggesting that addition capacity could have been obtained had the premature failure not occurred. Even with the rotation, the anchorage clearly provided an increase in capacity over the unanchored wall. The development of continuity between the FRP strengthening material and the surrounding boundary elements is key to increasing a walls out-of-plane strength and blast resistance for walls of similar slenderness ratios with arching action. Obviously this approach induces additional demands on the reinforced concrete frame and demands special attention to examine the overall system behavior of the framing system.

CONCLUSIONS

The objective of this research program was to evaluate the effectiveness of developing continuity between an FRP strengthened wall system and surrounding RC boundary elements. The effects of bond pattern and variable laminate strip width were also investigated. The conclusions drawn from this research are as follows:

- Additional capacity is gained by using all of the strengthening methods used in this research with exception to the case of unanchored NSM rods using a stacked bond. This reinforcement technique behaved much the same as the unreinforced wall.

- The development of continuity between the wall system and the surrounding frame provides additional capacity in the out-of-plane direction over the case where the strengthening material is not anchored to the boundary elements, both under static and blast testing.

- Bond pattern, stacked versus running, had limited effect on the out-of-plane strength of the walls.

- The laminate strips tend to hold the wall in tact as it fails, thereby reducing the scatter of debris under static out-of-plane testing. This reduces the risk to the inhabitants of the building.

- Increasing the width of the laminate strips provides additional capacity and allows the wall to fail as a unit, almost eliminating debris scatter under static out-of-plane testing. Laminates consisting of grid configurations may provide even a higher degree of limiting debris scatter.

- The bond characteristics of the various pastes used to apply NSM rods needs to be further investigated to properly evaluate the true strength of the anchorage details.

- The field blast test dynamically validate the laboratory results which suggest that the use of anchorage details or the development of continuity between the wall system and the surrounding RC frame provide additional capacity in the out-of-plane direction beyond that gained by strengthening alone.

ACKNOWLEDGEMENTS

The authors would to acknowledge the NSF/Industry sponsored Repair of Buildings and Bridges with Composites Cooperative Research Center (RB2C) at the University of Missouri-Rolla (USA) for supporting this research study.

REFERENCES

Angel, R., Abrams, D. P., Shapiro, D., Uzarski, J., and Webster, M. (1994). "Behavior of Reinforced Concrete Frames with Masonry Infills." Structural Research Series No. 589, Department of Civil Engineering, Univeristy of Illinois at Urbana-Champaign, IL.

El-Domiaty, K. A., Myers, J. J., Belarbi, A. (2002). "Blast Resistance of Un-Reinforced Masonry Walls Retrofitted with Fiber Reinforced Polymers." Center for Infrastructure Engineering Studies Report 02-28, University of Missouri – Rolla, Rolla, MO.

Galati, N., Tumialan, J. G., and Nanni, A. (2003) "Influence of Arching Mechanism in Masonry Walls Strengthened With FRP Laminates." Submitted to ACI International, January, 2003.

Hughes Brothers, Inc. (2001) "Glass Fiber Reinforced Polymer (GFRP) Rebar, Aslan 100." Seward, NE.

Masonry Standard Joint Committee (2002). "Building Code Requirements for Masonry Structures," ACI 530-02/ASCE 5-02/TMS 402-02. American Concrete Institute, American Society of Civil Engineers, and The Masonry Society, Farmington Hills, MI, Reston, VA, Boulder, CO.

Mays, G.C. and Smith, P.D. (1995) Blast Effects on Buildings. Thomas Telford, London.

Myers, J. J., Belarbi, A. El-Domiaty, K. A., "Blast Resistance of Un-reinforced Masonry Walls Retrofitted with Fiber Reinforced Polymers," The Masonry Society Journal, Boulder, Colorado, Vol.22, No.1, September 2004, pp. 9-26.

Shapiro, D., Uzarski, J., Webster, M., Angel, R., and Abrams, D. (1994). "Estimating Out-of-Plane Strength of Cracked Masonry Infills." Report, Department of Civil Engineering, University of Illinois, Urbanna, IL sponsored by the National Science Foundation, Arlington, VA.

Watson Bowman Acme Corp. (2002a). "Wabo®MBrace Composite Strengthening System, Engineering Design Guidelines." Amherst, NY.

Watson Bowman Acme Corp. (2002b). "Wabo®MBrace EG900, Unidirectional E-Glass Fiber Fabric." Amherst, NY.

Table 1 -- Experimental Test Matrix

Walls			FRP Sheets	Sheet Width	NSM FRP Rods	Anchorage	Stacked Bond	Running Bond
Phase I	Series I	#1					√	
		#2					√	
		#3	√	2.5"		√	√	
		#4	√	2.5"		√	√	
		#5			√	√	√	
		#6			√	√	√	
	Series II	#7	√	2.5"			√	
		#8			√		√	
		#9			√	√		√
		#10	√	2.5"		√		√
		#11	√	4.5"		√	√	
		#12	√	6.5"		√	√	
Phase II		#1	√	2.5"			√	
		#2	√	2.5"		√	√	

Conversion: 1in = 25.4 mm   Key: √ - includes detail

Table 2 -- Properties of GFRP Rebar (Hughes 2001)

Bar Size	Cross Sectional Area (in^2)	Nominal Diameter (in)	Tensile Strength (ksi)	Tensile Modulus of Elasticity (ksi)
#2	0.0515	0.25	120	5920

Conversions: 1 in = 25.4 mm, 1 in^2 = 645.2 mm^2, 1 ksi = 6.895 MPa

Table 3 -- Properties of GFRP Fabric (Watson 2002b)

Nominal Thickness (in)	Ultimate Tensile Strength (ksi)	Tensile Modulus of Elasticity (ksi)	Ultimate Rupture Strain	Ultimate Tensile Strength per Unit Width (k/in)
0.0139	220	10500	2.1%	3.06

Conversions: 1 in = 25.4 mm, 1 ksi = 6.895 MPa, 1 k/in = .175 kN/mm

Table 4 -- Properties of Application Materials ([1]Watson 2002a, [2]ChemRex® 2002)

	Primer[1]	Putty[1]	Saturant[1]	Paste[2]
Tensile Strength (psi)	2500	2200	8000	4000
Tensile Strain	0.40	0.06	0.03	0.01
Tensile Modulus (psi)	104,000	260,000	440,000	-
Poisson's Ratio	0.48	0.48	0.40	-
Compressive Strength (psi)	4100	3300	12,500	12,500
Compressive Strain	0.10	0.10	0.05	-
Compressive Modulus (psi)	97,000	156,000	380,000	450,000

Conversion: 1 psi = 6.895 kPa

Table 5 -- Summary of Failure Modes for Phase I

Wall	Initial Failure Mode	Brittle / Ductile
1	Bed Joint Failure	Brittle
2	Bed Joint Failure	Brittle
3	Delamination	Ductile
4	Delamination / FRP Rupture	Neutral
5	FRP Shear at Connections	Brittle
6	FRP Rupture	Neutral
7	Delamination	Ductile
8	Bed Joint Failure	Brittle
9	FRP Shear and Pullout at Connections	Neutral
10	Delamination	Ductile
11	Delamination	Ductile
12	Connection Pullout	Ductile

Table 6 -- Equivalent Uniform Pressures

Wall #	Experimental Pressure (psi)	Equivalent Uniform Pressure (psi)
1	5.3	3.5
2	6.6	4.4
3	12.0	8.0
4	11.4	7.6
5	10.2	6.8
6	11.0	7.3
7	9.6	6.4
8	4.8	3.2
9	9.4	6.2
10	11.0	7.3
11	12.6	8.4
12	15.2	10.1

Table 7 -- Experimental and Theoretical Results

Wall #	Experimental Pressure (psi)	Theoretical Pressure (psi)	Reinforcement Index (ω_f)
1	3.5	3.7	-
2	4.4	3.7	-
3	8.0	6.75	0.1556
4	7.6	6.75	0.1556
5	6.8	6.46	0.2631
6	7.3	6.46	0.2631
7	6.4	5.3	0.1556
8	3.2	3.9	0.2631
9	6.2	6.46	0.2631
10	7.3	6.75	0.1556
11	8.4	7.4	0.2787
12	10.1	7.85	0.4083

Conversion: 1 psi = 6.895 kPa

Table 8 -- Summary of Blast Events and Levels of Failure

Wall	Charge Weight (lb)	Standoff Distance (ft)	Level of Damage
Wall #1	2	6	Light Damage
	4	6	Failure
Wall #2	3	6	Light Damage
	4	6	Heavy Damage
	5	6	Failure

Conversions: 1 ft = 12 in =25.4 mm, 1 lb = 0.45 kg

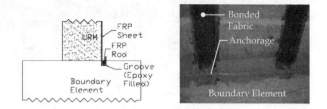

Figure 1 — FRP Laminate Detail

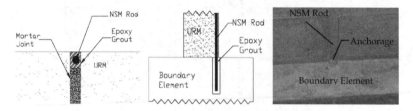

Figure 2 — NSM FRP Rod Detail

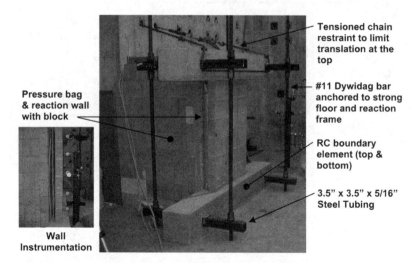

Figure 3 — Phase I Test Setup and Instrumentation (Laboratory)

Figure 4 — Phase II Test Set-up with Suspension of Charge (Field)

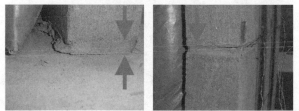

a) Formation of Arching Action: Base of Wall, Mid-height of Wall

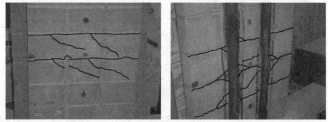

b) Crack Pattern Development in Strengthened Walls Prior to Failure: Wall #5, Wall #7

c) Debris Scatter: URM (Wall#2), NSM GRFP (Wall #8), Bonded Fabric (Wall #12)

Figure 5 — Wall Behavior of Various Systems as Noted

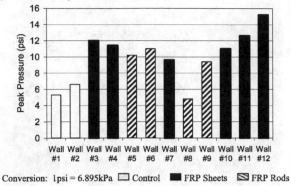

Conversion: 1psi = 6.895kPa ☐ Control   ■ FRP Sheets   ▨ FRP Rods

Figure 6 — Peak Out-of-Plane Pressure Results for Phase I Test Walls at Failure

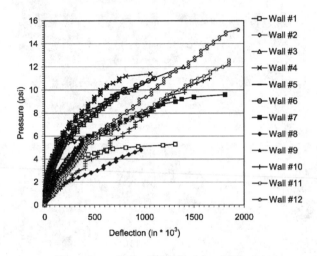

Conversions: 1 psi = 6.895 kPa, 1 in = 25.4 mm

Figure 7 — Pressure versus Displacement for Phase I Walls

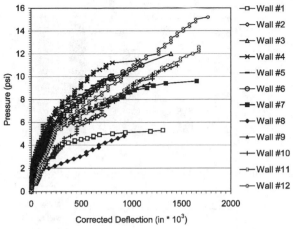

Conversions: 1 psi = 6.895 kPa, 1 in = 25.4 mm

Figure 8 — Pressure versus Modified Displacement

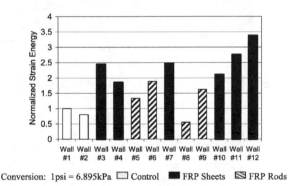

Conversion: 1psi = 6.895kPa   ☐ Control   ■ FRP Sheets   ▧ FRP Rods

Figure 9 — Normalized Strain Energy Ratio

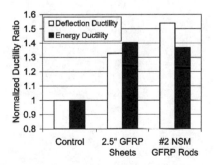

Figure 10 — Strengthening Scheme Effects on Normalized Ductility Ratio

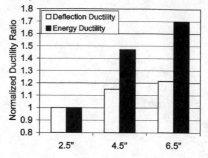

Figure 11 — Laminate Strip Width Effects on Normalized Ductility Ratio

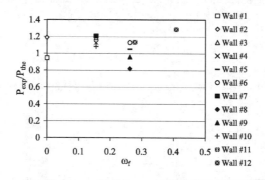

Figure 12 — Pressure Ratio versus Reinforcement Index

Out-of-Plane Bending of URM Walls Strengthened with FRP Strips – Modeling and Analysis

by E. Hamed and O. Rabinovitch

Synopsis: The modeling and analysis of un-reinforced masonry walls strengthened with externally bonded unidirectional composite laminates and subjected to out-of-plane loading is investigated. The model is based on variational principles, static equilibrium, and compatibility between the structural components, which include the masonry units (blocks), the mortar joints, the FRP strips, and the adhesive layers. The modeling of the masonry block and the mortar joints follows the small displacements and the Bernoulli-Euler beam theories. The strengthening FRP strips are modeled following the lamination theory and the adhesive layers are modeled as two dimensional linear elastic continuum. The effects of cracking of the mortar joints and the debonding of the composite laminates near the mortar joints are also considered. A numerical example is presented. The numerical results quantitatively describe some of the phenomena that govern the behavior of the strengthened wall and may lead to its failure. A summary and conclusions close the paper.

Keywords: analysis; cracking; debonding; FRP; masonry; out-of-plane bending; strengthening; theoretical modeling

Ehab Hamed is a Ph.D. student at the Faculty of Civil and Environmental Engineering at the Technion – Israel Institute of Technology.

Dr. Oded Rabinovitch is a senior lecturer and Horev fellow at the Faculty of Civil and Environmental Engineering at the Technion – Israel Institute of Technology.

INTRODUCTION

The structural strengthening and upgrading of existing masonry walls has recently gained much attention. One of the reasons that make this task challenging is the vulnerability of the masonry wall to seismic, blast, and wind forces. These forces are associated with in-plane and out-of-plane bending, which may yield loss of functionality, structural damage, injury to residents, and even loss of lives under earthquake events, blast, or explosions (Ehsani, Saadatmanesh, and Velazquez-Dimas 1999). In other cases, upgrading of masonry walls to become load carrying members is required in order to improve the overall resistance of the entire building. For example, masonry walls in buildings that were designed with little or no regard to seismic and wind forces, in-filled masonry frames in "soft stories", masonry walls in structures that exhibit high levels of inter-story drift, and others may require structural strengthening.

Strengthening and upgrading of masonry walls to resist out-of-plane or in-plane bending can be conducted in many ways. Kahn (1984) and Karantoni and Fardis (1992) discussed the application of shotcrete or ferrocement to the outer faces of the wall. Bhende and Ovadia (1994) investigated the addition of steel plates to the face of the wall. Dawe and Aridru (1993) studied the application of external prestressing, and Manzouri et al. (1996) examined the efficiency of filling the interior voids and cracks with grout and the introduction of steel reinforcement into the masonry wall. Although these methods offer a broad range of practical solutions, they are often time consuming. Furthermore, they add significant mass to the structure, encroach upon available working space, and adversely affect the aesthetics of the repaired area (Triantafillou 1998). One solution that overcomes these obstacles is the use of externally bonded composite materials. This technique uses strips, rods, sheets, or fabrics made of composite and fiber reinforced plastic (FRP) materials that provide the existing wall with additional tensile reinforcement (Schwegler 1995, Hamilton and Dolan 2001).

In general, masonry walls subjected to out-of-plane loading fail by bed-joint cracking at midspan followed by a sudden snap and total collapse (Hamoush et al. 2001). The application of the externally bonded composite layers modifies the bending behavior of the wall and converts it into a load carrying member. Experimental investigations have shown that the strength of the wall can be increased by a factor of up to 50 (Gilstrap and Dolan 1998). In addition, while externally bonded composite laminates tend to decrease the ductility of strengthened RC beams, the strengthened masonry wall exhibits a significant increase in ductility (Ghobarah and El Mandooh Galal, 2004).

Based on the experimental studies of Velazquez-Dimas and Ehsani 2000, Hamilton and Dolan 2001, Hamoush et al. 2002, and Kiss et al. 2002, the modes of failure of the strengthened masonry wall include rupture of the FRP strip or crushing of the masonry at the bed-joints, debonding (delamination) of the FRP composite at the edges and near the bed-joints, shear failure of the masonry blocks, and peeling of the composite near diagonal shear cracks. Additional modes of failure such as outward buckling/wrinkling of the compressed FRP strip (Kuzik, Elwi, and Cheng 2003), out-of-plane sliding of the blocks at the bed-joints (Tumialan, Galati, and Nanni 2003), punching shear under concentrated loads (Tan and Patoary 2004), flexural-shear cracking, and failure of the block between adjacent FRP strips (Albert, Elwi, and Roger Cheng 2001) have also been observed. Most of these modes of failure are unique to the strengthened wall and have not been observed in un-strengthened masonry walls or in other types of strengthened members such as RC beams, plates and slabs. However, in spite of the comprehensive experimental characterization of these effects, it appears that there are still no analytical models that are able to quantitatively describe and explain the behavior of the strengthened wall. The challenge of developing such model is addressed in this paper.

Among the approaches for the structural analysis of the strengthened wall, the strain compatibility approach has been widely used (Hamilton and Dolan 2001; Hamoush et al. 2002). This model provides an acceptable prediction of the ultimate load in cases of flexural failure by FRP rupture or crushing of the masonry blocks. However, the effect of the delamination, local phenomena in the vicinity of the mortar joints, and other modes of failure are beyond its capability. A similar model, which is also based on the strain compatibility approach while limiting the strains in the FRP strip at different loading stages, was proposed by Velazquez-Dimas and Ehsani (2000). The finite element (FE) method was used by Lee at al. (1996) for the analysis of un-strengthened masonry walls. However, the different length scales (thickness of the adhesive layer and the composite layer versus the depth of the masonry blocks), the differences in the order of magnitude of the mechanical properties of the materials involved, and the singularities and stress concentrations in critical locations make the FE analysis of the strengthened wall very complicated and computational effort consuming.

In this study, a model for the out-of-plane bending behavior of un-reinforced masonry walls strengthened with unidirectional composite strips is presented. The objective of the study is to quantitatively describe the out-of-plane bending behavior of the strengthened masonry wall with emphasis on the critical phenomena that have been experimentally observed. The theoretical model derived in the paper adopts a unidirectional (one-way) representation of the strengthened wall, and uses variational principles, static equilibrium, and compatibility requirements between the structural components (masonry units (blocks), mortar, FRP strips, adhesive layers). Each masonry unit is modeled as a Bernoulli-Euler beam with small deformations. The mortar joints are also modeled using a beam theory, but with a unique constitutive model that represents the cracking and the relatively low stiffness of the mortar joints. Following the concepts of the high order approach (Rabinovitch and Frostig 2000), the composite laminates are modeled using the lamination theory and the adhesive layers are modeled as 2D linear elastic continuum with shear and vertical normal rigidities only. The integrity of the wall is achieved by imposing compatibility requirements between the various structural components.

As reflected by many experimental investigations, the bending behavior of unidirectional strengthened masonry walls is characterized by unique phenomena. Among these, this paper focuses on the deformability of the adhesive layer and its influence on the behavior of the strengthened wall; the cracking of the mortar joint; and the development of stress concentrations and debonding near the mortar joints. The localized stress field near the corners of the masonry blocks; the development of compressive stresses in the composite strip; and the ability of the bonded composite strips to convert the masonry wall into a load carrying member are also investigated. The mathematical formulation, which includes the assumptions, the derivation of the equilibrium equations and the boundary/continuity conditions, the compatibility and debonding conditions, the stress and deformation fields of the adhesive layers, and the governing equations, is presented next. A numerical example that investigates the bending behavior of a strengthened masonry wall and examines the capabilities of the proposed model is also presented. A summary and conclusions close the paper.

RESEARCH SIGNIFICANCE

Masonry walls strengthened with composite materials have been widely studied through experimental researches. The small number and the limited capabilities of the theoretical and analytical models that deal with the complex and unique behavior of strengthened masonry walls triggers the development of a more inclusive and rigorous model. Such model would contribute to the understanding of the behavior of this family of strengthened members. This research presents a theoretical model for the out-of-plane bending behavior of strengthened masonry walls. Its significance is in the contribution to the establishment of a theoretical foundation for the understanding of the physical behavior of the strengthened wall and thus to its effective design and safe use.

MATHEMATICAL FORMULATION

The general layout of the strengthened masonry wall and the sign conventions for the coordinates, deformations, loads, stresses, and stress resultants appear in Figure 1. This layout, and especially the boundary conditions and the strengthening using one-way FRP strips, imply that the structural response of the strengthened wall is a unidirectional (one-way) one. Thus, the deformations and stresses in the un-strengthened direction of the masonry wall and their effects on the behavior of the strengthened direction are not considered. The derivation of the mathematical model assumes that the material behavior of the FRP, adhesive, and the masonry blocks is linear elastic. However, it is assumed that the mortar is linear elastic in compression but its tensile strength and stiffness are negligible. In addition, the longitudinal rigidity of the adhesive layer is neglected with respect to the in-plane stiffness of the masonry blocks and the FRP strips.

The assembly of the various components into a whole structure is based on compatibility between the block units and the mortar joints. Perfect bonding is also attributed to the adhesive-masonry and the adhesive-FRP interfaces. However, debonded (delaminated) regions that may form in the vicinity of the mortar joints are considered.

These debonded regions may be a result of two scenarios. The first is due to insufficient leveling of the mortar joints, which results in a potentially imperfect placement of the adhesive near the mortar joints. The second scenario is due to flexural cracking of the mortar or debonding between the mortar and the masonry unit. In this case, the vertically cracked surface cannot transfer shear stresses. Correspondingly, the horizontal shear stresses at the crack tip (i.e. at the mortar-adhesive interface) are also nullified and a localized debonded region is formed. Thus, the debonded interfaces are free of shear stresses and can only transfer vertical normal compressive stresses if vertical contact exists. Finally, it is assumed that the debonded regions exist before loading and do not grow under it, and that the external loads are applied to the block or the mortar only.

The general layout of the wall (Figure 1) indicates that the strengthened panel includes two main types of regions: un-strengthened regions near the external supports, and strengthened regions, in which the FRP strips are bonded on both faces of the masonry panel. The strengthened region may include two types of sub-regions, namely a fully bonded sub-region and a debonded sub-region. A further distinction between debonded sub-regions in which vertical contact exists, and debonded sub-regions without such contact, is made. In case vertical contact exists, the delaminated faces can freely slip with respect to each other (thus cannot resist shear stresses), but maintain vertical compatibility and resist vertical normal compressive stresses. In case vertical contact does not exist, the delaminated faces are free of shear and vertical normal stresses.

Variational principle and Kinematic relations

The equilibrium equations and the boundary/continuity conditions for the strengthened masonry panel are derived through the variational principle, which requires that:

$$\delta(U + V) = 0 \tag{1}$$

where U is the strain energy, V is the potential of the external loads, and δ is the variational operator. The first variation of the internal strain energy is

$$\delta U = \sum_{1}^{Nb} \int_{V_{block}} \sigma_{xx}^{block} \delta \varepsilon_{xx}^{block} dv_{block} + \sum_{1}^{Nm} \int_{V_{mortar}} \sigma_{xx}^{mortar} \delta \varepsilon_{xx}^{mortar} dv_{mortar} + \int_{V_t} \sigma_{xx}^t \delta \varepsilon_{xx}^t dv_t + \int_{V_b} \sigma_{xx}^b \delta \varepsilon_{xx}^b dv_b$$

$$+ \int_{V_{ta}} (\tau_{xz}^{ta} \delta \gamma_{xz}^{ta} + \sigma_{zz}^{ta} \delta \varepsilon_{zz}^{ta}) dv_{ta} + \int_{V_{ba}} (\tau_{xz}^{ba} \delta \gamma_{xz}^{ba} + \sigma_{zz}^{ba} \delta \varepsilon_{zz}^{ba}) dv_{ba} \tag{2}$$

where the superscripts *block, mortar, t, b, ta* and *ba* refer to the block, mortar, top FRP strip, bottom FRP strip, top adhesive layer, and bottom adhesive layer, respectively; σ_{xx}^i and ε_{xx}^i are the longitudinal normal stresses and strains in the masonry block *(i=block)*, the mortar joint *(i=mortar)*, the upper FRP strip *(i–t)*, and the lower FRP strip *(i=b)*; τ_{xz}^j and σ_{zz}^j *(j=ta,ba)* are the shear and vertical normal stresses in the upper and lower adhesive layers, respectively, γ_{xz}^j and ε_{zz}^j *(j=ta,ba)* are the shear angle and the vertical normal strain in the upper and lower adhesive layers, respectively, and Nb and Nm are the number of the masonry blocks and mortar joints, respectively.

The kinematic relations for the masonry blocks, the mortar joints, and the upper and lower FRP strips independently follow the Bernoulli-Euler assumption and the theory of small displacements as follows:

$$w_i(x,z_i) = w_i(x) \quad ; \quad u_i(x,z_i) = u_{oi}(x) - z_i w_{i,x}(x) \quad ; \quad \varepsilon_{xx}^i(x,z_i) = u_{oi,x}(x) - z_i w_{i,xx}(x) \quad (3a,b,c)$$

where w_i and u_{oi} are the vertical and longitudinal displacements at the reference line (mid-height) of the masonry block *(i=block)*, the mortar joint *(i=mortar)*, the upper FRP strip *(i=t)* or the lower FRP strip *(i=b)*, z_i is measured from the mid-height of each component downwards (Figure 1b), and $(\,)_{,x}$ denotes a partial derivative with respect to x.

Since both the masonry blocks and the mortar joints are modeled as Bernoulli-Euler beams, the equilibrium equations in terms of stress resultants and the boundary/continuity conditions for the block regions and the mortar joint regions are the same and the two components differ only in their constitutive behavior. For brevity, the superscripts *(block)* and *(mortar)* appear in Eq. (1) are replaced with the superscript *(c)*, where *(c=block)* refers to regions in which a block is sandwiched between two FRP strips and adhesive layers ("block regions", sections A-A, B-B in Figure 1a), and *(c=mortar)* refers to regions in which the mortar replaces the block ("mortar regions", section C-C in Figure 1a).

The kinematic relations for the adhesive are based on 2D linear elasticity as follows:

$$\varepsilon_{zz}^j(x,z_a^j) = w_{a,z}^j(x,z_a^j) \quad ; \quad \gamma_{xz}^j(x,z_a^j) = u_{a,z}^j(x,z_a^j) + w_{a,x}^j(x,z_a^j) \quad (4a,b)$$

where w_a^j and u_a^j are the vertical and longitudinal displacements of the upper *(j=ta)* and lower *(j=ba)* adhesive layers.

The first variation of the potential of the external loads equals:

$$\delta V = -\int_{x=0}^{x=L}\left((q_z\delta w_c + n_x\delta u_{oc} + m_x\delta w_{c,x} - \sum_{k=1}^{NC}\left(P_k\delta w_c(x_k) + N_k\delta u_{oc}(x_k) + M_k\delta w_{c,x}(x_k)\right)\delta_D(x-x_k)\right)dx \quad (5)$$

where q_z, n_x, m_x are the external distributed loads and moments exerted at the block or at the mortar, respectively, P_k, N_k and M_k are concentrated loads and bending moments at $x=x_k$, δ_D is the Dirac function, and NC is the number of the concentrated loads.

Compatibility and debonding conditions

The compatibility conditions at the fully bonded adhesive-bock, adhesive-mortar, and adhesive-FRP interfaces are

$$w_a^{ta}(x,z_a^{ta}=0) = w_t(x) \quad ; \quad u_a^{ta}(x,z_a^{ta}=0) = u_{ot}(x) - \frac{h_t}{2}w_{t,x}(x) \quad (6a,b)$$

$$w_a^{ta}(x,z_a^{ta}=c_a^{ta}) = w_c(x) \quad ; \quad u_a^{ta}(x,z_a^{ta}=c_a^{ta}) = u_{oc}(x) + \frac{h_c}{2}w_{c,x}(x) \quad (7a,b)$$

$$w_a^{ba}(x,z_a^{ba}=0) = w_c(x) \quad ; \quad u_a^{ba}(x,z_a^{ba}=0) = u_{oc}(x) - \frac{h_c}{2}w_{c,x}(x) \quad (8a,b)$$

$$w_a^{ba}(x,z_a^{ba}=c_a^{ba}) = w_b(x) \quad ; \quad u_a^{ba}(x,z_a^{ba}=c_a^{ba}) = u_{ob}(x) + \frac{h_b}{2}w_{b,x}(x) \quad (9a,b)$$

where h_c $(=h_{block}$ or $h_{mortar})$, h_t and h_b are the heights of the masonry block, the mortar joint, and the upper and lower FRP strips, respectively, c_a^{ta} and c_a^{ba} are the thicknesses of the upper and lower adhesive layers, respectively, and z_a^j *(j=ta,ba)* are measured from the upper interface of each adhesive layer downwards, see Figure 1b.

In the debonded sub-regions, the debonded interface cannot transfer shear or vertical tensile stresses. Hence, the requirements of compatible deformations are replaced with the conditions of stress free surfaces. For example, if the upper adhesive-block interface is debonded and vertical contact exists, Eq. (7b) is replaced with:

$$\tau_{xz}^{ta}(x, z_a^{ta} = c_a^{ta}) = 0 \tag{10}$$

while Eq. (7a) remains unchanged. If vertical contact does not exist, both compatibility conditions, Eqs. (7a) and (7b), are replaced with the following stress conditions:

$$\sigma_{zz}^{ta}(x, z_a^{ta} = c_a^{ta}) = 0 \quad ; \quad \tau_{xz}^{ta}(x, z_a^{ta} = c_a^{ta}) = 0 \tag{11}$$

Equilibrium equations

The equilibrium equations for the strengthened (bonded or debonded) regions are formulated using the variational principle (Eqs. 1,2,5), along with the kinematic relations (Eqs. 3,4), and the compatibility requirements (Eqs. 6-11), as follows:

$$N_{yy,x}^{t} + \alpha_{frp}^{t} b_t \tau_{xz}^{ta}(x, z_a^{ta} = 0) = 0 \tag{12}$$

$$N_{xx,x}^{c} - \alpha_c^t b_t \tau_{xz}^{ta}(x, z_a^{ta} = c_a^{ta}) + \alpha_c^b b_b \tau_{xz}^{ba}(x, z_a^{ha} = 0) = -n_x \tag{13}$$

$$N_{xx,x}^{b} - \alpha_{frp}^{b} b_b \tau_{xz}^{ba}(x, z_a^{ba} = c_a^{ba}) = 0 \tag{14}$$

$$M_{xx,xx}^{t} + \alpha_{frp}^{t} b_t \frac{h_t}{2} \tau_{xz,x}^{ta}(x, z_a^{ta} = 0) + \beta^t b_t \sigma_{zz}^{ta}(x, z_a^{t} = 0) = 0 \tag{15}$$

$$M_{xx,xx}^{c} + \alpha_c^b b_b \frac{h_c}{2} \tau_{xz,x}^{ba}(x, z_a^{ba} = 0) + \alpha_c^t b_t \frac{h_c}{2} \tau_{xz,x}^{ta}(x, z_a^{ta} = c_a^{ta}) + \beta^b b_b \sigma_{zz}^{ba}(x, z_a^{ba} = 0) \tag{16}$$

$$- \beta^t b_t \sigma_{zz}^{ta}(x, z_a^{ta} = c_a^{ta}) = -q_z + m_{x,x}$$

$$M_{xx,yy}^{b} + \alpha_{frp}^{b} b_b \frac{h_b}{2} \tau_{xz,x}^{ba}(x, z_a^{ba} = c_a^{ba}) - \beta^b b_b \sigma_{zz}^{ba}(x, z_a^{ba} = c_a^{ba}) = 0 \tag{17}$$

$$\tau_{xz,x}^{ta} + \sigma_{zz,z}^{ta} = 0 \tag{18}$$

$$\tau_{xz,z}^{ta} = 0 \tag{19}$$

$$\tau_{xz,x}^{ba} + \sigma_{zz,z}^{ba} = 0 \tag{20}$$

$$\tau_{xz,z}^{ba} = 0 \tag{21}$$

where N_{xx}^i and M_{xx}^i are the in-plane stress resultant and the bending moment in the masonry block or the mortar ($i=c=block$ or $mortar$), the upper FRP strip ($i=t$), and the lower FRP strip ($i=b$), respectively, b_i ($i=t,b$) is the width of the upper and lower FRP strips, α_i^j is a flag (=0 or 1) that reflects the bonding condition at the upper ($j=t$) and lower ($j=b$) adhesive-FRP ($i=frp$) and adhesive-block/mortar ($i=c$) interfaces, respectively, and β_i is a flag that reflects the debonding type (with or without contact). The different combinations of the bonding conditions and type, and the corresponding flags are summarized in Table 1. Note that Eqs. (12-21) are valid for both the block region ($c=block$) and the mortar joint region ($c=mortar$). The distinction between the two cases is achieved through the constitutive relations that are discussed next.

Constitutive relations

At the material point level, the constitutive relations for the blocks are:

$$\sigma_{xx}^{block} = E_{block} \varepsilon_{xx}^{block} \tag{22}$$

where E_{block} is the modulus of elasticity of the block. On the other hand, the material behavior of the mortar assumes linear elastic behavior in compression and negligible tensile strength and stiffness. Thus, the material point level constitutive law for the mortar is:

$$\sigma_{xx}^{mortar} = \begin{cases} E_{mortar}\,\varepsilon_{xx}^{mortar} & if \quad \varepsilon_{xx}^{mortar} \le 0 \\ 0 & if \quad \varepsilon_{xx}^{mortar} > 0 \end{cases} \tag{23}$$

where E_{mortar} is the modulus of elasticity of the mortar under compression.

At the cross section level, the constitutive relations are determined by introducing the kinematic relations of Eq. (3c) into Eqs. (22, 23), and using the definitions of the in-plane and bending internal stress resultants. Thus, in general, the constitutive laws for the block and the mortar cross section are:

$$N_{xx}^c = \int_{A_c} \sigma_{xx}^c(z_c)dA_c = A_{11}^c u_{o,x} - B_{11}^c w_{c,xx} \qquad (c=block/mortar) \tag{24}$$

$$M_{xx}^c = \int_{A_c} \sigma_{xx}^c(z_c)z_c dA_c = B_{11}^c u_{o,x} - D_{11}^c w_{c,xx} \qquad (c=block/mortar) \tag{25}$$

where, A_{11}^c, B_{11}^c and D_{11}^c are the extensional, extensional-flexural, and flexural rigidities of the masonry block (c=block) or the mortar joint (c=mortar). The rigidities of the block yield the traditional extensional and flexural rigidities of an elastic section as follows:

$$A_{11}^{block} = \int_{-h_{block}/2}^{h_{block}/2} b_{block}E_{block}\,dz_{block} = EA_{block} \quad ; \quad B_{11}^{block} = \int_{-h_{block}/2}^{h_{block}/2} b_{block}E_{block}z_{block}\,dz_{block} = 0 ; \tag{26a-c}$$

$$D_{11}^{block} = \int_{-h_{block}/2}^{h_{block}/2} b_{block}E_{block}z_{block}^2\,dz_{block} = EI_{block}$$

where b_{block}, EA_{block} and EI_{block} are the width, extensional stiffness, and flexural stiffness of the block section.

In the mortar joint, the negligible tensile strength and stiffness and the combined in-plane and bending tractions resulting from the external load or the composite action of the masonry panel and the FRP strips, require special consideration. The general stress distributions under the combined tractions appear in Figure 2a. Correspondingly, a distinction is made between the case where the thrust line is located within the middle-third of the mortar cross section and the case where it is located out of the of the middle-third, see Figure 2a and Heyman 1996. (A third case in which the mortar joint is fully detached may develop due to external tensile tractions, yet it is unlikely to occur under normal conditions). In the first case, the mortar joint is un-cracked, it is subjected to compressive stresses through its height, and it exhibits a linear elastic behavior. Therefore, the equivalent rigidities are given by Eq. (26) with the subscript (or superscript) "mortar" instead of "block". In case the thrust line is located outside of the middle-third, the mortar joint is cracked and the material points that undergo positive strain do not contribute to the stiffness of the cross section. Thus, the equivalent rigidities take the following form (see Figure 2b):

$$A_{11}^{mortar} = \int_{-h_{mortar}/2}^{z_o^{mortar}} b_{mortar}E_{mortar}\,dz_{morta} = E_{mortar}b_{mortar}\left(\frac{h_{mortar}}{2} + \varphi\,z_o^{mortar}\right) \tag{27}$$

$$B_{11}^{mortar} = \int_{-h_{mortar}/2}^{z_o^{mortar}} b_{mortar} E_{mortar} z_{mortar} dz_{mortar} = -\varphi \frac{E_{mortar} b_{mortar}}{2} \left(\left(\frac{h_{mortar}}{2} \right)^2 - \left(z_o^{mortar} \right)^2 \right) \tag{28}$$

$$D_{11}^{mortar} = \int_{z_o^{mortar}}^{h_{mortar}/2} b_{mortar} E_{mortar} z_{mortar}^2 dz_{mortar} = \frac{E_{mortar} b_{mortar}}{3} \left(\left(\frac{h_{mortar}}{2} \right)^3 + \varphi \left(z_o^{mortar} \right)^3 \right) \tag{29}$$

where b_{mortar} and z_o^{mortar} are the width and the height of the compression zone in the mortar cross section, $\varphi = 1$ in case the mortar section is locally subjected to axial compression combined with a positive bending moment, and $\varphi = -1$ if it is combined with a negative moment (see Figure 2b,c). The case in which the mortar joint is fully detached implies that the rigidities of the mortar joint are zero and the functionality of the strengthened wall depends on the FRP strips only. In general, the above formulation introduces a level of nonlinearity into the analytical model and requires the employment of nonlinear or iterative tools. Alternatively, the height of the compression zone may be assessed using simplified approaches. One of these approaches is based on the assumption of a linear strain distribution through the height of the entire strengthened section. Other simplified approach (which is adopted in the numerical study presented in this paper following Hasetline and Moore 1981) assumes that the depth of the compressed zone equals 0.2 the height of the mortar joint.

Note that by adopting the equivalent rigidities formulation (Eqs. 27-29) and by selecting the mid-height surface of the blocks as the reference line throughout the entire panel (blocks and mortar joints), the eccentricity of the cracked joints is introduced through the non-vanishing coupling rigidity B_{11}^{mortar}. This approach simplifies the assembly of the blocks and the mortar joints into a heterogeneous panel with a common reference line.

The constitutive relations for the FRP strips adopt the lamination theory and read

$$N_{xx}^i = A_{11}^i u_{oi,x} - B_{11}^i w_{i,xx} \quad ; \qquad M_{xx}^i = B_{11}^i u_{oi,x} - D_{11}^i w_{i,xx} \qquad (i=t,b) \tag{30a,b}$$

where A_{11}^i, B_{11}^i and D_{11}^i $(i=t,b)$ are the extensional, extensional-bending, and flexural rigidities (Vinson and Sierakowski 1986) of the upper and lower FRP strips, multiplied by their width. Finally, the constitutive relations for the adhesive are:

$$\sigma_{zz}^j = E_a^j \varepsilon_{zz}^j \qquad ; \qquad \tau_{xz}^j = G_a^j \gamma_{xz}^j \qquad (j=ta,ba) \tag{31a,b}$$

where E_a^j and G_a^j $(j=ta,ba)$ are the modulus of elasticity and the shear modulus of the upper and lower adhesive layers, respectively. Note that the constitutive relations assigned for the adhesive layer account for its compressibility in the vertical (out-of-plane) direction and its shear deformability. Due to the significantly lower elastic properties of the adhesive (with respect to the adjacent FRP strips, masonry blocks, or even the mortar joints) its deformability affects the structural behavior of the strengthened element and should be considered.

Adhesive layers - stress and displacement fields

The stress and displacement fields of the adhesive layers in the fully bonded regions are derived using Eqs. (18-21), along with the compatibility requirements (Eqs. 6-9) and the kinematic and constitutive relations (Eqs. 4,31), and they take the following form:

$$\tau^j_{xz,z}(x,z^j_a) = \tau^j_{xz}(x) = \tau^j_a \tag{32}$$

$$\sigma^j_{zz}(x,z^j_a) = -\frac{2z^j_a - c^j_a}{2}\tau^j_{a,x} + \frac{\lambda E^j_a(w_j - w_c)}{c^j_a} \tag{33}$$

$$w^j_a(x,z^j_a) = -\frac{z^{j2}_a - c^j_a z^j_a}{2E^j_a}\tau^j_{a,x} + \frac{\lambda(w_j - w_c)z^j_a}{c^j_a} + \frac{(1+\lambda)}{2}w_c + \frac{(1-\lambda)}{2}w_t \tag{34}$$

$$u^j_a(x,z^j_a) = \frac{\tau^j_a z^j_a}{c^j_a} + \frac{\tau^j_{a,xx}}{2E^j_a}\left(\frac{z^{j3}_a}{3} - c^j_a\frac{z^{j2}_a}{2}\right) - \frac{\lambda(w_{i,x} - w_{c,x})z^{j2}_a}{2c^j_a}$$

$$-\frac{(1+\lambda)}{2}\left(w_{c,x}\left(z^{ba}_a + \frac{h_c}{2}\right) + u_{oc}\right) + \frac{(1-\lambda)}{2}\left(w_{t,x}\left(z^{ta}_a + \frac{h_c}{2}\right) + u_{ot}\right) \tag{35}$$

where $\lambda = -1$ for the upper $(j=t)$ adhesive layer and $\lambda = 1$ for the lower $(j=b)$ layer. The stress fields at the delaminated adhesive layer (with or without contact) are:

$$\tau^j_{xz,z}(x,z^j_a) = \tau^j_{xz}(x) = \tau^j_a = 0 \quad ; \quad \sigma^j_{zz}(x,z^j_a) = \frac{\beta^j \lambda E^j_a(w_j - w_c)}{c^j_a} \tag{36,37}$$

Governing equations

The governing equations for the strengthened (fully bonded or debonded) regions are derived using Eqs. (12-21), the constitutive relations (Eqs. 24-31), the compatibility requirements (Eqs. 7b,9b), and the stress and deformation fields of the adhesive layers (Eqs. 32-37). The governing equations are stated in terms of the unknown displacements w_c, w_t, w_b, u_{oc}, u_{ot}, u_{ob}, and shear stresses τ^{ta}_a and τ^{ba}_a as follows:

$$A^t_{11}u_{ot,xx} - B^t_{11}w_{t,xxx} + \alpha^t_{frp}\alpha^t_c b_t \tau^{ta}_a = 0 \tag{38}$$

$$A^c_{11}u_{oc,xx} - B^c_{11}w_{c,xxx} - \alpha^t_{frp}\alpha^t_c b_t \tau^{ta}_a + \alpha^b_{frp}\alpha^b_c b_b \tau^{ba}_a = -n_x \tag{39}$$

$$A^b_{11}u_{ob,xx} - B^b_{11}w_{b,xxx} - \alpha^b_{frp}\alpha^b_c b_b \tau^{ba}_a = 0 \tag{40}$$

$$D^t_{11}w_{t,xxxx} - B^t_{11}u_{ot,xxx} - \alpha^t_{frp}\alpha^t_c(c^{ta}_a + h_t)\frac{b_t \tau^{ta}_{a,x}}{2} + \beta^t\frac{b_t E^t_a}{c^t_a}(w_t - w_c) = 0 \tag{41}$$

$$D^c_{11}w_{c,xxxx} - B^c_{11}u_{oc,xxx} - \alpha^b_{frp}\alpha^b_c(c^{ba}_a + h_c)\frac{b_b \tau^{ba}_{a,x}}{2} - \alpha^t_{frp}\alpha^t_c(c^{ta}_a + h_c)\frac{b_t \tau^{ta}_{a,x}}{2}$$

$$+ \beta^b\frac{b_b E^{ba}_a}{c^{ba}_a}(w_c - w_b) - \beta^t\frac{b_t E^{ta}_a}{c^{ta}_a}(w_t - w_c) = -q_z + m_{x,x} \tag{42}$$

$$D^b_{11}w_{c,xxxx} - B^b_{11}u_{oc,xxx} - \alpha^b_{frp}\alpha^b_c(c^{ba}_a + h_b)\frac{b_b \tau^{ba}_{a,x}}{2} - \beta^b\frac{b_b E^{ba}_a}{c^{ba}_a}(w_c - w_b) = 0 \tag{43}$$

$$\alpha^t_{frp}\alpha^t_c\left(u_{ot} - u_{oc} - \frac{(c^{ta}_a + h_t)}{2}w_{t,x} - \frac{(c^{ta}_a + h_c)}{2}w_{c,x} + \frac{\tau^{ta}_a c^{ta}_a}{G^{ta}_a} - \frac{\tau^{ta}_{a,xx}(c^{ta}_a)^3}{12E^{ta}_a}\right) = 0 \tag{44}$$

$$\alpha^b_{frp}\alpha^b_c\left(u_{oc} - u_{ob} - \frac{(c^{ba}_a + h_b)}{2}w_{b,x} - \frac{(c^{ba}_a + h_c)}{2}w_{c,x} + \frac{\tau^{ba}_a c^{ba}_a}{G^{ba}_a} - \frac{\tau^{ba}_{a,xx}(c^{ba}_a)^3}{12E^{ba}_a}\right) = 0 \tag{45}$$

Note that Eqs. (44) and (45) result from the compatibility conditions of the longitudinal deformations at the upper and lower adhesive layers (Eqs. 7b, 9b). Thus, they exist in the fully bonded sub-regions and vanish in the debonded ones. Likewise, the corresponding shear stresses, which are uniform through the thickness of each adhesive layer (see Eq. 32), also vanish in the debonded sub-regions and are not considered as unknowns. The distinction between the block and mortar regions (both governed by Eqs. 38-45) is made through the different elasto-geometric (rigidities) properties of the two regions (see Eqs. 26-29). For simplicity, it is assumed that due to the relatively small length of the mortar joint, these elasto-geometric properties are uniform through the length of the joint. The link between the various sub-regions (block, mortar, bonded, debonded, etc.) is achieved through the continuity conditions that are discussed next.

Continuity conditions

The continuity conditions at any point $x=x_k$ within the fully bonded sub-region are:

$$u_{oi}^{(-)} = u_{oi}^{(+)} \; ; \; w_i^{(-)} = w_i^{(+)} \; ; \; w_{i,x}^{(-)} = w_{i,x}^{(+)} \; ; \; w_a^{j(-)}(z_a^j) = w_a^{j(+)}(z_a^j) \quad (i=c,t,b) \qquad \text{(46a-d)}$$

$$N_{xx}^{i(-)} - N_{xx}^{i(+)} = \kappa N_k^i \qquad (i=c,t,b) \qquad \text{(47)}$$

$$\left(M_{xx,x}^{i(-)} + (1-\kappa)b_i h_i \tau_a^{i(-)} + \kappa \frac{h_c}{2}(b_t \tau_u^{t(-)} + b_b \tau_a^{b(-)}) - \kappa m_x^{(-)} \right)$$

$$-\left(M_{xx,x}^{i(+)} + (1-\kappa)b_i h_i \tau_a^{i(+)} + \kappa \frac{h_c}{2}(b_t \tau_a^{t(+)} + b_b \tau_a^{b(+)}) - \kappa m_x^{(+)} \right) = \kappa P_k \qquad (i=c,t,b) \qquad \text{(48)}$$

$$-M_{xx}^{i(-)} + M_{xx}^{i(+)} = \kappa M_k^i \qquad (i=c,t,b) \qquad \text{(49)}$$

$$\tau_a^{j(-)} = \tau_a^{j(+)} \qquad (i=c,t,b) \qquad \text{(50)}$$

where the *(-)* and *(+)* superscripts denote quantities left and right to the point $x=x_k$, respectively, and P_k, N_k and M_k are concentrated loads and bending moments at $x=x_k$. If the connection point $x=x_k$ is located within a debonded sub-region, the continuity conditions include Eqs. (46a-c,47,49) and Eq. (48) after dropping the terms that include the vanishing shear stress.

The continuity conditions at the connection points between the fully bonded sub-region and the debonded sub-region include Eqs. (46a-c,47-49) after dropping the vanishing shear stress at the debonded sub-region, and an additional condition that is applied only to the fully bonded sub-region. This condition requires that the shear stress, which is unknown only in the fully bonded region, equals zero at the connection point.

For brevity, the boundary conditions are not explicitly presented here. However, they can be obtained by degenerating the continuity conditions and selecting either a kinematic condition or a static one (i.e. Eq. 46a or 47, Eq. 46b or 48, Eq. 46c or 49, Eq. 46d or 50).

The governing equations of the various regions along with the boundary and continuity conditions are numerically solved using the Multiple Shooting method (Stoer and Bulirsch 1993). The determination of the type of debonded sub-regions (with or without contact) is conducted iteratively. Namely, one type is assumed and verified through the results of the analysis. If the results contradict the assumption, the assumed type of the debonded region is switched and the structure is re-analyzed.

NUMERICAL STUDY

The bending behavior of a masonry panel strengthened with unidirectional GFRP strips bonded to its upper and lower faces is investigated. The geometry, mechanical properties, and the lateral loads (self weight and two concentrated loads) follow the experimental study of Kiss et al. (2002) and appear in Figure 3. Two cases of bonding conditions are considered. At first, the FRP strips are assumed to be fully bonded through the length of the masonry panel. The second case assumes that debonded regions exist through the length of the mortar joints, plus a distance that is estimated as twice the thickness of the adhesive layer (see Figure 3c). These effects are considered in order to describe the behavior of the strengthened masonry wall at progressive levels of load, where cracking of the mortar joint, debonding between the mortar and the masonry unit, or potentially imperfect placement of the adhesive near the mortar joints may trigger the development of such debonded regions.

The response of the fully bonded masonry wall in terms of the deformations, internal forces, and stresses in the adhesive layers is described in Figure 4. The results show the slope change of the deflection line at the mortar joints (Figure 4a). They also show the amplified longitudinal deformations through the length of the joint (Figure 4b). This effect is attributed to cracking of the mortar and to its relative deformability. However, while the un-strengthened masonry wall fails after progressive cracking of one mortar joint near mid-span, the strengthened wall allows the development of cracks in more than one mortar joint along the wall. This effect, which is detected by the analysis, provides the strengthened wall with the ability to resist bending moments beyond the cracking point, by composite action in terms of axial forces in the masonry panel and the FRP strips.

Figures 4c and 4d quantitatively show that the reduced flexural stiffness of the mortar joints yield a reduction in the localized bending moment carried by the wall, and a corresponding increase in the axial tensile/compressive forces in the FRP strips and the wall. The shear forces distribution (see Figure 4e) shows that although the global shear force is null between the two concentrated loads ($380<x<520$ mm), self equilibrated shear forces develop in the masonry panel due to localized effects near the mortar joints. These localized effects are also observed in the distribution of the shear and vertical normal (peeling) stresses in the adhesive layers near the mortar joints (see Figure. 4f,g,h). The stress concentrations observed in these figures provide a quantitative explanation of the debonding failure mechanism that has been observed in experimental studies (Hamilton and Dolan 2001; Kiss et al. 2002). Peeling stress concentrations are also observed at the edges of the FRP strips, and explain the edge debonding failure that was also experimentally observed (Hamoush et al. 2001; Hamoush et al. 2002).

The distribution of the peeling stresses in the adhesive layers near the mortar joint at mid-span appears in Figure 5. The curves for the lower adhesive interfaces reveal that the adhesive-block interface is subjected to compression stresses at the edges of the mortar joint. These stresses result from the vertical interaction of the corners of the block, the adhesive layer, and the FRP strip. In other words, the lower corners of the masonry blocks are compressed against the FRP strip. On the other hand, the upper corners of the blocks pull the upper FRP strip downwards, result in the formation of tensile vertical stresses at the adhesive-block interface, and may lead to debonding failure. Within the mortar joint itself, the stresses at the adhesive-mortar interface diverge from the stresses at the adhesive FRP one. This effect is attributed to the shear gradient terms that appear in the vertical normal stress fields (Eq. 33). Yet, the average stresses through the thickness, which equal the stresses at $z^j_a = c^j_a/2$ and result from the vertical deformability of the adhesive (see Eq. 33), clarify that the FRP strips are compressed against the lower face of the joint and pulled down by the upper face. In both cases, the localized effects and stress concentrations as well as the cracking of the mortar material probably lead to debonding. The response of the structure under these conditions is examined next.

The response of the debonded masonry wall in terms of the vertical deflections and internal axial forces appear in Figure 6. The results show that the formation of the debonded regions decreases the flexural rigidity of the masonry wall and increases its vertical deflections. The distribution of the axial forces in the wall and the FRP strips reveals that the upper FRP strip is subjected to compression forces. These compression forces and the reduced or vanished lateral support of the adhesive layer may lead to local buckling of the strip and to further propagation of the debonded region (Rabinovitch 2004a)

The distribution of the shear and the vertical normal stresses in the adhesive layers in the vicinity of the mortar joint at mid-span is described in Figure 7. These results show that the formation of the debonded regions significantly increases the magnitudes of the shear and vertical normal stresses (see Figures 4f,g,h and 7). It is also seen that the magnitudes of these stresses are far beyond the strength of the cementicious materials involved and may lead to a localized or overall debonding failure. This failure mechanism, which is quantitatively characterized here has been observed in many experimental studies (see Kiss et al. 2002; Kuzik et al. 2003; Tumialan et al. 2003). Furthermore, the results presented in Figure 7 can be used for the evaluation of a fracture mechanics criterion for the propagation of the debonding failure (Rabinovitch and Frostig 2001, Rabinovitch 2004b).

SUMMARY AND CONCLUSIONS

A theoretical approach for the description of the out-of-plane bending behavior of un-reinforced masonry walls strengthened with unidirectional composite strips has been presented. The structural high order model is based on variational principles, static equilibrium, and compatibility requirements between the structural components (masonry blocks, mortar, FRP strips, adhesive layers). The Bernoulli-Euler beam assumptions and the theory of small displacements have been adopted for the modeling of the masonry blocks and the mortar joints. The strengthening composite laminates have been modeled using the lamination theory, and the adhesive layers have been considered as 2D linear elastic continua. Compatibility requirements at the interfaces of the adhesive layers and at the connection points between the various structure components have been used to assemble the various components into a whole structure.

The capabilities of the proposed approach have been examined through a numerical example. It has been shown that as expected and aimed by the strengthening process, the behavior of the strengthened masonry wall is similar to the behavior of a reinforced masonry wall, where cracks can develop in more than one critical mortar joint. This effect is well predicted by the analytical model. The numerical study has shown that the proposed model quantitatively describes the localized interaction between the FRP strips and the corners of the masonry block. This effect may lead to local shear failure of the FRP strips, and thus have to be considered in the design of strengthened masonry walls. The results have also revealed the development of high shear and vertical normal (peeling) stresses in the adhesive layers at the vicinity of the mortar joints as well as at the edges of the FRP strips. These stress concentrations provide a quantitative explanation of the debonding failure that has been observed in many experimental studies. In addition, the numerical study has quantitatively characterized the development of compressive forces in the upper FRP strip. This effect may lead to buckling or local wrinkling of the FRP strip, especially at the debonded zones near the mortar joints.

In conclusion, it is seen that the bending behavior of strengthened masonry walls is characterized by unique aspects that have not been observed in the un-strengthened wall or in other strengthened members. This study throws light on some of these aspects and sets a basis for further investigations of this unique structural member. Topics for ongoing and future investigation of strengthened URM walls include a detailed parametric study with emphasis on variables that are prescribed by the given properties of the wall (dimensions, materials, slenderness ratio, construction method, boundary conditions, etc.), and parameters that can be controlled by the designer of the strengthening system (FRP and adhesive materials and geometry, strengthening scheme, etc.). An enhancement of the analytical model to account for nonlinear effects that evolve at the higher load levels and a detailed comparison to experimental data are also under investigation. Finally, the challenge of developing a practical design protocol based on the theoretical tools and information gathered in the study is also considered for future research.

REFERENCES

Albert, M.L., Elwi, A.E., and Cheng, J.J.R., 2001, "Strengthening of Unreinforced Masonry Walls using FRPs", *Journal of Composite for Construction*, Vol. 5, No. 2, pp. 76-84.

Bhende, D. and Ovadia, D. 1994, "Out-of-Plane Strengthening Scheme for Reinforced Masonry Walls" *Concrete International* Vol. 16, No. 4, pp. 30-34.

Dawe, J.L. and Aridru, G.G., 1993, "Prestressed Concrete Masonry Walls Subjected to Uniform Out-of-Plane Loading", *Canadian Journal of Civil Engineering*, Vol. 20, No. 6, pp. 969-979.

Ehsani, M.R., Saadatmanesh, H. and Velazquez-Dimas, J.I., 1999, "Behavior of Retrofitted URM Walls Under Simulated Earthquake Loading", *Journal of Composites for Construction*, Vol. 3, No. 3, pp. 134-142.

Ghobarah, A. and El Mandooh Galal, K., 2004, "Out-of-Plane Strengthening of Unreinforced Masonry Walls with Openings", *Journal of Composites for Construction*, Vol. 8, No. 4, pp. 298-305.

Gilstrap, J.M. and Dolan, C.W., 1998, "Out-of-plane Bending of FRP-Reinforced Masonry Walls", *Composites Science and Technology*, Vol. 58, No. 8, pp. 1277-1284.

Hamilton, H.R. and Dolan, C.W., 2001, "Flexural Capacity of Glass FRP Strengthened Concrete Masonry Walls", *Journal of Composites for Construction*, Vol. 5, No. 3, pp. 170-178.

Hamoush, S.A., McGinley, M.W., Mlakar, P., Scott, D. and Murray, K., 2001, "Out-of-Plane Strengthening of Masonry Walls with Reinforced Composites", *Journal of Composites for Construction*, Vol. 5, No. 3, pp. 139-145.

Hamoush, S.A., McGinley, M.W., Mlakar, P., Terro, M.J., 2002, "Out-of-Plane Behavior of Surface-Reinforced Masonry Walls", *Construction and Building Materials*, Vol. 16, No. 6, pp. 341-351.

Hasetline, B.A. and Moore, J.F.A., 1981, *Handbook to BS5628: Structural use of Masonry, Part1: Unreinforced Masonry*. Brick Development Association Windsor, U.K

Heyman, J., 1996, *Arches, Vaults, and Buttresses.* Variorum, Ashgate Publishing company, Brookfiled, Vermont, USA.

Kahn, L.F., 1984, "Shotcrete Strengthening of Brick Masonry Walls", *Concrete International: Design and Construction*, Vol. 6, No. 7, pp. 34-39.

Karantoni, F.V. and Fardis, M.N., 1992, "Effectiveness of Seismic Strengthening Techniques for Masonry Buildings", *Journal of Structural Engineering*, Vol. 118, No. 7, pp. 1884-1902.

Kiss, R.M., Kollar, L.P., Jai, J. and Krawinkler, H., 2002, "Masonry Strengthened with FRP Subjected to Combined Bending and Compression, Part II: Test Results and Model Predictions", *Journal of Composite Materials*, Vol. 36, No. 9, pp. 1049-1063.

Kuzik, M.D., Elwi, A.E. and Cheng, J.J.R., 2003, "Cyclic Flexure Tests of Masonry Walls Reinforced with Glass Fiber Reinforced Polymer Sheets", *Journal of Composite for Construction*, Vol. 7, No. 1, pp. 20-30.

Lee, J.S., Pande, G.N., Middleton, J. and Kralj, B., 1996, "Numerical Modeling of Brick Masonry Panels Subject to Lateral Loading", *Computers and Structures*, Vol. 61, No. 4, pp. 735-745.

Manzouri, T., Schuller, M.P., Shing, P.B. and Amadei, B., 1996, "Repair and Retrofit of Unreinforced Masonry Structures", *Earthquake Spectra*, Vol. 12, No. 4, pp. 903-922.

Rabinovitch, O., 2004a, "Nonlinear (Buckling) Effects in RC Beams Strengthened with Composite Materials Subjected to Compression". *International Journal of Solids and Structures*, vol. 41, No. 20, pp. 5677-5695.

Rabinovitch, O., 2004b, "Fracture-Mechanics Failure Criteria for RC Beams Strengthened with FRP Strips - a Simplified Approach". *Composite Structures*, vol. 64, No. 3-4, pp. 479-492.

Rabinovitch, O. and Frostig, Y., 2000, "Closed-Form High-Order Analysis of RC Beams Strengthened with FRP Strips" *Journal of Composites for Construction*, Vol. 4, No. 2, pp. 65-74.

Rabinovitch, O. and Frostig, Y., 2001, "Delamination Failure of RC Beams Strengthened with FRP Strips – A Closed-Form High-Order and Fracture Mechanics Approach", *Journal of Engineering Mechanics*, Vol. 127, No. 8, pp. 852-861.

Schwegler, G., 1995, "Masonry Construction Strengthened with Fiber Composites in Seismically Endangered Zones". *10th European Conference on Earthquake Engineering*, Duma (Ed.), Balkema, Rotterdam, pp. 2299-2303.

Tan, K.H. and Patoary, M.K.H., 2004, "Strengthening of Masonry Walls against Out-of-Plane Loads using Fiber-Reinforced Polymer Reinforcement", *Journal of Composites for Construction*, Vol. 8, No. 1, pp. 79-87.

Triantafillou T.C., 1998a, "Strengthening of Masonry Structures using Epoxy-Bonded FRP Laminates", *Journal of Composites for Construction*, Vol. 2, No. 2, pp. 96-104.

Tumialan, J.G., Galati, N. and Nanni, A., 2003, "Fiber-Reinforced Polymer Strengthening of Unreinforced Masonry Walls Subjected to Out-of-Plane Loads", *ACI Structural Journal,* Vol. 100, No. 3, pp. 321-329.

Velazquez-Dimas, J.I. and Ehsani, M.R., 2000, "Modeling Out-of-Plane Behavior of URM Walls Retrofitted with Fiber Composites", *Journal of Composites for Construction*, Vol. 4, No. 4, pp. 172-181.

Vinson, J. R. and Sierakowski, R. L., 1986, *The Behavior of Structures Composed of Composite Materials*, Martinus-Nijhoff, Inc., Dordrecht, The Netherlands

Table 1 — Values of the α_i^j and β^j flags corresponding to the bonding conditions and type of delamination at the interfaces of the adhesive layers.

i	j	Location	Condition	α_i^j	β^j
frp	t	upper adhesive-FRP interface	Bonded	1	1
			Debonded with contact	0	1
			Debonded without contact	0	0
c	t	upper adhesive-block/mortar interface	Bonded	1	1
			Debonded with contact	0	1
			Debonded without contact	0	0
frp	b	lower adhesive-FRP interface	Bonded	1	1
			Debonded with contact	0	1
			Debonded without contact	0	0
c	b	lower adhesive-block/mortar interface	Bonded	1	1
			Debonded with contact	0	1
			Debonded without contact	0	0

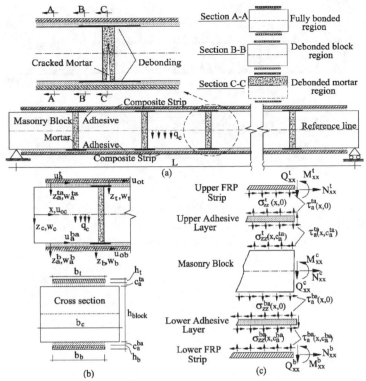

Figure 1 – Geometry, loads, sign conventions, and stress resultants: (a) Geometry and loads; (b) Coordinate systems and deformations; (c) Stresses and stress resultants.

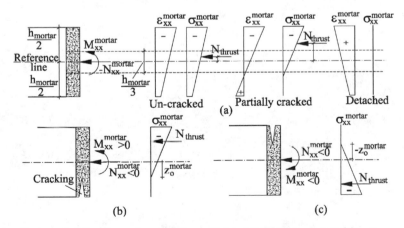

Figure 2 – Stress distributions through the height of the mortar joint: (a) influence of the location of thrust line; (b) Stress distribution in cracked mortar joint subjected to positive moment and axial compressive force; (c) Stress distribution in cracked mortar joint subjected to negative moment and axial compressive force.

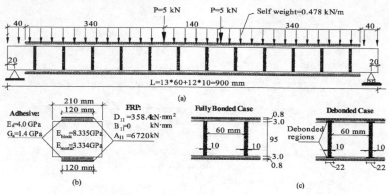

Figure 3 – Geometry, material properties, and load pattern: (a) Geometry and loading scheme; (b) Cross section and mechanical properties; (c) Full bonded and debonded case of the FRP strips near the mortar joints.

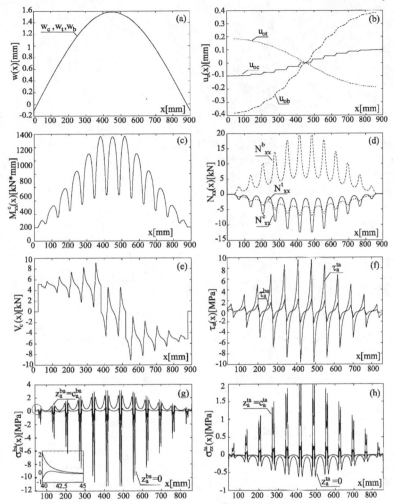

Figure 4 – Response of the fully bonded strengthened wall: (a) Vertical deflections; (b) Longitudinal deformations; (c) Bending moment in the wall; (d) Axial forces; (e) Shear forces in the wall; (f) Shear stresses at the adhesive layers; (g) Vertical normal stresses at the lower adhesive layer; (h) Vertical normal stresses at the upper adhesive layer.

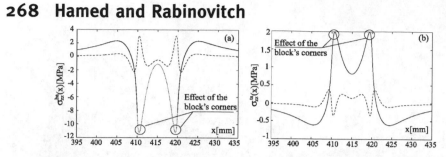

Figure 5 – Vertical normal stresses in the adhesive at the critical mortar joint (at mid-span) – fully bonded case: (a) Stresses at the lower adhesive layer; (b) Stresses at the upper adhesive layer. (Legend: —— Adhesive-block interface, ····Adhesive-mortar interface,— Adhesive-FRP interface)

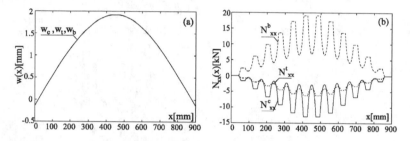

Figure 6 – Response of the wall debonded near the mortar joints: (a) Vertical deflections; (b) Axial forces.

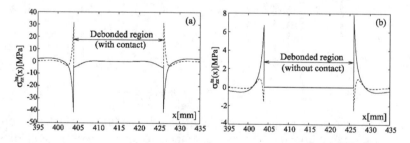

Figure 7 – Vertical normal stresses in the adhesive near the critical joint at mid-span – debonded wall: (a) Stresses at the lower adhesive layer; (b) Stresses at the upper adhesive layer. (Legend: –– Adhesive-block/ mortar interface, — Adhesive-FRP interface.)

Design Guidelines for Masonry Structures: Out of Plane Loads

by N. Galati, E. Garbin, G. Tumialan, and A. Nanni

<u>Synopsis:</u> Unreinforced masonry (URM) walls are prone to failure when subjected to out-of-plane loads caused by seismic loads or high wind pressure. Fiber Reinforced Polymers (FRP) in the form of laminates or grids adhesively bonded to the masonry surface with epoxy or polyurea based resins; or FRP bars used as Near Surface Mounted (NSM) reinforcement bonded to the masonry using epoxy or latex modified cementitious pastes, have been successfully used to increase flexural and/or shear capacity of URM walls. However, the practical application of FRPs to strengthen masonry structures is only limited to few research projects due to the limited presence of specific design guidelines. This paper describes provisional design guidelines for the FRP strengthening of masonry walls subject to out of plane loads. The proposed design methodology offers a first rational attempt for consideration by engineers interested in out-of-plane upgrade of masonry walls with externally bonded FRP systems.

Keywords: bar shapes; design; epoxy- or cementitious-based paste; FRP grids; FRP laminates; masonry; NSM FRP bars; out-of-plane; polyurea

Nestore Galati is a research engineer at the University of Missouri – Rolla. He obtained is PhD in Civil Engineering at the University of Lecce- Italy where he also received his B.Sc. in Materials Engineering. He obtained his M.Sc. degree in Engineering Mechanics at the University of Missouri-Rolla. His research has concentrated on the area of retrofitting of masonry and upgrade and in-situ load testing of RC structures.

Enrico Garbin is PhD candidate at University of Padua, Italy. His research interests include seismic behavior of RC and masonry structures, use of advanced materials for new construction and for retrofitting of existing structures, and use of innovative techniques for structural health monitoring.

Gustavo Tumialan is a staff engineer for Simpson Gumpertz and Heger in Boston, Massachusetts. He received his B.S. in Civil Engineering from Pontificia Universidad Catolica del Peru; and Ph.D. in Civil Engineering from the University of Missouri-Rolla. He is active in the field of rehabilitation of masonry and reinforced concrete structures. He is member of ACI - Committee 440 and the Existing Masonry Committee of TMS.

Antonio Nanni FACI, is the V & M Jones Professor of Civil Engineering at the University of Missouri – Rolla, Rolla, MO. He was the founding Chair of ACI Committee 440, Fiber Reinforced Polymer Reinforcement, and is the Chair of ACI Committee 437, Strength Evaluation of Existing Concrete Structures.

RESEARCH SIGNIFICANCE

A design methodology for the FRP strengthening of un-reinforced masonry walls subject to out of plane loads is presented. Non-Bearing and bearing walls are studied taking into account the influence of the boundary conditions. Different types of FRP strengthening are investigated: Glass Grid Reinforced Polyurea (GGRP), FRP laminates, and Near Surface Mounted (NSM) FRP bars.

INTRODUCTION

Unreinforced masonry (URM) walls are prone to failure when subjected to overstress caused by out-of-plane and in-plane loads. Externally bonded FRP laminates have been successfully used to increase the flexural and/or the shear capacity of the strengthening of unreinforced masonry (URM) walls subjected to overstresses (Schwegler and Kelterborn, 1996; Hamilton and Dolan, 2001; Tumialan et al., 2001). The effectiveness of Near Surface Mounted (NSM) FRP bars to increase both strength and ductility of URM walls subject to out-of-plane loads was proven by several researchers (Hamid, 1996; Galati et al., 2004). A field application on flexural strengthening with NSM FRP bars of cracked URM walls in an educational facility in Kansas City - Missouri, showed effectiveness and practicality of such technique (Tumialan et al., 2003). Glass Grid Reinforced Polyurea (GGRP) was successfully used by Yu et al., (2004), to strengthen URM walls subject to both, out-of-plane and in-plane loads.

This paper presents design guidelines for the strengthening of masonry structures strengthened either with Glass Grid Reinforced Polyurea (GGRP), FRP laminates, and Near Surface Mounted (NSM) FRP bars. Such design guidelines are based upon the present available literature and design codes. Both, non-bearing and bearing walls will be presented as well as strength limitations due to arching action.

The design procedures presented in this paper are based on principles of equilibrium and compatibility and the constitutive laws of the materials, and they written in the form of a design guideline in order to facilitate its immediate use.

DESIGN PHILOSOPHY

Strength design methodology

The design of FRP reinforcement for out-of-plane and in-plane loads is based on limit state principles. The design process for masonry walls requires investigating several possible failure modes and limit states (CNR-DT 200, 2004).

In this paper the strength design approach of reinforced masonry members is adopted, to assure consistency with the Building Code Requirements for Masonry Structures (ACI 530-02/ASCE 5-02/ TMS 402-02) and with other ACI document on masonry (ACI 530.1-02/ACSE 6-02/TMS 602-02 "Specification for Masonry Structures', ACI 530-02/ASCE 5-02/TMS 402-02 "Commentary on Building Code Requirement for Masonry Structures", ACI 530.1-02/ASCE 6-02/TMS 602-02 "Commentary on Specification for Masonry Structures").

The strength reduction factors given in Building Code Requirements for Masonry Structures (ACI 530-02/ASCE 5-02/ TMS 402-02) together with the load factors given in ASCE 7-98 "Minimum Design Loads for Building and Other Structures" are used, unless otherwise noted.

DESIGN MATERIAL PROPERTIES

The materials considered in this paper are the masonry and the FRP system. The masonry material properties should be obtained from the Building Code Requirements for Masonry Structures (ACI 530-02/ASCE 5-02/ TMS 402-02) or equivalent codes or as provided by the producers. For the FRP system, the materials properties are those provided by the manufacturers.

FRP Design Material Properties

The FRP material is considered linear elastic up to failure. The material properties guaranteed by the manufacturer should be considered as initial values that do not include the effects of long-term exposure to the environment. Because long-term exposure to various environments can reduce the tensile strength and creep rupture and fatigue endurance of the FRP system, the material properties used in design equations should be reduced based on the type and level of environment and loads exposure.

Equations (1) to (2) give the tensile properties that should be used for the design, taking into account the environment exposure. The design strength, f_{fu}, should be determined, according to ACI 440.2R-02 as:

$$f_{fu} = C_E\, f_{fu}^*$$ (1)

where: C_E is the environment reduction factor summarized in Table 1, and f_{fu}^* is the guaranteed tensile strength of FRP provided by the manufacturer. The design rupture strain should be determined as:

$$\varepsilon_{fu} = C_E \varepsilon_{fu}^*$$ (2)

where ε_{fu}^* is the guaranteed rupture strain of the FRP system. The design modulus of elasticity is assumed to be the same as the value reported by the manufacturer: $E_f = E_{f,ave}$.

Reduction for debonding at ultimate -- FRP debonding can occur if the force in the FRP cannot be sustained by the interface of the substrate. In order to prevent debonding of the FRP, a limitation should be placed on the strain level developed in the laminate. The debonding of FRP in flexure or shear is accounted through a parameter k_m. The effective design strength and strain, f_{fe} and ε_{fe}, of the FRP should be considered as:

$$f_{fe} = k_m f_{fu} = k_m C_E f_{fu}^*$$ (3)

$$\varepsilon_{fe} = k_m \varepsilon_{fu} = k_m C_E \varepsilon_{fu}^*$$ (4)

Table 2 summarizes values for k_m based on test results on un-reinforced masonry (URM) walls strengthened with GGRP, FRP laminates and NSM FRP bars (Tumialan et al., 2003-a, Galati et al.2004). It should be noted that in the case of GGRP it is reasonable to conservatively assume $k_m = 0.65$ as for the case of FRP laminates applied on puttied masonry.

Reductions for creep rupture at service -- Walls subjected to sustained load such as retaining or basement walls, creep rupture considerations need to be taken into account (ACI 440.2R-02, 2002). In such cases, for serviceability check, the designed admissible tensile stress, $f_{f,s}$, should not exceed the values presented in Table 3.

Masonry

Most masonry materials exhibit nonlinear behavior in compression, and a negligible tensile strength disregarded in the present guideline. The stress distribution for the part of masonry in compression should be determined from an appropriate nonlinear stress-strain relationship or by a rectangular stress block suitable for the given level of strain in the masonry. The stress block has dimensions $\gamma f_m'$ and γd. Expressions for β_1 and γ are given in equation (5) (Tumialan et al., 2003-a).

$$\beta_1 = 2 - \frac{4\left[\left(\dfrac{\varepsilon_m}{\varepsilon_m'}\right) - \tan^{-1}\left(\dfrac{\varepsilon_m}{\varepsilon_m'}\right)\right]}{\left(\dfrac{\varepsilon_m}{\varepsilon_m'}\right)\ln\left[1+\left(\dfrac{\varepsilon_m}{\varepsilon_m'}\right)^2\right]} \quad \text{and} \quad \gamma = 0.90\frac{\ln\left(1+\left(\dfrac{\varepsilon_m}{\varepsilon_m'}\right)^2\right)}{\beta_1\left(\dfrac{\varepsilon_m}{\varepsilon_m'}\right)} \tag{5}$$

where $\varepsilon_m' = \dfrac{1.71 f_m'}{E_m}$ and $\tan^{-1}\left(\dfrac{\varepsilon_m}{\varepsilon_m'}\right)$ is computed in radians. The strength and the

modulus of elasticity of the masonry can be computed as recommended in the Building Code Requirements for Masonry Structures (ACI 530-02/ASCE 5-02/ TMS 402-02) as $E_m = 700 f_m'$, for clay masonry and $E_m = 900 f_m'$, for concrete masonry. The maximum usable strain, ε_{mu}, at the extreme compressive side is assumed to be 0.0035 (in./in.) for clay masonry and 0.0025 (in./in) for concrete masonry. When masonry crushing failure occurs the parameters β_1 and γ can assume the values shown in Table 4.

DESIGN PROCEDURE

General considerations

The failure of masonry panels for out-of-plane loads could be due by earth pressure, seismic loads, dynamic vibrations, verticality flaw, wind pressure, and by arch thrust (CNR DT, 2004, Tumialan, 2003-a).

The failure modes of URM walls strengthened with FRP systems and subject to out-of-plane loads can be summarized as follow:

- FRP debonding: due to shear transfer mechanisms at the interface masonry/FRP, debonding may occur before flexural failure. Debonding starts from flexural cracks at the maximum bending moment region and develops towards the support. Since the tensile strength of masonry is lower than that of the resin, the failure typically occurs in the masonry for walls strengthened with FRP laminates or GGRP (Tumialan, 2003-a, Hamilton, 2001).

In the case of NSM FRP strengthening, since after cracking the tensile stresses at the mortar joints are taken by the FRP reinforcement, cracks can develop in the masonry units oriented at 45° or in the head mortar joints. Some of these cracks follow the embedding paste and masonry interface causing debonding and subsequent wall failure (Galati et al., 2004). In the case of smooth rectangular NSM FRP bars, the failure mode can be due to the sliding of the bar inside the epoxy (Galati et al., 2004). Finally, if deep grooves are used, debonding can also be caused by splitting of the embedding material (Galati et al., 2004).

- Flexural failure: after developing flexural cracks primarily located at the mortar joints, a failure can occur either by rupture of the FRP reinforcement or masonry crushing (Tumialan, 2003-a, Tumialan, 2003-b). Typically, flexural failure of masonry strengthened with FRPs is due to compressive crushing in walls strongly strengthened. FRP rupture is less desirable than masonry crushing being that the latter more ductile (Triantafillou, 1998). Both failure modes are acceptable in governing the design of out-of-plane loaded walls strengthened with FRP systems provided that strength and serviceability criteria are satisfied.

- Shear failure: cracking starts with the development of fine vertical cracks at the maximum bending region. Thereafter, two types of shear failure could be observed: flexural-shear or sliding shear. The first type is oriented at approximately 45°, and the second type occurs along bed joint, near the support, causing sliding of the wall at that location. The crack due to flexural-shear mode cause a differential displacement in the shear plane, which often results in FRP debonding (Tumialan, 2003-a, Hamoush, 2002).

The recommendations given in this section are only for members of rectangular cross-sections with strengthening applied to one side, as the experimental work has almost exclusively considered members with this shape and the FRP strengthening assumed to work only in tension, not in compression.

General Assumptions
The following assumptions and limitations should be adopted:
- The strains in the reinforcement and masonry are directly proportional to the distance from the neutral axis, that is, a plane section before loading remains plane after loading.
- The tensile strength of masonry is neglected.
- There is no relative slip between external FRP reinforcement and the masonry, until debonding failure.
- The wall can be assumed to behave under simply supported conditions (i.e. arching mechanism is not present).

The FRP design strength is adjusted for the effects of environmental exposure by means of the coefficient C_E as defined in ACI440.2R-02, and for the effects of debonding by the parameter k_m.

Flexural behavior of non-load bearing walls
The ultimate strength design criterion states that the design flexural capacity of a member must exceed the flexural demand (Eq. (6)).
$$\phi M_n \geq M_u \tag{6}$$
where ϕ is the strength reduction, which should be taken as 0.7 (Tumialan et al. 2003-a), and M_n and M_u represent the nominal and ultimate moment, respectively.

Computations are based on force equilibrium and strain compatibility. The distribution of strain and stress in the FRP reinforced masonry for a rectangular cross-section under out-of-plane load is shown in Figure 1, where the value of the stress block parameters γ and β_1 associated with a parabolic compressive stress distribution are given in Eq. (5).

The general equations to evaluate the nominal moment capacity, M_n, for a strip of masonry are given as:

$$\left(\gamma f_m^{'}\right)\left(\beta_1 c\right)b = A_f f_f \tag{7}$$

$$M_n = \left(\gamma f_m^{'}\right)\left(\beta_1 c\right)b\left(d - \frac{\beta_1 c}{2}\right) \tag{8}$$

$$\frac{\varepsilon_m}{c} = \frac{\varepsilon_f}{d-c} = \frac{\varepsilon_m + \varepsilon_f}{d} \tag{9}$$

where d is the distance from extreme compression fiber to the centroid of the tension reinforcement.

Failure mode -- The flexural capacity of FRP strengthened masonry subject to out-of-plane loads is dependent on whether the failure is governed by masonry crushing or FRP debonding or rupture. The failure mode can be determined by comparing the FRP reinforcement ratio for a strip of masonry to the balanced reinforcement ratio, defined as the ratio where masonry crushing and FRP debonding or rupture occurs simultaneously. The FRP reinforcement ratio for a strip of masonry is computed as:

$$\rho_f = \frac{A_f}{b\,t} \tag{10}$$

then, according to equilibrium and compatibility, the balanced reinforcement ratio is:

$$\rho_{fb} = \gamma\beta_1 \frac{f_m^{'}}{f_{fe}} \frac{\varepsilon_{mu}}{\varepsilon_{mu} + \varepsilon_{fe}} = \gamma\beta_1 \frac{f_m^{'}}{f_{fe}} \frac{E_f \varepsilon_{mu}}{E_f \varepsilon_{mu} + f_{fe}} \tag{11}$$

If the reinforcement ratio is below the balanced ratio ($\rho_f < \rho_{fb}$), FRP rupture or debonding failure mode governs. Otherwise, when $\rho_f > \rho_{fb}$, masonry crushing governs.

Nominal flexural capacity -- *Masonry crushing failure*: When $\rho_f > \rho_{fb}$, the failure is initiated by crushing of the masonry, and the stress distribution in the masonry can be approximated with a rectangular stress block defined by the parameters β_1 and γ that in this case assume the values shown in Table 4.

According to 440.1R-03 and based on the equations (7) to (9), the following equations can be derived:

$$M_n = \left(\gamma f_m^{'}\right)ab\left(d - \frac{a}{2}\right) = A_f f_f \left(d - \frac{a}{2}\right) \tag{12}$$

$$a = \beta_1 c = \frac{A_f f_f}{\gamma f_m^{'} b} \tag{13}$$

$$f_f = E_f \varepsilon_{mu} \frac{\beta_1 d - a}{a} \tag{14}$$

Substituting a from Eq. (13) into Eq. (14) and solving for f_f gives:

$$f_f = \left(\sqrt{\left(\frac{E_f \varepsilon_{mu}}{2} \right)^2 + \frac{\gamma \beta_1 f_m'}{\rho_f} \frac{d}{t} E_f \varepsilon_{mu}} - \frac{E_f \varepsilon_{mu}}{2} \right) \leq f_{fe} \tag{15}$$

The nominal flexural strength can be determined from Eq. (12), (13) and (14). Based on compatibility, the stress level in the FRP can be found from Eq. (15), and needs to be less or equal to f_{fe}.

FRP debonding or rupture: When $\rho_f < \rho_{fb}$, the failure of the wall is initiated by rupture or debonding of the FRP, and the equivalent stress block depends on the maximum strain reached by the masonry. In this case, an iterative process should be used to determine the equivalent stress block. The analysis incorporates four unknowns: the masonry compressive strain at the failure ε_m, the depth to the neutral axis c, and the parameters β_1 and γ.

Once the value of the four parameters have been found, the flexural capacity can be computed as shown in Eq. (16):

$$M_n = A_f f_{fe} \left(d - \frac{\beta_1 c}{2} \right) \tag{16}$$

For this type of failure, the upper limit of the product $\beta_1 c$ for balanced conditions is equal to $\beta_1 c_b$. Therefore, a simplified and conservative calculation of the nominal flexural capacity of the member can be based on Eq. (17) and (18):

$$M_n = A_f f_{fe} \left(d - \frac{\beta_1 c_b}{2} \right) \tag{17}$$

$$c_b = \left(\frac{\varepsilon_{mu}}{\varepsilon_{mu} + \varepsilon_{fe}} \right) d \tag{18}$$

Flexural behavior of load bearing walls

The ultimate strength design criterion states the design capacity of a member subject to flexural and axial load should be:

$$\frac{P_u}{\phi P_n} + \frac{M_u}{\phi M_n} \leq 1 \tag{19}$$

Again, ϕ is conservatively assumed to be 0.7 for flexure and/or axial loads. Computations are based on force equilibrium and strain compatibility. The geometry of the un-cracked cross-section is given in Figure 2. The distribution of strain and stress in the FRP reinforced masonry for a rectangular section under out-of-plane and axial loads are shown in Figure 1.

The nominal axial strength, P_n, for the masonry strip of width b should be evaluated according to the Building Code Requirements for Masonry Structures (ACI

530-02/ASCE 5-02/TMS 402-02), and shall not exceed the values given in Eq. (20) or Eq. (21).

(a) For members having $\dfrac{h}{r} \leq 99$:

$$P_n = 0.80\left[\gamma f_m' A_n\right]\left[1-\left(\frac{h}{140r}\right)^2\right] \tag{20}$$

(b) For members having $\dfrac{h}{r} > 99$:

$$P_n = 0.80\left[\gamma f_m' A_n\right]\left(\frac{70r}{h}\right)^2 \tag{21}$$

where, in this paragraph, r is the minimum radius of gyration of the uncracked cross-section of width l (Figure 2), A_n is the net cross-section area of the masonry strip of width b, and h the effective height of wall.

Using the stress distribution for a masonry section subject to flexural and axial load, the general equations of equilibrium and compatibility, written relative to the center of gravity, G, are given as:

$$\left(\gamma f_m'\right)(\beta_1 c)b - P_u = A_f f_f \tag{22}$$

$$M_n = \left(\gamma f_m'\right)(\beta_1 c)b\left(t - \frac{\beta_1 c}{2}\right) + A_f f_f\left(d - \frac{t}{2}\right) \tag{23}$$

$$\frac{\varepsilon_m}{c} = \frac{\varepsilon_f}{d-c} = \frac{\varepsilon_m + \varepsilon_f}{d} \tag{24}$$

The moment M_n can be also evaluated relative to the FRP reinforcement (Eq.(25)) or to the center of compression of the masonry (Eq.(26)).

$$M_n = \left(\gamma f_m'\right)(\beta_1 c)b\left(d - \frac{\beta_1 c}{2}\right) - P_u\left(d - \frac{t}{2}\right) \tag{25}$$

$$M_n = A_f f_f\left(d - \frac{\beta_1 c}{2}\right) + P_u\left(\frac{t}{2} - \frac{\beta_1 c}{2}\right) \tag{26}$$

Failure mode -- The flexural capacity of a FRP load bearing wall is dependent on failure mode. The failure mode can be determined by comparing the FRP reinforcement ratio (Eq. (10)) to the balanced reinforcement ratio Eq. (27).

$$\rho_{fb} = \frac{f_m'}{f_{fe}}\left[\gamma \beta_1 \frac{\varepsilon_{mu}}{\varepsilon_{mu} + \varepsilon_{fe}} - \frac{P_u}{b\,t\,f_m'}\right] = \frac{f_m'}{f_{fe}}\left[\gamma \beta_1 \frac{E_f \varepsilon_{mu}}{E_f \varepsilon_{mu} + f_{fe}} - \frac{P_u}{b\,t\,f_m'}\right] \tag{27}$$

If the reinforcement ratio is below the balanced ratio ($\rho_f < \rho_{fb}$), FRP rupture or debonding failure mode governs. Otherwise, ($\rho_f > \rho_{fb}$) masonry crushing governs.

<u>Nominal flexural capacity</u> -- *Masonry crushing failure:* When $\rho_f > \rho_{fb}$, the failure is initiated by crushing of the masonry, and the stress distribution in the masonry can be approximated with a rectangular stress block defined by the parameters β_1 and γ that in this case assume the values shown in Table 4. Based on equations (20) to (26), the following expressions can be derived:

$$M_n = \left(\gamma f_m'\right) a b \left(d - \frac{a}{2}\right) - P_u\left(d - \frac{t}{2}\right) = A_f f_f\left(d - \frac{a}{2}\right) + P_u\left(\frac{t}{2} - \frac{a}{2}\right) \tag{28}$$

$$a = \beta_1 c = \frac{\left(A_f f_f + P_u\right)}{\gamma f_m' b} \tag{29}$$

$$f_f = E_f \varepsilon_{mu} \frac{\beta_1 d - a}{a} \tag{30}$$

Considering equations from (28) to (30), in the case of masonry crushing, the following values for f_f and c can be obtained:

$$f_f = \left(\sqrt{\left(\frac{E_f \varepsilon_{mu}}{2} - \frac{P_u}{2 A_f}\right)^2 + \left(\frac{\gamma \beta_1 f_m'}{\rho_f}\frac{d}{t} - \frac{P_u}{A_f}\right)E_f \varepsilon_{mu}} - \left(\frac{E_f \varepsilon_{mu}}{2} + \frac{P_u}{2 A_f}\right)\right) \le f_{fe} \tag{31}$$

$$c = \frac{a}{\beta_1} = \frac{\rho_f t}{\beta_1 \gamma f_m'}\left[\sqrt{\left(\frac{E_f \varepsilon_{mu}}{2} - \frac{P_u}{2 A_f}\right)^2 + \frac{\beta_1 \gamma f_m'}{\rho_f}\frac{d}{t}E_f \varepsilon_{mu}} - \left(\frac{E_f \varepsilon_{mu}}{2} - \frac{P_u}{2 A_f}\right)\right] \tag{32}$$

FRP debonding or rupture: When $\rho_f < \rho_{fb}$, the failure of the wall is initiated by debonding or rupture of the FRP, and the equivalent stress block depends on the maximum strain reached by the masonry. In this case, an iterative process should be used to determine the equivalent stress block. The analysis incorporates four unknowns given the value of P_u: the masonry compressive strain at failure ε_m, the depth to the neutral axis c, and the parameters γ and β_1. Solving for this system of equations may be laborious.

Alternatively, according to the Building Code Requirements for Masonry Structures (ACI 530-02/ASCE 5-02/TMS 402-02, section 3.2.2) values of β_1 and γ equal to 0.80 can be assumed. Therefore, the following simplified equations can be used:

$$M_n = \left(0.80 f_m'\right)\left(0.80 c\right) b \left(d - \frac{0.80 c}{2}\right) - P_u\left(d - \frac{t}{2}\right) \tag{33}$$

$$c = \frac{\rho_f t}{0.80^2 f_m'}\left[\sqrt{\left(\frac{E_f \varepsilon_{mu}}{2} - \frac{P_u}{2 A_f}\right)^2 + \frac{0.80^2 f_m'}{\rho_f}\frac{d}{t}E_f \varepsilon_{mu}} - \left(\frac{E_f \varepsilon_{mu}}{2} - \frac{P_u}{2 A_f}\right)\right] \tag{34}$$

$$f_f = E_f \varepsilon_{mu} \frac{d - c}{c} \le f_{fe} \tag{35}$$

Shear Limitations

The nominal moment calculated for flexural behavior should be compared and, if necessary, limited by the one associated with shear failure. In fact, if a large amount of FRP is applied, the failure can be controlled by shear instead of flexure. The theoretical shear capacity of the FRP strengthened masonry should be evaluated according to the Building Code Requirements for Masonry Structures (ACI 530-02/ASCE 5-02/TMS 402-02). The shear strength capacity should exceed the shear demand, as shown in (36):

$$\phi V_n \geq V_u \tag{36}$$

Due to the fact that the FRP system is only bonded onto the masonry surface, its contribution can be neglected, and the nominal strength becomes:

$$V_n = V_m \tag{37}$$

The shear strength provided by the masonry, V_m, shall be computed using equation (38) for non-load bearing walls, and equation (39) for load bearing walls. The value of $\dfrac{M}{Vt}$ need not be taken greater than 1.0.

$$V_m = \left[4.0 - 1.75\left(\frac{M}{Vt}\right)\right] A_n \sqrt{f_m'} \tag{38}$$

$$V_m = \left[4.0 - 1.75\left(\frac{M}{Vt}\right)\right] A_n \sqrt{f_m'} + \frac{P}{4} \tag{39}$$

where M is the maximum moment at the section under consideration, V is the corresponding shear force, t the thickness of the masonry, A_n the net cross-section area of the masonry strip of width b, f_m' the specified compressive strength of masonry and P is the axial load.

The nominal shear capacity, V_n, shall not exceed the following limits:

(a) When $\dfrac{M}{Vt} \leq 0.25$:

$$V_n \leq 6 A_n \sqrt{f_m'} \tag{40}$$

(b) When $\dfrac{M}{Vt} \geq 1.00$

$$V_n \leq 4 A_n \sqrt{f_m'} \tag{41}$$

(c) For values of $\dfrac{M}{Vt}$ falling in the range 0.25 to 1.00, a linear interpolation can be used to determine the limiting value of V_n.

STRENGTHENING LIMITATIONS DUE TO ARCHING ACTION

When a wall is built between supports that restrain the outward movement, membrane compressive forces in the plane of the wall, accompanied by shear forces at the supports, are induced as the wall bends. The in-plane compression forces can delay cracking. After cracking, a so-called arching action can be observed. Due to this action, the capacity of the wall can be much larger than that computed assuming simply supported conditions. Experimental works (Tumialan et al., 2003-b, Galati, 2002-a, Galati, 2002-b, Carney, 2003), have shown that the resultant force between the out-of-plane load and the induced membrane force could cause the crushing of the masonry units at the boundary regions.

The arching mechanism must be considered in the quantification of the upgraded wall capacity to avoid overestimating the contribution of the strengthening. Three different modes of failure have been observed in walls exhibiting the arching mechanism:
- Flexural failure (i.e. rupture of the FRP in tension or crushing of the masonry)
- Crushing of masonry at the boundary regions
- Shear failure.

Figure 3 illustrates a comparison between the load-deflection curves obtained in the case of simply supported walls and walls with the end restrains, tested under four point bending (Galati et al., 2002-a). A significant influence of the boundary conditions in the wall behavior is observed. Figure 3 shows that the increase in the ultimate load for walls strengthened with 3 in. and 5 in. wide GFRP laminates were about 175 and 325%, respectively. If the .wall is restrained (i.e. arching mechanism is observed) the same effectiveness of the FRP reinforcement is not observed because crushing of the masonry units at the boundary regions controls the wall behavior. In this case, the increase in the out-of-plane capacity for strengthened specimens with 3 and 5 in. wide GFRP laminates was about 25%. It is to be stressed that capacity of an unstrengthened URM wall with end restrains is far superior to that of an identical simply supported wall with FRP strengthening.

Design Procedure

When a non load bearing wall is built solidly between supports capable of resisting an arch thrust with no appreciable deformation or when walls are built continuously past vertical supports (horizontal spanning walls), the lateral load resistance of the wall can benefit from the arching action if height to thickness ratio is less than 20. In such cases, the ultimate strength design criteria states the design ultimate load capacity of a member should be:

$$\phi q_n \geq q_u \tag{42}$$

where $\phi = 0.6$, and q_n and q_u have dimensions kN/m. The design procedure for unstrengthened and strengthened walls is presented herein. The design procedure presented herein allows determining the nominal resisting uniform force, q_n, for both unstrengthened and strengthened URM walls. The resisting force for loading conditions other than the uniform pressure can be derived from q_n.

The resisting force, Q_n, for a concentrated load at mid-height of the wall is given by equation (43):

$$Q_n = \frac{q_n h}{2} \qquad (43)$$

where h is the height of the wall. For a triangular distribution, the maximum resisting pressure $\bar{q}_n$ can be determined using the following equation:

$$\bar{q}_n = \frac{q_n}{2} \qquad (44)$$

Unstrengthened Masonry Walls -- Analysis may be based on a three-hinge arch, when the bearing of the arch thrust at the supports and at the central hinge should be assumed as 0.1 times the thickness of the wall, as indicated on Figure 4. If chases or recess occur near the thrust-lines of the arch, their effect on the strength of the masonry should be taken into account (Eurocode 6 Sec. 6.3.2).

The arch thrust should be assessed from knowledge of the applied lateral load, the strength of the masonry in compression, the effectiveness of the junction between the wall and the support resisting the thrust, and the elastic and time depending shortening of the wall. The arch thrust may be provided by a vertical load (Eurocode 6 Sec. 6.3.2). The resisting force, q_n, per width b of wall is given by equation (45):

$$q_n = 0.58 f_m' b \left(\frac{t}{h} \right)^2 \qquad (45)$$

where b, t and h are the width, thickness and height of the wall, respectively. If the clamping force per width b of the wall, C, is needed, it can be easily computed using equation (46).

$$C = 0.58 f_m' \frac{bt}{10} \qquad (46)$$

Strengthened Masonry Walls -- In addition to the general assumptions presented in the first part of the paper, the wall is also assumed cracked at mid-height, and that the two resulting segments can rotate as rigid bodies about the supports. With reference to Figure 5, the resisting force per unit area of wall is given by equation 8.44:

$$q_n = \frac{8}{h^2} \left(\gamma_1 \beta_{11} w_m b_1 f_m' a_c + A_f f_f a_f \right) \qquad (47)$$

where h is the height of the wall, A_f is the area of FRP reinforcement, w_m is the width of the wall, γ and β_1 define the stress block. The additional subscripts 1 or 2 for γ and β_1 has been used to single out the corresponding section. Finally, a_f and a_c define the arm of both the force in the FRP and of the clamping force, respectively. For small values of rotation of the wall θ, a_f and a_c can be determined as follows:

$$a_f = d - \frac{\beta_{12} \left(\varepsilon_{m2} \right) b_2}{2} \qquad (48)$$

$$a_c = a_f - \frac{\beta_{11} \left(\varepsilon_{m1} \right) b_1}{2} \qquad (49)$$

where b_1 and b_2 represent the bearing widths at the supports and at mid-height, respectively. It is important to notice that γ and β_1 are functions of the maximum compressive strain at the considered cross-section (ε_{m1} or ε_{m2}), as expressed in equations 7.5 and 7.6 in Chapter 7.

Equation 8.42, when accounting for equations 8.43 and 8.44, contains five unknowns: ε_{m1}, ε_{m2}, b_1, b_2, and f_f. Such unknowns can be determined using the procedure based on compatibility and equilibrium equations presented herein (Galati, 2002-a; Tumialan et al., 2003-b).

The free-body diagram shown in Figure 5 (b) can be derived analyzing the top segment of the masonry wall depicted in Figure 5 (a). From the equilibrium of forces in the vertical direction, the following relationship can be drawn:

$$C_2 = C_1 + T_f \tag{50}$$

where C_1 and C_2 are the clamping forces at top and mid-height of the wall, respectively, T_f is the force in the FRP strengthening. Considering the stress block distribution, the clamping forces by wall strip width, w_m, acting on the restrained ends of the wall can be calculated as:

$$C_1 = \gamma_1 \beta_{11} w_m b_1 f_m' \tag{51}$$

$$C_2 = \gamma_2 \beta_{12} w_m b_2 f_m' \tag{52}$$

where the additional subscripts 1 and 2 for γ and β_1 have been used to single out the corresponding cross-section. The tensile force developed by the FRP laminate is:

$$T_f = A_f f_f = A_f E_f \varepsilon_f \tag{53}$$

Based on equations (51) to (53), equation (50) can be re-written as:

$$\gamma_2 \beta_{12} w_m b_2 f_m' = \gamma_1 \beta_{11} w_m b_1 f_m' + A_f E_f \varepsilon_f \tag{54}$$

Equation (54) expresses the equilibrium of the forces. The compatibility of deformations is expressed with the following two equations (Tumialan et al., 2003)

$$t - b_1 - b_2 = \frac{h}{2} \cdot \frac{1 - \cos\theta}{\sin\theta} \cong \frac{1}{16} \frac{h^2}{b_1} \varepsilon_{m1} \tag{55}$$

$$\frac{b_2}{b_1} = \frac{\varepsilon_{m2}}{\varepsilon_{m1}} \tag{56}$$

Moreover, assuming that the deformation of the FRP occurs in an unbonded length, l_b, the strain in the FRP can be estimated using the equation:

$$\frac{\frac{t - b_2}{b_2} h \varepsilon_{m1}}{16 l_b} = \frac{\frac{t - b_2}{b_1} h \varepsilon_{m2}}{16 l_b} \tag{57}$$

To date, there is no experimental basis for the determination of l_b. Based on experimental observations (Tumialan et al., 2003-b) a suggested value for l_b is equal to 1.5 in (37.5 mm).

Given the failure mode (i.e. set the maximum value for ε_{m1} or ε_{m2} or ε_f), equations 8.49 to 8.52 can be iteratively solved for the remaining four unknowns out of the five (ε_{m1}, ε_{m2}, b_1, b_2, and f_f). Comparing the results of the first iteration with the ultimate values of ε_{m1}, ε_{m2} and ε_f, the actual failure mode of the wall can be determined and, therefore, a second iteration will give the actual value of all the unknowns.

Shear Limitations

The design shear strength for walls for which the arching action cannot be neglected, shall be in accordance to equation (38) after setting the ratio $\dfrac{M}{V t} = 1.00$.

CONCLUSIONS

The design methodology proposed in this paper offers a first rational attempt for consideration by engineers interested in out-of-plane upgrade of masonry walls with externally bonded FRP systems. Both, the case of bearing and not load bearing walls, and of infill and load-bearing walls for which the arching action cannot be neglected were considered. Additional experimental work as well as a reliability analysis is needed in order to determine a more comprehensive set of design factors.

ACKNOWLEDGMENTS

The authors would like to acknowledge the financial support of the National Science Foundation Industry/University Cooperative Research Center – Repair of Bridges and Buildings with Composites (RB2C) - located at the University of Missouri–Rolla and Bondo Corporation and Techfab for their financial support.

REFERENCES

ACI 440M Guide Draft-1 (2004), "Guide for the Design and Construction of Externally Bonded FRP System for Strengthening Unreinforced Masonry Structures", Reported by ACI Committee 440, 2004.

440.1R-03 (2003), "Guide for the Design and Construction of Concrete Reinforcement with FRP Bars", Reported by ACI Committee 440, 2003.

ACI 440.2R-02 (2002), "Guide for the Design and Construction of Externally Bonded FRP Systems for Strengthening Concrete Structures", Reported by ACI Committee 440, 2002.

ASTM E519-02 (2002), "Standard Test Method for Diagonal Tension (Shear) in Masonry Assemblages", ASTM International, www.astm.org

Carney, P. W. (2003), "Out of plane static and blast resistance of unreinforced masonry wall connections strengthened with fiber reinforced polymers," Thesis (M.S.), University of Missouri - Rolla

284 Galati et al.

CNR-DT 200/2004 (2004), "Istruzioni per la Progettazione, l'Esecuzione ed il controllo di Interventi di Consolidamento Statico mediante l'utilizzo di Compositi Fibrorinforzati, Materiali, strutture in c.a. e in c.a.p., strutture murarie", final Draft, Rome CNR 13 July 2004 (in italian)

Eurocode 6 (2004), "Design of masonry structures, Part 1-1: Common rules for reinforced and unreinforced masonry structures", CEN European Committee for Standardization, final draft prEN 1996-1-1, April 2004.

Galati N. (2002-a), "Arching effect in masonry walls reinforced with fiber reinforced polymer (FRP) materials," Thesis (M.S.), University of Missouri – Rolla

Galati N., Tumialan J.G., La Tegola A., and Nanni A. (2002-b), "Influence of Arching Mechanism in Masonry Walls Strengthened with FRP Laminates," ICCI 2002, San Francisco, CA, June 10-12, 10 pp. CD-ROM

Galati N., Tumialan J.G., Secondin S., and Nanni A. (2004), "Strengthening with FRP Bars of URM Walls Subject to Out-of-Plane Loads," International Conference Advanced Polymer Composites for Structural Applications in Construction" ACIC 2004, The University of Surrey, Guildford, Surrey, UK, CD-ROM 11

Hamid A.A., (1996) "Strengthening of Hollow Block Masonry Basement Walls with Plastic Reinforcing Bars", The Masonry Society Journal.

Hamilton, H.R. III, and Dolan, C. W., (2001) "Flexural Capacity of Glass FRP Strengthened Concrete Masonry Walls," J. of Comp. for Const., ASCE, Vol.5, No.3, August 2001, pp. 170-178.

Hamoush S., McGinley M., Mlakar P., Terro M.J. (2002), "Out-of-Plane behavior of surface-reinforced masonry walls", Construction and Building Materials, Vol. 16, Is. 6, pp. 341-351

Schwegler, G., and Kelterborn, P. (1996). "Earthquake Resistance of Masonry Structures strengthened with Fiber Composites," 11th World Conf. on Earthquake Eng., Acapulco, Mexico, 1996, 6 pp. CD-ROM.

Triantafillou T.C. (1998), "Strengthening of Masonry Structures Using Epoxy-Bonded FRP Laminates", Journal of Composites for Construction, Vol. 2, No. 2, May 1998, pp. 96-103

Tumialan J. G., Galati N., Nanni A. and Tyler D., (2003) "Flexural Strengthening of Masonry Walls in q High School Using FRP Bars," ACI, Special Pubblication 215-12, pp 413-428.

Tumialan, J. G., Morbin, A., Nanni, A. and Modena, C. (2001) "Shear Strengthening of Masonry Walls with FRP Composites," COMPOSITES 2001 Conv. and Trade Show, Composites Fabricators Assoc., Tampa, FL, October 3-6, 6 pp. CD-ROM.

Tumialan J.G., Galati N., Nanni A. (2003-a), "FRP Strengthening of UMR Walls Subject to Out-of-Plane Loads", ACI Structural Journal, Vol. 100, No. 3, May-June, pp. 312-329.

Tumialan J.G., Galati N., Nanni A. (2003-b), "Field Assessement of Unreinforced masonry Walls Strengthened with Fiber Reinforced Polymer laminates", Journal of Structural Engineeering, Vol. 129, No. 8, August 2003, pp. 1047-1055

Yu P., Silva P.F., Nanni A. (2004), "Application of Bondo Polyurea in Structural Strengthening of RC Beams and UMR Walls", Final Report, Report No. CIES 01-49, Center for Infrastructure Engineering Studies, University of Missouri-Rolla, August 2004.

Table 1 -- Environment Reduction Factors

Exposure condition	Fiber type	Environment reduction factor C_E
Masonry, interior exposition	Carbon	0.95
	Glass	0.75
	Aramid	0.85
Masonry, exterior exposition	Carbon	0.85
	Glass	0.65
	Aramid	0.75
Masonry, aggressive environment	Carbon	0.85
	Glass	0.50
	Aramid	0.70

Table 2 -- k_m Factors for Different Strengthening Systems

Strengthening System	Limitations	Resin Type	k_m
GGRP	-	Polyurea	0.65
FRP Laminates	If putty is used	Epoxy	0.65[1]
	If putty is not used	Epoxy	0.45[1]
NSM FRP Bars	FRP rectangular bars, Groove having the same height of the bar and width 1.5 times the one of the bar	Epoxy	0.65[4]
	FRP circular bars, Square groove 1.5 times the diameter of the Bar[4]	Epoxy	0.35[2]
	FRP circular bars, Square groove 2.25 times the diameter of the Bar	Epoxy / LMCG[3]	0.55[2]

[1] Based on Tumialan et al., 2003-a
[2] Based on Galati et al. 2004
[3] Latex Modified Cementitious Grout
[4] Latex Modified Cementitious Grout can not be used with a standard square groove having dimensions 1.5 times the diameter of the bar

Table 3 -- $f_{f,s}$ for Different Fiber Types

Fiber Type		
Carbon	Glass	Aramid
$0.55 f_{fu}$	$0.20 f_{fu}$	$0.30 f_{fu}$

Table 4 -- Stress Block Patameters β_1 and γ

Parameter	Concrete	Clay
β_1	0.805	0.822
γ	0.853	0.855

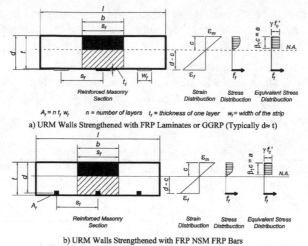

a) URM Walls Strengthened with FRP Laminates or GGRP (Typically d≈ t)

b) URM Walls Strengthened with FRP NSM FRP Bars

Figure 1 — Internal strain and stress distribution for a horizontal rectangular section of a strip of masonry under out-of-plane loads, without axial compression

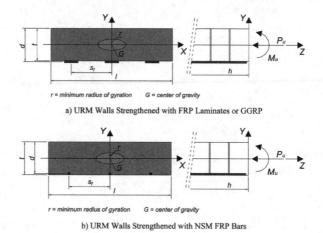

a) URM Walls Strengthened with FRP Laminates or GGRP

b) URM Walls Strengthened with NSM FRP Bars

Figure 2 — Geometric parameters of the uncracked section under out-of-plane loads with axial compression

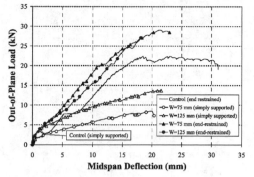
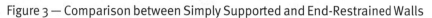

Figure 3 — Comparison between Simply Supported and End-Restrained Walls

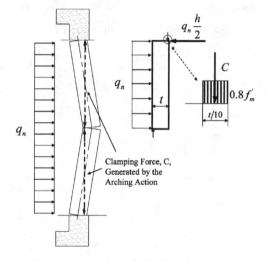

Figure 4 — Design of Rigid Arching Walls

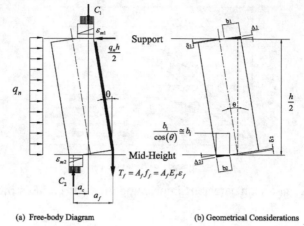

(a) Free-body Diagram   (b) Geometrical Considerations

Figure 5 — Half Part of Analyzed Wall

FRP Repair Methods for Unreinforced Masonry Buildings Subject to Cyclic Loading

by P.B. Foster, J. Gergely, D.T. Young, W.M. McGinley, and A. Corzo

Synopsis: Unreinforced masonry building specimens were evaluated under cyclic lateral loading. Various fiber reinforced polymer (FRP) composite configurations were used to repair and retrofit the masonry structures. In the first phase, three different composite systems were used to repair pre-damaged masonry structures. These systems included: a wet lay-up woven glass fabric; a near surface mounted (NSM) extruded carbon FRP plate; and a glass FRP grid attached via a high elongation polyurea resin. Retesting of the repaired structures revealed increases as much as 700% in terms of energy dissipation and 300% in terms of pseudo-ductility. The second phase involved retrofitting similar undamaged building specimens with FRP composites. Significant increases in strength, ductility and energy dissipation were observed. The seismic performance of each structure was increased with the addition of a minimal amount of composite material as compared to the unreinforced structures.

Keywords: FRP; near surface mounted; repair; retrofit techniques; seismic loading; unreinforced masonry; wet lay-up

Peter B. Foster is a doctorate research assistant at the University of North Carolina at Charlotte. His Masters' Thesis involved load testing a novel type of steel connection and has been involved in a multitude of research projects which included load testing bridges, structural insulated panels, FRP applications for concrete and masonry systems, as well as finite element studies.

Janos Gergely is an associate professor at the University of North Carolina at Charlotte. He has worked extensively with FRP composites, and their applications to concrete and masonry structures and components. He has also been involved in studies focused on bridge performance and analysis.

David T. Young is the chair of the Department of Civil Engineering at the University of North Carolina at Charlotte. He has been involved in a wide variety of research project involving masonry and concrete, FRP composites and structural insulated panels. His interests are in forensic engineering and NDE.

W. Mark McGinley is a professor in the Department of Architectural Engineering at North Carolina A & T State University. His interests are in the use of FRP composites to enhance the performance of masonry structures. He has conducted a number of research investigations on masonry materials, components and systems.

Anna Corzo is a graduate research assistant at North Carolina A & T State University and is working with Mark McGinley in developing an analytical model to predict the behavior of FRP reinforced masonry systems.

RESEARCH SIGNIFICANCE

The research presented herein describes a methodology for the repair and retrofit of unreinforced masonry (URM) buildings using fiber reinforced polymer (FRP) composites. This methodology can be directly applied to structures that have been damaged due to significant lateral loading or structures that are substandard and need to be retrofitted and, as seen in this paper, can significantly enhance the strength and performance characteristics of these structures. As compared to traditional repair/retrofit methods, this new methodology is appreciably more cost-effective as well as less intrusive.

INTRODUCTION

Fiber reinforced polymer (FRP) composites have been extensively investigated in the last decade and a half throughout the world. The interaction of FRP materials with concrete and masonry has been studied under a variety of loading conditions. With regard to masonry, FRP composites have been used to strengthen unreinforced masonry shear and flexural walls with resounding success. The extent of this research has been limited, with few exceptions, to component testing.

The seismic capacity of unreinforced masonry (URM) shear walls is minimal. Experimental studies (Eshani et al., 1996) proved that using FRP composites could enhance the seismic performance of unreinforced masonry shear walls. The dynamic response of URM shear walls has been investigated as well (Al-Chaar et al., 1999). In another study, the performance of URM shear walls subjected to cyclic lateral loads were greatly enhanced when CFRP composite laminates were used to retrofit the masonry components (Gergely et al., 2000).

The capacity of unreinforced masonry flexural walls has been shown to have drastically increased by using FRP composites (Albert, Elwi and Cheng 2001). In this investigation, both glass and carbon FRP laminates were used. In a similar investigation, flexural walls strengthened with FRP materials resulted in significantly increased flexural capacities, provided that shear was controlled at the support (Hamoush et al., 2001). In a similar study, the capacities of GFRP flexural walls were compared to design equations that resulted in an overprediction of no more than 20% (Hamilton and Dolan, 2001).

The performance of unreinforced masonry infill walls has been explored as well. The capacity and performance of these infill walls can be significantly enhanced by FRP composite retrofit (Silva et al., 2001). Infill walls, subjected to both on and off-axis loading, were shown to have moderate to significant strength increases when retrofitted with glass FRP (Hamid et al., 2005).

However, few investigations focused on the entire masonry building system subjected to quasi-static lateral loading. One such project evaluated a full-scale masonry building system with flexible diaphragms, that utilized several different FRP composite laminates, as well as post-tensioning, to enhance the performance of the system (Moon et al., 2002). It was found that the system-wide performance was greatly enhanced under cyclic lateral loading. It has also been shown that the dynamic capacity of URM walls reinforced with FRP composites can be significantly improved with failure occurring in the masonry units, rather than the mortar joints (Marshall, Sweeney and Trovillion, 1999).

In general, the experimental evaluation of full scale building models is quite expensive and testing of individual components seems to be a much more cost-effective solution. The results of these component tests, however, may be somewhat skewed in some instances and there may exist idiosyncrasies that may not be accurately captured by component testing. In a full scale building model, redistribution of lateral forces and changes in end conditions (i.e. fixed to cantilever) can be achieved when cracking occurs. These cannot be captured accurately with component testing. Also, with component testing, out-of-plane rotations may occur that would not normally occur in a full scale building model. It may be that component testing is applicable for obtaining general performance characteristics of a particular FRP system but, ultimately, the system must be implemented on a full scale structure to attain an accurate response.

The goal of the present research was to investigate several repairs and retrofit methods for existing unreinforced masonry structures, with rigid diaphragms, subjected to quasi-static lateral loading. In support of this effort, large-scale structural tests were performed at UNC Charlotte, and the small component tests were performed at NC A&T.

EXPERIMENTAL DESIGN

Several full-scale unreinforced masonry buildings were constructed. Due to size constraints that existed within the load frame, each structure was constructed with the dimensions 2.48 m (8') in height, 3.25 m (10'8") in width, and 4.47 m (14'8") in length (see Figure 1). The shear walls on each side were perforated with two 82 cm (32") wide standard doorway openings in two structures, resulting in a symmetric structural configuration. An asymmetric configuration was also tested, and in this case only one shear wall was perforated with the standard doorway opening.

A rigid roof diaphragm, designed as a deep beam, was used to prevent any independent movement that may occur in the walls with a flexible diaphragm. This roof diaphragm was placed atop a mortar bed and secured to the structure utilizing 32 anchorage points evenly spaced around the perimeter of the bond beam. Dywidag Threadbar® reinforcing bars were used to transfer the applied lateral load through the roof panel via a hydraulic piston. A gravity load simulator was used to apply a 667 kN (150 Kips) vertical load to the structure, simulating the weight of two additional floors. The loading mechanism was comprised of wide-flange steel members, and this simulator was designed such that the gravity load would be maintained under an applied cyclic lateral load, i.e. the entire gravity load system was allowed to translate, together with the building, under lateral load while maintaining the applied gravity load. This vertical load was monitored throughout the test both by the hydraulic system pressure transducer and by strain measuring instruments.

Each masonry building was constructed atop reusable concrete foundations. These foundations had removable dowel bars that extended 41 cm (16") above the top of the foundation, and these bars were grouted in the masonry wall. The dowel bars were used to prevent any overturning or sliding that may have occurred at the foundation level, and forced the failure to occur in the shear wall panels. In addition to the bottom 41 cm (16"), the top 20 cm (8") bond beam was also grouted with a 27.6 MPa (4000 psi) grout. This was done to allow anchorage for the roof diaphragm and foundation elements. The remainder of the masonry walls was ungrouted.

The testing procedure for each structure was identical: after the 667 kN (150 Kips) gravity load had been applied, the lateral load was applied at a rate of 3.5 kN (800 pounds) per second, with a load step of 44.5 kN (10 Kips). Each load step was comprised of three push-pull cycles. In the first phase, after failure of the URM building was achieved, the lateral load was removed, followed by the release of the gravity load. Following the removal of the instrumentation, the structure was evaluated and an appropriate FRP composite repair technique was selected. Following the FRP repair, the structure was retested. In the case of the retrofitted structures (Phase II), the FRP composite was applied without pre-testing the building. During the loading process, an array of instrumentation was used, which included displacement transducers, strain transducers, strain gages, linear potentiometers, pressure transducers and load cells.

EXPERIMENTAL RESULTS

CMU-Configuration I

Two symmetric CMU structures were fabricated and tested in this configuration. Both structures, as previously stated, were unreinforced. One structure was damaged and subsequently repaired using a glass FRP composite system and retested. The second structure was retrofitted using an identical composite system, although the layout was altered.

The performance of the unreinforced (baseline) structure remained linear until a lateral load level of approximately 267 kN (60 Kips) (see Figure 2). At this point, microcracking began to occur in the bed and head joints, which developed into a more extensive diagonal step cracking pattern as the load was increased. The cracking in the large pier was limited to the mortar joints, while in the small pier diagonal tension cracking was observed in the masonry units and the mortar joints. Failure occurred at an applied lateral load level of 400 kN (90 kips) with an associated lateral displacement of 6.35 mm (0.25"). The damage associated with this load level and displacement was not catastrophic, and the structure continued to withstand the gravity load at failure.

For both testing phases, the composite system used to strengthen the symmetric CMU building specimens was SikaWrap Hex 100G glass fiber woven fabric in conjunction with Hexcel 306 resin. This is an externally applied wet lay-up GFRP composite system that can be applied to one or both surfaces of a wall, depending on the site conditions and the performance requirements. In this project, all the FRP systems were applied only to the exterior surface of the shear walls.

The composite layout for the repaired structure was determined using a strut-and-tie model. The strut and tie model was developed from a simple free body diagram (FBD) as seen in Figure 3. In this figure, only the FBD for the large pier is shown. The axially applied load, P, and the applied lateral load, V, are known and this allows for the determination of the internal forces of the pier. The axial load, P, as previously specified was 667 kN (150 Kips) distributed evenly along the length of the shear walls. The applied lateral load was set to a level of 778.4 kN (175 kips) with an equal dispersion between the two shear walls. Distribution factors, based on rigidity calculations, were used to determine the lateral force applied to each pier. The internal shear forces, V1 and V2, characterize the shear strength of the GFRP laminate as determined through small component testing and the additional shear capacity achieved through friction and the applied axial load, respectively. The tensile forces, T and T2, represent solely the tensile strength of the GFRP laminate as the damaged unreinforced masonry substrate cannot hold significant tensile forces. From this, the quantity of GFRP composite needed to maintain rotational equilibrium was determined and yielded the number of 30 cm (12") layers needed to repair the structure. It was determined that two 1.016 mm (0.04") thick layers would be used throughout the majority of the structure, with the exception of the flexure-dominated small piers, where three vertical layers were used (see Figure 4), in addition to the diagonal laminates. The composite layout for the retrofitted symmetric CMU structure was determined using a similar procedure although the laminate configuration differed. It was comprised of only vertical layers, a somewhat less efficient but simpler layout to implement. The layout called for three 25 cm (10") strips in the large piers, evenly spaced across the pier, and two in the small piers. Each laminate was comprised of two layers of glass fiber.

As expected, the performance of the repaired structure was far superior to the unreinforced test. The lateral load reached a maximum level of 667 kN (150 kips), and the associated displacement in the push cycle had a magnitude of 22.35 mm (0.88"). At failure, the FRP laminate received little to no damage. The majority of the damage occurred in the large pier and was localized to the bottom of the doorway, as anticipated. The mode of failure was masonry substrate failure. The nature of the composite layout was such that significant forces from the flexural and shear reinforcement were transmitted through this region in a pull cycle, during which the structure failed. The tensile forces in the diagonal and vertical composite laminates created a moment about the dowel bar embedded with the CMU, a load which the masonry units could not withstand. A possible way to avoid this would be to wrap this region with the same fabric, or provide some other anchoring means to prevent the splitting that occurred. Later tests proved that when this region is protected, the structure will likely reach higher lateral loads.

The retrofitted CMU structure shown in Figure 5 performed well under the cyclic lateral load. The structure resisted approximately 35% more load than the URM baseline building, with a maximum applied load of 547 kN (123 kips). The displacement accompanying that load level was 12.2 mm (0.48") in push. Surprisingly, the stiffness of the structure did not significantly degrade until the lateral load reached a level of 400 kN (90 kips). The mode of failure observed was diagonal tension cracking in all piers .

Hysteretic and backbone curves were developed for the undamaged, repaired and retrofitted tests using guidelines set forth be the Federal Emergency Management Agency Document 356 (FEMA, 2000). The envelope created defines the general behavior of the structure in terms of applied load versus displacement. A comparison of the backbone curves from the unreinforced and the repaired tests shows that the FRP reinforced system had significantly more lateral capacity, but also had an increased pseudo-ductility (see Figure 6). Repairing the damaged CMU building with the glass FRP laminate increased the strength of the system by 67% beyond the capacity of the URM baseline building, and increased the energy dissipation (strain energy) of the system by nearly 550%, based on the compression quadrant of the curve. The repair of the damaged URM building with the GFRP laminate resulted in not only a stronger system, but also a much more ductile one as well. The performance of the retrofitted structure (using less composite material with a different layout), although not as "spectacular" as the repaired structure, did show remarkable increases in terms of both strength and energy dissipation. A 35% increase in strength and a 220% increase in energy dissipation were observed.

The maximum tensile strain in the GFRP laminate measured during the test of the repaired structure and had a magnitude of approximately 0.54%. This value was recorded in the flexure-dominated pier at the location of the lintel. The maximum strain measured in the retrofitted structure was approximately 0.8%. This occurred in the small pier at the level of the grouted section at the base of the structure. These strain levels are significantly less than the ultimate tensile strain of 2.1% of the GFRP laminate. Again, this indicates that the failure mode was based on the masonry substrate rather than the capacity of the composite laminate.

Brick-Configuration I

Similar to the previously discussed buildings, two symmetric two-wythe brick structures were evaluated for seismic performance. Near-surface mounted (NSM) systems were used for the brick buildings, provided by Hughes Brothers, Inc. Aslan 500 #2 carbon FRP bars were used in the repaired structure, while Aslan 100 #3 glass FRP rods was used in the retrofitted structure. The epoxy resin used in both structures was Unitex® Pro-Poxy 300, as recommended by the NSM system developer. The installation of the NSM rods involved cutting a vertical groove into the shear walls at the desired locations, placing an epoxy bed into the groove, placing the coated rod/bar into the groove and filling the remaining void with epoxy resin. The grooves were cut such that they passed through head joints, and from a short distance, the composite application resembled expansion joints.

The baseline brick building depicted a well-balanced behavior between the push and pull cycles. Also, the linear portion of the curve was seen until a lateral load level of 222 kN (50 Kips). At the peak failure load, the damage to the masonry system was nearly catastrophic (see Figure 7). The large piers exhibited massive diagonal tension cracks as well as bed joint sliding in several locations. The entire front half of the structure moved away from the remaining portions, leaving the structure in two pieces. As a result, the lateral stiffness of the structure was near zero, although the gravity load was maintained at failure.

Due to the damage incurred, it was debated whether there is a good reason to spend the effort repairing the structure. In a real repair project these walls would have been demolished, and rebuilt. However, for research purposes, it was decided to repair the building and retest it. Due to the extent of the movement, the mortar joints were repointed before the CFRP composite NSM bars were installed. A total of sixteen bars were placed vertically in the shear walls, with 5 rods installed in each of the large piers, and the remainder divided between the small piers. The response of the repaired structure to the laterally applied load was similar to the unreinforced test (see Figure 8). The stiffness of the structure began to diminish at the load level of 222 kN (50 kips). At the load level of 400 kN (90 kips), the original failure surface of the unreinforced building had reappeared and the CFRP NSM bars then carried the lateral load, mostly through tension and dowel action. Failure occurred at a load level of 490 kN (110 kips) with an associated lateral displacement of 16.51 mm (0.65"). The peak load level was nearly identical to the URM brick results, and the energy dissipation for the repaired building was significantly improved. This was a remarkable result, considering the fact that the URM building was catastrophically damaged and only minimal NSM reinforcement was used.

The retrofitted clay masonry structure received GFRP NSM rods, as compared to the CFRP bars used for the repaired building. Based on the results of small component tests performed, it was decided to use only fourteen rods in the retrofitted configuration, with four rods in each of the large piers and three in each of the small piers. Fewer rods were used to promote cracking in the shear walls and thereby increase ductility. The lateral load level achieved was 534 kN (120 kips), which represents an increase of approximately 10% over the unreinforced structure. The associated lateral displacement was 5.11 mm (0.201"), which represented only a nominal increase as compared to the URM building results.

There was a notable increase in the energy dissipation of the repaired system as compared to the unreinforced test, especially in the pull cycle. The added CFRP composite bars embedded within the shear walls not only fully returned the capacity of the shear walls, but the energy dissipation of the repaired structure were far superior to the unreinforced structure. The amount of energy that could be dissipated was increased by a factor of nearly 3, as compared to the baseline structure which had a calculated strain energy of 1652 Nm (14.62 kip-in) based on the tension quadrant of the backbone curve. The presence of the CFRP bars provided an increase in the ductility and also greatly enhanced the performance of the structure.

The retrofitted structure showed increases in terms of ductility and energy dissipation. As previously mentioned, the GFRP composite system used was based on small component testing as was anticipated to provide nominal gains. Despite this, a 70% increase in ductility and 100% increase in strain energy, based on the tension quadrant of the backbone curve.

The strain measured during the tests of both the repaired and retrofitted structures were such that the ultimate tensile strain of the GFRP rods and CFRP bars was reached at failure. Prior to failure, the strain measurements were approximately 0.5% and 0.3% in the CFRP bars and GFRP rods, respectively. The strain levels are below the ultimate capacities of 1.5% and 1.9% for the CFRP bars and GFRP rods, respectively. In the case of the repaired structure, the strain level of 0.5% was maintained, while the structure was degrading and a clear slip plane had developed. Significant dowel action was observed in the CFRP bars. The mode of failure for the retrofitted structure was shearing of the GFRP rods (see Figure 9).

CMU-Configuration II-Phase I

The asymmetric CMU structure was repaired with a GFRP grid system. One layer of TechFab MeC-Grid® G15000-BX1 glass grid, with an opening size of roughly 12.7 mm x 6.35 mm (0.5" x 0.25"), was placed in a polyurea matrix provided by Bondo Corporation on each shear wall. The layout of the GFRP grid system was of a strut-and-tie configuration.

The undamaged, unreinforced structure failed under an applied lateral load of 445 kN (100 kips). The associated lateral displacement was 6.35 mm (0.25"). Due to the asymmetry of the structure, torsion was induced. The torsional displacement was 3.81

mm (0.15") with a of 2.5 mm (0.1") displacement variant between the shear walls. The failure mechanism for the structure was diagonal tension cracking that occurred in all piers starting at an applied lateral load of 89 kN (20 kips). The tension failure was primarily localized in the mortar joints.

The GFRP grid system drastically enhanced the performance of the damaged structure. At the point of failure, the lateral displacement was three times higher and the lateral load had increased by 25%. The flexible nature of the polyurea resin allowed the structure to achieve even more rotation under the applied lateral load. At failure, extensive diagonal tension cracking occurred in a push cycle (see Figure 10). Composite failure was observed in one of the diagonal elements, where the diagonal tension force clearly exceeded the capacity of the glass FRP grid system.

A comparison of the backbone curves, for both the unreinforced and the repaired structures (see Figure 11) reveals that the repaired structure significantly outperformed the unreinforced structure with respect to both strength and deformation. The energy associated with the cyclic lateral force was drastically increased with only one layer of the GFRP composite grid system. The associated strain energies are quite impressive, with an increase of approximately 425%, based on the compression quadrant of the hysteretic curve.

SUMMARY OF EXPERIMENTAL RESULTS

Several masonry structures were load tested under a constant gravity load and increasing cyclic lateral load. It has been shown that a minimal amount of FRP composite can drastically improve the performance of unreinforced masonry systems in both repair and retrofit applications. GFRP wet lay-up, GFRP and CFRP near surface mounted rods and bars, and GFRP grid systems were tested. An increase in strength, pseudo-ductility and energy dissipation were observed in all cases. Due to the pre-damaged (and significantly cracked) nature of the repaired buildings, these structures outperformed the retrofitted ones in terms of energy dissipation. It is important to note that, although substantial damage was produced in all building specimens, the gravity load was not lost during any testing. Since the applied gravity load modeled two additional floors, it was imperative that it be maintained, even at failure.

The externally applied GFRP wet lay-up composite applied to the symmetric CMU structures resulted in a more ductile and stronger system, as compared to the unreinforced configuration. The addition of the GFRP laminate resulted in a 67% increase and a 541% increase in strength and energy dissipation calculated by integration of the backbone curves, respectively (see Table 1). Failure modes in all cases were diagonal tension cracking in all piers. Since masonry substrate failure governed in all cases, the level of strain measured was much lower than the ultimate tensile of the GFRP laminate, ranging from 0.5% to 0.8% and resulting in a safety factor of 3-4 against composite rupture.

The catastrophically damaged symmetric brick structure was repaired using

CFRP NSM bars and regained 100% of the strength of the system. The addition of the composite bars increased the lateral displacement of the system by a factor of nearly 3.5 (see Table 2), resulting in a 300% increase in strain energy dissipation. The retrofitted structure received GFRP NSM rods, which also increased the performance of the building specimen, although not to the same degree. As with the previous structure, the method of failure was diagonal tension cracking, present in all piers of the structure. The measured strain values in both the CFRP bars and GFRP NSM rods ranged from 0.3% for retrofitted system to 0.5% for the repaired system prior to masonry failure.

Similar results were observed for the asymmetric CMU structure as compared with symmetric CMU structure. Increases of 420% in energy dissipation and 270% were measured during the test of the repaired structure. The degree of rotation of the structure was increased due to the application of the polyurea-GFRP grid composite system. Similar to the other structures, diagonal tension was the observed failure mode. At failure, composite rupture was observed at one location where the tension force exceeded the capacity of the GFRP grid.

Future research may include optimization of the FRP systems used and exploration of new systems. Since only two building configurations were examined, the possibility for a broader investigation exists as well as development of an anchoring methodology, if a positive foundation connection does not exist.

CONCLUSIONS

The experimental program has shown that FRP composite laminates can restore and significantly enhance the seismic performance of unreinforced masonry structures, in a damaged and undamaged configuration. Multiple FRP systems were utilized and all were well below ultimate strain values for each respective laminate at failure of the masonry building specimens.

ACKNOWLEDGEMENTS

The authors would like to thank the National Science Foundation (Award Number CMS-0324548) for funding this project. Also, the authors would like to thank their composite and masonry industry partners, who have made significant material and labor contributions to the project. These included the Carolina Concrete Masonry Association (CCMA), the Brick Industry Association (BIA), the North Carolina Masonry Contractors Association (NCMCA), Sika and Unitex Chemicals.

REFERENCES

Albert, M., Elwi, A. and Cheng, J.J., 2001, "Strengthening of Unreinforced Masonry Walls Using FRPs," *Journal of Composites for Construction*, May, V. 5, No. 2, pp. 76-84.

Al-Chaar, G.K. and Hassan, H.A., 1999, May 10-12, "Seismic Testing and Dynamic

Analysis of Masonry Bearing and Shear Walls Retrofitted with Overlay Composite," *Proceeding of ICE '99*, Cincinnati, OH, 1999 (Session 4-B), pp. 1-10.

Ehsani, M.R. and Saadatmanesh, H., 1996, "Seismic Retrofit of URM Walls with Fiber Composites," *The Masonry Society*, December, V. 14, No. 2, pp. 63-72.

Federal Emergency Management Agency, November 2000, *Prestandard and Commentary for the Seismic Rehabilitation of Buildings*, FEMA 356, Section 2.8.3 pg 2-28, Washington, D.C.

Gergely, J., and Young, D.T., October 2001, "Masonry Wall Retrofitted with CFRP Materials," *Proceedings of the 1st International Conference on Composites in Construction*, Porto, Portugal, 2001, Figueiras et al (eds), Swets and Zeitlinger, pp. 565-569.

Hamid, A.A., El-Dakhakni, W.W., Hakam, Z.H. and Elgaaly, M., 2005, "Behavior of Composite Unreinforced Masonry – Fiber-Reinforced Polymer Wall Assemblages Under In-Plane Loading," *Journal of Composite for Construction*, V. 5, No. 1, pp. 73-83.

Hamilton III, H.R. and Dolan, C.W., 2001, "Flexural Capacities of Glass FRP Strengthened Concrete Masonry Walls," *Journal of Composites for Construction*, V. 5, No. 3, pp. 139-145.

Hamoush, S., McGinley, M., Mlakar, P., Scott, D. and Murray, K., 2001, "Out-of-plane Strengthening of Masonry Walls with Reinforced Composites," *Journal of Composites for Construction*, August, V. 5, No. 2, pp. 139-145.

Marshall, O.S, Sweeney, S.C. and Trovillion, J.C., 1999, "Seismic Rehabilitation of Unreinforced Masonry Walls," *Proceedings of the Fourth International Symposium on Fiber Reinforcement for Reinforced Concrete Structures*, Batlimore, MD, 1999 (SP188-26), American Concrete Institute, pp. 287-295.

Moon, F.L., Yi, T., Leon, R.T. and Kahn, L.F., 2002, September 9-13, "Seismic Strengthening of Unreinforced Masonry Structures with FRP Overlays and Post-tensioning," *Proceeding of the 12th European Conference on Earthquake Engineering*, London, England, 2002 (Paper Reference 123), pp. 10.

Silva, P., Myers, J., Belarbi, A., Tumialan, G., El-Domiaty, K., and Nanni, A., 2001, July 4-6, "Performance of Infill URM Wall Systems Retrofitted with FRP Rods and Laminates to Resist In-Plane and Out-of-Plane Loads", *Proceeding of Structural Faults and Repairs*, London, UK, 2001, pp. 1-12.

Table 1 – Energy dissipation (ε-energy) comparisons throughout experimental testing. (Units: N-m [k-in])

Structure	Side	Unreinforced	Repaired	% Increase	Retrofitted	% Increase
Brick-Symm	Comp.	1365 [12.08]	3384 [29.95]	148	1845 [16.33]	35
	Tension	1652 [14.62]	6538 [57.87]	296	3342 [29.58]	102
CMU-Symm	Comp.	1711 [15.14]	10960 [97.0]	541	5511 [48.78]	222
	Tension	1249 [11.05]	6843 [60.57]	448	2616 [23.15]	110
CMU-Asymm*	Comp.	1685 [14.91]	8779 [77.7]	421	---	---
	Tension	2281 [20.19]	9711 [85.95]	326	---	---

*Based on displacement measurements taken from perforated shear wall.

Table 2 – Pseudo-ductility comparisons of unreinforced, repaired and retrofitted structures. (Units: mm [in])

Structure	Side	Unreinforced	Repaired	% Increase	Retrofitted	% Increase
Brick-Symm	Comp.	4.43 [.18]	8.63 [.34]	95	5.29 [.21]	19
	Tension	5.02 [.20]	17.29 [.68]	244	8.63 [.34]	72
CMU-Symm	Comp.	6.05 [.24]	22.34 [.88]	269	12.28 [.48]	101
	Tension	4.65 [.18]	14.27 [.56]	207	7.21 [.28]	58
CMU-Asymm*	Comp.	5.49 [.22]	20.40 [.80]	272	---	---
	Tension	7.09 [.28]	20.0 [.79]	182	---	---

*Based on displacement measurements taken from perforated shear wall.

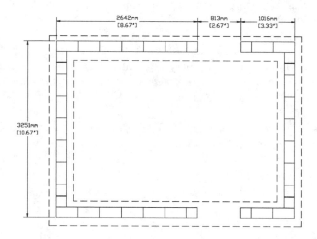

Figure 1 – Plan view of the symmetric configuration (asymmetric not shown).

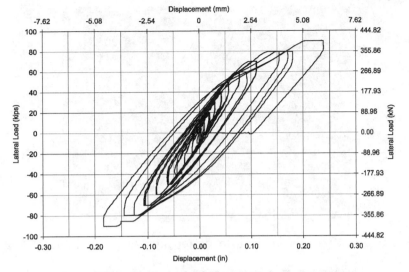

Figure 2 – Hysteretic curve for the unreinforced, symmetric CMU structure.

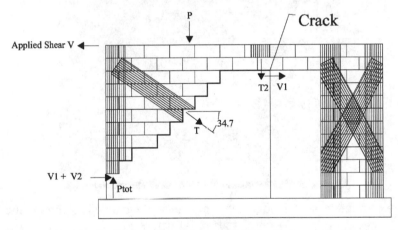

Figure 3 – Free body diagram from determination of composite layout using a strut and tie configuration.

Figure 4 – GFRP composite layout for the repaired symmetric CMU structure.

Figure 5 – Damage sustained during experimental load testing of the symmetric retrofitted CMU structure.

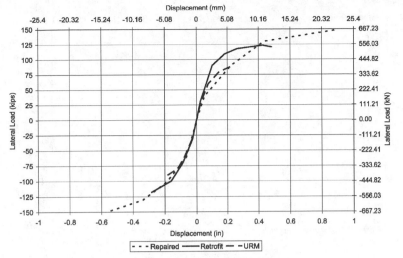

Figure 6 – Idealized backbone curve for the unreinforced, repaired and retrofitted symmetric CMU structures.

Figure 7 – Damaged incurred during experimental testing of the symmetric unreinforced brick structure.

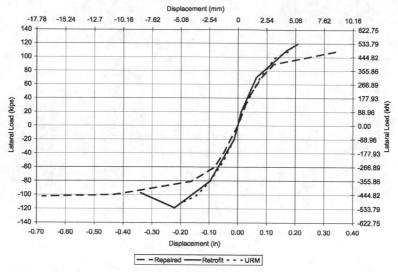

Figure 8 – Backbone curves for unreinforced, repaired and retrofitted symmetric brick structures.

Figure 9 – Diagonal tension failure and GFRP NSM rod rupture of the symmetric retrofitted brick structure.

Figure 10 – Damaged induced by an applied 556 kN (125 kips) lateral force to a repaired asymmetric CMU structure.

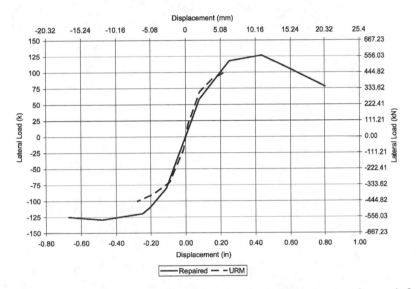

Figure 11 – Backbone curve comparison for asymmetric CMU structure in unreinforced and repaired condition.

In-plane Strengthening of Unreinforced Masonry Walls with Prestressed GFRP Bars

by P. Yu, P.F. Silva, and A. Nanni

Synopsis: An innovative mechanical device was developed and successfully used in this research program to apply low levels of prestressing force to glass fiber reinforced polymer (GFRP) bars. In addition, this device contains special features that allow for the application of GFRP bars to the retrofit of unreinforced masonry (URM) walls using the technique typically known as near surface mounted (NSM) reinforcement. Experimental studies were undertaken to investigate the feasibility of using this device for the retrofit of URM walls. A theoretical model was developed to calculate the shear capacity of URM walls retrofitted with prestressed NSM-GFRP bars. Numerical results matched well with the experimental results, which are described and presented in this paper.

Keywords: glass fiber-reinforced polymer; in-plane load; masonry walls; model; prestressing; strengthening

Piyong Yu is a graduate student at the Center for Infrastructure Engineering Studies (CIES) at the University of Missouri – Rolla (UMR). His research interests include the rehabilitation of masonry and reinforced concrete structures.

Pedro Franco Silva is an assistant professor of Civil Engineering in CIES at UMR. He received his PhD from the University of California-San Diego, Calif. His research interests include earthquake engineering and the seismic performance evaluation of bridge structures.

Antonio Nanni, FACI, is the V&M Jones Professor of Civil Engineering and Director of CIES at UMR. He is chair of ACI 437, Strength Evaluation of Existing Concrete Structures; and founder of 440, Fiber Reinforced Polymer Reinforcement; 544, Fiber Reinforced Concrete; 549, Thin Reinforced Cementitious Products and Ferrocement; the Concrete Research Council, and Joint ACI-ASCE-TMS Committee. His research interests include construction materials, their structural performance, and field application.

INTRODUCTION

It is well documented in the literature that unreinforced masonry (URM) structures have demonstrated poor performance during earthquakes. Consequently, effective retrofit techniques are needed to upgrade masonry walls.[1,2] Some of the methods used to strengthen masonry walls include: (a) filling of cracks and voids with injected grout; (b) stitching of cracks with metallic or fiber reinforced polymer (FRP) materials; (c) external jacketing; (d) external or internal post-tensioning with steel ties; (e) externally bonding of FRP laminates; and (f) installation of Near Surface Mounted (NSM) glass FRP (GFRP) bars.[1,2,3,4,5]

External post-tensioning with steel or carbon FRP ties has been applied on many historic masonry or stone structures to prevent expansion of existing cracks.[1] Figure 1 shows one such application using steel ties to provide the necessary horizontal confinement and strengthening for the building. However, this method is not very attractive because it alters the architectural integrity of the historic building, as clearly seen in Fig. 1. Other disadvantages such as steel corrosion have led researchers to investigate alternative techniques.

A strengthening scheme for URM walls using prestressed GFRP bars was studied in this program. The GFRP bars were installed inside grooves cut along the horizontal joints according to the NSM technique. Low levels of prestressing were then induced on the GFRP bars using a promising innovative system.[6] Since the bars are now hidden inside the joints of the walls and bonded with epoxy, which can be made to match closely the color of the existing walls, the architectural integrity of historic buildings can now be preserved. Furthermore, the bars are now protected from the environment.

In order to validate this retrofit technique six concrete URM walls were tested under a diagonal compressive loading test setup. Except for a control unit, one wall was strengthened with GFRP bars without prestress and the other four walls were strengthened with prestressed GFRP bars using the prestressing technique presented in this paper. This prestressing device was successfully used to apply low level of prestressing to GFRP bars. Different layouts of the bars and prestressing levels were studied, and test results show that the shear capacity of the strengthened walls was increased compared to the control wall. A failure theory, developed by Mann and Müller[7], was adopted to predict the shear capacity of the strengthened URM walls. Analysis indicates that this modified theory predicted the shear capacity of the tested walls within a reasonable degree of accuracy.

EXPERIMENTAL PROGRAM

Test matrix and specimens

Six URM walls with overall dimensions of 162.6 x 162.6 x 15.2 cm (64 x 64 x 6 in.) were built with concrete masonry blocks following a running bond pattern. The strengthening schemes of the five walls are described in Fig. 2 and Table 1, and are discussed next.

Wall W1 was constructed with no retrofit scheme and was used as the control unit to establish a baseline for performance. Wall W2 was strengthened with seven No. 6 (No. 2) GFRP bars, and the remaining walls were retrofitted with different amounts of No. 6 (No. 2) GFRP bars and at different prestressing levels. The walls were only strengthened on one side, because most likely in real retrofit applications this will be the case. Strain gages were attached to the GFRP bars to draw the strain profiles along the rods.

The walls were tested in a closed loop fashion, as shown in Fig. 3. Two hydraulic jacks were connected in parallel and positioned at one end of the walls to apply the desired load, which was measured by two load cells. The diagonal load was applied through two high strength steel rods positioned on both sides of the walls. The four corners of the walls were grouted to prevent crushing of the units in contact with the steel shoes (Fig. 2).

Prestressing of GFRP bars and strengthening of URM walls

The GFRP bars were bonded to the structure by adopting the retrofit technique designated in the literature as NSM method.[5] The bars were placed inside horizontal square grooves previously made and filled with epoxy. The grooves had dimensions of approximately 10 x 10 mm (3/8 x 3/8 in.) along the bed joints. The prestressing system was then fixed to the structure by tightening one steel nut with a regular wrench. The anchorage between the bars and the walls was achieved with a conventional prestressing steel chuck. Using a torque wrench and strain gages installed in the bar, the applied load was controlled to the desired strain level. The completed system after prestressing and with the anchors still in place is depicted in Fig. 4. Once the epoxy inside the grooves was properly cured, the bars were cut outside of the walls where PVC pipes are located, and the anchors removed for other applications. Further details can be found elsewhere.[6]

When exposed GFRP bars are stressed, a distinctive stress corrosion reaction results in the formation of sharp cracks causing significant material strength degradation. Many industrial failures of GFRP composites have been attributed to stress corrosion due to acid attack.[8] However, in the test and potential in situ applications, the prestressed GFRP bars are/will be protected from the environment. Therefore, stress corrosion is not a concern within this type of research application.

Material properties

GFRP bars – The average tensile strength of the bars were 1023.9 MPa (148.5 ksi).

Epoxy – The epoxy used for bonding the GFRP bars to the URM walls was a two-component general purpose adhesive with 100% solids and a non-sag consistency. The manufacturer specified properties were: compressive strength 86.2 MPa (12.5 ksi), tensile strength 27.6 MPa (4.0 ksi), elastic modulus 3.1 GPa (450 ksi), and elongation at break 1%.

Mortar – Type N mortar (ASTM) was used to build the walls and was available in bags that contained a dry premixed composition of cement and sand. The compressive strength, determined by five specimens with nominal dimensions of 51 x 51 x 51 mm (2 x 2 x 2 in.), was 5.7 MPa (823 psi).

Concrete blocks – The blocks used for the walls had nominal dimensions of 152 x 203 x 406 mm (6 x 8 x 16 in.). The net area compressive strength was 16.8 MPa (2.4 ksi), averaged from four samples. The modulus of elasticity was calculated at 15.1 GPa (2.2 Msi) according to the Masonry Standard Joint Committee Specification (MSJC).[9] All of the material properties are shown in Table 2.

EXPERIMENTAL RESULTS

Stress of the GFRP bars

Prestress losses obtained from strain gages mounted on the GFRP bars are shown in Fig. 5, where the left and right vertical axis corresponds to the ratio of the prestress remained in the GFRP bars over the tensile strength of the bars and the ratio of the prestress over the initial applied prestress, respectively. Referring to this figure it is clear that more than 90% of the initial stress was maintained in the bars. This observation was typical for recorded results throughout all strain gages in all walls.

Failure modes of the walls

In all walls the ultimate condition was reached in a brittle mode shortly after the first crack was observed. For wall W1, a typical shear friction failure mode was observed running through the entire diagonal direction of the wall (Fig. 6). Similarly, in all the strengthened walls a shear friction failure mode was observed, but the diagonal crack pattern varied from wall to wall. Part of the strengthened walls at ultimate condition is shown in Figs. 7, 8, and 9. In addition, the average crack width of the strengthened side was smaller than that on the unstrengthened side. This type of failure is well documented in the literature. [7,10]

Failure loads of the walls

The applied force, F, at ultimate and the corresponding calculated shear capacity, V, in all walls are summarized in Table 3. The experimental and calculated shear capacity is further shown in Fig. 10. Compared with the control wall, the shear capacity of the strengthened walls increased significantly, and is further discussed in the next section.

THEORETICAL ANALYSIS OF THE SHEAR CAPACITY

Failure mechanism and shear capacity of the plain URM walls: V_m

Prevalent failure modes of URM walls typically include shear friction, shear slide, and compression failure. [7,10,11]

<u>Shear friction failure along the bed and head joints</u> -- $V_{s,f}$ –Mann and Müller [7] developed a failure theory to explain the behavior of masonry walls subjected to shear and compression stresses, which were assumed to be uniformly distributed along the entire length of the joints. In addition, the strength of the head joints was neglected and the shear stress was transferred only through the bed joints. According to Mann and Müller, the horizontal shear stress, τ_m, acting on the blocks along the bed joint creates a moment and needs to be balanced by a vertical couple, as shown in Fig. 11a. The vertical stress distribution is then modified by this stress couple. Referring to Fig. 11a, the horizontal surface of the block is equally divided into two parts namely: I and II. The modified compressive stresses on half of parts I and II are, respectively [7]

$$\sigma_{1,2} = \sigma_y \pm \frac{2h\tau_m}{b} \tag{1}$$

Where, h and b are the height and the width of masonry unit, respectively. According to friction theory, the first crack initiates in segment II along the bed joints, where the vertical stress σ_2 is lower. The shear strength of the bed joints is further expressed as [7]

$$\tau_m = \tau_m^0 + \mu_m \sigma_y \tag{2}$$

Where

$$\tau_m^0 = \frac{\tau_0}{1 + 2\mu\frac{h}{b}} \text{ and } \mu_m = \frac{\mu}{1 + 2\mu\frac{h}{b}} \tag{3}$$

Where, τ_0 is the shear bond strength of the mortar joint and μ is the coefficient of shear friction.

According to the test setup (see Fig. 3), a typical compression strut develops along the diagonal direction of the walls (Fig. 12). Within the strut, the vertical and horizontal stresses were assumed to be uniformly distributed and one typical block inside the shaded region is shown in Fig. 11b, where the compressive stress at the head joints due to the diagonal load is shown as well. In this paper, the shear bond strength of the mortar joint τ_0 was estimated at 3% of the masonry compressive strength f'_m, and the coefficient of shear friction μ was equal to 0.3, as proposed by Paulay and Priestley.[11]

In Fig. 12, the width of the compressive strut W is estimated by $\dfrac{L}{4\cos\theta}$,[11] where L is the width of the masonry walls and θ is the angle between the diagonal and the horizontal direction. The horizontal projection of this strut is (Fig. 12)

$$D = \frac{L}{2\sin 2\theta} \leq L \tag{4}$$

The vertical compressive stress σ_y on the blocks inside the strut is (Figs. 12 and 13)

$$\sigma_y = \frac{V\tan\theta}{tD} \tag{5}$$

Where, t is the net thickness of the URM walls, and V is the horizontal component of the diagonal load F.

Similar to Mann and Müller's analysis, the first crack occurs when the shear stress in segment II reaches the critical value of τ_m inside the strut in the test (Fig.12). After the first horizontal crack occurs, more cracks may occur in segment II inside the strut and develop in a stepped form, passing through both of the bed and head joints (as shown in Figs. 6, 7, 8, and 9).

A free body diagram of the URM walls is shown in Fig. 13. When the tensile strength of the head joints is neglected, the horizontal force equilibrium at ultimate condition is given by (Fig. 13)

$$V_{s,f} = \tau_m Lt \tag{6}$$

Where, τ_m is defined in Eq.2. Therefore, the ultimate shear capacity based on the shear friction failure is

$$V_{s,f} = \frac{\tau_m^o LDt}{D - \mu_m L\tan\theta} \tag{7}$$

<u>Shear slide failure of bed joints</u>: $V_{s,s}$ – The horizontal crack, originated at segment II of the bed joints, propagates to the bed joints outside of the diagonal strut at the same level. The wall then separates and slides along the cracked joints. The horizontal force equilibrium is

$$V_{s,s} = \tau Dt + \tau_0(L - D)t \tag{8}$$

Where, τ is the average shear stress inside the diagonal strut and is estimated as

$$\tau = \tau_0 + \mu\sigma_y \tag{9}$$

Therefore, the ultimate shear capacity based on the shear slide failure is

$$V_{s,s} = \frac{\tau_0 Lt}{1 - \mu \tan\theta} \qquad (10)$$

<u>Compression failure of the diagonal strut</u>: V_c – For masonry infilled frames, when compression failure of the infill masonry wall occurs the diagonal load is [11]

$$C = \frac{2}{3} Z t f'_m \sec\theta \qquad (11)$$

Where, Z is the contact length of the concrete frame to the infill walls. Therefore, the ultimate shear capacity based on diagonal compression failure is

$$V_c = \frac{2}{3} Z t f'_m \qquad (12)$$

The ultimate shear capacity estimated for the tested walls was then governed by the minimum failure load of these three failure modes

$$V_m = \min(V_{s,f}, V_{s,s}, V_c) \qquad (13)$$

The normalized shear capacity V_m/Ltf'_m is shown in Fig. 14. For walls with lower aspect ratio H/L, the walls fail with shear friction mode and the shear capacity is determined by Eq.7. With an increase in the aspect ratio, the shear slide failure tends to govern and the shear capacity is computed with Eq.10. In addition, the failure of crushing of the masonry walls along the diagonal compression strut was evaluated in terms of Eq.12 for two values of Z/H (0.10 and 0.20). It is shown that crushing of the compression strut only occurs when Z/H is near or below 0.10 (Fig. 14), which is not practical in most cases.

Contribution of GFRP bars to shear capacity of the strengthened URM walls: V_F

Assuming the strengthened walls fail based on the shear friction mode, the increase of the shear capacity due to the prestressed GFRP bars can be divided into two components: (1) contribution of the prestress of the bars $V_{F,P}$; (2) conventional contribution of the bars $V_{F,N}$.

<u>Contribution of the prestress</u>: $V_{F,P}$ –The prestress increases the horizontal pressure along the head joints and the load to separate the adjacent units in the horizontal direction. Consequently, the shear capacity of the walls increases. The contribution of the prestress to the shear capacity $V_{F,P}$ is the summation of the prestress of all bars that remained after prestress losses

$$V_{F,P} = mnA_f f_{fu} \qquad (14)$$

Where, m and n are the prestress after losses (in percentage) and quantity of the prestressed GFRP bars, respectively.

Conventional contribution: $V_{F,N}$ – According to MSJC the contribution of reinforcement to the shear strength of beams, piers, and column is [9]

$$V_{F,N} = 0.5nA_f f_{fu} \tag{15}$$

In the test, the highest increase of the recorded strain of the GFRP bars during testing was approximately 20% of the ultimate strain. The coefficient in Eq.15 was then modified to be 0.2 in this research. Therefore, the shear contribution of the prestressed GFRP bars is

$$V_F = (m + 0.2)nA_f f_{fu} \tag{16}$$

Shear capacity of URM walls strengthened with prestressed GFRP bars: V_W

If the shear friction failure mode governs the failure of plain URM walls, then

$$V_m = V_{s,f} = \min(V_{s,f}, V_{s,s}, V_c) \tag{17}$$

Correspondingly, the shear capacity of strengthened walls is the summation of the shear force taken by the plain URM walls and the GFRP bars

$$V_W = V_m + V_F \le V_c \tag{18}$$

However, if the shear slide or the compression failure mode governs the failure of the unstrengthened walls, the overall failure mode will not be changed due to the strengthening schemes. For conservative considerations in those two cases, V_F is taken as zero and the shear capacity is

$$V_W = V_m \tag{19}$$

Validation of the analytical model with the experimental results

The following is a list of values that were used in the analytical prediction of the shear capacity of the tested URM walls in this program. The masonry block width b was 40.6 cm (16 in.) and the height h was 20.3 cm (8 in.). The URM walls had a net thickness t of 7.6 cm (3 in.), a width L of 293.0 cm (64 in.), and a height H of 293.0 cm (64 in.). The angle θ between the diagonal and the horizontal directions was equal to 45°. In the test, the contact length between the infill wall and frame Z was defined to be vertical projection of the diagonal strut at the corners of the walls $\dfrac{W}{2\cos\theta}$, and it was equal to 40.6 cm (16 in.) (see Fig. 12).

Wall W1 (plain masonry wall) – The shear capacity is evaluated in terms of
1. $V_{s,f}$ =89.0 kN (20.0 kips)
2. $V_{s,s}$ =89.0 kN (20.0 kips)
3. V_c =345.9 kN (77.8 kips)
4. V_m = 89.0 kN (20.0 kips)

These values indicate that this unit may fail with shear friction failure along the diagonal or shear slide along the bed joints, because $V_{s,f}$ is equal to $V_{s,s}$. In the test, the failure occurs with stepped cracks along the diagonal (Fig. 6).

Walls W2 to W6 (strengthened masonry walls) – Since the failure mode of plain URM walls is shear friction, the contribution of the GFRP bars needs to be considered. The shear capacity of the strengthened walls was:

1. V_F =46.3 kN (10.4 kips), $V_W = V_{s,f} + V_F$ =135.2 kN (30.4 kips) for wall W2

    V_F =43.1 kN (9.7 kips), $V_W = V_{s,f} + V_F$ = 132.1 kN (29.7 kips) for wall W3

    V_F =41.8 kN (9.4 kips), $V_W = V_{s,f} + V_F$ =130.8 kN (29.4 kips) for wall W4

    V_F =81.8 kN (18.4 kips), $V_W = V_{s,f} + V_F$ =170.8 kN (38.4 kips) for wall W5

    V_F =92.5 kN (20.8 kips), $V_W = V_{s,f} + V_F$ =181.5 kN (40.8 kips) for wall W6

2. V_c =345.9 kN (77.8kips)

The theoretical shear capacities are compared with the experimental values in Figs. 10 and 15 and Table 3.

CONCLUSIONS

The following conclusions were drawn from this research

1. Prestress losses of the NSM GFRP bars were relatively low, as more than 90% of the initial strains were maintained.

2. Analytical model predicted that the aspect ratio, H/L, can have a significant impact on the failure modes of both unstrengthened and strengthened URM walls, but this must be further validated.

3. The experimental shear capacity of the URM walls strengthened with the prestressed GFRP bars increased significantly.

4. The theoretical model gave a reasonable agreement of the shear capacity of the URM walls with the test results. However, further research must be undertaken.

ACKNOWLEDGEMENTS

The financial support provided by the National Science Foundation Industry/ University Research Center on Repair of Buildings and Bridges with Composites (RB2C) based at the University of Missouri-Rolla is greatly appreciated.

NOTATION

A_f = cross-sectional area of one GFRP bar

b = length of one concrete block

C = failure load of the walls when the diagonal compression failure occurs

D = horizontal width of the diagonal compression strut

F = diagonal compressive load of the URM wall

f_{fu} = ultimate strength of one GFRP bar

f_p = prestress remained in the GFRP bars after relaxation

f_{pu} = the initial prestress of the GFRP bars before relaxation

f'_m = compressive strength of the masonry

h = height of one concrete block

H = height of the URM walls

L = width of the URM walls

m = prestress over tensile strength of one GFRP bar (in percentage)

n = quantity of the GFRP bars

t = net thickness of the URM walls

V = horizontal component of the diagonal compressive load F

V_c = failure load of the walls when the compression failure occurs

V_F = contribution of the GFRP bars to shear capacity of the walls

$V_{F,N}$ = conventional contribution of the GFRP bars to shear capacity of the walls

$V_{F,P}$ = contribution of prestress of prestressed GFRP bars to shear capacity of the walls

$V_{s,f}$ = failure load of the walls when the shear friction failure occurs

$V_{s,s}$ = failure load of the walls when the shear slide failure occurs

V_m = shear capacity of the plain URM walls without strengthening

V_W = shear capacity of the strengthened URM walls with the GFRP bars

W = width of the diagonal compression strut

Z = contact length of the concrete frame to the infill walls

θ = the angle between the diagonal load F and the horizontal direction

μ = coefficient of internal friction of the blocks

μ_m = modified coefficient of internal friction of the blocks

σ_1 = vertical compressive stress on the segment I of the blocks inside the diagonal strut

σ_2 = vertical compressive stress on the segment II of the blocks inside the diagonal strut

σ_x = average horizontal compressive stress on blocks inside the diagonal strut

σ_y = average vertical compressive stress on blocks inside the diagonal strut

τ = shear strength of the mortar joints

τ_o = shear bond strength of the mortar joints

τ_m = modified shear strength of the mortar joints

τ_m^0 = modified shear bond strength of the mortar joints

REFERENCES

1. Triantafillou,T.C., and Fardis, M.N, 1997, "Strengthening of historic masonry structures with composite materials", *Material and Structures*, V. 30, pp. 486-496

2. EI-Dakhakhni, W.W, Hamid, A.A, and Elgaaly, M., 2004, "Seismic Retrofit of Concrete-Masonry-Infilled Steel Frames with Glass Fiber-Reinforced Polymer Laminates", *Journal of Structural Engineering*, Vol. 130, pp. 1343-1352.

3. Tumialan, J. G., Galati, N., and Nanni, A., 2003, "Fiber-Reinforced Polymer Strengthening of Unreinforced Masonry Walls Subjected to Out-of-plane Loads", *ACI Structural Journal,* V.100, No.3, May-June, pp. 321-329.

4. Tumialan, G., Morbin, A., Nanni, A., and Modena, C., 2001, "Shear Strengthening of Masonry Walls with FRP Composites", *Composites 2001 Convention and Trade Show*, Composites Fabricators Association, Tampo, FL, October 3-6, CD-ROM.

5. De Lorenzis, L., Nanni, A., and La Tegola, A., 2000, "Bond of Near Surface Mounted FRP rods in Concrete Masonry Units", *Proceedings of the Seventh Annual International Conference on Composites Engineering (ICCE/7)*, Denver, Colorado, pp. 3-4.

6. Yu, P., Silva, P.F., and Nanni, A., 2004, "Innovative Mechanical Device for the Innovative Mechanical Device for the Post-Tensioning of GFRP Bars for Masonry Type Retrofit Applications", *Experimental Mechanics,* V. 44, No.3, June, pp. 272-277.

7. Mann, W., and Müller, H., 1982, "Failure of Shear-Stressed Masonry-An enlarged Theory, Tests and Application to Shear Walls", *Proceedings of the British Ceramic Society*, 30, pp. 223-235.

8. Hong, P.J., 1990, "A Model for Stress Corrosion Crack Growth in Glass Reinforced Plastics", *Composites Science and Technology*, 38 (1990), pp. 23-42.

9. Masonry Standard Joint Committee (MSJC) Specification, 2002, "Building code requirement for masonry structures (ACI 530-02/ASCE 5-02/TMS 402-02).''

10. Crisafulli, F.J., Carr, A.J., and Park, R., 1995, "Shear Strength of Unreinforced Masonry Panels", *Pacific Conference on Earthquake Engineering*, Australia, 20-22.

11. Paulay, T., and Priestley, M.J.N., 1992, "Seismic Design of Reinforced Concrete and Masonry Buildings", John Wiley & Sons, Inc., New York.

Table 1 – Summary of strengthening schemes

Wall	W1	W2	W3	W4	W5	W6
Quantity of GFRP bars	None	7	2	3	4	7
Prestress value (%)	--	0	45	22	42	20
Location of GFRP bars	--	1-7	3,5	2,4,6	1,3,5,7	1-7

*Refer to Fig. 2

Table 2 – Material properties

	Compressive strength f'_m MPa (ksi)	Tensile strength f_{fu} MPa (ksi)	Elastic modulus GPa (Msi)	Ultimate strain (%)
Mortar	5.7 (0.8)	--	--	--
Concrete block	16.8 (2.4)	--	15.1 (2.2)	--
Epoxy	86.2 (12.5)	27.6 (4.0)	3.1 (0.45)	1
GFRP bar	--	1023.9 (148.5)	--	--

Table 3 – Comparison of theoretical and calculated shear capacity V

Wall	Failure mode	Theoretical shear capacity V kN (kips)	Calculated shear capacity V^* kN (kips)	Experimental ultimate load F kN (kips)	$\dfrac{V}{V^*}$	Strengthened V^* / Control V^*
W1	Shear friction	89.0 (20.0)	98.3 (22.1)	138.8 (31.2)	0.90	1.00
W2	Shear friction	135.2 (30.4)	158.5 (35.7)	224.2 (50.4)	0.85	1.62
W3	Shear friction	132.1 (29.7)	138.4 (31.1)	195.7 (44.0)	0.95	1.41
W4	Shear friction	130.8 (29.4)	149.1 (33.5)	210.8 (47.4)	0.88	1.52
W5	Shear friction	170.8 (38.4)	149.1 (33.5)	210.8 (47.4)	1.15	1.52
W6	Shear friction	181.5 (40.8)	166.1 (37.4)	234.9 (52.8)	1.09	1.69

*This force was calculated from the experimental diagonal load F

Fig.1 – Externally strengthening of a stone building with steel ties

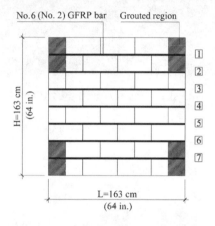

Fig.2 – Locations of GFRP bars in strengthened walls

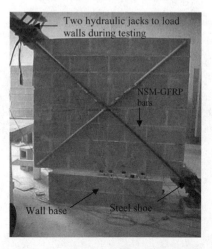

Fig.3 – Test setup of the URM walls

Fig.4 – Masonry wall strengthened with prestressed GFRP bars (W5)

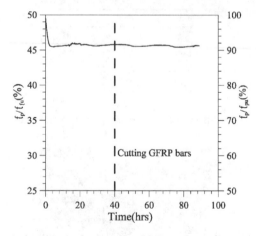

Fig.5 – Relaxation of prestress of GFRP bars (from Wall W3)

Fig.6 – Failure mode of wall W1

Fig.7 – Failure mode of wall W3

Fig.8 – Failure mode of wall W4

Fig.9 – Failure mode of wall W6

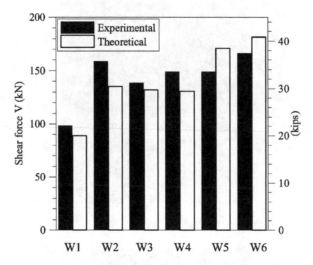

Fig.10 – Comparison of theoretical and calculated shear capacity *V*

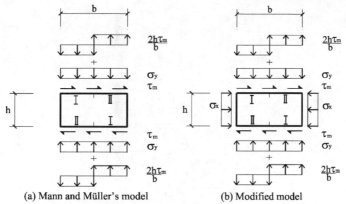

(a) Mann and Müller's model   (b) Modified model

Fig.11 – Stress analysis of one typical masonry block

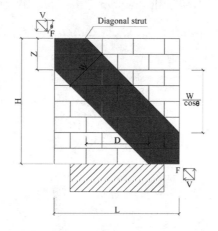

Fig.12 – Diagonal compression strut of the walls

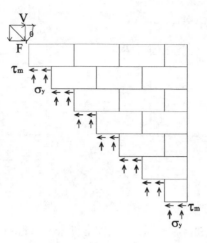

Fig.13 – Free body diagram of the plain URM walls

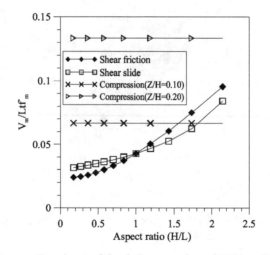

Fig.14 – Envelope of the failure modes of URM walls

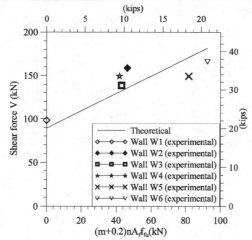

Fig.15 – Predicted shear capacity *V* versus experimental results

Cyclic In-Plane Shear of Concrete Masonry Walls Strengthened by FRP Laminates

by M.A. Haroun, A.S. Mosallam, and K.H. Allam

Synopsis: Cyclic in-plane shear tests were conducted on six full-scale walls built from reinforced concrete masonry units and strengthened by unidirectional composite laminates. Carbon/epoxy, E-glass/epoxy and pre-cured carbon/epoxy strips were placed on one or both sides of the walls. Each wall sample was loaded with a constant axial load simulating the gravity load, and incremental cyclic lateral shear loads were applied in accordance with the Acceptance Criteria (AC-125) of the International Code Council Evaluation Services (ICC-ES 2003). Displacements, strains and loads were continuously monitored and recorded during all tests. Evaluations of the observed strength and ductility enhancements of the strengthened wall samples are made and limitations of such retrofit methods are highlighted for design purposes.
Results obtained from current tests indicated that the limit-state parameter influencing strength gain of the FRP retrofitted walls was the weak compressive strength of the masonry units, especially at the wall toe where high compression stresses exist. Despite such a premature failure caused by localized compression damage of the masonry at the wall toe, notable improvement in their behavior was achieved by applying the FRP laminates to either one or two sides of the walls. However, it should be cautioned that available theoretical models may significantly overestimate the shear enhancement in the FRP strengthened walls, if other limiting failure modes are not considered.

Keywords: FRP laminates; in-plane cyclic loads; masonry walls; retrofit and repair

Medhat A. Haroun is Dean and AGIP Professor, School of Sciences and Engineering, American University in Cairo, Egypt, and Professor Emeritus, University of California, Irvine. His research deals with theoretical and experimental modeling of seismic behavior of structural systems: tanks, bridges, and buildings. Honors include UCI Distinguished Faculty Award, AUC Excellence in Teaching Award, and ASCE Huber Research Prize.

Ayman S. Mosallam joined the faculty of the University of California, Irvine, after serving as Research Professor at California State University, Fullerton. He is a registered Professional Engineer and has over than twenty-five years of experience in structural engineering with a particular interest in polymer composites, reinforced concrete, large-scale testing, seismic repair and rehabilitation, and code development.

Khaled H. Allam is a Registered Professional Civil Engineer at KPFF Consulting Engineers in Orange County, California. He obtained M.S degree from University of California, Irvine. His research interest is innovative retrofit systems for structures.

INTRODUCTION AND RESEARCH SIGNIFICANCE

Polymer composites have been accepted by the construction industry worldwide as structurally-efficient and cost-effective repair and rehabilitation systems. For the past decades, fiber reinforced polymer (FRP) composites have been successfully used to strengthen seismically-deficient and corroded reinforced concrete members, as well as masonry and wood members.

The 1994 Northridge, California, earthquake caused many structures to shift over the first story due to lack of adequate shear strength. Since then, there has been a significant research interest to study the seismic shear behavior of both reinforced and un-reinforced masonry walls. (Ehsani et al, 1999, Haroun and Ghoneam, 1997, Mosallam and Haroun, 2003, Mosallam et al, 2001). Reinforced masonry wall design is generally based on an elastic approach that focuses, mainly, on the linear behavior of the wall with no consideration of the effect of energy dissipation that occurs in the non-linear range. In addition, the majority of building codes have mainly concentrated on calculating the required vertical wall reinforcement to resist flexural moments. The small inadequate ratio of horizontal steel was shown to be incapable of resisting the seismic shear that occurred in recent seismic activities. The horizontal reinforcement considered for lateral load is not sufficient to resist high frequency earthquakes and provide ductility against the possibility of collapse. Consequently, there has been a need for utilizing more effective alternative techniques to increase the shear strength of these walls and to improve their ductility performance.

WALL SAMPLES AND TEST SETUP

The scope of this research program was to evaluate the in-plane shear behavior of masonry walls externally reinforced with FRP composite laminates. Six full-scale wall samples were tested under a combination of constant axial load with incremental lateral (push-pull) cyclic loads. As shown in Figure 1 and 2, each wall specimen was 72 in. (183 mm) high and 72 in. (183 mm) long, and constructed from one wythe of 6 in. x 8 in. x 16 in. (152 mm x 203 mm x 406 mm) hollow concrete blocks. Each wall has a base footing and a top loading reinforced concrete beam. The walls were fully grouted and detailed with five vertical reinforcing bars placed uniformly in the wall. These bars were continuous from the footing base to the top beam without any lap splice, and were strain gauged at the base-wall intersection level to capture the first yield of the steel bars. All wall specimens had a vertical steel reinforcement ratio of 0.54% with no horizontal reinforcement in the direction of the applied shear force to simulate a deficient and/or old wall construction. Four short dowels were distributed between the vertical steel bars at each interface between the wall and both the top loading beam and the footing.

As noted above, the wall specimens were built with a height-to-length aspect ratio of 1:1 to promote a shear dominated behavior under in-plane loading. Wall specimen number 1 was used as a control wall (as-built) whereas wall number 2 was cracked first and used for investigating repair techniques. The remaining four wall specimens were retrofitted with unidirectional carbon/epoxy laminates on one or two sides, E-glass/epoxy laminates on two sides, and carbon strips overlay on one side of the wall (Table 1).

All wall samples were built at the same time, and shared materials from the same batch. The reinforcing bars were grade 60 and were tested, according to ASTM standards, to measure the tensile strength as displayed in Table 2. Strength tests at 28 days on masonry prisms, grout cylinders, and mortar cylinders yielded 485 psi (3.34 MPa), 2750 psi (18.96 MPa), and 2120 psi (14.62 MPa), respectively. For the carbon/epoxy and E-glass/epoxy laminates, a specimen from each batch, 12 in. x 12 in. (304 mm x 304 mm), was fabricated and tested to ensure the same quality for all retrofitted specimens. All such specimens were tested to obtain their ultimate strength, modulus at yield and strain at ultimate strength as listed in Table 3.

The in-plane wall displacement was monitored by three displacement potentiometers located at height 24 in. (610 mm), 48 in. (1219 mm), and 71.5 in. (1816 mm), and the loading beam displacement was measured at 82 in. (2082 mm), all from the wall base level. For all tests using FRP laminates, strain gages were bonded on the external surface of the FRP to monitor the tensile/compressive strain.

GENERAL OBSERVATIONS

For all tested specimens, each wall was cycled laterally (in-plane) following a specified load-control regime. Once yielding has been achieved, a displacement control regime is adopted. At each load or displacement level, the wall is cycled for three identical cycles.

As-Built Wall

The control as-built wall was cyclically tested to failure and demonstrated a pure shear mode. The failure of the specimen was initiated by diagonal shear cracks and developed a diagonal strut action resulting in the crushing of the wall edge boundaries under compressive stresses. The wall cracking is displayed in Figure 3 and the hysteretic loops of the wall are shown in Figure 4 illustrating the sudden degradation of strength and stiffness of the as-built wall specimen at a low ductility level.

Repaired Wall

To demonstrate the use of composite laminates in the possible repair of a cracked wall, specimen number 2 was cyclically pre-cracked and then repaired and retested. In the pre-cracked wall specimen, two major localized damages appeared: diagonal shear cracks across the wall and local compression failure of the wall toe on one side. The diagonal cracks and the wall toe were first repaired with high strength epoxy resin, and then a single layer of carbon/epoxy composite laminate was applied on each side of the wall. In addition, a U-shaped laminate was applied at the pre-damaged toe.

The failure of the repaired wall was dominated by a combination of shear and flexural modes contrary to the single shear mode of the as-built wall. The primary mode of failure was attributed to the exceedance of the compressive strength at the end elements of the wall. This failure led the wall to lose its overall stiffness and become unable to resist any further lateral load. The addition of a U-shaped laminate at the cracked wall toe resulted in increasing the strength at this location, and consequently, it pushed the failure to occur at the un-reinforced toe on the other side of the wall. When the performance of the repaired wall is compared with that of the as-built wall, it becomes clear that the repair technique has improved the strength and energy dissipation of the wall. It not only succeeded in restoring the capacity of the original wall, but also increased it to a level 120 % of that of the original wall capacity. The energy dissipation observed for repaired specimen was also increased to 167 % of that of the control wall.

Retrofitted Walls

The predominant mode of failure in all single-side strengthened wall specimens was in the form of shear failure of the un-strengthened side of the wall. This shear failure was a combination of diagonal tension cracks as well as step cracks initiated at the base of the un-strengthened face. However, unlike the as-built specimen, single-side strengthened wall specimens suffered from another mode of localized failure in the form of a compression crushing of one of the wall toes. In fact, this localized failure mode at the wall toes was the controlling factor in determining the ultimate capacity of the single-side strengthened wall specimens.

The common mode of failure of all two-side FRP strengthened wall specimens was also compressive failure of the masonry units at the bottom ends (toes) of the wall specimens (Figure 5). The application of the composite laminates to the two sides of the wall specimens contributed an appreciable stiffness gain which was evident from the displacement profiles of such specimens. However, the overall usable strength gain was limited by the masonry compression properties rather than the ultimate tensile strength of the unidirectional FRP laminates.   This applies to all FRP strengthening systems evaluated in this study, including E-glass/epoxy wet lay-up laminates, carbon/epoxy wet lay-up laminates, and pre-cured unidirectional carbon/epoxy strips.   The premature compression failure of the wall toes resulted in appreciable shear and flexural stiffness degradations that was amplified by the loss of the grout confinement leading to local buckling of the vertical steel bars near the ends of the walls. It is recommended to develop optimized techniques to enhance the properties of masonry at the wall toes. One simple technique, which was adopted and was proven effective in the repair application in this study, is to apply an FRP U-laminate at the bottom ends and through the thickness of the wall.

One important observation noted from all retrofitted test specimens, at high stress level and just before the failure, was the development of white lines parallel to the fibers at the neighborhood of the mortar lines. This can be attributed to the development of large shear deformation at the mortar lines that resulted in fracture of the non-structural cross-stitches holding the unidirectional fibers, parallel to the applied in-plane loads. This can serve as an indicator for damages of either the masonry and/or the mortar hidden under the composite laminates of the FRP strengthened walls.

Unlike the wet lay-up strengthening systems, cohesive debonding of the ends of the pre-cured carbon/epoxy strips was observed at higher stress levels following the large deformation caused by the compression failure of the masonry units at the wall toes and the local buckling of the unconfined bottom length of the vertical steel bars near the damaged masonry.

SUMMARY OF EXPERIMENTAL RESULTS

Despite the premature failure caused by localized compression failure of the masonry at the wall toes, notable gains in strength, stiffness and ductility were achieved by applying the FRP laminates to either one or two sides of the walls. Adding a single carbon layer at each side of the pre-damaged wall specimen resulted in 20% strength gain as compared to capacity of the as-built specimen that was tested to failure. For a retrofitted wall specimen strengthened with a single ply of carbon/epoxy at both sides of the wall (Figure 6), the capacity was 130% of the ultimate capacity of the as-built specimen, while the capacity of the single-side carbon/epoxy retrofitted wall was 115% as compared to the control specimen. The E-glass/epoxy double-side retrofitted wall achieved a slightly less strength gain, where the ultimate capacity was 128% of the as-built wall capacity. For the wall specimen retrofitted with carbon/epoxy strips applied to a single side, the ultimate capacity was 118% of the as-built ultimate capacity. Table 4 presents a summary of the ultimate strength of all specimens tested in this study. A strength comparison between the different wall specimens is also presented in Figure 7.

The ductility of the carbon/epoxy repaired specimen was 1.7 times that of the as-built specimen. For the retrofitted specimens, the enhancement in the ductility ranged from 3.4 times that of the as-built in case of double-side carbon/epoxy retrofit to 6 folds in the case of pre-cured carbon/epoxy strips.

ANALYTICAL EVALUATION OF WALL CAPACITY

The shear capacity of a masonry wall strengthened by FRP laminates can be divided into two components

$$V = V_m + V_f \tag{1}$$

where V is the total shear capacity of the strengthened wall, V_m is the shear strength of the masonry wall alone, and V_f is the shear strength contributed by the FRP laminates.

According to Pauley and Priestley (1992), the shear strength of the masonry wall can be estimated using the following relation

$$V_m = v_m\, d_w\, t_w \tag{2}$$

where d_w is the effective length of the wall taken as 0.8 of the actual length L_w of the wall, t_w is the wall thickness, and v_m is the masonry shear stress which empirically depends on the masonry crushing strength, the applied axial load, and the wall's gross cross sectional area. Application of Equation 2 shows that the shear strength contributed by the masonry only for all walls under consideration is 50.5 kips (225 kN).

The shear strength of the carbon/epoxy laminate was calculated based on AC-125 ICC-ES Acceptance Criteria (2003). Accordingly, the shear strength enhancement for a rectangular wall section of length L_w in the direction of the applied shear force, with a laminate thickness, t_f, on two sides or one side of the wall at an angle, θ, to the wall axis is calculated from the following relations

For a <u>two-sided</u> retrofit $\qquad V_f = 2.0\, t_f\, \sigma_f\, L_w\, sin^2\theta \tag{3a}$

For a <u>single-sided</u> retrofit $\qquad V_f = 0.75\, t_f\, \sigma_f L_w\, sin^2\theta \tag{3b}$

in which $\qquad\qquad\qquad\qquad \sigma_f = 0.004\, E_f \leq 0.75\, \sigma_{uf} \tag{4}$

where σ_f is the stress developed in the laminates; E_f is the longitudinal modulus of elasticity of the FRP composite material; and σ_{uf} is the ultimate tensile strength of the laminates. The above equations may only be applied to FRP laminates. In the case of carbon/epoxy strips, the equation proposed by Zhao et al (2002) may be used. Table 5 shows the theoretically computed shear strength of the tested wall samples, if only shear enhancement is taken into consideration. These are clearly much higher than those observed experimentally. However, if flexural behavior of the wall is also considered (Allam 2002), lower and upper bounds of the lateral load that can be resisted by the walls were calculated at 71.5 kips (318 kN) and 120 kips (534 kN), respectively, which are more consistent with test observations.

CONCLUSIONS

The main purpose of this study was to evaluate the gain in the shear strength of reinforced masonry walls when repaired or retrofitted by FRP laminates. However, results obtained from current tests indicated that the limit-state parameter influencing the strength gain of the FRP retrofitted walls are the weak compressive strength properties of the masonry units, especially at the wall toes where high compression stresses exist.

Despite the premature failure caused by localized compression failure of the masonry at the wall toes, notable gains in strength, stiffness and ductility were achieved by applying the FRP laminates to either one or two sides of the walls. However, it should be cautioned that available theoretical models significantly overestimate the shear enhancement in the FRP strengthened walls. For this reason, serious modifications to these equations must be made to reflect the actual performance of the strengthened walls in shear and to consider other major limiting factors on the strength gain of such walls.

RECOMMENDATIONS

It is recommended to develop optimized techniques to enhance the properties of masonry at the wall toes. One simple technique, which was adopted and was proven effective in the repair application in this study, is to apply an FRP U-laminate at the bottom ends and through the thickness of the wall. For field applications, a slit can be made at the ends of the walls for about 1 to 2 feet (0.3 to 0.6 meters) above the footing or the floor level, where a thin wet lay-up laminate can be applied in a U-shape on both sides of the wall and through the wall thickness. In order to validate this concept, both experimental and analytical studies should be conducted.

ACKNOWELDGEMENTS

This project is part of a comprehensive ICC-ES evaluation program of the TufLam™ FRP composite strengthening systems. The authors would like to acknowledge the financial support of the Structural Composites Construction Incorporated (SCCI). Thanks are due to Dr. A. Hassan, Dr. H. Elsanadedy, Mr. J. Kiech and Mr. R. Kazanjy for their contributions to the experimental program.

REFERENCES

Allam, K. H. (2002), "In-Plane Cyclic Behavior of Concrete Masonry Walls Enhanced by Advanced Composite Laminates," Master Thesis, University of California, Irvine, USA.

Ehsani, M., Saadatmanesh, H., and Velazquez, J.L. (1999), "Behavior of Retrofit URM Walls under Simulated Earthquake Loading," Journal of Composites Construction, Vol. 3, pp. 134-142.

Haroun, M.A., and Ghoneam, E.H. (1997), "Seismic Performance Testing of Masonry Infilled Frames Retrofitted by Fiber Composite," Proceedings of the International Modal Analysis Conference, Vol. 2, Orlando, Florida, pp. 1650-1656.

International Code Council – Evaluation Services ICC-ES (2003), "Acceptance Criteria for Concrete and Reinforced and Unreinforced Masonry Strengthening Using Fiber Reinforced Polymer (FRP) Composite Systems," AC-125, Whittier, California.

Mosallam, A.S., and Haroun, M.A. (2003), "Seismic Strengthening of Unreinforced Thick Masonry Walls Using Polymer Composites," Proceedings of 10th Structural Faults and Repair Conference, London, United Kingdom (CD).

Mosallam, A., Haroun, M., Almusallam, T., and Faraig, S. (2001), "Experimental Investigation on the Out-of-Plane Response of Unreinforced Brick Walls Retrofitted with FRP Composites," Proceedings of 46th International SAMPE Symposium, Vol. 2, California, pp. 1364-1371.

Paulay, T., and Priestley, M. J. N., Seismic Design of Reinforced Concrete and Masonry Buildings, John Wiley & Sons, 1992.

Zhao, T., Zhang, C.I., and Xie, J. (2002), "Study and Application on Strengthening Cracked Brick Walls with Continuous Carbon Fiber Sheet," Proceedings of 1st International Conference on Advanced Polymer Composites for Structural Applications in Construction, Southampton, United Kingdom, pp. 309-316.

Table 1 -- Test Matrix

Test #	Wall #	Test Code	Test ID
1	1	WU1	Control (ultimate)
Description: As-built.			
2	2	WU2	Control (cracked)
Description: As-built to be repaired with carbon/epoxy laminates.			
3	2	WU2-C-R	Carbon/epoxy repair
Description: Single layer of carbon/epoxy laminate on each side of the repaired wall.			
4	3	W3-C-RT	Carbon/epoxy retrofit
Description: Single layer of carbon/epoxy laminate on one side of the retrofitted wall.			
5	4	W4-C-RT	Carbon/epoxy retrofit
Description: Single layer of carbon/epoxy laminate on each side of the retrofitted wall.			
6	5	W5-E-RT	E-glass/epoxy retrofit
Description: Two layers of E-glass/epoxy laminate on each side of the retrofitted wall.			
7	6	W6-CS-RT	Carbon strips retrofit
Description: Horizontal strips spaced 4 in. (101 mm), on center, on one side of the retrofitted wall.			

Table 2 -- Properties of Reinforcing Steel

Bar size	Yield Stress, ksi (MPa)	Ultimate Strength, ksi (MPa)
# 6	60 (414)	94 (648)

Table 3 -- Properties of FRP Composite Materials

Type	Thickness (t) inch (mm)	Ultimate Strength ksi (MPa)	Strain at Ultimate (μ strain)	Modulus of Elasticity ksi (GPa)
Carbon/epoxy	0.045 (1.14)	154 (1,061)	0.012	14×10^3 (96.5)
E-glass/epoxy	0.045 (1.14)	74 (510)	0.022	3.5×10^3 (24.2)
Carbon strips	0.047 (1.19)	420 (2,896)	0.018	22×10^3 (151.7)

Table 4 -- Summary of Ultimate Strength of Tested Wall Samples

Sample ID	Description	Ultimate Strength, kips (kN)
WU1	Control (ultimate)	83 (369.18)
WU2	Control (cracked)	62 (275.78)*
WU2-C-R	Carbon/epoxy repair (two sides)	100 (444.8)
W3-C-RT	Carbon/epoxy retrofit (single side)	95 (422.6)
W4-C-RT	Carbon/epoxy retrofit (two sides)	108 (480.38)
W5-E-RT	E-glass/epoxy retrofit (two sides)	106 (471.49)
W6-CS-RT	Carbon strips retrofit	98 (435.9)

*Maximum load at which the cracking test was stopped.

Table 5 -- Theoretical Shear Strength of Tested Wall Samples

Sample ID	Masonry Strength kips (kN)	FRP Strength kips (kN)	Total Strength kips (kN)
W3-C-RT		134.1 (596)	184.6 (821)
W4-C-RT	50.5 (225)	357.7 (1,591)	408.2 (1,816)
W5-E-RT		181.4 (807)	231.9 (1,032)

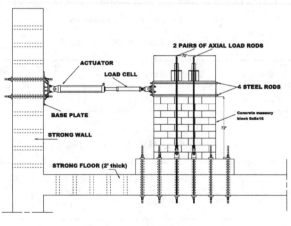

Figure 1 — Test Set-Up.

Figure 2 – Gravity and Cyclic Shear Loading on Typical Wall.

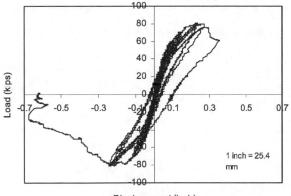

Figure 3 — Shear Failure of the As-built Wall Sample.

Figure 4 — Hysteretic Loops of the As-built Wall Sample.

Figure 5 — Premature Compression Failure of Masonry Units at Wall Toes
(a) Crack Initiation   (b) Local Buckling of the Far-End Vertical Steel Rebar.

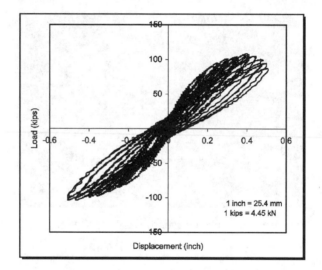

Figure 6 – Hysteretic Loops of Retrofitted Wall Sample W4-C-RT.

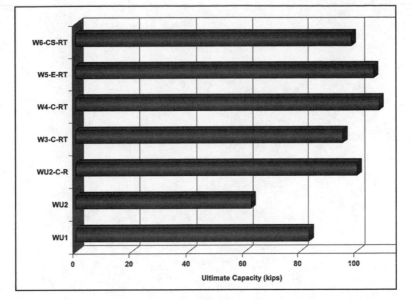

Figure 7 — Strength Comparison of Tested Wall Samples.

Seismic Performance of Masonry Infill Walls Retrofitted With CFRP Sheets

by M. Saatcioglu, F. Serrato, and S. Foo

Synopsis: A significant portion of existing building stock that was constructed prior to the enactment of modern seismic design provisions consists of gravity-load-designed reinforced concrete frames, infilled with unreinforced masonry walls. These structures are susceptible to extensive seismic damage when subjected to strong earthquakes and require retrofitting in order to comply with the provisions of current building codes. Experimental investigation of gravity-load-designed reinforced concrete frames, infilled with concrete block masonry, has been conducted to develop a seismic retrofit strategy that involves the use carbon fiber reinforced polymer (CFRP) sheets. Two half-scale concrete frames, infilled with masonry walls were tested with and without seismic retrofitting. The retrofit technique consisted of CFRP sheets, surface bonded on the masonry wall, while also anchored to the surrounding concrete frame by means of specially developed CFRP anchors. The frame-wall assemblies were tested under constant gravity loads and incrementally increasing lateral deformation reversals. The results indicate that infilled frames without a seismic retrofit develop extensive cracking in the walls and frame elements. The elastic rigidity reduces considerably resulting in softer structure. The failure may occur in non-ductile frame elements, especially in columns. Retrofitting with CFRP sheets controls cracking and increases lateral bracing, improving the elastic capacity of overall structural system. The retrofitted specimen tested in the current investigation showed approximately 300% increase in lateral force resistance, promoting elastic response to earthquake loads as a seismic retrofit strategy. Experimental observations and results are presented in the paper.

Keywords: concrete frames; ductility; earthquakes; fiber reinforced polymer (FRP); infill walls; masonry walls; retrofitting; seismic retrofit

342 Saatcioglu et al.

Murat Saatcioglu, FACI, is a professor of structural engineering and university research chair at the University of Ottawa, Ottawa, Canada. He is a member of ACI Committees 374, Performance Based Seismic Design of Concrete Buildings; 441, Concrete Columns; and 369, Seismic Repair and Rehabilitation, and Canadian Standards Association's (CSA) Committee S806-02 on FRP Structural Components and Materials for Use in Buildings.

Fabio Serrato, is a structural design engineer at Carl Walker Inc. Kalamazoo, Michigan. He received his M.A.Sc. degree in Civil Engineering from the University of Ottawa, Otawa, Canada.

Simon Foo, MACI, is a risk management specialist at the Innovations and Solutions Directorate of Public Works and Government Services Canada. He is a member of CSA Technical Committee on Seismic Risk Reduction and ACI 369 on Seismic Repair and Rehabilitation.

INTRODUCTION

Unreinforced masonry (URM) elements are used extensively as infill wall panels in reinforced concrete and steel frame structures. URM infills fulfill architectural and other functional requirements, such as forming a significant portion of building envelop, partitioning, temperature and sound barriers, while also providing adequate compartmentalization against fire hazard. Masonry is a locally available building material that has a long history of successful use in the construction industry.

The role of URM in resisting seismic forces has been somewhat misunderstood or ignored in the past due to the brittle nature of the material. Lack of knowledge on its performance under seismic loading has discouraged engineers from relying on the interaction of URM-infills with the enclosing structural system. Therefore, it has become a common practice to ignore the participation of infills in resisting lateral loads. The brittle behavior of masonry elements during past earthquakes has discouraged the use of URM as a seismic resistant structural component, justifiably so. However, previous research has shown the beneficial effects of the interaction between URM infills and structural elements for seismic performance of existing frame buildings. Researchers have concluded that proper use of URM infills in frames could result in significant increases in the strength and stiffness of structures subjected to seismic excitations (Klingner and Bertero 1978, Mehrabi et al. 1996, Bertero and Brokken, 1983). However, the locations of infills in a building must be carefully selected to avoid or minimize torsional effects. Architectural restrictions have to be considered when assigning these locations. Whenever infills are placed in an existing building, the increase in strength demand must be compared with the increase in strength capacity, since infilled frames may attract higher forces as compared to bare frames.

A large number of buildings, particularly those constructed prior to the enforcement of ductile design philosophies of the 1970's, have utilized URM-infilled concrete frames. The concrete frames were primarily designed and detailed to resist gravity loads. It has been a common practice to consider URM infill walls as non-structural elements, though they often interact with the enclosing frame, sometimes creating undesirable effects. These elements may contribute to the increase in strength and stiffness of the overall building, though they are not capable of improving inelastic deformability. Any seismic improvement to be expected from URM infills is limited to the elastic range of material. The elastic behavior may not be ensured during a strong seismic excitation, and the subsequent brittle failure may lead to disastrous consequences. Buildings that were designed for higher seismic forces than those corresponding to the elastic threshold of URM need to be protected during seismic response. One technique to achieve this goal is to strengthen the URM by means of epoxy-bonded carbon fiber reinforced polymer (CFRP) sheets.

Two half-scale reinforced concrete frame-concrete block infill assemblies were tested under simulated seismic loading to develop a seismic retrofit strategy for these structures. The first specimen was built to reflect the majority of existing buildings constructed prior to 1970's, with a gravity-load designed frame. The second specimen was retrofitted with CFRP sheets for improved seismic resistance. The experimental program and the results are presented in the paper.

EXPERIMENTAL PROGRAM

Test specimens

Two identical specimens were constructed at the Structures Laboratory of the University of Ottawa for testing. The frames were designed following the requirements of ACI 318-1963 to represent an older building prior to the enactment of modern seismic design requirements. The columns had a 250 mm square section and 8-#15 (16 mm diameter) deformed bars, resulting in 2.56% reinforcement ratio. The longitudinal column reinforcement was lap spliced just above the foundation, with 24 bar-diameter lap lengths, which conformed to the requirements of ACI 318-1963 for compression members. Column ties consisted of 6.35 mm (#2 US) diameter smooth wire with 90-degree bends with 6 bar diameter extensions. The ties were spaced at 125 mm (1/2 the column dimension), providing virtually no confinement to column concrete. The beams, 250 mm wide and 350 mm deep, were reinforced with 3-#15 top bars and 2-#15 bottom bars, resulting in 0.77% and 0.54% reinforcement ratios, respectively. The beam stirrups were in the form of closed hoops with 90-degree bends, following the same detailing as that for column ties with 125 mm spacing. Figures 1 and 2 illustrate the details of the frame members.

Concrete was supplied by a local ready-mix company for both frames. Once the foundation of each specimen was cast, the reinforcement cage for frame was assembled. The formwork for frames consisted of 19 mm plywood boards wrapped with plastic to preserve the moisture in concrete during curing, and to facilitate removal of the formwork. The formwork was removed after 3 days. The specimens were moist cured for another 3 to 4 days after the removal of formwork. A professional contractor was hired to build the masonry infills to implement the actual practice in industry. Figure 3 illustrates the construction process. The infill walls were not connected to the frames by means of anchors. Masonry prisms were prepared and mortar samples were taken to verify that it met the requirements of Mortar Type N with compressive strength of at least 5.2 MPa at 28 days (Drysdale et al., 1994).

One of the specimens was retrofitted after testing as the as-built specimen, and its performance was used to assess the retrofit requirements for the subsequent specimen. Two aspects of the retrofit strategy employed were of concern; i) the amount and arrangement of CFRP sheets and ii) the possibility of delamination of sheets and measures against them. After carefully examining the results of the first test, it was decided to use one sheet per face parallel to each of the two diagonals, resulting in two sheets per wall face. The CFRP sheets were placed diagonally to increase their efficiency since their primary function was to resist diagonal tension.

Specially designed CFRP anchors were developed and used to minimize/eliminate the delamination of CFRP sheets from the surface of the wall. This was done by drilling holes in frame members adjacent to the wall, with approximately 45-degree inclination towards the centre of the frame elements, and inserting the anchors to be epoxy glued into concrete. The anchors were produced in the Laboratory by twisting strips of CFRP sheets and folding into two, as illustrated in Figure 4. A hammer drill was used to make approximately 50 mm deep, 12 mm diameter holes in columns and beams for the FRP anchors (Figure 5). The anchors were placed and epoxy glued after the application of surface bonded CFRP sheets. Wooden pieces were inserted into the anchor holes during the placement of CFRP sheets, as guides, and also to avoid filling of the holes with epoxy. This is illustrated in Figure 6.

The recommendations of the supplier were followed to place the sheets as described below:

- Masonry substrate was prepared by removing loose concrete and dust to provide good adhesion between CFRP and masonry substrate.
- Epoxy primer coat was applied by using a roller.
- Epoxy putty filler was applied on the surface to fill mortar joints as much as possible, as well as any imperfections on the surface. This was done using a trowel. Figure 3(b) shows a wall specimen after the application of epoxy primer and filler putty.
- The first coat of epoxy resin (saturant) was applied diagonally over the primed and puttied surface using a roller.
- Fibers were saturated with epoxy and the first layer of fibers (sheet) was placed over the masonry surface.
- The second coat of saturant was applied over the first layer of sheet.
- The second saturated layer of sheet was placed over the wall with fibers running diagonally, perpendicular to the fibers of the first layer.
- Final coat of saturant was applied as illustrated in Figure 3(c).
- The CFRP anchors were carefully embedded through two layers of fiber sheets into the holes in the frame and bonded to the CFRP sheets with epoxy resin as illustrated in Figure 7.

- A ribbed steel roller was used to remove excess saturant from the sheets and to improve surface bond by removing entrapped air pockets.

Material properties

CFRP Composite – A carbon fiber composite system was used to retrofit the masonry wall, consisting of carbon fibers and epoxy resin. The thickness of carbon fiber sheet was 0.165 mm/ply, which was increased to 0.8 mm/ply when impregnated with epoxy resin. The stress-strain relationship of composite material was established by coupon tests and showed linear-elastic behavior up to the rupturing strength in tension. The capacity of the composite material was established to be approximately 785 MPa based on the fiber content corresponding to 0.8mm/ply. This observation was found to be in agreement with the 3800 MPa fiber strength reported by the manufacturer. The maximum tensile strain of 1.67% specified by the manufacturer was also in agreement with the values recorded during coupon tests. The above values translate into the elastic modulus of 47,000 MPa for the composite material, which is in line with 227,000 MPa reported for the modulus of elasticity of carbon fibers alone.

Concrete and Reinforcing Steel – Concrete was cast in two different batches. The 28-day concrete strengths for unretrofitted and retrofitted frames were established by standard cylinder tests to be approximately 42 MPa, and 38 MPa, respectively.

The stress-strain relationships of reinforcement were determined by performing standard coupon tests. The yield strength of transverse reinforcement (bar diameter = 6.35 mm) used in both frames was approximately 400 MPa. The stress-strain relationship for longitudinal reinforcement (#15 re-bar) in both frames indicated a yielding strength of approximately 425 MPa.

Test setup, instrumentation, and loading program

The specimens were secured to the laboratory strong floor using four threaded steel rods to provide full fixity. Vertical loads were applied on the columns and top beam to simulate gravity loading. This was done by means of externally post-tensioned cables anchored to steel loading assemblies. The loading assemblies were built with hollow structural steel (HSS) sections (75mm x 75mm x 9.5mm). Three HSS sections were welded together to apply axial loads on columns. These assemblies were positioned at the top of columns and underneath the foundation. Each cable was loaded using a hydraulic jack to strain to a value corresponding to 100 kN, generating a total load on each column of 400 kN. The vertical loads on the beam were applied in a similar manner. However, a single HSS was sufficient for each concentrated vertical load on the beam. Three concentrated loads were applied at an equal spacing on the beam, each having a magnitude of 40 kN to simulate distributed gravity load. An MTS hydraulic actuator, with a force capacity of 1000 kN and a stroke capacity of 500 mm was used to apply cyclic in-plane lateral loading. The actuator was connected to a steel reaction frame at one end and to the top beam of the infilled frame at the other end to apply horizontal load during the pushing phase of loading. Four high-strength Dywidag bars were placed externally along the top beam and connected to a loading plate at the far end of the top beam to facilitate pulling of the specimen during lateral load reversals. The test setup is illustrated in Figure 8.

An MTS temposonic LVDT, with a range of 250 mm, was used to measure lateral displacements of the frame along the longitudinal centre-line of the beam. The LVDT was mounted on a metal frame that was connected to the foundation of the specimen. Electric resistance strain gages were placed on re-bars inside the concrete frame. Strain gages were also placed on the post-tensioning cables used to apply vertical loads. These gages were all connected to the same data acquisition system. The MTS actuator was controlled with a computer system that recorded the magnitude of the lateral load, displacements from the actuator's internal LVDT, and displacement readings from the temposonic LVDT.

Lateral loading was applied by the MTS actuator in deformation control mode, as the increments of lateral drift ratios were increased. Three cycles of lateral displacements were applied at each deformation level until failure. The failure was assumed to occur when the strength of specimen was reduced by more than 50% of its ultimate resistance.

TEST RESULTS

The infilled frames tests indicated a high level of participation of the walls to frame response. The walls were able to stiffen the frames significantly, especially during the initial stages of loading. The unretrofitted wall experienced gradual stiffness degradation under increasing levels of deformation reversals. Progressive cracking of masonry units and the mortar joints led to the dissipation of energy, without affecting the strength of frame. The lateral drift was controlled by the stiffening effect of wall. Of particular interest was the simultaneous degradation of strength and stiffness of the wall and the frame, contrary to the common design assumption that the masonry walls would disintegrate early in seismic response, leaving the frames as the only structural system to resist earthquakes. This may be the characteristics of the wall system tested in the current investigation, as the walls were constructed to be fully in contact with the frames. Nevertheless, this observation does indicate that the two materials, i.e., reinforced concrete and infill masonry are compatible in resisting lateral deformations. During loading, the initial resistance was provided mostly by the walls, and the load resistance was gradually transferred to the frames through progressive cracking and softening of the walls. The eventual failure of the unretrofitted specimen was caused by the hinging of columns within the reinforcement splice regions near the ends, while a significant portion of the cracked infill wall remained intact. Figure 9 shows the hysteretic force-lateral drift relationship for the unretrofitted frame-wall assembly. The relationship indicates that a peak load of 273 kN was attained in the direction of first load excursion at approximately 0.25% lateral drift ratio and remained constant up to about 1% drift, though there was substantial stiffness degradation during each cycle of loading. Gradual strength decay was observed after 1% drift, and the assembly failed due to the failure of reinforced columns in their reinforcement splice region at about 2% lateral drift.

The retrofitted specimen showed a substantial increase in elastic rigidity and strength. The initial slope of the force-deformation relationship was high, corresponding to that of uncracked wall stiffness and remained at the uncracked stiffness value until the specimen approached its peak resistance. It was clear that the CFRP sheets controlled cracking and helped improve the rigidity of the wall. The specimen resisted a peak load of 784 kN in the direction of first load excursion, indicating an improvement of approximately a factor of 3. The peak load was attained at approximately 0.3% lateral drift ratio. The CFRP sheets maintained their integrity until after the peak resistance was reached. There was no delamination observed throughout the test. However, the CFRP sheets started to rupture gradually, near the opposite corners in diagonal tension. This resulted in strength decay. By about 0.5% lateral drift, approximately 25% of the peak load resistance was lost. The load resistance continued to drop during subsequent deformation reversals and the resistance dropped down to the level experienced by the unretrofitted specimen at approximately 1% drift ratio. The behavior beyond this level was similar to that of the earlier unretrofitted specimen, and the failure occurred at 2% drift ratio. The hysteretic relationship recorded during the test is illustrated in Figure 10. Further details of the experimental program, including additional data can be found elsewhere (Serrato and Saatcioglu 2004).

CONCLUSION

The following conclusions can be drawn from the experimental research reported in this paper:

- Unreinforced masonry infill walls in reinforced concrete frame structures can provide significant lateral stiffness during seismic response provided that they are not isolated from the surrounding frame elements. These walls can lead to sufficient drift control until after their elastic limit is exceeded. Beyond this level, the walls suffer from significant strength and stiffness deterioration. The overall strength of infilled frames is governed by that of the bare frame.

- Surface bonded CFRP sheets placed parallel to wall diagonals and sufficiently anchored to the surrounding frame can control cracking and resulting stiffness deterioration in unreinforced infill masonry walls. This results in continued in-plane lateral bracing of the wall during seismic response.

- Surface bonded CFRP sheets placed parallel to wall diagonals and sufficiently anchored to the surrounding frame can result in significant increase in the strength of overall structural system. However, they do not contribute to ductility. Upon failure of CFRP, the wall strength reduces to a level that is equivalent to that of unretrofitted wall. The strength of the specimen tested in this investigation increased by a factor of three due to the use of a single layer of CFRP sheet on each side of the wall parallel to each diagonal.

- The CFRP anchors used in securing the sheets to the surrounding frame can function effectively in preventing the delamination of sheets, and promote frame-infill interaction during seismic response.

348 Saatcioglu et al.

- Seismic retrofit design of CFRP covered masonry walls should be based on elastic force limits rather than the principles of ductility and energy dissipation.

ACKNOWLEDGEMENTS

The research project reported in this paper was funded jointly by Public Works and Government Services Canada (PWGSC) and Natural Sciences and Engineering Research Council of Canada (NSERC). Their contributions are greatly acknowledged.

REFERENCES

ACI Committee 318, "Building Code Requirements for Reinforced Concrete," American Concrete Institute, Detroit, 1963.

Bertero, V., and Brokken, S. "Infills In Seismic Resistant Building," Journal of the Structural Engineering, ASCE, Vol. 109, No 6, June, 1983, pp 1337-1361.

Drysdale, R. G., Hamid, A. A., Baker, L. R., "Masonry Structures: Behavior and Design", Prentice Hall, Englewood Cliffs, New Jersey, 1994.

Klingner, R. E., and Bertero, V. V. "Earthquake Resistance of Infilled Frames," Journal of the Structural Division, ASCE, Vol. 104, No ST6, June, 1978, pp. 973-989.

Mehrabi, A. B., Shing, P. B., Schuller, M. P., and Noland, J. N. "Experimental Evaluation of Masonry-Infilled RC Frames", Journal of Structural Engineering, ASCE, Vol. 122, No. 3, March 1996, pp 228-237.

Serrato, F. and Saatcioglu, M. "Seismic Retrofit of Masonry Infill Walls in Reinforced Concrete Frames," Research Report No: OCCERC 04-31, The Ottawa-Carleton Earthquake Engineering Research Centre, Department of Civil Engineering, the University of Ottawa, Ottawa, Canada, Nov. 2004, pp. 105.

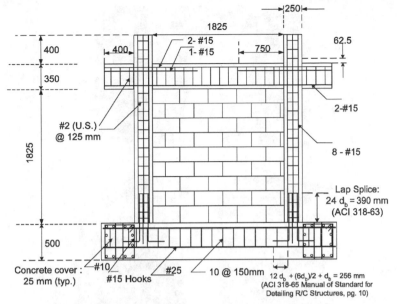

Figure 1 Details of the frame-wall assemblies tested

(a)Column cages   (b)Spliced bars   (c)Frame after casting

Figure 2 Frames during construction

(a) Masonry wall under construction

(b) Application of filler putty   c) Wall retrofitting with FRP sheets

Figure 3 Infill walls under construction

Figure 4 CFRP anchor

Figure 5 Anchor holes

Figure 6 Wooden guides

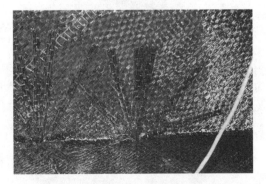

Figure 7 CFRP anchors in place

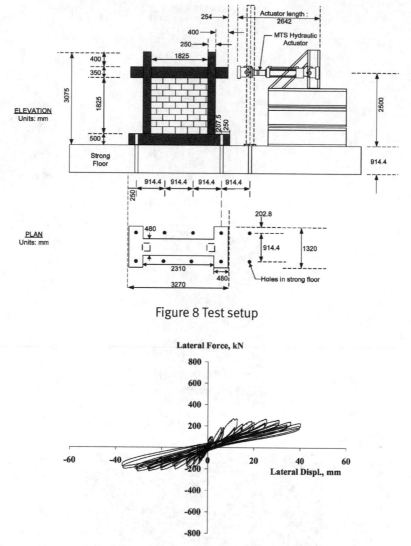

Figure 8 Test setup

Figure 9 Hysteretic force-displacement relationship for unretrofitted specimen

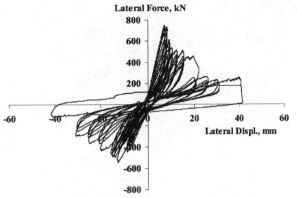

Figure 10 Hysteretic force-displacement relationship for retrofitted specimen

Chapter 4

Bond of FRP Bars, Sheets, Laminates, and Anchorages to Concrete

The Effect of Adhesive Type on the Bond of NSM Tape to Concrete

by C. Shield, C. French, and E. Milde

Synopsis: Near surface mounted (NSM) reinforcement is becoming a viable material for strengthening existing concrete structures in flexure and shear. As this method moves from the laboratory testing phase to implementation in the field, designers will need to be able to predict the capacity of the strengthening scheme. Bond of NSM reinforcement to concrete will likely be a key factor in determining the strengthened capacity. There are a number of parameters that affect bond between steel rebar and concrete, where only two different materials are present. In the case of bond of NSM reinforcement, a third material, adhesive, is present, further complicating the analytical description of bond of NSM reinforcement. This paper presents the results of two types of bond tests that were used to determine the effect of epoxy adhesive material properties on the bond of NSM tape to concrete. From the testing, it was found that even for epoxy adhesives with similar shear strengths, the bond strength and bond failure mechanisms differ significantly.

Keywords: bond; FRP; NSM

356   Shield et al.

Carol Shield is an Associate Professor in the Department of Civil Engineering at the University of Minnesota. She serves as secretary of ACI Committee 440.

Cathy French is a Professor in the Department of Civil Engineering at the University of Minnesota. She is member of ACI Committees 352 and 318 and Chairs subcommittee 318B.

Emily Milde is a Master's Degree candidate in Structural Engineering at the University of Minnesota. She is currently an Intern Structural Engineer at Opus Architects and Engineers.

INTRODUCTION

Several different types of fiber reinforced polymers (FRP) are available for use in strengthening applications. The most common are externally bonded sheets and plates. Near surface mounted (NSM) reinforcement is a recent development that has been gaining attention. Near surface mounted FRP is installed by first cutting a groove in the surface of the concrete. The groove is then filled with an appropriate adhesive, and the NSM reinforcement is placed in the groove. Round FRP rods have been studied by researchers for use as NSM reinforcement (De Lorenzis, Nanni, and La Tegola 2000; De Lorenzis and Nanni 2001, 2002; De Lorenzis, Lundgren and Rizzo 2004; Valerio and Ibell 2003). Rectangular FRP tape is emerging as an alternative NSM reinforcement, but is not yet widely available.

Externally bonded (EB) and NSM FRP each have their own advantages and drawbacks. Externally bonded FRP sheets can easily form to the shape of the structural member. This makes them more suitable for curved, I-shaped, T-shaped or otherwise oddly shaped members. A disadvantage of the EB FRP is the possibility of premature debonding. NSM FRP cannot usually form to the shape of a beam, but it can be anchored into flanges or adjacent members. The NSM FRP is much less sensitive to surface preparation than the EB FRP. NSM FRP also has more consistent quality because it is pre-cured and pultruded. It is also better protected after installation because it is embedded in the concrete.

Tensile bond tests conducted by Blaschko and Zilch (1999) indicated that applying FRP in slits greatly increased the ultimate load over applying FRP to the surface of the concrete. Hassan and Rizkalla (2002) compared the cost and effectiveness of rectangular NSM reinforcement to round NSM bars and EB sheets. Externally bonded sheets and rectangular NSM strips were found to be the most cost effective repair methods when cost of materials, equipment and labor were analyzed for a 30% increase in flexural strength of an existing bridge. Carolin, Nordin and Talisten (2001) conducted tests on seven beams strengthened in flexure with one or two square (10 x 10 mm [0.4 x 0.4in]) NSM reinforcement, and found that the strength of the beam was increased by 57% when one bar was added and between 93 and 141% when two bars were added, depending on the bonding material used.

De Lorenzis, Lundgren and Rizzo (2004) conducted a bond test on round NSM bars using groove filling material (epoxy paste and expansive cement) as a variable. It was found that tests with epoxy-filled grooves had higher strength than tests with expansive cement-filled grooves, attributed to the higher tensile strength of the epoxy. Carolin, Nordin and Talisten (2001) tested beams strengthened on the underside in flexure with square NSM rods and recommended using cement mortar instead of epoxy for improvement of health and environmental conditions on site.

Surface treatment of the bars was a variable in bond tests conducted by De Lorenzis Lundgren and Rizzo (2004). Both CFRP and GFRP bars were tested. Specimens containing bars with a spirally wound surface deformation pattern had higher bond strengths than specimens containing bars with a ribbed surface. De Lorenzis and Nanni (2002) found that deformed bars were more likely to cause a splitting failure of the surrounding material, while sand coated rods were more likely to pull out.

Surface preparation of the groove was also a variable in bond tests conducted by De Lorenzis, Lundgren and Rizzo (2004). The specimens with rougher grooves made by saw cutting achieved higher loads than specimens with smooth formed grooves.

Groove width was a variable in bond tests performed by De Lorenzis and Nanni (2001, 2002). They found that increasing the size of the groove increased the strength until the increase in groove width changed the mode of failure from splitting of the epoxy cover to longitudinal and inclined cracking of the surrounding concrete. De Lorenzis, Lundgren and Rizzo (2004) found that increasing the groove size increased the bond strength by delaying the splitting failure.

Several researchers investigated the effects of different bond lengths. Yan et al. (1999) using pullout tests, De Lorenzis and Nanni (2001) using a Rilem type hinged beam bond test and De Lorenzis, Lundgren and Rizzo (2004) using pullout tests with round No. 3 NSM FRP bars, found that increasing the bonded length of the NSM reinforcement over the range of $4d_b$ to $9d_b$ significantly increased the ultimate load, while a further increase in bonded length to $14d_b$ showed only modest increases in ultimate load. Hassan and Rizkalla (2003) conducted beam tests using rectangular NSM tape (25 x 1.2mm [1 x 0.05 in]) to strengthen beams in flexure, where the length of the NSM tape was a variable. They found that tape lengths of less than 500 mm (19.7 in) did not result in a significant load increase over the control beam with no FRP due to premature debonding of the tape. Tape lengths between 500 mm (19.7 in) and 1700 mm (66.9 in) resulted in a load increase of approximately 15 to 45% over the control beam with no NSM, but eventually failed by debonding of the FRP tape. Tape lengths greater than 1700 mm (66.9 in) had the highest loads, with a load increase of approximately 53% over the control beam, and failed by rupture of the strips. In contrast to the previously mentioned work on round NSM bars, Hassan and Rizkalla's work indicated that the development length for the rectangular NSM tape on the beams in the study was approximately 850 mm (33.5 in), the length of tape on either side of the central flexural crack. The large discrepancy in development length may be due to the significantly different types of testing and not necessarily due to the different cross-sectional shape.

Hassan and Rizkalla also performed a parametric analytical study which indicated that development length increased with increasing internal steel reinforcement ratio.

It is difficult to directly compare the results of the tests on NSM FRP because several different types and shapes of bars were used in the tests, and there were very few test results available for rectangular NSM FRP tape. To further gain an understanding of the performance of NSM tape for strengthening, it was included in an investigation at the University of Minnesota regarding shear strengthening of bridge pier caps.

One of the first issues of importance discovered in the tests was the choice of adhesive to bond the NSM reinforcement to the concrete. To this end, a study was conducted to investigate the effect of adhesive on NSM bond using small-scale bond tests in which the concrete was highly confined. Larger scale bond tests were then used to better predict bond strength in field applications and to investigate parameters such as fiber orientation to crack and vibration during cure.

SMALL-SCALE ADHESIVE TESTS

Some NSM tapes are not commercially sold as a system. Consequently, appropriate choice of adhesive is a critical factor to the success of the system. As part of the research investigation, seven different adhesives were chosen for evaluation. Six specimens of each adhesive were tested. The adhesive that exhibited the best behavior along with one of the more popularly used adhesives were used for the larger scale NSM FRP bond tests.

Test set-up

The FRP system chosen was Aslan 500 CFRP tape manufactured by Hughes Brothers. Aslan 500 Tape was chosen because it was the most readily available rectangular near surface mounted tape in the US. The rectangular profile is desirable because it increases the bondable surface area for a given cross-sectional tape area. It was made from 4.8 GPa (700 ksi) carbon fiber, 60% by volume in a bisphenol epoxy vinyl ester resin matrix. The tape was rectangular with a 2 x 16 mm (0.079 x 0.63 in) cross section and area of 32 mm^2 (0.05 in^2). The manufacturer specified guaranteed tensile strength of the tape was 2.1 GPa (300 ksi) and the modulus was 130 GPa (19,000 ksi). The manufacturer specified ultimate strain of the tape was 1.7%. The measured tensile strength was determined in the laboratory to be 2.5 GPa (360ksi) with a coefficient of variation of 4.5% from an average of six tensile tests.

The adhesives chosen for testing were Sikadur Anchorfix-3, Master Builders/Chemrex Concresive 1420, 3M DP600NS, 3M DP460NS, Sonneborn Epofil, Sikadur 35 Hi-Mod LV, and Sikadur 32 Hi-Mod. Recommended cure time, viscosity, tensile strength, tensile modulus, elongation to failure, and shear strength of these adhesives are given in Table1. All information in the table was provided by each manufacturer except the elongation to failure for 3M DP460NS, which was determined in previous research at the University of Minnesota (Nozaka, 2002). A dash in the table indicates that the information was not available.

The small-scale bond test setup is shown in Figure 1. Concrete blocks measuring approximately 152 x 152 x 203 mm (6 x 6 x 8 in) were cast in modulus of rupture beam molds. Plywood spacers were set in the molds to divide the beams into three pieces approximately 203 mm (8 in) in length. The plywood spacers moved slightly during casting, causing the lengths of the blocks to vary between 150 and 250 mm (6 and 10 in). All of the blocks had a 152 x 152 mm (6 x 6 in) cross section which were the dimensions of importance to the test. Due to volume limitations of the mixer, the blocks were cast in two batches on consecutive days. Compressive strengths and split tensile strengths, measured as the average of three tests per ASTM C39 and ASTM C469 respectively using 102x203 mm (4 x 8 in) cylinders at the beginning of the adhesive tests, are shown in Table 2.

A 6.4 x 19 mm (0.25 x 0.75 in) groove was saw-cut into one of the 152 mm (6 in) high sides of the concrete blocks using a tuck pointing blade on a circular saw. The groove was cut perpendicular to the top side of the block. The grooves were then brushed clean and high pressure air was used to remove any remaining dust or debris. The FRP tape was cut to a length of 406 mm (16 in), then lightly sanded and wiped with acetone to remove any dust or grease from the surface.

The procedure for applying the adhesive differed depending on the consistency of the adhesive. Epofil, Sikadur 32, and Sikadur 35 were two-component liquid adhesives. To apply them to the specimens, the FRP tape was inserted into the center of the groove. Small pieces of tape, placed so as to not interfere with flow of the adhesive, were used as spacers at the top and bottom of the FRP to keep the FRP in the center of the groove and slightly forward of the back of the groove. The front of the groove was then sealed with duct tape, leaving the top of the groove open. For Sikadur 32 and Sikadur 35, the two components of the adhesive were measured and mixed thoroughly using a drill fitted with a mixing attachment. The adhesive was then transferred to a squeeze bottle and dispensed into the groove from one side of the top until the entire groove was full. After filling, the front of the tape was tapped to dislodge any trapped air. Post-test inspection did not reveal any voids in the adhesive. Epofil was a two-component cartridge that was mixed with a static mixing nozzle and dispensed into the top of the groove using a manual gun. Sikadur Anchorfix-3, Concresive 1420, 3M DP600NS and 3M DP460NS were non-sag two-component adhesives that were supplied in cartridges. They were mixed with static mixing nozzles and applied using manual guns. The grooves were not taped for application of the non-sag adhesives. The grooves were filled from the front with the adhesive. The FRP tape was then coated with the adhesive and inserted into the center of the groove. The surface of the adhesive was smoothed with a trowel.

After application of the adhesive, the FRP tape was clamped to a steel tube and angle seated on top of the block. The angle kept the FRP tape perpendicular to the top of the block while the adhesive cured.

Approximately 24 hours prior to testing, aluminum tabs were epoxied to the top end of the FRP tape. The 16 x 102 mm (0.625 x 4 in) tabs were cut from 3.2 mm (0.125 in) thick aluminum. They were lightly sanded and wiped with acetone before bonding them to the FRP tape with a fast-curing epoxy.

After the tab epoxy had cured, the specimen was inserted into an MTS 900 kN (200 kip) test frame. The test setup is shown in Figure 1. Shims were used under the concrete block to position the top of the block tightly against the top plate. The top plate kept the block from rotating as the tensile force was applied eccentrically to the block through the FRP tape, which was located on one side of the block. The plate also put the concrete in a state of compression to promote an adhesive failure rather than a concrete failure to facilitate comparison of the different adhesives. The crosshead was lowered until the grips covered approximately 76 mm (3 in) of the tabs. The hydraulic grips were closed and the testing began. The loading was displacement-controlled at a rate of 0.25mm/min (0.01 in/min). Testing continued until the load dropped suddenly due to failure of bond between the tape and the adhesive, failure of the concrete, failure of the adhesive, or fracture of the FRP tape. If the FRP tape had not broken, it was pulled out of the concrete block before both parts were removed from the test frame.

Some of the blocks were re-used for a second round of testing. A groove was cut in the face opposite the first groove. The same procedure was followed for preparation and testing.

Small-scale adhesive test results

There were several different types of failures observed in the adhesive tests. Most of the adhesive tests using Anchorfix-3, Concresive 1420, Sikadur 32, Sikadur 35, and Epofil failed by a loss of bond between the FRP tape and the adhesive. This usually occurred in a very sudden fashion. In some of these tests, shock from the abrupt loss of bond immediately caused the FRP tape to break. In these cases, the FRP broke off cleanly with no fraying of the fibers. The tests using 3M DP600NS all failed by loss of bond between the adhesive and the concrete. The tests using 3M DP460 NS failed in various ways including loss of bond between the tape and the adhesive, failure in the adhesive, failure in the concrete, and fracture of the FRP tape. When failure occurred by fracture of the FRP tape, it was not a clean break. The fibers began to fray and break off individually until the tape completely deteriorated. Figure 2 shows a photograph of a typical failure for each adhesive type.

Table 3 shows the maximum loads reached in each of the tests, as well as the age of adhesive cure at the time of testing. A dash in the table indicates that the test was not valid because of a failure due to the gripping method. Figure 3 shows a graph comparing the average and highest load of the six tests conducted for each of the seven adhesives. The tests using Sika Anchorfix-3 had the lowest ultimate strengths with an average load of 30 kN (6.8 kips) and a maximum load of 34 kN (7.7 kips). The tests using 3M DP460NS had the highest ultimate strengths with an average load of 73 kN (16.5 kips) and a maximum load of 81 kN (18.3 kips). The other adhesives all performed comparably, with maximum loads between 50 and 60 kN (11.3 and 13.6 kips).

No strong correlation between any of the manufacturer provided adhesive properties and the results of the small-scale bond tests was observed. Adhesive tensile strength and shear strength correlated the least well (r^2 values of approximately 0.07 and 0.02 respectively), with tension modulus and elongation at failure correlating better (r^2 values of 0.53 and 0.42). Clearly the results of these tests indicate that picking an adhesive solely based on the manufacturer provided shear strength will not lead to the best system. Nozaka (2002) found that the bond strength between FRP and steel was strongly dependent on the ductility of the adhesive. Long lengths of adhesive are required to yield in shear to build up enough force through redistribution to fracture the FRP. It is possible that the development length of NSM tape in highly confined concrete is also highly dependent on the ability of the adhesive to deform and yield in shear, which cannot typically be determined from the manufacturer provided data.

LARGE-SCALE BOND TESTS

Large-scale bond tests that more typically represent the state-of-stress of concrete in NSM applications were also conducted. The test setup is shown in Figure 4. Two concrete blocks were placed adjacent to each other to simulate a crack between them, with two strips of FRP NSM tape connecting the two blocks on each side of every specimen. The blocks were loaded in a direction perpendicular to the crack to simulate crack opening. An advantage of this test setup was the ability to test the FRP without gripping it directly, which eliminated a possible stress concentration due to mechanical grips. This test setup also allowed tensile load to be applied to FRP on both sides of the block at one time while keeping the moment in the tape strips to a minimum.

Parameters

Variables addressed in this part of the study are summarized in Table 4 and included tape orientation, adhesive type, and vibration during cure of the adhesive, through the use of six specimen. For each adhesive type, two repeats were performed with the tape oriented at 90 degrees to the simulated crack. Tape orientation was a variable for the NSM tape because it was anticipated that NSM tape may be used for shear as well as flexural strengthening. In shear strengthening applications, it is quite possible that the tape would not be placed at 90 degrees to the crack, so a worst case scenario, with the tape strips placed at 45 degrees to the crack was investigated. Specimen naming convention consisted of the manufacturer-strip orientation-repetition number (e.g. Specimen Sika-90-2 was the second test [repeat] using Sika Anchorfix-3 with the tape oriented 90 degrees to the simulated crack.

Although adhesive 3M DP460NS performed best in the small-scale adhesive tests, it was among the most expensive adhesives and had very little reported data in the literature regarding its use between FRP and concrete. On the other hand, Sika Anchorfix-3 was a commonly used adhesive to bond FRP to concrete. Therefore both 3M DP460NS and Sika Anchorfix-3 were chosen for study in the large-scale specimen to represent a promising adhesive and a more typically used adhesive.

One specimen with NSM FRP was loaded with a small amplitude cyclic load while the adhesive was curing. This was done to investigate possible effects of service load vibrations occurring during cure of the adhesive. If it is to be applied in the field under traffic loading, the effect of cyclic loading on bond strength may be important.

Specimen Design and Construction

Each half specimen was a rectangular shape, 610 x 305 x 305 mm (2 x 1 x 1 ft). The two halves joined together to form a square shape 610 x 610 x 305 mm (2 x 2 x 1 ft). Specimen drawings can be found in Figure 5. A square shape was chosen to allow investigation of different tape orientations.

Sufficient Grade 60 reinforcement was used in each specimen half to prevent cracks from occurring elsewhere in the specimen. Four Grade B7 threaded rods embedded into the specimen attached the specimen to the loading frame. There was not enough room inside the half specimen to anchor the rods through bond, so a nut was placed in the middle of the half specimen to serve as mechanical anchorage.

Two unbonded 32 mm (1.25 in) diameter steel rods extended through snug-tight greased PVC sleeves in both halves of the specimen. This system acted similarly to a linear bearing to ensure that the crack opened evenly with no rotation of the blocks.

The concrete was delivered from a ready-mix plant. The mix design specified normal weight concrete with a compressive strength of 28 MPa (4000 psi) at 28 days. Compressive strengths measured per ASTM C39 at the beginning and end of the large-scale tests with 102x203 mm (4x8 in) cylinders and split tensile strengths measured per ASTM C496, along with their coefficients of variation and the concrete age at the time of testing are given in Table 5.

Instrumentation and Data Acquisition

A computer data acquisition system was used to record time, displacement, and load from every test. Data was taken at a rate of 1 Hz. All of the tests were displacement-controlled to allow data to be collected for FRP on both sides of the specimen (i.e., as each strip failed, the test was continued in a controlled manner until all of the strips had failed). In every test, four linear variable differential transformers (LVDTs) located near the corners were used to measure the crack opening on each corner of the block to ensure that the blocks were mainly translating and not rotating as the test progressed.

Procedure

The specimen was placed into the load frame and adjusted so that the top half of the specimen could slide freely on the steel rods running between the two specimen as the crosshead was raised and lowered. Once the top and bottom halves were sufficiently aligned with each other and with the steel beams, the blocks were grouted to the steel beams. Nuts were then placed on the threaded rods and fully torqued.

Two 6.4 x 19 mm (0.25 x 0.75 in) grooves 76 mm (3 in) on either side of center were cut into the full length of the specimens on each side with a diamond tuck pointing blade on a circular saw. The grooves were then brushed and blown out with pressurized air to remove any dust or debris. A layer of petroleum jelly was placed on the horizontal surfaces between the two blocks to ensure the two blocks would not be bonded together by adhesive that might migrate into the interface.

The adhesives used came in two-part cartridges that were mixed using a static mixing nozzle and dispensed with a manual or pneumatic gun. A small amount of adhesive was discarded at the beginning of each cartridge to ensure that only well-mixed adhesive was applied. The FRP tape was cut to 610 mm (24 in) and one piece of tape was coated with adhesive. Prior to filling, the edges of the grooves were masked with tape to prevent excess adhesive on the surface of the concrete. One groove was then filled with adhesive and the tape was inserted so that it was centered in the groove. Any remaining voids were filled with adhesive and the excess adhesive was removed with a trowel. The surface was smoothed, and the masking tape was removed. The same procedure was then followed for the other three grooves.

The hydraulics were turned on and the load was set to 4.5 kN (1 kip) in compression in order to maintain a constant position while the FRP cured. The load was held for at least 48 hours until testing began. Keeping the load constant was more successful than keeping the displacement constant, possibly due to shrinkage in the grout.

One specimen had cyclic loading applied while the adhesive was curing. For that specimen, the load was set to cycle between 4.5 and 22 kN in compression at a frequency of 1 Hz. This load range was determined from an analysis of overhang of the pier cap of MN Bridge 19855/56 with the bridge loaded with passenger cars. The maximum flexural stress on the face of the pier cap overhang, assuming uncracked sections was calculated. The 17.5 kN load applied to the large-scale test specimen resulted in this same stress, The stress was then applied to the bond test specimen in compression. It was not possible to apply any tension to the specimen while the adhesive was curing because there was nothing to resist tension holding the two halves of the specimen together. The loading began as soon as the FRP had been applied to the specimen, and continued for 3 days until the specimen was tested.

Testing of the six specimens was conducted approximately 48 hours after bonding, with the exception of Specimen Sika-90-1 which was tested approximately 24 hours after bonding, and Specimen 3M-90-3V, which was tested approximately 72 hours after bonding to accommodate the cyclic loading during cure. In all cases, full cure should have been reached after 24 hours (Table 1). The tests were monotonic with a crosshead loading rate of 0.00127 mm/sec (0.00005 in/sec). The tests continued until failure (loss of load carrying capability) or a crosshead displacement of 1.25 mm (0.5 in).

Large-Scale Bond Test Results

Table 4 gives the results of the tests for the six large-scale specimen, including, maximum load reached in the test, load drops associated with the failure of each individual tape strips and the order of strip failure (the superscripts on the failure loads). A superscript "s" indicates that two or more different tape locations failed simultaneously and the total load was divided equally among the tape strips.

For each of the specimens using Sika Anchorfix-3, the failure began with cracking of the adhesive that extended into the surrounding concrete, followed by a shear failure in the concrete near the interface between the two blocks, often resulting in the loss of a large piece of concrete from the block as shown in Figure 6a. Ultimately the specimen using Sika Anchorfix-3 lost bond between the FRP tape and the adhesive, with the tape pulling out of the adhesive. During this phase, load carrying capacity was provided by friction between the tape and the adhesive. All of the FRP tapes that had been bonded using Sika Anchorfix-3 showed little to no concrete attached to the surface and significant damage to the matrix during post-test inspection as shown in Figure 7a. The variability in bond between the first two tests using Sika Anchorfix-3 was high. The tests were identical with the exception of cure time, but Specimen Sika-90-1showed indications of poorly cured adhesive in two locations upon post test inspection, even though the manufacturer recommended installation procedure was followed. It was unclear why these two locations showed poor cure, while the rest of the adhesive had cured well. The second test, Sika-90-2, was a repeat of the first test with an ultimate load of 169 kN (38.1 kips), and was taken as more predictive of a well cured bond between FRP and concrete using Sika Anchorfix-3.

The specimens using 3M DP460NS exhibited progressive cracking in the concrete directly surrounding the groove, ultimately failing when large pieces of concrete were broken off between the two blocks as shown in Figure 6b. None of the tapes fractured during the test. There was no failure between the FRP and adhesive or in the adhesive as shown by the large amount of concrete remaining bonded to the tape after the conclusion of the test in Figure 7b. The two identical tests of the specimen using the 3M DP460-NS (Specimens 3M-90-1 and 3M-90-2) did not exhibit high variability.

Effect of Adhesive Type for NSM FRP -- The specimens using 3M DP460NS exhibited much higher loads than those using Sika Anchorfix-3 even though the manufacturer's reported shear strengths of both adhesives were similar, with Sika Anchorfix-3 having a slightly larger value (Table1). The maximum load for the 90 degree specimen with Sika Anchorfix-3 was 169 kN (38.1 kips). The average maximum load for the three 90 degree specimens with 3M DP460-NS was 246 kN (55.3 kips), with a coefficient of variation of 2.0%. The specimen using 3M DP460NS had an increase in ultimate load of 45% over the specimen using Sika Anchorfix-3, simply by changing the adhesive type with all other parameters kept constant.

Although these differences in strength were quite large, they were not as large as those observed between these two adhesives in the highly confined small-scale tests (143%). The difference in strength in the large-scale tests was mainly attributable to the difference in failure modes due to the different adhesive types. Specimens using 3M DP460NS had higher strengths because they were able to take better advantage of the strength of the FRP tape. The maximum loads recorded for the specimens made with 3M DP460NS were close to the load expected to fracture the four tapes when using the manufacturer's guaranteed strength for the Aslan 500 tape (266 kN [59.7 kips] based on a guaranteed strength of 2.1 GPa [300 ksi]). According to Table 1, 3M DP460NS had a larger elongation to failure, perhaps indicating that it was a more ductile adhesive in shear. From the results of the large-scale tests, it appeared that a tougher, less brittle adhesive may be more effective in bonding the dissimilar substrates of FRP and concrete. The failure surfaces for typical NSM specimens with both adhesives are shown in Figures 6a and 6b.

Effect of Fiber Orientation for NSM FRP -- Tape orientation was taken as a variable to examine potential shear strengthening cases. In a shear crack, the crack opens as well as slides. As a consequence, it may be desirable to place the reinforcement at a 45 degree angle to the crack. For this investigation, it was also of interest to see if this orientation could demonstrate good ductility while still carrying a considerable amount of load. Thus, the third test using Sika Anchorfix-3 (Specimen Sika-45-1) was conducted with the NSM reinforcement oriented 45 degrees to the crack.

As expected, this specimen had a much lower ultimate load than the specimen using Sika Anchorfix-3 with the reinforcement oriented 90 degrees to the crack (Specimen Sika-90-2). The 45 degree test reached a load of 99 kN (22.3 kips), 59% of the strength of the 90 degree specimen. This configuration was less stiff than the tests with fibers oriented 90 degrees to the crack, and had a larger displacement at peak load. The 45 degree specimen had approximately twice the displacement of the 90 degree specimen with the same adhesive at peak load (6.1 mm [0.24 in] for Sika-45-1, 3.0 mm [0.12 in] for Sika-90-2) The load vs. displacement plot for the 45 degree Sika Anchorfix-3 specimen (Sika-45-3), one of the 90 degree specimens using Sika Anchorfix-3 (Sika-90-2) and one of the 90 degree specimens using 3M DP460NS (3M-90-2), are shown in Figure 8.

Failure of Specimen Sika-45-1 began with progressive cracking of the concrete and epoxy on the surface, followed by shear failure of the concrete near the crack and a loss of bond between the FRP tape and the adhesive. The failure surface of the 45 degree specimen, shown in Figure 6c, lost much less concrete than the 90 degree specimen. The FRP tape did not fracture at the crack, but its flexibility allowed it to bend as the crack opened. There was significant damage at failure in the matrix of the tape near the crack.

Effect of Vibrations During Cure of Adhesive on NSM FRP -- The test of the specimen (3M-90-3V) that was cyclically loaded during curing of the 3M DP460NS adhesive had a maximum capacity of 242 kN (54.5 kips). The average of the tests using the same parameters, but no cyclic loading during cure (3M-90-1 and 3M-90-2) was 246 kN (55.3 kips). This shows that there was an insignificant change in strength when cyclic loading was applied to the specimen while the adhesive was curing. The failure surface of the specimen subject to vibrations was similar to that of the control specimens with the 3M adhesive. The failure surfaces are shown in Figure 6d.

CONCLUSIONS AND RECOMMENDATIONS

Bond strength of the NSM tape was highly dependent on both the adhesive as well as the state of stress in the concrete surrounding the tape. The dependence did not correlate to the manufacturer supplied shear strength of the adhesive. For most adhesives, even in the small scale tests with highly confined concrete, the failure was in the concrete or adhesive. Only the specimens using 3M DP460NS were able to achieve strengths large enough to break the NSM. In the more realistic large-scale tests neither adhesive could develop enough bond to break the NSM tape, with a maximum force of 98% of the strip strength using 3M DP460NS. The improvement in ultimate strength for the specimens using 3M DP460NS was likely attributable to the larger ductility of this adhesive.

However, there was no good correlation between any manufacturer provided adhesive material properties and the strength and ductility of the test specimen. Indicating that other material properties such as ultimate shear strain and shear modulus of the adhesives may play an important role in determining the amount of bond that can be developed. The limited scope of the work reported herein should be expanded to determine the role of concrete strength, NSM tape type, other forms of adhesives, and the effect of flexure in the tape, with the long term goal to develop a design equation for bond of NSM that accounts for all the important physical parameters.

REFERENCES

Blaschko, M., and Zilch, K., 1999, "Rehabilitation of Concrete Structures with Strips Glued into Slits," *Proceedings of the Twelfth International Conference on Composite Materials*, Paris, France, July 5-9.

Carolin, A., Nordin, M., and Taljsten, B., 2001, "Concrete Beams Strengthened with Near Surface Mounted Reinforcement of CFRP," *Proceedings of the International Conference on FRP Composites in Civil Engineering*, pp. 1059-1066.

De Lorenzis, L., Lundgren, K., and Rizzo, A., 2004, "Anchorage Length of Near-Surface Mounted Fiber-Reinforced Polymer Bars for Concrete Strengthening-Experimental Investigation and Numerical Modeling," *ACI Structural Journal*, March-Apr, V. 101, No. 2, pp. 269-278.

De Lorenzis, L., and Nanni, A., 2001, "Characterization of FRP Rods as Near-Surface Mounted Reinforcement," *ASCE Journal of Composites for Construction*, May, V. 5, No. 2, pp. 114-121.

De Lorenzis, L., and Nanni, A., 2002, "Bond between Near-Surface Mounted Fiber-Reinforced Polymer Rods and Concrete in Structural Strengthening," *ACI Structural Journal*, Mar/Apr, V. 99, No. 2, pp. 123-132.

De Lorenzis, L., Nanni A., and La Tegola, A., 2000, "Flexural and Shear Strengthening of Reinforced Concrete Structures with Near Surface Mounted FRP Rods," *Proceedings of the Third International Conference on Advanced Composite Materials in Bridges and Structures*, Aug, pp. 521-528.

Hassan, T., and Rizkalla, S., 2002, "Flexural Strengthening of Post-Tensioned Bridge Slabs with FRP Systems," *PCI Journal*, V. 47, N. 1, pp. 76-93.

Hassan, T., and Rizkalla, S., 2003, "Investigation of Bond in Concrete Structures Strengthened with Near Surface Mounted CFRP Strips," *ASCE Journal of Composites for Construction*, V. 7, N. 3, pp. 248-257.

Nozaka, K., 2002, "Repair of Fatigued Steel Bridge Girders with Carbon Fiber Strips," PhD Thesis, Department of Civil Engineering, University of Minnesota, 174 pp.

Valerio, P. and Ibell, T. J., 2003, "Shear Strengthening of Existing Concrete Bridges," *Structures and Buildings*, Feb, V. 156, No. 1, pp. 75-84.

Yan, X., Miller, B., Nanni, A., and Bakis, C., 1999, "Characterization of CFRP Rods Used as Near-Surface Mounted Reinforcement," *Proceedings of the Eighth International Structural Faults and Repair Conference,* M. D. Forde Editor., Engineering Technics Press, Edinburgh, CD-ROM version.

Table 1 Properties of Adhesives

Adhesive	Recommended Cure Time	Viscosity	Tensile Strength (MPa)	Tensile Modulus (MPa)	Elongation to Failure (%)	Shear Strength (MPa)
3M DP600NS	1 hour	Non-sag	-	-	-	24.7
Sikadur Anchorfix-3	24 hours	Non-sag	32.4	830	1.2	33.8
MBT/Chemrex Concresive 1420	24 hours	Non-sag	27.6	-	1.0	-
Sikadur 32 Hi Mod	7 days	2800 cps	35.2	2200	1.8	40.7
Sikadur 35 Hi Mod LV	> 7 days	375 cps	61.4	2800	5.4	35.2
Sonneborn Epofil	> 7 days	750 cps	50.3	3100	8.0	60.7
3M DP460NS	24 hours	Non-sag	35.2	2500	2.1	31.0

Table 2 Concrete Strengths for Adhesive Tests

Specimen Set (age of concrete)	Compressive Strength (MPa)	COV (%)	Split Tensile Strength (MPa)	COV (%)
Set 1 (58 days)	55.7	6.8	4.21	8.6
Set 2 (59 days)	64.5	6.3	4.55	5.0

Table 3 Maximum Loads for Adhesive Tests

Adhesive	Cure Time (days)	Test 1 Load (kN)	Test 2 Load (kN)	Test 3 Load (kN)	Test 4 Load (kN)	Test 5 Load (kN)	Test 6 Load (kN)	Ave. Load (kN)	Max. Load (kN)
Anchorfix-3	4	29.8	34.2	30.2	26.7	29.4	30.2	30.2	34.2
Concresive 1420	3		42.3	37.8	50.3	42.7	34.2	37.8	40.9	50.3
Epofil	18	58.7	-	52.9	53.8	59.2	59.2	56.9	59.2
Sikadur 35	17	-	-	52.5	53.8	49.8	53.4	52.5	53.8
Sikadur 32	16	53.4	53.4	55.2	-	60.5	56.5	55.6	60.5
DP600NS	4	43.6	54.3	50.7	41.8	48.0	48.0	47.6	54.3
DP460NS	4	76.1	62.7	81.4	76.1	79.6	65.4	73.4	81.4

Table 4 Large Scale Bond Tests

Test	Adhesive	Fiber Orientation to Crack (deg)	Date of test	Peak Load (kN)	Load Drops[a] East (kN)	Load Drops[a] West (kN)
Sika-90-1	Sika	90	11/13/03	123	N 35[1] S NA	N NA S NA
Sika-90-2	Sika	90	11/20/03	169	N 44[3] S 18[4]	N 13[1] S 32[2]
Sika-45-1	Sika	45	11/23/03	99	N 21[1s] S 21[1s]	N 34[2s] S 34[2s]
3M-90-1	3M	90	5/30/04	242	N 81[3] S 60[2]	N 59[1s] S 59[1s]
3M-90-2	3M	90	6/3/04	250	N 64[4] S 64[1]	N 59[3] S 59[2]
3M-90-3V	3M	90	6/7/04	242	N 51[3] S 68[4]	N 67[2] S 67[1]

[a] Superscripts on loads indicate the order of failure

Table 5 Concrete Strengths for Bond Tests

Test Date (age of concrete)	Compressive Strength (MPa)	COV (%)	Tensile Strength (MPa)	COV (%)
11/20/03 (224 days)	54.1	2.3	4.41	10.0
6/20/04 (437 days)	60.0	3.1	4.55	7.3

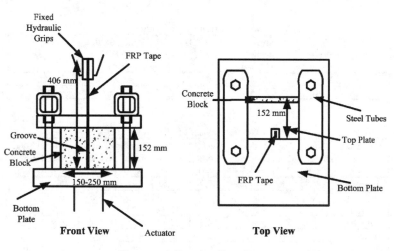

Figure 1 Adhesive Test Setup

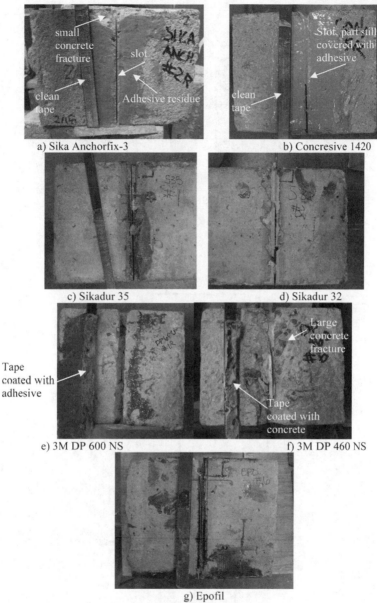

a) Sika Anchorfix-3

b) Concresive 1420

c) Sikadur 35

d) Sikadur 32

e) 3M DP 600 NS

f) 3M DP 460 NS

g) Epofil

Figure 2 Adhesive Test Specimens after Failure

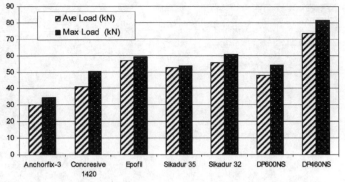

Figure 3 Adhesive Test Load Comparison Graph

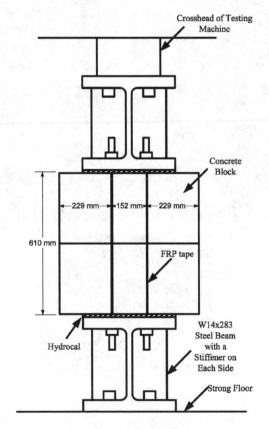

Figure 4 Large-ScaleTest Setup for 90 Degree Tests

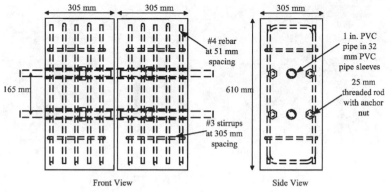

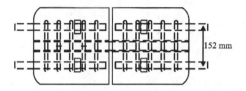

Figure 5 Bond Test Specimen Drawings

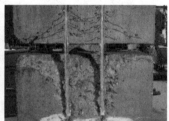

a) Specimen Sika-90-2   b) Specimen 3M-90-2

c) Specimen Sika-45-1   d) Specimen 3M-90-3V

Figure 6 Comparison of NSM Failure Surfaces

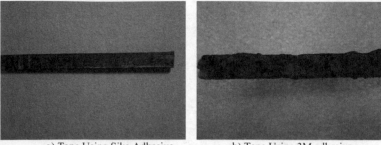

a) Tape Using Sika Adhesive   b) Tape Using 3M adhesive

Figure 7 Comparison of Tapes after Failure

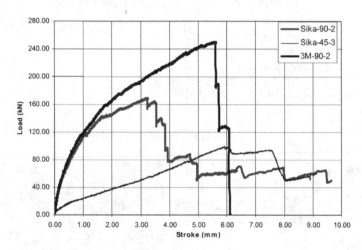

Figure 8 Load vs Deflection Plots

Modeling of FRP-Concrete Bond Using Nonlinear Damage Mechanics

by C.A. Coronado and M.M. Lopez

Synopsis: In this study, experimental and numerical procedures are proposed to predict the debonding failure of concrete elements strengthened with fiber-reinforced polymers (FRP). The experimental tests were designed to obtain the tensile softening curve and fracture energy of the *concrete-epoxy interface*. Results indicate that the fracture energies of plain concrete and *concrete-epoxy interface* are different in magnitude. The numerical simulations were conducted using a plastic-damage model. In this approach, the damage is defined using the tensile softening curve. Numerical results were validated against experimental results obtained from pull-off bond tests. The numerical models were capable of predicting the experimentally observed load-strain, strain distributions, failure load, and failure mechanism of the pull-off specimens.

Keywords: concrete-epoxy interface; finite element modeling; fracture energy; FRP laminate; numerical modeling; plastic damage; pull-off bond test; softening curve

Carlos A. Coronado is a Ph.D. candidate and research assistant at Penn State University. He graduated with a Master's of Science in structural engineering from the Universidad del Valle, Colombia. His current research interests include the numerical modeling and experimental characterization of the interface formed between concrete and different types of repair materials.

Maria M. Lopez is an Assistant Professor at Penn State University. She is active on several ACI technical committees including committee 544, fiber-reinforced concrete; 543, Fracture mechanics; and 440, fiber-reinforced polymer reinforcements. Dr. Lopez's research interests include the use of fiber-reinforced polymer materials for civil infrastructure applications.

INTRODUCTION

Bonded concrete-repairs are widely used in civil engineering applications. These repairs often govern the behavior and failure of the structures where they occur. Such is the case with concrete overlays and reinforced concrete beams strengthened in flexure with bonded plates (ACI 440.2R, 2002; Granju, 2001). In these applications, debonding of the concrete-repair is the single most important factor affecting the structural performance. Therefore, appropriate methods of characterizing and modeling the bond of the concrete-repair interface are required.

In this study, experimental and numerical procedures are proposed in order to predict the bond behavior between concrete and FRP. The experiments were designed to obtain the softening curve of the *concrete-epoxy interface*. The numerical simulations are conducted using a plastic-damage model. In this approach, the damage is defined using the softening curve. The numerical simulations were validated against experimental results obtained from pull-off bond tests. The numerical models are capable of predicting the experimentally observed load-strain, strain distributions, and failure mechanism of these specimens.

RESEARCH SIGNIFICANCE

Debonding is the single most important factor affecting the performance of *bonded concrete repairs*. Experimental observations indicate that the debonding failure of these systems is due to a crack (or group of cracks) propagating along the *concrete-repair* interface. In order to improve the understanding of bonded concrete repairs, appropriate methods of characterizing and modeling the bond of the *concrete-repair interface* (CEI) are required.

The experimental methodology presented in this article was used to obtaining the parameters required for modeling the debonding failure of concrete members strengthened with FRP. The use of these parameters is illustrated and validated. For this purpose, numerical simulations of pull-off bond tests (SL) are conducted. The frameworks of nonlinear fracture and damage mechanics are used during the experimental characterization and numerical modeling.

OBJECTIVES

The main objectives of this research were:

- Propose a comprehensive approach for the characterization and modeling of the debonding failure of concrete members strengthened with FRP.

- Determine the mechanical properties governing the *debonding-failure* of the interface formed after bonding an FRP laminate to a concrete substrate.

- Use the mechanical properties of the *concrete-epoxy interface* to predict the debonding failure of FRP-strengthened specimens.

THEORETICAL FRAMEWORK

The framework of damage mechanics can be used for modeling and explaining the *debonding failure* of members strengthened with FRP. For instance, consider the FRP-strengthened beam shown in Figure 1. Before any microcracking occurs, the beam's behavior is virtually elastic and no damage has taken place (i.e., the *damage variable, d,* is zero). Once the interfacial stresses reach a *threshold value* (i.e., the *tensile strength, f't,* of the interface), microcracks start to form close to the *concrete-epoxy interface*. At this point, the damage of the interface starts to increase ($0 < d < 1$). Under further loading, the microcracks start to coalesce and macrocracks form. These macrocracks result in debonding of the FRP and total loss (*damage*) of the *concrete-epoxy* bond ($d = 1$).

It is the authors' assumption that the damage of the *concrete-epoxy interface* is due to high *interfacial* stresses. Under this assumption, the *damage variable, d,* can be defined using the concept of softening curve, which is borrowed from nonlinear fracture mechanics, Figure 2. The area under this curve (G_F) is considered a material property and can be determined experimentally as discussed later (Elices et al., 2002). In this context, G_F represents the total external energy supply, per unit of area, required to create and propagate a crack along the *concrete-repair interface*. From the softening curve, the damage variable can be defined following Lubliner et al. (1989):

$$d = \frac{1}{G_F} \int_0^w \sigma(w)dw \qquad\qquad 0 \le d \le 1 \qquad\qquad (1)$$

$$\sigma = f_t(d) \qquad\qquad (2)$$

The previous damage definition has been implemented in several constitutive models (e.g., Lee and Fenves, 1998; Llubliner et al. 1989). In this study, the behavior of concrete and *concrete-epoxy interface* is approximated using a plastic-damage model. For this purpose, the authors assume that most of the damage of the *concrete-epoxy interface* occurs in the concrete side. The basic parameters required by this plastic-damage model are listed in Table 1 and described in the following sections.

EXPERIMENTAL PROGRAM

The experimental program conducted in this study was aimed to provide the material properties required during the numerical modeling of concrete members strengthened with FRP, Table 1. For such purpose, material characterization procedures and validation experiments were conducted. The intention of this approach is to demonstrate that the material properties determined in the characterization phase suffice to predict the global and local behavior observed during the validation experiments.

Material Properties

The mechanical properties of the materials used during the experimental program were determined using standard testing procedures. These properties are presented in Table 2. Two concrete mixes were used: PC-1 and PC-2, respectively. The laminate properties shown in Table 2 were calculated using the nominal fiber thickness given by the manufacturer (i.e., 0.165 mm/ply). Figure 3 shows the epoxy behavior under uniaxial tension and pure shear. As can be seen, the stress-strain behavior of the epoxy is parabolic under axial tension, whereas it can be assumed as linear under shear; these behaviors were accounted for during the numerical modeling.

In addition to the mechanical properties presented in Table 2, the stress-strain behavior of concrete in compression is required. It is approximated here using Todeschini's model, which is defined by Eqs. 3 and 4, recommended by MacGregor (1997). The experimental procedures used in determining the tensile softening curve of concrete and *concrete-epoxy interface* are discussed in the following sections.

$$\varepsilon_0 = 1.71 \frac{f'_c}{E_c} \tag{3}$$

$$\sigma_c = \frac{1.8 f_c (\varepsilon / \varepsilon_0)}{1 + (\varepsilon / \varepsilon_0)^2} \tag{4}$$

Tensile-softening curve

In the context of damage mechanics, the softening curve is the principal material property required in predicting the tensile failure of concrete and the *concrete-epoxy interface*. This curve can be approximated using linear or bilinear models as shown in Figure 4. Three properties are worth mentioning for the bilinear approximation shown in this figure: the tensile strength (f'_t), the size-effect fracture energy (G_f), and the cohesive fracture energy (G_F).

Tensile Strength f'_t – the tensile strength is the property controlling the initiation of microcracking and damage. In this study, the splitting tensile test was used to determine this material property. Following Rocco et al. (2001), the peak load of the split-cylinder test (P_{uc}) can be expressed as a function of the unknown "*true*" tensile strength (f'_t) and the characteristic length (l_1) in the following form,

$$f_{st} = \frac{2P_{uc}}{\pi D_c L_c} = \frac{f'_t}{-5.544 + 21.526\dfrac{D_c}{l_1}} + 1.0318 f'_t \qquad (5)$$

$$l_1 = \frac{E_c w_1}{2 f'_t} \qquad (6)$$

where f_{st} is the tensile strength from the splitting test; D_c (102 mm) and L_c (204 mm) are the length and diameter of the cylinder, respectively; and w_1 is the intercept of the initial tangent to the softening curve, as indicated in Figure 4. It must be noted that w_1 is not known a priori but can be determined with the technique described in the next section. Once w_1 is known, the ratio D_c/l_1 can be computed *a posteriori* and the value of f'_t determined. Since the value of w_1 is a function of f'_t, the process of determining w_1 can be iterated to obtain better estimates of both f'_t and w_1. It must be emphasized that this procedure is followed here since, otherwise, the results from the splitting test are size-dependent, as shown by Rocco et al. (2001).

In this study, two specimen types were used to obtain the tensile strength, f'_t, of plain concrete and the *concrete-epoxy interface* respectively. The plain concrete (PC) specimens were tested following ASTM C496. A different procedure was used in determining the tensile strength of the *concrete-epoxy interface*. For such a purpose, concrete cylinders 28 days old were cut in half along the longitudinal direction and glued back together using epoxy. The epoxy was then left to cure at least 7 days. Before testing, the cylinder edges were ground in order to avoid shear stresses due to an uneven application of the load. The splitting test was conducted following ASTM C496.

The geometry, failure mode, and specimen characteristics of the splitting test specimens are shown in Figure 5. It is interesting to note in Figure 5c that the failure plane oscillated along the bond line of the two cylinder halves. The tensile test results are summarized in Table 3. The corrected, *true*, tensile strength, f'_t, is presented in Table 4. For the epoxy used in this study, the tensile strength of the concrete and *concrete-epoxy interface* are similar and within the experimental range of error.

Size-effect fracture energy G_f – the size-effect fracture-energy, G_f, represents the area under the initial tangent of the softening curve; this tangent is shown in Figure 4. In this study, the fracture energy G_f was determined using the procedure outlined by Elices et al. (2002). For such purposes, three-point bending tests of notched specimens were conducted. The specimens' geometry is shown in Figure 6 and Table 3. During the testing, the peak load, P_{max}, the crack mouth opening displacement (CMOD), and the mid-span deflection were recorded. From the peak load, P_{max}, and beam's geometry, the characteristic length, l_1, can be calculated using Eq. (7) (Elices et al., 2002). Once l_1 is known, w_1 and G_f can be obtained from Eqs. 6 and 9, respectively.

$$l_1 = D(1 - \alpha_0^{1.7})\left[\frac{11.2}{\left(x^2 - 1\right)^2} + \frac{2.365}{x^2}\right] \quad ; \qquad x = \frac{f'_t}{f_p} \qquad (7)$$

$$f_p = \frac{P_{max}S}{2Bb^2} \tag{8}$$

$$G_f = \frac{f_t w_1}{2} \tag{9}$$

Cohesive Fracture energy G_F – the *cohesive* fracture energy, G_F, is the energy required to create and fully break a crack of unit length. This property can be determined using the work of fracture method (Bazant and Planas,(1998). In this study, G_F was approximated as $G_F = 2.5G_f$, following the findings of Elices and Planas (1990). With G_F and f_t known, the shape of the softening curve was approximated, as shown in Figure 4. The fracture mechanics properties obtained following this procedure are presented in Table 4. As can be seen, the fracture energies of plain concrete, PC, and *concrete-epoxy interface,* CEI, have different values. These results indicate that this particular epoxy introduces a toughening effect, i.e., crack deflection which results in a higher fracture energy, G_F, of the *concrete epoxy interface* when compared to the fracture energy, G_F, of plain concrete.

The mentioned toughening effect is illustrated in Figure 4c, in which the deflection of the *interfacial* crack is evident. This crack deflection indicates that the damage process, which results in debonding of the *concrete epoxy interface*, occurs in a *band* (*damage band*) or region in the neighborhood of the *bond surface*. From a practical point of view, this experimental observation implies that the *damage band* should be modeled during the simulation of debonding failures.

Validation experiments

The single shear lap test (Pull-off test) is used in order to validate the experimental and numerical procedures presented in this study. During this test, the concrete-epoxy interface is under mixed-mode loading; i.e., normal and shear stresses are acting along the bond line and its surroundings, as shown by Tong and Steven (p. 56, 1999). The geometry and principal characteristics of the single shear lap specimens (SL-PC2) are shown in Figure 7. The concrete mix PC-2 was used for these specimens. Samples SL-PC2-3 and SL-PC2-4 were instrumented, as shown in Figure 7.

Load, displacement, and strain were recorded continuously during the testing. Displacement control was used for the entire loading process. The peak load was reached within five (5) minutes in order to minimize rate effects and subcritical crack growth. Experimental results and failure mode are shown in Table 5 and Figure 8. As can be seen, the failure is due to crack(s) propagating in a band of about one-inch (1") width. Within this band, the crack(s) change(s) several times in direction due to the mixed nature of the loading being applied. This results in pull-off of the concrete cover at different locations along the bond line.

It must be mentioned that the failure mode and load of the pull-off test are influenced by the location of the reaction plate at the top of the concrete block. Such influence has been investigated experimentally by Yao et al. (2005). Accordingly, in their findings, the bond strength is reduced with the height of the free concrete edge (c_c),

Figure 7. Such reduction was found to be on the order of 10%, when c_c varied from 120 mm to 5 mm (Yao et al., 2005). For the purposes of this study, c_c was selected as 25 mm, since this is the typical value of the concrete cover used in reinforced concrete structures. Furthermore, it must be emphasized that the pull-off tests conducted in this study serve as a validation procedure rather than as a method for determining the bond strength. In other words, any value of c_c could have been chosen in order to validate the numerical simulations.

NUMERICAL MODELS

The finite-element program ABAQUS (Hibbitt et al., 2004) was used during the numerical analyses conducted in this study. The typical finite-element mesh is shown in Figure 9. The concrete, epoxy, and FRP were modeled using plain strain elements with reduced integration (CPE4R). Refined meshes were employed for modeling the *damage-band*, epoxy, and FRP. Bonded (tied) contact was used for connecting meshes with different densities. In this approach, each of the nodes on the fine mesh has the same displacement as the point on the coarse mesh to which it is closest. This allows for the modeling of normal and shear stresses along the entire *tied surfaces*.

The material properties used during the simulations are shown in Table 2 and Table 4. The damage of the *concrete-epoxy interface* was assumed to occur mostly in the concrete side, as mentioned before. Therefore, the fracture properties of the *concrete-epoxy interface* (CEI-2) were assigned only to the *damage band,* shown in Figure 9. The geometry of this band was selected following Bazant and Planas' (1998, p. 227) recommendations. They report that in concrete cracking, localization occurs in a crack band width $h_c = 3d_a$, where d_a is the maximum aggregate size.

Numerical results are shown in Figure 10. In this figure, the load-strain curves of the instrumented specimens SL-PC2-3 and SL-PC2-4 are compared with the numerical simulations. Excellent agreement is observed for the entire load-strain curves of specimen SL-PC2-3. The model also accurately predicts the load-strain distribution of specimen SL-PC2-4. Very good agreement is also obtained for the predicted peak load. During the experiments, the average peak load was 15.2 kN, as shown in Table 5, while the numerical prediction is 15.4 kN.

The failure mode was also accurately predicted by the numerical simulations. Figure 11 shows the model predictions. In this figure, the damage distribution is plotted for models with different mesh densities. As can be seen, the failure mode coincides with experimental results shown in Figure 8. Such results are independent of the mesh choice, as discussed in the next section.

Mesh sensitivity

Numerical simulations must be objective. Results should not depend on subjective aspects such as the choice of mesh or element size. In this study, three mesh configurations were used, as shown in Figure 11. The obtained load-displacement and load-strain curves are shown in Figure 12. All three simulations predict almost identical results. Slight differences, within less than 7%, are observed in the peak loads, displacements, and strains. These results confirm the objectivity of the numerical solution.

CONCLUSIONS

Experimental and numerical procedures were proposed in order to predict the debonding failure of concrete elements strengthened with fiber-reinforced polymers (FRP). The framework of damage mechanics was used during these procedures. Conclusions derived from the analysis are as follows:

The experimental procedure followed in this study was effective in determining the concrete and *concrete-epoxy interface* fracture energies. It was found that the fracture energies of plain concrete and *concrete-epoxy interface* are different in magnitude.

It was found that for the epoxy used in this study, the tensile strength of concrete and *concrete-epoxy interface* are similar and within the experimental range of error.

The epoxy used in this study introduces a toughening effect, i.e., crack deflection, which results in a higher fracture energy of the *concrete epoxy interface* when compared to the fracture energy of plain concrete.

The crack deflection, induced by the epoxy, occurs in a *band* (*damage band*) or region in the neighborhood of the *bond surface*. From a practical point of view, this experimental observation implies that the *damage band* should be modeled during the simulation of debonding failures.

The finite element model used in this study predicts, with very good accuracy, the strain distributions, failure load, and the failure mechanisms of the single shear lap test. The materials properties required during the simulations can be obtained using the experimental procedures outlined in this article.

ACKNOWLEDGEMENT

The research activities described in this paper have been supported by a CAREER grant from the National Science Foundation (CMS-0330592). This support is gratefully acknowledged.

REFERENCES

ACI 440.2R-02 (2002). "Guide for the Design and Construction of Externally Bonded FRP Systems for Strengthening Concrete Structures." Reported by ACI Committee 440.

ASTM C496. "Standard Test Method for Splitting Tensile Strength of Cylindrical Concrete Specimens". Annual Book of ASTM Standards, Vol 04.02.

Bazant, Z. P. and J. Planas (1998). "Fracture and size effect in concrete and other quasibrittle materials". Boca Raton, CRC Press.

Elices, M., G. V. Guinea, J. Gómez, and J. Planas. (2002). "The cohesive zone model: Advantages, limitations and challenges." Engineering Fracture Mechanics 69(2): 137-163.

Granju, J.-L. (2001). "Debonding of thin cement-based overlays." Journal of Materials in Civil Engineering 13(2): 114-120.

Hibbitt Karlsson and Sorensen. (2004). ABAQUS: Theory manual. Pawtucket, RI, Hibbitt Karlsson & Sorensen.

Lee, J., and L. G. Fenves (1998). "Plastic-damage concrete model for earthquake analysis of dams" Earthquake Engineering & Structural Dynamics 27(9): 937-956.

Lubliner, J., J. Oliver, S. Oller, and E. Onate (1989). "Plastic-damage model for concrete." International Journal of Solids and Structures 25(3): 299-326.

MacGregor, J. G. (1997). "Reinforced concrete: mechanics and design." Upper Saddle River, NJ, Prentice Hall.

Planas, J., and M. Elices (1990). "Fracture criteria for concrete. Mathematical approximations and experimental validation." Engineering Fracture Mechanics International Conference on Fracture and Damage of Concrete and Rock and Special Seminar on Large Concrete Dam Structures, Jun 4-6 1988 35(1-3): 87-94.

Rocco, C., G. V. Guinea, B., J. Planas, and M. Elices. (2001). "Review of the splitting-test standards from a fracture mechanics point of view." Cement and Concrete Research 31(1): 73-82.

Tong, L., and G. P. Steven (1999). "Analysis and design of structural bonded joints." Boston, Kluwer Academic.

Yao, J., Teng, J. G., and Chen, J. F. (2005). "Experimental study on FRP-to-concrete bonded joints." Composites Part B: Engineering 36(2): 99-113.

Table 1 – Material constitutive models

Material	Constitutive Model	Required Material Properties
FRP	Linear Elastic	Modulus of Elasticity (E_{FRP}) Poisson ratio (ν_{FRP}) Tensile strength (F_{uFRP})
Epoxy	Elastoplastic Von Mises material with isotropic hardening	Modulus of Elasticity (E_e) Poisson ratio (ν_e) Tensile strength (F_{ue}) Stress strain curve in tension
Plain Concrete (PC)	Concrete Damaged plasticity	Modulus of Elasticity (E_c) Poisson ratio (ν_c) Compressive strength (f'_c)
Concrete-epoxy Interface (CEI)		Tensile strength (f_t) Fracture energy (G_F) Stress strain curve in compression Softening curve in tension

Table 2 – Mechanical properties for the materials used during the experimental program

Properties	Concrete PC-1	Concrete PC-2	Epoxy WS	CFRP
Modulus of elasticity, E (GPa)	21.7 ± 0.7	27.7 ± 2.1	3.0±0.35	207.9 ± 8.6
Poisson ratio	0.2	0.2	0.38	--
Yield strength, f_y (MPa)	--	--	20.1 ± 7.1	--
Tensile strength, f_{tu} (MPa)	3.2 ± 0.15	4.0 ± 0.2	42.2 ± 1.6	4667 ± 390
Compressive strength, f'_c (MPa)	32.2 ± 1.6	48.8 ± 3.3	--	--
Ultimate strain (%)	--	--	--	2.05 ± 0.1

Table 3 – Geometry and material properties of three-point bending specimens.

Sample	f'_c [MPa]	f_{st} [MPa]	P_{max} [N]	B [mm]	D [mm]	f_p [MPa]	α_0
CEI-1-A[1]	32.2	2.9	8804	154.9	145.9	1.36	0.33
CEI-1-B	32.2	2.9	8524	153.5	146.3	1.30	0.32
CEI-2-A	48.8	4.1	10583	150.1	148.3	1.62	0.33
CEI-2-B	48.8	4.1	11867	146.7	153.5	1.60	0.30
CEI-2-C	48.8	4.1	10866	152.6	145.4	1.73	0.33
PC-1-A	32.2	3.2	8245	144.6	153.8	1.25	0.34
PC-1-B	32.2	3.2	8261	152.8	146.0	1.21	0.31
PC-1-C	32.2	3.2	8234	148.6	153.1	1.21	0.33
PC-2-A	48.8	4.0	10635	149.2	153.3	1.56	0.33
PC-2-B	48.8	4.0	10759	154.0	147.3	1.55	0.31

[1] Concrete epoxy interface (CEI); Plane concrete (PC)

Table 4 – Fracture mechanics properties of plain concrete and concrete epoxy interface.

Sample	f_t [MPa]	l_l [mm]	w_l [mm]	G_f [N/m]	G_F [N/m]	w_{ch} [mm]
CEI-1	2.51	291.89	67.53	84.76	211.90	0.084
CEI-2	3.74	149.01	40.27	75.35	188.37	0.050
PC-1	2.96	115.81	31.55	46.63	116.57	0.039
PC-2	3.68	120.01	31.92	58.80	146.99	0.040

Table 5 – Failure load of single shear lap specimens.

Test	P_{max}
SLPC2-1	14.66
SLPC2-2	15.84
SLPC2-3	14.72
SLPC2-4	15.61
Average	14.77
Coefficient of variance	4%

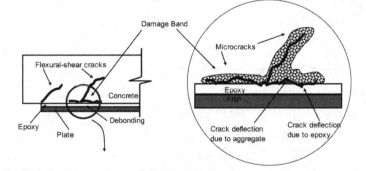

Figure 1 – Damage of the concrete-epoxy interface during debonding failure.

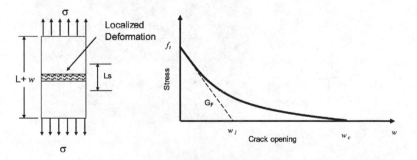

Figure 2 – Softening curve for the concrete-epoxy interface under uniaxial tensile loading.

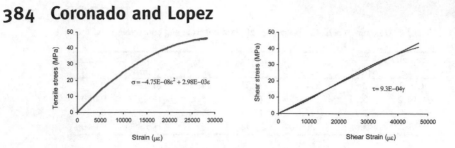

Figure 3 – Epoxy stress-strain behavior. (a) Axial tension, (b) Shear.

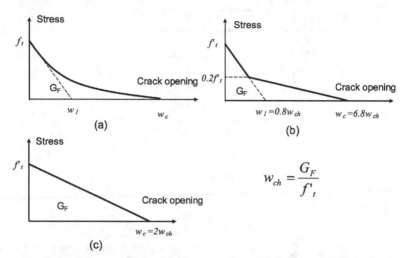

Figure 4 – Concrete Softening function and bilinear approximation.

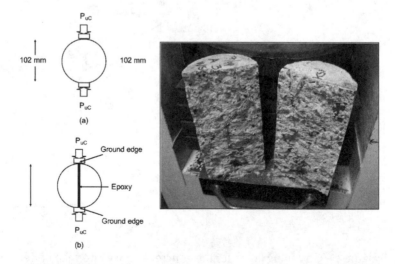

Figure 5 – Splitting tensile test specimens. (a) Plain concrete (PC);
(b) Concrete-epoxy interface (CEI); (c) Typical failure of the CEI.

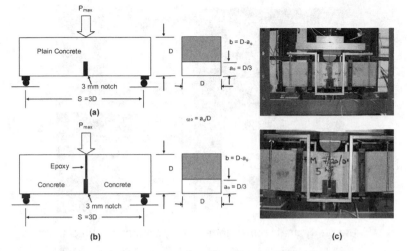

Figure 6 — Three-point bending test. (a) Plain concrete (PC) specimens; (b) Concrete epoxy interface (CEI) specimens; (c) Failure mode of CEI specimens.

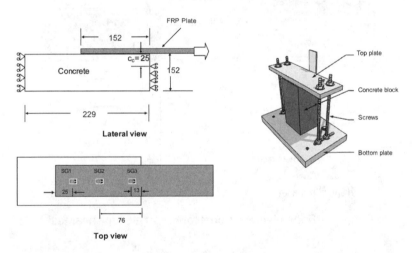

Figure 7 – Geometry of the single shear lap specimens. Dimensions in millimeters (mm).

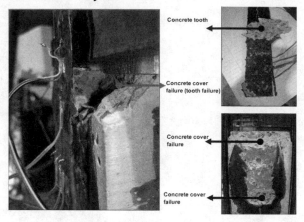

Figure 8 – Typical failure mode of single shear lap specimens.

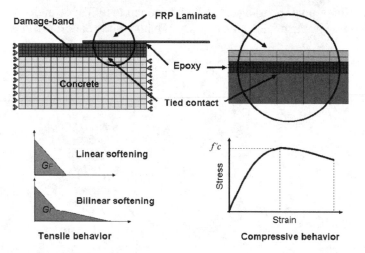

Figure 9 – Finite element model for the single lap shear test.

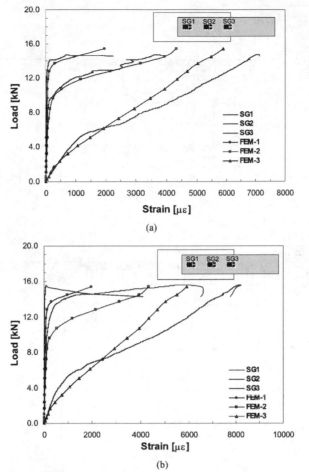

Figure 10 – Single shear lap test. Compared load-strain results
(a) specimen SL-PC2-3, (b) specimen SL-PC2-4.

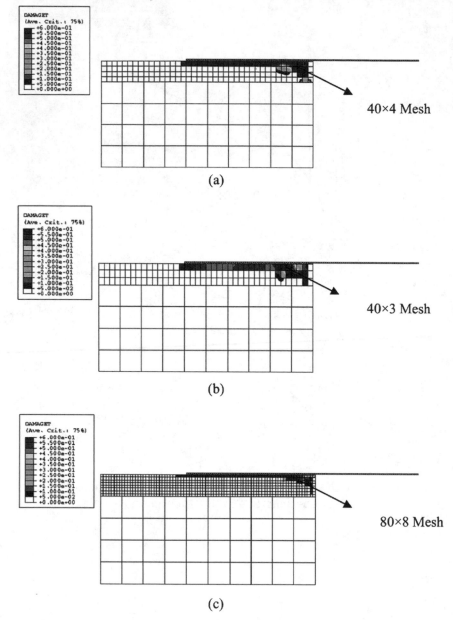

Figure 11 – Numerical prediction of the failure mode. (a) 40×4 Mesh,
(b) 40×3 Mesh, and (c) 80×8 Mesh.

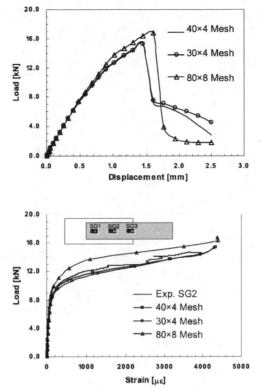

Figure 12 — Mesh Sensitivity. (a) Global load displacement response.
(b) Local load strain response.

Experimental Study of Short NSM–FRP Bar Anchorages

by D.G. Novidis and S.J. Pantazopoulou

Synopsis: In this paper bond tests conducted on short NSM-bar anchorages are presented and results are evaluated to determine the effect of design variables on development capacity. Specimens were designed so as to minimize spurious secondary stresses along the bonded length. Parameters of the study were the groove dimensions and the embedment length. Bars were CFRP reinforcement with sandblasted surface characteristics. Analytical expressions are derived from first principles so as to interpret the observed experimental trends. The parameters of the analytical bond stress - slip model are calibrated to the experimentally derived bond-slip relationships in order to establish a limit-state model for post-installed reinforcement, for practical upgrading applications.

Keywords: analytical model; bond; carbon fiber reinforcement polymer bars; mechanical tests; near surface mounted bars; slip

D. G. Novidis is PhD candidate at Demokritus University of Thrace (DUTh), Greece. He holds a Civil Engineering Diploma and MSc. Degrees from the same University. His research interests are in bond and development of FRP reinforcement, and in the use of FRPs in repair/upgrading of reinforced concrete structures.

S. J. Pantazopoulou, F.ACI, is Professor in the Dept. of Civil Engineering, DUTh, Greece. Prior to her current appointment she was tenured Assoc. Professor of Civil Engineering at the Univ. of Toronto, Canada. She holds PhD and MS degrees in Structural Engineering from the Univ. of California at Berkeley, and an Engineering Diploma from NTUA, Greece. She is member of several ACI Committees and *fib* TGs.

INTRODUCTION

With the advent of FRP materials alternative uses for upgrading/strengthening of existing structures with these new materials have been explored. Increasing or replacing damaged reinforcement in concrete members by post-installing rebars in surface grooves cut on the tension side or the web is known as near-surface mounting technology. This method had been tried in the past with steel rebars with the help of a cement-based mortar for groove filler. However, as with externally bonded laminates performance of the strengthened element was fraught with a number of problems arising usually near the cutoff points, owing mostly to the stress concentrations at these points and the tendency of the post-installed bars to delaminate. Today this technology has received renewed attention because the variety of FRP products available (bars having a lower modulus than steel, jacketing sheets, and enhanced mortars) provide solutions to some of the problems encountered in upgrading substandard or damaged reinforced concrete members.

Added reinforcement is usually mounted with the help of the filler material that is placed inside the surface groove, which, after hardening functions as the cover of the post-installed bar. Performance of the bar largely depends on its development capacity, which is controlled by its surface characteristics, its extensional stiffness and by the fracture properties of the filler. Once placed, near-surface mounted bars (NSM) are subjected to similar displacement conditions as the existing reinforcement. If force transfer can be supported through bond, then the bar may effectively function as the existing reinforcement, developing strain and its force capacity, thereby contributing to the member resistance. In the absence of bond it is not possible to develop a strain gradient along the bar as would be required by statics, and thus no tension stiffening may be mobilized. Thus, bond of post-installed reinforcement is essential in securing composite action with the existing members. Bond also controls the strength and the eventual failure mode of the upgraded member.

Stress transfer and interaction occurs along two interfaces, namely the cylindrical contact surface between the NSM bar and the surrounding filler, and along the rectangular boundary of the groove, between groove-filler and the surrounding concrete. In both cases force transfer is achieved through frictional bond if the surfaces in contact are smooth. Mechanical interlocking between bar and filler may also occur if the bar used has surface deformations; ribs add the possibility of a splitting failure of the filler over the anchorage length, a failure mode that is less likely in sand-coated bars with or without indentations. When using FRP rods for NSM reinforcement, the dependable bond strength is controlled entirely by the surface characteristics and rod stiffness, the groove dimensions and the shear behavior of the filler material. Considering the variety of commercially available products the breadth of the available international experimental database is limited as of yet. Extensive experimentation is still needed in order to expand the available data on bond and development of NSM bars, so as to fully characterize a general design model for practical applications. Laboratory studies and field applications for bond of composite bars are reported by Ehsani et al. (1996), Consenza et al. (1997), Guo and Cox (2000), Focacci et al. (2000), Malvar et al. (2003), Cox and Bergeron-Cochran (2003), Thiagarajan (2003), Pilakoutas and Achillides (2004). Studies specifically addressing bond and development capacity of NSM – FRP bars have been conducted by De Lorenzis (2000), De Lorenzis and Nanni (2001, 2002), De Lorenzis et al. (2004).

In this paper the bond properties of FRP rods used as NSM reinforcement are explored through testing of twenty-four short-anchorage concrete specimens having various bonded lengths and prefabricated groove dimensions. To this end the familiar pullout specimen was pertinently modified so as to minimize any spurious diagonal compression stress fields owing to the loading setup that may affect the attained bond strength, (Figures 1,2,3). An analytical model for bond of composite reinforcing bars used as NSM post-installed reinforcement is also developed and calibrated against the experimental data.

RESEARCH SIGNIFICANCE

The objective of this paper is to evaluate the modified familiar pullout specimen and to quantify the essential parameters that characterize the bond-slip relationship for this NSM reinforcement as a function of the groove dimensions, groove filler material properties and bar diameter and stiffness. Such a relationship is an essential tool for detailing post-installed reinforcement in retrofit solutions aiming to upgrade the flexural and shear resistance of substandard or damaged existing reinforced concrete elements.

EXPERIMENTAL PROGRAM

A total of twenty-four (24) specimens were tested. The typical specimen comprised a 150.0 mm square, 300.0 mm long concrete block with the test bar located in a plane of symmetry of the block near a free surface. Surface grooves having a square cross-section were preformed and the bar was installed using epoxy paste.

To achieve almost uniform conditions over the anchorage so as to quantify bond strength short anchorage specimens were used. The embedded length extended from the top of the specimen down by $3D_b$, $5D_b$, $7.5D_b$ or $10D_b$. Note that bond tests of NSM bars are hampered by the same problems as bond tests of conventional reinforcement. For this reason, a specially designed eccentric-pullout specimen was used in the study, in order to minimize unintended stress conditions in the anchorage zone. Thus, compression-field breakers were used in order to interrupt the diagonal compression field that rises due to friction at the bearing plate. To avoid tension-cracking specimens were reinforced in tension by externally bonded CFRP sheets placed lengthwise and extending downwards to the end of the specimen. The sheets terminated at a transverse distance of 50.0 mm from the bar axis, so as to avoid interference with the shear stress-state normally developing along the anchored NSM-bar.

Test variables were the bonded length; the size of the groove and the passive confining pressure exerted by externally bonded (EB)-CFRP jackets placed transversely over the anchorage. Commercially supplied C25/30 concrete was used, having an average 28-day cylinder compressive strength of 30.0 MPa (34.6 MPa at the time of testing). NSM-bars were 12 mm diameter CFRP rods having a sandblasted surface and winding helical lengthwise indentations (product commercially known as Aslan 200 CFRP bar). Nominal bar properties were as follows: modulus of elasticity of 124.0 GPa, and 1.7% ultimate strain. Specimens were allowed to cure for 28 days and then the epoxy paste was prepared by mixing the two components (resin and hardener) in 3:1 proportion (Figure 4). The mechanical properties of the epoxy paste as specified by the manufacturer were 30.0 MPa tensile strength and 3.0 GPa Modulus of Elasticity. To control experimental scatter specimens were tested in replicas of three; thus values listed in Table 1 are the average of three tests. Character J in the specimen identification code denotes anchorages confined with a single EB-CFRP layer for confinement. Nominal properties of the jacket material as reported by the manufacturer were, modulus of elasticity of 230.0 GPa and 1.5% ultimate strain.

Specimens were instrumented with linear-variable differential transducers. Furthermore, the strain field in the concrete surface over the anchorage was mapped with digital imaging of a grid of optical targets throughout the test. Relative displacements and strains were obtained in pixels after processing of successive digital images and were subsequently converted to mm using as a reference the size in pixels of the target diameter, which was 10.0 mm. This method of mapping the strain field is similar to that using a Zurich gauge between metallic targets, whereby normalized changes in the size of the grid (including diagonals) can be used as a strain rosette to calculate principal strains. The accuracy of the method depends on the analysis capacity (number of pixels per square mm) and distance of the camera from the surface monitored. The loading setup is depicted in Figures 3 and 5. The free end of the test bar was inserted in a 150.0 mm long, 60.0 mm diameter and 5mm thick steel pipe filled with epoxy paste that was bearing against the steel plate of the loading frame. This method of loading was meant to eliminate local crushing of the FRP bar in the transverse direction owing to the hydraulic pressure exerted by the conventional machine grips. Tensile load was applied to the NSM rod using a 200.0 KN piston in several stages, monotonically increasing to specimen failure. Failure occurred by splitting and successive pullout along one of the two interfaces mentioned in the preceding.

Table 1 outlines the variables of the experimental program. Specimens were tested in replicas of three so as to control experimental scatter. Based on the combination of these variables, eight groups of specimens are identified (group number corresponds to the first letter in the specimen acronym, A:H); the numeral that follows identifies the specimen number in the group of replicas it belongs, whereas the numeral after the period denotes the anchorage length (in multiples of the bar diameter, D_b). The numeral after the letter G identifies in mm the depth of the square grooves: either 25.0 mm or 20.0 mm grooves were used. The last character in the specimen identification code is the letter J denoting, when used, anchorages jacketed with a single EB-CFRP layer for confinement, Figure 6. Nominal properties of the jacket material as reported by the manufacturer were, modulus of elasticity of 230.0 GPa and 1.5% ultimate strain.

THE STATE OF STRESS ALONG AN NSM BAR

The frictional mechanism is the basic physical construct that underlies all design models for bond. In attempting to develop the FRP bar in a near-surface groove the effort is to mobilize frictional forces along the critical interfaces, i.e. at the contact between the bar and the groove filler, and at the contact between the groove filler and the surrounding concrete. A common drawback of some commercially available FRP bars is that they do not possess a ribbed profile so as to mobilize significant normal pressures with slip. Thus, weaker friction may develop in these cases. By establishing equilibrium in polar coordinates for the ring of the cover-filler surrounding the bar, (Figure 7), it may be shown that normal pressures generated in the radial direction due to bar slip attenuate with increasing radius. In this regard the normal pressure σ_c acting at the concrete-filler (CF) interface is at a ratio of d_g/D_b with respect to the corresponding pressure on the filler-bar (FB) interface, σ_b, (Figure 7). Assuming comparable magnitudes for the coefficient of friction at the two interfaces, it is more likely that frictional resistance is less at the CF contact and hence this surface is more susceptible to shear failure. This point is supported by the available experimental evidence. Of the alternative modes of failure by far the most commonly observed is debonding at the CF interface; debonding at the FB interface is less common although it has been seen in some cases. Depending on the controlling failure mode, the corresponding interface bond stress identifies the strength of the respective mechanism. Near failure the two mechanisms of shear transfer are competitive, with slip localizing along the failed interface as evidenced by digital image processing of bond specimens during loading. In these cases, the associated slip magnitude in the non-failing interface was insignificant.

For short anchorages, the average bond stresses are obtained from the developed bar stress, f_B, and the contact surface area, which equals the product of the anchorage length L_b and the contact perimeter, p_b and p_g for bar and groove respectively:

$$f_b^b = \frac{D_B}{4} \cdot \frac{f_B}{L_b} \quad ; \quad f_b^c = \frac{D_B}{4} \frac{f_B}{L_b} \cdot \frac{\pi D_B}{p_g}; \text{ where } p_g = 3d_g; \text{ thus, } f_b^c \approx f_b^b \frac{D_B}{d_g} \qquad (1)$$

thus the bond demand at the CF joint, f_b^c, is at a ratio of D_B/d^g with the bond demand at the FB joint, f_b^b. (Subscript B in (1) refers to the bar properties; subscript b refers to bond and development.) Filler properties appear to have great influence on the failure mode and development capacity of the post-installed NSM bar: considering average bond stresses along the two contact surfaces, it may be shown from equilibrium of an infinitesimal bar length dx, that the filler material is under a state of shear and radial compression, with zero resultant longitudinal stress. Thus diagonal tension failure is anticipated, with the angle of inclination of principal stresses in the longitudinal direction resembling that which occurs in the cover of a ribbed bar. In such a state of stress, brittleness of the filler material is non-conducive to efficient bar development through redistribution of stress in the filler. Thus it is more likely that a more compliant material with residual post-cracking resistance will produce a more ductile response in the bar at a lower development capacity, whereas a brittle but stiff filler will lead to higher loads and instantaneous debonding with a quick release of energy upon fracture. The basic differential equation governing equilibrium of an elementary segment of length dx over the anchorage is (Filippou Popov and Bertero 1982, Cosenza et al 1997, De Laurenzis et al 2001):

$$\frac{df_B(x)}{dx} = \frac{p_{crit}}{A_B} \cdot f_b^{crit}(x) \tag{2}$$

Subscript "crit" refers to the critical contact surface (i.e., the contact surface where failure is anticipated either along the bar-epoxy or along the epoxy-concrete joint.) Since the filler material is in a state of shear with no longitudinal tension on average, the slip value of the bar is defined with reference to the concrete as:

$$\frac{ds_{b,c}}{dx} = \varepsilon_B \tag{3}$$

There is no differential slip between filler and surrounding concrete ($ds_{f,c} \approx 0$); rather, a rigid body slip occurs along the CF joint, equal to:

$$s_{f,c} = \gamma_f^{ave} \cdot c_f \quad \text{where the filler shear distortion equals} \quad \gamma_f^{ave} = (1 + v_f) \cdot \frac{\left(f_b^b + f_b^c\right)}{E_f} \tag{4}$$

and c_f is the filler thickness over the bar. If the post-installed bar is elastic (usually the case with FRP bars) then bar stress is in linear relation with associated strain (i.e., $\varepsilon_B(x) = \sigma_B(x)/E_B$). A bilinear bond-slip relationship with an initial elastic branch and a yield plateau for slip values ranging between s_y and the slip at failure, s_u, is assumed for the weaker of the two contact surfaces (Figures 8, 9). For a given bar strain at the critical section, $\varepsilon_{B,o}$ the distribution of bar strain and slip over the anchorage length L_b, is obtained considering (2) and (3): here it is important to identify whether the bond-slip law has entered the stage of plastification over the bar. This occurs when the bar strain exceeds the milestone value, $\varepsilon_B^{b,y}$, which corresponds to an associated slip equal to s_y. Thus, for the two cases solution of the differential equations listed above leads to the following results:

(a)   For bar strains below the strain limit $\varepsilon_B^{b,y}$:

$$\varepsilon_{B,o} \leq \varepsilon_B^{b,y}: \quad \varepsilon_B(x) = \varepsilon_{B,o} e^{-\omega x \left[\dfrac{1-e^{2\omega(x-L_b)}}{1-e^{-2\omega L_b}}\right]} ; \qquad f_B(x) = E_B \cdot \varepsilon_B(x)$$

$$\text{(5)}$$

$$\text{and,} \quad s_B(x) = \dfrac{\varepsilon_{B,o}}{\omega} e^{-\omega x \left[\dfrac{1+e^{2\omega(x-L_b)}}{1-e^{-2\omega L_b}}\right]} ; \qquad f_b^{crit}(x) = K_{crit} \cdot s_B(x)$$

In (5) $\varepsilon_{B,o}$ is the strain at the critical section; variable ω^2 is defined as $\omega^2 = p_{crit} K_{crit}/E_B A_B$. The value $\varepsilon_B^{b,y}$ is calculated from:

$$\varepsilon_B^{b,y} = \omega \cdot s_y \dfrac{1 - e^{-2\omega L_b}}{1 + e^{-2\omega L_b}} \tag{6}$$

(b) For bar strains above the strain limit $\varepsilon_B^{b,y}$:

$$\varepsilon_{B,o} > \varepsilon_B^{b,y}: \quad \varepsilon_B(x) = \varepsilon_{B,o} - x \cdot \dfrac{p_{crit} \cdot f_{b,max}}{A_B E_B} \quad ; \quad f_B = E_B \cdot \varepsilon_B(x) \tag{7}$$

Whereas the length l_p of bar over which bond is in the plastic range and the corresponding slip s:

$$l_p = (\varepsilon_{B,o} - \varepsilon_B^{b,y}) \dfrac{D_B E_B}{p_{crit} \cdot f_{b,max}} \tag{8}$$

$$s = s_y + 0.5 \cdot l_p \cdot (\varepsilon_{B,o} - \varepsilon_B^{b,y}) \leq s_u \quad ; \quad f_b^{crit}(x) = f_b \tag{9}$$

Clearly, bar strains and stresses, as well as the bond and slip decay exponentially with distance from the critical section over the anchorage length, L_b, for the segment where bar strains are below the limit $\varepsilon_B^{b,y}$. From the above expressions it follows that slip and the corresponding bond stress is nonzero at the free end of the bar (at the end of anchorage):

$$\varepsilon_{B,o} \leq \varepsilon_B^{b,y}: \quad s_{L_b} = \dfrac{2\varepsilon_{B,o} e^{-\omega L_b}}{\omega(1 - e^{-2\omega L_b})} \neq 0 \tag{10}$$

$$\varepsilon_{B,o} > \varepsilon_B^{b,y}: \quad s_{L_b} = \dfrac{2\varepsilon_B^{b,y} e^{-\omega(L_b - l_p)}}{\omega(1 - e^{-2\omega(L_b - l_p)})} \neq 0 \tag{11}$$

The average bond stress, f_{crit}^{ave}, is estimated from the applied bar load distributed over the anchorage length:

$$f_{crit}^{ave} = \frac{1}{L_b}\left[\int_0^{l_p} f_{b,max}\,dx + \int_{l_p}^{L_b} f_b(x)\,dx\right] \Rightarrow f_{crit}^{ave} = \left[\frac{l_p}{L_b} + \frac{\varepsilon_y}{s_y \cdot L_b \cdot \omega^2}\right] \cdot f_{b,max} \quad (12)$$

Clearly, the anchorage cannot sustain bond stresses exceeding the limit value $f_{b,max}$. Plastification initiates when peak bond stress reaches the bond strength, $f_{b,max}$; failure occurs by debonding when the slip at the critical section exceeds the limit value s_u. This is the point when bar axial strain reaches the value:

$$\varepsilon_{B,o} = 2\frac{(s_u - s_y)}{l_p} + \varepsilon_B^{b,y} \quad (13)$$

As debonding proceeds bond stresses tend to even out over the remaining bonded length, till the stress distribution becomes practically uniform and equal to the strength. When this occurs there are no reserves to the anchorage and failure is abrupt. To simulate this failure process, which is marked by progressive debonding from the critical section towards the free end of the bar, bar strain at the critical section is kept constant, whereas peak bond stress is taken equal to strength. By progressively increasing the debonded length, l_r, so that the residual anchorage length is $L_{b,r} = L_b - l_r$, (Figure 10), equilibrium is sought for the anchorage: thus, the values of $x - l_r$, $L_b - l_r$ and $l_p - l_r$ are used in equations (5)-(13) in place of x, L_b and l_p. For comparison with experimental values, the slip at the critical section (i.e., at $x=0$) is obtained by adding the amount of bar elongation that occurs over the debonded length to the slip at the end of the debonded length (from equation 9). Thus,

$$s(x = 0) = s_u + \varepsilon_{B,o} \cdot l_r \quad , \quad s_u(x = l_r) = s_u \quad (14)$$

TEST RESULTS

Test results in terms of ultimate pullout load, ultimate slip and failure mode, are summarized in Table 1. Also listed are the average bond strengths calculated at the two interfaces, namely f_b^c between the concrete and the groove filler (epoxy paste), and f_b^b between the CFRP rod and the groove filler. Two alternative failure modes were observed in this experimental study, namely either pullout at the interface between concrete and groove filler or pullout at the interface between CFRP rod and groove filler. Figures 11, 12 and 13 plot for all specimens the ultimate load, and the f_b^c and f_b^b values versus the associated interface slip values (s_b^c and s_b^b represent the relative slip between concrete and groove filler, and the relative slip between rod and groove filler, respectively). Depending on the controlling failure mode, the corresponding interface bond stress identifies the strength of the respective mechanism. From digital image processing slip localization was observed to occur along the failed interface whereas slip in the other interface was negligible.

Figures 14, 15 highlight the effect of L_b and d_g on average bond strength; for a given groove size the development capacity in terms of total load carried by the joint increases with the bonded length, L_b. However, the average bond strength decreases after a critical bonded length value, due to the non-uniform distribution of the bond stresses along the bonded length (Figure 14). As in the case of high strength concrete, specifying development capacity in terms of an average stress fails as a concept when extrapolated to groove filler owing to the brittleness of the material as compared with conventional normal strength concrete.

For a given bonded length, L_b, the ultimate load increases with increasing groove size, but the parametric sensitivity of the average bond strength depends on the failure mode as observed in Figure 15. Thus, f_b^c reduces as the groove size increases while f_b^b rises. For the specimens that failed at the concrete – epoxy interface, which was observed in the majority of specimens tested, actual failure comprised a combination of adhesion breakdown and frictional sliding between epoxy and concrete. The latter component depends on the amount of normal stresses at the epoxy – concrete interface. This in turn depends on the normal stresses exerted by the CFRP rod on the surrounding material at a given slip magnitude, being a function of the surface pattern of the rod. Since the surface pattern of the CFRP rod used in the present study is sandblasted with lengthwise winding helical indentations the normal stresses exerted by the CFRP rod are relatively small and attenuate at a linear rate with distance, reaching the minimal value at the epoxy - concrete interface as illustrated from the thick-ring analogy depicted in Figure 7.

In the present study it was illustrated that confinement by means of CFRP jacketing over the anchorage length of NSM bars may delay failure at the concrete – epoxy interface promoting failure at the CFRP rod – epoxy contact. In general, all jacketed specimens (identified as B-5G25-J in Table 1) resulted in higher average bond strength values as compared to the other specimens.

Figures 16, 17 plots results of the above procedure applied to specimens E-3G20 and F-7.5G25 having development lengths of $3D_b$ (=36mm) and $7.5D_b$ (=90mm) respectively, at the onset of specimen failure (associated bar strains at the critical section were $\varepsilon_{o,r}$=0.00051 and 0.0013 for the two cases). Note that upon first attainment of the critical bar strain bond distribution is highly nonlinear. Deviation from the nominal average bond stress value is also given in the plot. Calculated response of specimen F-7.5G25 is plotted in Figure 17: as debonding proceeds bond stresses tend to even out over the remaining bonded length, till the stress distribution becomes practically uniform and equal to the strength (average bond stress of 3.9 MPa for $L_{b,r}$ of 63 mm, whereas the bond strength is 4MPa). When this occurs there are no reserves to the anchorage and failure is abrupt. The slip reported from the tests corresponds to that final point; in the example considered slip upon first attainment of the limiting bar strain at the critical section was 0.54 mm, whereas at failure it had increased to 0.75 mm, Figure 17.

CALIBRATION OF THE MODEL WITH THE EXPERIMENTAL RESULTS

The analytical expressions derived in the preceding are general and represent the behavior in three stages: linear elastic-bond slip resistance, plastification of the bond mechanism, and eventual rupture of bond and debonding. The stage of plastification is insignificant for short anchorages because the system carrying shear stresses (contact interfaces) has no redundancy and cannot support redistribution. Plastification is seen in extended anchorages (L_b exceeding $12D_b$), (Novidis D., et al, 2005).

When using the model to calibrate the experimental results, measured values were corrected for the rigid body displacement of the bar that was owing to epoxy distortion, as indicated by the term $s_{f,c}$ in Eqn. (4). The corrected values are listed in Table 1. The principal unknowns of the correlation were, the bond strength of the critical interface, f_{crit}^{max}, and the associated slip at the characteristic points needed to completely define the bond-slip curve, s_y, s_r. Application of the model to the entire family of the test results yielded the following values for these two variables (Figure 10):

(a) For Failure at the Bar-Epoxy Joint: f_{crit}^{max} =8MPa, s_r = {0.4-0.65}mm
(b) For Failure at the Epoxy Paste – Concrete Joint: f_{crit}^{max} =4MPa, s_r = {0.5-0.75}mm
(Note: 25.4 mm = 1 in. , 1 MPa = 145 psi or 6.895 MPa = 1 ksi.)

For the short anchorages in general no plastification could be identified from the correlation, a result consistent with the experimental observation. Thus for the present tests, s_y = s_r. Note that a range of values was obtained for slip, as this variable demonstrates much greater variability than stress, owing to the many factors influencing the test measurement (e.g. the exact point in the process of failure to which the recorded value corresponds, the relatively small magnitude of slip as compared to the accuracy of the recording instruments, etc).

CONCLUSIONS

In this paper bond of NSM – FRP bars in concrete was examined analytically and experimentally. A total of 24 short anchorage specimens were tested in order to establish design values for the characteristic points of a bilinear elastic-plastic limit state bond-slip model for post-installed FRP reinforcement with epoxy groove filler. The familiar eccentric pullout specimen was modified so as to minimize any spurious diagonal compression stress fields owing to the loading setup that may affect the attained bond strength. From the experimental study it was found that the modified specimen is an effective device for investigation of bond, producing reliable data while maintaining a manageable specimen size.

Two alternative failure modes were observed in the tests, namely either pullout at the interface between concrete and epoxy paste and pullout at the interface between CFRP rod and epoxy paste. It was found that by increasing the groove size, higher average bond strengths might be obtained when failure is controlled by pullout at the interface between rod and groove filler. When failure occurs at the interface between concrete and groove filler the variation of the groove size either decreases the average bond strength or has no significant effect, (Figure 15). Similarly, increasing the bonded length, L_b for a given groove size produces a commensurate increase in development capacity of the joint (in terms of total load carried). However, the average bond strength decreases after a critical bonded length, due to the non-uniform distribution of the bond stresses along the bonded length as shown in Figure 14.

The anchorage length was evaluated by solving the differential equation of bond considering the local bond stress-slip relationship of the NSM rods. The experimental results were correlated with the analytical model developed in the paper, to calibrate design values for bond strength and limiting slip at rupture. Depending on the mode of failure, bond strength was evaluated at 4 MPa (for failure at the epoxy-concrete joint) and 8 MPa (for failure at the bar-epoxy joint). A range of values between 0.5 and 0.75 mm was obtained for the associated slip at anchorage failure.

ACKNOWLEDGEMENTS

This investigation was conducted in the Laboratories of the Department of Civil Engineering, Democritus University of Thrace (DUTH), Xanthi - Greece under a research program funded by the Hellenic General Secretariat for Research and Technology (PENED2001). MAC BETON HELLAS S.A. generously donated the epoxy-pastes used in the experimental program. FRP reinforcement was purchased from Hughes Brothers (USA).

REFERENCES

ACI Committee 440 (2001). "Guide for the Design and Construction or Concrete Reinforced with FRP Bars (*ACI 440.1R-01*)", American Concrete Institute, Farmington Hills, Michigan, 42pp.

CEB (1991) "CEB-FIP Model Code 1990, Design Code", Comite Euro-International du Beton, Lausanne, Switzerland.

Consenza E., Manfredi G., Realfonzo R. (1997). "Behavior and Modeling of Bond of FRP Bars to Concrete", *ASCE J. of Composites for Construction*, V. 1, No.2, pp. 40-51.

Cox J. V. and Bergeron-Cochran K. (2003). "Bond between carbon fiber reinforced polymer bars and concrete. ii: computational modeling", *ASCE J. of Composites for Construction*, V. 7, No. 2, pp. 164–171.

De Lorenzis, L. (2000). "Strengthening of RC structures with near surface mounted FRP rods", MS thesis, Dept. of Civ. Engrg., The University of Missouri–Rolla, Rolla, Mo.

De Lorenzis L., Lundgren K., and Rizzo A. (2004). "Anchorage Length of Near-Surface Mounted FRP Bars for Concrete Strengthening—Experimental Investigation and Numerical Modeling", *ACI Structural J.*, V. 101, No. 2, pp. 269-278.

De Lorenzis L. and Nanni A. (2001). "Characterization of FRP rods as near-surface mounted reinforcement", *ASCE J. of Composites for Construction*, V. 5, No. 2, pp. 114–21.

De Lorenzis L. and Nanni A. (2002). "Bond of Near-Surface Mounted FRP Rods to Concrete in Structural Strengthening", *ACI Structural J.*, V. 99, No. 2, pp. 123-132.

De Lorenzis L., Rizzo A. and La Tegola A. (2002). "A modified pull-out test for bond of near-surface mounted FRP rods in concrete", *Elsevier, Composites: Part B*, V. 33, pp. 589–603

Ehsani M.R., Saadatmanesh H., Tao S. (1996). "Design Recommendations for Bond of FRP Rebars to Concrete", *ASCE J. of Structural Engineering*, V.122, No. 3.

Eligehausen R., Popov E.P., Bertero V.V. (1983). "Local Bond Stress-Slip Relationships of Deformed Bars under Generalized Excitations", Report No.83/23, Earthquake Engineering Res. Cnl. (EERC), University of California, Berkeley, Calif.

Focacci F., Nanni A., Bakis CE. (2000). "Local Bond-Slip Relationship for FRP Reinforcement in Concrete", *ASCE J. of Composites for Construction*, V.4, No.1, pp. 1-9.

Guo J. and Cox. J. V. (2000). "An Interface Model for the Mechanical Interaction between FRP Bars and Concrete", *Elsevier, Journal of Reinforced Plastics and Composites*, V. 19, No. 1, pp. 15-33.

Malvar L. J., Cox J. V. and K. Bergeron Cochran (2003). "Bond between Carbon Fiber Reinforced Polymer Bars and Concrete. I: Experimental Study", *ASCE J Composites for Construction*, V. 7, No.2, pp. 154–163.

Novidis D., Pantazopoulou S. J. and Tentolouris E. (2005). "Experimental study of bond of NSM – FRP reinforcement", under review, *Elsevier Construction and Building Materials*.

Pilakoutas K. and Achillides Z. (2004). "Bond Behavior of Fiber Reinforced Polymer Bars under Direct Pullout Conditions", *ASCE J. of Composites for Construction*, V. 8, No.2, pp. 173–181.

Thiagarajan G. (2003). "Experimental and Analytical Behavior of Carbon Fiber-Based Rods as Flexural Reinforcement", *ASCE J. of Composites for Construction*, V.7, No. 1, pp. 64–72.

Table 1 – Description of specimens and test results (averages of 3 specimens)

Specimen Code	P_{max}, KN	Groove d_g, mm	Actual D_b, mm	d_g/D_b, k	L_b in D_b	S_b^c, mm, Eq(4)	S_b^b, mm, Eq(4)	$f_{b\ ave}^c$, MPa	$f_{b\ ave}^b$, MPa	Failure mode*
A1-5G25										
A2-5G25	15.20	25	12.0	2.08	5		0.67	3.38	**6.72**	2
A3-5G25										
B1-5G25-J										
B2-5G25-J	16.92	25	12.0	2.08	5		0.38	3.76	**7.48**	2
B3-5G25-J										
C1-5G20										
C2-5G20	12.05	20	12.0	1.67	5	**0.41**		3.35	5.33	1
C3-5G20										
D1-3G25										
D2-3G25	7.96	25	12.0	2.08	3	**0.65**		2.95	5.87	1
D3-3G25										
E1-3G20										
E2-3G20	7.17	20	12.0	1.67	3	**0.50**		3.32	5.28	1
E3-3G20										
F1-7.5G25										
F2-7.5G25	18.20	25	12.0	2.08	7.5	**0.78**		2.70	5.36	1
F3-7.5G25										
G1-10G25										
G2-10G25	21.60	25	12.0	2.08	10	**0.97**		2.40	4.78	1
G3-10G25										
H1-10G20										
H2-10G20	19.40	20	12.0	1.67	10	**0.91**		2,69	4.29	1
H3-10G20										

* 1 = failure at the epoxy–concrete interface; 2 = failure at the bar-epoxy interface.
Note: 25.4 mm = 1 in. , 1 MPa = 145 psi or 6.895 MPa = 1 ksi , 1KN = 224.8 lb

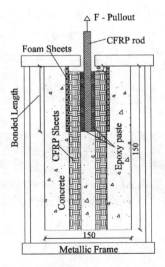

Figure 1— Disposition of specimens.

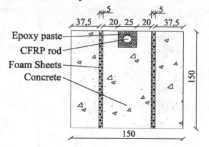

Figure 2 — Plan view of specimens.

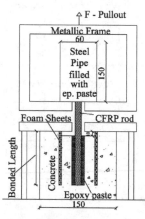

Figure 3 — Disposition of applied pullout load.

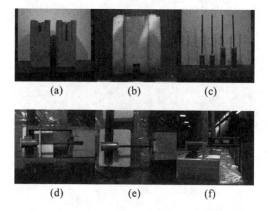

(a)   (b)   (c)

(d)   (e)   (f)

Figure 4 — Preparation of specimens (a), (b), (c), (d), (e) and (f).

Figure 5 — Instrumentation of the two LVTDs.

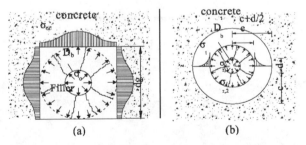

Figure 6 — B Specimens with CFRP jacket.

(a)   (b)

Figure 7 — (a) Internal pressures around the bar perimeter and at the filler – con. Interface: σ_o, $\sigma_{e,c}$, (b) Internal pressure around the bar perimeter σ_n.

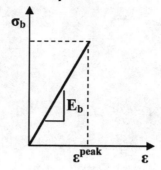

Figure 8 — A linear model of the stress-strain diagram for FRP bars.

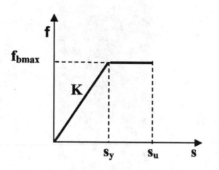

Figure 9 — A bilinear model of the bond-slip diagram.

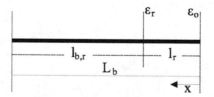

Figure 10 — Critical section of the bar whereas bond stresses decay from the peak value with increasing distance x.

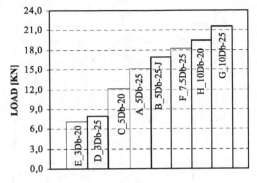

Figure 11 — Ultimate load for all specimens.

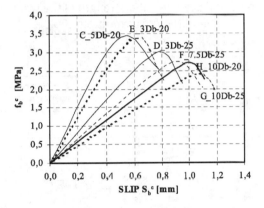

Figure 12 — Average bond strength between concrete and epoxy–versus–slip.

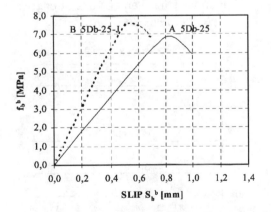

Figure 13 — Average bond strength between CFRP rod and epoxy–versus–slip, (B specimens have been confined with CFRP jacket).

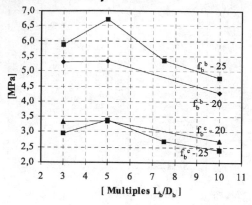

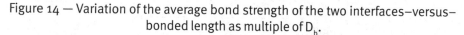

Figure 14 — Variation of the average bond strength of the two interfaces—versus—bonded length as multiple of D_b.

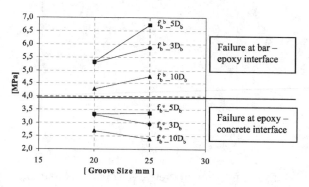

Figure 15 — Variation of the average bond strength of the two interfaces—versus—groove size.

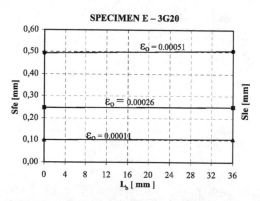

Figure 16 — Results of the procedure that described in eq. (5-13) applied to specimen E-3G20 having development length of $3D_b$ (=36mm), at the onset of specimen failure (associated bar strain at the critical section were $\varepsilon_{o,r}$=0.00051).

SPECIMEN F – 7.5G25

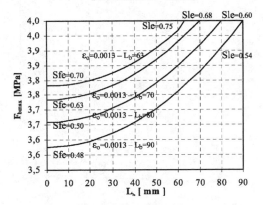

Figure 17 — Results of the procedure that described in eq. (5-13) applied to specimen F-7.5G25 having development length of $7.5D_b$ (=90mm), at the onset of specimen failure (associated bar strain at the critical section were $\varepsilon_{o,r}$=0.0013).

Debonding in FRP-Strengthened Flexural Members with Different Shear-Span Ratios

by Z. Wu and S. Hemdan

<u>Synopsis:</u> The use of fiber reinforced polymers (FRP) as external reinforcement to the tension face of concrete members has bean accepted as an effective mean of strengthening and retrofitting aging and deteriorated structures. One of the problems which limit the full utilization of the material strength is the premature failure due to debonding. In this study a special attention is paid to investigate the effect of the shear-span ratio on the debonding due to flexural cracks of concrete in FRP-retrofitted beams. Based on finite element analysis and nonlinear fracture mechanics, the debonding behavior of FRP-retrofitted beams with different shear-span ratios was investigated. The initiation of macro debonding of each beam was recorded and main factors affecting the effective bonding length were discussed. It is concluded that shear-span ratio has an insignificant effect on the initiation of macro debonding while it greatly affects the effective bonding length.

<u>Keywords:</u> debonding; effective bonding length; fiber reinforced polymers (FRP); shear-span ratio

Zishen Wu is a professor in the department of Urban & Civil Engineering at Ibaraki University, Japan. He received his BS and MS from the Southeast University, China in 1983 and 1985, respectively and his PhD from Nagoya University, Japan in 1990. His research interests include computational fracture and failure mechanics of composite materials and structures, maintenance engineering, structural health monitoring, intelligent and security engineering.

Said Hemdan is a PhD candidate in the department of Urban & Civil Engineering at Ibaraki University, Japan. He received his BS and MS from Assiut University, Egypt in 1996 and 2001 respectively. His research interests include nonlinear finite modeling and optimization design of reinforced concrete members retrofitted by fiber reinforced polymers.

INTRODUCTION

The use of fiber reinforced polymers (FRP) as an innovative method for retrofitting concrete members in both shear and flexure has been developed during the last two decades to replace the conventional methods such as steel plates. Although fiber reinforced polymers (FRP) have many advantages such as high efficiency, low weight/strength ratio, and ease of application, they still have their own drawbacks. One of the most popular problems accompanied with the use the bonded FRP sheets is the premature failure due to debonding which limits the complete utilization of the material strength. Over the last several years, great efforts have been paid to understand the debonding phenomena and to develop methods on evaluating the debonding load-carrying capacities.

Two different types of debonding were observed, the first type is debonding at the cut-off point of FRP caused by stress concentration as investigated by many researchers such as (Roberts, 1989; Ziraba et al., 1994; Malek et al., 1998; Ahmed et al., 2000). This type of debonding is dominant in case of FRP/steel plates. The another type of debonding which is observed to be dominant in the case of using FRP sheets with relatively thin thickness is that debonding initiates from the end of flexural cracks near the region of the maximum bending moment which is usually called intermediate crack-induced debonding. The intermediate crack-induced debonding has attracted many researchers to investigate it (e.g. Triantafillou, 1992; Täljstin, 1996; Wu et al. 1997; Bizindavyi et al., 1999; Yoshizawa et al., 2000; Sato et al., 2000; Lorenzis et al., 2001; Chen & Teng., 2001; Niu et al. 2002; and Wu & Yin, 2002). Täljstin (1997) presented the use of a linear and nonlinear fracture mechanics approach for the plate bonding technique. The expressions on stress transfer and mode II fracture are derived analytically for a simplified shear stress-deformation curve. Yoshizawa et al. (2000) found that the local shear stress distribution, effective transfer length, initiation and propagation of debonding could be well represented by the bi-linear model shown in Figure 4. They also, identified an average bond-slip relationship with local bond strength of 8 MPa and fracture energy of 1.2 MPa mm. Niu et al. (2002) performed a parametric study to illustrate the effect of the adhesive properties on the behavior of the FRP retrofitted beams; they concluded that the interfacial fracture energy is the only parameter to govern the ultimate load-carrying capacity. Chen and Teng (2001) assessed the performance of some of the available bond models and proposed a new model defined by combining fracture mechanics analysis with experimental evidence. Their model is a modification of the model proposed by Yuan and Wu (1999). Chen and Teng (2001) observed that the shear-slip behavior of plate concrete bonded joints may be well represented by a triangular shear-slip model similar to that shown in Figure 4. Because the relative slip at the peak shear is relatively too small when compared to the relative slip at failure, Chen and Teng (2001) suggested that a linear decreasing shear-slip model may be used. Their model uses the compressive strength of concrete instead of the shear-slip properties because accurate measurements of the later are practically difficult. Also, they took into consideration the width ratio of the bonded plate to the concrete member.

An important feature observed in many literatures is that there is a certain length beyond which any increase in the FRP bond length cannot increase the bond strength and hence can not contribute in the increase of the load-carrying capacity, this length is regarded as the effective bonding length. Extensive work has been done by Wu and his research group to illustrate the phenomena of debonding and the effect of cracks on the effective bonding length in a flexural concrete member. They concluded that the effective bonding length may be increased due to existence of cracks as compared with the effective transfer length determined without any crack in the simple shear test.

Based on the pre-mentioned studies and various relevant field applications, code provisions have been also issued in various countries: In the USA, ACI 440.2R-02 *Guide for the Design and Construction of Externally Bonded FRP Systems for Strengthening Concrete Structures* (ACI 2002) suggests a simple method for avoiding premature failure due to debonding in the anchorage zone and through all the interface length by limiting the allowable strain in the FRP to a value smaller than the FRP rupture strain. The method of the ACI is empirically and derived mainly from the experimental data.

Based on the analogy between the simple shear test without cracking and the RC beam behavior with cracking, the Japanese Code of Standards (JSCE Recommendations 2001) introduces an explicit method for checking the FRP strengthened beams against the premature failure due to loss of bonding. A double check procedure is suggested to ensure that no debonding occurs neither in the anchorage zone nor along the adhesive interface. JSCE suggested a procedure to calculate the flexural capacity and axial load-carrying capacity of members that fail due to peeling of the continuous fiber. Two main parameters are essential for the JSCE method: fracture energy which may be measured from the simple shear tests and the generalized effective bonding length in RC beam with cracking. Although JSCE suggests some equations for calculating the effective bonding length, these equations need to be developed to take into account the effect of more parameters such as shear-span ratio.

Fib bulletin 14 (2001) has a similar approach to the JSCE and states that stress check against delamination between two adjacent cracks can be carried out by limiting the FRP stress difference between the two cracks to a limited value. Also, Fib bulletin 14 (2001) provides relationships for estimating the average distance to be assumed between flexural cracks. The mechanical meaning of the Fib method is still unclear.

Evaluation of the aforementioned code provisions has demonstrated that no one of them is completely valid and can be used to describe debonding accurately (C. Faella at al. 2004). In addition, there are still some difficulties in estimating some of the needed parameters involved by the suggested equations such as crack spacing, fracture energy of the adhesive and effective bonding length. So, we still need to develop such equations to obtain a reliable one. In order to find a suitable relation for estimating the debonding failure capacity, basic factors affecting the effective bonding length as well as the effect of shear-span ratio on the debonding behavior are discussed in this paper.

RESEARCH SIGNIFICANCE

Debonding due to flexural cracks of concrete in FRP-retrofitted beams has been reported as a dominant mode of failure. Code provisions have been issued in various countries to give suggestions and recommendations to overcome this premature failure mode. JSCE suggested a procedure to calculate the flexural capacity and axial load-carrying capacity of members that fail due to peeling of the continuous fiber. Two main parameters are essential for the JSCE method: fracture energy which may be measured from the simple shear tests and the generalized effective bonding length in RC beam with cracking. Although JSCE suggests some equations for calculating the effective bonding length, these equations need to be developed to take into account the effect of more parameters such as shear-span ratio.

FINITE ELEMENT MODELING

A schematic drawing for the details of the simulated beams is shown in Figure 1. As it is clear, every beam is composed of three main parts: concrete, reinforcing bars and FRP sheets. For all cases, the beam cross section and reinforcement ratio were held constant, while the clear span was varied as which will be discussed later. Because we are using FRP sheets not the thicker FRP plates, so debonding of FRP sheets from the end of flexural cracks is expected to be the dominant failure mode. Therefore, to simulate the real response accurately, it is necessary to establish appropriate models for considering the crack propagation in concrete, bond-slip behavior between reinforcing bars and concrete and bonding behavior along the adhesive layer.

The discrete crack model is adopted here to simulate the initiation and propagation of cracks. It is assumed that flexural cracks are vertical along the whole depth of the beam and can develop only in the prescribed locations, no shear cracks are taken into consideration. According to Hillerborg et al. (1976), the linear softening curve shown in Figure2 is considered to model the mod I tension softening behavior of concrete. In this model, the cohesive crack, where forces are followed by a given softening curve, is assumed to initiate if the tensile stress reaches the tensile strength ft. The macro crack, where no shear stress is transferred along the interface of crack, is formed when the energy required to create one unit area of crack is met. The fracture energy G_f^c is represented by the area below the curve. Unloading and reloading are modeled by a secant path. The values of $ft = 3.0$ MPa and $G_f^c = 0.12$ MPa mm were employed in this study. The bond-slip model proposed by Morita et al. (1967) is adopted to simulate the interfacial behavior between concrete and the deformed steel bars as shown in Figure 3. Unloading and reloading are modeled by a secant path, which means upon a slip reversal, a straight line back to the origin is followed.

A linear softening model shown in Figure 4 is adopted to simulate the mode-II fracture of the adhesive. This model is considered valid for simulating the bond behavior regardless whether debonding occurs within the interfacial concrete or through the adhesive layer. In this model, when the local bond stress attains the local bond strength τ_f, micro-debonding is initiated and followed by a decrease in the local bond stress until it becomes zero where macro-debonding is formed . The slope of the ascending branch represents the interfacial stiffness k_s, while the area under the curve represents the mode II fracture energy G_f^b. According to Yoshizawa et al. (2000), the values of τ_f, k_s, and G_f^b are chosen to be 8.0 MPa, 160 MPa/mm, and 1.2 MPa mm respectively as a set of reference values.

As described above, the tensile behavior of concrete is modeled using a discrete cracking model. On the other hand, the response of concrete in compression is modeled by Drucker-Prager perfect plasticity, where the internal friction angle is taken as $10°$. The compressive strength and the Young's modulus of concrete were chosen to be 40 MPa and 34.2 GPa respectively.

Steel reinforcing bars are considered as a linear elastic-perfectly plastic material, as shown in Figure 5. The slope of the curve represents the modulus of elasticity Es and the maximum stress represents the yield stress f_y. The elastic-perfectly plastic behavior is modeled by Von Mises yield criterion. The values of Es and f_y are chosen to be 210 GPa and 364 MPa, respectively. FRP sheets in general behave in linear elastic manner up to rupture. The modulus of elasticity and the tensile strength are chosen to be 230 GPa and 4.1 GPa, respectively.

The simply supported beam shown in Figure1 was simulated numerically. The small deformation theory is adopted, large deformations are not considered in the simulation. Due to symmetry, only half of the beam was solved with appropriate boundaries. Beams were solved under three point loading using displacement control. The concrete beam is modeled by 4-node plane stress elements, steel bars and FRP sheets are modeled by 2-node linear truss elements connected to concrete by zero-thickness line interface elements. The flexural cracks are modeled by zero-thickness line interface elements at a spacing of 40mm from the mid-span to the support of the beam. The crack spacing was chosen to be 40mm to cover a wide range of crack distributions, especially in the case of relatively longer spans.

NUMERICAL SIMULATION

As stated above, we are aiming mainly at clarifying the effect of the shear-span ratio on the debonding behavior in addition to investigating the parameters that affect the effective bonding length. To investigate the effect of the shear-span ratio on the debonding behavior, 5 beams of clear spans of 1800, 2400, 3000, 3600, and 4800 which are corresponding to shear-span ratios of 4.5, 6.0, 7.5, 9.0, and 12.0 respectively were solved. Each beam was retrofitted by 2 layers of CFRP of thickness of 0.111 mm/layer. The interfacial stiffness k_s of the adhesive was chosen to be 160 MPa/mm and held constant in this case. The load at initiation of macro debonding and the corresponding FRP strain were recorded. The effective bonding length of each beam was identified.

To investigate the parameters which affect the effective bonding length, the following main parameters were taken into consideration:

1. The interfacial stiffness k_s of the adhesive in which five values of 40, 80, 160, 320, and 640 MPa/mm were investigated. The interfacial stiffness was changed by changing the relative slip at the peak shear stress.
2. The reinforcing stiffness $(E_{frp}.t_{frp})$ where t_{frp} had the values of 0.111, 0.222, and 0.333 mm while E_{frp} held constant of 230 GPa which corresponds to $E_{frp}.t_{frp}$ of 25.53, 51.06, and 76.59 GPa mm respectively.
3. The fracture energy of the adhesive G_f^b in which the values of 0.6, 1.2, 1.8 were investigated. The fracture energy was changed by changing the ultimate relative slip, while the interfacial stiffness ks and the peak shear stress were held constant.
4. Shear-span ratio as discussed above in the first case.
5. Span length at constant shear-span ratio, where three beams of spans 2400, 4800, and 7200 mm having a shear-span ratio of 6.0 were investigated. The steel reinforcement ratio of each beam was held constant.

A beam of clear span of 2400 was selected to investigate the first three parameters. The effective bonding length of each case was identified.

RESULTS AND DISCUSSIONS

The effect of the shear-span ratio on the debonding behavior

Look at Figure 6, one can observe that for all investigated beams of shear-span ratios of 4.5, 6.0, 7.5, 9.0, and 12.0, the bending moment at the initiation of macro debonding is approximately constant which in return means that shear-span ratio has no significant effect on mode-II debonding initiation. That is an acceptable result because mode-II debonding is mainly dependant on the axial stress of FRP. Figure 7 exhibits the FRP stress distributions at the initiation of macro debonding for the different beams. Irregular behavior of the stress distribution curves may be attributed to crack spacing of 40mm, but this has insignificant effect on the accuracy of the results. Because we employed a constant cross section and a constant reinforcement ratio for all the beams in this case, the axial stress of FRP is dependent only on the bending moment. It is worth mentioning that shear-span ratio may have significant effect on mode-I debonding, but this is still under investigation and out of scope of this paper.

Basic parameters affecting the effective bonding length

It is well known that the effective bonding length for a joint in the simple shear test is defined as the length beyond which no further increase in failure load can be achieved. Also, the initiation of macro debonding is considered as a type of failure, even if there is some gain in the ultimate strength after initiation of macro debonding, that gain will be insignificant. Based on these two points and without lack of generality, the effective bonding length of the investigated beams may be considered as shown in Figure 8 after approximating the FRP stress distributions to bilinear curve.

Mesh sensitivity -- To check the sensitivity of the identified effective bonding length to mesh size, three different mesh sizes were investigated, 5*5, 10*10, and 20*20 mm. Referring to Figure 9, it is obvious that mesh size has no significant effect on the identified effective bonding length and hence mesh size of 10*10 mm will be adopted through this study.

As stated above, to investigate the parameters which affect the effective bonding length, the following main parameters were taken into consideration:

1. The interfacial stiffness k_s.
2. The reinforcing stiffness ($E_{frp} \cdot t_{frp}$).
3. Fracture energy of the adhesive layer.
4. Shear-span ratio.
5. Span of the beam.

In the following section, every parameter will be discussed in detail.

Interfacial stiffness K_s -- It is well known that if the interfacial stiffness of the adhesive is relatively high, this will cause rapid transfer of stresses from concrete to FRP and consequently results in delaying yielding of steel reinforcements, Niu and Wu (2002). Figure 10 shows the FRP stress distributions for the investigated beams having Ks values of 40, 80, 160, 320, and 640 MPa/mm. The identified effective bonding lengths are 4.55, 360, 330, 320, and 330 respectively. It is clear that for relatively small values of interfacial stiffness, the effective bonding length decreases as the interfacial stiffness increases. For values of interfacial stiffness higher than 160 MPa/mm, one can observe that interfacial stiffness has no significant effect on the effective bonding length. So, using adhesive having low interfacial stiffness will cause an increase to the effective bonding length which in return means relieving the stress concentration in the FRP and consequently delaying the debonding failure.

Effect of ($E_{frp} \cdot t_{frp}$) -- To study the effect of the FRP reinforcing stiffness ($E_{frp} t_{frp}$) on the effective bonding length, three RC beams of clear span of 2400mm retrofitted with one, two, and three layers of FRP were investigated. The corresponding values of FRP reinforcing stiffness ($E_{frp} t_{frp}$) are 25.53, 51.06, and 76.59 GPa mm respectively. Figure 11 shows the FRP stress distributions for the investigated beams. It is clear that as the FRP reinforcing stiffness ($E_{frp}.t_{frp}$) increases, the effective bonding length increases.

Effect of of the fracture energy of the adhesive G_f^b -- Figure 12 shows the FRP stress distributions for the investigated beams having G_f^b of 0.6, 1.2, and 1.8. Using the method described previously in Figure 8, effective bonding length of 240, 330, and 430 mm respectively are obtained. As shown in Figure 13, as the fracture energy increases the effective bonding length increases.

Effect of shear-span ratio -- Figure 7 shows the FRP stress distributions for the investigated beams having clear spans of 1800, 2400, 3000, 3600, and 4800 which are corresponding to shear-span ratios of 4.5, 6.0, 7.5, 9.0, and 12.0 respectively. Adopting the method described previously for identifying the effective bonding length after approximating the FRP distribution curve to a bilinear one, effective bonding lengths of 230, 330, 410, 480, and 690 mm respectively can be obtained easily. It is clear that as the shear-span ratio increases, the effective bonding length increases. Figure 14 illustrates this relation which may be approximated to a straight line.

Effect of span -- Another factor which may affect the effective bonding length is the span at constant shear-span ratio. Keeping the shear-span ratio constant of 6.0, three beams of spans of 2400, 4800, and 7200 were solved. Figure 15 shows the FRP stress distributions for the investigated beams. Effective bonding lengths of 230, 350, and 610 may be obtained respectively. It is clear that as the span increases, the effective bonding length increases as indicated in Figure 16.

CONCLUSIONS

Based on finite element analysis and nonlinear fracture mechanics, the effect of shear-span ratio on the debonding behavior as well as factors affecting the effective bonding length of FRP-retrofitted RC beams was investigated. The following conclusions may be drawn:

1. The shear-span ratio has no significant effect on mode-II debonding initiation.
2. As the shear-span ratio increases, the effective bonding length increases. Increasing the span length while holding the shear-span ratio constant causes an increase in the effective bonding length. This has an implication that beams having larger spans and larger shear-span ratios after initiation of macro debonding may gain more loads before final failure.
3. Using adhesive having low interfacial stiffness will cause an increase to the effective bonding length which in return means relieving the stress concentration in the FRP and consequently delaying the debonding failure.
4. Using adhesive having high fracture energy may increase the effective bonding length and consequently delays the final failure after the initiation of macro debonding.
5. As the FRP reinforcing stiffness ($E_{frp}t_{frp}$) increases, the effective bonding length increases.

REFERENCES

ACI Committee 440.2 R-02 (2002), Guide for the Design and Construction of Externally Bonded FRP Systems for Strengthening Concrete Structures.

Ahmed, O., Van Gemert, D. and Vandeewall, L. (2000), *Improved Model for Plate-End Shear of CFRP Strengthened R.C. Beams*, Journal of Cement and Concrete Composites.

Bizindavyi, L. and Neale, K. W. (1999), Transfer lengths and bond strengths for composites bonded to concrete, *ASCE Journal of Composites for Construction*, 3(4), pp. 153-160.

Chen, J. F. and Teng, J. G. (2001), *Anchorage Strength Models for FRP and Steel Plates Attached to Concrete,* Journal of Structural Engineering, ASCE, Vol. 127, No. 7, pp. 784-791.

Faella C., Martinelli E., and Nigro E. (2004). Debonding in FRP strengthened RC beams: comparison between code provisions, *FRP Comosites in Civil Engineering-CICE 2004,* pp. 189-197.

fib: Bulletin d'information n. 14 (2001), *Externally bonded FRP reinforcement for RC structures.*

Hillerborg, A.; Modeer, M.; and Petersson, P.E., (196), *Analysis of Crack Formation and Crack Growth in Concrete by Means of Fracture Mechanics and Finite Elements,* Cement and Concrete Research, pp. 773-782.

DIANA-8.1 User's Manual, TNO Building and Construction Research, Lakerveld b.v., The Hague, 2003.

JSCE (2001), Recommendation for upgrading of concrete structures with use of continuous fiber sheets, Concrete Engineering Series 41.

Lorenzis, L. D., Miller, B. and Nanni, A. (2001), Bond of FRP laminates to concrete, *ACI Materials Journal*, 98(3), pp. 256-264.

Malek, A. M., Saadatmanesh, H. and Ehsani, M. R. (1998), Prediction of failure load of R/C beams strengthened with FRP plate due to stress concentration at the plate end, *ACI Structural Journal*, 95(1), pp. 142-152.

Morita, S.; Muguruma, H.; and Tomita, K. (1967), *Fundamental Study on Bond between Steel and Concrete,* Transaction of AIJ, 131(1), pp.1-8.

Nakaba, K., Kanakubo, T., Furuta, T. and Yoshizawa, H. (2001), Bond behavior between fiber-reinforced polymer laminates and concrete, *ACI Structural Journal*, 98(3), pp. 359-367.

Niu, H. and Zhishen Wu (2002), *Strengthening Effect of RC Flexural Members with FRP Sheets Affected by Adhesive Layers,* Journal of Applied Mechanics, Vol. 5, pp. 887-897.

Roberts, T. M. (1989), Approximate analysis of shear and normal stress concentrations in the adhesive layer of plated RC beams, *The Structural Engineer*, 67(12), pp. 229-233.

420   Wu and Hemdan

Sato, Y., Asano, Y. and Ueda, T. (2000), Fundamental study on bond mechanism of carbon fiber sheet, *JSCE Journal of Material, Concrete Structures and Pavements*, 47(648), pp. 71-87 (in Japanese).

Täljsten, B. (1997), Strengthening of beams by plate bonding, *ASCE Journal of Materials in Civil Engineering*, 9(4), pp. 206-212.

Triantafillou, T. C., and Plevris, N., (1992), *Strengthening of RC Beams with Epoxy-Bonded FRP Composite materials,* Materials and Structures, V. 25, pp. 201-211.

Wu, Z. S. Matsuzaki, T., and Tanabe, K., (1998), *Experimental Study on Fracture Mechanism of FRP-Reinforced Concrete Beams,* Proceeding of JCI Symposium on Non-metallic FRP Reinforcement for Concrete Structures, pp. 119-126.

Wu, Z. S. and Yin, J. (2002), *Numerical analysis on interfacial fracture mechanism of externally FRP-strengthened structural member,* JSCE Journal of Material, Concrete Structures and Pavements, 55(704), pp. 257-270.

Yaun, H. and Wu, Z.(1999), *Interfacial Fracture Theory in Structures Strengthened with Composite of Continuous Fiber,* Proceedings of Symposium of China and Japan, Science and Technology of 21st Century, Tokyo, Japan, pp. 142-155.

Yoshizawa, H., Wu, Z. S., Yuan, H. and Kanakubo, T. (2000), *Study on FRP-concrete interface bond performance*, JSCE Journal of Material, Concrete Structures and Pavements, 49(662), pp. 105-119.

Ziraba Y.N., Baluch M.H., Basunbul I.A., Sharif A.M., Azad A.K. and Al-Sulaimani G.J. (1994), *Guidelines toward the Design of Reinforced Concrete Beams with External Plates. ACI Structural Journal* 91(6), pp. 639-646.

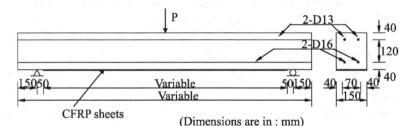

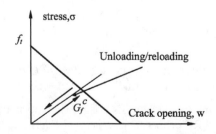

(Dimensions are in : mm)

Figure 1—Details of the simulated beam.

Figure 2 —Linear tension softening model for concrete.

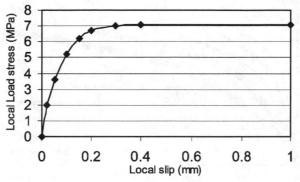

Figure 3—Bond-slip model for steel-concrete interface (Morita et al., 1967).

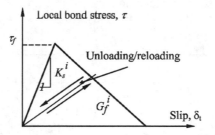

Figure 4—Linear softening model for FRP-concrete interface.

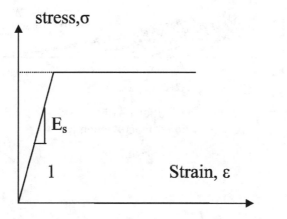

Figure 5—Elastic-perfectly plastic model for steel reinforcement.

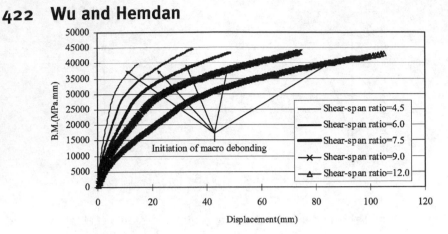

Figure 6—Mid-span displacement versus mid-span B.M. for different shear-span ratios.

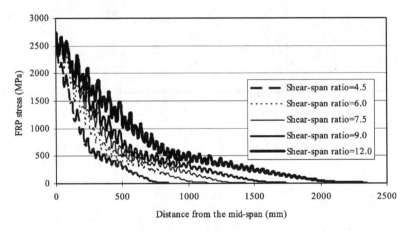

Figure 7—FRP stress distribution for different shear-span ratios at initiation of macro debonding.

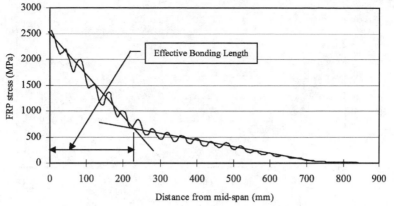

Figure 8—Suggested method for identifying the effective bonding length.

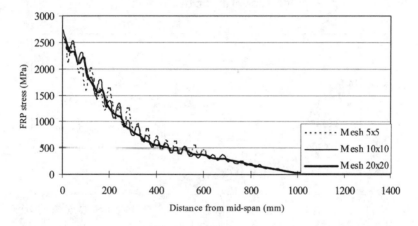

Figure 9—Sensitivity of the effective bonding length to the mesh size.

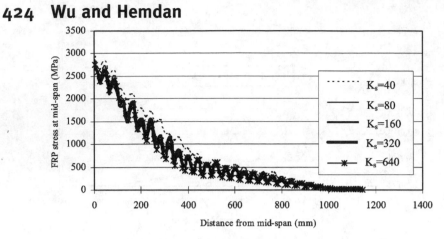

Figure 10—FRP stress distribution for different interfacial stiffness at initiation of macro debonding.

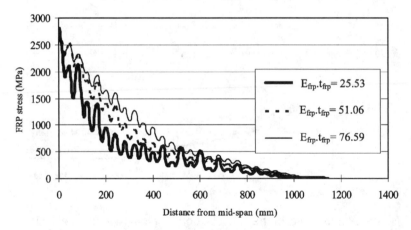

Figure 11—FRP stress distribution for different values of FRP reinforcing stiffness $(E_{frp}.t_{frp})$ at initiation of macro debonding.

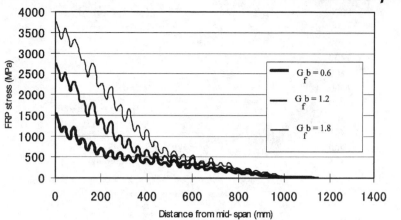

Figure 12—FRP stress distribution for different values of fracture energy of adhesive at initiation of macro debonding.

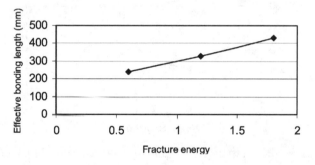

Figure 13—Relation between fracture energy of adhesive and effective bonding length.

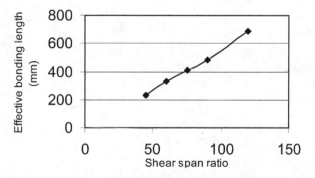

Figure 14—Relation between shear-span ratio and effective bonding length.

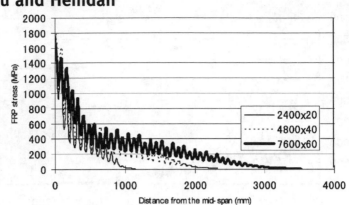

Figure 15—FRP stress distribution for different values of spans with a constant shear-span ratio at initiation of macro debonding.

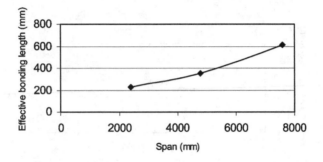

Figure 16—Relation between span and effective bonding length.

Enhanced End Anchorage of Bonded FRP Repairs

by M.J. Chajes, W.W. Finch, Jr., and H.W. Shenton

Synopsis: When strengthening concrete structures using externally bonded composite material plates, the bond between the concrete and plate typically fails well before the plate's tensile capacity is achieved. To enhance the bond performance, using composite fabric between the plate and the concrete was explored. Six tension development specimens were tested to determine the maximum tensile load that could be applied to a CFRP plate before it debonded. Three specimens had a Sika CarboDur strip bonded directly to the concrete, while three others had the CarboDur strip bonded to two plies of Sikawrap Hex 230C fabric material. The fabric was used to distribute the load over a larger bond area. The specimens without fabric failed at an average load of 60.5 kN while the specimens with the CFRP fabric failed at an average load of 92.6 kN. These failure loads represent 18 % and 27.5% of the plate capacity respectively.

Keywords: adhesives; bonding; composite materials; concrete; development length; force transfer

Michael J. Chajes is Professor and Chair of the Department of Civil and Environmental Engineering Department at the University of Delaware. He is a registered professional engineer and is a member of ACI. His research interests include bridge evaluation and rehabilitation including the use of advanced composite materials.

William W. Finch, Jr. is President of Structural Testing Incorporated, a firm that specializes in both bridge testing and also the evaluation of advanced composite material applications. He is a registered professional engineer and is a member of ACI. He is also the owner of WF Construction.

Harry W. Shenton, III is an Associate Professor in the Department of Civil and Environmental Engineering Department at the University of Delaware. His research interests include structural health monitoring and condition assessment, in-service and long-term monitoring of bridges, sensors and instrumentation for civil infrastructure systems, structural dynamics and earthquake engineering.

INTRODUCTION

As a nation's infrastructure ages, the number of deficient structures continues to grow. Because of the prohibitive cost of replacing all of the sub-standard structures, innovative techniques for rehabilitating deteriorating structures are needed.

Numerous researchers have studied strengthening of existing structures, in particular concrete structures, through the use of externally bonded steel and composite material plates.[1-18] With regard to steel plates, the studies have focused on problems found at the concrete-steel interface including interface corrosion and peel stresses.[2-9] Other studies have shown that bonding prestressed and non-prestressed composite-material plates and fabrics to concrete beams can lead to significant increases in flexural and shear capacity.[10-19] Possibly the single most critical aspect in strengthening a concrete structure through externally bonded plates is the bond between the plate and the concrete. Studies such as that by Chajes et al.[1] have evaluated the bond strength and have shown that a composite plate will typically debond well before it reaches its ultimate tensile capacity. Some more recent studies have looked at ways to improve the bond using FRP wraps placed over the bonded plates.[20] This study will investigate the ability to improve bond behavior by placing composite fabric between the bonded plates and the concrete substrate.

RESEARCH SIGNIFICANCE

To more fully utilize the capacity of externally bonded composite plates, improved anchorage is needed. This paper presents research aimed at improving the bond in the anchorage zone so as to more fully utilize the tensile capacity of the bonded plate.

A simple and relatively effective method for improving end anchorage involving the bonding of a CFRP fabric directly to the concrete surface at the anchorage area, and then bonding the composite plate to the fabric has been studied.

SPECIMEN FABRICATION

A total of nine tension development test specimens were fabricated (six to be used for testing, and three to be used as back up specimens in case any problems in the testing procedure were encountered). Six specimens (three for testing and three back-ups) were fabricated as shown in Figure 1 (specimens nos. 1, 3, 4, 5, 6 & 9), and three specimens (all for testing) were fabricated as shown in Figure 2 (specimens nos. 2, 7 & 8). The specimens shown in Figure 1 had a 102 mm wide CarboDur strip[21] bonded directly to the 254 mm by 305 mm by 610 mm concrete block (610 mm bond length), whereas the specimens shown in Figure 2 had the CarboDur bonded directly to two plies of SikaWrap Hex 230C material[22] arranged at a plus/minus 45 degree layout. The purpose of the plus/minus 45 degree carbon fiber sheet was to distribute the load over a wider concrete bond area. Please note that for the specimens with the fabric, the fabric and plate were bonded at the same time, and the fabric extended beyond the width of the 102 mm plate all the way out to the edges of the 305 mm wide block. As such, the bonded fabric area was much greater than the bonded area of just the plate alone. After fabrication, specimens 1, 3, and 9 were set aside as extra samples, and specimens 2, 4, 5, 6, 7 & 8 were instrumented with strain gages as shown in Figure 3 and subsequently tested. The back-up specimens (1, 3, and 9) were not needed and will not be discussed further.

The concrete blocks were cast using 27.6 MPa concrete with a 2% non-chloride accelerator added. In addition to the concrete development specimens, 6 standard concrete cylinders were cast for compression testing. These concrete development specimens were moist-cured for 1-½ weeks and then the carbon fiber materials were applied (once the specimens had dried out. The cylinders were cured under the same conditions at the specimens and were tested at the same time as the development tests were conducted. Ten compression test cubes, 51 mm x 51 mm x 51 mm, were fabricated from the SikaDur 30 epoxy that was used to bond the CarboDur plate to the test specimens. It should be noted that in addition to the strain gages bonded to the CarboDur, strain gages were bonded to each piece of vertical rebar at the mid-height of the specimen. The application of both types of carbon fiber reinforcing systems was performed per the Sika Corporation's recommended procedures.

The 102 mm wide by 1.2 mm thick CFRP CarboDur plates used (Type 2-101) had a tensile modulus of 165 GPa, a strain to failure of 1.69%, and an ultimate tensile strength of 336 kN. The SikaWrap Hex 230C fabric had a weight per square yard of 1.9 N, and carbon fibers having a tensile strength of 3.45 GPa and a tensile modulus of 230 GPa. These are the manufacturer's values.

TEST PROCEDURE

A universal testing machine was used to load the specimens as shown in the photograph in Figure 4, and the schematic in Figure 5. The specimens were placed into the testing machine using an overhead crane. They were tested under displacement control at a slow rate of loading until well beyond debonding of the CarboDur from the concrete specimen. A set of rollers were used to stabilize the specimen and the set-up appeared to work well and not cause a clamping effect. Strains were recorded during testing on a continuous basis using a PC based data acquisition system. Six specimens were tested, three with no fabric (specimens 4, 5, and 6) and three with fabric (2, 7, and 8).

TEST RESULTS

The six concrete compressive test cylinders and the SikaDur 30 epoxy compressive test cubes were tested on the same dates as the tension development specimens. The average compressive failure stress of the cylinders was 34.0 MPa. Ten compressive cubes of SikaDur 30 were tested and the average compressive failure stress of the cubes was 79.6 MPa.

Load versus microstrain plots for all six test specimens are shown in Figures 6 through 11. The first three plots, Figures 6, 7, and 8, are plots of the specimens that were fabricated with the CarboDur plate bonded directly to the concrete. Figures 9, 10, and 11 are plots of the specimens that were fabricated with the CarboDur plate bonded to the SikaWrap Hex 230C carbon fiber fabric that was in turn bonded to the concrete surface. Failure was defined as the load at which the CarboDur starts to debond from the test specimen. This load is indicated by the vertical dashed line in the figures. The average load at which the CarboDur debonded when bonded directly to the concrete surface was 60.5 kN and the average load at which the CarboDur debonded when bonded to the SikaWrap Hex 230C was 92.6 kN.

The test specimens with the CarboDur bonded directly to the concrete failed in the concrete with a thin layer of concrete bonded to the complete surface of the CarboDur plate. The test specimens with the CarboDur bonded to the SikaWrap carbon fabric failed between the SikaDur 30 adhesive and CarboDur plate through either a debonding of the adhesive to the plate or a surface failure of the CarboDur plate. This explains the higher failure load because the weaker concrete surface bond no longer governs.

The failure loads represented approximately 18% of the ultimate strength of the CarboDur strip being developed when bonded directly to the concrete surface and 27.5% of the CarboDur strip ultimate tensile strength being developed when bonded to the Hex 230C carbon fabric. In the plots one can see that after the bond breaks, the plate was carrying the entire load. Prior to failure, the interior gages on the plate showed a reduced tension because some of the force has transferred into the concrete specimen (now being carried by the rebar).

Figure 12 shows a comparison of the strain distribution along the bonded plates (both with and without fabric) at a load just before failure. One can easily see that the plate with the fabric beneath it (specimen 8) carries significantly more load (i.e. has more strain at all points along the length) than the plate that is bonded directly to the concrete (specimen 6). Figures 13 (specimen 6) and 14 (specimen 8) show plots of the strain distribution along the length of the CarboDur plate at 70%, 90% and 100% of the ultimate load that causes debonding of the CarboDur plate. The zero point on the x-axis is the starting point of the bonded CarboDur plate and 610 mm on the x-axis is the end of the bonded length of the CarboDur plate. These figures illustrate the transfer of the force in the CarboDur plate to the concrete specimen. Figure 13 (the CarboDur bonded directly to the concrete) shows that only approximately 100 to 125 mm of bond length are utilized to transfer the force into the concrete. Figure 14 shows that improved development length was achieved when the CarboDur plate was bonded to the SikaWrap Hex 230C carbon fiber material. The Hex 230C material effectively distributed the load to a wider area of concrete.

CONCLUSIONS

A simple method for improving the bond of CFRP plates to concrete has been presented. The application of CFRP fabric between the plate and the concrete in which the fabric extends beyond the footprint of the plate has shown to improve the force transfer and increase the ultimate bond strength. For the six tension development test specimens tested, the three specimens without the fabric failed at an approximate average failure load of 60.5 kN while the three specimens with the CFRP fabric failed at an average load of 92.6 kN. This represents a 53% increase in ultimate load. Furthermore, the failure load increased from approximately 18 % of the ultimate strength of the CFRP plate to 27.5% of the plate's ultimate strength. The results are based on tests of single lap specimens. Future work involving full-scale beams that simulate realistic field applications, as well as analytical modeling of the bond behavior, would be very useful.

ACKNOWLEDGMENTS

The authors would like to thank the Sika Corporation for helping to support this research. We would like to thank Doug Baker and Michael Davidson for their assistance with test equipment and fixtures.

REFERENCES

1. Chajes, M.J., Finch, W.W., Januszka, T.F., and Thomson, T.A. "Bond and Force Transfer of Composite-Material Plates Adhered to Concrete," *Structural Journal*, ACI, 93(2), 1996, 208-217.

2. Swamy, R.N., Jones, R., and Bloxham, J.W. "Crack Control of Reinforced Concrete Beams Through Epoxy Bonded Steel Plates." *Adhesion Between Polymers and Concrete*, RILEM 86, 1986, pp. 542-555.

3. Jones, R., Swamy, R.N., and Ang, T.H. "Under- and Over-reinforced Concrete Beams With Glued Steel Plates." *The International Journal of Cement Composite and Lightweight Concrete*, 1(1), 1982, pp. 19-32.

4. Johnson, R.P., and Tait, C.J. "The Strength in Combined Bending and Tension of Concrete Beams With Externally Bonded Reinforcing Plates." *Building and Environment*, 16(4), 1981, pp. 287-299.

5. Macdonald, M.D., and Calder, A.J.J. "Bonded Steel Plating for Strengthening Concrete Structures." *International Journal of Adhesion and Adhesives*, (4), April 1982, pp. 119-127.

6. Solomon, S.K., Smith, D.W., and Cusens, A.R. "Flexural Tests of Steel-Concrete-Steel Sandwiches." *Magazine of Concrete Research*, 28(94), March 1976, pp. 12-20.

7. Hamoush, S.A., and Ahmad, S.H. "Debonding of Steel Plate-Strengthened Concrete Beams." *Journal of Structural Engineering*, 116(2), February 1990, pp. 356-371.

8. Oehlers, D.J., and Moran, J.P. "Premature Failure of Externally Plated Reinforced Concrete Beams." *Journal of Structural Engineering*, 116(4), April 1990, pp. 978-995.

9. Jones, R., Swamy, R.N., and Charif, A. "Plate Separation and Anchorage of Reinforced Concrete Beams Strengthened by Epoxy-Bonded Steel Plates." *The Structural Engineer Part A*, 66(12), June 1988, pp. 85-94.

10. Chajes, M.J., Thomson, T.A., Finch, W.W., and Januszka, T.F. "Flexural Strengthening of Concrete Beams Using Externally Bonded Composite Materials," *Construction and Building Materials*, 8(3), 1994, pp. 191-201.

11. Ritchie, P.A, Thomas, D.A., Lu, L.W., and Connelly, G.M. "External Reinforcement of Concrete Beams Using Fiber Reinforced Plastics." *ACI Structural Journal*, 88(4), July-August 1991, pp. 490-500.

12. Saadatmanesh, H., and Ehsani, M.R. "RC Beams Strengthened With GFRP Plates. 1: Experimental Study." *Journal of Structural Engineering*, 117(11), November 1991, pp. 3417-3433.

13. Meier, U., Deuring, M., Meier, H., and Schwegler, G. "Strengthening of Structures With CFRP Laminates: Research and Applications in Switzerland." *Advanced Composite Materials in Bridges and Structures*, CSCE, Sherbrooke, Canada, 1992, pp. 243-251.

14. Karam, G.N. "Optimal Design For Prestressing With FRP Sheets in Structural Members." *Advanced Composite Materials in Bridges and Structures*, CSCE, Sherbrooke, Canada, 1992, pp. 277-285.

15. Triantafillou, T.C., and Plevris, N. "Post-Strengthening of R/C Beams With Epoxy-Bonded Fiber Composite Materials." *Advanced Composites Materials in Civil Engineering Structures*, ASCE, Las Vegas, Nevada, 1991, pp. 245-256.

16. Triantafillou, T.C., and Meier, U. "Innovative Design of FRP Combined With Concrete." *Advanced Composite Materials in Bridges and Structures*, CSCE, Sherbrooke, Canada, 1992, pp. 491-500.

17. Triantafillou, T.C., Deskovic, N., and Deuring, M. "Strengthening of Concrete Structures With Prestressed Fiber Reinforced Plastic Sheets." *ACI Structural Journal*, 89(3), May-June 1992, pp. 235-244.

18. Triantafillou, T.C., and Deskovic, N. "Innovative Prestressing with FRP Sheets: Mechanics of Short-Term Behavior." *Journal of Engineering Mechanics*, 117(7), July 1991, pp. 1652-1672.

19. Chajes, M.J., Januszka, T.F., Mertz, D.R., Thomson, T.A., and Finch, W.W. "Shear Strengthening of Reinforced Concrete Beams Using Externally Applied Composite Fabrics," *ACI Structural Journal,* ACI. 92(3), 1995, 295-303.

20. Hamad, B., Soudki, K, Harajli, M., and Rteil, A. "Experimental and Analytical Evaluation of Bond Strength of Reinforcement in Fiber-Reinforced Polymer-

Wrapped High-Strength Concrete Beams." *ACI Structural Journal,* ACI. 101(6), 2004, 747-754.

21. Sika CarboDur, Product Data Sheet, Edition 9.2003, ID No. 332, Sika Corporation, Lyndhurst, NJ, 07071.

22. SikaWrap Hex 230C, Product Data Sheet, Edition 7.2003, ID No. H33230, Sika Corporation, Lyndhurst, NJ, 07071.

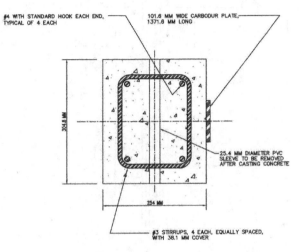

Figure 1 — Typical Section of CarboDur Test Specimen with SikaWrap and No Fabric

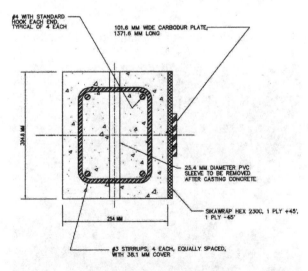

Figure 2 — Typical Section of CarboDur Test Specimen with SikaWrap Fabric

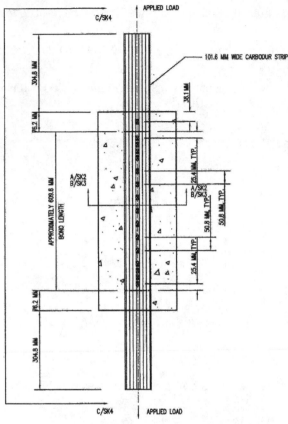

Figure 3 — Elevation View of Instrumented of Test Specimen

Figure 4 — Photograph of Specimen During Testing

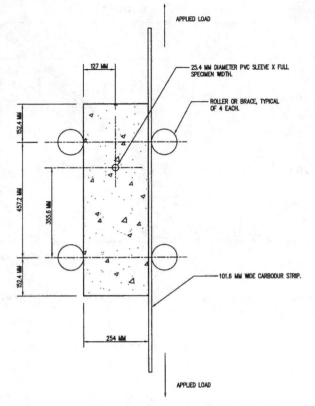

APPLIED LOAD

127 MM

25.4 MM DIAMETER PVC SLEEVE X FULL SPECIMEN WIDTH.

ROLLER OR BRACE, TYPICAL OF 4 EACH.

152.4 MM

457.2 MM

355.6 MM

152.4 MM

101.6 MM WIDE CARBODUR STRIP.

254 MM

APPLIED LOAD

Figure 5 — Schematic of Specimen in Braced Test Fixture

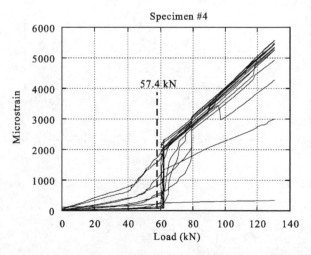

Figure 6 — Load versus Strain Plot for Specimen 4 (No Fabric)

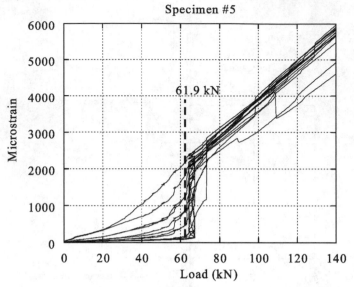

Figure 7 — Load versus Strain Plot for Specimen 5 (No Fabric)

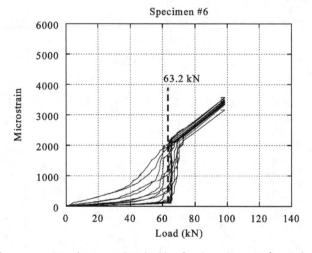

Figure 8 — Load versus Strain Plot for Specimen 6 (No Fabric)

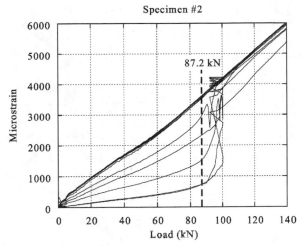

Figure 9 — Load versus Strain Plot for Specimen 2 (With Fabric)

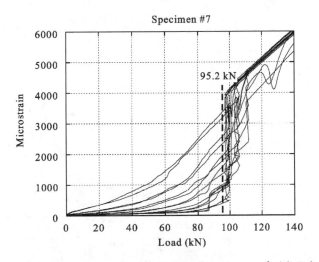

Figure 10 — Load versus Strain Plot for Specimen 7 (With Fabric)

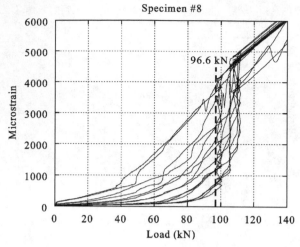

Figure 11 — Load versus Strain Plot for Specimen 8 (With Fabric)

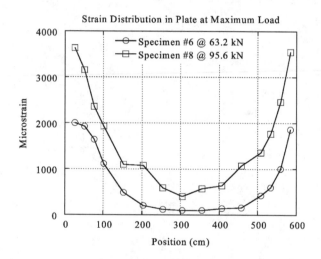

Figure 12 — Comparison of Strain Distribution at Ultimate Load

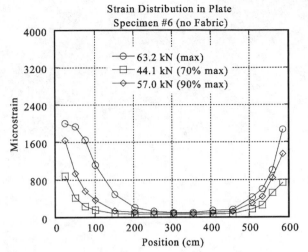

Figure 13 — Strain Distribution for No Fabric Specimen at 70, 90, and 100 % of Load

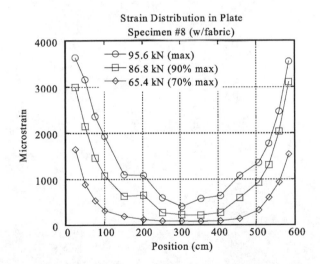

Figure 14 — Strain Distribution for Fabric Specimen at 70, 90, and 100 % of Load

FRP-Concrete Bond Behavior: A Parametric Study Through Pull-Off Testing

by B.M. McSweeney and M.M. Lopez

Synopsis: The sensitivity of the FRP-concrete bond failure load to changes in geometric and material parameters is described, and initial comparisons to predictions from existing bond models are made. To accomplish this, load and strain data from a series of single-lap pull-off tests is analyzed, in which carbon fiber reinforced polymer (CFRP) strips of varying width, thickness, and bonded length were pulled from concrete blocks of varying concrete strength. It was found that the concrete compressive strength had limited effects on the bond failure load, and longer bonded lengths increased the time up to failure load. Changes to the bonded width and FRP thickness had a significant impact on the bond failure load. Failure load predictions produced by three studied bond models were found to be strongly influenced by the material properties used as input, and were occasionally insensitive to the parameters varied.

Keywords: bond behavior; bond modeling; CFRP; concrete; fiber-reinforced composite; pull-off test

Brian M. McSweeney graduated with a Master's Degree in Civil Engineering from Penn State University in 2005, and is now a structural engineer at Linton Engineering in Vienna, VA. He earned his B.S. in Civil Engineering from the University of Virginia in 2003. His research is focused on the bond behavior between fiber-reinforced polymer composites and a concrete substrate.

Maria M. Lopez is an Assistant Professor at Penn State University. She is active on several ACI technical committees including committee 544, fiber-reinforced concrete; 543, fracture mechanics; and 440, fiber-reinforced polymer reinforcements. Dr. Lopez's research interests include the use of fiber-reinforced polymer materials for civil infrastructure applications.

INTRODUCTION

Repair and rehabilitation methods for civil infrastructure in the United States has become a topic of great interest to engineers. In Pennsylvania alone, 40% of the 5,788 concrete bridges are either structurally deficient or functionally obsolete[1]. One relatively new option for the flexural strengthening of concrete members is the use of externally-bonded fiber-reinforced polymer (FRP) laminates. These composite materials consist of an epoxy matrix with fiber reinforcement of either glass, carbon, or aramid materials (GFRP, CFRP, or AFRP, respectively). FRP composites have the benefits of a high strength-to-weight ratio, corrosion resistance, and ease of fabrication. FRP repairs for flexural members can be prefabricated and bonded to the concrete using a separate adhesive, or can be bonded and cured in-situ directly through the epoxy[2] in a process often called wet layup. This study focuses on carbon FRP (CFRP) bonded to concrete using wet layup.

While the use of FRP laminates or sheets bonded to concrete is becoming more common for the repair of concrete flexural members, the behavior of the bond is still in question. The ACI440.2R-02 design guidelines intentionally avoid bond failure between an FRP flexural repair and a concrete substrate as a failure mode. The guidelines incorporate an additional reduction factor into design equations "to arrive at a strain limitation to prevent debonding"[3]. No design equation to calculate bond failure load (given material and geometric properties for the FRP and substrate) has yet been included in these guidelines, but extensive work has been done to propose a number of bond models.

Research Significance

The objective of this study is to conduct a series of experimental tests to collect data with which three proposed bond models can be evaluated. The study uses the maximum load carried by the bond prior to failure in a pull-off test as an indicator of bond performance, and compares that load to the predicted load from three existing bond models.

EXPERIMENTAL PROGRAM

To achieve the objective, two sets of (21) 127x127x254mm concrete block specimens were fabricated; one set having an ASTM C39[4] compressive strength (f'_c) of 35MPa (Mixture A) and the other set 46MPa (Mixture B). CFRP strips were bonded to the blocks via wet layup after preparing the concrete bonding surfaces by grinding away the laitance. The parameters of interest to this study were the concrete strength, the bonded length of the FRP (76.2, 152.4, and 203.2mm), the bonded width of the FRP (25.4, 50.8, and 76.2mm), and FRP thickness (1, 2, and 3 plies). Table 1 summarizes the parameters and specimens, and Figure 1 illustrates the specimen geometries. Note in Figure 1 that a "narrowing extension" had to be wet-laid with the FRP strip when the bonded width was greater than 25mm so that the strip would fit within the grips, which had a maximum useable width of 25mm. All FRP strips were oriented with the fibers in the longitudinal direction (0°).

A restraint frame was designed and constructed for the 49kN load frame used during testing, as shown in the sketch in Figure 2. Placing a plate (the "crack shim") between the top of the concrete block and the top plate of the restraint with a 25mm setback created a free edge for the concrete. The effects of this free edge were studied during testing.

Three specimens were tested for each parameter of interest, as indicated in Table 1. Out of each set of three, one specimen was instrumented with 120Ω strain gages at the center of the bonded width. Load-displacement data was collected for all specimens, and strain data was available for each parametric variation. Although the maximum load transferred by the FRP-concrete bond was the key factor in the comparison to existing bond models, the strain data aided in evaluating the bond behavior. Figure 3 depicts the strain gage layout on the specimens.

TESTING AND OBSERVATIONS

Each specimen was carefully secured in the restraint frame and checked for alignment with the grips to prevent normal stresses from forming along the bonded length. Normal (peeling) stresses have been shown to have a significant effect on the bond behavior[5]. Sandpaper was placed between the self-tightening grips and the CFRP strips to ensure that slip would not occur.

Testing was conducted on the specimens under a displacement-controlled rate of 0.381mm per minute. Using this rate, the specimens reached failure within 7 to 15 minutes on average, and the propagation of the crack front through the CFRP-concrete bond was found to be stable. Data was collected at 20 points / second (20Hz). When an instrumented specimen was tested, the gages were connected to a ¼ Wheatstone bridge to collect strain data.

Propagation of Failure Through the Bond

During the initial portion of the test procedure (the first few minutes), the specimens did not display any signs of distress. After a few minutes, cracking and small popping sounds could be heard as the load applied to the strip (and the bond) continued to increase. Near the end of each test, as the final bond failure approached, cracking sounds became louder and occurred at much shorter intervals. On occasion, just prior to complete failure, the cracking came in a continuous string, analogous to an unzipping sound. This has been observed in other tests, such as those conducted by Lopez in 2000[6].

As bond failure progressed, small diagonal cracks began to form in the edges of the CFRP strips, as shown in Figure 4. These cracks appeared at regular intervals, and never propagated significantly inwards from the edges of the strip. The angle of the cracks suggests an edge effect, affecting the crack front propagation, seen clearly in the crack front of a GFRP strip bonded to steel in Figure 4, tested by the author prior to this study. Each new crack in the CFRP strips appeared at a time approximately corresponding to a shift in the crack front as a portion of the FRP-substrate bond fails.

As tests progressed, 86% of the 42 specimens developed cracks in the concrete substrate that ran from the edge of the FRP strip up to a point at the leading edge of the concrete block. Figure 5 depicts these cracks, which resulted in a tooth of concrete at the top of the block. The angle, α, tended to increase as the bonded width of the strip increased, but the distance, x, remained the same, leading to a wider tooth of concrete at the free edge. Rarely did the tooth approach the edge of the shim plate used in the restraint setup – thus, the unrestrained edge distance of 25mm between the shim plate and the bonding surface was sufficient to prevent an unrealistic restraint of the concrete against failure. The tooth did not displace visibly during testing, but upon final bond failure, the tooth remained bonded to the FRP strip. Concrete cover failure as part of overall bond failure has been noted by several researchers, including Lopez in 2003[7] and Buyukozturk in 1998[8].

Final Bond Failure

The final failure of the CFRP-concrete bond was extremely brittle, and accompanied by a large release of energy. Some of the FRP strips were damaged by the energy release at final failure, exhibiting splitting in the longitudinal (fiber) direction, while the epoxy of some other strips had transverse cracking post failure. Fragments of concrete were torn from the block when the strip broke free.

Increased bonded length, width, and thickness parameters led to longer time-to-failure, but changes to the bonded width had the greatest impact on time-to-failure. For the shortest bonded length (76mm), there was little warning before final failure. All other FRP strips provided some warning, prior to final bond failure, in the form of the cracking sounds.

While a few specimens experienced pull-out of an additional tooth of concrete at the end of the bonded length during failure, the vast majority only lost the concrete tooth at the leading edge of the bond as discussed previously. The remainder of the bonded region past the leading edge on the concrete block was free of epoxy after failure, and the FRP strips did not exhibit a significant amount of paste left clinging to the strip. However, the concrete surface was rough, with numerous microcracks oriented in a manner indicative of the direction of bond failure progression. A pattern of regularly-spaced microcracks was visible throughout the bonded area, with small pieces of paste peeled back by the propagation of the bond failure as shown in Figure 6.

EXPERIMENTAL RESULTS

The data obtained from experimentation can be subdivided into three categories. These are the maximum load data, which was compared to failure load predictions from existing bond models; load-displacement data; and strain data along the bonded region. The "base specimen" for comparison is defined as a block with a 1-ply FRP strip, a 25mm bonded width, and a 152mm bonded length. The average failure load for this specimen is defined as P^*.

Maximum Load Data

Results for the maximum load carried by the CFRP-concrete bond (P_{max}) are summarized in Figures 7(a) and (b). The data had relatively little scatter, with an average coefficient of variance (CV) of 7.83%. The limited scatter is a strong indication of the validity of the data presented in this report.

From these figures and the values obtained for P_{max}, it can be seen that concrete strength did not have a significant impact on the maximum load carried by the bond for this type of test. Maximum loads either increased slightly by about 1% or decreased slightly by around 7% – this variation can be associated with differences in specimen fabrication or inherent material variability in the concrete.

Figures 8(a) and (b) show the changes in the normalized failure load ratio (P_{max} / P^*) as the parameters of study change. The P_{max} value used here for the comparisons is the average from each group of three specimens from both Mixture A and B. When the bonded length changed from 76mm to 152mm, the failure load ratio increased by about 17%; beyond 152mm bonded length, the failure load did not increase significantly, and in fact decreased by about 4%. This behavior is believed to be related to the fact that the bonded length may have exceeded the effective bonded length as defined by Täljsten[9] after 152mm. Thus, the additional bonded length only increased the time to failure for the system. This, then, indicates that an increase in bonded length beyond the length needed to transfer stresses between the FRP and concrete can provide some warning time prior to the FRP-concrete system failure. However, the extra length will not significantly impact the system failure load.

On average, increasing the bonded width had more impact on the bond failure load (an increase of approximately 95% from 25mm bonded width to 76mm bonded width) than an increase in the FRP thickness (which caused a total failure load increase of approximately 55% from 1 to 3 plies). Increasing bonded width, while shown here to be effective at adding to the load capacity of the FRP-concrete system, may not be practical due to limitations imposed by the width of the member being repaired. Extra plies of FRP added to the load capacity of the system as well, so when increasing bonded width is not an option, adding to the repair thickness will still increase the failure load substantially.

The relationship between changes in each parameter and the resulting failure load increase was investigated. It was found that the failure loads did not increase by the same ratio as the FRP thickness or width – that is to say, the failure load did not double when the bonded width or FRP thickness doubles. Moreover, the incremental changes in failure load for changes in each parameter were not necessarily linear for the data obtained in this study, as seen for the FRP thickness. The increase in the ratio of Figures 8(a) and (b) from one to two plies was approximately 25%, while the increase in the ratio from two to three plies was 31%. However, the change was linear for additional bonded width, with an increase of 48% from 25mm to 51mm of bonded width and 47% from 51mm to 76mm of bonded width. More testing may need to be done to determine if the nonlinearity for the changes in FRP thickness is due to inherent variability in the data. Because the load-carrying capacity of the system does not increase in a one-to-one manner for added width or quantity of plies, it may be necessary to vary multiple parameters at once when designing an FRP flexural strengthening solution to obtain the desired amount of additional load capacity for the system.

Strain Data

The strain data collected supports observations by researchers such as Täljsten[9] and Miller and Nanni[10]. Figures 9, 10 and 11 for the Mixture A specimens, respectively, show that as the load was initially applied to the specimens, the strains on the FRP near the free edge of the concrete increased steadily until bond failure began. At that point, if strain redistribution was able to occur along the bonded length (i.e., additional bonded length was available for crack propagation to occur prior to bond failure), the strain near the free edge of the bonded length reached a peak value and remained near that value for the remainder of the test. Meanwhile, strain began to increase at the next gage on the bonded length, and so forth as the crack propagated until complete debonding failure occurred. Debonding failure is believed to result from a condition where the remaining bonded length is insufficient to transfer stresses between the FRP and the concrete substrate.

Specimens with longer bonded length, greater width, and greater thickness experienced this strain redistribution at higher loads. Also, strains in specimens with greater thickness or bonded width tended to be smaller in magnitude than the strains in the specimens with one ply of carbon fiber and 25mm of bonded width for the same load. Longer bonded length did not reduce the magnitude of strain at the location of the strain gages, but it did ensure that strain redistribution could occur (again, increasing the time up to failure of the system). For the 76mm bonded length, strain redistribution did not occur to any significant degree prior to complete bond failure.

Plots for the Mixture B specimen strain data were similar in nature to those presented for Mixture A. As with the load data, the difference in concrete compressive strength between Mixtures A and B had little impact on the strain data.

Load-Displacement Data

Additional information about the bond behavior can be obtained from the crosshead load and displacement data. Figure 12 illustrates the load-displacement relationships for the specimens of Mixture A.

After an initial adjustment period, and prior to the beginning of debonding, the slopes of the load-displacement curves exhibited similar slopes for each set of three specimens. For those specimens with varied bonded length, the slopes of all specimens matched quite well for each mixture (A and B). However, as bonded width and FRP thickness increase, the slopes increase. This indicates that while an increase in bonded length for the FRP strip does not change the overall stiffness of the FRP-concrete system, an increase in bonded width or FRP thickness can strongly influence the stiffness of the system. Future research should consider other types of FRP material (such as GFRP) bonded to concrete and examine the changes in stiffness for that system as the same parameters are varied. Specimens with a greater bonded width or FRP thickness experience less deformation for a given load. Increasing the bonded length from 76mm to 152mm increases the displacement of the FRP-concrete system in proportion to the increase in maximum load. However, as also seen in the maximum load data, an increase from 152mm to 203mm of bonded length results in little change in maximum load and displacement values.

At the higher bonded lengths, it can be noted that the load-displacement curve appears to reach a plateau where the load continues to increase and drop repeatedly within a narrow range of load values, until complete bond failure. This supports the concept of an effective or active strain transfer region[9] that shifts as the bond failure propagates. When the effective strain transfer region reaches its maximum capacity for strain transfer, bond failure occurs in that region, forcing the region to shift down the bonded length. The load drops momentarily, reducing the strains; then, the strains increase again in the new transfer region until they reach the maximum capacity once again, and so forth.

Edge Effects

By examining the load-displacement and normalized failure load ratio (P_{max} / P^*) data for those specimens with bonded width variant, it can be seen that the aforementioned edge effects may influence system failure load results for specimens less than 51mm in bonded width. The increase in the normalized failure load ratio is greater between 25mm and 51mm than between 51mm and 76mm for both Mixture A and B. Similarly, the change in crosshead displacement at failure is greater between 25mm and 51mm of bonded width than between 51mm and 76mm. This data, in conjunction with the observed crack patterns, supports the theory that the bond behavior for narrow FRP strips is influenced by the U-shaped crack front – the apparent FRP-concrete system capacity may be less than expected. For wider strips, the influence of these edge effects becomes negligible relative to the overall width of the strip. Further study is required to confirm this, as it has significant implications for the validity of results obtained in previous studies by other researchers, which often used strips 25mm in width.

ANALYTICAL STUDY

The experimental results are compared to maximum load predictions from three existing bond models, as proposed by Maeda et al. in 1997[11] (Model 1), Miller and Nanni in 1999[10] (Model 2), and Chen and Teng in 2001[12] (Model 3). These three models were chosen because the models incorporate equations to find effective bond length and also to find the maximum load the FRP-concrete system can carry. Model 2 is a variation on Model 1.

The input to these models includes the actual material properties obtained during the course of experimentation and through coupon testing, as well as published material properties for the CFRP system. Table 2 summarizes the necessary input values for each model.

Two different modulus of elasticity values were used for the FRP as input. The first modulus, called the *nominal* modulus of elasticity (*Nom. E*), is based upon the manufacturer's published fiber modulus. This is used in conjunction with the published FRP thickness (again, based upon the fiber cloth thickness) and the other input values to generate the nominal modulus line on each plot of failure load output. The second modulus, called the *actual* modulus (*Act. E*), is based on coupon testing data of CFRP strips. This is used in conjunction with the actual measured specimen thickness values and the other input values to generate the actual modulus line.

The failure load output for the models is plotted as a continuous curve over the range for each parameter studied, with the experimental data shown on the same plot. This output is shown in Figures 13, 14, and 15 for Mixture A.

The plots in Figure 13 show that Models 1 and 2 are insensitive to changes in the bonded length, while Model 3 follows the trend of the data but is too conservative for longer bonded lengths. Models 1 and 2 do not predict the decrease in failure load for the FRP-concrete system as bonded length becomes shorter than 152mm, but remain conservative within the data range of 76 – 203mm bonded length when the nominal manufacturer's values for E_{FRP} and FRP thickness are used.

From the plots in Figure 14, it can be seen more clearly that the models were calibrated using published (nominal) values for E_{FRP} and FRP thickness, as these values give better predictions of the failure loads. All three models follow the trend in the P_{max} values; Models 1 and 2 result in a linear trend, and Model 3 in a power curve trend. Model 2 has a steeper slope than Model 1, and may overestimate the failure load for the FRP-concrete system at higher values of bonded width.

The plots of Figure 15 indicate that all models predict changes in the failure load for increasing FRP thickness with changes in slope as plies are added. The failure load predictions from Model 2 best correlate to the shape of the trend in the data for increasing FRP thickness. It is conservative with respect to the data by a consistent amount. The other two models seem to predict a greater change in failure load from two to three plies than from one to two plies – this does not match the trends in the experimental data. Moreover, all three models are conservative with respect to the failure loads. Since adding layers of FRP is one of the primary methods used to obtain extra capacity in externally-bonded FRP repairs, this particular lack of sensitivity to trends in the data will require further study and analysis. As with Figure 14, these plots indicate that using the actual composite properties as input (as opposed to nominal properties) has a negative impact on the failure load predictions for these three particular models, rendering them completely insensitive to thickness changes.

Output for Mixture B was quite similar to the output for Mixture A both in trends and magnitudes of P_{max} for variations in the parameters of interest.

CONCLUSIONS AND RECOMMENDATIONS

This study has created a new database of experimental data for modeling the bond behavior of the FRP-concrete system.

Conclusions

The failure load of the FRP-concrete system depends primarily upon the width of the bonded region and the thickness of the FRP repair. Increasing the bonded width had the greatest effect on increasing the failure load of the system, but may be limited by the geometry of the member being repaired. Increases in FRP thickness also significantly increased the failure load of the system.

Bond failure occurs in stages, provided there is sufficient bonded length to develop the full capacity of the bond in more than one region. The strain data and load-displacement data both support this observation.

Failure of the bond at a free edge, such as at a crack in a beam, will most likely result in the pull-out of a tooth of concrete before the bond begins to fail at the epoxy-concrete interface. Final catastrophic failure of the FRP-concrete bond is extremely brittle. The energy release is sufficient to also fail the FRP itself in some cases, and to tear out teeth of concrete.

Edge effects were observed during testing that may influence the results obtained using specimens with FRP strips less than 51mm in bonded width, which may have negatively impacted results obtained in previous studies of the FRP-concrete bond.

All of the studied bond models appear to be calibrated to nominal material properties (modulus of elasticity and thickness of the FRP). If actual composite properties obtained from material testing are used, the models may not be conservative in the predictions of failure load.

The models from Maeda et al.[11] and Miller and Nanni[10] are both insensitive to changes in failure load as bonded length changes. While this is acceptable when the bonded length exceeds the effective bonded length, it may be unconservative for bonded lengths shorter than the effective length.

All three models are relatively accurate in their predictions for failure loads as bonded width varies. Miller and Nanni's[10] model may over-predict the failure load for bonded widths greater than those studied despite having the closest correlation to the failure loads in the range of the data gathered.

The models studied underestimate the performance gain from increased FRP thickness. The model developed by Miller and Nanni[10] shows a closer trend to the experimental failure loads with increasing FRP thickness.

Overall, Miller and Nanni's model[10] fit the data more closely than the other two. However, the trends in the experimental data are most accurately modeled by Chen and Teng's bond model[12].

Recommendations

Results from this study show that further study must be devoted to developing and extensively testing a unified bond model for external FRP reinforcement bonded to concrete. While it is believed that other FRP materials (such as glass FRP) bonded to concrete may exhibit similar results, testing should be done on such systems to determine the influence of the change in FRP elastic modulus on the test results, as compared to the CFRP results presented here. Once a complete bond model for FRP-concrete systems is developed and tested, it can be integrated into the ACI 440 design guidelines.

ACKNOWLEDGEMENT

The author thanks the National Science Foundation for its support of the research activity described in this study (grant CMS 0330592).

REFERENCES

[1]   Federal Highway Administration, "Material Type of Structure by State" Bridge Technology (2004), Excel File, http://www.fhwa.dot.gov/bridge/mat03.xls (accessed February 25, 2004).

[2] Bakis, C.E., et al., "Fiber-Reinforced Polymer Composites for Construction – State-of-the-Art Review." Journal of Composites for Construction, V6 (2002), pp. 73-87.

[3] American Concrete Institute. ACI440.2R-02: Guide for the Design and Construction of Externally Bonded FRP Systems for Strengthening Concrete Structures. Farmington Hills, MI, ACI, 2002.

[4] American Society for Testing of Materials. "ASTM C39 / C39 M: Standard Test Method for Compressive Strength of Cylindrical Concrete Specimens." Annual Book of ASTM Standards. Conshohocken: ASTM, 2001.

[5] Karbhari, V.M., et. al. "On the Durability of Composite Rehabilitation Schemes for Concrete: Use of a Peel Test." Journal of Materials Science 32 (1997), pp. 147 - 156.

[6] Lopez, M.M. "Study of the Flexural Behavior of Reinforced Concrete Beams Strengthened by Externally Bonded Fiber Reinforced Polymeric (FRP) Laminates." Ph.D. Thesis University of Michigan, 2000.

[7] Lopez, M.M., and A.E. Naaman. "Concrete Cover Failure or Tooth Type Failure in RC Beams Strengthened with FRP Laminates." Fiber-Reinforced Polymer Reinforcement for Concrete Structures, Proc. of the Sixth International Symposium on FRP Reinforcement for Concrete Structures, 8-19 Jul. 2003, Singapore. Ed. Kiang Hwee Tan. River Edge: World Scientific Publishing Co., 2003., pp. 317 - 326.

[8] Buyukozturk, Oral, and Brian Hearing. "Failure Behavior of Precracked Concrete Beams Retrofitted with FRP." Journal of Composites for Construction 2 (1998): pp. 138-144.

[9] Täljsten, Björn. "Defining Anchor Lengths of Steel and CFRP Plates Bonded to Concrete." Journal of Adhesion and Adhesives Great Britain: Elsevier Sci. Ltd, 1998.

[10] Miller, Brian, and Antonio Nanni. "Bond Between CFRP Sheets and Concrete." Proc., ASCE 5th Materials Congress, 10-12 May 1999, Cincinnati. Ed. L.C. Bank, pp. 240-247.

[11] Maeda, Toshiya, et al. "A Study on Bond Mechanism of Carbon Fiber Sheet." Non-Metallic (FRP) Reinforcement for Concrete Structures. Proc. of the Third International Symposium, Vol. 1, Oct. 1997, Japan: Japan Concrete Institute, pp. 279-286.

[12] Chen, J.F., and J.G. Teng. "Anchorage Strength Models for FRP and Steel Plates Bonded to Concrete." Journal of Structural Engineering 127 (2001), pp. 784-791.

Table 1 – Test Matrix

	Mixture A						
	Varying b_{FRP}			Varying L_b		Varying t_{FRP}	
Specimen No.	1,2,3	4,5,6	7,8,9	10,11,12	13,14,15	16,17,18	19,20,21
Plate Thickness (# of plies)	1	1	1	1	1	2	3
Bonded Length (mm)	152	152	152	76	203	152	152
Bonded Width (mm)	25	51	76	25	25	25	25
	Mixture B						
Specimen No.	22,23,24	25,26,27	28,29,30	31,32,33	34,35,36	37,38,39	40,41,42
Plate Thickness (# of plies)	1	1	1	1	1	2	3
Bonded Length (mm)	152	152	152	76	203	152	152
Bonded Width (mm)	25	51	76	25	25	25	25

Table 2 – Input values needed for each model

Model	Concrete Width	f_c	FRP Thickness	Bonded Width	Bonded Length	E_{FRP} (Elastic Modulus)	Strain Gradient (constant)
1			X	X	X	X	X
2			X	X	X	X	X
3	X	X	X	X	X	X	

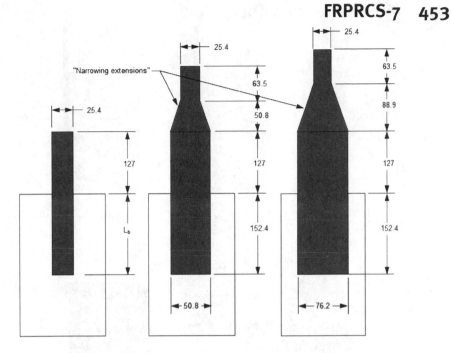

Figure 1 – Sketch of specimen geometries (not to scale)

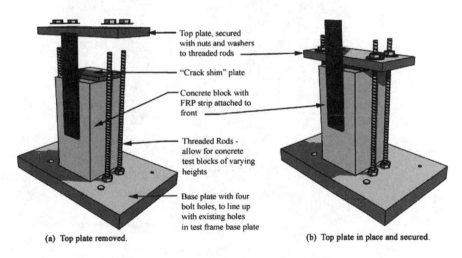

(a) Top plate removed.

(b) Top plate in place and secured.

Figure 2 – Sketch of the restraint frame for specimens

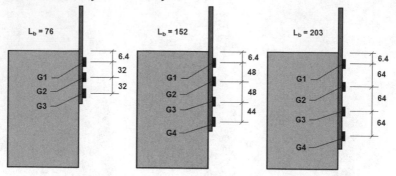

Figure 3 – Sketch of specimen instrumentation along the bonded length (not to scale).

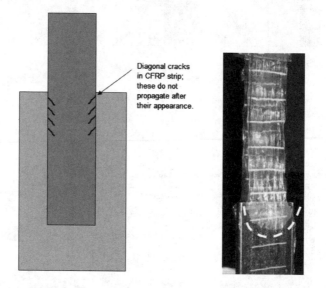

Figure 4 – Exaggerated sketch of edge cracking in the CFRP during testing, compared to edge effects observed in GFRP strips bonded to steel.

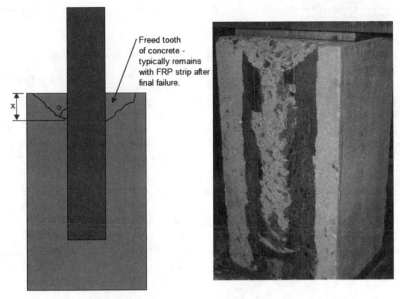

Figure 5 – Free edge failure of the concrete.

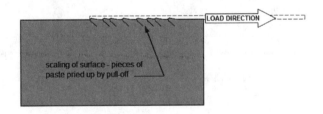

Figure 6 – Concrete surface scaling due to pull-off of the FRP strip.

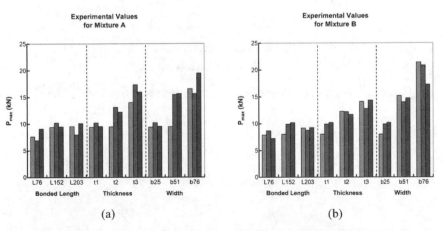

Figure 7 – (a) P_{max}- results for Mixture A (b) P_{max} results for Mixture B.

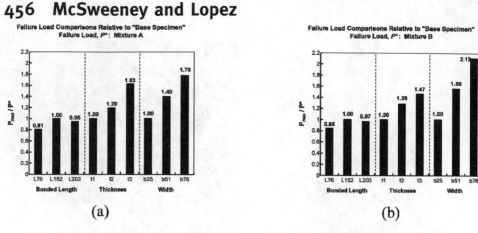

Figure 8 – (a) P_{max} / P^* for Mixture A (b) P_{max} / P^* for Mixture B.

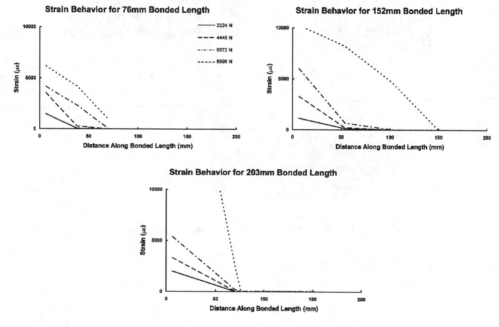

Figure 9 – Strain data as bonded length varies: Mixture A specimens.

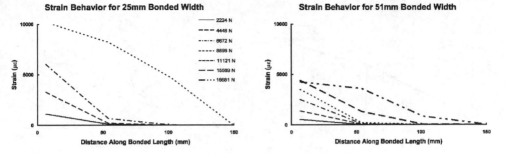

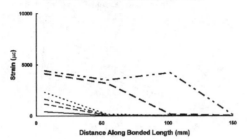

Figure 10 – Strain data as bonded width varies: Mixture A specimens.

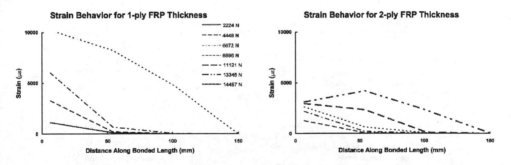

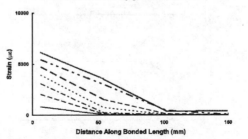

Figure 11 – Strain data as FRP thickness (number of plies) varies:
Mixture A specimens.

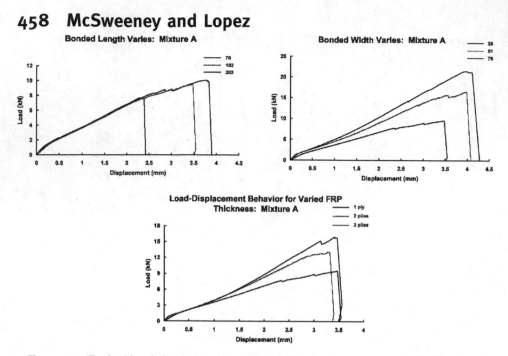

Figure 12 – Typical load-displacement results for the parameters studied: Mixture A.

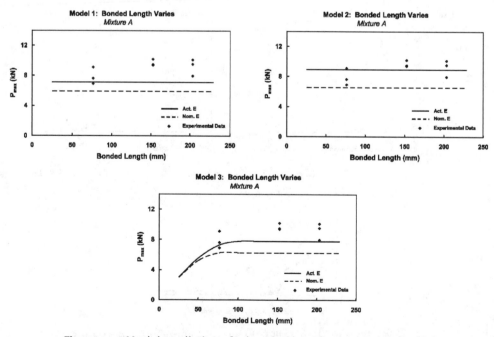

Figure 13 – Model predictions for bonded length variant: Mixture A.

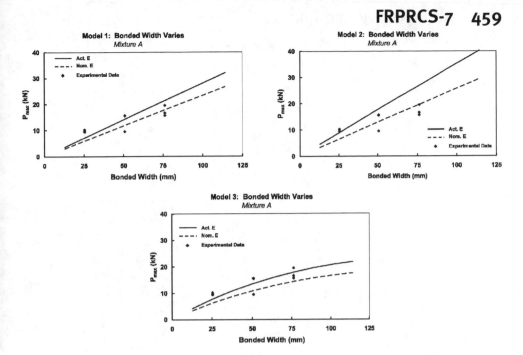

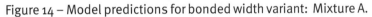

Figure 14 – Model predictions for bonded width variant: Mixture A.

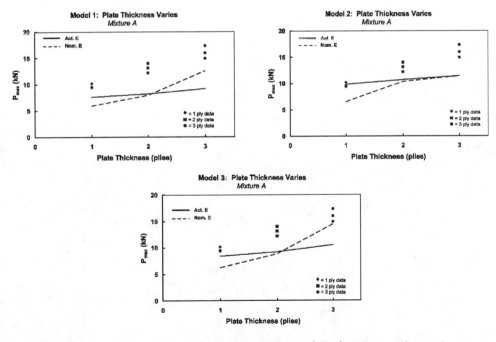

Figure 15 – Model predictions for FRP thickness (plies) variant: Mixture A.

Modelling of Debonding Failures in FRP-Strengthened Two-Way Slabs

by W.E. El Sayed, U.A. Ebead, and K.W. Neale

Synopsis: With the development of the technology of strengthening existing concrete structures using externally bonded FRP composites, a number of issues related to the structural behaviour of such structures require investigation. One of the most important issues is the interfacial behaviour between the FRP composite and the concrete as this often controls the failure mode of the strengthened member. Several studies have progressed on the development of appropriate interfacial behaviour (bond–slip) models and on understanding the debonding phenomena. This work focuses on the numerical modelling of FRP-strengthened two-way reinforced concrete slabs where the predominant mode of failure is debonding at the FRP-concrete interface. An appropriate finite element model that represents the interfacial behaviour for FRP-strengthened two-way slabs is presented. The proposed model successfully simulates the different flexural failure modes, particularly those related to the debonding of the FRP sheets. The ultimate load carrying capacity and load–deflection relationships are predicted with reasonable accuracy.

Keywords: bond-slip model; debonding; FRP composites; interfacial behaviour; two-way slabs

W.E. El Sayed, M.Sc., is a Ph.D student, Department of Civil Engineering, Université de Sherbrooke, Sherbrooke, Quebec, Canada. He received his M.Sc. from Helwan University, Cairo, Egypt in 2001. His research interests include the strengthening of reinforced concrete structures using composite materials and the finite element modelling of FRP-strengthened structures.

U.A. Ebead, Ph.D., P.Eng. is an NSERC Postdoctoral Fellow, Department of Civil Engineering, Université de Sherbrooke, and Assistant Professor at Helwan University, Cairo, Egypt. He received his Ph.D. from Memorial University of Newfoundland, Canada in 2002. His research interests include the strengthening of concrete structures and the finite element analysis of concrete structures.

K.W. Neale, Ph.D., Eng. is Canada Research Chair in Advanced Engineered Material Systems, Department of Civil Engineering, Université de Sherbrooke. His research interests include the rehabilitation and strengthening of structures using composite materials, micromechanics, solid mechanics, and materials engineering.

INTRODUCTION

FRP composites are deemed very promising materials for use as external reinforcements for reinforced concrete structural elements. As a result, there has been a numerous amount of research work conducted on the use of FRP composites in strengthening reinforced concrete structures. Several studies have been conducted on the use of FRPs for the shear strengthening of concrete slabs. In some of these studies, slabs were strengthened either with unidirectional FRP sheets around the column stub[1,2,3] or with bidirectional FRP sheets bonded over the whole slab width[4]. Another study used L-shaped FRP plates anchored to the slab[5]. In addition, FRP strengthening materials have been used for the flexural strengthening of one-way[6,7] and two-way slabs[8,2,7]. In other studies, the slabs were strengthened with equally spaced FRP sheets in both directions over the whole slab width[9,10]. Another investigation studied prestressed FRP sheets for the strengthening of two-way slabs[11]. There are limited theoretical investigations on the behaviour of FRP-strengthened slabs. The present study attempts to provide theoretical insight on the efficiency of using FRP composites for such applications.

Finite element analysis is undoubtedly one of the most effective numerical methods for modelling the behaviour of reinforced concrete two-way slabs strengthened with FRP composites. Reitman and Yankelevsky[12] developed a new technique for the nonlinear analysis of reinforced concrete slabs under various loading conditions. Their technique used a nonlinear finite element analysis of a grid representing the slab based on yield-line theory. Seim et al.[7] used a beam analogy to provide an approximate simulation of the overall response of FRP-strengthened one-way slabs. The moment-curvature relationship and a hypothetical three-stage load–displacement curve have been used. The hypothetical three-stage curve represented the uncracked response, the response before yielding of the steel, and the yielding of the steel and failure of either the concrete or the FRP composites. Other researchers used finite element packages to investigate the structural behaviour of concrete slabs, both unstrengthened[13] and strengthened[1,10]. In another study, the strengthened slabs were designed as three-layered plates using a simplified laminated plate model[9]. The bottom layer represented the FRP strip, the middle layer represented the steel reinforcement, and the upper layer represented the compressive concrete. Usually, for the FRP-strengthened slabs, a full-bond was assumed between the concrete and the FRP composites.

From the available literature on the FRP strengthening of concrete slabs, it is clear that the amount of theoretical research is very limited compared to the experimental work. It is obvious that further analytical research is required. In this study, we attempt to enrich the theoretical research on the use of the FRPs as a strengthening technique for two-way slabs. This study introduces a nonlinear finite element analysis to simulate the behaviour of these applications. This analysis takes into account the material nonlinearity and also the FRP-concrete interfacial behaviour. The main advantage of the present numerical model is its ability to accurately represent the entire load–deflection behaviour, including crack development, and all possible failure modes such as debonding and the classical flexural failures.

FRP-CONCRETE INTERFACE

The interfacial bond between the concrete and the FRP composites has a significant effect on the overall performance of the strengthened members. Experimental studies showed that the overall behaviour is brittle for externally bonded FRP beams due to mechanisms such as delamination and/or peeling-off. Such a premature failure leads to an inefficient usage of the FRP materials and prevents the strengthened members from reaching their full capacities. Observations showed that there are two different debonding types that occur in experiments, both of which result in the delamination of the FRP sheets from the concrete[14,15,16]. The first type of debonding takes place in the adhesive layer, where the interfacial shear strength is exceeded. The other type is represented by diagonal micro-cracks of the concrete at the interface. Existing studies suggest that the main failure mode for FRP-to-concrete bonded joints in direct shear tests is a concrete failure due to shear, occurring generally at a few millimetres from the adhesive layer inside the concrete[17,18].

Generally, there are two approaches for the simulation of the debonding of FRP-strengthened reinforced concrete structures using a nonlinear FE model. In the first approach, debonding is directly simulated by modelling the cracking and failure of concrete elements adjacent to the adhesive layer[19,20]. To simulate the debonding behaviour using this approach, a very fine finite element mesh with element sizes being one order smaller than the thickness of the fracture layer of the concrete is used[20]. This approach has the simplicity of the fixed angle crack model rather than using the rotating angle crack model and has the capability of tracing the paths of the cracks as deformation progresses. This approach is referred to as meso-scale modelling. The second approach is to use interface elements between the FRP composites and the concrete where the failure of these elements represents debonding[21,15]. In this approach an appropriate bond–slip model needs to be employed for the interface elements. Unfortunately, the meso-scale model is difficult to be implemented in the case of three-dimensional modelling due to the large computational demands. Thus, in this study, interface elements are used to represent the interfacial behaviour between the concrete and the FRP composites.

NONLINEAR FINITE ELEMENT MODEL

The finite element package ADINA[22] is used to simulate the structural behaviour of FRP-strengthened reinforced concrete two-way slabs. This analysis considers the complexities of the interfacial behaviour, concrete nonlinearity, and the different failure modes for the concrete. A three-dimensional model is developed. A comparison between the theoretical results and the available experimental data is discussed.

MATERIAL MODELLING

Plain Concrete

A plasticity-based model is used to represent the concrete in compression and adopts the smeared crack approach for the tensile behaviour of the cracked concrete[22]. The stress–strain relationship of the concrete in compression exhibits a nearly linear elastic response up to about 30% of the concrete compressive strength. This is followed by a plastic behaviour until the concrete compressive strength is reached. Beyond the compressive strength, the concrete stress–strain relationship exhibits strain softening until crushing. Figure 1-a shows the idealized uniaxial stress–strain curve for the concrete, and Figure 1-b shows the biaxial failure surface of the concrete.

The stress–strain relationship for the concrete in tension is assumed to follow a linear ascending branch with a slope that is equal to the concrete modulus of elasticity, E_c, until the maximum tensile stress, σ_t, is reached as shown in Figure 1-a. In this study the smeared crack model is used, in which it is assumed that a plane of failure is developed perpendicular to the corresponding principal stress direction. The normal and shear stiffness across the plane of failure are reduced and plane stress conditions are assumed to exist at the plane of tensile failure[22].

Steel Reinforcement and FRP Composites

Typical stress–strain relationships for reinforcing steel bars used in concrete are obtained from tests. For all practical purposes, steel exhibits the same stress–strain relationship in compression as in tension. This relationship can appropriately be represented by using an elastic-plastic constitutive model as shown in Figure 2-a. For the FRP composite, a linear elastic material behaviour up to failure is assumed as shown in Figure 2-b.

Bond–Slip Model

The mechanical behaviour of the FRP-concrete interface can be modelled as a relationship between the local shear stress, τ, and the relative displacement, S, between the FRP sheet and the concrete substrate. A nonlinear bond–slip model is employed in the present study[20]. This model is a nonlinear relationship between the shear stress, τ, and the relative slip, S, as shown in Figure 3. The area under the τ–S curve presents the interfacial fracture energy, G_f, which is defined as the energy required for complete debonding per unit bond area. The τ–S relationship is as follows:

$$\tau = \tau_{max}\left(\sqrt{\left(\frac{S}{S_o - S_e} + \left\{\frac{S_e}{2(S_o - S_e)}\right\}\right)^2} - \frac{S_e}{2(S_o - S_e)}\right) \qquad \text{if } S \leq S_o \qquad (1)$$

$$\tau = \tau_{max} \cdot \exp\left[\left(\frac{\tau_{max} S_o}{G_f - G_f^a}\right)\left(1 - \frac{S}{S_o}\right)\right] \qquad \text{if } S > S_o \qquad (2)$$

The maximum bond strength, τ_{max}, is governed by the tensile strength of the concrete, f_t, and the width ratio, β_w, as follows:

$$\tau_{max} = 1.5\beta_w f_t \qquad (3)$$

where the width ratio, β_w, depends on the FRP plate width, b_f, and the concrete member width, b_c, as follows:

$$\beta_w = \sqrt{\frac{2.25 - b_f / b_c}{1.25 + b_f / b_c}} \qquad (4)$$

S_o is the slip corresponding to the maximum bond strength, τ_{max}, as shown in Figure 3. S_o has an elastic and plastic component, S_e and S_p, respectively as follows:

$$S_o = S_e + S_p = \frac{\tau_{max}}{K_o} + 0.0195\beta_w f_t, \qquad (5)$$

K_o is the equivalent initial stiffness which depends on the elastic shear stiffness of the concrete, K_c, and that of the adhesive, K_a, as follows:

$$K_o = \frac{K_a K_c}{K_a + K_c}, \qquad (6)$$

$$K_c = \frac{G_c}{t_c}, \qquad K_a = \frac{G_a}{t_a}, \qquad (7)$$

Here G_c and G_a are the shear moduli for the concrete and the adhesive, respectively, t_a is the thickness of the adhesive layer, and t_c is the effective thickness of the interfacial concrete layer. The thickness of this layer depends strongly on the concrete fracture energy and can be assumed to be 5 mm[23]. The total fracture energy, G_f, can be expressed as:

$$G_f = 0.308\beta_w^2 \sqrt{f_t}, \qquad (8)$$

while the fracture energy of the ascending branch G_f^a, can be expressed as:

$$G_f^a = \tau_{max} S_o\left[\frac{2A}{3}\left(\frac{1 + B^2 A}{A}\right)^{3/2} - B - \frac{2}{3}B^3 A\right] \qquad (9)$$

where $A = \frac{(S - S_o)}{S_o}$, $B = \frac{S_e}{2(S_o - S_e)}$ $\qquad (10)$

Due to the various applications of the bond–slip model to FRP-strengthened slabs with different configurations, it is obvious that the interfacial fracture energy is less affected by the FRP stiffness; however, it is affected by the mechanical properties of the concrete and to a lesser extent by those of the adhesive. Further details of this model can be found in[20].

GEOMETRICAL MODELLING

Due to the geometrical and loading symmetry of the slabs, only one quarter of the slab was modelled. Symmetrical boundary conditions were placed along the two axes of symmetry. Figure 4 shows the configurations of the different test specimens that are investigated in this study. Figure 5 shows a FE representation of one of these specimens, tested by Ebead and Marzouk[1]. For the two-way slab specimens, the strengthening materials were attached to the tension side of the slab in both directions. The orthotropy of each FRP strip was accounted for in the constitutive material properties of the corresponding elements depending on the fibre orientation. In the proposed model, in order to represent the interface elements, two sets of nodes were employed. One set of nodes was used for the concrete elements, while the second set of nodes was used for the FRP elements. These two sets of nodes were connected together using the interface elements that allow relative horizontal displacements (slip) to take place between the FRP composites and the concrete surface.

3-D quadrilateral solid elements were used to represent the concrete. Using such elements satisfies the shear and bending deformations due to their quadratic interpolation functions. The element employed has 27 nodes; each node has three degrees of freedom.

Thin shell elements were used to represent both the unidirectional and bidirectional FRP composites. The thin shell element has nine nodes; each node has three degrees of freedom.

Nonlinear translational spring elements were employed to represent the FRP-concrete interface. Each element has two nodes; each node has three degrees of freedom. These elements were employed in the same direction as that of the unidirectional fibres and in both directions for the bidirectional fibres. A force–displacement relationship was developed based on the bond–slip model used. The interface elements have only stiffness in the horizontal direction. Constraint equations were applied in the vertical direction between the concrete elements and the FRP elements.

Steel reinforcement elements embedded in the concrete are best described as one-dimensional elements. Truss elements are very appropriate to represent the steel reinforcement. Two-node truss elements were generated and connected to the 27-node 3-D solid elements. The truss element has two nodes; each node has three translational degrees of freedom. The assemblage of the various elements is shown in Figure 5.

NUMERICAL RESULTS

In order to calibrate the proposed model, a comparison was first made between the theoretical and the experimental results for one-way reinforced concrete slabs strengthened with FRP composites. Tables 1 and 2 show the material properties for the different specimens and the different configurations are described in Table 3 and Figure 4. The comparisons are made in terms of the ultimate load carrying capacity and the load–deflection relationship. In addition, the slip profiles along the FRP plate are given.

Ultimate Load Carrying Capacity

The proposed model proved its ability to predict the ultimate load carrying capacity of the FRP-strengthened slabs with a good accuracy. Table 4 shows the comparison between the experimental and the theoretical results for the different sets of specimens. The model is able to predict the load capacity of one-way slabs as well as that for the two-way slabs. Besides, the model can predict the load capacity of both flexural and shear strengthened slabs with an accuracy of about 90%. In addition, the proposed model can successfully simulate the structural behaviour of the post-strengthened slabs including the load sequences and the application of the strengthening materials. For all the cases, FRP debonding was the common failure mode.

Load–Deflection Relationships

Comparisons between experimental and numerical results in terms of the load–deflection relationships are important to evaluate the accuracy of the model in predicting the overall behaviour and stiffness characteristics of the slabs. Figures 6 and 7 show the predicted load–deflection relationships along with the experimental results of Seim et al.[7] and Mosallam and Mosalam[10], respectively.

For the specimens of Seim et al.[7], the load–deflection relationship follows the yield plateau of the unstrengthened slab in the case where a small amount of FRP strengthening material is employed. This was obtained for Specimen S43 that has the shortest strengthening plate length of 1090 mm. For Specimens S57.5 and S72 that have plate lengths of 1460 and 1830 mm, respectively, the load–deflection relationship stopped abruptly as shown in Figure 6. The maximum deflection value of Specimen S43 was about 20% higher than those of Specimens S57.5 an S72. The same yield plateau was obtained for Specimen SA10 of Harajli and Soudki[2]. This specimen has the smallest width of bonded FRP of 100 mm as described in Table 3. For Specimens SA20 and SA30 (that have the strengthening plate widths of 200 and 300 mm, respectively), the load–deflection relationship stopped abruptly at about 65% of the maximum deflection of the corresponding control specimen. In addition, it was found that there is no significant difference in the response between Specimens SA20 and SA30. The maximum deflection value of Specimen SA10 was about 30% higher than those of Specimens SA20 and SA30. Based on the aforementioned discussion, it is evident that using a small amount of FRP material does not significantly affect the ductile behaviour of the strengthened slabs. By using a larger amount of FRP, the overall behaviour changes to a more brittle behaviour. Furthermore, as the strengthening plate length increases, the specimen strength increases. However, there is an optimum width of the strengthening plate beyond which no significant improvement is achieved. This agrees with other observations[8].

Figure 7 shows the load–deflection relationships for the specimens tested by Mosallam and Mosalam[10]. It shows the comparison between the experimental and theoretical results for the unreinforced and the reinforced specimens before and after strengthening. These specimens were subjected to a uniform load using a high-pressure water bag underneath the specimen. The effect of the strengthening is obvious in the stiffer behaviour observed for the strengthened specimens. Specimen Ret-85 was loaded up to 85% of the ultimate load of the control specimen. Then the strengthening materials were applied before the load continued up to failure. This was successfully simulated in the FE model by activating the FRP material at the specified load level. A good agreement was obtained with the experimental results, as shown in Figure 7-c.

Slip Profiles along the FRP-Concrete Interface

Based on the good agreement between the numerical and the experimental results using the proposed model in terms of the load capacity and deformational characteristics, some other quantities can be predicted that are difficult to measure in the laboratory. These quantities include the slip distribution along the FRP-concrete interface that will help understanding the interfacial behaviour between the FRP composites and the concrete.

Figure 8 shows the slip profiles along the FRP-concrete interface at different load levels for Specimen S43 of Seim et al.[7]. The slip profiles indicated that the main failure mode was characterized as debonding of the FRP plate; this is in agreement with the experimental observations. For Specimen S43, which has the shortest strengthening length, the final slip values at the plate end were 30% higher than those at the central zone of the slab. This explains the end plate debonding failure mode that was observed experimentally. On the other hand, the slip distribution for Specimens S57.5 and S73 is almost uniform at the final load stages as a result of the longer length of the attached FRP plate. It can be concluded that a bonded length of at least 65% of the specimen length is necessary to prevent the end plate debonding failure mode.

The specimens tested by Harajli and Soudki[2] experienced punching shear failure associated with a debonding of the FRP plates. It was found that increasing the FRP plate width resulted in a reduction in the slip values. This is because the wider the width of the FRP plate the more uniform the transferred shear stresses between the FRP plate and the concrete substrate, which in turn leads to a less interfacial slip. Figure 9 shows the slip profiles over the FRP plate length of Specimen SA30, which has the largest plate width of 300 mm. The slip profile shows that the slip values at the centre of the slab were significantly higher than those at the plate end at all load levels. This indicates that the delamination started at the centre of the slab then propagated towards the plate end.

The specimens tested by Mosallam and Mosalam[10] were strengthened using equally-spaced FRP plates in both directions, as indicated in Table 3. The slip profiles had the same shape for the reinforced specimen, R-Str, and the unreinforced specimen, Un-Str. For the unreinforced Specimen, Un-Str, the slip values were 60% smaller than those of the reinforced specimen, R-Str. Furthermore, due to the intersection of the FRP plates in the two directions, the slip values were 50% lower at the overlapping zone than those at the adjacent zone on the curve shown in Figure 10. This means that the transverse strengthening plate contributes strongly to reducing the interfacial slip.

The numerical results for the specimens of Ebead and Marzouk[1] showed a rapid increase in the interfacial slip at the centre of the slab followed by an almost uniform distribution along the plate length until the plate end. This means that the debonding occurred under the applied load then propagated towards the plate end. Figure 11 shows the slip profiles for Specimen SGF-0.35. It was found that increasing the FRP plate stiffness leads to less interfacial slip between the FRP and the concrete. Also, the steel reinforcement ratio has no effect on the interfacial slip between the FRP and the concrete. Moreover, by comparing Figures 9 and 11, unlike Specimen SA30 of Harajli and Soudki[2], the slip values for Specimen SGF-0.35 of Ebead and Marzouk[1] were relatively high because this specimen failed mainly in flexure. On the other hand, the specimen of Harajli and Soudki had small slip values because it failed mainly due to punching shear.

Failure Characteristics

Debonding failure starts to occur when the slip value, S_o, that corresponds to the maximum shear stress defined in the bond slip model is reached at any point. The failure starts around the point of load application and moves towards the plate end. When there is a sufficient bond length, the failure occurs in the zone around the load point such as in the case of Specimen SGF-0.35 as indicated in Figure 11. In this figure, it is obvious that the slip values at failure are maximum 2.5 mm next to the load point. On the other hand for Specimen S43 where there is insufficient bond length, the debonding occurs at the plate ends as Figure 8 indicates from the slip distribution along the bonded length. From the slip profiles, it can be concluded that the load at which the slip values reach S_o, where the debonding starts to occur, is 15% higher than that at which intermediate flexural cracks appear. This means that debonding usually occurs due to intermediate flexural cracks (except for Specimen SA30 that failed in shear).

CONCLUSION

The use of appropriate interface elements in the FE analysis enabled very good simulations of the interfacial behaviour between the FRP and the concrete in FRP-strengthened two-way slabs. The proposed model can capture the debonding phenomena and predict the possible failure modes associated with the delamination of the FRP composites off the concrete. Also, this model has the capability to predict the slip profiles along the FRP-concrete interface. The slip profiles are useful for a better understanding of the expected debonding behaviour. By increasing the FRP plate width, more uniform stresses are transferred between the FRP plate and the concrete substrate. This results in lower slip values at the FRP-concrete interface.

ACKNOWLEDGEMENTS

This research was funded by the Natural Sciences and Engineering Research Council of Canada (NSERC) and the Canadian Network of Centres of Excellence on Intelligent Sensing for Innovative Structures (ISIS Canada). KWN is Canada Research Chair in Advanced Engineered Material Systems and the support of this program is gratefully acknowledged.

REFERENCES

1. Ebead, U.A. and Marzouk, H. (2004): Fiber-Reinforced Polymer Strengthening of Two-Way Slabs. *ACI Structural Journal*, **101**(5), 650-659.

2. Harajli, M.H. and Soudki, K.A. (2003): Shear Strengthening of Interior Slab-Column Connections Using Carbon Fibre-Reinforced Polymer Sheets. *Journal of Composites for Construction, ASCE,* **7**(2), 145-153.

3. Sharaf, M.H. and Soudki, K.A. (2004): Strengthening of Slab-Column Connections With Externally Bonded CFRP Strips. *Advanced Composite Materials in Bridges and Structures, ACMBS IV*, M. El-Badry and L. Dunasazegi, Eds., Canadian Society for Civil Engineering, 8 p.

4. Rochdi, E.H., Bigaud, D., Ferrier, E. and Hamelin, P. (2004): Strengthening of Slab-Column Connections With Externally Bonded CFRP Strips. *Advanced Composite Materials in Bridges and Structures, ACMBS IV*, M. El-Badry and L. Dunasazegi, Eds., Canadian Society for Civil Engineering, 8 p.

5. Biddah, A. (2004): Punching Shear Strengthening of Reinforced Concrete Slabs Using CFRP L-Shaped Plates. *Advanced Composite Materials in Bridges and Structures, ACMBS IV*, M. El-Badry and L. Dunasazegi, Eds., Canadian Society for Civil Engineering, 8 p.

6. Erki, M.A. and Heffernan, P.J. (1995): Reinforced Concrete Slabs Externally Strengthened With Fibre-Reinforced Plastic Materials. Proceedings of the Second International RILEM Symposium *FRPRCS-2*, L. Taerwe, Ed., University of Ghent, Belgium, 509-516.

7. Seim, W., Horman, M., Karbhari, V. and Seible, F. (2003): External FRP Post-Strengthening of Scaled Concrete Slabs. *Journal of Composites for Construction, ASCE*, **5**(2), 67-75.

8. Ebead, U.A. (2002): Strengthening of Reinforced Concrete Two-Way Slabs. *Ph.D. Thesis*. Memorial University of Newfoundland. St. John's, NF, Canada.

9. Limam, O., Foret, G. and Ehrlacher, A. (2003): Reinforced Concrete Two-way Slabs Strengthened with CFRP Strips: Experimental Study and a Limit Analysis Approach. *Journal of Composite Structures*, **60**, 467-471.

10. Mosallam, A.S. and Mosalam, K.M. (2003): Strengthening of Two-Way Concrete Slabs With FRP Composite Laminates. *Construction and Building Materials*, **17**, 43-54.

11. Longworth, J., Bizindavyi, L., Wight, R.G. and Erki, A. (2004): Prestressed CFRP Sheets for Strengthening Two-Way Slabs in Flexure. *Advanced Composite Materials in Bridges and Structures, ACMBS IV*, M. El-Badry and L. Dunasazegi, Eds., Canadian Society for Civil Engineering, 8 p.

12. Reitman, M.A. and Yankelevsky, D.Z. (1997): A New Simplified Method for Nonlinear Reinforced Concrete Slabs Analysis. *ACI Structural Journal*, **94**(4), 399-408.

13. Marzouk, H. and Chen, Z.W. (1993): Nonlinear Analysis of Normal and High-Strength Concrete Slabs. *Canadian Journal of Civil Engineering*, **20**, 696-707.

14. Nguyen, D., Chan, T. and Cheong, H. (2001): Brittle Failure and Bond Development Length of CFRP-Concrete Beams. *Journal of Composites for Construction, ASCE,* **5**(1), 12-17.

15. Wu, Z.S. and Yin, J. (2003): Fracturing Behaviour of FRP-Strengthened Concrete Structures. *Journal of Engineering Fracture Mechanics*, **70**, 1339-1355.

16. Thomsen, H., Spacone, E., Limkatanyu, S. and Camata, G. (2004): Failure Mode Analyses of Reinforced Concrete Beams Strengthened in Flexure with Externally Bonded Fibre-Reinforced Polymers. *Journal of Composites for Construction ASCE*, **8**(2), 123-131.

17. Bizindavyi, L. and Neale, K.W. (1999): Transfer Lengths and Bond Strengths for Composites Bonded to Concrete. *Journal of Composites for Construction, ASCE*, **3**(4), 153-160.

18. Chen, J.F. and Teng, J.G. (2001): Anchorage Strength Models for FRP and Steel Plates Bonded to Concrete. *Journal of Structural Engineering, ASCE*. **127**(7): 784-791.

19. Wu, Z.S. (2003): Element-Level Study on Stress Transfer Based on Local Bond Properties. *Technical Report of JCI Technical Committee on Retrofit Technology*, 44-56.

20. Lu, X.Z., Teng, J.G., Ye, L.P. and Jiang, J.J. (2004): Bond-Slip Models for FRP Sheet/Plate-to-Concrete Interfaces. *Proceeding of 2nd International Conference of Advanced Polymer Composites for Structural Applications in Construction (ACIC)*, 10 p.

21. Wong, R.S.Y. and Vecchio, F.J. (2003): Towards Modeling of Reinforced Concrete Members With Externally Bonded Fiber-Reinforced Polymer Composite. *ACI Structural Journal*. **100**(1), 47-55.

22. ADINA (2003): Theory Manual, Version 8.1. *ADINA R & D, Inc. Watertown, USA*.

23. Lu, X.Z., Jiang, J.J., Teng, J.G. and Ye, L.P. (2004): Finite Element Models for FRP-to-Concrete Bonded Joints. *Construction and Building Materials*, (accepted for publication).

Table 1 — Concrete and steel reinforcement properties

Specimen	Concrete		Steel	
	f_c' (MPa)	E_c (GPa)	f_y (MPa)	f_u (MPa)
Seim et al.[7]	33.2	30	462	765
Harajli & Soudki[2]	34.5	26.4	487	745
Mosallam and Mosalam[10]	31.9	25.2	401.4	720
Ebead & Marzouk[1]	33.8	26.1	450	660

Table 2 — FRP composite and adhesive properties

Specimen	FRP type	FRP			Adhesive	
		f_t (MPa)	ε_u %	E (GPa)	f_t (MPa)	E (GPa)
Seim et al.[7]	CFRP	2270	1.2	198	24.8	2.69
Harajli & Soudki[2]	CFRP	3500	1.5	230	25	3.2
Mosallam & Mosalam[10]	CFRP	1209	1.2	101	22.2	2.65
Ebead & Marzouk[1]	CFRP	2800	1.7	170	24.8	4.5
	GFRP	600	2.2	26.1	72.4	3.1

Table 3 — Configurations of the analyzed specimens

Set	Specimen	Configuration	FRP dimensions (mm)		
			length	width	thick.
Seim et al.[7]	S43	43" FRP plate length	1090	50	1.19
	S57.5	57.5" FRP plate length	1460		
	S72	72" FRP plate length	1830		
Harajli & Soudki[2]	SA10	100 mm FRP plate width	670	100	1.0
	SA20	200 mm FRP plate width		200	
	SA30	300 mm FRP plate width		300	
Mosallam & Mosalam[10]	Un-Str	Unreinforced-Strengthened	2640	420	1.74
	R-Str	Reinforced-Strengthened	2640	420	1.74
	Ret-85	Repaired after 85% of P_u			
Ebead & Marzouk[1]	SGF-0.35	GFRP plate, $\rho = 0.35$ %	1900	300	1.0
	SCF-0.35	CFRP plate, $\rho = 0.35$ %			1.2
	SGF-0.5	GFRP plate, $\rho = 0.50$ %	1900	300	1.0
	SCF-0.5	CFRP plate, $\rho = 0.50$ %			1.2

Table 4 — Comparison between the analytical and the experimental results

Set	Specimen	Experimental		Theoretical		$\dfrac{P_{exp}}{P_{theo}}$
		Max. Defl. (mm)	Ultimate Load (kN)	Max. Defl. (mm)	Ultimate Load (kN)	
Seim et al.[7]	Cont.	211.80	21.75	190.70	25.75	0.84
	S43	42.16	32.40	38.86	33.80	0.95
	S57.5	31.50	40.0	38.10	37.81	1.05
	S72	28.19	41.2	33.78	48.40	0.85
Soudki & Harajli2	Cont	26	49	25	46	1.06
	SA10	17.5	47	21.2	46.3	1.02
	SA20	13.5	64	16.0	66.0	0.96
	SA30	12	64	14.9	68.8	0.93
Mosallam & Mosalam[10]	Un-Cont	19.55	8.14	22.1	11.25	0.72
	Un-Str	59.69	47.89	63.5	42.29	1.13
	R-Cont	74.93	31.61	86.86	30.65	1.03
	R-Str	73.15	58.9	71.12	64.17	0.91
	Ret-85	59.69	61.78	68.58	56.03	1.10
Ebead & Marzouk[1]	Ref-0.35	48.5	248.5	38.9	260	0.96
	SGF-0.35	31.47	338	30.8	313	1.08
	SCF-0.35	22.37	360	26.8	345.8	1.04
	Ref-0.5	40.65	320.1	42.0	307.4	1.04
	SGF-0.5	26.7	414.6	26.5	369.9	1.12
	SCF-0. 5	24.4	448.1	21.7	405.1	1.10

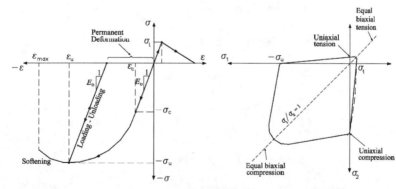

(a) Idealized uniaxial stress—strain curve for concrete.   (b) Biaxial failure surface of concrete.

Figure 1—Constitutive model for concrete

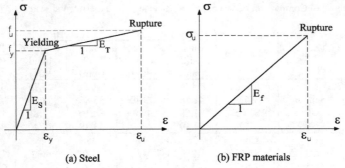

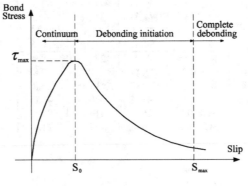

Figure 2—Typical stress-strain curves for steel and FRP materials

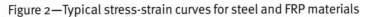

Figure 3—Bond-slip model

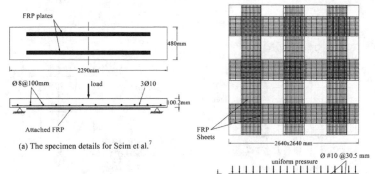

(a) The specimen details for Seim et al.[7]

(b) The specimen details for Mosallam and Mosalam[10]

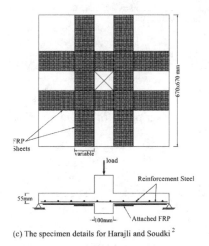

(c) The specimen details for Harajli and Soudki[2]

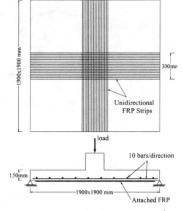

(d) The specimen details for Ebead and Marzouk[1]

Figure 4—Configurations of the test specimens

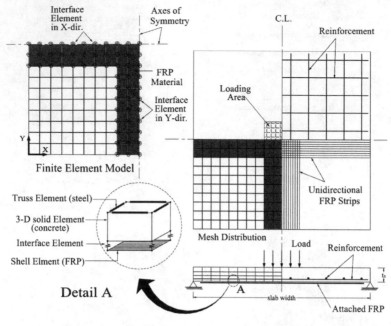

Figure 5—Typical specimen details and model description for Ebead and Marzouk[1]

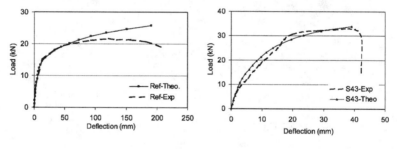

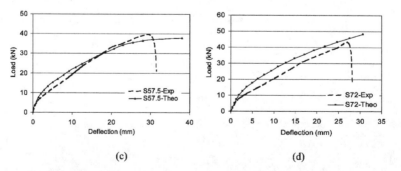

(a)

(b)

(c)

(d)

Figure 6—Comparison with test results of Seim et al. specimens[7]

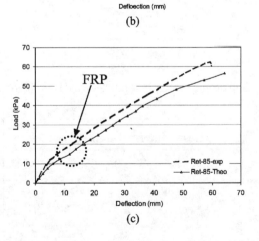

Figure 7—Comparison with test results of Mosalam et al. specimens[10].

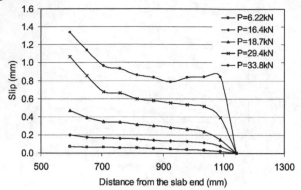

Figure 8—Slip profiles for Specimen S43 of Seim et al.[7]

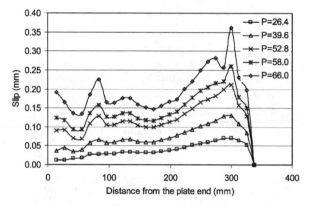

Figure 9—Slip profiles for Specimen SA30 of Harajli and Soudki[2]

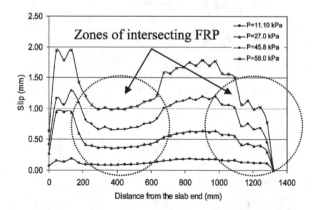

Figure 10—Slip profiles for Specimen R-Str of Mosallam and Mosallam[10]

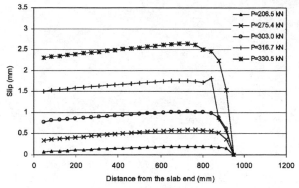

Figure 11—Slip profiles for Specimen SGF-0.35 of Ebead and Marzouk[1]

Chapter 5

Strengthening of Existing Concrete Structures with FRP Systems, Part 1

Design Guidelines for the Strengthening of Existing Structures with FRP in Italy

by L. Ascione, A. Benedetti, R. Frassine, G. Manfredi, G. Monti, A. Nanni, C. Poggi, and E. Sacco

<u>Synopsis:</u> A regulatory document was issued by the National Research Council (CNR) of Italy on the use of FRP for strengthening structures: 'Instructions for Design, Execution and Control of Strengthening Interventions by Means of Fibre-reinforced Composites' (2004). Emphasis is also given to specific requirements for seismic applications.

This document, described in more details in the paper, sets for the first time in Italy some standards for production, design and application of FRP for reinforced concrete and masonry constructions. It is also conceived with an informative and educational spirit, which is crucial for the dissemination, in the professional sphere, of the mechanical and technological knowledge needed for an aware and competent use of such materials.

The document is the result of a remarkable joint effort of almost all professors and researchers involved in this emerging and promising field, from 15 universities, of the technical managers of major production and application companies, and of the representatives of public and private companies that use FRP for strengthening artifacts. Thus, the resulting FRP code naturally incorporates the experience and knowledge gained in ten years of studies, researches and applications of FRP in Italy.

<u>Keywords:</u> design guidelines; fiber-reinforced polymers; masonry structures; reinforced concrete structures; seismic strengthening

Luigi Ascione is Full Professor of Structural Mechanics at the University of Salerno, Italy. His research activities have covered various aspects in the mechanics of structures and materials such as stability and post-critical behaviour of elastic structures, experimental analysis of materials and structures, mechanics with unilateral constraints, composite materials and structures. He is the general co-ordinator of the Task Group that prepared the regulatory document CNR-DT 200/2004.

Andrea Benedetti is Professor of Special Structural Problems at the University of Bologna, Italy. His research activity is focused on problems where the non linear behaviour of materials is essential: masonry, concrete, externally bonded FRP reinforcement, and fire damage assessment. His work is contained in more than 80 publications.

Roberto Frassine is Full Professor of Material Science at the Politecnico of Milan, Italy. He received his PhD in Materials Engineering in 1989 and is currently Associate Professor in Polymers and Composites Materials at the Politecnico of Milano (Italy). His research activities are in the fields of viscoelastic properties, time-dependent yield and fracture, processing-structure and structure-properties relationships and impact, fatigue and dynamic fracture of polymers and composites.

Gaetano Manfredi is Full Professor of Structural Engineering at the University of Naples "Federico II", Italy. He is member of *fib* WG 7.1 "Seismic Commission – Assessment of Existing Structures", WG 7.2 "Seismic Commission – Displacement Based Design" and WG 9.3 "FRP Reinforcement". His research interests include earthquake engineering and the use of advanced composites in civil structures.

Giorgio Monti is Full Professor of Structural Engineering at the University "La Sapienza" of Rome. He is member of the Commission Design of *fib* (fédération internationale du béton), in the groups "Seismic Concrete Design", "Computer-based Modelling and Design" and "Design of Structures Reinforced by FRP". His research interests span from reliability to the assessment and retrofitting of existing structures in seismic zones.

Antonio Nanni, FACI, is the V & M Jones Professor of Civil Engineering at the University of Missouri – Rolla, Rolla, MO and Professor of Structural Engineering at the University of Naples Federico II, Naples, Italy. He was the founding Chair of ACI Committee 440, Fiber Reinforced Polymer Reinforcement, and is the Chair of ACI Committee 437, Strength Evaluation of Existing Concrete Structures.

Carlo Poggi is Full Professor of Structural Engineering at the Department of Structural Engineering of Politecnico of Milano. His research activities have covered various aspects in the mechanics of structures and materials such as stability of structures, plastic collapse of structures, mechanical properties of composite materials and composite structures for aerospace, marine and civil applications.

Elio Sacco is Full Professor of Structural Mechanics at the University of Cassino, Italy. His research activity is focused on problems of fracture mechanics, of constitutive equations for masonry and for shape memory alloys, and of non-linear structural analysis by FEM.

INTRODUCTION

The peculiar situation of Italy with regards to the preservation of existing constructions results from the combination of two aspects: a) seismic hazard over the whole of the national territory, recently refined by a new seismic zonation, with medium-high intensity over a large portion of it, the highest expected PGA being 0.35g for a 475 years return period, and b) extreme variety of the built environment, perhaps with no comparison in the entire world.

Construction typology in Italy encompasses examples reckoned as Country's (and world's) historical, architectural and cultural heritage – which include buildings of various function and importance, such as palaces, temples, churches, cloisters, theatres, thermae, memorials, city walls, castles, simple dwellings, civil engineering works as bridges harbours and aqueducts – dating back to more than 2000 years ago, throughout the ancient- middle- modern- and contemporary ages, down to those built in the 20th century. The former are largely made of masonry, although under this name again a great quantity of techniques and materials are indicated, from those using stone of various natures, squared or not, regularly placed or loose, or clay bricks of different quality, or combinations of them, and binders extremely different in nature, in application ways and in ageing conditions. Instead, the latter mainly consist of reinforced concrete constructions, if not uniform, at least more homogeneous. This has motivated the growth of two clearly distinct fields of research and application of fiber-reinforced polymers (FRP): one for (generally old) masonry and one for (relatively recent) reinforced concrete constructions. The first one is more peculiar, apart the complexity of the subject, as masonry constructions have less alternatives of strengthening means and have received less applications and studies.

It goes without saying that for the historical, cultural and architectural heritage, the issue of structural safety is only one aspect of the wider concepts of restoration, preservation and conservation. In this respect, it should be underlined that these concepts do not allow a systematic use of innovative materials, such as FRP, for strengthening purposes, unless it is demonstrated that they comply with the strict requirements regarding formal and material compatibility. These essential considerations have such complex and articulated implications that they would deserve deeper consideration that is beyond the scope of this paper.

With the distinction in the two above described main fields of research on FRP, namely, masonry and reinforced concrete, the first studies have started in the early 90's by some pioneering groups that were striving at finding new solutions for increasing the safety of existing constructions, that could compete with the more developed and usual ones of concrete jacketing, steel plating, base isolation, and dissipative bracings.

In the last ten years the interest has spread so widely and rapidly that now FRP has become one of the most active and prolific research fields throughout the country. The most important testimony of the intense activity in Italy in the field of FRP is the recently issued regulatory document CNR-DT 200/2004 [1]: 'Instructions for Design, Execution and Control of Strengthening Interventions by Means of Fibre-reinforced Composites' (2004), under the auspices of the Research National Council (CNR).

THE NEW FRP CODE IN ITALY

The CNR-DT 200/2004: 'Instructions for Design, Execution and Control of Strengthening Interventions by Means of Fibre-reinforced Composites' (2004) is composed of the following chapters:
- Materials (with Annex),
- Basic concepts of FRP strengthening and special problems,
- Strengthening of reinforced concrete and prestressed concrete structures,
- Strengthening of masonry structures.

Materials

The chapter on materials has a prevailing informative character and contains the fundamental information needed to obtain a basic knowledge of the composite materials, of their components (fibres, matrices, and adhesives) and of their physical and mechanical properties. It also includes an annex describing the most usual production techniques and some basic notions on the mechanical behavior of composites.

The most notable aspect is that a possible classification of composites usually adopted for structural strengthening is proposed, and some appropriate criteria for product qualification and acceptance are introduced. Moreover, the concept is introduced of FRP as a strengthening *system*, enforcing all applicators to sell fiber-reinforced material and bonding agent as a certified package.

It is widely recognised that the design of a FRP strengthening system is a critical process. The various components (fibers, resin and the support) have different mechanical properties and roles but must be selected and designed to work together in a unique system. Therefore the properties of the components, their interactions and the properties of the final FRP must be well known and defined. The chapter on materials provides both general information on the mechanical, physical and chemical properties of FRP materials and indications for the qualification of the components and the systems on use in the reinforcement of civil engineering structures.

Specific sections of the chapter are dedicated to the main components, namely the fibers and the textiles, the resin and the adhesives. For each of them the main properties are discussed and some examples of the technical data sheets that should be provided with the products are reported. For all the mechanical, physical and chemical properties that must be determined or verified, reference is made to the appropriate testing procedures and the relevant European and American standards. The terms and quantities that are commonly used in the textile or chemical fields and are not familiar to civil engineers are properly explained.

An informative section is dedicated to the different reinforcing systems. The main aim is to clarify that the material properties can be referred to the total cross-sectional area for the prefabricated strips. On the contrary when in-situ resin impregnated systems are used, the final FRP thickness varies with the amount of resin and cannot be known in advance. For this reason the calculations may be based on the properties of the bare fibers but a reduction factor should be included to account for the efficiency of the system and for other detrimental variables such as the textile architecture or possible misalignments of the fibers.

It is known that conventional materials used in the civil engineering field are covered by standard specifications that both ensure the properties of the materials and provide standard procedures for the tests. The CNR –DT 200 document suggests two levels of qualification for the FRP materials that imply a set of mechanical and physical tests for the definition of short-term or long-term material properties respectively. The complete systems are also classified in two categories. In both cases all the basic components of the FRP must be tested and certified while a series of tests on the complete system in full scale and with the proper substrate must be performed for class A systems. Certified systems of this class have the advantage of being subject to less severe safety factors.

Basic Concepts

It is stated that the design of the FRP strengthening intervention must meet with the requirements of strength, serviceability and durability. In case of fire, the strengthening resistance must be adequate to the prescribed exposure time. The design working life of the strengthened structure is taken equal to that of new structures. This implies that the design actions to be considered are those of the current design codes for new constructions.

The safety verifications are performed for both the serviceability and the ultimate limit states. The format is that of the partial factor method. The design properties of both the materials and the products are obtained from the characteristic values, divided by the appropriate partial factor.

A rather innovative point (following the indications of EN 1990) is that the design properties X_d of the existing materials in the structure to be strengthened are obtained as a function of the number of tests performed to acquire information on them:

$$X_d = \frac{\eta}{\gamma_m} \cdot m_X \cdot (1 - k_n \cdot V_X), \tag{1}$$

where η is a conversion factor, lower than 1, accounting for special design problems (related to environmental conditions and long duration phenomena), γ_m is the material partial factor, m_X is the mean value of the property X resulting from the number n of tests, the value k_n is given as a function of the number n and the coefficient of variation

V_X is supposed known. The latter can be assumed equal to 0.10 for steel, to 0.20 for concrete and to 0.30 for masonry and timber.

The partial factor γ_m for FRP at the ultimate limit states (γ_f) is taken as 1.10 under quality control and as 1.25 in other situations. Similarly, the partial factor γ_m for delamination at the ultimate limit state ($\gamma_{f,d}$) is taken as 1.20 under quality control and as 1.50 in other situations.

The design capacity is given as:

$$R_d = \frac{1}{\gamma_{Rd}} \cdot R\{X_{d,i}; a_{d,i}\} \, ,$$
(2)

where $R\{\cdot\}$ is the function describing the relevant mechanical model considered (e.g., flexure, shear, anchorage, etc.) and γ_{Rd} is a partial coefficient accounting for the uncertainties in the above capacity model (equal to 1.00 for flexure, 1.20 for shear and 1.10 for confinement). The function arguments are, in general, a set of mechanical and geometrical properties, of which $X_{d,i}$ and $a_{d,i}$ are the design value and the nominal value of the i-*th* quantity, respectively.

An essential and innovative aspect is related to the safety verifications in the presence of fire. It is suggested that the load combination for exceptional situations, where E_d is the design value of the indirect thermal action due to fire, refers to the following situations:

- Exceptional event in the presence of strengthening (with E_d), in case the strengthening was designed for a predefined fire exposure time. In this case, the service actions of the frequent combination are to be considered. The elements capacity, appropriately reduced to account for the fire exposure time, should be computed with the partial coefficients relevant to the exceptional situations;
- After the exceptional event (without E_d), in the absence of strengthening. In this case, the service actions of the quasi-permanent combination are to be considered. The elements capacity, appropriately reduced to account for the fire exposure time, should be computed with the partial coefficients relevant to the service situations.

Reinforced Concrete Structures

Debonding -- Two different collapse modes for debonding are considered: end debonding (mode I) and intermediate debonding for flexural cracking (mode II).

The optimal anchorage length of FRP strip (Figure 1) is given as (length units in mm):

$$l_e = \sqrt{\frac{E_f \cdot t_f}{2 \cdot f_{ctm}}} \, ,$$
(3)

where E_f is the modulus of the FRP overlay in the fibers direction, t_f is the thickness of FRP and f_{ctm} is the concrete mean tensile strength.

The design debonding strength for end debonding (Mode I) is:

$$f_{fdd} = \frac{1}{\gamma_{f,d} \cdot \sqrt{\gamma_c}} \cdot \sqrt{\frac{2 \cdot E_f \cdot \Gamma_{Fk}}{t_f}} \quad , \tag{4}$$

with $\Gamma_{Fk} = 0.03 \cdot k_b \cdot \sqrt{f_{ck} \cdot f_{ctm}}$ (forces in N, lengths in mm) , $\tag{5}$

where Γ_{Fk} is the characteristic value of fracture energy of bond between concrete and FRP, k_b is a scale/covering coefficient ≥ 1, $\gamma_{f,d}$ is the delamination partial factor, and f_{ck} is the concrete characteristic strength.

The design debonding strain for intermediate debonding is:

$$\varepsilon_{fdd} = \frac{k_{cr} \cdot f_{fdd}}{E_f} , \tag{6}$$

where k_{cr} is a coefficient assumed equal to 3.0.

Flexure -- The flexural capacity is attained when either the concrete compressive strain reaches its ultimate value or when the FRP tensile strain reaches its ultimate value $\varepsilon_{fd} = \min(\eta_a \varepsilon_{fu}/\gamma_f, f_{fdd}/E_f)$ where the first value corresponds to failure and the second to the design debonding as previously defined. The flexural capacity is then given as (notation in Figure 2):

$$M_{Rd} = \frac{1}{\gamma_{Rd}} \cdot \left[\psi \cdot b \cdot x \cdot f_{cd} \cdot (d - \lambda \cdot x) + A_{s2} \cdot \sigma_{s2} \cdot (d - d_2) + A_f \cdot \sigma_f \cdot d_1 \right], \tag{7}$$

where the neutral axis x is found by solving:

$$0 = \psi \cdot b \cdot x \cdot f_{cd} + A_{s2} \cdot \sigma_{s2} - A_{s1} \cdot f_{yd} - A_f \cdot \sigma_f , \tag{8}$$

in which σ_{s2} is the stress in the superior compressed re-bars, σ_f is the tensile stress in the FRP reinforcement, f_{cd} is the concrete design strength, f_{yd} is the yield stress in the inferior re-bars, ψ and λ are non-dimensional coefficients representing the intensity and the position of the compressive concrete resultant, respectively. However, the strengthened capacity cannot be considered as greater than the 60% of initial capacity.

Flexure in the presence of axial force – The flexural capacity in the presence of an axial force N_{sd} can be evaluated by means of eqns. (7 and 8), substituting the first member of eqn. (8) by N_{sd}. Longitudinal fibers must be accurately confined in order to avoid their debonding as well as the spalling of the support material.

Shear and Torsion -- Shear strengthening configurations can be in the form of side bonded, U-jacketed and wrapped FRP strips/sheets. The design shear strength of the strengthened element is given as:

$$V_{Rd} = \min\left\{ V_{Rd,ct} + V_{Rd,s} + V_{Rd,f}, V_{Rd,max} \right\}, \tag{9}$$

where $V_{Rd,ct}$, $V_{Rd,s}$ and $V_{Rd,f}$ are the concrete, transverse steel and FRP contribution, respectively, while $V_{Rd,max}$ is the shear producing collapse in the compressed diagonal concrete strut.

The FRP contribution to the overall strength is given based on the chosen strengthening configuration. For side bonding (see Figure 3 for notation):

$$V_{Rd,f} = \frac{1}{\gamma_{Rd}} \cdot \min\{0.9 \cdot d, h_w\} \cdot f_{fed} \cdot 2 \cdot t_f \cdot \frac{\sin\beta}{\sin\theta} \cdot \frac{w_f}{p_f}, \tag{10}$$

where the partial safety factor γ_{Rd} is equal to 1.20, while for U-jacketing and wrapping:

$$V_{Rd,f} = \frac{1}{\gamma_{Rd}} \cdot 0.9 \cdot d \cdot f_{fed} \cdot 2 \cdot t_f \cdot (\cot\theta + \cot\beta) \cdot \frac{w_f}{p_f}, \tag{11}$$

where f_{fed}, termed effective debonding strength, is given, in the case of side bonding, as:

$$f_{fed} = f_{fdd} \cdot \frac{z_{rid,eq}}{\min\{0.9 \cdot d, h_w\}} \cdot \left(1 - 0.6 \cdot \sqrt{\frac{l_{eq}}{z_{rid,eq}}}\right)^2, \tag{12}$$

with:

$$z_{rid,eq} = z_{rid} + l_{eq}, \quad z_{rid} = \min\{0.9 \cdot d, h_w\} - l_e \cdot \sin\beta, \quad l_{eq} = \frac{s_f}{f_{fdd}/E_f} \cdot \sin\beta, \tag{13}$$

where l_e is the optimal anchorage length given in (3), s_f is the ultimate delamination slip assumed as 0.2 mm and E_f is the elastic modulus of FRP reinforcement in the fiber direction.

In the case of U-jacketing and wrapping, respectively, it is given by:

$$f_{fed} = f_{fdd} \cdot \left[1 - \frac{1}{3} \cdot \frac{l_e \cdot \sin\beta}{\min\{0.9 \cdot d, h_w\}}\right], \tag{14}$$

$$f_{\text{fed}} = f_{\text{fdd}} \cdot \left[1 - \frac{1}{6} \cdot \frac{l_e \cdot \sin\beta}{\min\{0.9 \cdot d, h_w\}} \right] + \frac{1}{2} (\phi_R \cdot f_{\text{fd}} - f_{\text{fdd}}) \cdot \left[1 - \frac{l_e \cdot \sin\beta}{\min\{0.9 \cdot d, h_w\}} \right], \quad (15)$$

where f_{fd} is the FRP design strength, and:

$$\phi_R = 0.2 + 1.6 \cdot \frac{r_c}{b_w}, \qquad 0 \le \frac{r_c}{b_w} \le 0.5, \qquad (16)$$

is a coefficient depending on the rounding radius R with respect to the beam web width b_w.

With regards to strengthening in torsion, this is obtained through the application of wrapping strips/sheets at an angle of 90° to the element axis. The design torsional strength of the strengthened element is given as:

$$T_{\text{Rd}} = \min\left\{ T_{\text{Rd,s}} + T_{\text{Rd,f}}, T_{\text{Rd,max}} \right\}, \qquad (17)$$

where $T_{\text{Rd,s}}$ and $T_{\text{Rd,f}}$ are the transverse steel and FRP contribution, respectively, while $T_{\text{Rd,max}}$ is the torque producing collapse in the compressed diagonal concrete strut. The FRP contribution to the torsional strength is given as:

$$T_{\text{Rd,f}} = \frac{1}{\gamma_{\text{Rd}}} \cdot 2 \cdot f_{\text{fed}} \cdot t_f \cdot b \cdot h \cdot \frac{w_f}{p_f} \cdot \cot\theta, \qquad (18)$$

where f_{fed} is given by (14) and γ_{Rd} is equal to 1.20.

Confinement -- This aims both at increasing the ultimate strength in elements under axial load, and the ductility in FRP-confined elements under axial load and flexure. In case of elements with circular cross-section of diameter D, the confined/unconfined concrete strength ratio is:

$$\frac{f_{\text{ccd}}}{f_{\text{cd}}} = 1 + 2.6 \cdot \left(\frac{f_{\text{l,eff}}}{f_{\text{cd}}} \right)^{2/3}, \qquad (19)$$

that depends on the effective confinement pressure exerted by the FRP sheet, given as:

$$f_{\text{l,eff}} = k_{\text{eff}} \cdot f_l, \qquad (20)$$

where k_{eff} is an effectiveness coefficient (≤ 1), equal to the ratio between the volume of confined concrete $V_{\text{c,eff}}$ and the total volume of concrete element V_c.

The confinement pressure is:

$$f_1 = \frac{1}{2} \cdot \rho_f \cdot E_f \cdot \varepsilon_{\text{fd,rid}},$$ (21)

$$\varepsilon_{\text{fd,rid}} = \min\{\eta_a \cdot \varepsilon_{fk} / \gamma_f;\ 0.004\},$$ (22)

where η_a is the conversion factor related to environmental conditions, γ_f is the confinement partial safety factor equal to 1.10 and $\varepsilon_{\text{fd,rid}} = 0.004$ is the FRP conventional ultimate strain, corresponding to an unacceptable degradation of concrete. The geometrical percentage of reinforcement for circular section is:

$$\rho_f = \frac{4 \cdot t_f \cdot b_f}{D \cdot p_f},$$ (23)

where t_f and b_f are the thickness and the height of FRP strips, p_f is the distance of the strips and D the diameter of circular section (Figure 4).
In the case of continuous wrapping, ρ_f is equal to:

$$\rho_f = \frac{4 \cdot t_f}{D}.$$ (24)

However, the strengthened capacity cannot be considered as greater than the 60% of the initial capacity.

For the case of rectangular sections with dimensions $b \times d$, with corners rounded with a radius $r_c \geq 20$ mm, the geometrical percentage of reinforcement can be computed by:

$$\rho_f = \frac{4 \cdot t_f \cdot b_f}{\max\{b,d\} \cdot p_f}.$$ (25)

In the case of continuous wrapping, ρ_f is equal to:

$$\rho_f = \frac{4 \cdot t_f}{\max\{b,d\}}.$$ (26)

With regards to the ductility increase, the sectional ultimate curvature can be evaluated, in a simplified way, by adopting the classical parabola-rectangle law for concrete (Figure 5), with the ultimate concrete strain given by:

$$\varepsilon_{ccu} = 0.0035 + 0.015 \cdot \sqrt{\frac{f_{l,\text{eff}}}{f_{cd}}}.$$ (27)

Masonry Structures

The application of FRP on masonry walls has the primary aim of increasing their strength and, secondarily, of increasing their collapse displacements. The objectives of FRP strengthening in masonry structures are: a) transmission of stresses either within the structural elements or between adjacent elements, b) connection between elements, c) in-plane stiffening of slabs, d) limitation of cracks width, e) confinement of columns in order to increase their strength. It is again underlined that the choice of the strengthening FRP material should avoid any incompatibility, both physical and chemical, with the existing masonry.

The strengthening intervention can include: a) increase of strength in walls, arches or vaults, b) confinement of columns, c) reducing the thrust of thrusting elements, d) transformation of non structural elements into structural elements, e) stiffening of horizontal slabs, f) application of chains in the building at the slabs and roof levels.

The masonry walls can be FRP-strengthened to prevent the out-of-plane collapse modes due to: overturning, vertical flexure, and horizontal flexure (Figure 6).

In these cases, the design of the FRP strengthening is performed through simple equilibrium between the acting forces and the resisting force of FRP strips located on top of the wall to restrain its rotation.

With regards to the in-plane collapse modes, these are due to flexure or shear. The wall shear strength is given by the sum of the masonry and the FRP shear strengths:

$$V_{Rd} = \min\left\{ V_{Rd,m} + V_{Rd,f}, V_{Rd,max} \right\}. \tag{28}$$

When the FRP strips are parallel to mortar joints the expression of $V_{Rd,m}$ and $V_{Rd,f}$ are given by:

$$V_{Rd,m} = \frac{1}{\gamma_{Rd}} \cdot d \cdot t \cdot f_{vd}, \tag{29a}$$

$$V_{Rd,f} = \frac{1}{\gamma_{Rd}} \cdot \frac{0.6 \cdot d \cdot A_{fw} \cdot f_{fd}}{p_f}, \tag{29b}$$

where γ_{Rd} = partial safety factor (in this case 1.20), d = steel depth (if any), t = wall thickness, f_{vd} = design shear strength of masonry, A_{fw} = FRP strip area, p_f = FRP strip spacing and f_{fd} = FRP design strength.

In the same case, the expression of $V_{Rd,max}$ is given by:

$$V_{Rd,max} = 0.3 \cdot f_{md}^h \cdot t \cdot d, \tag{30}$$

where f_{md}^h is the masonry compressive resistance in the horizontal direction, it is parallel to the mortar joints.

When strengthening elements with either single (barrel vaults, in Figure 7) or double (groin and cross vaults, in Figure 8) curvature, the FRP strips should contrast the relative rotation at the hinge zones that develop where the limited tensile strength of masonry is attained. Thus, application of FRP strips over the outer (inner) surface of the vault thickness can prevent the formation of hinges on the opposite inner (outer) surface.

The FRP strengthening of arches includes two possible structural schemes: a) arch on fixed restraints, and b) arch supported by columns. The aim is to avoid the formation of four hinges, which would imply collapse. The FRP-strengthening is applied on either (preferably) the outer or the inner surface, in the form of fabrics that adapt better to a curved shape than prefab strips.

The FRP strengthening of domes should increase the capacity of both the membrane and the flexural regimes. For the former, FRP strips should be applied circumferentially around the dome base (Figure 9) , while for the latter, FRP strips should be applied along the meridians.

The load bearing capacity of masonry columns can be increased by confining them through FRP. The confining system can consists in an external overlay and/or in internal bars. The confined strength (which cannot be taken as greater than 1.5 the initial strength) can be computed as:

$$f_{mcd} = f_{md} + k' \cdot f_{l,eff} , \tag{31}$$

where f_{md} is the initial masonry strength and k' is an effectiveness coefficient that can be assumed equal to:

$$k' = \frac{g_m}{1000} , \tag{32}$$

where g_m (kg/m^3) is the mass density of masonry, when there isn't specific experimental evaluations.

The effective confining pressure $f_{l,eff}$ is evaluated as:

$$f_{l,eff} = k_{eff} \cdot f_l = k_H \cdot k_V \cdot f_l , \tag{33}$$

$$\text{with } f_l = \frac{1}{2} \cdot (\rho_f \cdot E_f + 2 \cdot \rho_b \cdot E_b) \cdot \varepsilon_{fd,rid} , \tag{34}$$

where E_f is the modulus of the FRP external overlay in the fibers direction, E_b is the longitudinal elastic modulus of the FRP internal bars, $\varepsilon_{f,rid} = 0.004$, ρ_f and ρ_b are the FRP external overlay and the FRP internal bars ratios, respectively.

FRP Strengthening in Seismic Zones

The above described chapters on strengthening also contain specific indications regarding constructions in seismic zones. These follow the approach of the most recent Italian and International codes, with regards to: assessment techniques, safety requirements (limit states), seismic protection levels, analysis methods, and verification criteria (distinction between ductile and brittle elements).

Reinforced Concrete Buildings -- FRP strengthening is regarded as a selective intervention technique aiming at: a) increasing the flexural and shear capacity of deficient members, b) increasing the ductility (or the chord rotation capacity) of critical zones through confinement, c) improving the performance of lap splice zones through confinement, d) prevent longitudinal steel bars buckling through confinement, and e) increase the tensile strength in partially confined beam-column joints through application of diagonal strips.

A relevant innovation concerns the definition of the inspiring principles of the intervention strategies: a) all brittle collapse mechanism should be eliminated, b) all "soft story" collapse mechanism should be eliminated, and c) the global deformation capacity of the structure should be enhanced, either: c1) by increasing the ductility of the potential plastic hinge zones without changing their position, or c2) by relocating the potential plastic hinge zones by applying capacity design criteria. In this latter case, the columns should be flexure-strengthened with the aim of transforming the frame structure into a high dissipation mechanism with strong columns and weak beams.

Failure of brittle mechanisms such as shear, lap splicing, bar buckling, and joint shear should be avoided. For shear, the same criteria apply as for the non-seismic case, with the exception that side bonding is not allowed and FRP strips/sheets should only be applied orthogonal to the element axis. For lap splices of length L_s, adequate FRP confinement should be provided, having thickness:

$$t_f = \frac{\max\{b,d\}}{2 \cdot E_f} \cdot \left(1000 \cdot \frac{f_l}{k_H} - E_S\right), \tag{35}$$

where E_s = steel modulus, and f_l = confinement pressure:

$$f_l = \frac{A_s \cdot f_y}{\left[\dfrac{u_e}{2 \cdot n} + 2 \cdot (d_b + c)\right] \cdot L_s}, \tag{36}$$

where u_e= perimeter of the cross section inscribed in the longitudinal bars, of which n are spliced, and c = concrete cover. For bar buckling, adequate FRP confinement should be provided, having thickness:

$$t_f = \frac{10 \cdot n \cdot \max\{b, d\}}{E_f},$$ (37)

where n = total number of longitudinal bars under potential buckling.

Masonry Buildings -- Starting from the same principles as for RC buildings, when FRP-strengthening a masonry building one should also consider that: a) masonry walls inadequate to resist vertical and horizontal actions should be strengthened or rebuilt, b) orthogonal and corner walls should be adequately connected, c) slab/wall and roof/wall connections should be ensured, d) thrusts from roofs, arches and vaults should be counter-reacted by appropriate structural elements, e) slabs should be in-plane stiffened, f) vulnerable elements that cannot be strengthened should be eliminated, g) irregularity of buildings cannot be corrected by FRP applications, h) local ductility increase should be pursued whenever possible, and i) the application of local FRP strengthening should not reduce the overall structural ductility.

Quality Control

A series of *in situ* checks and operations are specified in order to validate the quality level of the applications of composite materials: check and preparation of the substrate, evaluation of the substrate degradation, removal and reconstruction of the substrate with possible treatment of steel bars.

A series of requirements for a correct application are also given with regards to: humidity conditions, environmental and substrate temperature, construction details and rules. The quality control of the application is then based on semi-destructive and non-destructive tests.

CONCLUSIONS

The peculiarity of Italy, highly seismic and endowed with a built environment unique in the world, extremely various and rich of cultural value, renders all research in this field a continuous and challenging task.

This nationwide effort has resulted in a first regulatory document (CNR-DT 200/2004), that was conceived both for regulating a rapidly growing professional and technical market, as well as for an informative and educational purpose. The document is deemed of great importance for the dissemination, in the professional sphere, of the physical and technological knowledge necessary to conscious and competent use of FRP in strengthening.

A version in English of the document is under preparation and will be available in summer 2005.

REFERENCES

[1] CNR-DT 200/2004: 'Instructions for Design, Execution and Control of Strengthening Interventions by Means of Fibre-reinforced Composites' (2004)

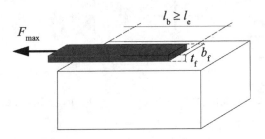

Figure 1 - Notation for anchorages.

Figure 2 - Notation for flexural strengthening.

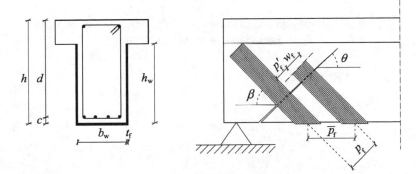

Figure 3 - Notation for shear strengthening (in lack of specific evaluation, it can assume $q = 45°$).

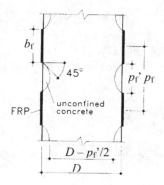

Figure 4 - Notation for confinement (column vertical section).

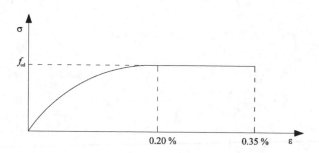

Figure 5 – Parabola-rectangle law (f_{cd} = concrete design strength).

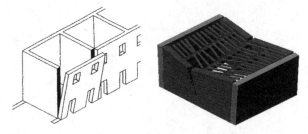

Figure 6 - Collapse modes of masonry walls: overturning (left) and horizontal flexure (right).

Figure 7 - FRP strengthening of a masonry barrel vault.

Figure 8 - FRP strengthening of a masonry cross vault.

Figure 9 - FRP strengthening of a masonry dome.

Finite Element Modelling of RC Beams Retrofitted with CFRP Fabrics

by H.B. Pham and R. Al-Mahaidi

Synopsis: In this paper, non-linear finite element modelling of debonding failure of rectangular reinforced-concrete beams strengthened with externally-bonded Carbon Fiber Reinforced Polymer (CFRP) fabrics under bending is presented. Smeared crack models were used to simulate concrete cracking. Interface elements were used to model the bond between concrete and reinforcement. The model proved to be able to simulate the beams' behaviour, predicting the failure modes, the failure loads and the reinforcement strain distributions relatively well. The parametric study using the FEA clarified the effect of several important factors on the capacity of the beams.

Keywords: CFRP; concrete; debonding; modelling

Huy Binh Pham is a PhD student investigating strengthening concrete structures using FRP.

Riadh Al-Mahaidi is the Head of the Structures Group in the Department of Civil Engineering at Monash University in Melbourne, Australia. He received his MSc and PhD degrees from Cornell University. His research interests include finite element modelling of concrete structures, strength assessment of concrete bridges and rehabilitation of concrete and steel structures using FRP composites. He is a member of ACI-ASCE Committee 447 and ACI-440.

INTRODUCTION

Even for a simply supported rectangular reinforced concrete beam retrofitted with FRP, there is much complication regarding debonding failure. One reason is that the behaviour of those beams is determined by a large number of parameters such as the beam dimensions, the amount of steel reinforcement and the FRP dimensions. As a consequence, even though there have been a large number of experimental studies, the failure mechanisms are still not fully understood and the influences of several parameters are not yet proved clearly.

Nonlinear finite element analysis can provide a powerful tool to study the behaviour of concrete structures. There have been a number of numerical studies of retrofitted beams. Linear elastic analysis has been used to verify the stress concentration at the ends of laminates or at a tip of a flexural crack (Malek et al. 1998; Teng et al. 2002). To be able to pick up the concentration, very fine meshes were required. Non-linear analysis using smeared cracks has been also reported. Two-dimensional models have been built by Arduini et al. (1997), Ross et al. (1999), David et al. (1999), Hearing (2000), Rahimi and Hutchinson (2001) and Yin and Wu (2001). Three-dimensional models were reported in Arduini et al. (1997), Fanning and Kelly (2000), Zarnic and Bosiljkov (2001), Wong et al. (2001) and Shokrieh and Malevajerdy (2001). Most studies were able to predict the overall behaviour of retrofitted beams giving close load versus deflection curves. However, most models were not able to capture local debonding failures. Cracking pattern was rarely reported. Several studies showed the general crack bands but the debond cracks were not observed. There was an attempt to simulate debonding using a new interface element (Wong et al. 2001). However, in this model, delamination surface location was assumed to be the bond line between FRP and concrete with a known interface constitutive relation. In brief, all linear elastic finite element analysis and nonlinear finite element analysis using smeared cracks have not been able to capture clearly debonding failure modes.

In the present study, another attempt was made using non-linear finite element code, DIANA (de Witte and Kikstra, 2003). The tasks are to build a model which can represent the beam crack pattern and predict the ultimate loads reasonably and to use the model to carry out a parametric study.

EXPERIMENTAL STUDY

To investigate the failure mechanisms of RC beams with FRP bonded on soffit, a thorough literature review and an experimental program has been carried out (Pham and Al-Mahaidi, 2004a; Pham and Al-Mahaidi, 2004b). Prediction models have also been proposed (Pham and Al-Mahaidi, 2004c). The experimental program included testing of 18 reinforced concrete beams. The typical beam dimensions material properties are shown in Figure 1 and Table 1. The varied parameters are shown in Table 2.

From these studies, two main flexural debond modes have been identified for retrofitted beams having sufficient shear strength (Figure 2). *End debond* is the failure that originates near the plate end and propagates in the concrete along the tension steel reinforcement. It is the result of the tensile force pulling on the end of FRP. This leads to a high shear stress level at the weakest layer near tension reinforcement level, inducing a longitudinal debonding crack there. *Intermediate span debond* is the failure that originates either from a wide flexural crack (*flexure crack debond*) or a wide flexure shear crack (*flexure-shear crack debond*). It then propagates to the laminate ends parallel to the bonded plate near the adhesive/concrete interface. This debond mode is due to the fact that concrete at the tip of a main flexure-shear crack or flexural crack is subjected to a high tensile force in the CFRP near the crack tip.

FINITE ELEMENT MODELLING

Finite element idealisation

A typical finite element mesh and boundary conditions are shown in Figure 3. The element size was maintained at approximately 12x12 mm. The aspect ratios (length over height) ranged from 1.0 to a 1.2. Since the geometry of the beams, loading and boundary conditions were symmetrical about the centreline, only half a beam was modelled. The model was supported vertically at the base and horizontally at the beam's centreline with roller supports. Load was applied to a single node in the steel plate element.

The concrete was modelled using four-node quadrilateral isoparametric plane stress element (Q8MEM) as shown in Figure 4a. Each element has eight degrees of freedom (dof) with two displacements, u_x and u_y, at each node. A 2 x 2 Gaussian integration scheme was used. The steel and CFRP reinforcement in the beams were modelled as beam element (L7BEN). This is a two-node, two-dimensional beam element. Basic variables are the translations u_x and u_y and the rotation ϕ_z in the nodes (Figure 4b). The element is numerically integrated over their cross-section and along their axis. Two-point and three-point Gauss integration scheme are used along the bar axis and in the bar cross-section respectively. The reinforcement elements are connected to concrete elements by 2+2 nodes structural interface elements (L8IF) as shown in Figure 4c. Three-point Newton-Cotes integration scheme and linear interpolation are used. The nodes for the steel reinforcement were superimposed on top of the concrete nodes.

To avoid stress-locking problem, a rotating crack model was used for the concrete elements below the tension reinforcement. The concrete in this region behaved like plain concrete. The shear strength of cracked surfaces due to aggregate interlock would be very small and could be ignored. The concrete portion above the tension steel was reinforced with steel bars and should exhibit certain shear strength after cracking. It was modelled with a fixed crack model with a shear retention factor. It was assumed that the shear retention factor was constant. The formulations for these two models are reported in de Witte and Kikstra (2003).

Concrete constitutive relation in tension was described using a nonlinear tension softening stress-strain relationship proposed by Hordijk (1991) and shown in Figure 5a. The peak tensile strength of the concrete was determined accordance to the Comite Euro-International du Beton (1991) (Formula 2.1-4). The area under the stress-strain curve is given by G_f/h, where G_f is the Mode-I fracture energy and h is the crack band width of the element. The crack band width is related to the area of the 2D element, which was determined in DIANA as the square root of the total area of the element. The fracture energy was calculated using the expression by Trunk and Wittmann (1998)

$$G_f = a \cdot d_a^n \tag{1}$$

where d_a is the maximum aggregate size, a = 80.6 and n = 0.32. G_f is in N/mm and d_a is in mm. The shear stiffness of diagonal cracks was assumed to be 5 % of the uncracked concrete.

The behaviour of the concrete in compression is described by the function of Thorenfeldt et al. (1987), which can describe the hardening and softening behaviour of concrete (Figure 5b). More details can be found in de Witte and Kikstra (2003).

The CFRP composite was assumed to be an isotropic material and modelled as linear elastic. The CFRP elastic modulus was taken to be 209000MPa as tested. The Poisson ratio of the CFRP composite was assumed to be 0.30. The adhesive elastic modulus was taken as 8500MPa. A bond-slip relationship, which was established from shear-lap tests (Pham and Al-Mahaidi, 2004d), was assumed for the interface (Figure 5c).

At the tensile reinforcement level, there is reduction of concrete volume due to the presence of the bars. To account for that, the thickness of the concrete at the level of the reinforcement was reduced. This was achieved by reducing the thickness of the concrete elements just below the tensile beam elements such that the concrete area lost was equivalent to the area of tension reinforcement. The influence of concrete area lost associated with compression reinforcement and stirrups was ignored

Verification of finite element model

To verify that the finite element base models simulated the behaviour of the experimental control beam properly, three items from the experimental and numerical results were compared. They were crack patterns at failure, the load displacement behaviour and the strain development in CFRP.

The crack pattern after peak load for the control beam model is illustrated in Figure 6. For clarity, only the cracks, in which the crack strain is large enough so that the crack tensile stress has reduced by 50 % from maximum tensile strength, are shown. The model indicated that the beam experienced main vertical flexural cracks in the constant moment region. It failed by opening of those cracks and crushing of concrete fibre at the top. This is the typical ductile failure of reinforced concrete beams.

For retrofitted beams, the NLFEA indicated two failure modes. The crack pattern developments are illustrated in Figures 7 to 9 for three example beams, E1, S2 and S1, respectively. Failure modes indicated are end span debond, intermediate span debond and a combination of end and intermediate span debond, respectively. The patterns are very similar to those observed in the experiments (Figure 10). Similar results are obtained for all other beams.

The peak loads are compared in Figure 11. The predicted values by FEA agree relatively well with the experimental measurements. The loading curves for E1, S2 and S1 are compared in Figure 12. For S2 and S1 beams, two FE models were built with and without clamping and labelled as 'S2C' and 'S2N', respectively. Again, the predictions and the measurement are in good agreement.

Figure 13 compares the experimental and numerical CFRP strain distributions along the bonded joint for the three beams. The distributions were compared at several load levels. The experimental and numerical load levels were selected to be as close as possible to each other. The distributions match reasonably accurately both for these two examples and for all other beams in the test series as well. Since the debonding failure occurred in a very brittle manner, the experimental CFRP strain distributions during propagation of debonding was not available. Therefore, comparison was not possible after peak.

PARAMETRIC STUDY

Figure 14a plots the predictions of the ultimate shear load capacity for the beam with similar dimension as E1 (E1-type beam) bonded with different CFRP thicknesses. When the FRP thickness is small, the failure mode is intermediate span debond. As the number of plies increases from 1 to 3, the capacity increases to a peak of around 79 kN. As the number of plies increases further, failure modes shifted to end debond and the load capacity reduces relatively.

Figure 14b lists the FEA results for beams with different tensile reinforcement amounts for E1-type beams. Shifting of failure mode from intermediate span debond to end debond (E1-type beams) is also observed as the steel amount increases.

Figure 15 shows the effect of varying CFRP bond length in the shear span. The failure mode is end debond for all cases. As expected, the debonding capacity reduces slightly with the bond length.

CONCLUSIONS

This paper has presented the FEA of a series of retrofitted beams with different parameters. The following findings are drawn from this work:

1) The FE model was able to simulate the beams' behaviour. It predicted the ultimate capacity, the crack patterns as well as the CFRP strain distributions relatively well for all 16 cases. Three failure modes were observed as end debond, intermediate span debond and combination of both end and intermediate span debond.

2) The parametric study was able to clarify the trends as the CFRP thickness, tensile reinforcement amount and CFRP bond length were varied.

REFERENCES

Arduini, M., Di Tommaso, A. and Nanni, A. (1997), "Brittle failure in FRP plate and sheet bonded beams", *ACI Structural Journal*, Vol. 94, No. 4, pp. 363-370.

Comite Euro-International du Beton (1991). CEB-FIP Model Code 1990. London, Great Britain, Thomas Telford.

David, E., Djelal, C., Ragneau, E. and Bodin, F. B. (1999), Use of FRP to strengthen and repair RC beams: experimental study and numerical simulations. Proceedings of the Eighth International Conference on Advanced Composites for Concrete Repair, London, UK.

de Witte, F. C. and Kikstra, W. P. (2003). DIANA User's Manual. Delft, The Netherlands, TNO DIANA BV.

Fanning, P. J. and Kelly, O. (2000), "Smeared crack models of RC beams with externally bonded CFRP plates", *Journal of Computational Mechanics*, Vol. 26, No. 4, pp. 325-332.

Hearing, B. (2000) Delamination in reinforced concrete retrofitted with fiber reinforced plastics. Deparment of Civil and Environmental Engineering, Massachusetts Institute of Technology.

Hordijk, D. A. (1991) Local approach to Fatigue of Concrete, Delft University of Technology.

Malek, A. M., Saadatmanesh, H. and Ehsani, M. R. (1998), "Prediction of failure load of R/C beams strengthened with FRP plate due to stress concentration at the plate end", *ACI Structural Journal*, Vol. 95, No. 2, pp. 142-152.

Pham, H. B. and Al-Mahaidi, R. (2004a), "Assessment of Available Prediction Models for the Strength of FRP Retrofitted RC Beams", *Composite Structures*, Vol. 66, No. 1-4, pp. 601-610.

Pham, H. B. and Al-Mahaidi, R. (2004b), Bonding Characteristics between Fibre Fabrics Bonded to Concrete Members Using Wet Lay-up Method. The Second International Conference on FRP Composites in Civil Engineering, Adelaide, Australia, A. A. Balkema, pp. 407-412.

Pham, H. B. and Al-Mahaidi, R. (2004c), "Experimental Investigation into Flexural Retrofitting of Reinforced Concrete Bridge Beams using FRP Composites", *Composite Structures*, Vol. 66, No. 1-4, pp. 617-625.

Pham, H. B. and Al-Mahaidi, R. (2004d), Predicting models for debonding failures of CFRP retrofitted RC beams. The Second International Conference on FRP Composites in Civil Engineering, Adelaide, Australia, A. A. Balkema, pp. 531-539.

Rahimi, H. and Hutchinson, A. (2001), "Concrete Beams Strengthened with Externally Bonded FRP Plates", *Journal of Composites for Construction*, Vol. 5, No. 1, pp. 44-56.

Ross, C. A., Jerome, D. M., Tedesco, J. W. and Hughes, M. L. (1999), "Strengthening of reinforced concrete beams with externally bonded composite laminates", *ACI Structural Journal*, Vol. 96, No. 2, pp. 212-220.

Shokrieh, M. M. and Malevajerdy, S. A. M. (2001), Strengthening of reinforced concrete beams using composite laminates. FRP composites in civil engineering, Hong Kong, Elsevier, pp. 507-515.

Teng, J. G., Zhang, J. W. and Smith, S. T. (2002), "Interfacial stresses in reinforced concrete beams bonded with a soffit plate: a finite element study", *Construction & Building Materials*, Vol. 16, No., pp. 1-14.

Thorenfeldt, E., Tomaszewicz, A. and Jensen, J. J. (1987), Mechanical properties of high-strength concrete and applications in design. Proc. Symp. Utilization of High-Strength Concrete (Stavanger, Norway), Tapir.

Trunk, B. and Wittmann, F. H. (1998), Experimental investigation into the size dependence of fracture mechanics parameters. Third international conference of fracture mechanics of concrete structures, D-Freiburg: Aedificatio Publ., pp. 1937-1948.

Wong, W. F., Chiew, S. P. and Sun, Q. (2001), Flexural strengthening of RC beams strengthened with FRP plate. FRP composites in civil engineering, Hong Kong, Elsevier, pp. 633-640.

Yin, J. and Wu, Z. (2001), Structural performances of steel-fiber reinforced concrete beams with externally bonded CFRP sheets. FRP composites in civil engineering, Hong Kong, Elsevier, pp. 641-648.

Zarnic, R. and Bosiljkov, V. (2001), Behaviour of Beams Strengthened with FRP and Steel Plates. The 2001 Structural Congress and Exposition, Washington, D.C.

NOTATIONS

a	Shear span
A_f, E_f	Area and Young's modulus of FRP
A_s, E_s	Area and Young's modulus of tension steel
d_a	maximum aggregate size
f_{cm}	mean value of the compressive strength of concrete at the relevant age
f_{ct}	mean value of the tensile strength of concrete
f_{sy}	steel yield stress
G_f	Mode-I fracture energy
L_f	Bonded length of CFRP in shear span

Table 1 - Mechanical properties of materials used

Material	E (MPa)	f_t (MPa)	f_{sy} (MPa)	f_{cm}(MPa)
CFRP fabrics	208000	3800*	-	-
Concrete	-	- - -	-	54
Steel (Y12)	205000	651	551	-
Steel (Y10)	204000	483	334	-
Steel (Y6)	238000	576	423	-

* given by the manufacturer

Table 2 - Variables in the experimental program

Design	n_s x d_b	$d_{b,sv}$ - s	Cover (mm)	n_f x $t_{f,0}$	L_0 (mm)
Control	3 x 12	10 - 125	25	N/A	N/A
E1	3 x 12	10 - 125	25	6 x 0.176	150
E2	3 x 12	10 - 125	25	6 x 0.176	350
E3	2 x 12	10 - 125	25	6 x 0.176	150
E4	3 x 12	10 - 125	45	6 x 0.176	150
E5	3 x 12	10 - 125	25	9 x 0.176	150
S1	3 x 12	06 - 125	25	2 x 0.176	150
S2	3 x 12	06 - 90	25	2 x 0.176	150
S3	2 x 12	06 - 125	25	2 x 0.176	150

n_s x d_b: number of tension steel bars x bar diameter (mm)
$d_{b,sv}$ - s: stirrup diameters (mm) – spacing (mm)
cover: concrete cover to stirrup
n_f x $t_{f,0}$: number of plies x ply thickness (mm)
L_0: distance from end of FRP to nearest support

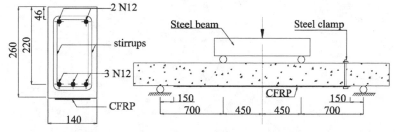

All dimensions are in mm.

Figure 1 - Experiment set-up

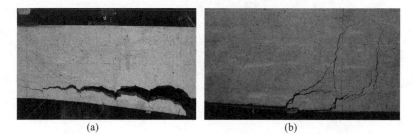

(a)   (b)

Figure 2 – End cover peeling (a) and mid-span debond (b)

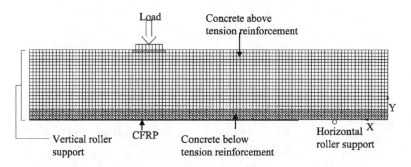

Figure 3 - 2D Mesh of beams loaded in 4-point bending

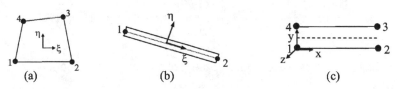

(a)   (b)   (c)

Figure 4 - Q8MEM element (a), L7BEN element (b) and L8IF element (c)

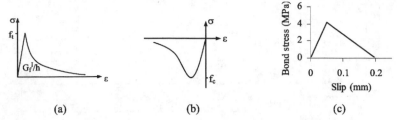

(a)   (b)   (c)

Figure 5 - Concrete in tension (a), concrete in compression (b) and
interface bond-slip behaviour (c)

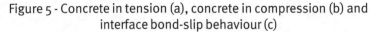

Figure 6 – Crack pattern of C1

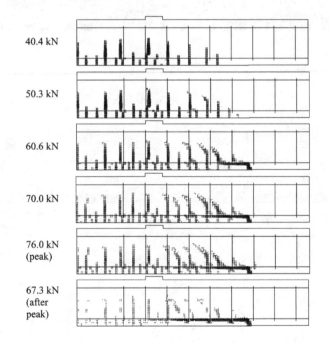

40.4 kN

50.3 kN

60.6 kN

70.0 kN

76.0 kN
(peak)

67.3 kN
(after
peak)

Figure 7 – Crack pattern development of model E1 (unclamped side)

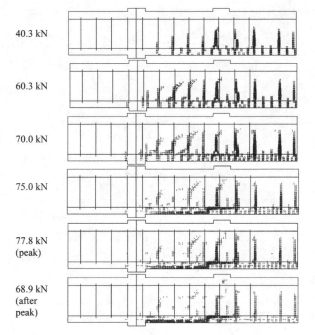

40.3 kN

60.3 kN

70.0 kN

75.0 kN

77.8 kN
(peak)

68.9 kN
(after
peak)

Figure 8 – Crack pattern development of model S2 (clamped side)

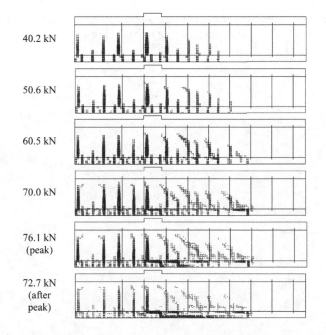

40.2 kN

50.6 kN

60.5 kN

70.0 kN

76.1 kN
(peak)

72.7 kN
(after
peak)

Figure 9 – Crack pattern development of model S1 (unclamped side)

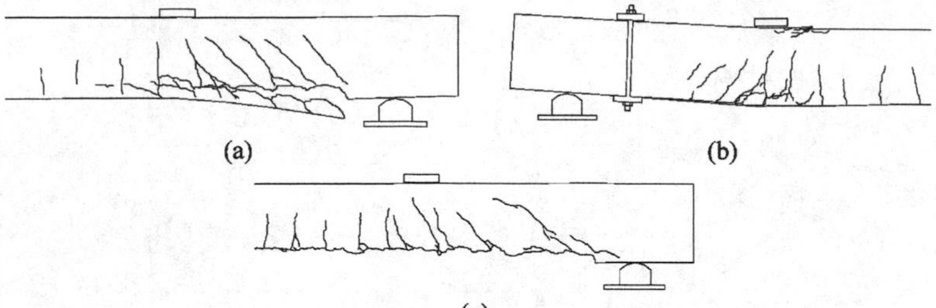

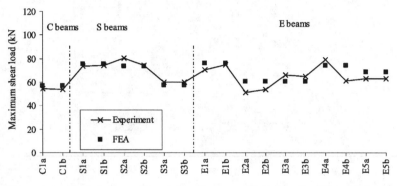

Figure 10 – Observed crack pattern of beams E1 (a), S2 (clamped side) (b) and S1 (unclamped side) (c)

Figure 11 – Correlation of peak loads as predicted by the FE model and measured in the experiments

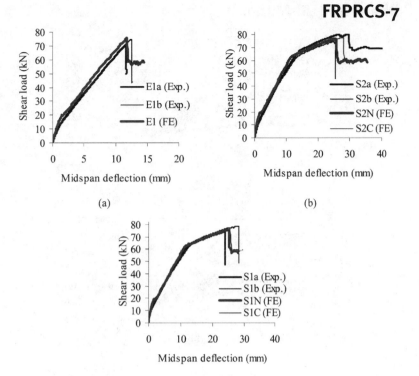

(a)

(b)

(c)

Figure 12 – Comparison of the load-deflection curves for beams E1 (a), S2 (b) and S1 (c)

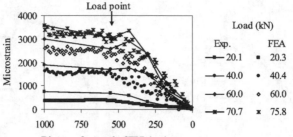

(a)

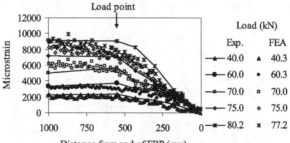

(b)

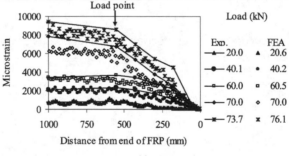

(c)

Figure 13 – Comparison of CFRP strain distributions for beams E1a (debonded side) (a), S2a (undebonded side) (b) and S1a (debonded side) (c)

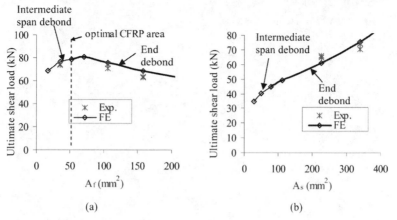

(a)   (b)

Figure 14 – Effect of varying CFRP thickness (a) and tensile reinforcement amount (b)

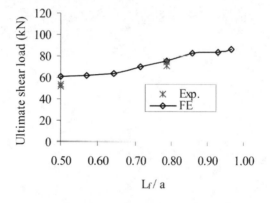

Figure 15 – Effect of varying CFRP bond length in shear span

Behavior of RC Beams Strengthened with Externally Post-Tensioning CFRP Strips

by K.-S. Choi, Y.-C. You, Y.-H. Park, J.-S. Park, and K.-H. Kim

Synopsis: Experimental study has been performed in order to investigate the behavior of RC beams strengthened with externally post-tensioning CFRP (Carbon Fiber Reinforced Polymer) strips. A total of 11 specimens have been manufactured of which specimens strengthened with bonded or unbonded CFRP strips considering the level of post-tensioning as experimental variable, and a specimen with simply bonded CFRP strips. The following phenomena have been observed through the experimental results. The specimen with simply bonded CFRP strips failed below 50% of its tensile strength due to premature debonding. On the other hand, all the specimens strengthened with post-tensioning CFRP strips showed sufficient strengthening performance up to the ultimate rupture load of the CFRP strips. Also, it was observed that the cracking loads and yield loads of the strengthened beams were increased proportionally to the post-tensioning level, but the ultimate loads were nearly equal regardless of the post-tensioning level and bonded or unbonded system. In addition, the yield loads and cracking loads of the beams with bonded post-tensioning systems were increased about 20% compared with those of the unbonded post-tensioning systems. However, the beams strengthened with unbonded post-tensioning CFRP strips showed ductile behaviour with large deflections resulting from the reduced member stiffness.

Keywords: bonded; CFRP strips; debonding failure; external post-tensioning; strengthening; unbonded

Ki-Sun Choi received his MS degree in architectural engineering from Hannam University, Korea. He is a researcher at the Building Research Department of the Korea Institute of Construction Technology.

Young-Chan You received his MS degree and Ph.D. in architectural engineering from Hanyang University, Korea. He is a chief researcher at the Building Research Department of the Korea Institute of Construction Technology.

Young-Hwan Park received his MS degree and Ph.D. in civil engineering from Seoul National University, Korea. He is a chief researcher at the Structure Research Department of the Korea Institute of Construction Technology.

Jong-Sup Park received his MS degree in civil engineering from Myongji University, Korea. He is a senior researcher at the Structure Research Department of the Korea Institute of Construction Technology.

Keung-Hwan Kim received his MS degree in structural engineering from Colorado State University and his PhD from University of Texas. He is vice-president of the Korea Institute of Construction Technology.

INTRODUCTION

Since CFRP (Carbon Fiber Reinforced Polymer) strips are manufactured by pultrusion process, their quality is superior to CFRP sheets and their construction process on job site can be shortened. CFRP strips factory manufactured are produced normally with smaller width and larger thickness than CFRP sheets because of the necessity to standardize their production. For this reason, bonding force between CFRP strips and concrete substrate is not sufficient to sustain tensile force in CFRP strips. Therefore premature debonding failure cannot be avoided when strengthening is done by simply bonding the CFRP strips (Seim, W., et al. 2001; Camata, G., et. al. 2003). The unused strength of CFRP strip caused by premature debonding can be fully utilized by post-tensioning CFRP strips at such a degree (El-Hacha, R., et. al. 2004; Zou, X. W. 2003).

There is much necessity to develop hardware and software technologies about strengthening with post-tensioning CFRP strips in order to activate its exploitation. Hardware technology includes basic elements such as development of high strength CFRP strips, anchorage system and jacking device. Software technology comprises the investigation on the behavior of strengthened beams with externally post-tensioning CFRP strips with respect to the specific parameters.

The researchers involved in this study have already developed high strength CFRP strips, anchorage system and jacking device (KICT, 2005). Using these basic technologies, the behavior of RC beams strengthened with externally post-tensioning CFRP strips has been investigated with respect to the level of post-tensioning, the bonded or unbonded post-tensioning systems.

EXPERIMENTAL PROGRAM

Post-tensioning systems

Strengthening with externally post-tensioning CFRP strips can be classified into bonded or unbonded systems. After introduction of post-tensioning, the bonded system permanently bonds the CFRP strips to the concrete member. On the other hand, the unbonded system just clamps the CFRP strips by means of mechanical anchorages after introduction of post-tensioning.

Level of post-tensioning

In bonded post-tensioning system, classical bending theory based on the perfect composite behavior can be applied if the level of post-tensioning is determined to prevent debonding until the CFRP strip reaches its ultimate rupture strength. However, the level of post-tensioning may be limited considering creep rupture of CFRP strips due to long-term loading and durability consideration against environmental condition. The failure mechanism of a beam strengthened with bonded CFRP strips with respect to the level of post-tensioning can be classified as illustrated in Fig. 1.

Zone 1: $\varepsilon_{fpi} > \varepsilon_{fu} - \varepsilon_{fd}$

Zone 1 corresponds to the case where the introduced post-tensioning exceeds the unused strain (rupture strain – debonding strain) of the CFRP strips after debonding. In such case, the ultimate load may be attained while maintaining composite behaviour without debonding of the CFRP strips.

Zone 2: $0 \le \varepsilon_{fpi} < \varepsilon_{fu} - \varepsilon_{fd}$

Zone 2 corresponds to the case where the introduced post-tensioning runs below the unused strain of the CFRP strips. Progressive debonding of the CFRP strip occurs as the load is increased. And, the bonded system switches into unbonded state after complete debonding and then reaches its ultimate load. In such case, bending theory based on strain compatibility conditions of the section cannot be applied since the CFRP strips and the member does not sustain complete composite behaviour at the end.

Zone 3: zero post-tensioning or simply bonded

Zone 3 corresponds to the case where post-tensioning is not introduced or the CFRP strip is simply clamped by means of a mechanical anchorage after bonding. The general behaviour of strengthened beams is similar to that of the beams with simply bonded CFRP strips. After the CFRP strip reaches its debonding strain, the system turns into unbonded state and then reaches the ultimate rupture load. The system with simply bonded CFRP strip is not expected to exhibit any strengthening performance in Zone 4.

Material properties

A design compressive strength of concrete of 18MPa has been planned considering the degradation state of the existing structure to be strengthened. D10 and D13 have been used for tension and compression rebars, respectively. D10 has also been selected for the shear rebar. The CFRP strips adopted in this study are the unidirectional CFRP strips developed by KICT in Korea. Table 1 and Table 2 summarize the material properties of the CFRP strips and epoxy resin.

Design of the specimens and test set-up

The specimens are small-scaled models manufactured with the shape and dimensions depicted in Fig. 2. Strengthening has been done by installing the CFRP strip in the tension zone at the soffit of the specimen with mechanical anchorage as shown in Fig. 3. Three-point load has been applied.

The specimens consist of both bonded and unbonded systems with specific levels of post-tensioning. The levels of post-tensioning were selected with reference to the tensile strength of the CFRP strips, that is 0%, 20%, 40% and 60%. The bonded system has been manufactured by bonding the CFRP strip with the specimen by means of epoxy resin after the introduction of post-tensioning. A control specimen and a specimen with simply bonded CFRP strip have also been manufactured to compare the structural performances of each post-tensioning system. Table 3 lists the parameters of each specimen.

RESULTS

Failure mode

As shown in Fig. 4, the specimen strengthened with simply bonded CFRP strip failed by the abrupt debonding of CFRP strips starting from the center to the extremity. The strain of the CFRP strip measured at debonding failure reached 6,852µ, which corresponds to nearly 50% of its tensile strength.

All the specimens strengthened with bonded CFRP strips which are post-tensioned below 60% of their tensile strength turned into unbonded states after debonding. Final failure modes were governed by CFRP rupture at the mid-span (Fig. 5(a)). However, the specimen post-tensioned with 70% of the tensile strength reached the ultimate state by the rupture of CFRP strips without any occurrence of debonding (Fig. 5(b)).

From the experimental results, it was observed that the failure mode of the beam strengthened with bonded system was affected by the level of post-tensioning that has been introduced in the CFRP strips. On the contrary, all the specimens strengthened with unbonded system failed by the rupture of the CFRP strip regardless of the level of post-tensioning as shown in Fig. 6.

Effects of strengthening

Table 4 summarizes the experimental results. As mentioned above, the specimen strengthened with simply bonded CFRP strip showed debonding failure at 50% of the tensile strength of the CFRP strip without developing sufficient strengthening. On the other hand, the anchorage system used in this study was able to clamp the CFRP strips up to their rupture strength. Therefore, it was verified that all the specimens strengthened with post-tensioning CFRP strips using the anchorage system reached the rupture strength of the CFRP strip prior to concrete crushing. The ultimate loads were nearly equal regardless of the level of post-tensioning.

Fig. 7 shows the losses of the CFRP strain for some length of time. As shown in Fig 7, it is clear that the post-tensioned CFRP strain rapidly decreases with anchorage set, but the decreasing rate is reduced and converged to a certain value. In this post-tensioning system, total losses of post-tensioned CFRP strain for 300 hours are about 5% of its initial strain.

Load-deflection curve

Fig. 8 and 9 show the load-deflection relation of the specimens with bonded and unbonded systems with respect to the level of post-tensioning. Observation of the curves reveals that the cracking load, yield load and debonding load increase with the level of post-tensioning, but the deformation capacity after debonding decreases.

The general load-deflection curves of specimen strengthened with post-tensioning level below 60% of the CFRP strength are characterized by the transient diminution of the load caused by the sequential first and second debonding from mid-span to beam end. These diminutions imply that bonded system is switched into unbonded system. After this, deflection is increased rapidly up to the ultimate load.

The specimen strengthened with post-tensioning level of 70% of the CFRP strength maintained consistent stiffness as a composite section since debonding did not occur. As a result, the deflection and crack width were reduced and the maximum loads were increased.

On the other hand, the stiffness of the specimens strengthened with unbonded system become lower because of the non-composite action between members and CFRP strips. The large deflection caused by reduced member stiffness resulted in ductile behavior up to failure state.

Stiffness of the beam

Fig. 10 compares the load-deflection curves of the beam with bonded or unbonded CFRP strips. The initial stiffness 'A' effective for the whole section and stiffness 'B' of the cracked section do not show any significant dependence on either bonding or unbonding status of the CFRP strip. After yielding of the steel rebar, the beams with bonded system maintained stiffness 'C' until debonding as a composite section with increased load carrying capacity, but the beam with unbonded system maintained stiffness 'D' as a non-composite section from the beginning. The beam with bonded system being switched into non-composite section after debonding, its stiffness degrades to become stiffness 'D'.

Longitudinal strain gradient of CFRP strips

Strain gauges were installed at spacing of 150 mm all along the length in order to survey the longitudinal strain gradient of CFRP strip. Measurement results are plotted in Fig. 11 and 12.

In a beam with bonded system, the strain measured in CFRP strip exhibited linear distribution along the length at first, the strain gradient was increased proportionally to the load sharing rate of the CFRP strip after the yield of steel rebar. After debonding of the CFRP strips, the slope of the strain distribution was drastically reduced as the section switched into a non-composite section. However, the strain distributions of the CFRP strip in the unbonded zone were different due to local friction force between CFRP strips and concrete.

In a beam with unbonded system, the strain of the CFRP strip exhibited identical distribution all over the loading zone. Since the stress is redistributed along the length of the CFRP strip, smaller strain is measured at identical measuring points compared with bonded system. Moreover, the strain of the CFRP strip at ultimate state appeared to be decreased by about 10% compared with bonded system. Table 5 shows the ultimate strains of the CFRP strip in bonded or unbonded systems

CONCLUSIONS

Tests have been performed in order to investigate the flexural behaviour of RC beams strengthened with externally post-tensioning CFRP strips with respect to the level of post-tensioning and the bonding or unbonding of CFRP strip. The following conclusions are drawn from the test results.

1) The specimen strengthened with simply bonded CFRP strip failed by debonding at about 50% of the tensile strength of CFRP strip.

2) In the case of the specimens strengthened with bonded CFRP strip post-tensioned below 60% of its tensile strength, final failure was governed by the rupture of CFRP strip after being switched into unbonded state with debonding of CFRP strips.

3) Strengthening with bonded CFRP strip post-tensioned exceeding 70% of its tensile strength can lead to rupture of CFRP strip without debonding.

4) The cracking and yield loads of the strengthened beams were increased proportionally to the level of post-tensioning, but the ultimate loads were nearly identical regardless of the post-tensioning level and bonded or unbonded system.

5) The beams strengthened with unbonded CFRP strips showed ductile behavior with large deflections resulting from the reduced member stiffness.

REFERENCES

Camata, G., Spacone, E. and Saouma, V., 2003, "Debonding Failure of RC Structural Members Strengthened with FRP Lamimates" *Proc., 6th Int. Symp. on Fiber-Reinforced Polymer Reinforcement for Concrete Structures(FRPRCS-6)*, pp. 267-276.

El-Hacha, R., Wight, R. G. and Green, M. F., 2004, "Prestressed Carbon Fiber Reinforced Polymer Sheets for Strengthening Concrete Beams at Room and Low Temperatures" *J. Compos. for Constr.*, ASCE, V. 8, No. 1, pp. 3-13.

Seim, W., Hörman, M., Karbhari, V. and Seible, F., 2001, "External FRP poststrengthening of scaled concrete slabs" *J. Compos. for Constr.*, ASCE, V.5, No. 2, pp. 67-75.

KICT, 2005. *Development of New Technology on Strengthening RC Structures with Externally Post-tensioning CFRP Strips,* Korea Institute of Construction Technology.

Zou, X. W., 2003, "Flexural Behavior and Deformability of Fiber Reinforced Polymer Prestressed Concrete Beams" *J. Compos. for Constr.*, ASCE, V. 7, No. 4, pp. 275-284 .

Table 1— Mechanical properties of CFRP strip

Thickness (mm)	Width (mm)	Tensile strength (MPa)	Modulus of elasticity (GPa)	Ultimate strain (%)
1.5	50	2,167	173	1.25

Table 2— Mechanical properties of epoxy resin

Tensile strength (MPa)	Tensile shear strength (MPa)	Pot life (min)
33.5	4.3	26

Table 3— Summary of the test beams

Specimen	Strengthening Details	Strengthening system	Post-tensioning, f_{fpi}/f_{fu} *(%)
Control	-	-	-
NB1		Simply bonded	-
PB1-0R			0
PB1-2R	CFRP Width : 50mm		20
PB1-4R		Bonded post-tensioning system	40
PB1-6R	CFRP THK : 1.5mm		60
PB1-7R			70
PU1-0R	Bond length : 1,900mm		0
PU1-2R		Unbonded post-tensioning system	20
PU1-4R			40
PU1-6R			60

* f_{fpi} : Post-tensioning stress, f_{fu} : Ultimate tensile strength of CFRP strip

Table 4— Test results

Beam	P_{cr} (kN)	P_y (kN)	P_d (kN)	P_{max} (kN)	ε_{fpi} (μ)	ε_{fd} (μ)	ε_f (μ)	ε_{fu} (μ)	Failure mode[*]
Control	18.2	40.4	-	-	-	-	-	-	A
NB1	13.7	56.3	77.0	77.0	-	6,852	6,852	6,852	B
PB1-0R	24.5	55.4	80.5	121.5	0	7,002	12,218	12,218	C
PB1-2R	26.4	71.6	105.0	123.0	2,367	8,309	10,317	12,684	C
PB1-4R	42.4	85.2	120.1	125.2	5,011	6,882	7,239	12,250	C
PB1-6R	51.8	100.5	119.6	122.8	7,410	6,023	6,098	13,508	C
PB1-7R	61.9	115.5	-	126.5	8,069	-	4,987	13,056	D
PU1-0R	18.9	43.0	-	115.0	0	-	10,655	10,655	D
PU1-2R	33.5	56.4	-	119.8	2,540	-	8,662	11,202	D
PU1-4R	47.0	75.9	-	120.7	5,200	-	6,257	11,457	D
PU1-6R	54.1	83.6	-	122.5	7,402	-	4,469	11,871	D

* A: concrete crushing after steel yield,
  B: debonding failure of CFRP strip after steel yield,
  C: rupture of CFRP strip after debonding of CFRP strip,
  D: rupture of CFRP strip after steel yield.

P⌐
  cr   : crack load
  y   : steel yield load
  d   : CFRP debonding load
  max : maximum load

ε⌐
  fpi : CFRP initial post-tensioning strain
  fd : CFRP debonding strain
  f   : CFRP measured strain
  fu : CFRP total ultimate strain ($\varepsilon_{fpi}+\varepsilon_f$)

Table 5— Comparison of CFRP ultimate strains in bonded or unbonded post-tensioning systems

Post-tensioning	Bonded system-		Unbonded system-		
	ε_{fu} (μ)	$\varepsilon_{fu}/\varepsilon_{fu}$	ε_{fu} (μ)	$\varepsilon_{fu}/\varepsilon_{fu}$	$\varepsilon_{fu}/\varepsilon_{fu}$
Simple bond	6,852	0.55	-	-	-
0%	12,218	0.98	10,655	0.85	0.87
20%	12,684	1.01	11,202	0.89	0.88
40%	12,250	0.98	11,457	0.91	0.94
60%	13,508	1.08	11,871	0.95	0.88
70%	13,056	1.04	-	-	-

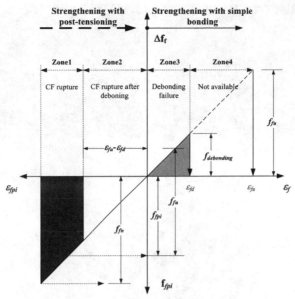

Figure 1— Failure mechanism of beams strengthened with bonded CFRP strips with respect to the level of post-tensioning

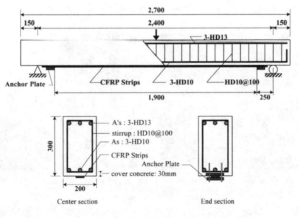

Figure 2— Dimensions and details of test beam

(a) Anchorage system   (b) Jacking device

Figure 3—Anchorage and jacking device for post-tensioning CFRP strips

Figure 4—Typical failure mode of beam with simply bonded CFRP strips

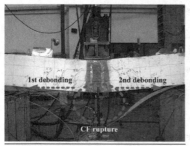

(a) below 60% of tensile strength   (b) more than 70% of tensile strength

Figure 5—Typical failure mode of beams strengthened with bonded
post-tensioning CFRP strips

Figure 6—Typical failure mode of beam with unbonded post-tensioning CFRP strips

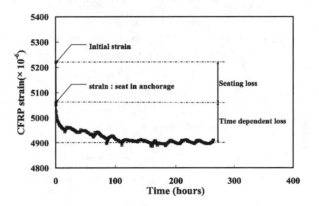

Figure 7—Losses of post-tensioned CFRP strain

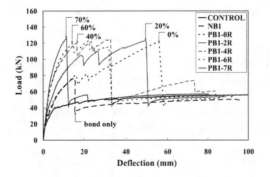

Figure 8—Load-deflection curves of beams strengthened with bonded
post-tensioning systems

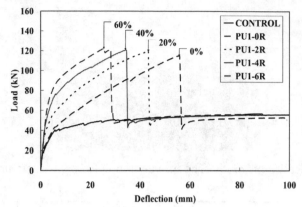

Figure 9—Load-deflection curve of beams strengthened with unbonded post-tensioning systems

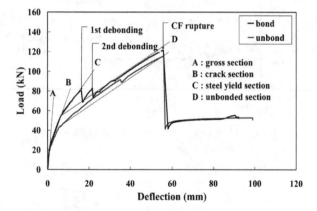

Figure 10—Typical load-deflection curves of beams with bonded or unbonded post-tensioning systems

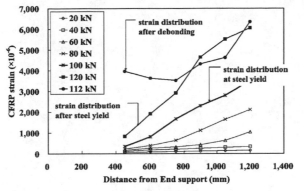

Figure 11—Typical strain distribution of the bonded CFRP strips along the length

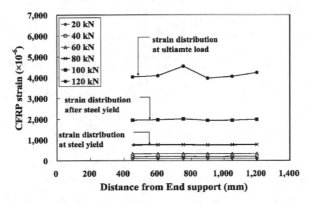

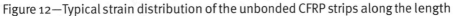

Figure 12—Typical strain distribution of the unbonded CFRP strips along the length

Influence of Bond Behavior on the Cross Sectional Forces in Flexural RC Members Strengthened with CFRP

by G. Zehetmaier and K. Zilch

Synopsis: In strengthened RC members the distribution of cross sectional tensile forces is affected by the significant differences in bond behavior of the reinforcement layers. As the tensile forces in the externally bonded reinforcement is the essential input value for bond verification for example at the end anchorage, the detailed knowledge of the distribution of forces in cracked sections is of fundamental importance. In this paper the common models to describe the interactions in tensile members are summarized and an advanced numerical model based on nonlinear bond stress-slip relationships for strengthened flexural members is presented. On the basis of experimental results combined with parametric studies, the effects of various parameters – for example the axial stiffness of CFRP, the diameter of internal rebars or the concrete compressive strength – on the interactions between the different reinforcement layers are examined. For practical design bond coefficients for a simplified calculation of cross sectional tensile forces are proposed.

Keywords: bond behavior; bond coefficients; cross-sectional forces; externally bonded reinforcement; flexural members; numerical model; tension stiffening

Dipl.-Ing. Gerhard Zehetmaier, born in 1973, received his engineering degree in 1999. Currently he is research assistant at the Department of Concrete Structures of the Technical University Munich. His research work comprises strengthening with externally bonded and near surface mounted reinforcement as well as bridge construction and fatigue verification. He is member of fib TG 9.3.

Prof. Dr.-Ing. Konrad Zilch, born in 1944, graduated at the Technical University Darmstadt in 1969. Following post graduate studies at the UCB and at the Boundary Layer Wind Tunnel Laboratory (University of Western Ontario) he worked at T.Y Lin International and the STRABAG AG. Since 1993 he is head of the Department of Concrete Structures at the Technical University Munich. He is member of several national and international committees, e.g. CEN TC250/SC2 – EUROCODE 2.

INTRODUCTION

For flexural members the calculation of sectional forces assuming an even distribution of strains according to the *Bernoulli*-hypothesis is common practice. In general the *Bernoulli*-hypothesis leads to satisfying results if the bond behavior of the different reinforcement layers is equivalent. Tests on RC members strengthened with externally bonded reinforcement (EBR) show that there are significant discrepancies between measured tensile forces in the EBR and calculated values according to the *Bernoulli*-hypothesis (Ueda et al. 2002; Zehetmaier and Zilch 2003). These discrepancies are caused by the considerably different bond behavior of internal and externally bonded reinforcement. The resulting distribution of forces depends on the interaction of the different reinforcement layers. Like in RC members strengthened with EBR, in prestressed concrete members two reinforcement layers with different bond behavior are combined. According to design regulations, e.g. CEB FIP Model Code 1990 (CEB 1993) or EN 1992-1-1 (CEN 2004), the effects of the different bond characteristics have to be taken into consideration in serviceability and fatigue verification by the use of simplified bond coefficients.

RESEARCH SIGNIFICANCE

The tensile forces of the reinforcement layers are the essential input values e.g. for bond verification at the end anchorage or in the regions subjected to shear forces as well as for serviceability verification e.g. the calculation of crack widths. Therefore a detailed knowledge of the distribution of cross sectional tensile forces is of fundamental importance. In strengthened flexural members the interactions of internal and externally bonded reinforcement governs this distribution. In this paper the common models to describe interactions are summarized, an advanced model for flexural members strengthened with EBR is presented and proposals for bond coefficients with regard to a simplified calculation of cross sectional tensile forces are made.

INFLUENCE OF BOND ON CROSS SECTIONAL FORCES

Bond Behavior

As the interactions of different reinforcement layers are predominantly governed by the specific bond behavior, in preliminary studies bond models for externally bonded CFRP as well as embedded steel rebars have been established.

Externally bonded reinforcement – To represent the bond behavior of CFRP EBR a bilinear bond stress-slip relationship acc. to Fig. 1 has been derived from bond tests with double lap specimens. The basic parameters – the elastic energy G_e corresponding with the linear elastic behavior of bond, the interfacial fracture energy G_f and the bond shear strength τ_{f1} – have been determined by regression analysis of measured ε_f-s_f-relations using an approach presented in (Niedermeier and Zilch 2001) which in general is equivalent to the method published in (Dai et al 2005).

Internal rebars – To describe the bond behavior of embedded ribbed rebars numerous models are available. Based on the results of own bond tests the basic formulation acc. to (Eligehausen et al. 1983) was modified to represent splitting failure in case of small concrete cover. For smooth bars a rigid-plastic bond stress-slip relationship is assumed (Fig. 1). Effects influencing the bond behavior, e.g. the position during casting, are taken into account by means of reduction factors.

Influence of Bond on Tensile Members / Ties

The basic principles of the interaction of reinforcement layers with different bond characteristics may be illustrated on the basis of tensile members Fig. 2 a). The distribution of strains (ε_{fr}, ε_{sr}) or tensile forces (F_{fr}, F_{sr}) in cracked sections is strongly dependent on the bond behavior of the reinforcement layers and the specific ratio of axial stiffness to bond-effective circumference described with the shape coefficients $c_{s,f}$ and $c_{s,s}$ (Eq. 1). In addition to the mentioned parameters the stage of cracking and the crack spacing have significant impact on the distribution of cross sectional forces. A mechanical description of the interactions between different reinforcement layers can be established by applying equilibrium and compatibility conditions to a differential element. The resulting system of coupled differential equations represents a second order boundary value problem. Neglecting the influence of concrete deformations on the reinforcement strains and relative displacements, two independent differential equations Eqs. (2) and (3) are derived.

$$c_{s,f} = \frac{1}{E_f t_f}; \qquad c_{s,s} = \frac{4}{E_s d_s} \qquad (1)$$

$$s_f'' - \tau_f(s_f) \cdot c_{s,f} = 0 \qquad (2)$$

$$s_s'' - \tau_s(s_s) \cdot c_{s,s} = 0 \qquad (3)$$

To solve the boundary value problem, bond stress-slip relationships $\tau_f(s_f)$, $\tau_s(s_s)$ and boundary conditions regarding strains or relative displacements (slip) are necessary.

The mechanical or numerical modeling of tensile members – commonly referred to as "tension stiffening models" – makes use of the symmetries for strain and slip distribution in the center between two cracks (Fig. 2 a) and the basic compatibility condition for the slip in cracked sections acc. to Eq. (4),

$$s_{sr} = s_{fr} = \frac{w_{cr}}{2} \tag{4}$$

where s_{sr} and s_{fr} are the relative displacements in the cracked section. As the crack width w_{cr} is supposed to be constant throughout the cross section, the usually assumed uniform crack spacing leads to a crack width that equals twice the relative displacement (Eq. 4). At present various models for strengthened tensile members exist. In (Holzenkämpfer 1994, Rostásy et al. 1996) an analytical model based on simplified rigid-plastic bond stress-slip relationships was presented. Numerical models taking into account nonlinear bond stress-slip relationships were published e.g. in (Sato et al 2002; Ulaga 2003; Pecce and Ceroni 2004).

Influence of Bond on Flexural Members

Tie models in combination with an effective area in tension may be used to describe the behavior of flexural members, but the so-called "tension chord models" are only valid for specific boundary conditions. Due to the assumption of symmetry between two cracks in conjunction with the compatibility of relative displacements at the cracks according to Eq. (4), the tension chord models may be appropriate to describe the interactions in regions with constant bending moment. The assumptions do not apply to regions subjected to shear forces (Fig. 2 b). In general, tension chord models assume equal inner lever arms for the different reinforcement layers. For the calculation of crack width or crack spacing, where only the total tensile force is of importance, the difference is negligible. As far as the distribution of tensile forces is concerned, tension chord models result in too small EBR forces, as usually the inner lever arm of EBR exceeds that of internal rebars. It is obvious that the differences in inner lever arms decrease with increasing member depth. But especially in case of slender slabs – at least in Germany the majority of strengthening applications - the effects of different distances to the neutral axis should be taken into account.

FLEXURAL MEMBERS – A NUMERICAL MODEL

To investigate the effects of bond behavior on cross sectional tensile forces in particular, a numerical model based on a modular program system has been developed. The model follows a discrete crack approach and allows the iterative calculation of strains (ε_f, ε_s) and relative displacements (s_f, s_s) along the entire length of the reinforcing elements on the basis of realistic nonlinear constitutive laws and bond models. Flexural members are represented by a sequence of macro-elements which idealize discrete segments e.g. between two adjacent cracks (Fig. 3).

Model Components

Element level (Fig. 3 b) – The behavior of strengthened flexural members at element level is characterized by material laws of concrete and reinforcement as well as bond models for internal and externally bonded reinforcement. The concrete compression zone $(\rightarrow F_c^{i,j})$ is described using a nonlinear material law acc. to (CEB 1993) implemented in a layer model. For reinforcing bars and CFRP laminates elastic-plastic and linear-elastic constitutive laws are used respectively. The bond behavior of externally bonded CFRP reinforcement and internal rebars is described with the already introduced nonlinear bond models. Local effects, e.g. local bond degradation in the immediate vicinity of initial flexural cracks due to the formation of diagonal cracks caused by the EBR or the cone shaped cracks surrounding the embedded bar have an impact on the interactions. To account for local effects, correction factors for the compatibility conditions (see Eq. 9) depending on d_s, t_f and f_{cm} were derived from 56 tests on strengthened RC tensile members (Zehetmaier and Zilch 2003). This allows the use of global bond models in numerical calculations as well as in analytical solutions for bond coefficients. The bond models do not include hysteretic behavior. Effects resulting from the reversal of relative displacement between reinforcement and concrete due to the formation of a new crack or the incremental shift of the points of zero slip ($s_f = 0$ and $s_s = 0$ respectively, see Fig. 6) with increasing mean strain are not considered.

System level (Fig. 3 a) – In contrast to tensile members, for flexural members no boundary conditions at element level – expressed e.g. with a fixed relation between s_f and s_s – exist. Boundary conditions may only be formulated at system level for specific cross sections, e.g. at the system symmetry axis (Eq. 5) or at the ends of the reinforcing elements (Eq. 6) (see Fig. 3 for notation)[a].

$$s_{sr}^{n,2} = s_{sr}^{n+1,1} \qquad s_{fr}^{n,2} = s_{fr}^{n+1,1} \tag{5}$$

$$s_s(x_s = 0) \neq 0 \qquad \varepsilon_s(x_s = 0) = 0$$
$$s_f(x_f = 0) \neq 0 \qquad \varepsilon_f(x_f = 0) = 0 \tag{6}$$

The discrete macro-elements are linked with equilibrium and compatibility conditions. As the Bernoulli-hypothesis is not valid, there are three independent constraints referring to cracked sections (i.e. the elements boundaries):

- equilibrium of sectional forces and action effects acc. to Eq. (7)

$$\sum N^{i,j} \Rightarrow F_{fr}^{i,j} + F_{sr}^{i,j} + F_c^{i,j} - N^{i,j} = 0$$
$$\sum M^{i,j} \Rightarrow F_{fr}^{i,j} z_{fr}^{i,j} + F_{sr}^{i,j} z_{sr}^{i,j} - M^{i,j} = 0 \tag{7}$$

- compatibility of strains acc. to Eq. (8)

---

[a] The first superscript (i) denotes the element (element number $1 \leq i \leq n$), the second superscript denotes the element boundary ($j = 1,2$) (see Fig. 3)

$$\varepsilon_{fr}^{i,2} = \varepsilon_{fr}^{i+1,1} \qquad \varepsilon_{sr}^{i,2} = \varepsilon_{sr}^{i+1,1}$$

$$(8)$$

- compatibility of relative displacements acc. to Eq. (9)

$$\left(s_{sr}^{i,2} + s_{sr}^{i+1,1}\right) \cdot k_x^{i,i+1} = s_{fr}^{i,2} + s_{fr}^{i+1,1}$$

$$(9)$$

In Eq. (9) $k_x^{i,i+1}$ takes into account the influence of different distances to the neutral axis (i.e. the influence of curvature) and includes the correction factors to consider local bond effects. Eq. (9) may be interpreted as a compatibility condition of crack widths at the z-coordinates of the reinforcement layers (Fig. 3 c). The mentioned independent constraints are used to control the numerical calculation and to formulate convergence conditions respectively.

As mentioned above, the distribution of cross sectional forces depends on the crack pattern and particularly on the spacing between two adjacent cracks. To reduce complexity in the established numerical model a predefined crack pattern with constant crack spacing is assumed. The mean crack spacing equals 1.4 l_t where l_t is the transfer length resulting from the cracking moment; l_t is calculated based on the presented bond models. The factor 1.4 was obtained from the evaluation of stabilized crack patterns observed in tests with strengthened tensile and flexural members and takes into account the differences between the calculated initial crack spacing ($=s_{cr,max} = 2\, l_t$) and the mean crack spacing at stabilized cracking.

Comparison with experimental results

To identify effects of bond behavior on the distribution of cross sectional tensile forces an extensive experimental program covering tests on strengthened ties and flexural members was conducted at the Technical University Munich (Zehetmaier and Zilch 2003). The comparison of measured FRP strain in a predefined crack and calculated strain assuming even cross sections (*Bernoulli*-hypothesis) displayed in Fig. 4 a demonstrates the significant influences of bond behavior. Especially at low loading range (serviceability level) the measured strains exceed the values calculated on the basis of the *Bernoulli*-hypothesis by 80%. In contrast to the *Bernoulli*-hypothesis the F-ε_{fr}-relation in eight predefined cracks calculated with the presented numerical model is in good agreement with measured values (Fig. 4 a, b). Of course the distribution of tensile forces is fixed if the internal reinforcement yields (Fig. 4 a). Various tests on strengthened flexural members reported in literature, where reinforcement strains in discrete cracks were available, were recalculated. Although in absence of a predefined crack pattern successive cracking occurred, in general the calculated values based on the mean crack spacing are in good accordance with measured data.

Due to the general material laws and bond models, the complete loading range from serviceability to ultimate limit state may be followed. In general bond failure of EBR is initiated when the required increase in tensile force between two cracks exceeds the bond capacity. In Fig. 5 the distribution of strains of internal and externally bonded reinforcement along half of the system is displayed for the last convergent load increment. The bond failure of EBR will initiate in element 12 as there the bond capacity is reached. It should be noted that the location where bond failure starts according to the numerical model corresponds to observations in tests. In Fig. 6 the calculated distribution of strains and relative displacements for four different load steps in case of an element in the shear span is displayed. The lacking symmetries at element level as well as the shift of the points of zero slip are obvious.

BOND COEFFICIENTS

Background

Like in the design of PC members, the effects of different bond behavior on tensile forces in cracked sections may be taken into consideration by means of simplified bond coefficients. As the coefficients should be independent from specific reinforcement ratios or distances to the neutral axis a formulation according to Eq. (10) was chosen. Obviously δ_f equals 1 if the *Bernoulli*-hypothesis is accurate.

$$\delta_f = \frac{\varepsilon_{fr}}{\varepsilon_{sr}} \cdot \frac{d-x}{d_f - x} \tag{10}$$

Due to the formulation the bond coefficients will be referred to as "strain ratio". From equilibrium of internal forces a relationship between ε_{fr} and ε_f^{II} is derived (Eq. 11), where ε_f^{II} is the EBR strain in the cracked section calculated acc. to the *Bernoulli*-hypothesis. With predetermined bond coefficients δ_f the realistic strain ε_{fr} can be determined with Eq. (11). For deep beams or ties where $(d_f-x)/(d-x)$ approaches 1, the resulting expression – Eq. (11-right) – corresponds to the equation given in (CEB 1993) for the design of PC members.

$$\varepsilon_{fr} = \frac{\left(1 + \dfrac{E_f A_f}{E_s A_s} \cdot \dfrac{z_f}{z_s} \cdot \dfrac{d_f - x}{d - x}\right) \cdot \delta_f}{1 + \dfrac{E_f A_f}{E_s A_s} \cdot \dfrac{z_f}{z_s} \cdot \dfrac{d_f - x}{d - x} \cdot \delta_f} \cdot \varepsilon_f^{II} \quad \Rightarrow \quad \varepsilon_{fr} = \frac{\left(1 + \dfrac{E_f A_f}{E_s A_s}\right) \cdot \delta_f}{1 + \dfrac{E_f A_f}{E_s A_s} \cdot \delta_f} \cdot \varepsilon_f^{II} \tag{11}$$

Parameters governing tensile forces

The numerical model can be used to estimate the influence of various parameters on the distribution of tensile forces in cracked sections. The results of parametric studies displayed in Fig. 7 are based on the simply supported slab acc. to Fig. 6. The resulting strain ratio δ_f at midspan is plotted versus the corresponding strain ε_f^{II}. The following general tendencies may be derived (note: $\delta_f > 1$ means that the EBR tensile strain exceeds the value calculated acc. to the *Bernoulli*-hypthesis):

- Due to the softening behavior of EBR bond the strain ratio decreases with increasing load. The pronounced maximum refers to single cracking, i.e. the idealized transfer lengths of internal and external reinforcement are smaller than half the crack spacing;

- With increasing bar diameter or decreasing axial stiffness of FRP EBR the strain ratio δ_f increases (Fig. 7 a, b);
- The impact of the concrete compressive strength on the strain ratio is limited to service loads (Fig. 7 c) although there is a dominant correlation between compressive or tensile strength and bond capacity;
- The crack spacing is governing the initiation of debonding as well as the redistributions of forces to the internal reinforcement, but it is only of little importance regarding the maximum strain ratio (Fig. 7 d).

The results of parametric studies in Fig. 7 demonstrate that, in contrast to the design of PC members, constant bond coefficients will only be a roughly approximate representation of the real interaction behavior. The complex interaction is mainly due to the softening behavior of EBR bond.

Concept for Bond Coefficients

As a consequence of the complex interaction behavior, bond coefficients may only be determined for specific boundary conditions e.g. the single cracking stage. Due to the used global bond models analytical expressions can be established depending on the parameters of the bond stress-slip relationships and geometrical variables. In Fig. 8 two different types of bond coefficients for the combination of externally bonded CFRP strips and ribbed reinforcing bars are displayed. The coefficient $\delta_{f,max}$ (Fig. 8 a) corresponds to single cracking and should be used e.g. for fatigue verification of bond at serviceability level. For bond verification in ULS the coefficients $\delta_{f,e}$ (Fig. 8 b) are appropriate.

CONCLUDING REMARKS

In this paper the fundamental characteristics of a numerical model are summarized which allows to examine RC members strengthened in flexure with externally bonded CFRP. Comparisons with test results confirm the reliability of calculations. Based on numerical results the dominant parameters influencing the interaction of the different reinforcement layers are identified and bond coefficients to simplify the calculation of cross sectional tensile forces are established. As the distribution of strains and in particular the relative displacements at the cracked sections are connected to the element curvature, the numerical model also allows to calculate the flexural deformations of strengthened members. Currently improvements to take into account crack formation and crack spacing are implemented in the program system.

REFERENCES

CEB 1993: CEB FIP Model Code 1990. Comité Euro-International du Béton, Lausanne.

CEN 2004: EN 1992-1-1:2004-04 Design of Concrete Structures – Part 1-1: General Rules and Rules for Buildings (EUROCODE 2). Comité Européen de Normalisation.

Dai, J., Ueda, T., Sato, Y., 2005, "Development of the Nonlinear Bond Stress-Slip Model of Fiber Reinforced Plastic Sheet-Concrete Interfaces with a Simple Method," *Journal of Composites for Construction*, Vol. 9, No. 1, pp. 52-62.

Eligehausen, R., Popov, V., Bertero, V., 1983, „Local Bond Stress-Slip Relationship of Deformed Bars Under Generalized Excitations, *Report UCB/EERC 82/23*, Berkeley, 169 pp.

Holzenkämpfer, P, 1994,. *Ingenieurmodelle des Verbundes geklebter Bewehrung für Betonbauteile (Engineering Bond Models for Externally Bonded Reinforcement),* Ph.D-thesis, Technical University Braunschweig, 217 pp. (in German)

Niedermeier, R., Zilch, K., 2001, "Zugkraftdeckung bei klebearmierten Biegebauteilen („Verification of the Envelope Line of Tensile Forces for Flexural Members Strengthened with Externally Bonded Reinforcement") ," *Beton- und Stahlbetonbau*, Vol. 96, No. 12, pp. 759-770. (in German)

Pecce, M., Ceroni, F., 2004, "Modeling of Tension-Stiffening Behavior of Reinforced Concrete Ties Strengthened with Fiber Reinforced Plastic Sheets," *Journal of Composites for Construction*, Vol. 8, No. 6, pp.510-518.

Rostasy, F.S., Holzenkämpfer, P., Hankers, Ch., 1996, "Geklebte Bewehrung für die Verstärkung von Betonbauteilen," ("Externally Bonded Reinforcement for the Strengthening of Concrete Members") In *Betonkalender 1996*. Berlin, Ernst & Sohn. 30 pp. (in German)

Sato, Y., Ueda, T., Yamaguchi, R., Shoji, K., 2002, "Tension Stiffening Effect of Reinforced Concrete Member Strengthened by Carbon Fiber Sheet," Proc. *Bond in Concrete – From Research to Standards, Budapest, Hungary,2002*, G. Balazs et al. (eds.), pp. 606-613.

Ueda, T., Yamaguchi, R., Shoji, K., Sato, Y., 2002, "Study on Behavior in Tension of Reinforced Concrete Members Strengthened by Carbon Fiber Sheets," *Journal of Composites for Construction*, Vol. 6, Nr. 3, pp. 168-174.

Ulaga, T., 2003, *Betonbauteile mit Stab- und Lamellenbewehrung: Verbund- und Zuggliedmodellierung (Concrete Members with Internal and Externally Bonded*

Reinforcement: Bond and Tie Models). Ph.D.-thesis, Swiss Federal Institute of Technology, Zürich, 161 pp. (in German)

Zehetmaier, G., Zilch, K., 2003, "Interaction Between Internal Bars and External FRP Reinforcement in RC Members" Proc., FRPRCS 6, K.H. Tan (ed.), Vol. 1, pp. 397-406.

NOTATION

The following symbols are used:

A_f	cross section (EBR)	h	depth
A_s	cross section (rebars)	k_x	curvature coefficient
E_c	Young's modulus (concrete)	s_{cr}	crack spacing
E_f	Young's modulus (EBR)	s_f	slip (EBR)
E_s	Young's modulus (rebars)	s_s	slip (rebars)
F_c	concrete compressive force	t_f	thickness of EBR
F_{fr}	EBR force (cracked section)	w_{cr}	crack width
F_{sr}	Rebar force (cracked section)	x	depth of compression zone
G_e	elastic bond energy (EBR)	z_f	inner lever arm (EBR)
G_f	bond fracture energy (EBR)	z_s	inner lever arm (rebars)
M	bending moment	δ_f	bond coefficient ("strain ratio")
N	axial force	ε_f	strain (EBR)
$c_{s,f}$	shape coefficient (EBR)	ε_{fr}	strain in the cracked section (EBR)
$c_{s,s}$	shape coefficient (rebars)	ε_f^{II}	strain acc. to Bernoulli-hypothesis
d	effective depth (rebars)	ε_s	strain (rebars)
d_f	effective depth (EBR)	ε_{sr}	strain in the cracked section (rebars)
d_s	rebar diameter	τ_f	bond stress (EBR)
f_{cm}	concrete compressive strength	τ_{fl}	bond strength (EBR)
f_{ctm}	concrete tensile strength	τ_s	bond stress (rebars)

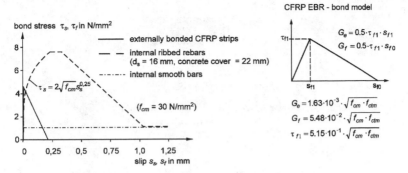

Figure 1 – Bond models for internal rebars and externally bonded reinforcement

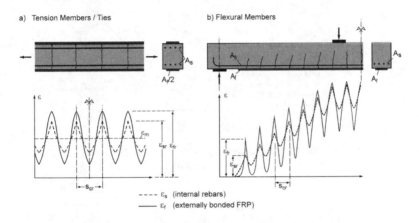

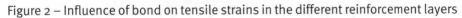

Figure 2 – Influence of bond on tensile strains in the different reinforcement layers

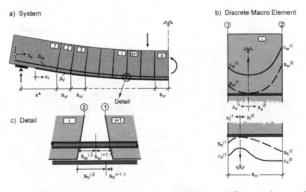

Figure 3 – Numerical model for strengthened flexural members

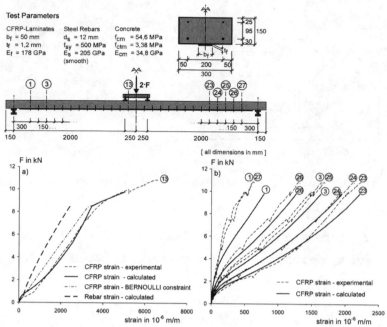

Figure 4 – Calculated reinforcement strains vs. measured strains in predefined cracks
(slab No. B2-08-B2C)

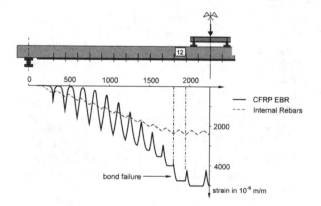

Figure 5 – Calculated reinforcement strains prior to bond failure
(last convergent increment) (slab No. B2-08-B2C)

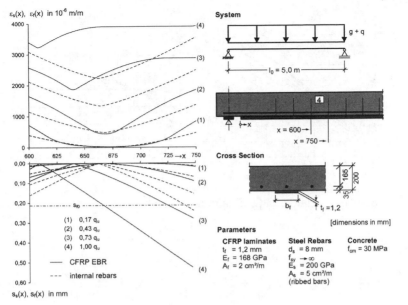

Figure 6 – Reinforcement strains and related slip in element No. 4 for 4 different load steps

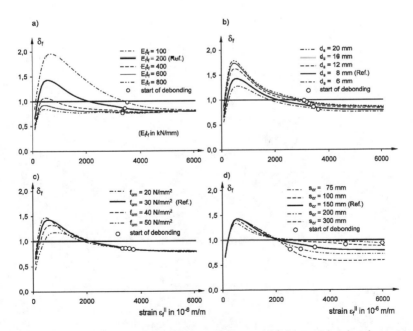

Figure 7 – Development of strain ratio d_f at midspan with increasing load, parameters affecting the distribution of cross sectional forces (Reference system acc. to Fig. 6)

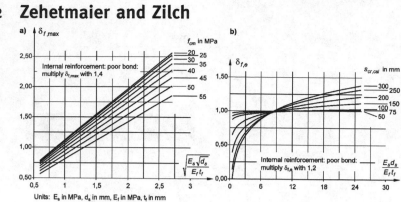

Figure 8 – Bond coefficients for CFRP-laminates combined with ribbed reinforcing bars

FRP-Strengthening in Shear:
Tests and Design Equations

by G. Monti and M.A. Liotta

Synopsis: This paper presents the results of an experimental/analytical study aiming at obtaining a clear understanding of the underlying mechanisms of the shear strengthening of reinforced concrete beams with fibre reinforced polymers (FRP). Through the definition of the generalised constitutive law of a bonded FRP sheet, of the compatibility imposed by the shear crack opening, and of the appropriate boundary conditions depending on the strengthening configuration, analytical expressions of the stress field in the FRP sheet crossing a shear crack are obtained. These expressions allow to easily define closed-form equations for the effective strength of FRP strips/ sheets used for shear strengthening, as function of both the adopted strengthening configuration and some basic geometric and mechanical parameters. The FRP contribution is then added to those of concrete and steel. The equations accuracy has been verified through correlation studies with experimental results obtained from the literature and from laboratory tests on purposely under-designed real-scale beam specimens, strengthened with different FRP schemes.

Keywords: FRP; shear design equations; shear strengthening; shear tests

544 Monti and Liotta

Giorgio Monti is a Full Professor of Structural Engineering at the University "La Sapienza" of Rome. He is a member of the Commission Design of *fib* (fédération internationale du béton), in the groups "Seismic Concrete Design", "Computer-based Modelling and Design" and "Design of Structures Reinforced by FRP". His research interests span from reliability to the assessment and retrofitting of existing structures in seismic zones.

Marc'Antonio Liotta is a PhD student in Structural Engineering at the Structural and Geotechnical Engineering Department of the Università La Sapienza of Rome. His main interests are in FRP strengthened reinforced concrete members, behaviour of reinforced concrete structures and application of prestressed FRP fabrics on concrete members.

1. INTRODUCTION

In the development of practical and reliable design equations for shear strengthening of reinforced concrete elements with FRP composite materials, three aspects still remain not perfectly understood. The first regards the shear resisting mechanism that develops when FRP strips/sheets are side bonded, rather than U-jacketed or wrapped, to the element; in this case, a different mechanism than the Mörsch truss activates, that is, a "crack-bridging" mechanism, similar in nature to those of aggregate interlock, dowel effect and concrete tooth. The second aspect regards the evaluation of the contribution of the FRP transverse strengthening to the shear capacity: FRP is subjected to a variable tensile stress along the crack profile, which is conveniently expressed as an effective stress whose intensity is usually given through diagrams and not through closed-form equations. The third aspect regards the evaluation of the relative contributions to the shear capacity of concrete, steel and FRP at ultimate; it is not guaranteed that both concrete and stirrups can exploit their maximum strength when in the presence of FRP strengthening. The clarification of these aspects is the object of the present work, where they are treated from both the experimental and the analytical standpoint.

2. EXPERIMENTAL TESTS

Twenty-four beam specimens, purposely designed as under-reinforced in shear, were tested with a 3-point bending scheme (Monti et al. 2004). The concrete mean compressive cubic strength was $R_{cm} = 13.3$ MPa and the steel rebars had mean yield strength $f_{ym} = 500$ MPa. The geometric dimensions of the beams were: span 2.80 m, cross-section width 250 mm and depth 450 mm. The longitudinal reinforcement was made of 4ϕ20 bottom and of 2ϕ20 top, while shear stirrups ϕ8/400 mm were used. In view of the external strengthening application the bottom corners of the beam were rounded with 30 mm radius. All strips/sheets of the external strengthening were in a single layer of CFRP, having thickness 0.22 mm and elastic modulus $E_f = 390$ GPa. Figure 1 shows both the specimen's dimensions and loading scheme. The nomenclature used for each typology is represented in Table 1.

2.1   Tests description and results

REF1: Reference specimen, unstrengthened. Formation of the first cracks at 100 kN load. Progressive opening and formation of further cracks until failure. The failure was reached due to rupture of the second stirrup, at 550 mm from the left beam end, at 210 kN. Figure 3

REF2: Reference specimen, unstrengthened. Formation of the first cracks at 110 kN load. Failure reached due to rupture of the third stirrup, at 900 mm from the left beam end, at 187 kN.

SS90: Beam with S-strengthening, with CFRP strips 150 mm wide at $\beta = 90°$, with 300 mm spacing. The first cracks were observed at the load of 120-130 kN. The beam failure was reached at 200 kN. The reinforcement seemed to strengthen the beam very little, because the principal crack crossed the strips close to their end.

SS45: Beam with S-strengthening, with CFRP strips 150 mm wide at $\beta = 45°$, with 300 mm spacing, measured along the beam axis. The first cracks were observed at the load of 120/130 kN. At 170 kN the strip at right of midspan started to debond at the bottom. The beam failure was reached at 202 kN for complete delamination of the lower part of the second and third strip at the left of the beam.

SSVA: Beam with S-strengthening, with CFRP strips 150 mm wide with variable inclination (30°-45°-60°) and with spacing as in Figure 7. At 90 kN of load the first flexural cracks were noted. Around 140 kN debonding of the third strip from left of the beam started. Beam failure at 210 kN due to failure with complete debonding from the top of the 30°-strip.

SF90: Beam with S-strengthening, with CFRP sheets at $\beta = 90°$. At 208 kN debonding occurs at the beam midspan. At 213 kN first shear cracks observed. At 225 kN beam failure with rupture of the stirrup at 900 mm from the beam end.

US90: Beam with U-strengthening, with CFRP strips 150 mm wide at $\beta = 90°$, with 300 mm spacing. Failure was due to the rupture of the third stirrup, after debonding of the second strip from left occurred. The failure load of 190 kN was close to the unstrengthened beam because the strips were not activated.

US60: Beam with U-strengthening, with CFRP strips 150 mm wide at $\beta = 45°$, with 300 mm spacing, measured orthogonally to the strips. Formation of the first shear cracks at 135 kN. The third strip from left started to debond from the top at 165 kN. Debonding also started from the bottom at 199 kN, probably because of a crack at the beam soffit. Specimen failure at 222 kN, apparently without stirrup rupture.

USVA: Beam with U-strengthening, with CFRP strips 150 mm wide with variable inclination (30°-45°-60°) as in SSVA. Vertical flexural cracks at midspan around 100 kN of load. Slightly inclined cracks at midspan around 110 kN. Formation of shear cracks located between the strips at 30° and 45°. Specimen failure at 240 kN for debonding from the top of the 30° strip.

USV+: Beam with U-strengthening, with CFRP strips 150 mm wide with variable inclination (30°-45°-60°) as in USVA with a further bottom collaboration strip on the beam sides. Shear cracks around 170 kN. Debonding of the mid-span strip at the beam bottom. Specimen failure at 270 kN without stirrup rupture.

US45+: Beam with U-strengthening, with CFRP strips 150 mm wide at $\beta = 45°$, with 300 mm spacing, measured along the beam axis. At 100 kN first flexural cracks were observed. At 167 kN first shear cracks were observed. At 223 kN debonding of the second strip from left started. At 232 kN debonding of the second strip from left started. Beam failure at 251 kN with complete debonding of the second strip.

US90(2): Beam with U-strengthening, with CFRP strips 150 mm wide at $\beta = 90°$, with 300 mm spacing. At 90 kN first flexural cracks were observed. At 127 kN first shear cracks were observed. At 135 kN the third strip from left started to debond. At 166 kN debonding of the second strip from left. Failure at 179 kN, accompanied by opening of the stirrups hooks.

UF90: Beam with U-strengthening, with CFRP sheets at $\beta = 90°$. First crack at 178 kN. Debonding starts at 206 kN. At 215 kN reinforcement buckling at the beam top, probably due to the upper concrete compression. Fabric failure around 250 kN and specimen failure at 260 kN with stirrups rupture.

An information emerged from this first series of tests regards the limitation of the strips spacing. It has been verified that the strip spacing should be sufficiently close to avoid the formation of cracks that do not cross at least one strip. From Figure 16 it can be seen that, thinking to "condense" the strips in an "equivalent stirrup" on the strip axis, having the same height of the strip minus the effective bond length L_e from both ends in case of Side-bonding and only from one in case of U-Jacketing, shear cracks can develop that do not cross excessively spaced strips in the effective zone.

In fact, in the case of side bonding there exists a field, shown in Figure 8, where the crack, represented in its minimum and maximum inclination, can freely pass in between strips, without crossing and activating them. From the figure, it can be seen that such field reduces its extension passing from Side-bonding to U-jacketing and increasing the fibre inclination. This conclusion is supported by the observation of tests and from correlation of theoretical and experimental results shown in section 4 and suggests to adopt the following limitations in case of strip strengthening: the width w_f and the spacing p_f of the strips, measured (in mm) orthogonally with respect to the angle β of the fibre direction, should respect the following limitations: 50 mm $\leq w_f \leq$ 250 mm, $p_f \leq$ min$\{0.5\ d,\ 3\ w_f,\ w_f + 200$ mm$\}$ and $p_f \geq 2\ w_f$. Obeying to these limitations a second series of beam tests was carried out.

US45++: Beam with U-strengthening, with CFRP strips 50 mm wide at $\beta = 45°$, spacing 150mm along the beam axis. First crack at 184 kN. Failure of first strip at 261 kN and shear failure of beam at 267 kN.

WS45++: Beam with W-strengthening, with CFRP strips 50 mm wide at $\beta = 45°$, spacing 150 mm along the beam axis. First crack at 182 kN. Failure of first strip at 291 kN and shear failure of beam at 317 kN.

Ref3 and Ref4: Two more reference specimens to ensure the concrete strength of the second casting to be the same as the one in the first series of beams. Shear failure at 187 and 200 kN of load, respectively.

US45+ "A": Beam with U-strengthening, with CFRP strips 150 mm wide at $\beta = 45°$ with 225 mm spacing, measured along the beam axis. First crack at 184 kN. First shear cracks at 197 kN. Failure at 334,2 kN with stirrup overlap opening.

US45++ "B": Beam with U-strengthening, with CFRP strips 150 mm wide at $\beta = 45°$ with 225 mm spacing, measured along the beam axis. First crack at 184 kN.

US45++ "C": Beam with U-strengthening, with CFRP strips 150 mm wide at $\beta = 45°$ with 225 mm spacing, measured along the beam axis. First shear cracks appear at 184 kN. First strip breaks at 288 kN. Strengthening failure around 364 kN and specimen failure at 366kN without stirrups rupture.

US45++ "F": Beam with U-strengthening, with CFRP strips 150 mm wide at $\beta = 45°$ with 300 mm spacing, measured along the beam axis. Formation of first shear cracks between 204 and 212 kN. Shear failure at 300 kN, with stirrup overlap opening at one side and stirrup failure on the other side.

US45++ "E": Beam with U-strengthening, with CFRP strips 150 mm wide at $\beta = 45°$ with 300 mm spacing, measured along the beam axis. Formation of first shear cracks at 210 kN. Shear failure at 327 kN, with stirrup overlap opening at one side and stirrup failure on the other side.

US45++ "D": Beam with U-strengthening, with CFRPstrips 150 mm wide at $\beta = 45°$ with 300 mm spacing, measured along the beam axis. First crack at 210 kN. Shear failure at 229 kN, with stirrup overlap opening at one side and stirrup failure on the other side.

3. DESIGN EQUATIONS FOR FRP SHEAR STRENGTHENING

This section tries to provide a coherent analytical framework to describe the behaviour of RC elements FRP-strengthened in shear, following previous efforts made by other authors (Täljsten 1997, Triantafillou 1998, Khalifa et al. 1998). The developed theory aims at describing the FRP stress distribution $\sigma_{f,cr}(x)$ along a shear crack (as qualitatively sketched in Figure 27) through closed-form equations, as opposed to regression-based formulas (Triantafillou & Antonopoulos 2000). Once this is correctly defined, the FRP resultant across the crack can be computed and the FRP contribution to the resisting shear be found. The analytical developments arrive at defining three predictive equations for: Side Bonding (S), U-jacketing (U) and Wrapping (W).

The obtained expressions of the strength are given in terms of readily available geometrical and mechanical quantities of both the FRP strengthening and the RC beam and are then used to compute the FRP contribution to the overall shear strength, together with that of concrete and transverse reinforcement. These equations have been adopted in the new Code for FRP strengthening recently issued by the Italian Research Council (CNR 2005)

In the following developments, the following hypotheses are made (notation in Figure 28):

- Shear cracks are evenly spaced along the beam axis, and inclined with angle θ,
- At the ULS the cracks depth is equal to the internal lever arm $z = 0.9 \, d$,
- In the case of U-jacketing (U) and wrapping (W), the resisting shear mechanism is based on the Moersch truss, while in the case of side bonding (S), because the Moersch truss cannot form as the tensile diagonal tie is missing, a different resisting mechanism of "crack-bridging" is considered to develop.

In order to fully characterize the physical phenomenon, the following aspects must be analytically defined: a) the failure criterion of an FRP strip/sheet bonded to concrete, b) the stress-slip constitutive law, c) the compatibility equations (*i.e.*, the crack opening), and d) the boundary conditions (*i.e.*, the available bonded lengths on both sides of the crack depending of the different configurations).

3.1 Generalised failure criterion of an FRP strip/sheet bonded to concrete

The criterion includes the two cases of: a) straight strip/sheet, and b) strip/sheet wrapped around a corner. Two quantities are introduced: the *effective bond length* l_e and the *debonding strength* $f_{fdd}(L)$, expressed as function of the available bond length l_b.
The effective bond length (optimum anchorage length) is given as:

$$l_e = \sqrt{\frac{E_f t_f}{2 f_{ctm}}} \quad \text{[length in } mm\text{]} \tag{1}$$

where: E_f = FRP sheet elastic modulus, t_f = sheet thickness, $f_{ctm} = 0.27 \cdot R_{ck}^{2/3}$ = concrete mean tensile strength (with R_{ck} = concrete characteristic cubic strength).
The specific rupture energy Γ_{Fk} of the concrete-strngthening bond can be expressed as:

$$\Gamma_{Fk} = 0.03 \cdot k_b \cdot \sqrt{f_{ck} \cdot f_{ctm}} \quad \text{,[units: } N, \, mm\text{]} \tag{2}$$

where f_{ck} is the concrete characteristic cubic strength and k_b = covering/scale coefficient (Brosens and Van Gemert 1999), given as:

$$k_b = \sqrt{\frac{2 - w_f / p_f}{1 + w_f / 400}} \geq 1 \tag{3}$$

where, for strips: w_f = width measured orthogonally to β, p_f = spacing measured orthogonally to β; while for sheets $k_b = 1$. Note however that w_f should not exceed $\min(0.9d, \, h_w) \cdot \sin(\theta + \beta)/\sin\theta$, with d = beam effective depth, h_w = beam web depth, β = angle of strip/sheet to the beam axis, θ = crack angle to the beam axis.
The debonding strength is given as:

$$f_{fdd} = \frac{0.80}{\gamma_{f,d}} \sqrt{\frac{2 E_f \Gamma_{Fk}}{t_f}} \quad \text{units: } [N, \, mm] \tag{4}$$

where $\gamma_{f,d}$ is a partial safety factor depending on the application accuracy. In case the available bond length l_b is lower than the optimum anchorage length, l_e, the design strength should be reduced to the value $f_{fdd,rid}$ given as:

$$f_{fdd,rid} = f_{fdd} \cdot \frac{l_b}{l_e} \left(2 - \frac{l_b}{l_e} \right). \tag{5}$$

The ultimate strength of the FRP strip/sheet, which includes the case when it is wrapped around a corner rounded with a radius r_c, is:

$$f_{fu}(l_b, \delta_e, r_c) = f_{fdd}(l_b) + \langle \phi_R \cdot f_{fu} - f_{fdd}(l_b) \rangle \cdot \delta_e, \quad \text{where:}$$

$$\delta_e = \begin{cases} 0 & \text{free end} \\ 1 & \text{end around a corner} \end{cases} \tag{6}$$

where it can be seen that the debonding strength depends on the available bonded length l_b and $\langle \cdot \rangle$ denotes that the bracketed expression is zero if negative. It is noted that the sheet wrapped around a corner attains a fraction ϕ_R of the ultimate strength f_{fu} of the FRP sheet depending on the coefficient ϕ_R as function of the rounding radius r_c with respect to the beam width b_w (Campione and Miraglia 2003):

$$\phi_R = 0.2 + 1.6 \frac{r_c}{b_w}, \qquad 0 \le \frac{r_c}{b_w} \le 0.5 \tag{7}$$

When $l_b \ge l_e$, the expression for the ultimate strength of the FRP strip/sheet, wrapped around a corner with a radius r_c, becomes:

$$f_{fu,W}(r_c) = f_{fdd} + \left\langle \phi_R \cdot f_{fu} - f_{fdd} \right\rangle \tag{8}$$

3.2   Generalised stress-slip constitutive law
The generalised stress-slip law σ_f (u, l_b, δ_e) of FRP strips/sheets bonded to concrete, including both cases of free end or wrapped around a corner, is shown in Figure 29.

3.3   Compatibility (crack width)
Considering a reference system with the origin fixed at the upper limit of the shear crack and with abscissa x along the crack itself (Figure 31), the crack width (normal to the crack axis) along the shear crack can be expressed as $w = w(x)$.

3.4   Boundary conditions (available bond length)
The boundary conditions refer to the available bond length $L(x)$ on both sides of the shear crack and should be defined according to the strengthening scheme adopted: either S=Side bonding, U=U-jacketing, W=Wrapping (Figure 31).

3.5   FRP stress profile along the shear crack
In order to obtain the stress profile in the FRP sheet along the crack as a function of both the crack opening and the available bond length on both sides of the crack itself, one has to substitute into the constitutive law σ_f (u, l_b, δ_e): a) the compatibility equation $u = u(\alpha, x)$, b) the boundary condition $l_b = l_b(x)$ given according to the strengthening configuration, and c) the end constraint given by the appropriate value of δ_e. Figure 32 qualitatively depicts the $\sigma_{f,cr}(x)$ profiles along the crack for the three different strengthening configurations considered, when sheets are used. In the configuration S, the stress profile is truncated towards the end of the crack, where the available length tends to zero. In the configuration U, the stress profile remains constant where the available length allows the full debonding strength to be developed throughout the crack length. In the configuration W, the stress profile rises towards the end of the crack, where, after

complete debonding, the sheet is restrained at both ends and subjected to simple tension up to its tensile strength.

3.6   Determination of FRP contribution to the shear strength

The objective is to obtain the maximum contribution of the FRP strips/sheet to the shear strength. This means to identify, among all possible shapes of the FRP stress profile $\sigma_{f,cr}[u(\alpha,x),l_b(x)]$, which changes with the crack opening α, the one offering the maximum contribution.

3.6.1 Effective stress in the FRP sheet -- To this aim it is expedient to define an effective stress in the FRP sheet, inclined to an angle β as the FRP fibres, as the mean FRP stress field $\sigma_{f,cr}(x)$ along the shear crack length $z/\sin\theta$:

$$\sigma_{fe}(\alpha) = \frac{1}{z/\sin\theta} \cdot \int_0^{z/\sin\theta} \sigma_{f,cr}[u(\alpha,x),l_b(x)]\,dx \tag{9}$$

which might be regarded as an equivalent constant FRP stress block along the shear crack.

3.6.2 Effective debonding strength -- The maximum of the FRP effective stress, which is termed the effective debonding strength f_{fed}, is found by imposing:

$$\frac{d\sigma_{fe}[x_u(\alpha)]}{d\alpha} = \frac{d\sigma_{fe}(x_u)}{dx_u} \cdot \frac{dx_u(\alpha)}{d\alpha} = 0 \tag{10}$$

where the chain rule has been used. Solution of (10) allows to determine the FRP stress profile with the maximum area, that is, the effective strength of the FRP shear strengthening.

In the case of side-bonding (however, not allowed for seismic strengthening):

$$f_{fed} = f_{fdd} \cdot \frac{z_{rid,eq}}{\min\{0.9\,d, h_w\}} \cdot \left(1 - 0.6\sqrt{\frac{l_{eq}}{z_{rid,eq}}}\right)^2 \tag{11}$$

where :

$$z_{rid,eq} = z_{rid} + l_{eq}, \quad z_{rid} = \min\{0.9\,d, h_w\} - l_e \cdot \sin\beta, \quad l_{eq} = \frac{s_f}{f_{fdd}/E_f} \cdot \sin\beta \tag{12}$$

and it is observed that: z_{rid} is equal to the minimum between the effective depth of the section (that is equal to the vertical projection of the crack) minus the bottom part where there is not enough bond length and the beam web in case of T sections, l_{eq} is the bonded length projected vertically that would be necessary if the fabric strain $\varepsilon_{fdd} = f_{fdd}/E_f$ was uniform and s_f is the slip at debonding.

In the case of U-jacketing:

$$f_{fed} = f_{fdd} \cdot \left[1 - \frac{1}{3} \frac{l_e \sin \beta}{\min \{0.9\,d, h_w\}} \right] \tag{13}$$

In the case of wrapping:

$$f_{fed} = f_{fdd} \cdot \left[1 - \frac{1}{6} \frac{l_e \sin \beta}{\min \{0.9\,d, h_w\}} \right] + \frac{1}{2} (\phi_R \cdot f_{fd} - f_{fdd}) \cdot \left[1 - \frac{l_e \sin \beta}{\min \{0.9\,d, h_w\}} \right] \tag{14}$$

where f_{fd} is the design ultimate strength of the FRP to be evaluated as in

$$f_{fu,W}(r_c) = f_{fdd} + \left\langle \phi_R \cdot f_{fu} - f_{fdd} \right\rangle \tag{15}$$

In the previous equation the second term should be considered only when positive.

3.7   Shear capacity with FRP

In case the reinforcement type is U or W, the Moersch resisting mechanism can be activated and the shear carried by FRP is expressed as:

$$V_{Rd,f} = \frac{1}{\gamma_{Rd}} \cdot 0.9\,d \cdot f_{fed} \cdot 2\,t_f \cdot (\cot \theta + \cot \beta) \cdot \frac{w_f}{p_f} \tag{16}$$

while for side-bonding (S) the FRP role is that of "bridging" the shear crack, so that:

$$V_{Rd,f} = \frac{1}{\gamma_{Rd}} \cdot \min \{0.9\,d, h_w\} \cdot f_{fed} \cdot 2\,t_f \cdot \frac{\sin \beta}{\sin \theta} \frac{w_f}{p_f} \tag{17}$$

with d = beam effective depth, f_{fed} = design effective strength of the FRP shear strengthening, given either by (11) for side bonding or by (13) for U-jacketing or by (14) for wrapping, t_f = thickness of FRP strip/sheet (on single side) with angle β, θ = crack angle, s_f, w_f = strip spacing and width, respectively, measured orthogonally to the fibre direction β.

Assuming cracks inclined of an angle $\theta = 45°$ with respect to the vertical and strips/sheets vertically aligned at $\beta = 90°$, the two previous equations become:

$$V_{Rd,f} = \frac{1}{\gamma_{Rd}} \cdot 0.9\,d \cdot f_{fed} \cdot 2\,t_f \cdot \frac{w_f}{p_f} \tag{18}$$

$$V_{Rd,f} = \frac{1}{\gamma_{Rd}} \cdot \min \{0.9\,d, h_w\} \cdot f_{fed} \cdot 2\,t_f \cdot \sqrt{2} \frac{w_f}{p_f} \tag{19}$$

The shear verification should be performed by comparing the design acting shear with the shear capacity, given by:

$$V_{Rd} = \min \{ V_{Rd,ct} + V_{Rd,s} + V_{Rd,f}, V_{Rd,\max} \} \tag{20}$$

where $V_{Rd,ct}$ is the concrete contribution, given by (*e.g.*, EC2 (CEN 1991), not accounted for):

$$V_{Rd,ct} = \frac{0.18}{\gamma_c} b_w \cdot d \cdot \min\left\{1+\sqrt{\frac{200 \text{ mm}}{d}}, 2\right\} \cdot \sqrt[3]{100 \cdot \min\{0.02, \rho_{sl}\} \cdot f_{ck}} \tag{21}$$

and $V_{Rd,s}$ is the steel contribution, given by:

$$V_{Rd,s} = 0.9 d \cdot f_{yd} \frac{n_{st} \cdot A_{st}}{s_{st}} (\cot\theta + \cot\beta_{st}) \sin\beta_{st} \tag{22}$$

where $f_{ctd} = 0.7 f_{ctm}/\gamma_c$ = concrete tensile strength, $\gamma_c = 1.5$ = concrete partial coefficient, b_w = web section width, ρ_{sl} = longitudinal geometric ratio, f_{ck} = concrete characteristic cylindrical strength, f_{yd} = design steel yield strength, n_{st} = transverse reinforcement arm number, A_{st}, s_{st} = area (one arm) and spacing of traverse reinforcement, e β_{st} = stirrups angle.

In (20), $V_{Rd,max}$ is the strength of the concrete strut, given by (*e.g.*, EC2):

$$V_{Rd,max} = 0.9 d \cdot b_w \cdot v \cdot f_{cd} \cdot (\cot\theta + \cot\beta_{st})/(1 + \cot^2\theta) \tag{23}$$

with

$$v = 0.6[1 - f_{ck}/250] \quad [\text{in MPa}]. \tag{24}$$

VALIDATION OF DESIGN EQUATIONS

The results obtained with the above presented equations are both applied to the case of the specimen beams tested in the lab and to other authors' test specimens; the results are shown in Figure 33, also in comparison with the results obtained following the ACI 440 code (ACI 440.2R-02).. Partial coefficients were set to 1 for the prediction of experimental results, and material properties were considered with their mean values. In the equations for the variable inclination reinforcements a mean value of the strips inclinations is considered, while spacing is the effective one. The shear capacity of the reference beam was computed as the mean between the two tested unstrengthened specimens. Please note that in the specimen SS90, SS45, and US90, the contribution of FRP strengthening was not considered, as it was recognised that the diagonal shear cracks did not cross the strips.

It can be observed that the mean error on the predictions that activated the FRP strengthening is 7%, with a peak of 15% for the configurations US60 and UF90. Such an error can be considered as acceptable. Further tests are being carried out to validate the proposed equations on different reinforcing schemes.

The work presented here addressed some of the still unsolved aspects in previous analytical treatments of shear strengthening of beams with composite materials (FRP) and proposes possible solutions for them. In particular, closed-form analytical expressions for the effective strength of FRP strips/sheets crossing the shear crack were found, which are then introduced in design equations for the contribution of FRP to the shear strength of RC elements. In this respect, it has been clarified that the FRP contribution to the shear strength should be computed for U and W configurations with equation (16), based on the formation of the Moersch truss, while for S configurations equation (17) should be used instead, which considers the "bridging" of cracks. The equations developed showed good correlation with purposely carried out experimental tests. The equations matched the shear capacity increase with a more than acceptable error.

4. ACKNOWLEDGEMENTS

The authors wish to thank Interbau srl company of Milan, Italy, for the beam specimens preparation and the CFRP application.

REFERENCES

ACI 440.2R-02 (2002). Guide for the Design and Construction of Externally Bonded FRP Systems for Strengthening Concrete Structures. *American Concrete Institute, Committee 440*

Brosens, K., and Van Gemert, D. (1999). Anchorage design for externally bonded carbon fiber reinforced polymer laminates. *Proc. 4th Int. Symposium on FRP Reinforcement for Concrete Structures*, Baltimore, USA, pp. 635-645.

Campione, G., and Miraglia, N. (2003). "Strength and strain capacities of concrete compression members reinforced with FRP". Cement and Concrete Composites, Elsevier, 25, 31-41.

CEN (1991). Eurocode 2: Design of concrete structures – Part 1-1: General rules and rules for buildings. ENV 1992-1-1, Comité Européen de Normalisation, Brussels, Belgium.

CNR (2005). Instructions for Design, Execution and Control of Strengthening Interventions with FRP. *Consiglio Nazionale delle Ricerche, Roma, Italy.*

fib (2001). Design and Use of Externally Bonded FRP Reinforcement (FRP EBR) for Reinforced Concrete Structures. *Bulletin no. 14, fib Task Group 9.3 'FRP Reinforcement for Concrete Structures'.*

Khalifa, A., Gold, W. J., Nanni, A. and Aziz, A. M. I. (1998). Contribution of externally bonded FRP to shear capacity of rc flexural members. *ASCE Journal of Composites for Construction*, 2(4), 195-202.

Monti, G., Santinelli, F., and Liotta, M.A. (2004). Shear strengthening of beams with composite materials. *Proc. 2nd International Conference on FRP Composites in Civil Engineering CICE 2004*, Adelaide, Australia, December.

Täljsten B. (1997). Strengthening of concrete structures for shear with bonded CFRP-fabrics. *Recent advances in bridge engineering, Advanced rehabilitation, durable materials, nondestructive evaluation and management*, Eds. U. Meier and R. Betti, Dübendorf, 57-64.

Triantafillou, T. C. (1998). Shear strengthening of reinforced concrete beams using epoxy-bonded FRP composites. *ACI Structural Journal*, 95(2), March-April, 107-115.

Triantafillou, T. C. and Antonopoulos, C. P. (2000). Design of concrete flexural members strengthened in shear with FRP. *ASCE Journal of Composites for Construction*, 4(4), 198-205.

Table 1. Typology, nomenclature and experimental shear strength of the tested beams.

STRENGTH'G APPLICATION	STRENGTH'G TYPE	FIBRES ANGLE	NAME	STRENGTHENING CONFIGURATION	EXP. SHEAR STRENGTH (kN)
	NONE	-	REF		95.0
SIDE BONDING	STRIPS width 150 mm spacing 300 mm	90°	SS90*		100.0
		45°	SS45		101.0
		60°, 45°, 30°	SSVA		105.0
	SHEETS	90°	SF90		112.5
U-JACKETING	STRIPS width 150 mm spacing 300 mm	90°	US90*		95.0
		60°	US60		111.0
		60°, 45°, 30°	USVA		120.0
		60°, 45°, 30°	USVA+		135.0
		45°	US45+		126.0
		90°	US90(2)*		90.0
	SHEETS	90°	UF90		125.0
	STRIPS, width 50mm spacing 100 mm		US45++		133.5
	SHEETS		UF45+ A		158.5
			UF45++B		167.0
			UF45++C		172.0
	STRIPS width 150 mm spacing 225 mm		US45++F		182.85
			US45++E		150.15
			US45+ D		163.45
WRAPPING	STRIPS width 50mm spacing 100 mm		WS45++		114.5

* In these tests, the shear cracks did not fully activate the FRP strips, which then did not contribute to the shear strength.

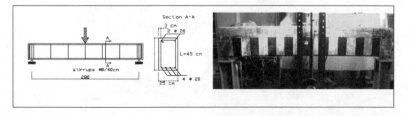

Figure 1 — Reference specimen dimensions and loading scheme (left) and representative picture of a test (right).

Figure 2 — Reference specimen REF1

Figure 3 — Reference specimen REF2

Figure 4 — Specimen SS90

Figure 5 — Specimen SS45

Figure 6 — Specimen SSVA

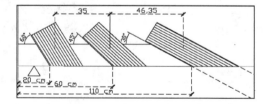

Figure 7 — Configuration of SSVA strengthening.

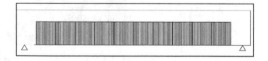

Figure 8 — Specimen SF90

Figure 9 — Specimen US90

Figure 10 — Specimen US60

Figure 11 — Specimen USVA

Figure 12 — Specimen USV+

Figure 13 — Specimen US45+

Figure 14 — Specimen US90

Figure 15 — Specimen UF90

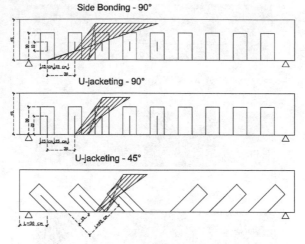

Figure 16 — Crack formation fields with inadequate strip spacing.

Figure 17 — Specimen US45+

Figure 18 — Specimen WS45+

Figure 19 — Specimen REF3

Figure 20 — Specimen REF4

Figure 21 — Specimen US 45+ "A"

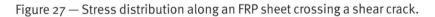

Figure 22 — Specimen US45++"B"

Figure 23 — Specimen US45++"C"

Figure 24 — Specimen US45++"F"

Figure 25 — Specimen US45++"E"

Figure 26 — Specimen US45++"D"

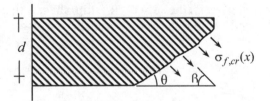

Figure 27 — Stress distribution along an FRP sheet crossing a shear crack.

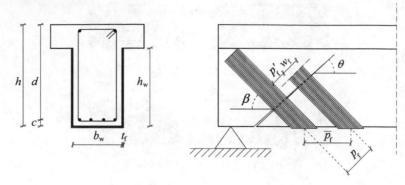

Figure 28 — Geometry notation.

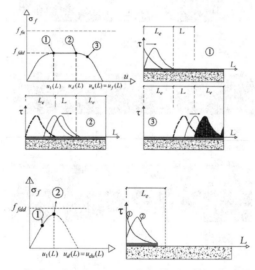

Figure 29 — Stress-slip law for the case of FRP strip/sheet with free end and with sufficient bond length (Top), and with small bond length (Bottom).

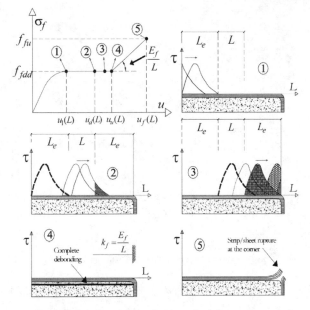

Figure 30 — Stress-slip law for the case of FRP strip/sheet wrapped around a corner.

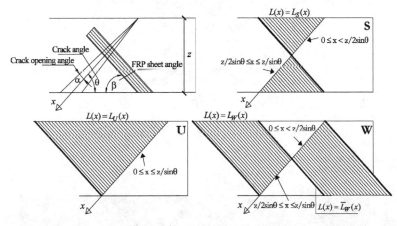

Figure 31 — Boundary conditions (available bond length) for three strengthening configurations: S = Side bonding, U = U-jacketing, and W = Wrapping.

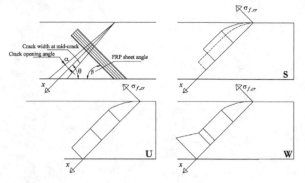

Figure 32 — Typical stress profiles in FRP sheets along the shear crack for three strengthening configurations: S = Side bonding, U = U-jacketing, and W = Wrapping.

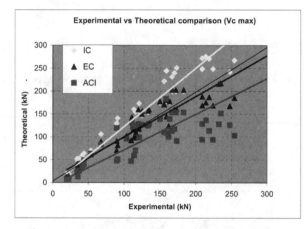

Figure 33 — Prediction-test results comparison.

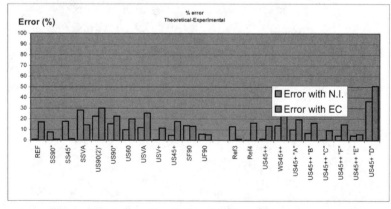

Figure 34 —Prediction-test results error.

Post-Repair Performance of Corroded Bond Critical RC Beams Repaired with CFRP

by B.C. Craig and K.A. Soudki

Synopsis: Presented in this paper is an investigation on the ability of externally applied fiber reinforced polymer (FRP) laminates to maintain bond of steel reinforcement in concrete members subjected to corrosion. Specimens were transversely confined with CFRP laminates in the bond zone after being subjected to various degrees of corrosion ranging from 2 to 10% theoretical mass loss. Some specimens were further corroded after repair to assess the effects of further structural deterioration. Control beams were subjected to minor amounts of corrosion and tested to failure without repair. Test results showed that CFRP wrapping was able to confine the corrosion cracking resulting in an overall flexural failure for all repaired specimens unlike the brittle bond splitting failures of the control specimens. CFRP confinement proved less effective at higher levels of pre-repair corrosion. Initial amounts of post-repair corrosion enhanced the performance of the CFRP repair by increasing the confining pressure; however the concrete rapidly deteriorated as the corrosion increased. In general, CFRP confinement was found to provide superior bond performance with respect to the unrepaired members indicating the potential future use in field applications of bond repair.

Keywords: bond; CFRP; confinement; corrosion; repair

Brent C. Craig is a Civil Engineer within the Central Hydro Division at Acres International in Niagara Falls, Canada. He received his Master of Applied Science in Civil Engineering at the University of Waterloo in 2002. His graduate research focused on the use of externally applied fiber reinforced polymer laminates in the structural rehabilitation of concrete members.

Khaled Soudki is the Canada Research Chair Professor in Innovative Structural Rehabilitation at the University of Waterloo, Canada. He is a member of ACI Committees 440, Fiber Reinforced Polymer Reinforcement; 222, Corrosion of Metals; 546, Repair of Concrete; 550, Precast Concrete Structures. His research interests include corrosion, durability, rehabilitation, and strengthening of concrete structures using fiber reinforced polymer composites.

INTRODUCTION

The corrosion of steel reinforcement is a major concern in reinforced concrete (RC) infrastructure. Incidences of chloride-induced corrosion commonly lead to structural deficiencies and dangerous public safety concerns in highway bridges, parking garages, marine structures, and various other infrastructure. The option to replace structures is not always practical and performing repairs that are not durable may be costly to the owner and users. It is therefore important to find innovative methods to repair and/or strengthen deteriorated structures in an efficient, durable, and cost effective manner.

A primary influence of corrosion on the performance of flexural members is the degradation of bond interaction between the reinforcing steel and concrete, potentially resulting in catastrophic failures. Numerous studies have investigated the effects of corrosion on flexural members and shown that as the degree of corrosion increased, the loss of tensile steel area was not directly proportional to the loss in flexural capacity.[1,2,3] At higher levels of corrosion, the loss of bond strength between the reinforcing steel and the concrete becomes more detrimental.[1,2,3]

As stress levels increase in the tension reinforcement, bond forces radiate from the bar developing hoop stresses in the surrounding concrete.[4] Bond forces are transferred from the steel to the concrete through chemical adhesion, friction and/or mechanical interlock.[5] The interaction of each transfer mechanism depends on the configuration of the bar. However, the use of deformed bars allows mechanical interlock to be the most predominant form of transfer.[5]

Bond failure typically occurs in two distinct ways relating to the confinement of the reinforcing bar. In the presence of sufficient cover and or transverse reinforcement, a bond pullout failure is expected.[6] In a bond pullout failure, concrete is able to withstand the radial tensile forces generated by the bond transfer force. However, failure is caused by localized shearing of the concrete between the reinforcing bar lugs, thus allowing the steel to pull though the concrete. When there is insufficient confinement, bond failure occurs when the tensile strength of the concrete is exceeded causing the development of longitudinal cracks parallel to the reinforcing bar.[7,8] Cracking causes a loss of confinement and mechanical interlock resulting in a bond splitting failure.

Corrosion of the reinforcing steel bar reduces the bond strength in several fashions. The development of corrosion products around the bar initially eliminates the chemical adhesion to the surrounding concrete. As corrosion progresses, frictional forces are reduced as the products may act as a lubricating layer.[9] The expansive nature of the corrosion products develops hoop stresses in the surrounding concrete similar to those developed from bond forces. As the radial forces exceed the tensile capacity of the concrete, cracks develop resulting in a diminished mechanical interlock due to the reduced confining pressure. To a lesser degree, corrosion reduces steel rib cross section which has an negative impact of the mechanical interlock.[9]

It is important to note that low levels of corrosion have proven to be beneficial to bond forces.[9,10,11] Prior to the tensile forces being exceeded in the concrete, the expansive pressure generated by the corrosion products increase the confining pressure around the bar, thus reducing slip potential. In addition, the initial formation of corrosion products increases the bar surface roughness and therefore increase the frictional component of the bond interaction.[9]

Confinement has a beneficial effect on the capacity of anchorage bond.[12,13] By providing additional cover or transverse reinforcement, crack openings are minimized, thus better maintaining mechanical interlock. It still is important however to realize that inadequate development length could still lead to bond pullout failure regardless of the confinement provided.

Fiber reinforced polymer (FRP) reinforcement has emerged as viable solution to enhancing the bond capacity in existing RC members. FRP are light weight materials that provide excellent strength and durability. Literature has shown that the addition of externally applied FRP sheets does enhance the bond capacity in situations where adequate confinement is not provided.[14,15,16,17,18]

To the author's knowledge, minimal research has been conducted on the potential use of FRP as a confining wrap in order to prevent the deterioration of bond in members subject to corrosive environments. The focus of this study is the investigation of the post-repair effects of FRP laminates on corrosion damaged, bond critical, RC members. Carbon FRP (CFRP) was used as a confining wrap in the bond zones to counteract the effects of corrosion and assist in maintaining mechanical interlock, thereby maintaining/enhancing the load carrying capabilities.

EXPERIMENTAL PROGRAM

The test program consisted of twenty-two medium-scale reinforced concrete bond-beam specimens. Beams were designed to investigate the effect of externally applied CFRP laminates with respect to maintaining the bond interaction between reinforcing steel and concrete in corrosion damaged flexural members. The test matrix for the study is provided in Table 1. Bond lengths were varied to examine the effectiveness of CFRP confinement for beams designed to fail either by flexure or bond. Bond lengths were controlled at 450 mm (series L1, l_d/d_b = 28), 350 mm (series L2, l_d/d_b = 22), and 250 mm (series L3, l_d/d_b = 16). Only series L3 with the 250 mm bond length was designed to fail by bond splitting in the uncorroded state.

To effectively investigate post-repair behavior, the specimens were subjected to initial degrees of corrosion prior to the application of the CFRP laminates. In some cases, specimens were further corroded to represent situations where the deterioration was not fully arrested. Four corrosion levels were selected to represent various degrees of deterioration (2, 5, 10 and 15% theoretical mass loss). Series L2, with the intermediate bond length of 350 mm, was corroded to the highest degree of post-repair corrosion to investigate section behavior at higher levels of mass loss. For each specified bond length, a control specimen was left uncorroded without any application of CFRP.

Specimen Design

The beam specimen measured 150 mm wide, 250 mm high and 2000 mm long as illustrated in Figure 1. Beams were reinforced to provide a ductile failure in the event that an adequate bond length was provided. Reinforcement consisted of two No. 15M, Grade 400, deformed bars to achieve a reinforcement ratio (ρ_s) of 0.012. The tension reinforcement was placed such that 25 mm of clear cover was maintained to the side and bottom of the bar (c/d_b = 1.56). The length of the reinforcing bar was selected such that a minimum of 100 mm was exposed on either end of the beam to allow for instrumentation. Shear zones were reinforced with 6 mm diameter stainless steel stirrups at 100 mm spacing to ensure adequate shear resistance. The 28-day concrete compressive strength was 42 MPa and the yield strength of the tension reinforcement was 440 MPa.

Reinforcing bars were debonded from the concrete outside of the bond zone using low-density polyethylene tube. Pockets were formed in the concrete tension zone near the midspan of the beam, outside of the bond zone, to allow for easy instrumentation of the tensile reinforcement during testing.

Induced Corrosion

Specimens were deteriorated using accelerated corrosion by means of an impressed current. The corrosion setup is shown in Figure 2. To facilitate the corrosion process, an 8 mm diameter hollow stainless steel tube was cast within the beam to act as an internal cathode. Furthermore, the concrete in the bottom third of the beam was contaminated with NaCl in order to produce a concrete mixture with 2.3% chlorides by mass of cement.

Specimens were connected in series by wires through the stainless steel bar and the tensile reinforcement to a power source. A constant current was applied such that the stainless steel bar acted as the cathode, and the tensile reinforcement as the anode in the corrosion reaction. The polyethylene debonding sleeves acted as insulator around the tension reinforcement ensuring that all corrosion occurred in the bond zones. Current was applied in such a manner that the impressed current density of 105 $\mu A/cm^2$ was maintained for all specimens.

Theoretical mass loss was calculated using Faradays Law, which specifies the amount of mass loss expected at a specific current density over, a specific time frame (Jones, 1996). Specimens were corroded at room temperature in a humidity tent where they were subjected to cycles of 2.5 days at 100% relative humidity and 1 day dry for the duration of the corrosion period (Figure 2).

Wrapping Scheme

Deteriorated bond zones were confined using single unidirectional CFRP sheets. The CFRP sheets measured 0.11mm thick with a tensile strength of 2450 MPa, and an elastic modulus of 160 GPa. A single wrapping scheme was implemented for all bond regions as shown in Figure 1. CFRP laminates were transversely placed in the concrete-steel bond zone such that the fiber direction was oriented perpendicular to the direction of main reinforcement. Fiber sheets were fully wrapped around the beam to provide full confinement of the concrete section, thus preventing premature failure due to debonding of the CFRP sheet. It was assumed that in real world applications where CFRP sheets could not be wrapped around the entire beam, adequate anchorage or development length of the sheet would be applied through other means.

Test Setup

Specimens were tested to failure in four-point bending using the configuration shown in Figure 3. The specimens were simply supported over a span of 1800 mm, with a constant moment region measuring 300 mm.

Overall beam performance was monitored using a load cell and a displacement transducer (LVDT) located at the midspan of the beam. Slip displacement of the tensile steel was measured using LVDT's mounted to the free ends of the reinforcing bar. The LVDT's determined the free end slip by measuring relative displacement between the reinforcing steel and concrete surface as the bar was pulled through the concrete. The tensile steel stress was measured using 5 mm strain gauges attached to the bars within the pre-cast pockets at four locations.

Load was applied under displacement control at a rate of 1 mm/min until failure was achieved. Failure was considered to occur when concrete crushed in the compression zone (ultimate condition), or until the slip of the main reinforcing bar at one end reached a value of 10 mm (bond failure).

RESULTS AND DISCUSSION

A summary of test results is provided in Table 2. A significant difference in results could be observed for each bond length. For a complete understanding of the bond interaction, both the load-deflection and bond slip-stress curves must be examined. When calculating the bond stress, values from the strain gauges on the tension bars were used and averaged out over the length of the bar in order to simplify the analysis. When comparing specimens with different bond lengths, the bond stress is normalized by dividing the value by the ultimate theoretical bond stress for that particular bond length.

The actual mass loss values for the corroded reinforcing bar were determined after testing in accordance with ASTM G1-90. Results of the mass loss are provided in Table 2. For the purposes of analysis, the theoretical mass loss values are referred to in the text.

For all bond lengths, the control specimen failed by bond splitting action. For Series L1 and L2, this deviated from the designed mode of failure. The two percent theoretical mass loss provided enough loss of confinement to cause both specimens to fail abruptly. Series L1, with the longer bond length, was still able to achieve yielding of the reinforcing bar prior to loss of load carrying capabilities. Visually, the effect of the bond splitting is shown in Figure 4 where longitudinal cracks were developed along the length of the bar through the bond zone. These cracks were partially developed from existing corrosion cracks. Lack of confinement around the reinforcing bar is evident as the concrete is cracked and displaced in the bond region. At the time of failure, a sudden loss in load carrying capability for all three members can be seen in the form of an instantaneous drop on the load deflection curve (Figure 5). None of the members were able to achieve their full flexural capacity. Similar behavior may be observed for all beams with respect to their bond slip behavior (Figure 6). It is evident that the series with the longer bond length was able to develop bond stresses closer to the ultimate capacity. The following sections discuss the behaviour of the different test series.

Series L1– 450 mm Bond Length

Six beams were corroded and tested with a 450 mm bond length with five of the specimens subject to repair. Three beams were repaired at 2% mass loss with two beams being further corroded to 5 and 10% mass loss. The two remaining beams were repaired at 5% theoretical mass loss, with one beam being further corroded to 10% mass loss. The control beam was tested without repair after being subjected to 2% mass loss.

It is apparent from the load-deflection plots in Figure 7 that the post-repair beams were able to outperform the control beam. The repaired beams were able to achieve a ductile flexural failure. Analysis of the slip data indicated that unlike the control specimen, there was zero or negligible free end slip in the reinforcement for all repaired specimens. This indicates that none of the specimens provided any evidence of potential premature failure due to bond degradation. Therefore, for this series, the wrapping improved the performance of the corrosion damaged beams up to of 10% theoretical mass loss.

Series L2 – 350 mm Bond Length

This series incorporated ten beams to examine the post-repair behavior. Four beams were repaired at 2% mass loss with three being further corroded to 5, 10 and 15% mass loss. Three beams were repaired at 5% mass loss with two specimens further corroded to 10 and 15% mass loss. Two beams were repaired at 10% mass loss with one further corroded to 15% mass loss. The remaining specimen acted as the control specimen and was tested without repair after being corroded to 2% mass loss.

Beams that were repaired at 2% theoretical mass loss were able to achieve a ductile flexural failure (Figure 8a). No additional deterioration was observed as the post-repair corrosion increased. Slip gauges indicate that little to no slip was measured up to failure of the specimen. At 10% mass loss, a slip of 0.04 mm was measured at the free end of only one bar indicating a potential weakening of the concrete around one of the tension bars. Slip initiation occurred around 3.66 MPa bond stress, which is 73% of the maximum predicted bond capacity. The slip was localized to one area of the tensile reinforcement and did not continue to propagate. The measurement is small and could be linked to experimental error. However, the possibility remains that under sustained load, the slip could continue eventually leading to failure. The high level of corrosion may have increased the confinement pressure beyond the compression strength of the concrete causing local crushing around the ribs leading to weakening of the bond strength. In general, the results indicate that beams in the L2 series confined at 2% mass loss were able to re-establish a ductile flexural failure with minimal indication of bond degradation. This result exhibits a significant improvement over the control specimen, which failed by bond splitting.

In terms of overall behavior, beams repaired at 5% mass loss show a similar trend to those repaired at 2% mass loss. Figure 8b shows that the load-deflection plots are relatively consistent with only slight differences most likely due to variations in material properties. However, when examining the local free end slip behaviour of the reinforcing bar, there is evidence that slip occurred in specimen W5C15, which was subjected to the highest level of post-repair corrosion. Specimen W5C10, corroded to 10% mass loss, behaved similarly to specimen W5C5 at 5% mass loss with little to no slip in the reinforcing bar at the time of failure. The additional post-repair corrosion damage induced in specimen W5C15 developed enough pressure to further deteriorate the concrete surrounding the reinforcement, thus weakening the mechanical interlock component of the bond. A free end slip of 0.07 mm was recorded at the time of failure. Slip initiation occurred at a bond stress of 3.60 MPa, which demonstrates a 28% reduction in bond stress at slip initiation relative to the theoretical maximum bond stress. This reduced level of bond stress at slip initiation is an indication of deterioration in the concrete, which could lead to a premature bond failure under sustained load conditions.

Specimens W10C10 and W10C15 exhibited load-deflection and load-slip behavior similar to the other flexural failures (Figure 8b). No significant degradation in bond strength was observed since no slip was recorded at the free ends of the tension steel. This indicates that specimen W10C10 out performed the other repaired beams corroded to 15 % mass loss (W2C15 and W5C15) due possibly to the initial bond enhancement created by small increases in the degree of corrosion. The concrete around the reinforcing bar was significantly cracked prior to repair, indicating that even though corrosion around the bar had occurred, some of the mechanical interlock could still be maintained. The small amount of corrosion after CFRP repair (2.2% experimental mass loss) has the potential of increasing the confining pressure in the bond zone since the expansion forces are restrained by the CFRP wrap. It is theorized that further increasing the post-repair corrosion levels would increase the internal pressure, locally exceeding the concrete strength around the reinforcing bar, and thus decreasing the bond strength.

Series L3 – 250 mm Bond Length

Similar to series L1, a total of six beams were examined in this series. Five beams were repaired at 2% and 5% mass loss with three beams being further corroded to higher levels up to 10% mass loss to examine the post-repair behavior. The sixth beam was tested without repair after attaining 2% mass loss.

Based on the load-deflection plot in Figure 9, the failure mode for all corroded and repaired specimens had a ductile flexural failure, where as the control specimen was subjected to a brittle bond splitting failure. The overall load-deflection behavior of specimens W2C5, W2C10 and W5C10 was similar to the beams that were not subjected to any further corrosion after repair (W2C2 and W5C5). This shows that even after additional corrosion, the overall performance of the CFRP repaired specimens was superior to the control specimen, U2C2, which failed by bond splitting.

Examining the bond-slip curves, several observations can be made. The beams that were repaired at 2% mass loss had small amounts of free end slip at failure (Figure 10). The slip values of the steel reinforcing bar at failure were 0.06, 0.02, and 0.03 mm for specimens W2C2, W2C5 and W2C10, respectively. CFRP confinement to the concrete at the low corrosion levels preserved the integrity in the bond zone. Further corrosion after repair even up to high degrees of mass loss, up to 10%, did not damage the steel ribs and the surrounding concrete to a condition that would result in a bond failure. The internal pressure developed from the CFRP confinement allowed the reinforcing steel to continue to interact with the concrete. In turn, the added pressure reduced the overall slip by increasing the stiffness in the bond zone. Some degradation was observed at 10% mass loss with a 2% reduction in the slip initiation bond stress over the other two specimens. However, this change is very small and could be a result of other factors including experimental error.

Different results were observed for the beams repaired at 5% mass loss. When tested directly after wrapping, W5C5 exhibited a free end slip of 0.08 mm at failure. When the corrosion was continued after repair up to a mass loss of 10%, the slip value for specimen W5C10 significantly increased to 1.03 mm (Figure 11). The free-end slip at yield was 0.07 mm whereas specimen W5C5 had slip initiation after yielding of reinforcement bar. The bond stress at slip initiation for W5C10 was 7% lower than the ultimate bond stress, exhibiting noticeable degradation compared to other repaired specimens. The large increase in slip at ultimate denotes that a bond pullout failure was in progress. The residual bond force caused by friction due to confinement was sufficient to lead to a flexural failure under static load conditions. However, under sustained load or fatigue testing, a premature bond pullout failure would be expected.

Effect of Bond Length

In general, the overall effect of the confinement on the bond strength was similar for all bond lengths. The repaired specimens out performed the control specimens as their failure mode was effectively altered from a bond splitting failure to a more ductile flexural failure. Similarly, regardless of bond length, all members were able to achieve the full bond stress.

It was found that when the initial bond length was adequate, confinement at low degrees of corrosion was able to reinstate the full flexural capacity of the members. Where specimens were designed with inadequate bond lengths, FRP confinement was limited to preventing bond splitting failures.

The importance of sufficient bond length is evident when examining the free end slip and the slip initiation bond stress. In the cases of ample bond length, reinforcing bars maintained bond behavior in the concrete up to high levels of corrosion. The L2 series showed slight potential for bond pullout failures at corrosion levels of 15% mass loss. In the case of the bond-deficient length (series L3), significant bond slip was evident in most post-repair conditions. This was more evident when confinement was applied after significant pre-repair corrosion had occurred.

In summary, post repair analysis shows that the application of the CFRP wrap must be implemented prior to excessive damage of the concrete.

CONCLUSION

Post-repair performance testing demonstrated the confining ability of CFRP wrap. CFRP repair may lead to higher bond strengths but under conditions of increased corrosion after repair, bond failure is still possible in the form of a pullout failure as opposed to the bond splitting failure witnessed in the unconfined corroded specimens. Corrosion plays a devastating role in reducing the structural capacity of unconfined flexural members. CFRP confinement of the bond zone in bond-beam specimens serves to maintain steel-concrete bond interaction. CFRP resisted the expansion forces caused by corrosion, thus reducing crack growth and maintaining the interlock between the reinforcing steel and concrete. As post-repair corrosion progressed, cracks were unable to expand due to the presence of CFRP sheets. In turn, CFRP developed stresses, which increased the internal confining pressure around the reinforcing bar that counteracted the expansion stresses due to corrosion.

The effect of CFRP on the confinement of corrosion-damaged members varies depending on whether the member has adequate bond or is bond deficient. For those members with inadequate bond length, the added CFRP confinement improved the performance of bond-deficient corroded members allowing them to outperform the unconfined specimen. It is important to understand the nature of failure of the CFRP confined specimens. Since no cracks were visible with the CFRP wrap in place, there are no indications of failures. Even under conditions of high ultimate bond stresses, the presence of low slip initiation bond stresses indicates that failure could potentially occur prematurely by bond pullout in the case of sustained loading or creep.

Confinement was found to be more effective when applied prior to excessive corrosion of the specimens. Typically, small amounts of post-repair corrosion were found to have no effect or in some instances helped increase bond strength as a result of increased confining pressures. However, as the post-repair corrosion levels increased, the bond strength deteriorated.

The overall structural performance of beams wrapped with CFRP was enhanced. However, caution and engineering judgement must be used in the application of this repair method since abrupt failure of the member due to bond pullout failure could occur without warning if repair is performed at high corrosion levels or if members were initially designed with inadequate bond. The confining wrap may increase the bond strength, but as with all repairs, this should not be used as a band-aid solution, and the cause of deterioration must be addressed to prevent further corrosion and deterioration.

ACKNOWLEDGEMENTS

This research is conducted with financial assistance from ISIS Canada Network of Centres of Excellence and Natural Sciences and Engineering Research Council of Canada.

REFERENCES

1. Uomoto, T., Tsuji, K., and Kakizawa, T. (1984) "Deterioration Mechanism of Concrete Structures Caused by Corrosion of Reinforcing Bars," Transaction of the Japan Concrete Institute, Vol. 6. pp. 163-170.

2. Capozucca, R. (1995) "Damage to Reinforced Concrete due to Reinforcement Corrosion," Construction and Building Materials, Vol. 9, No. 5, pp. 295-303.

3. Mangat, P.S. and Elgarf, M.S. (1999) "Flexural Strength of Concrete Beams with Corroding Reinforcement," ACI Structural Journal, Vol. 96, No. 1, pp. 149-158.

4. Tepfers, R. (1979) "Cracking of Concrete Cover along Anchored Deformed Reinforcing Bars," Magazine of Concrete Research, Vol. 31, No. 106, pp. 3-12.

5. Lutz, L.A. and Gergely, P. (1967) "Mechanics of Bond and Slip of Deformed Bars in Concrete," ACI Journal, Vol. 64, No. 11, pp. 711-721.

6. Cairns, J. and Abdullah, R.B. (1996) "Bond Strength of Black and Epoxy-Coated Reinforcement – A Theoretical Approach," ACI Materials Journal, Vol. 93, No. 4, pp. 362-369.

7. Ferguson, P.M. (1966) "Bond Stress – The State of the Art," ACI Committee 408, Journal of the American Concrete Institute, November, pp. 1161-1180.

8. Pillai, S.U., Kirk, D.W., and Erki, M.A. (1999) "Reinforced Concrete Design: Third Edition," McGraw-Hill Ryerson Limited, Ontario, Canada.

9. Almusallam, A.A., Al-Gahtani, A.S., Aziz, A.R., and Rasheeduzzafar (1996) "Effect of Reinforcement Corrosion on Bond Strength," Construction and Building Materials, Vol. 10, No. 2, pp. 123-129.

10. Al-Sulaimani, G.J., Kaleemullah, M., Basunbul, I.A., and Rasheeduzzafar (1990) "Influence of Corrosion and Cracking on Bond Behavior and Strength of Reinforced Concrete Members," ACI Structural Journal, Vol. 87, No. 2, pp. 220-231.

11. Cabrera, J.G. (1996) "Deterioration of Concrete due to Reinforcement Steel Corrosion," Cement & Concrete Composites, Vol. 18, pp. 47-59.

12. Soroushian, P., Choi, K., Park, G., Aslani, F., (1991) "Bond of Deformed Bars to Concrete: Effects of Confinement and Strength of Concrete," ACI Materials Journal, Vol. 88, No. 3, pp. 227-232.

13. Giuriani, E. and Plizzari, G.A. (1998) "Confinement Role in Anchorage Capacity," American Concrete Institute SP-180, MI, USA, pp. 171-193.

14. Kono, S., Inazumi, M., And Kaku, T. (1998) "Evaluation of Confining Effects of CFRP Sheets on Reinforced Concrete Members," Proceedings: Second International Conference on Composites in Infrastructure, Tucson, AZ, pp. 343-355.

15. Hamad, B.S., Soudki, K.A., (2001) "GFRP Wraps for Confinement of Bond Critical Regions in Beams," Proceedings: Composites in Construction International Conference, 10-12 October, Porto, Portugal.

16. Hamad, B., Rteil, A., Selwan, B., Soudki, K.A. (2004) "Behavior of Bond Critical Regions Wrapped with FRP Sheets in Normal and High Strength Concrete," *ASCE Journal of Composites for Construction,* Vol. 8, No. 3, 248-257.

17. Hamad, B., Rteil, A., Soudki, K.A., 2004. "Tension Lap Splices in High-Strength Concrete Beams Strengthened with GFRP Wraps," *ASCE Journal of Composites for Construction,* Vol. 8, No. 1, pp. 14-21.

18. Soudki, K.A. and Sherwood, E.G., (2003) "Bond Behaviour of Corroded Steel Reinforcement in Concrete Wrapped with Carbon Fibre Reinforced Polymer Sheets," *ASCE Journal of Materials in Civil Engineering*, Vol. 15, No. 4, pp. 358-370.

Table 1 – Experimental Test Matrix

Strenthening Scheme	Theoretical Degree of Corrosion (% mass loss)			
	2	5	10	15
Unconfined	U2C2			
Wrapped @ 2%	W2C2	W2C5	W2C10	W2C15*
Wrapped @ 5%		W5C5	W5C10	W5C15*
Wrapped @ 10%			W10C10*	W10C15*

* Specimens apply to series L2 only

Table 2 – Experimental Test Results

Specimen	Experimental Mass Loss (%)	Failure				Slip Initiation
		Mode	Load (kN)	Bond Stress (MPa)	Slip (mm)	Bond Stress (MPa)
Series L1 - 450 mm Bond Length						
U2C2	4.0	Bond Splitting	87.5	3.91	0.24	3.66
W2C2	4.5	Flexure	96.4	3.91	0.01	N/A
W2C5	6.0	Flexure	96.1	3.89	0.00	N/A
W2C10	7.2	Flexure	98.6	3.76	0.00	N/A
W5C5	6.7	Flexure	98.5	3.91	0.00	N/A
W5C10	10.2	Flexure	98.1	3.91	0.00	N/A
Series L2 - 350 mm Bond Length						
U2C2	3.2	Bond Splitting	87.9	4.99	0.13	3.96
W2C2	4.4	Flexure	94.1	5.03	0.01	N/A
W2C5	5.7	Flexure	106.9	5.03	0.00	N/A
W2C10	8.5	Flexure	98.7	4.89	0.03	3.66
W2C15	10.3	Flexure	97.1	5.03	0.01	N/A
W5C5	7.6	Flexure	98.7	5.03	0.03	4.19
W5C10	8.5	Flexure	96.5	4.97	0.00	N/A
W5C15	11.5	Flexure	96.2	5.03	0.07	3.6
W10C10	10.4	Flexure	96.6	5.03	0.01	N/A
W10C15	11.6	Flexure	97.2	5.03	0.01	N/A
Series L3 - 250 mm Bond Length						
U2C2	3.6	Bond Splitting	65	5.53	0.16	4.48
W2C2	4.6	Flexure	94.5	7.04	0.06	7.04
W2C5	6.5	Flexure	94.8	7.04	0.02	7.04
W2C10	8.2	Flexure	97.8	7.04	0.03	6.9
W5C5	7.4	Flexure	98.4	7.04	0.08	7.04
W5C10	11.5	Flexure	103.5	7.04	1.03	6.56

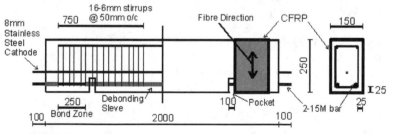

All dimensions are in mm

Figure 1 – Test Specimen Reinforcement and CFRP Layout

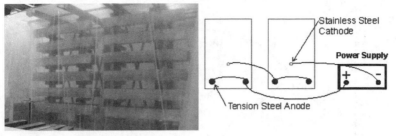

Figure 2 – Corrosion Tent and Electrical Setup

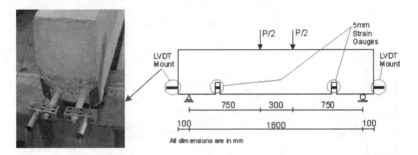

Figure 3 – Test Setup and Instrumentation

Figure 4 – Bond Splitting Failure Crack Pattern, Series L1

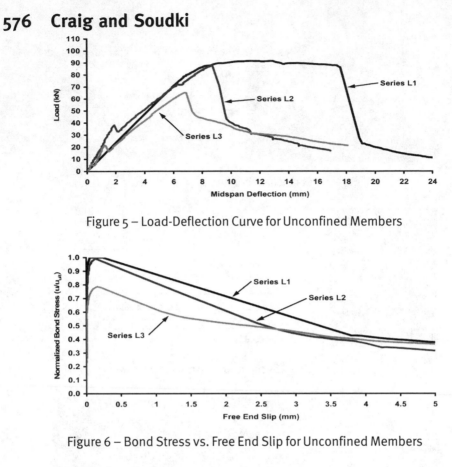

Figure 5 – Load-Deflection Curve for Unconfined Members

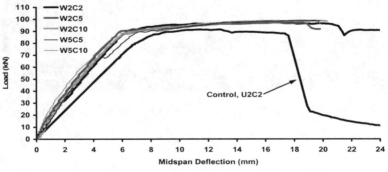

Figure 6 – Bond Stress vs. Free End Slip for Unconfined Members

Figure 7 – Load-Deflection Behavior, Series L1

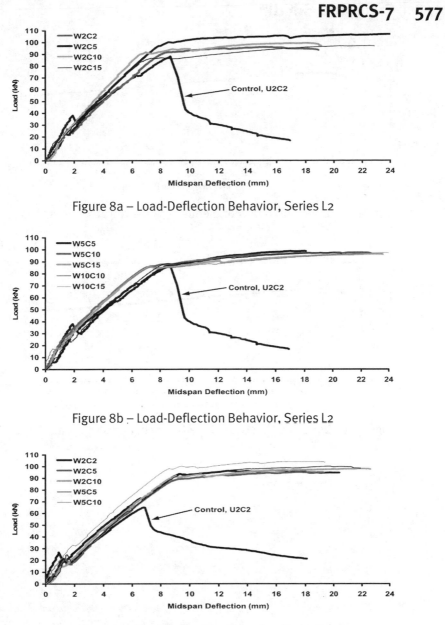

Figure 8a – Load-Deflection Behavior, Series L2

Figure 8b – Load-Deflection Behavior, Series L2

Figure 9 – Load-Deflection Behavior, Series L3

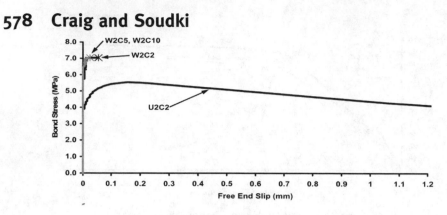

Figure 10 – Bond Slip Behavior, Series L3, Repaired at 2% Mass Loss

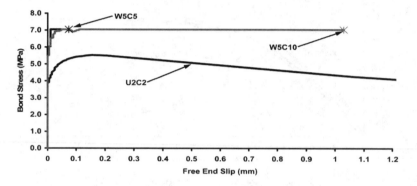

Figure 11 – Bond Slip Behavior, Series L3, Repaired at 5% Mass Loss

Bond Strengthening of Steel Bars Using External FRP Confinement: Implications on the Static and Cyclic Response of R/C Members

by M.H. Harajli

<u>Synopsis:</u> The effect of external confinement using fiber reinforced polymer (FRP) laminates on the bond strength and bond stress-slip response of steel bars under static and cyclic loading is experimentally investigated. Beam specimens with spliced bars at midspan were tested. The test variables included the diameter of the steel bars, the ratio of concrete cover to bar diameter and the area of FRP confinement. Without confinement, the spliced reinforcement suffered significant deterioration of bond strength, accompanied by considerable stiffness degradation of the beam specimens in the initial few load cycles following bond splitting. Confining the concrete with a relatively small area of carbon FRP sheets increased the bond strength of the steel bars, reduced the bond deterioration of the specimens under cyclic loading, and resulted in superior seismic performance. The characteristics of the bond stress-slip response of steel bars embedded in FRP confined concrete in comparison with unconfined concrete are discussed, and a general relationship for the local bond stress-slip response of reinforcing bars embedded in FRP confined concrete and subjected to monotonically increasing load in tension is presented. Also, a rational design expression is developed for estimating the area of external FRP confinement required for bond strengthening.

<u>Keywords:</u> bond stress; cyclic loading; fiber-reinforced polymer; seismic strengthening; splitting strength

580 Harajli

ACI member **Mohamed H. Harajli** is a professor in the Department of Civil and Environmental Engineering, American University of Beirut, Beirut, Lebanon. He is a member of ACI Committee 408, Bond and Development of Reinforcement. His research interests include design and behavior of concrete structures and structural strengthening.

INTRODUCTION

Ductility, or the ability to absorb and dissipate energy by post-elastic deformation is one of the most important characteristics of reinforced concrete (R/C) structures in regions of seismic hazard. Without ductile behavior, in the event of strong earthquake ground shaking, the structural elements may experience significant stiffness degradation and loss in load resistance which could lead to structural failure. One of the most effective measures for improving structural ductility is confinement of the concrete at the critical regions of the structure where plastic hinging is most likely to develop. For newly designed and constructed structures, concrete confinement can best be achieved by using closely spaced internal steel hoops. For existing structures in which the internal steel confinement is inadequate, the structure can be confined externally using traditional techniques such as concrete or steel jackets, or by using advanced composites such as fiber reinforced polymers (FRP) laminates. Compared to the traditional methods, in addition to the advantage of being light weight and non-corrosive material, the use of FRP for concrete confinement is much easier to apply, does not require expensive equipment or labor, and does not change the appearance of the structure. Another important advantage of FRP wraps compared to concrete or steel jackets is that FRP wraps do not increase the flexural stiffness of the structural members, which could be critical in seismic rehabilitation. Applications of FRP reinforcement, and guidelines for design and construction of externally bonded FRP systems for upgrading concrete structural elements under different types of load application including flexure, axial compression or tension, and one-way shear in beams are reported by ACI Committee 440[1] and ISIS Canada[2].

One critical zone in R/C structures where confinement becomes necessary for improving the structural performance under seismic loading is the base of columns or bridge piers in gravity load-designed buildings or bridges, respectively. Because these columns or piers are designed to withstand gravity load only, their longitudinal reinforcement is normally spliced at the base using compression lap splices. Since compression lap splices are relatively short and in most cases closely spaced along the plan dimension, the spliced reinforcement may develop splitting bond failure before acquiring their full tension capacity when these vertical elements are subjected to bending moment, particularly when insufficient internal confinement, in the form of hoops or ties, is provided. In other words, whenever these building structures or bridges are subjected to lateral earthquake ground shaking, the columns may not be able to develop adequate flexural capacity or ductility demanded by the earthquake, which could lead to substantial damage or failure of the structures. As pointed out by Saadatmanesh et al.[3], some of the most common causes of bridge failures following major earthquakes are substandard detailing of bridge piers associated with inadequate starter bar lap lengths and insufficient lateral confinement.

A recent experimental investigation of the effect of confinement on the seismic behavior of gravity load-designed concrete columns in which the longitudinal reinforcement is spliced at the column base[4] reported significant improvement in the response when the concrete is confined externally with FRP sheets. The same study also emphasized that the bond behavior and the splitting bond resistance of the spliced column bars are the main aspects that control the strength and ductility of the columns under cyclic loading. According to ACI Committee 440[1], and Priestly et al.[5], external confinement by FRP inhibits potential splitting of the reinforcement tension lap splice, leading to reduction in the required splice length of steel bars, and effectively clamps lap splices of compression reinforcement, thereby improving ductility. Nevertheless, experimental and analytical studies or guidelines for the use of externally bonded FRP systems for upgrading concrete structural elements when the potential mode of failure is by bond splitting of the steel bars are very limited.

In this paper, test results of the effect of FRP confinement on the bond strength and slip response between steel bars and concrete under monotonic and cyclic loading are presented and discussed. A general model of the bond stress-slip response of steel bars subjected to monotonically increasing load in tension is described, and a comprehensive equation for bond strengthening using external FRP confinement is generated.

RESEARCH SIGNIFICANCE

This research helps in better understanding of the effect of external FRP confinement on the bond strength and bond stress-slip response of steel bars under static and cyclic loading. Using the aspect of bond between steel bars and concrete, a general design equation is derived for estimating the area of external FRP confinement required for bond strengthening of steel reinforcement at the locations of the structure where bond failure of the steel bars could potentially lead to inferior seismic performance.

EXPERIMENTAL PROGRAM

Six beam specimens with spliced reinforcement at midspan were tested under cyclic loading. Typical plan dimensions, cross section and reinforcement details at the splice location of the specimens are shown in Fig. 1. The splice length was selected at $5d_b$, where d_b is the bar diameter, which is short enough to simulate "local" bond conditions, yet large enough to allow a reasonable number of bar ribs to participate in resisting the bar force through bond.. Two bar sizes were selected for the splice, namely, 20 mm, and 25 mm. Two top bars of 16 mm and 20 mm diameter were provided along with the bottom 20 mm, and 25 mm spliced bars, respectively, to resist the negative bending moment induced by cyclic loading. The transverse reinforcement in the shear span for all specimens, provided outside the splice region, consisted of deformed ϕ6mm closed ties spaced at 100 mm. The bottom cover, side cover and ½ clear spacing between the splices were kept identical for all specimens. The width of the specimens of 200 mm, the number of splices (two), and the bar sizes were all selected to produce two values of concrete cover (or ½ clear distance between the splices) to bar diameter c/d_b of 1.0, and 1.5, respectively. Without external confinement, all specimens were designed to experience a splitting mode of bond failure.

A summary of specimen designation and test parameters is given in Table 1. Provided also in Table 1 is a summary of specimens designations for almost identical specimens tested earlier by the author et al.[6] under static loading. The specimens were grouped into two groups (IIW and IIIW) depending on the diameter of the spliced bars.

The FRP reinforcement consisted of carbon fiber reinforced polymer flexible sheets (SikaWrap Hex-230C). The properties of the fiber fabric recommended by the manufacturer for design purposes are as follows: t_f = 0.13 mm; modulus of elasticity E_f = 230,000 MPa; tensile strength = 3500 MPa, and strain at break of the fibers = 1.5%. The corners of the beam section were first rounded to about 10 mm radius for FRP application, and then the epoxy saturated FRP laminates were wrapped around the perimeter of the section along the full length of the splice (see Fig. 1) with fiber orientation along the circumference and with 50 mm overlap at the top face of the beam. Specimens with the same reinforcement and concrete cover parameters were confined with one or two layers of FRP reinforcement.

The spliced bars consisted of Grade 60 steel. The clear distance between the bar ribs was measured at 9.0 mm and 13 mm for the T20 mm and T25 mm bars, respectively. The concrete mixes were prepared using ordinary Portland cement, washed sand and graded crushed limestone aggregate having 10 mm maximum size. Every two specimens were cast using one concrete batch. The concrete compressive strength f'_c of the various specimens, determined using three standard 150 x 300 mm cylinders taken for each batch, are given in Table 1.

The specimens were loaded in 4-point bending (Fig. 1) using the cyclic load history shown in Fig. 2. The load history was modified slightly for some specimens during the course of the test by reducing the maximum imposed deflection in a given cycle, depending on the extent of bond degradation experienced in each specimen. The distance between the externally applied concentrated loads for all specimens was taken equal to the splice length L_s, plus the height of member h. For the purpose of strain and slip measurements, the bars were exposed at the splice ends using two 70 mm wide by 70 mm deep notches. The steel strains were measured using two electric resistance strain gages attached to the bars at the notch location. Slip measurements were taken for each spliced bar using linear variable differential transformers (LVDTs). The body of the LVDT was attached to the spliced bars at the notch location, while the LVDT core was attached to the specimen at midspan (see Fig. 3). Deflections of the specimens were measured at midspan using one LVDT. All data were recorded automatically using computerized data acquisition system.

DISCUSSION OF TEST RESULTS

General behavior and mode of failure

Tensile flexural cracks tended to initiate in all specimens within the constant moment region at the notch locations close to the ends of the splices. With increasing load or number of cycles beyond flexural cracking, all specimens experienced splitting mode of bond failure (as opposed to pullout bond failure in well confined concrete). Splitting failure in the control unconfined specimens was clearly evident from the formation of side and bottom splitting cracks. A typical example of the splitting modes of bond failure is shown for the unconfined control specimens in Fig. 3.

Confining the concrete by FRP sheets at the spliced location improved the bond resistance of the spliced reinforcement in that region and resulted in a more ductile and stable behavior of the specimens. None of the beam specimens encountered delamination or bond failure of the FRP sheets themselves throughout the response. Examination of the concrete within the FRP wraps after stripping the sheets at the end of the test showed clear evidence of splitting failure.

Bond stress-slip response

Because the splice length is relatively short, all specimens failed in splitting mode long before yielding of the steel bars. The average local bond stress u within the splice length is converted into bond stress by considering equilibrium between the bond force and bar force ($A_b f_s = u \pi d_b L_s$):

$$u = \frac{A_b f_s}{\pi d_b L_s} = \frac{f_s d_b}{4 L_s} \tag{1}$$

where A_b is the area (taken equal to $\pi d_b^2 / 4$) and L_s is the splice length ($5d_b$) of the bar. The bar stress f_s was obtained directly from the experiment by multiplying the measured steel strain by the modulus of elasticity of steel.

Results of the cyclic response of the local bond stress, normalized to $\sqrt{f'_c \ (MPa)}$, versus slip for all specimens are shown in Fig. 4. The results of static local bond stress versus slip response obtained using a similar size specimen and identical test parameters[6] but with slightly different specimen configuration (see Table 1) are plotted against the cyclic response for comparison. Note that because the spliced bars were constrained against compression slip by the concrete above the notch, the slip of the bars during the cyclic response under negative bending was negligible in comparison with the tension slip mobilized under positive bending. Consequently, most of the discussion that follows applies to the response of the spliced bars in tension.

Figure 4 clearly shows that for all the specimens tested, the peak bond strengths obtained earlier under statically increasing load[6] are consistent with the peak bond strengths measured from the envelope (or backbone) curve of the current cyclic test results. Also, despite some discrepancy, the agreement between the static bond stress-slip response and the envelope of the cyclic response is quite reasonable.

Under statically increasing load in tension, the local bond stress-slip response of steel bars confined externally with FRP undergoes four distinct stages of behavior[6]: In the early stage of the response, the bond stress-slip behavior is quite identical to that of steel bars embedded in well confined concrete (see Eq. 2) with pullout mode of bond failure. With further increase in bond stress, internal tensile cracks develop under the influence of the radial pressure exerted by the rib bearing forces on the surrounding concrete. The development of these internal cracks, shortly before splitting, leads to a gradual softening of the bond stress, accompanied by an increase in the slip rate as compared to the response corresponding to pull-out mode of bond failure. During this stage, as the bond stress and slip increase, the internal cracks propagate gradually toward the concrete surface causing splitting failure.

Splitting of concrete causes a drop in the bond stress until a state of equilibrium is achieved between the radial component of the rib bearing force and the post-splitting tensile strength of the concrete matrix surrounding the splice. For unconfined concrete, the post-splitting tensile strength is attributed to the tension softening phenomenon associated with the interlocking of the rough concrete surfaces on either side of the crack. The presence of external FRP confinement reduces the width of the splitting cracks and increases the tensile resistance of the concrete matrix after splitting, leading to a larger post-splitting bond strength in comparison with plain unconfined concrete. With further increase in slip beyond splitting, the bar ribs tend to ride up the concrete keys between the ribs leading to a progressive widening of the splitting cracks and, consequently, a gradual degradation in bond resistance until the bond stress is transformed completely into pure friction between the top surface of the bar ribs and the surrounding concrete. The bond resistance reaches its lowest value when the ribs have traveled a distance equal approximately to the rib spacing.

Comparing the results of the various specimens in Fig. 4, it is clearly evident from the static behavior and the envelope curve of the cyclic bond stress-slip response that confining the concrete with FRP improves substantially the peak local bond strength in the initial stage of the bond stress-slip response. Using one layer of FRP increased the normalized bond strength $u / f'_c{}^{1/2}$ as compared to the unconfined control specimens by about 74% for the $\phi20$ mm splice (from 0.77 for specimen B20 to 1.34 for specimen B20FP1) and by 35% for the $\phi25$mm splice (from 0.7 for specimen B25 to 0.95 for specimen B25FP1). Increasing the area of FRP from one layer to two layers did not produce a noticeable increase in bond strength, but resulted in significant improvement in the ductility of bond–slip behavior in the post splitting stage. Note that because the bond strength increases with increase in the ratio of concrete cover to bar diameter c / d_b (Orangun et al.[7, 8], Zu and Darwin[9]), the bond strengths of the FRP confined or unconfined specimens with $\phi20$ mm splices (specimens in series IIW) having $c / d_b = 1.5$ were larger than the bond strengths of the specimens with $\phi25$ mm splices (specimens in series IIIW) having $c / d_b = 1.0$.

The results of the cyclic response in Fig. 4 also show that in the absence of confinement, the control specimens B20 and B25 experienced a dramatic and sudden loss in bond strength and considerable degradation in bond resistance in the initial few cycles following the formation of splitting cracks. For instance, beyond a relatively small slip of about 1.5 mm, the bond strength of the spliced reinforcement in the unconfined specimens dropped to about 33% and 10% of the peak bond strength for specimens B20 and B25, respectively. On the other hand, as a result of the limited crack growth provided by the FRP reinforcement, in addition to improved bond capacity, the degradation of local bond strength with increase in the number of loading cycles after splitting was substantially less in the FRP confined specimens as compared to the unconfined ones, leading to considerable improvement of seismic performance. Comparing the cyclic response of the specimens with different areas of FRP, it can be observed that while increasing the area of FRP from one layer to two layers did not produce increases in bond strength, it did lead to hardening of the bond stress-slip response immediately after splitting.   Consequently the specimens that were wrapped with two layers of FRP experienced less bond strength degradation with number of loading cycles in comparison with the specimens confined with only one layer. Note that since the splitting of large concrete area around the reinforcing bar demands more resistance from the FRP sheets to compensate for the loss of concrete tensile strength immediately after splitting, specimens having $c/d_b = 1.5$ (series IIIW) encountered less degradation in bond strength after splitting as compared to the specimens having $c/d_b = 1.0$ (series IIW).

Load-deflection response

Typical cyclic load-deflection responses of the FRP confined specimens in comparison with the unconfined control specimens are shown in Fig. 5. Because the load resistance and deflection of the specimens in tension are more or less related to the strength and slip of the spliced bars at the splice location, the load-deflection responses of the specimens under cyclic loading, particularly in the tension cycle, were intrinsically similar to the local bond stress-slip responses of the spliced bars. Therefore in the same manner that the FRP confinement enhanced the peak bond strength and reduced the bond degradation of the spliced reinforcement under cyclic loading, it enhanced also the peak load resistance and improved the ductility of the load-deflection response of the specimens.

LOCAL BOND STRESS - SLIP RELATIONSHIP

It is possible to predict the bond characteristics of reinforcing bars based on knowledge of the local bond strength and slip behavior. Several analytical studies, undertaken by incorporating local bond laws into numerical solution schemes of the bond problem[10-14], were able to very accurately predict experimental data of development/splice strength covering both pull-out and splitting modes of bond failure, and allowed for better understanding of the influence of various parameters on bond strength.

Based in part on the static bond stress-slip results presented in this paper, Harajli et al.[6] proposed a four-stage relationship to generate the local bond stress-slip response of steel bars embedded in concrete confined with flexible FRP sheets. Described schematically in Fig. 6, the relationship is composed of an initial ascending portion from zero bond stress up to αu_{max}; a stage of a linearly increasing bond strength from αu_{max} up to the splitting bond strength u_{max}; a drop in bond resistance immediately after splitting down to a post-splitting bond strength u_{ps} (for FRP confined concrete) or βu_{max} for plain unconfined concrete; and finally, a stage of progressively diminishing bond strength. The corresponding relationship and the characteristic parameters of the response, including the peak bond strength u_{max} described below, were generated using a wide range of test variables consisting of four different bar diameters of 16 mm, 20 mm, 25 mm, and 32 mm, respectively, and a variety of ratios of concrete cover to bar diameter c/d_b of 0.56, 0.88, 1.0, 1.34, 1.5, and about 2.0.

The relation between the bond stress and slip in the initial stage (from $u = 0$ up to $u = \alpha u_{max}$) coincides with the monotonic curve of the local bond-slip response of steel bars embedded in well confined concrete in which the mode of bond failure is by pull-out. The corresponding relationship is described using the following experimentally based equation derived earlier[15] using a test specimen similar to the one used by Eligehausen et al.[10]:

$$u = u_1 \, (s/s_1)^{0.3} \tag{2}$$

where u is the bond stress in MPa and s is the slip in mm; u_1 is the maximum bond stress corresponding to pull-out mode of bond failure, given as $u_{1\,(MPa)} = 2.57\sqrt{f'_{c(MPa)}}$; and s_1 is equal to $0.15c_0$, where c_0 is the clear distance between the ribs of the reinforcing bar, or equal to 1.5 mm if no information is available about the bar rib pattern.

The local splitting bond strength u_{max} (MPa) for steel bars embedded in plain unconfined concrete or concrete confined with carbon FRP sheets of the type used in the current investigation is calculated using the following (in SI units):

$$u_{max} = 0.78\sqrt{f'_c} \left(\frac{c + 32 A_{frp} \, / \, sn_s}{d_b} \right)^{2/3} \tag{3}$$

in which A_{frp} is the area of FRP sheets within a spacing s, crossing the potential plain of splitting. Whenever the FRP is wrapped along the full splice length, $A_{frp} \, / s = 2n_f t_f$, where n_f is the number of FRP layers and t_f is the design thickness of the fabric. n_s is the number of splices being developed along the plane of splitting. For FRP material having a modulus of elasticity E_f different from the one used in developing Eq. (3) (i.e.

$E_f = 230,000$ MPa), the coefficient "32.0" in Eq. (3) should be modified by multiplying it by the ratio $(E_f / 230,000)$.

Referring to Fig. 6: $\alpha = 0.7$ and s_1 is as defined in Eq. (2); $s_2 = 0.35c_o$ and $s_3 = c_o$, where c_o is the clear distance between the ribs of the reinforcing bar; $u_f = 0.35u_1$; and $u_{fr} = 0.30u_{ps}$. In the absence of information about the bar rib geometry, s_2, and s_3 can be taken equal to 3.5 mm, and 10 mm, respectively. For unconfined concrete, $\beta = 0.65$. The slip s_{max} at which u_{max} is mobilized, is calculated as the sum of the slip along the monotonic pull-out curve corresponding to u_{max} (Eq. 2) and the additional slip resulting from softening of the bond response shortly before splitting:

$$s_{max} = s_1 e^{(1/0.3)Ln(\frac{u_{max}}{u_1})} + s_o Ln(\frac{u_1}{u_{max}}) \tag{4}$$

where $s_o = 0.15$ mm for unconfined concrete and 0.20 mm for concrete confined with FRP sheets. For unconfined concrete, the descent in the bond stress-slip $(u - s)$ relationship beyond splitting is assumed to follow the expression:

$$u = \beta u_{max} (s / s_{max})^{-0.5} \tag{5}$$

For concrete confined with FRP sheets, the local post-splitting bond resistance is expressed as:

$$u_{ps} = u_{max}\left(0.5 + 22\frac{A_{frp}}{sn_s c}\right) \leq u_{max} \tag{6}$$

PROPOSED EXPRESSION FOR FRP BOND STRENGTHENING

Based on a combination of experimental data and the results of an analytical evaluation of the effect of FRP confinement on the bond strength of developed/spliced steel bars in tension, conducted using the local bond-slip law presented above, Harajli et al.[6] proposed the following expression for evaluating the increase in average bond strength U_{frp} (MPa) relative to plain unconfined concrete that can be attained when the concrete is confined externally with FRP sheets along the full splice/development length :

$$\frac{U_{frp}}{f'_c{}^{1/2}} = \frac{2n_f t_f E_f}{8000 n_s d_b} \leq 0.40 \tag{7}$$

The average bond strength U_c (MPa) at bond failure of reinforcing bars embedded in unconfined concrete and subjected to tension stress can be calculated using the equation proposed by Orangun et al.[7,8]:

$$\frac{U_c}{f'_c{}^{1/2}} = 0.1 + 0.25(c/d_b) + 4.15(d_b/L_d) \tag{8}$$

in which c is the smaller of side cover, bottom cover, or ½ the clear distance between the bars; and L_d is the development (or splice) length. The total average bond stress U (MPa) at bond failure of spliced or developed bars in tension when the concrete is confined externally with FRP sheets along the full splice or development length can be obtained by adding Eqs. (7) and (8):

$$\frac{U}{f'_c{}^{1/2}} = \frac{U_c + U_{frp}}{f'_c{}^{1/2}} = 0.1 + 0.25(c/d_b) + 4.15(d_b/L_d) + \frac{2n_f t_f E_f}{8000 n_s d_b} \tag{9}$$

Using equilibrium between the bar force and bond stress along the splice length ($A_b f_s = U \pi d_b L_d$) leads to:

$$\frac{A_b f_s}{f'_c{}^{1/2}} = 0.785 L_d (c + 0.4 d_b) + 16.6 A_b + \frac{n_f t_f E_f L_d}{1273 n_s} \tag{10}$$

One possible criterion that could be used for FRP bond strengthening of spliced steel bars at the critical hinging regions for the purpose of improving the seismic performance of concrete members (ex. base of gravity load designed columns or bridge piers where the longitudinal reinforcement is spliced with starter bars) would be to require that the reinforcement at the splice location be capable of developing a minimum steel stress $f_s = \kappa f_y$ where $\kappa \geq 1$, before experiencing splitting bond failure. Implementing this criterion by replacing the value of $f_s = \kappa f_y$ in Eq. (10) leads to the following proposed expression for estimating the required thickness $n_f t_f$ (mm) of the FRP reinforcement applied along the full splice length for bond strengthening:

$$n_f t_f = \frac{1000 n_s d_b}{E_f (L_d/d_b)} \left(\left(\frac{\kappa f_y}{\sqrt{f'_c}} - 16.6 \right) - \frac{L_d}{d_b} \left(\frac{c}{d_b} + 0.4 \right) \right) \tag{11}$$

Note that because of the limit imposed on the maximum increase in average bond strength U_{frp} in Eq. (7), and in the absence of experimental data to justify a larger upper limit, there is theoretically a maximum value of the steel stress beyond which bond splitting failure will occur irrespective of the area of FRP confinement used. The corresponding steel stress f_s can be estimated directly from Eq. (11) by replacing κf_y with f_s and the value of $n_f t_f$ on the left hand side of the equation by its upper limit of

$$n_f t_f = \frac{1600 n_s d_b}{E_f} \text{ calculated from Eq. (7).}$$

Variation of steel stress $f_s = \kappa f_y$ with FRP thickness $n_f t_f$ predicted using Eq. (11) (or Eq. 10) for different ratios of splice/development length to bar diameter L_d / d_b assuming a constant value of $c / d_b = 1.0$, and for different values of c / d_b assuming a constant value of $L_d / d_b = 30$, are shown in Fig. 7(a) and 7(b), respectively. The results were obtained using a reinforced concrete member (ex. column) in which the critical outermost tension reinforcement is composed of 4ϕ25 mm spliced bars (n_s = 4), $f'_c = 30$ MPa, and assuming E_f = 230,000 MPa and f_y = infinite. Note that the magnitude of the steel stresses for unconfined concrete ($n_f t_f = 0.0$) in Fig 7 are essentially those predicted using the Orangun equation (Eq. 8).

Fig. 7 shows that confining the concrete members at the splice locations with a relatively small area of FRP reinforcement leads to significant increases in the steel stress that can be mobilized at bond failure as compared to unconfined concrete. Fig. 7 also shows that as c / d_b or L_d / d_b decreases, larger areas of FRP are required to achieve some desired level of steel stress ($f_s = \kappa f_y$) before bond failure. The leveling-off of the curves above a value of $n_f t_f$ of about 0.75 in this particular example is attributed to the limitation imposed on U_{frp} in Eq. (7). Note that Eq. (11) also predicts (not shown in Fig. 7) that the effectiveness of the FRP in improving the bond strength increases as the modulus of elasticity of the FRP material increases.

SUMMARY AND CONCLUSIONS

The effect of FRP confinement on the local bond stress-slip response under static and cyclic loading was experimentally investigated. Beam specimens with spliced reinforcement at midspan were tested. The test parameters included the diameter of the spliced bars, the ratio of concrete cover to bar diameter, and the area of external FRP confinement. Without confinement all specimens were designed to experience splitting mode of bond failure. Based on the results of this investigation, the following conclusions can be drawn:

1. External confinement of concrete using FRP laminates leads to substantial improvement of the bond strength between spliced or developed bars and concrete.

2. Without confinement, the spliced bars suffered significant bond deterioration and loss in bond capacity in the first few loading cycles immediately after splitting.

3. Confining the spliced reinforcement with external FRP sheets improved the peak bond strength under cyclic loading, reduced considerably the bond strength degradation with increase in the number of loading cycles after splitting, and resulted in significant improvement of the cyclic bond stress-slip or load-deflection response.

4. While increasing the thickness of the FRP sheets from one layer two layers did not influence much the peak bond strength, it did produce hardening in the bond stress–slip response after splitting, and therefore significant improvement in ductility under cyclic loading.

A general relationship of the local bond stress-slip response for FRP confined concrete under monotonically increasing load in tension is presented. Also, a comprehensive design expression is generated for estimating the area of FRP wraps that should be used at the critical regions of the structure where bond strengthening becomes critically important for improving the seismic performance of the structural member.

ACKNOWLEDGMENTS

The author wishes to thank the Lebanese National Council for Scientific Research for their support and the American University of Beirut for providing the test facilities.

REFERENCES

1. ACI Committee 440, 2002, "Guide for the Design and Construction of Externally Bonded FRP Systems for Strengthening Concrete Structures," ACI 440.2R-02, American Concrete Institute, Det., MI.

2. ISIS Canada., 2001, "Strengthening Reinforced Concrete Structures With Externally Bonded Fibre Reinforced Polymers (FRPs)," *Design Manual No. 4*, K. Neale, ed.

3. Saadatmanesh, H., Ehsani, M. R., and Li, M. W., 1994, "Strength and Ductility of Concrete Columns Externally Confined with Fiber Composite Straps," *ACI Struct. J.*, 94(1), pp. 434-447.

4. Harajli, M. H., and Rteil, A.A., 2004, "Effect of Confinement Using FRP or FRC on Seismic Performance of Gravity Load-Designed Columns," *ACI Struct. J.*, 101(1), pp. 47-56.

5. Priestley, M.J.N., Seible, F., and Calvi, M., 1996, "Seismic Design and Retrofit of Bridges," John Wiley & Sons, Inc., New York, NY.

6. Harajli, M. H., Hamad, B. S., and Rteil, A., 2004, "Effect of Confinement on Bond Strength Between Steel Bars and Concrete," *ACI Struct. J.*, 101(5), September-October, pp. 595-603.

7. Orangun, C. O., Jirsa, J. O., and Breen, J. E., 1975, "Strength of Anchored Bars: A Reevaluation of Test Data on Development Length and Splices," *Research Report* No. 154-3F, Center for Highway Research, Univ. of Texas at Austin, 78 pp.

8. Orangun, C. O., Jirsa, J. O., and Breen, J. E., 1977, "Reevaluation of Test Data on Development Length and Splices," *ACI Journal, Proceedings* 74 (3), Mar., pp. 114-122.

9. Zuo, J. and Darwin, D., 2000, "Splice Strength of Conventional and High Relative Rib Area Bars and High-Strength Concrete," *ACI Struct. J.*, 97(4), July-Aug. pp. 630-641.

10. Eligehausen, R., Popov, E.P., and Bertero, V. V., 1983, "Local Bond Stress-Slip Relationships of Deformed Bars Under Generalized Excitations," Rep. UCB/EERC-83/23, Univ. of Calif., Berkeley.

11. Filippou, F. C., Popov, E. P., and Bertero, V. V., 1983, "Effects of Bond Deterioration on Hysteretic Behavior of Reinforced Concrete Joints," Rep. UCB/EERC-83/19, Univ. of Calif., Berkeley.

12. Esfahani R., and Rangan, V., 1998, "Bond Between Normal Strength and High-Strength Concrete (HSC) and Reinforcing Bars in Splices in Beams," *ACI Struct. J.*, 95(3), May-June, pp. 272-280.

13. Harajli, M. H., and Mabsout, M.E., 2002, "Evaluation of Bond Strength of Steel Bars in Plain and Fiber Reinforced Concrete," *ACI Struct. J.* , 99(4) July-Aug., pp. 509-517.

14. Harajli, M. H., 2004, "Comparison of Bond Strength of Reinforcing Bars in Normal and High-Strength Concrete," *Journal of Materials in Civil Engrg., ASCE,* 16(4), Jul./Aug. pp. 365-374.

15. Harajli, M. H., Hout, M., and Jalkh, W., 1995, "Local Bond Stress-Slip Relationship of Reinforcing Embedded in Fiber Reinforced Concrete," *ACI Materials J.* 92 (4), Jul-Aug., pp. 343-354.

Table 1 – Summary of test parameters

Series	Specimen Notation	Load Type	Bar size d_b (mm)	$c_s = c_b$ (mm)	c/d_b (a)	# of steel bar splice	Splice length (width of CFRP) (mm)	# of FRP plies	f'_c (MPa)
IIW	B20	Cyclic	20	30	1.5	2	100	None	35.8
	B20FP1	Cyclic	20	30	1.5	2	100	1	35.6
	B20FP2	Cyclic	20	30	1.5	2	100	2	35.6
	B2W[(b)]	Static	20	30	1.5	2	100	None	28.7
	B2W-CF1[(b)]	Static	20	30	1.5	2	100	1	41.9
	B2W-CF2[(b)]	Static	20	30	1.5	2	100	2	40.6
IIW	B25	Cyclic	25	25	1.0	2	125	None	35.8
	B25FP1	Cyclic	25	25	1.0	2	125	1	28.8
	B25FP2	Cyclic	25	25	1.0	2	125	2	28.8
	B3W[(b)]	Static	25	25	1.0	2	125	None	32.9
	B3W-CF1[(b)]	Static	25	25	1.0	2	125	1	37.4
	B3W-CF2[(b)]	Static	25	25	1.0	2	125	2	41.5

(a) c is the smaller of the clear concrete cover (c_b) or half the clear spacing between splices (c_s); in this study $c = c_s = c_b$

(b) Harajli et al. [6]

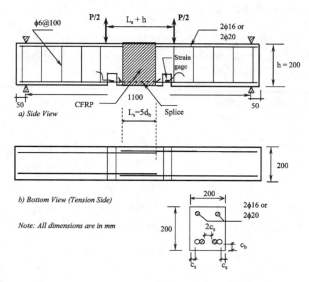

Figure 1 – Specimen dimensions and splice details

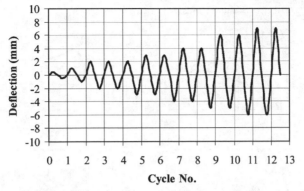

Figure 2 – History of cyclic loading used in the experiment

Figure 3 – Typical view of test setup and failure modes

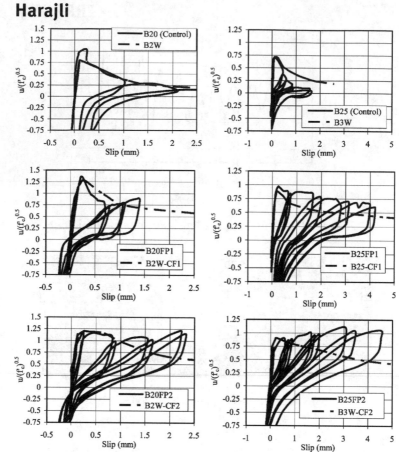

Figure 4 –Cyclic responses of normalized local bond stress versus slip

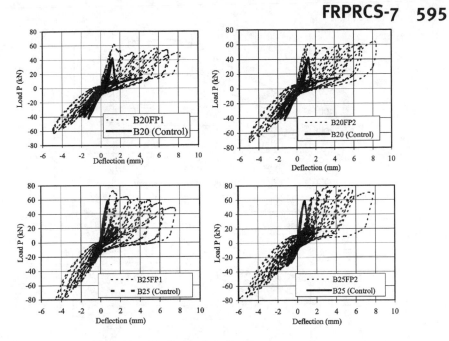

Figure 5 – Cyclic load-deflection responses

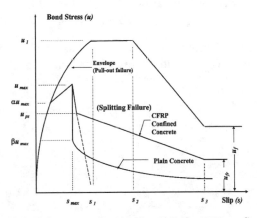

Figure 6 –Local bond stress-slip relationship for FRP confined concrete[6]

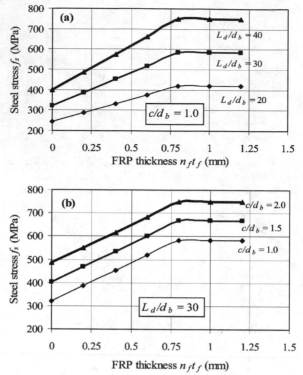

Figure 7 – Variation of steel stress with thickness of FRP confinement: (a) for different values of L_d/d_b, and (b) for different values of c/d_b

Chapter 6

Serviceability of FRP Reinforced Concrete Structures

Fatigue Behavior of Prestressed Concrete Bridge Girders Strengthened with Various CFRP Systems

by O.A. Rosenboom and S.H. Rizkalla

Synopsis: Carbon Fiber Reinforced Polymer (CFRP) materials provide a solution for the classical challenge facing bridge maintenance engineers: the upgrading of bridges whose service life has been exceeded but due to evolving demands must stay in service. Of particular concern are rural short span prestressed concrete bridges that may be required to carry loads above the initial design value. The CFRP strengthening systems presented in this paper have the potential to increase the ultimate capacity of such bridge girders. This research project is aimed at investigating the fatigue performance of CFRP strengthening systems for prestressed concrete bridge girders. Five 9.14 meter prestressed concrete bridge girders were tested under fatigue loading conditions: one as a control specimen and four strengthened with various CFRP systems including near surface mounted (NSM) bars and strips, and externally bonded sheets and strips. Results show that CFRP strengthened prestressed concrete bridge girders can withstand over two million cycles of fatigue loading equivalent to a 20 percent increase in live load with little degradation.

Keywords: bridge girder; externally bonded; fatigue behavior; fiber reinforced polymers; flexural behavior; near surface mounted; prestressed; strengthening

Owen Rosenboom is a PhD candidate at North Carolina State University.

Dr. Sami Rizkalla is a Distinguished Professor of Civil Engineering and Construction, Director of the Constructed Facilities Laboratory (CFL) and Director of the NSF I/UCRC - RB2C Center at North Carolina State University.

INTRODUCTION

Research objectives

Through examination of various parameters, this research aims to provide the prestressed/precast concrete industry as well as transportation agencies with economical and effective strengthening systems for prestressed concrete using CFRP materials. The feasibility of using carbon fiber reinforced polymers (CFRP) strengthening systems to upgrade the load carrying capacity of 40 year old prestressed concrete bridges is investigated. Although there is an ever-expanding research database of reinforced concrete structures strengthened with different CFRP systems, information on various strengthening techniques for prestressed concrete structures is very limited. The first phase of the research began was an investigation of the static behavior of prestressed concrete bridge girders strengthened with CFRP which can be found elsewhere[1].

Background and literature review

Many bridges in the United States do not conform to current design standards. If they are not strengthened they may need to be replaced to accommodate the increase in load. One of the major concerns to departments of transportation is short-span prestressed concrete bridges in rural areas which have exceeded their design life but due to evolving industry demands may be required to carry loads above the initial design value. CFRP systems have the potential for cost-effective retrofitting of prestressed concrete bridges by increasing the load-carrying capacity thus extending their service life.

The use of externally bonded and NSM CFRP systems to repair or strengthen reinforced concrete beams in flexure has been well researched[2-4]. Takacs and Kanstad[5] showed that prestressed concrete girders could be strengthened with externally bonded CFRP plates to increase their ultimate flexural capacity. Reed and Peterman[6] showed that both flexural and shear capacities of 30 year-old damaged prestressed concrete girders could be substantially increased with externally bonded CFRP sheets. Reed and Peterman also encouraged the use of CFRP U-wraps as shear reinforcing along the length of the girder in externally bonded systems to delay debonding failure. The use of NSM CFRP in prestressed concrete bridge decks was explored by Hassan and Rizkalla[3] and found to be a viable alternative to externally bonded systems.

Debonding of externally bonded FRP systems has been noted by many researchers often at the termination point of the FRP plate/sheet for members with a short span, and at the midspan section for long span members. Many models have been proposed to predict the failure loads of FRP strengthened reinforced concrete members due to plate-end debonding[7,8], yet the midspan debonding mechanism has not been as extensively researched[9,10]. U-wrap CFRP reinforcement has been recommended for use at the termination point of the main CFRP strengthening system, but the benefits of providing this extra reinforcement throughout the length of the girder is not known[9]. One of the benefits of NSM FRP strengthening is to reduce the propensity for debonding failure. Models to predict this debonding load have been characterized from earlier plate-based work[2].

The fatigue behavior of reinforced concrete beams strengthened with externally bonded CFRP systems has been investigated[11, 12], yet no work has been done on prestressed concrete members strengthened with CFRP and tested in fatigue. The fatigue behavior of prestressed concrete was extensively examined in the 1960's and 1970's[13, 14] with results showing that very little fatigue degradation occurs if the girder remains uncracked. If the fatigue load is above the cracking load and the stress range in the prestressing strands is high, failure will be due to rupture of the prestressing strands.

EXPERIMENTAL PROGRAM

Test girders

As part of a research program funded by the North Carolina Department of Transportation, ten prestressed concrete C-Channel type bridge girders were tested at the Constructed Facilities Laboratory at North Carolina State University. Five of the girders were tested under fatigue loading conditions: one as a control specimen (F0), two strengthened with near surface mounted CFRP bars (F1) and strips (F2) and two strengthened with externally bonded CFRP strips (F3) and wet lay-up sheets (F4). Five identical girders were tested under static loading conditions and details of the results of these experiments can be found elsewhere [1]. Table 1 shows the properties of the CFRP systems used determined from coupon tests as well as the concrete strength found from the testing of core samples. All girders were 9.14 m long C-Channel type prestressed concrete bridge girders (see Fig. 1). The girders were taken from a decommissioned bridge in Carteret County, NC, USA, which was erected in 1961. Each girder was prestressed with ten 1725 MPa seven-wire stress relieved, 11 mm prestressing strands (five in each web) and had a 125 mm deck with minimal reinforcing. The measured camber of the girders at midspan, due to prestressing, ranged from 32 to 38 mm. Details of the various strengthening systems are provided in Figure 1.

Design of strengthened girders

The design of the strengthened girders proceeded after testing the control girder under static loading conditions[1]. The objective of the strengthening was to achieve a 20 percent increase in the ultimate load carrying capacity with respect to the control girder, except for F4 which was designed for a 60 percent increase in the ultimate load carrying capacity to examine the behavior at a significantly higher range. Each strengthened girder was designed using a cracked section analysis program, Response 2000[15]. For the design, the manufacturer's properties were used to model the FRP materials. The prestressing steel and concrete material properties were taken from the provided specifications.

Flexural failure, defined as rupture of the FRP or crushing of the concrete in compression, was the desired mode of failure. The shear stresses in the concrete at the plate end were significantly lower than the shear strength of the concrete (according to equations proposed by Malek, et al[16]) so debonding at the plate end was not a concern. However, to delay FRP delamination-type failures along the length of the girder, 150 mm wide U-wraps were installed at 900 mm spacing for all externally bonded strengthened girders. This arrangement was selected to simulate typical anchorage details commonly used by the construction industry for reinforced concrete members strengthened with FRP.

Test setup, loading and instrumentation

All girders were tested using a 490 kN MTS hydraulic actuator. The actuator was mounted to a steel frame placed at the midspan of the girder. Typical prestressed concrete bridges constructed in the early 1960s usually have a substructure of wooden piles that is difficult to mimic in the laboratory. However, in order to simulate small displacements at the supports, the girder was supported at both ends on a 64 mm neoprene pad which in turn rested on a 25 mm steel plate. Hydrostone was used at the supports for leveling purposes. The width of the neoprene pad was 216 mm which provided a clear span of 8710 mm for the tested girder.

The behavior of the girders during testing was monitored using three sets of string potentiometers, placed at quarter spans, and two linear potentiometers to measure vertical displacement over the supports. The compressive strain in the concrete was measured using a combination of PI gauges (a strain gauge mounted to a spring plate) and electrical resistance strain gauges located beside and between the loading tires. PI gauges were also placed at the level of the lowest prestressing strand to measure the crack width and determine the strain profile through the section. The tensile strain in the CFRP reinforcement was measured using six strain gauges: two at midspan, two at 150 mm from midspan and two at 305 mm from midspan as shown in Figure 2. A picture of the test setup is provided in Figure 3.

The loading sequence for the tested girders started by increasing the applied load to a load level slightly higher than the cracking load, unloading, and then reloading again at a rate of 2.5 mm/min up to the load at which the crack at midspan reopens. This loading sequence was selected to determine the effective prestressing force in the girders by observing the re-opening of the flexural cracks[17]. The girders were then cycled between two load values at a frequency of 2 Hz. The dead load for all the girders was the same, 8.9 kN. The live load used for the control specimen was 40 kN. This was based on the service load the original girder was designed for (HS15 type loading) and includes the appropriate distribution and impact factors. For three of the strengthened girders tested in fatigue (F1, F2, F3), the live load was increased 20 percent to 49 kN and for F4 it was increased 20 percent for one million cycles and 60 percent for the next one million cycles.

FATIGUE TEST RESULTS AND DISCUSSION

Control girder

Cracking of the control specimen occurred at a load of 55.6 kN. After unloading and reloading, the flexural crack at midspan reopened at a load of 45.4 kN, which shows an approximate loss of prestress of 6.7 percent. After the initial loading cycles, the girder was cycled between 8.9 kN and 49 kN as described above. The load deflection behavior of the control girder is shown in Fig. 4. The stress range in the lower prestressing strand, which can be defined as

$$SR = \frac{f_{ps2} - f_{ps1}}{f_{pu}}$$

where f_{ps2} and f_{ps1} are the upper and lower stress in prestressing subjected to cyclic loading conditions and f_{pu} is the ultimate strength of the prestressing, is shown in Fig. 5 versus the number of cycles. The control girder survived 2 million cycles with very little degradation. The girder was then loaded to failure, which occurred at a load of 142 kN, a 3.64 percent decrease from the ultimate strength achieved in the static test of the control girder.

Near surface mounted CFRP strengthened girders

One girder strengthened with NSM bars and another strengthened with NSM strips (identical to two girders tested under static loading conditions) were tested in fatigue, F1 and F2 respectively. After the initial loading the strengthened girders were tested at a frequency of 2 Hz between 9 kN and 57.6 kN. The cracking load of the NSM bars and strips strengthened girders occurred at loads of 54 kN and 51 kN respectively. For both girders, the largest degradation in stiffness occurred between the secondary loading sequence (used to determine the prestress losses) and 10,000 cycles. Even with a substantial increase in stress range (see Figure 5) compared to the control girder, after 2 million cycles there was very little degradation in the load versus deflection plot for either girder. The load versus deflection plots for both girders are very similar – the one for the girder strengthened with NSM bars is shown in Figure 6.

After 2 million cycles the girders were tested up to failure and showed little difference between the girders tested under static loading conditions. For the NSM bars strengthened girder tested in fatigue, failure was due to crushing of the concrete at a load of 178 kN, compared to the statically tested girder which failed at a load of 181 kN (see Figures 7 and 8). The NSM strip strengthened girder tested in fatigue also failed due to crushing of concrete at a load of 163 kN, compared to the statically tested girder which failed at a load of 180 kN (40.6 k).

Externally bonded CFRP strengthened girders

Two girders strengthened with externally bonded CFRP systems were tested in fatigue: one girder strengthened with externally bonded strips (F3) and another strengthened with externally bonded sheets designed for a 60% increase in capacity compared to the control (F4). These were identical to two girders tested under static loading conditions[1].

The girder strengthened with externally bonded strips had a cracking load of 54 kN. After 10,000 cycles, the behavior of the strengthened girder changed markedly from that of the NSM strengthened girders. Whereas the stress range and the maximum deflections of the other girders stabilized, F3 showed a steady increase in both these quantities. At 625,000 cycles a large crack was noticed near midspan which led to localized delamination of the CFRP sheets from the concrete substrate. This crack was due to rupture of a prestressing strand. The test was continued and catastrophic failure occurred at 908,000 cycles due to progressive rupture of the prestressing strands followed by debonding of the CFRP strips. The rupture of the first prestressing strand constitutes failure of the girder, since the deflection under service loading exceeded the limits specified for this type of girder. Gradual degradation of the bond between the concrete and FRP due to the fatigue loading could have resulted in a loss of girder stiffness therefore increasing the stress range in the lower prestressing strands precluding failure. Although very little corrosion was detected in the prestressing strands in an examination after failure, this could also have been a cause of the early fatigue rupture. Load versus deflection behavior of this girder is shown in Figure 9.

A girder strengthened with externally bonded CFRP sheets designed to achieve a 60% increase in the ultimate load carrying capacity of the control girder was tested in fatigue (F4). The cracking load of this girder was measured to be 70 kN, much greater than any of the previously tested girders due to a higher effective prestressing force. Due to the uncertainty in calculating the stress range in the lower prestressing strand for the girder, it was decided to test this girder identically to the other strengthened girders. Very little degradation was noticed between the initial loading sequences up until 1 million cycles. At this stage it was decided to cycle between the load range of 8.9 kN to 72.7 kN, representing a 60 percent increase in live load. A static test performed at one million cycles showed very little change in stiffness up to 72.7 kN. Another 250,000 cycles degraded the girder so that a secondary stiffness can be seen after reopening of the crack as can be seen in the load versus deflection plot in Fig. 10.

The stress range in the prestressing strands, as shown in Fig. 5, was lower than that of the control girder up to one million cycles (due to the higher prestressing force), but increased dramatically after the 60 percent increase in live load was applied. Between 1.25 million cycles and 2 million cycles very little change was observed in the cracking pattern or the load-deflection behavior. After 2 million cycles the girder was tested to failure. Like the girder tested under static loading conditions, failure was due to rupture of the CFRP sheets followed by crushing of concrete. Due to the higher prestressing force observed in this girder, the girder tested in fatigue failed at a load of 245 kN, greater than the ultimate load observed for girder S6.

CONCLUSIONS

Five 40-year-old 9.14m long prestressed concrete girders have been tested under fatigue loading conditions. One was tested as a control specimen and four were strengthened with various CFRP systems. The cyclic loading was designed to simulate loads on an actual bridge girder, from the dead load to the dead load plus live load. Based on the results, the following conclusions can be drawn:

1. Prestressed concrete girders strengthened with NSM CFRP systems to achieve a 20 percent increase in ultimate load carrying capacity can withstand over two million cycles of a loading equivalent to a 20 percent increase in live load.
2. Girders strengthened with externally bonded CFRP sheets to achieve a 60 percent increase in ultimate load carrying capacity can withstand over one million cycles of loading equivalent to a 60 percent increase in live load.
3. The girder strengthened with externally bonded CFRP strips performed worse under fatigue loading conditions than either the NSM systems or the externally bonded sheet strengthened systems, although this performance could be due to other circumstances such as corrosion of the prestressing strands. More testing needs to be done to corroborate these findings.
4. The influence of the CFRP U-wraps placed along the length of the girder for preventing fatigue initiated debonding is not known and requires further research.

Future fatigue testing of similar prestressed concrete girders will involve the testing of an externally bonded CFRP sheets girder strengthened to achieve a 40 percent increase in strength, as well as a girder strengthened to achieve a 20 percent increase in strength using externally bonded high modulus sheets.

ACKNOWLEDGMENTS

The authors would like to acknowledge the contribution of Dr. Amir Mirmiran in the initiation of this study which is supported by the North Carolina Department of Transportation through Project 2004-15. The authors would also like to thank Jerry Atkinson, the technician at the Constructed Facilities Laboratory.

REFERENCES

1. Rosenboom, O.A., and Rizkalla, S.H., "Static behavior of 40 year-old prestressed concrete bridge girders strengthened with various FRP systems. *Proceedings of the Second International Conference on FRP Composites in Civil Engineering*, Adelaide, Australia, Dec. 2004.

2. Hassan T., and Rizkalla, S., "Investigation of Bond in Concrete Structures Strengthened by Near Surface Mounted CFRP Strips," *ASCE, Journal of Composites for Construction*, V. 7, No. 3, 2003, pp. 248-257.

3. Hassan T., and Rizkalla, S., "Flexural Strengthening of Post-Tensioned Bridge Slabs with FRP Systems," *PCI Journal*, V. 47, No. 1, 2002, pp. 76-93.

4. De Lorenzis, L., and Nanni, A., "Characterization of FRP Rods as Near-Surface Mounted Reinforcement," *Journal of Composites for Construction* V. 5, No. 2, 2001, pp. 114-121.

5. Takács, P. F., and Kanstad T., "Strengthening Prestressed Concrete Beams with Carbon Fiber Reinforced Polymer Plates," *NTNU Report R-9-00*, Trondheim, Norway, 2000.

6. Reed, C. E., and Peterman, R. J., "Evaluation of Prestressed Concrete Girders Strengthened with Carbon Fiber Reinforced Polymer Sheets," *ASCE, Journal of Bridge Engineering*, V. 9, No. 2, 2004, pp. 185-192.

7. Smith, S. T., and Teng, J. G., "FRP Strengthened RC Beam I – Review of Debonding Strength Models," *Engineering Structures* V. 24, No. 4, 2002, pp. 385-395.

8. Smith, S. T., and Teng, J. G., "FRP Strengthened RC Beam II – Assessment of Debonding Strength Models," *Engineering Structures* V. 24, No, 4, 2002, pp. 397-417.

9. Teng, J.G., Chen, J.F., Smith, S.T., Lam, L., "FRP Strengthened RC Structures", John Wiley and Sons, New York, NY, 2002.

10. Oehlers, D.J., Seracino, R., "Design of FRP and Steel Plate Structures" Elsevier, New York, NY, 2004.

11. Heffernan, P.J., Erki, M.A., "Fatigue Behavior of Reinforced Concrete Beams Strengthened with Carbon Fiber Reinforced Plastic Laminates", *Journal of Composites for Construction,* V. 8, No. 2, 2004, pp. 132-140.

12. El-Hacha, R., Filho, J., Rizkalla, S., and Melo, G., "Static and Fatigue behavior of Reinforced Concrete beams strengthened with different FRP Strengthening techniques", *Proceedings of the 2nd International Conference on FRP Composites in Civil Engineering* (CICE 2004), Adelaide, Australia, December 8-10, 2004.

13. Overman, T.R., Breen, J.E., and Frank, K.H., "Fatigue of pretensioned concrete girders', *Report 300-2F*, Center for Transportation Research at the University of Texas at Austin, 1984.

14. Rao, R. and Frantz, G.C., "Fatigue tests of 27-year-old prestressed concrete bridge box beams". *PCI Journal*, 1996, V. 41, No. 5, pp. 74-83.

15. Bentz, E. C., "Section Analysis of Reinforced Concrete Members," Ph.D. Thesis, University of Toronto, 2000.

16. Malek, A. M., Saadatmanesh, H., and Ehsani, M. R., "Prediction of Failure Load of RC Beams Strengthened with FRP Plate Due to Stress Concentration a the Plate End," *ACI Structural Journal,* V. 95, No. 1, 1998, pp. 142-152.

17. Zia, P., and Kowalsky, M. J., "Fatigue Performance of Large-Sized Long-Span Prestressed Concrete Girders Impaired by Transverse Cracks," *Federal Highway Administration Report, FHWA/NC/2002-024,* 2002.

Table 1. Material Properties of Tested Girders

Girder Designation	F0	F1	F2	F3	F4
Strengthening Technique	none	NSM Bars	NSM Strips	EB Strips	EB Sheets
FRP Details	--	1 #3 per web	2 strips per web	1 strip per web	4 plies per web
A_{FRP} (mm^2)	--	130	129	119	1960
Ultimate Tensile Strength of FRP (MPa)	--	2070	2070	2800	340
E_{FRP} (GPa)	--	124	124	165	46.5
FRP rupture strain	--	0.0167	0.0158	0.0170	0.00734
f'$_c$ (MPa)	Not available	66.7	48.0	61.4	67.8

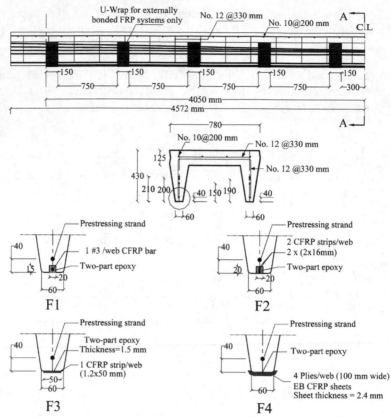

Figure 1 – Reinforcement and strengthening details of the C-Channel girders

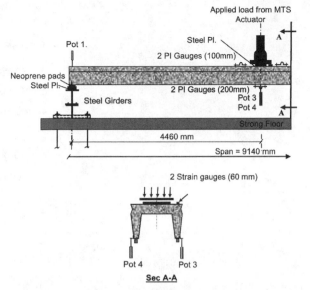

Figure 2 – Test setup for girders tested under fatigue loading conditions

Figure 3 – Typical test setup for girders tested in fatigue

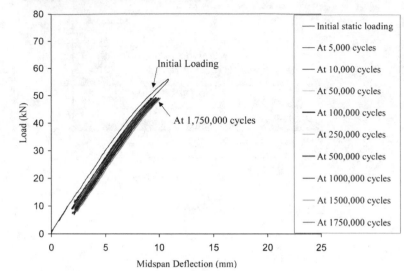

Figure 4 – Load versus deflection for control girder (F0)

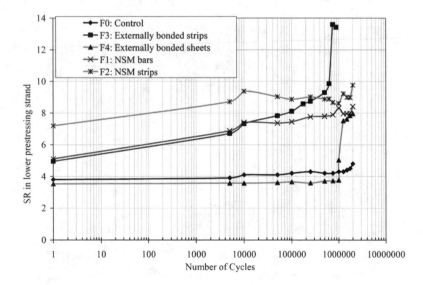

Figure 5 – Stress range versus the log of the number of cycles for girders tested in fatigue

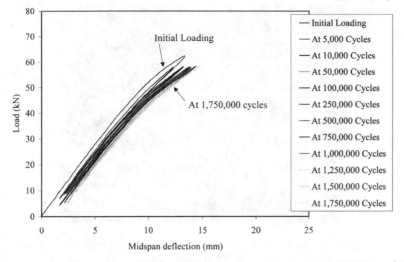

Figure 6 – Load versus deflection for girder strengthened with NSM bars (F1)

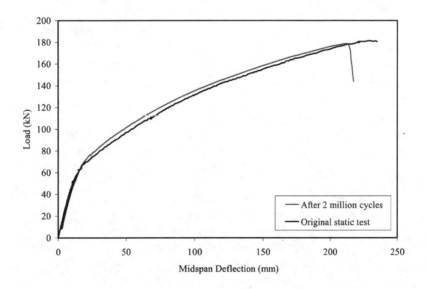

Figure 7 – Static test after two million cycles of girder F1

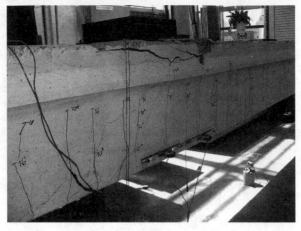

Figure 8 – Failure of girder F1 during final static test after two million cycles of fatigue loading

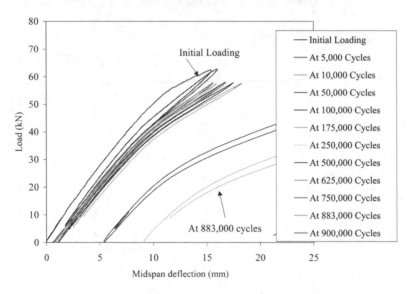

Figure 9 – Load versus deflection for girder strengthened with externally bonded CFRP strips (F3)

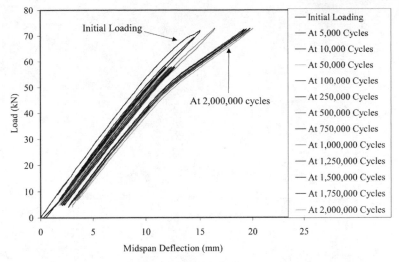

Figure 10 – Load versus deflection for girder strengthened with
externally bonded sheets (F4)

Fatigue Behavior of Reinforced Concrete Beams Strengthened with Different FRP Laminate Configurations

by R. Gussenhoven and S.F. Breña

Synopsis: This paper presents testing results of thirteen small-scale beams strengthened using carbon fiber-reinforced polymer composites tested under repeated loads to investigate their fatigue behavior. The beams were strengthened using different thicknesses and widths of composite laminates to identify parameters that would generate different failure modes. Two predominant fatigue failure modes were identified through these tests: fatigue fracture of the steel reinforcement with subsequent debonding of the composite laminate, or fatigue fracture of the concrete layer below the tension reinforcing steel (concrete peel off). Test results indicate that peak stress applied to the reinforcing steel in combination with composite laminate configuration are the main parameters that affect the controlling failure mode. Tests on large-scale components are required to verify the results presented in this paper.

Keywords: carbon fiber-reinforced polymers; fatigue loading; reinforced concrete; strengthening

Maj. Richard Gussenhoven is an Instructor in the Department of Mathematical Sciences and Assistant Director for the Center of Data Analysis and Statistics in the U.S. Military Academy at West Point. He received his MS degree at the University of Massachusetts Amherst in 2004.

Sergio F. Breña is an Assistant Professor at the University of Massachusetts Amherst. He is a member of ACI Committee 440 – Fiber Reinforced Polymer, ACI 374 – Performance Based Seismic Design, ACI-ASCE 445 – Shear and Torsion, and ACI 369 Repair and Rehabilitation.

INTRODUCTION

Strengthening existing reinforced concrete structures using composites is an accepted technology that is being increasingly used in a wide variety of applications. This technology is particularly promising for bridge applications because of the large number of existing bridges that need repair or replacement. Additionally some bridge strengthening applications are required because of an increase in vehicle weight over the years (Breña et al. 2003). As a result, old bridges designed using lower live-load models require strengthening to meet current load standards.

Carbon fiber-reinforced polymer (CFRP) composites have been investigated to strengthen existing older bridges. Many of the studies available in the literature have been conducted on specimens loaded statically with comparatively few studies performed using fatigue loading. An understanding of the factors affecting the fatigue performance of strengthened beams is required to develop appropriate design guidelines for bridge applications.

RESEARCH OBJECTIVES

The primary objective of this research study was to identify the failure modes that occur in beams strengthened using CFRP composites and loaded under repeated loads. Parameters including laminate width and thickness, amplitude of cyclic loading, and pre-existing damage in the beams were investigated.

DESCRIPTION OF LABORATORY SPECIMENS

Thirteen small-scale beams were fabricated and tested to investigate the effects of various parameters including composite laminate configuration on their fatigue performance. Two of these specimens were used as control specimens, and the rest were strengthened using carbon fiber-reinforced polymer (CFRP) laminates. The CFRP laminate configuration varied on the strengthened specimens within their respective specimen group. Details of the strengthening configurations corresponding to each specimen group are given in the following sections.

Specimen Geometry and Reinforcement

The laboratory specimens had a 102 by 102 mm cross-section with a 914 mm total length. Longitudinal reinforcement consisted of 2 – #10M reinforcing bars with a nominal yield stress of 410 MPa on the tension and compression sides of the beam. Transverse reinforcement consisted of 16 stirrups at a 51 mm spacing placed within the shear spans constructed using gage D4 deformed wire. The transverse reinforcement was designed to avoid shear failures of the beams after flexural strengthening. Concrete clear cover was 13 mm from the exterior face of the beam to the transverse reinforcement on all faces (Figure 1).

SPECIMEN GROUPS AND STRENGTHENING CONFIGURATIONS

Specimens were fabricated in three groups depending on casting date. The first two groups (A and B) consisted of five beams, and the third group (group C) consisted of three beams. Details of each specimen group are discussed below. Table 2 presents a summary of the experimental parameters varied in the three groups. The primary experimental parameter in each group was different as discussed in the following sections. Eleven of the specimens were strengthened using a commercially available CFRP strengthening system (Master Builders 1998 MBrace C-130). Measured material properties used in this research are presented in Table 1.

The CFRP laminates were bonded on the tension face of the beams and terminated 25 mm from the face of each support, resulting in a total laminate length of 762 mm. The laminates were formed by wet layup and cured at ambient temperature for at least seven days before testing. The laminates were either 89 or 51 mm wide, and were formed using one or two plies depending on the specimen group. No supplemental anchorage was used along the composite laminates to allow the possibility of debonding as a failure mode after repeated loading. The typical strengthening configuration is shown in Figure 2.

Group A

Strengthened beams in this group were strengthened using a one-ply, 89-mm wide laminate. The primary variable for these specimens was the loading amplitude so all specimens had the same laminate configuration. A high load amplitude, close to the expected fatigue limit of the reinforcing steel, was used when testing these specimens to investigate any favorable effects that the CFRP laminates might have in the fatigue life of the beams as a result of increased stiffness, reduction of crack width, and larger crack spacing at service loads. All these attributes have been reported in beams strengthened with composite laminates.

Two of the five beams in group A were used as control specimens: one unstrengthened specimen and one strengthened specimen. The unstrengthened specimen was tested using repeated loading that generated peak stresses in the tension steel equal to

80% of the yield stress. The strengthened control beam contained the same composite configuration as the rest of the specimens in this group, and was tested statically to failure to provide a static-load baseline for the repeated load tests.

Group B

Beams in group B were strengthened using CFRP laminates of equal width (51 mm) but different thickness. The effect of composite thickness on the fatigue behavior of the specimens was the primary experimental parameter in this group. The applied load was increased in beams strengthened using 2-ply CFRP laminates in order to generate the same peak stress in the reinforcing steel as beams with only 1-ply laminates. Laminate area in beams strengthened with a 1-ply laminate was approximately 57% of the area in group A specimens, and beams with 2-ply laminates had approximately 14% more area than beams in group A. The laminate surface contact area in these beams was only equal to 57% of the area provided in group A specimens. Interface stresses in beams with 1-ply laminates in group B would be theoretically equal to those in beams in group A. Because of higher thickness and smaller contact surface in specimens with 2-ply laminates, interface stresses would be twice as large as those generated in beams in group A at equal laminate strains.

Group C

The three beams in group C were subjected to 500,000 cycles of repeated loads generating a stress equal to 50% of the yield stress in the tensile reinforcement prior to strengthening. These beams were subsequently strengthened using identical CFRP configurations to those used in beams with 1-ply laminates from group B. The loading protocol after strengthening was designed to produce the same stress levels as those generated in the companion specimens in group B.

TEST SETUP, INSTRUMENTATION, AND LOADING PROTOCOL

All members were subjected to 4-point bending using an 8500 Instron testing machine. Load was measured using a 50 kN load cell mounted to the machine piston. The specimens were supported on plates and rollers, which simulated a simply supported condition and rested on a stiff beam mounted on the testing machine load frame. A picture of the experimental setup is shown in Figure 3.

Beam deflections were measured using linear potentiometers positioned at mid-span and at support centerlines to account for flexibility of the supporting beam. A linear variable displacement transducer (LVDT) integral to the hydraulic actuator was used to measure the deflection at the load points. In order to measure the strain profile across the cross section, each beam was instrumented at the mid-span with strain gages to measure the strain on the tension reinforcement, the compression face, and the CFRP laminate for the strengthened beams. Strain gages were applied to both reinforcing bars on the tension side of the beam with a protective coating to prevent damage during concrete placement. Two strain gages were positioned on the compression face at mid-span to measure concrete strains. In beams strengthened with CFRP laminates, two strain gages were applied at mid-span 13 mm from the exterior edge of the CFRP laminate (Figure 4).

Repeated Load Protocol and Data Acquisition

Repeated loads were applied using a sinusoidal variation between minimum and maximum values. A minimum load of 2.7 kN was selected to avoid shifting of the beams with cycling while generating stresses that could be considered representative of dead loads in beams. The maximum load was calculated to generate specified stress levels in the tension reinforcement as a fraction of the nominal yield stress (60, 70, or 80% of nominal yield stress). A peak stress in the reinforcement equal to 60% of yield was intended to represent service-load stress conditions that would be typically encountered in practice. A peak reinforcing bar stress of 80% of yield represents the upper strengthening limit recommended by ACI Committee 440 (2002). The load frequency used for all repeated-load tests was approximately 4-Hz.

Loading was stopped periodically after reaching predetermined numbers of cycles to inspect beams for damage and acquire data during a low-frequency cycle. The data acquisition cycle consisted of a full cycle of loading between the minimum and maximum load values while acquiring deformation and strain data.

EXPERIMENTAL RESULTS

Damage progression in the specimens was monitored visually by examining crack growth and formation of new cracks during testing. Most of the new cracks or growth of existing cracks occurred during the early cycles of loading (up to 20,000). Subsequently, observed damage in the beams did not increase substantially. In many cases, no indications of imminent failure were observed, especially in beams where the reinforcing bars failed by fatigue fracture. The observed failure modes and the associated damage progression are described in the next sections. Deflection and strain measurements as additional damage indicators are also presented afterward.

Observed Failure Modes

The observed failure modes were dependent on the selected peak stress levels in the tension reinforcement and the configuration of the CFRP laminates. Specimens that withstood at least 2 million cycles of loading without failing were considered to have an infinite fatigue life and were subsequently tested statically to failure. These specimens did not exhibit any strength degradation compared with the control specimen after cycling. Table 3 lists the steel reinforcement peak stress, the applied stress range, the number of cycles to failure, and the associated failure modes observed during the tests.

Specimens cycled at peak stresses equal to 70 or 80% of the yield stress exhibited one of two failure modes depending on stress amplitude and CFRP laminate configuration. Failure modes in these beams were characterized by fatigue fracture of the reinforcement with subsequent debonding of the CFRP laminate or failure of the concrete cover along the bottom reinforcement. Figure 5 shows a specimen that exhibited steel fracture and subsequent debonding of the CFRP laminate. In this type of failure mode, after the reinforcing bars fractured the excess force generated in the CFRP laminate triggered loss of bond between the laminate and concrete surface. These beams showed no visible indication of distress prior to fracture of the reinforcement, as the cracks had stabilized and ceased to grow in length.

Beams that exhibited failure along the concrete cover were characterized by visible indications of damage as loading progressed. Typically a crack formed at the end of the CFRP laminate during early loading cycles (Figure 6a and Figure 7a) and later propagated along a plane just below the tension reinforcing bar level. The end crack propagated sometimes in the form of diagonal cracks within the beam shear span causing severe crack widening to occur within this region (Figure 6b).

Global Behavior – S-N Plots

Stress range is widely recognized as having a major influence in the fatigue life of metals. A number of fatigue life models for steel reinforcing bars used in concrete applications have been published in the past. From a series of reinforcing bars repeated tension tests, Helgason and Hanson (1974) proposed the following relationship to determine fatigue life of reinforcing bars:

$$\log N = 6.969 - 0.0055 f_r \tag{1}$$

where N is the number of cycles to failure and f_r is the applied stress range in MPa. Fatigue fracture of reinforcing bars embedded in concrete typically occurs in the proximity of concrete cracks that form as a result of loading. As a result of a study to examine the effects of concrete on fatigue life of reinforcement, Moss (1982) developed a fatigue life model based on the results of bending tests of reinforcing bars embedded in concrete. The following relationship was developed from a regression analysis of test data:

$$N f_r^m = K \tag{2}$$

where m represents the inverse slope of the stress range – number of cycles curve (S-N curve) = 8.7, and K is a constant calibrated to the test data = 0.59 x 10^{27} to represent the mean minus two standard deviations curve. ACI Committee 215 (1997) recommends that the maximum stress range in reinforcing bars in non-prestressed flexural members be limited to the following relationship to achieve an infinite fatigue life:

$$f_r = 161 - 0.33 f_{min} \tag{3}$$

where f_{min} is equal to the minimum (permanent) stress acting in the reinforcement.

Test data points of the strengthened beams in this study are plotted along with Equations 1 to 3 in the S-N plot shown in Figure 8. It can be observed that the majority of the data points generated in this study fall close to the Helgason-Hanson fatigue life model. Only two data points fall significantly to the left of Helgason-Hanson model, indicating that these specimens had a lower fatigue life than predicted by this model. Both specimens failed by concrete cover failure, which explains the poor correlation with the model. The apparent slope of data points from the three beam series tested in this research study appears to follow the slope of the model proposed by Moss (1982). This result should not be surprising since the majority of the specimens were controlled by fatigue fracture, which would be consistent with the failure mode captured by this model.

Test results reported by Papakonstantinou et al. (2001) are also shown in Figure 8. These researchers conducted tests of unstrengthened beams and strengthened beams using glass fiber-reinforced composites. The reported failure mode for all their tests consisted of steel fatigue fracture. It should be pointed out, however, that for all the strengthened beams in their study the FRP laminates extended past the supports so FRP debonding was restrained by clamping from support plates near the beam ends. Data points from the tests by Papankonstantinou consistently reached higher fatigue lives than the beams reported in this paper. Data from unstrengthened specimens consistently lie to the right of the Helgason-Hanson model, and there is no clear evidence of an increase in fatigue life from the use of composite laminates. This result is in contrast with observations from other researchers that reported that fatigue life of FRP-strengthened concrete beams was higher than life of companion unstrengthened specimens (Barnes and Mayes 1999).

The influence of laminate geometric parameters (width and thickness) on fatigue life was investigated by plotting the ratio of laminate force to laminate width or laminate thickness as a function of number of cycles to failure. Only specimens that failed in fatigue were included in this study. Only two specimens were strengthened using twice the laminate thickness than the rest (Table 2), and only one of those failed in fatigue. Because of this, conclusive evidence of any effect of laminate thickness on fatigue life could not be obtained in this research.

Laminate width, on the other hand, seemed to have a definite influence on fatigue life of the specimens. Figure 9 shows plots of laminate force as a function of number of cycles to failure. The test data was separated into two groups depending on the laminate width used: 89 mm or 51 mm. As Figure 9 (a) indicates, the two specimens where the highest laminate force was developed during the first loading cycle eventually failed by concrete clear cover peel off. Furthermore, the data follow two distinct trends depending on laminate width, as shown by best-fit lines drawn through the data points. The ratio of laminate force to laminate width was used to generate the data shown in Figure 9 (b). It can be seen that, to achieve comparable increases in fatigue life, a higher decrease in laminate force per unit width was required in specimens with 51-mm laminates compared with those with 89-mm laminates. Fatigue life in specimens strengthened using 51-mm wide laminates increased approximately 3 times (from 150,000 to 440,000 cycles) for a laminate unit force decrease of approximately 0.13 kN/mm, while specimens strengthened with 89-mm laminates had approximately a six-fold increase in fatigue life (from 130,000 to 779,000) for a laminate unit force decrease of only 0.03 kN/mm. This behavior could be attributed to wider laminates being able to restrain crack opening more efficiently than narrower laminates, therefore delaying fatigue failure of reinforcing steel.

Deflection and Strain Variation with Cycling

Increase in deflection was used as another measure of damage accumulation with cycling. Mid-span deflection increased markedly during the first 10 to 20 thousand cycles as new cracks formed and existing cracks extended. Crack formation and growth resulted in a reduction in effective moment of inertia of the specimens, which caused an increase in deflections. After approximately 20 thousand cycles deflections did not increase significantly for the majority of the specimens except those failing by concrete cover peel off. Figure 10 shows the midspan deflection comparison between a specimen that failed by steel fatigue fracture (specimen B-1-2-70) and one that failed by concrete peel off (specimen B-2-2-70). Mid-span deflection in specimen B-2-2-70 increased significantly after 100 thousand cycles of loading as a result of initiation of cover peel off. Contrastingly there was only a slight increase in deflection in specimen B-1-2-70 after initial cycling up to approximately 20 thousand cycles. The behavioral trends observed in these two specimens are also representative of behavior observed in other specimens in this research.

Strains in the specimens were expected to follow a similar variation as deflection. Figure 11 shows the strain variation in the concrete, steel, and CFRP laminates as a function of number of load cycles in specimen B-1-2-70. In this figure negative values represent compressive strains (concrete strains) and positive values represent tensile strains (steel and CFRP strains). The figure indicates that concrete and steel strains follow the expected variation, that is, significant strain increase during the first few thousand cycles of loading with little increase afterwards. Contrastingly one of the CFRP strain gages exhibited a marked reduction, particularly after 150 thousand cycles of loading. These types of variations were not uncommon in other specimens so definitive conclusions about damage progression could not be made using strain gage data. From these results, it was found that displacement measurements provided the most reliable indicator of damage progression in these tests.

SUMMARY AND CONCLUSIONS

Thirteen small-scale beams were strengthened using CFRP composites and tested under fatigue loading. The effect of composite configuration on the fatigue failure mode was investigated through these tests. Three values of peak stress applied to the tension reinforcing steel, which generated different stress ranges during application of repeated loads, were studied. Two primary fatigue failure modes were identified in these studies: fatigue fracture of reinforcing steel for beams subjected to moderate peak stresses (up to 70% of yield), or fatigue fracture of the concrete cover below the reinforcing steel for beams subjected to high steel stresses (between 70 and 80% of yield) and stiff composite laminates. The fatigue test results could be predicted reasonably well with an existing fatigue life model proposed for reinforced concrete beams. Wider laminates were more effective than narrower laminates to increase fatigue life of strengthened beams. Beam deflection was considered a more reliable indicator of damage progression than measured strains in the beams. Additional tests in larger beams are needed to verify these tests results and identify other parameters that might affect fatigue behavior (shear span, existing reinforcement ratio).

REFERENCES

ACI Committee 215 (1997). *Considerations for Design of Concrete Structures Subjected to Fatigue Loading*, American Concrete Institute, ACI 215R-74 (Reapproved 1997), Farmington Hills, MI, 24 pp.

ACI Committee 440 (2002). *Guide for the Design and Construction of Externally Bonded FRP Systems for Strengthening Concrete Structures*, American Concrete Institute, ACI 440.2R-02, Farmington Hills, MI, 45 pp.

Barnes, R. A. and Mayes, G. C. (1999). "Fatigue Performance of Concrete Beams Strengthened with CFRP Plates." *Journal of Composites for Construction*, Vol. 3, No. 2, May, pp. 63-72.

Breña, S.F., Wood, S.L., and Kreger M.L. (2003). "Full-scale Tests of Bridge Components Strengthened using Carbon Fiber-Reinforced Polymer Composites." *ACI Structural Journal*, Vol. 100, No. 6, November-December 2003, pp. 775-784.

Helgason T. and Hanson J.M. (1974). "Investigation of Design Factors Affecting Fatigue Strength of Reinforcing Bars – Statistical Analysis." *Abeles Symposium on Fatigue of Concrete*, SP-41, American Concrete Institute, Farmington Hills, MI, pp. 107-138.

Master Builders, Inc. (1998). *MBrace^{TM} Composite Strengthening System-Engineering Design Guidelines,* Cleveland, OH, 1998.

Moss, D.S. (1982). "Bending Fatigue of High-yield Reinforcing Bars in Concrete." *TRRL Supplementary Rep. 748*, Transport and Road Research Laboratory, Crowthome, U.K.

Papakonstantinou, C.G., Petrou, M.F., and Harries, K.A. (2001). "Fatigue Behavior of RC Beams Strengthened with GFRP Sheets." *Journal of Composites for Construction*, ASCE, Vol. 5, No. 4, pp. 246-253.

Table 1 — Measured material properties

Parameter	Value
Concrete compressive strength, MPa	Group A = 28.3, Group B = 29.9, Group C = 29.0
Steel yield stress, MPa	420
CFRP tensile strength, MPa	3800
CFRP modulus of elasticity, GPa	260
CFRP rupture strain	0.0146

Table 2 — Specimen Matrix

Specimen	No. of Plies	CFRP Width	CFRP thickness[†]	f_s/f_y[*]
		(mm)	(mm)	
A-1-4-Mono	1	89	0.17	--
A-Control	0	0	0	--
A-1-4-80	1	89	0.17	0.80
A-1-4-70	1	89	0.17	0.70
A-1-4-60	1	89	0.17	0.60
B-1-2-80	1	51	0.17	0.80
B-1-2-70	1	51	0.17	0.70
B-1-2-60	1	51	0.17	0.60
B-2-2-70	2	51	0.34	0.70
B-2-2-60	2	51	0.34	0.60
C-1-2-80	1	51	0.17	0.80
C-1-2-70	1	51	0.17	0.70
C-1-2-60	1	51	0.17	0.60

[†] Thickness based on fiber properties.
[*] Ratio of applied tension steel stress to nominal yield stress.

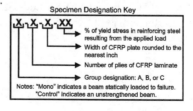

Specimen Designation Key

X-X-X-XX

% of yield stress in reinforcing steel resulting from the applied load
Width of CFRP plate rounded to the nearest inch
Number of plies of CFRP laminate
Group designation: A, B, or C
Notes: "Mono" indicates a beam statically loaded to failure. "Control" indicates an unstrengthened beam.

Table 3 — Summary of load parameters and failure modes

Specimen	Peak Stress, MPa	Stress Range, MPa	No. of cycles to failure	Failure mode
A-Control	368	310	183,674	Fatigue of Reinforcement
A-1-4-80	332	295	131,619	Concrete Clear Cover Peel Off
A-1-4-70	316	272	287,594	Fatigue of Reinforcement followed by CFRP Debonding
A-1-4-60	265	224	778,734	Fatigue of Reinforcement and Concrete Cover Peel Off
B-1-2-80	347	300	290,307	Fatigue of Reinforcement followed by CFRP Debonding
B-1-2-70	309	268	336,873	Fatigue of Reinforcement followed by CFRP Debonding
B-1-2-60	283	236	4,000,000+	Statically Loaded to Failure after 4 Million Cycles
B-2-2-70	328	270	150,000	Concrete Clear Cover Peel Off
B-2-2-60	256	199	2,000,000+	Statically Loaded to Failure after 2 Million Cycles
C-1-2-80U	220	175	500,000	No Failure
C-1-2-70U	229	182	500,000	No Failure
C-1-2-60U	193	137	500,000	No Failure
C-1-2-80S	320	326	326,775	Fatigue of Reinforcement followed by CFRP Debonding
C-1-2-70S	298	226	440,193	Fatigue of Reinforcement followed by CFRP Debonding
C-1-2-60S	227	189	4,000,000+	Statically Loaded to Failure after 4 Million Cycles

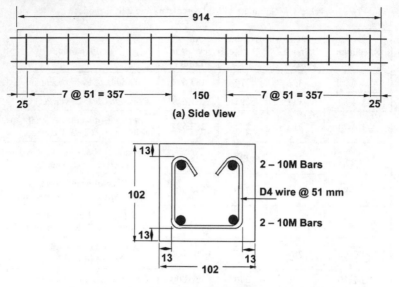

(a) Side View

(b) Cross Section

2 – 10M Bars

D4 wire @ 51 mm

2 – 10M Bars

Figure 1 — Specimen geometry and reinforcement

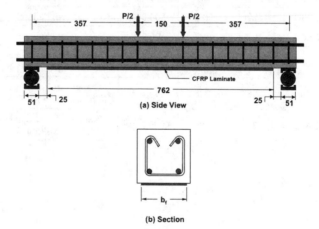

(a) Side View

CFRP Laminate

(b) Section

Figure 2 — CFRP strengthening configuration

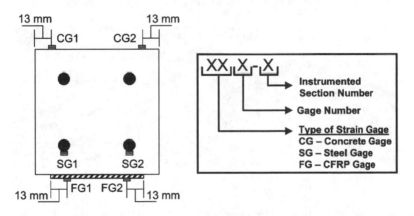

Figure 3 — Experimental setup used in the tests

Figure 4 — Specimen Instrumentation

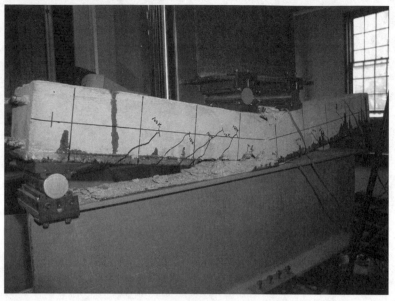

Figure 5 — CFRP debonding after fatigue fracture of reinforcing steel
(Specimen A-1-4-70)

(a)   (b)

Figure 6 — Damage progression in specimen A-1-4-80; (a) Formation of crack at end of
CFRP laminate, and (b) extension of end crack along concrete cover and shear span

(a)   (b)

Figure 7 — Damage progression in specimen B-2-2-70; (a) crack initiation at end of
CFRP laminate, and (b) concrete cover failure

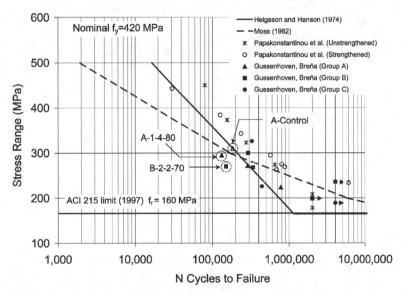

Figure 8 — Comparison between fatigue life models and results of tests

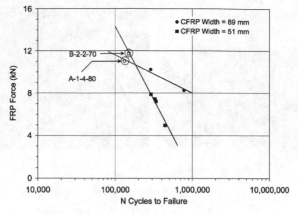

(a) Fatigue life as a function of CFRP laminate force in first cycle

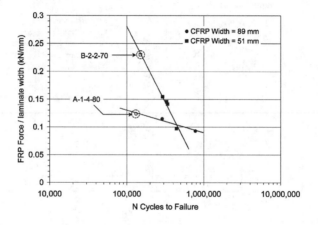

(b) Fatigue life as a function of CFRP laminate force/width in first cycle

Figure 9 — Effect of laminate width on fatigue life

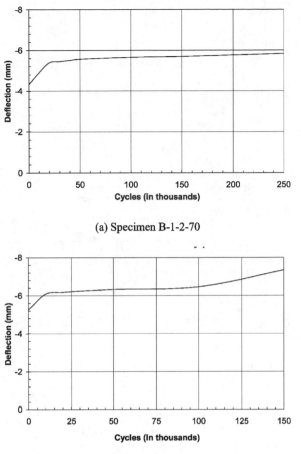

(a) Specimen B-1-2-70

(b) Specimen B-2-2-70

Figure 10 — Variation of mid-span deflection with number of applied cycles

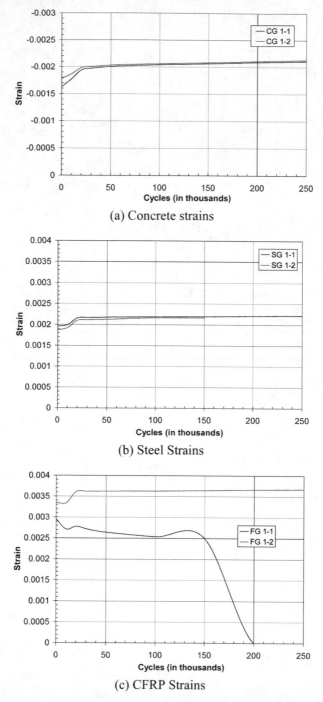

(a) Concrete strains

(b) Steel Strains

(c) CFRP Strains

Figure 11 — Strain variation with cycling (Specimen B-1-2-70)

Steel-Free Hybrid FRP Stiffened Panel-Concrete Deck System

by L. Cheng and V.M. Karbhari

Synopsis: This paper presents a design, analysis, and characterization of a hybrid deck system incorporating a thin Fiber Reinforced Polymer (FRP) stiffened hat-panel configured stay-in-place formwork that serves as flexural reinforcement with steel-free concrete poured on top. Quasi-static tests were conducted to first investigate the flexural behavior of the system. To understand the deck performance under traffic loads that induce repetitive stress cycles during the service life, a two-span continuous deck specimen (1.22 m wide) was tested by subjecting it to a total of 2.36 million cycles of load that simulates an AASHTO design truck with inclusion of the impact factor at both low and high magnitudes. The concrete-panel interfacial response due to the presence of sand and interlocking ribs was characterized by performing a series of 610 mm wide deck section tests, the results of which were used to calibrate a finite-element (FE) based analytical model. The effect of the shear span-to-depth ratio, carbon fiber reinforcement ratio, and rib spacing were then evaluated by performing a parametric study using the calibrated nonlinear FE model. A simplified design approach is also proposed.

Keywords: analysis; bridge deck; characterization; fiber reinforced polymer (FRP); finite element (FE); steel-free

Lijuan (Dawn) Cheng is a graduate research assistant in the Department of Structural Engineering at the University of California, San Diego. She is currently a PhD candidate and also a member of ASCE, SAMPE and ACI.

Vistasp M. Karbhari is a professor in the Department of Structural Engineering at the University of California, San Diego.

INTRODUCTION

As a result of the large percentage of concrete bridges in the United States that are structurally deficient due to corrosion of the embedded steel reinforcement and attendant concrete degradation in the bridge deck (Plecnik 1991), rapid replacement or rehabilitation methods are strongly desired. Developing new modular systems that consist of readily replaceable deck modules and that utilize high performance materials, such as advanced fiber reinforced polymer composites, is a potential solution to accomplish this. FRP composites are beneficial in bridge construction due to their high strength-to-weight ratio, low maintenance cost, and greatly increased durability with higher corrosion resistance compared to conventional steel reinforced concrete (RC) decks. However, in order to have any new system using innovative material stand up in the construction field, the evaluation of their application in prototype structures is essential so as to demonstrate their construction and performance capacity will achieve the designer's objectives.

Research on the application of FRP composites in bridge decks has advanced promisingly during the past twenty years. They have been considered as replacements for steel reinforcement in the concrete slabs in the form of tendons by Hassan et al. (2000), as 2D or 3D continuous grids and gratings by Yost and Schmeckpeper (2001), and as flat panels by Reising et al. (2001). Bridge decks consisting of pure polymer composites have also been studied by Karbhari et al. (2000). Due to the fact that the strength of a highway bridge deck is sensitive to repeated stressing, the fatigue behavior of bridge decks has also been extensively studied by Karbhari et al. (2000), GangaRao et al. (1999), Matsui et al. (2001), and Kwon et al. (2003). This paper is an attempt in the same direction, wherein the static and dynamic behavior of a recently developed hybrid modular deck system that is free of steel reinforcement and combines FRP composites and concrete material was evaluated. In this system, a steel-free concrete slab is cast on top of a carbon fiber reinforced FRP stiffened deck panel (as shown in Figure 1) that serves as both the flexural reinforcement and stay-in-place form for the concrete slab. To enhance the slab-deck interaction, the panel surface is sand treated and installed with discrete shear ribs. The objective of this study is to fully characterize the structural response of this prototype FRP-concrete deck system via a combination of analytical simulation and full-scale experimental investigation on the static flexural-shear response and the fatigue behavior under moving traffic loads. A finite element based analytical model was developed that demonstrated good agreement with the static test results; this was further utilized to perform a general parametric study of the system considering the critical design variables such as the shear span-to-depth ratio and the amount of carbon fiber reinforcement. The major results of this research work are discussed herein.

STATIC RESPONSE

Materials and setup

The composite deck panel, as shown in Figure 1, consists of a bottom plate (2.254 m long and 6.3 mm thick) with end hooks for slab-girder connection and adhesively bonded rectangular stiffeners filled with foam. The bottom plate is composed of 8 layers of unidirectional carbon fabric with an areal weight of 305 g/m^2 and 4 layers of E-glass chopped strand mat with an areal weight of 458 g/m^2, in a symmetric lay-up scheme of $[C/E/C_2/E/C]_S$. The rectangular stiffeners contain primarily unidirectional carbon fabric and foam core and are adhesively bonded onto the bottom plate. To enhance the shear interaction between the concrete slab and deck panel, the top surface of the panel is sand treated and additionally installed with shear ribs made of sand-epoxy paste. Five 610 mm wide testing specimens (SF1–SF5) were constructed from a steel-free concrete slab and such deck panel with different spacing for rectangular stiffeners and interfacial ribs (as summarized in Table 1). All the specimens were constructed with casting steel reinforced concrete blocks at both ends to simulate the condition of fixity to the supporting girders. The test setup is illustrated in Figure 2, where the specimen was simply supported by a roller at one end, a pin at the other, and quasi-statically loaded at mid-span from the top under a double-rod hydraulic actuator through an elastomeric loading pad up to different service and strength levels in relation to the AASHTO wheel load (98 kN), which is the AASHTO's HS-20 truck wheel load of 73 kN with the consideration of an impact factor of 33%. A load factor of 1.75 was used for STRENGTH I design level according to AASHTO (2004).

Observations and results

The displacement and strain response, crack pattern, and failure mode of each specimen were monitored during the test. The first three specimens (SF1–SF3) had a similar structural behavior with a stiffener spacing of 305 mm on center in each. Flexural cracks were found to first grow vertically near the bottom of slab at mid-span when the load reached 98 kN. As the load increased, additional cracking occurred at the locations of the interfacial ribs and further propagated diagonally toward the load point, as shown in Figure 3, due to the combined effect of flexural and shear stresses, i.e. principal tensile stresses. A sudden flexure-shear crack occurred in the slab at failure at a load carrying capacity similar among the three specimens, regardless of the spacing of the ribs. This is mainly due to the existence of the concrete end blocks at the ends, which provided a restraining effect from the ends such that the shear force could still be transferred from the slab to the deck even without the ribs at the interface. The ultimate capacity of SF4 with a stiffener spacing of 610 mm and SF6 without stiffeners were found to be about 17% and 43%, respectively, lower than that of the control panel SF1, indicating a significant effect of the spacing of the stiffeners on the global flexural capacity of the entire deck. Both specimens failed in a very similar flexure-shear mode as before. These cracks were found to propagate into the concrete end blocks in all the specimens except for SF1. The load vs. mid-span displacement response is shown in Figure 4 and Figure 5, where a largely linear-elastic behavior can be seen up to the ultimate load level that reached approximately twice the factored AASHTO wheel load demand for all the specimens (STRENGTH I, as represented by the top dashed line in the figure). The compressive strains in the concrete slab and tensile strains in the carbon fiber composite panel were all found to satisfy the code specified limit of 0.003 for concrete (ACI 1995) and the design specified allowable of 0.005 for carbon fiber reinforced composite at strength level (simplified recommendation for CFRP in the current design with the consideration of a reasonably large safety factor).

FATIGUE BEHAVIOR

Test configuration

The test specimen was constructed from two 1220 mm wide, 2254 mm long, and 6.3 mm thick FRP composite deck panels placed side-by-side, with a 196 mm thick steel-free concrete slab cast on top of the panels. The double span was selected for the specimen construction in order to include the continuity effect in the slab as would be experienced in the field application. A layer of carbon fiber reinforced mesh/grid was embedded in the concrete close to the top surface of the section at the middle one-third span region, serving as the tensile reinforcement at the negative bending moment region. The specimen was simply supported and loaded in a sinusoidal waveform by two 1828.8 mm apart patch loads of 84 kN via two double-rod hydraulic actuators to simulate one axle of the AASHTO truck wheel load while considering the dynamic load allowance factor. The setup configuration and the loading protocol are shown in Figure 6 and Figure 7, respectively. The specimen underwent 2.1 million cycles of fatigue service load and 250,000 cycles of doubled and 10,000 cycles of tripled fatigue load, after which the specimen was monotonically loaded up to failure. Deflections and strains were monitored using linear potentiometers and lead attached strain gages.

Results and discussion

During the fatigue stage, no tensile cracks were observed on the vertical sides of the specimen. Hairline cracking only formed on the top surface of the slab above the middle support at the end of the 2 million cycles of fatigue service load. The crack width satisfied the serviceability limit state per the code requirement. The maximum deflection of the structure under fatigue service load was found to be within the deflection-to-span ratio limit per the design code specification, satisfying the serviceability limit state with respect to the deflection limit. The structure had suffered no stiffness degradation during the first 2 million cycles of fatigue service load based on the small variation in the observed structural response. However, a substantial degradation of 37.6% was found during the subsequent 250,000 cycles of doubled fatigue service load and 44% during the further 10,000 cycles of tripled load (see Figure 8), indicating the higher the magnitude of the wheel load, the larger the amount of degradation in the system. The residual displacement in the system under all the fatigue load conditions was found to be insignificant and display a largely elastic and stable manner as seen in Figure 9, indicating no slippage at the slab-deck interface. The tensile strain and compressive strain experienced in the FRP composites and concrete material were found to be well below the design allowables. The existence of the carbon fiber composite mesh/grid that was embedded in the middle support of the negative bending moment region was found to be beneficial in controlling the crack width and post cracking behavior of the concrete slab. Therefore, a minimum amount of fiber mesh is highly recommended and should be designed based on the concrete tensile stress demand in those regions.

PARAMETRIC STUDY

Objectives

The inclined flexural-shear type of failure as discussed in the previous static tests is mainly caused by the combined shear and flexure, which is affected by many principal variables including the shear span-to-depth ratio, the amount of carbon fiber reinforcement, the spacing of the shear ribs, the tensile strength of concrete, the size of deck, the axial forces applied on the section, etc. Extensive experimental tests on these effects are desired but oftentimes not economically allowed due to the high cost of composite panels and the time-consuming nature of the lab tests. In addition, the flexure-shear failure load is hard to predict using conventional reinforced concrete theories since the stress state in the cracked section varies as the crack propagates from flexure to shear, which causes a stress concentration near the head of the crack. For this reason, therefore, a semi-analytical/semi-empirical approach is developed in this study with the major objectives of: a) experimentally characterizing the sand-rib interfacial property through shear bond-slippage tests; b) developing a refined analytical model that is validated by experimental results; c) performing a parametric study using the refined analytical model to evaluate the effect of selected design parameters.

Shear bond experiment

Three deck sections (SB1, SB2 and SB3), that each had the same slab width as previous but a slightly shorter span length (2.024 m instead of 2.254 m), were constructed without casting the reinforced concrete blocks at the ends so as to allow for the slippage at the interface between the slab and deck panel. Additionally, in order to examine the spacing effect of the shear ribs, the three specimens were designed to have a different rib spacing of 152 mm, 305 mm, and infinity (i.e. no ribs at all), respectively (see Table 1). A similar setup configuration to those for the static flexure tests was adopted here. Quasi-static cycles were included in the loading protocol during the tests in an attempt at detecting the interface debonding using the so-called modal vibration device. Measurements on the mid-span displacement, strain levels in composites and concrete, and slippage along the span length of the specimens were taken during the tests. Similar flexural-shear type of cracking was observed in SB1 and SB2, but specimen SB3 failed in a more flexure like mode. Horizontal cracking or debonding was observed at the interface between the concrete slab and the deck panel and it occurred much earlier in SB3 than the other two specimens. The ultimate capacities of SB1 and SB2 were found to be close to each other while SB3 showed a much lower capacity of about 37% (Figure 10). This is mainly attributed to the existence of the shear ribs that proved to be sufficient to provide the shear transferring between the slab and the panel. The displacement and strain levels experienced in SB3 are hence smaller than the other two, as seen in Figure 11.

Analytical model

An analytical model was created using the general-purpose finite element analysis software ABAQUS (2003). Figure 12 illustrates the assembled components and the mesh scheme of a half-span deck model utilizing symmetry. The composite deck panel was modeled with 4-node doubly curved general-purpose shell elements with reduced integration points (S4R) and 8-node linear brick elements (C3D8) were used for the modeling of the concrete slab. Linear elastic orthotropic properties were assumed for the composites. The concrete damaged plasticity model in ABAQUS was used to model the nonlinear behavior of concrete. This model uses the concepts of isotropic damaged elasticity in combination with isotropic tensile and compressive plasticity to represent the inelastic behavior of concrete.

Two mechanical characteristics exist at the interface between the concrete slab and the composite deck panel, i.e., the sand-bond surface and the shear rib feature. The former introduced a friction effect at the interface and the latter results in an interlocking or dowel type of mechanism that provides a much larger contribution than friction. To model the friction, the basic classical Coulomb friction model was selected in ABAQUS combined with the definition of surface interaction, where the two contacting surfaces are assumed to carry shear stresses up to a certain magnitude across their interface before they start sliding relative to one another. This stress is defined as critical shear stress τ_{crit}, at which sliding of the surfaces starts as a fraction of the contact pressure p between the surfaces ($\tau_{crit}=\mu p$), where μ is known as the coefficient of friction and was experimentally measured as 0.562 for this specific type of sand-bond surface. Figure 13a gives a two-dimensional representation for this model. The shear ribs at the interface were modeled with spring elements, which act between the node of panel and the node of slab, with its line of action being the line joining the two nodes. The relative displacement along this line stands for the relative slippage between the two surfaces. The behavioral property of the springs was represented by a proposed bond strength-slippage relationship as illustrated in Figure 13b. The springs are assumed to behave linear elastically before reaching their ultimate capacity. After that, a sudden failure occurs in them with a sudden load drop and the spring stiffness (i.e., force per relative displacement) goes down to zero. The spring stiffness during the elastic range, k, is defined by the shear bond strength of the interface, which was assumed to be 50 MPa/mm based on the experimental data for CFRP-concrete interface from the reference by Yoshizawa et al. (2000). The ultimate shear strength τ_{ult} is governed by the nominal shear strength of a typical epoxy resin. The failure modes considered in the analysis include the followings: (a) tensile failure in the carbon FRP composite in the deck panel when the strain level reaches the maximum allowable design limit that is defined as 50% of the maximum level (i.e., $\varepsilon_{cf}=0.5\times0.01=5000$ $\mu\varepsilon$); (b) compressive failure due to crushing in concrete at the top-of-slab when the compressive strain in concrete reaches the code specified ultimate level, i.e., $\varepsilon_{cu}=-3000$ $\mu\varepsilon$ (ACI 1995); (c) inclined flexural-shear failure when the diagonal shear crack forms, i.e., the maximum principal stress along the path of the crack, σ_{max}, exceeds the tensile strength of concrete, f_{ct}; (d) anchorage failure in ribs when the shear stress in springs exceeds the ultimate level as aforementioned.

Figure 14 and Figure 15 compare the analytically obtained mid-span displacement and maximum strain response with the experimental data in SB1, which shows a fairly good correlation between them but with a larger analytical load capacity than the test. This is mainly due to the fact that multiple sequences were introduced into the loading protocol through the entire test up to failure, which resulted in large stiffness degradation in the system because of the accumulated damages, and in turn, hindered the specimen from reaching its ultimate capacity, as it should have if monotonic load was applied up to failure. Due to the use of the concrete damaged plasticity model that is not based on the smeared crack approach, the notion of cracks developed in the slab was not visualizable. However, by introducing the concept of effective crack direction, a graphical visualization of the cracking patterns can be obtained, the results of which on SB1 is compared in Figure 16. The direction of the vector normal to the crack plane is parallel to the direction of the maximum principal plastic strain and the length of the colored vector is proportional to the amount of cracking. Concrete crushing was analytically obtained at ultimate strain, at which stage the potential formation of inclined crack was seen from the principal stress contour as shown in Figure 17. The maximum shear stress in the interfacial ribs was found to be much smaller than the ultimate strength. The developed analytical model was further verified via comparisons with test results from SB2. The load-displacement response and crack pattern are displayed in Figure 18 and Figure 19.

Design parameters

The inclined cracking behavior of the FRP-concrete deck is affected by many variables of which only a subset is studied herein, including the tensile strength of concrete (f_{ct}), the shear span-to-depth ratio (a/d), the amount of carbon fiber reinforcement (ρ_c), and the spacing of the ribs (s). The other geometrical and material parameters, such as the depth of the slab and the modulus of composites, had the same values as the current design and remained constants during the parametric study. It is known that the tensile strength of concrete (f_{ct}) can be directly related to the compressive strength (f_c') and also the typical range of f_c' for bridge deck application is from 16.5 MPa to 68.9 MPa, according to AASHTO. Two boundary cases and another typical design with f_c' equal to 43.9 MPa were selected here. The shear span-to-depth ratio (a/d) in concrete members without vertical stirrups has been found to likely determine the type of failure experienced in the member (MacGregor 1997). For this reason, a range from 2.5 to 6.5 was selected for a/d, in which cases the inclined flexural-shear type of failure typically occurs. The number of carbon fiber reinforcement layers at the bottom of the deck panel was allowed to vary from 4 layers to 12 layers, resulting in a plate total thickness within [4.37−7.42 mm]. The symmetry design in the laminate was maintained for all the cases. Two types of rib spacing were considered, namely, 152 mm and 305 mm. The deck system without shear ribs at the panel surface is not recommended based on the poor structural response observed in the test.

Figure 20 demonstrates the effect of different concrete strengths on the load-displacement response of the system. It can be seen that the load capacity can be increased 68.3% by changing the concrete strength from 16.5 MPa to 43.9 MPa and about 25.9% from 43.9 MPa to 68.9 MPa, indicating a larger gain in capacity as concrete strength increases at lower strength levels. The influence of the carbon fiber amount is illustrated in Figure 21 and Figure 22 for different rib spacings. The load capacity increases very linearly after the carbon fiber amount is increased up to 6 layers. Figure 23 and Figure 24 show the effect of shear span-to-depth ratio on the load capacity, which decreases quite uniformly as the deck gets slender. Slight differences in the load capacity can be seen between the deck with a rib spacing of 152 mm and the one with 305 mm, but is less substantial compared to the effect of the other variables.

DESIGN APPROACH

Composite deck panel

During the construction stage before the concrete sets, the composite deck panel itself should be designed to carry the concrete deck load, the composite deck dead load, and the construction live load. The maximum stress and strain levels in the deck for bending and shear can be computed by elastic theory from its sectional properties and moment and deflection coefficients determined from static analysis based on classical beam theory.

Shear-bond strength

As previously mentioned, the inclined shear crack loads are very difficult to predict due to the change in stress state within the cracked section. Therefore, a semi-empirical equation combining analytical simulations and statistical regression analysis is developed in this study to calculate the inclined shear capacity for the deck system. The ACI Building Code (1995) provides a shear equation in a form similar to the modified equation adopted herein as equation (1), which is based on the hypothesis that failure is initiated by the diagonal tension cracking,

$$\frac{V_u s}{b d} = \frac{m \rho d}{L'} + k \sqrt{f_c'} \tag{1}$$

In equation (1), V_u is the shear load, b is the deck width, and d is the distance from the extreme compressive fiber to the centroid of the deck. The reinforcement percentage, ρ, is given by A_s/bd. L' is the shear span length and f_c' is the compressive strength of concrete. The parameters, m and k, are the constants that are typically determined from an empirical fit to testing data. The term s accounts for the spacing of the shear transferring ribs. In this study, the specimen cases (60 in total) developed for the parametric study, where the analytical model was finely validated from the experimental tests, were used in deriving the shear design equation. A plot is made of the parameter, $V_u s/(bd\sqrt{f_c'})$, as ordinates X and $\rho d/(L'\sqrt{f_c'})$ as abscissas Y (see Figure 25). Two separate linear regressions are then performed with respect to the two rib spacing cases to determine the slope, m, and the zero intercept, k. In order to account for the insufficient specimen cases and variations that occur in the numerical results, a reduced regression line obtained by reducing the slope and intercept, respectively, of the original regression by 15% is proposed. The design equations as such obtained are given as follows,

$$\begin{cases} \dfrac{V_u s}{b d'} = \dfrac{19.983 \rho_c d'}{L'} + 0.0139\sqrt{f_c'} & \text{(for rib spacing } s=152 \text{ mm)} \quad (2a) \\[3mm] \dfrac{V_u s}{b d'} = \dfrac{44.509 \rho_c d'}{L'} + 0.0206\sqrt{f_c'} & \text{(for rib spacing } s=305 \text{ mm)} \quad (2b) \end{cases}$$

The confidence level of both of the regression analyses was about 95% and the R^2 statistic or the coefficient of determination for these equations is about 0.63 and 0.88, respectively. By combining dimensional analysis techniques with statistical regression analysis, alternative design formula by Zsutty (1968) uses an expression involving an exponent of one-third of the parameters $(f_c'\rho_c d'/L')$ in the following form

$$\frac{V_u s}{bd'} = m\left(\frac{f_c'\rho_c d'}{L'}\right)^{1/3} + k \qquad (3)$$

Following the similar procedure as before, a plot was made of the parameter $(f_c'\rho_c d'/L')^{1/3}$ as abscissa X and $V_u s/(bd')$ as ordinate Y, and the design equations thus obtained are described by Eq. (5-6) with better R-square values of 0.92 and 0.91 for cases with a rib spacing of 152 mm and 305 mm, respectively. These equations should be recommended for the use in the design.

$$\begin{cases} \dfrac{V_u s}{bd'} = 0.750\left(\dfrac{f_c'\rho_c d'}{L'}\right)^{1/3} - 0.366 \quad \text{(for rib spacing } s=152 \text{ mm)} \qquad (4a) \\[3mm] \dfrac{V_u s}{bd'} = 1.511\left(\dfrac{f_c'\rho_c d'}{L'}\right)^{1/3} - 0.770 \quad \text{(for rib spacing } s=305 \text{ mm)} \qquad (4b) \end{cases}$$

Flexural strength

The flexural capacity of the deck system can be obtained from the compatibility of strains and the equilibrium of internal forces along the section (see Figure 26), following the similar principle as that for the reinforced concrete structures. The controlling strain herein is either the maximum compressive strain of 0.003 in concrete or the allowable tensile strain of 0.005 in carbon fiber composites. The nominal moment strength, M_u, is obtained as a simple summation of internal moments of all the tensile and compressive forces on the cross section that is considered. Other design considerations such as deflection limitation should also be included.

SUMMARY

An extensive testing program has been conducted on a new deck system, which was based on the concept of using hybrid FRP-concrete materials with modular construction. The static flexure-shear response and the fatigue behavior of the system have been fully characterized, and the results were found to meet the code specified limits and requirements. A finite element based analytical approach was developed and the validity as well as the accuracy of this approach has been well demonstrated through the comparisons of the results with the experimental data. A parametric study primarily concerned with the design variables of concrete strength, shear span-to-depth ratio, carbon fiber amount, and shear rib spacing was performed.

Since only a very thin layer of CFRP composite plate (6 mm) is used in the 203 mm thick concrete slab, the material cost in the proposed system is comparable to the conventional RC system and other currently available steel-free concrete deck systems. Also considering the cost savings from installing and demolding the formwork for the concrete slab and the associated cost-consuming labor, as the construction of conventional RC deck would have, the overall market cost of this system is very competitive. Additional cost savings can be gained in the long-term maintenance cost during the service-life of the structure due to the significantly improved durability in the proposed system.

Based on the findings from the current study, the following notes shall be recommended for the design of this deck system: (a) The rectangular stiffeners on the deck panel should not be designed to carry the construction load only, but also should take into account the flexural strength of the entire FRP-concrete deck system; (b) Based on the fatigue behavior of the deck system, the fatigue limit state was found to be not a governing state in the design as long as a minimum amount of composite fiber mesh was provided at regions where concrete tensile stress exists; (c) Eq. (4) can be utilized to design the shear-bond strength for the FRP-concrete deck system.

The range of the parametric study herein was selected for the current research interests. Extension of this range to other existing variables, such as the concrete strength, and inclusion of further parameters, such as the depth of the slab, can be included in future work. Other potential effort can be focused on the long-term behavior of the deck system due to creep and shrinkage, and the temperature and moisture environment.

REFERENCES

AASHTO LRFD Bridge Design Specifications (2004), 3rd Edition, American Association of State Highway and Transportation Officials, Washington, D.C.

ABAQUS/Standard User's Manual (2003), Version 6.4, ABAQUS Inc.

ACI 318-95 (1995), Building Code Requirements for Structural Concrete (ACI 318-95) and Commentary (ACI 318R-95), American Concrete Institute, Farmington Hills, MI.

GangaRao, H.V.S., Thippeswamy, H.K., Shekar, V., and Craigo, C. (1999), "Development of Glass Fiber Reinforced Polymer Composite Bridge Deck," *SAMPE Journal,* 35(4): 12-24.

Hassan, T., Abdelrahman, A., Tadros, G., and Rizkalla, S. (2000), "Fibre reinforced polymer reinforcing bars for bridge decks," *Can. J. Civ. Eng.,* 27:839-849.

Karbhari, V.M., Seible, F., Burgueno, R., Davol, A., Wernli, M. and Zhao, L. (2000), "Structural Characterization of Fiber-Reinforced Composite Short-and Medium-Span Bridge Systems," *Applied Composite Materials*, 7 (2/3), pp. 151-182.

Kwon, S.C., Dutta, P.K., Kim, Y.H., and Lopez-Anido, R. (2003), "Comparison of the fatigue behaviors of FRP bridge decks and reinforced concrete conventional decks under extreme environment conditions," *KSME International Journal*, 17(1): 1-10.

MacGregor, J.G. (1997), *Reinforced Concrete: Mechanics and Design*, 3rd edition, Prentice-Hall Inc., New Jersey.

Matsui, S., Ishizaki, S. and Kubo, K. (2001), "An Experimental Study on Durability of FRP-RC Composite Deck Slabs of Highway Bridges," *Proceedings of the 3rd International Conference on Concrete Under Severe Conditions: Environment & Loading,* Vancouver, Canada, pp. 933-940.

Plecnik, J.M., and Azar, W.A. (1991), "Structural components, highway bridge deck applications," *International Encyclopedia of Composites*, 6: 430-445.

Reising, R.M.W., Shahrooz, B.M., Hunt, V.J., Lenett, M.S., Christopher S., Neumann, A.R., Helmicki, A.J., Miller, R.A., Kondury, S., and Morton, S. (2001), "Performance of five-span steel bridge with fiber-reinforced polymer composite deck panels," *Transportation Research Record 1770*, Paper No. 01-0337, pp.113-123.

Yoshizawa, H., Wu, Z.S., Yuan, H., and Kanakubo, T. (2000), "Study on FRP-Concrete Interface Bond Performance," *Journal of Materials, Concrete Structures and Pavements, JSCE*, 49:662, pp. 105-119.

Yost, J.R. and Schmeckpeper, E.R. (2001), "Strength and serviceability of FRP grid reinforced bridge deck," *Journal of Bridge Engineering*, 6(6):605-612.

Zsutty, T.C. (1968), "Beam Shear Strength Prediction by Analysis of Existing Data," Proceedings, *Journal of the American Concrete Institute*, November, 65(11): 943-951.

Table 1 – Summary of static tests

Specimen Number	Stiffener Spacing on panel (mm)	Rib Spacing on panel (mm)	P_{ult} (kN)	δ at Failure (mm)	Failure Mode	Crack in End Block
SF1[*†]	305 (2 stiffeners)	152	311.6	16.8	Combined flexural and diagonal shear	None
SF2	305 (2 stiffeners)	305	316.4	17.9		Some
SF3	305 (2 stiffeners)	No ribs	302.6	17.2		Some
SF4	610 (1 stiffener)	152	258.2	13.6		None
SF5	No stiffener	152	179.2	15.1		Some
SB1[*†]	305 (2 stiffeners)	152	321.4	14.5		No end block constructed
SB2	305 (2 stiffeners)	305	327.8	14.3		
SB3	305 (2 stiffeners)	No ribs	202.1	9.28	Flexure	

[*] Control panel as originally designed. [†] SF – Flexural specimen; SB – Shear bond specimen

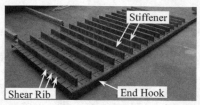

Figure 1 – Geometry of FRP composite deck panel

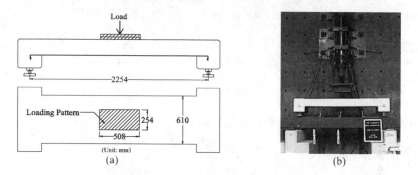

Figure 2 – (a) Geometry; (b) Configuration of static test setup

Figure 3 – Typical crack pattern for static test

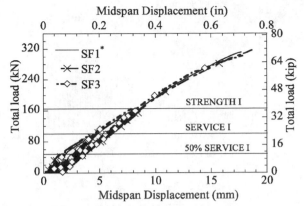

Figure 4 – Load vs. displacement response for SF1, SF2 and SF3

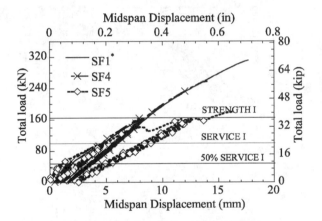

Figure 5 – Load vs. displacement response for SF1, SF4 and SF5

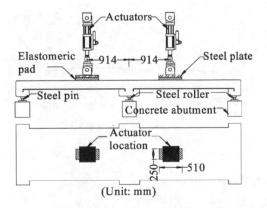

Figure 6 – Fatigue test setup

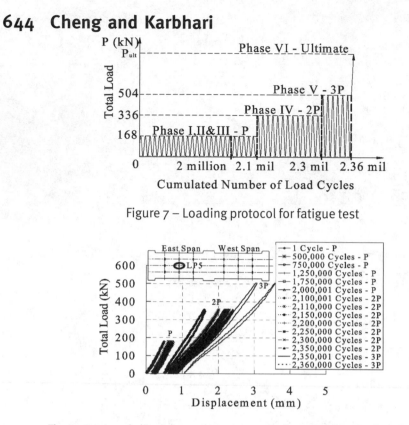

Figure 7 – Loading protocol for fatigue test

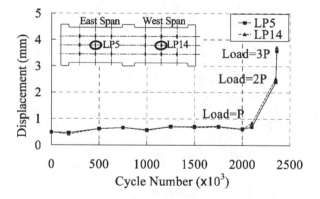

Figure 8 – Load-displacement response before ultimate for fatigue test

Figure 9 – Mid-span displacement history during fatigue test

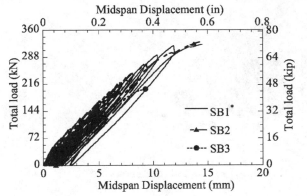

Figure 10 – Load-displacement response of shear-bond test

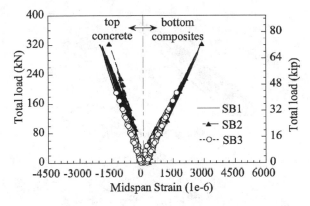

Figure 11 – Load-strain response of shear-bond test

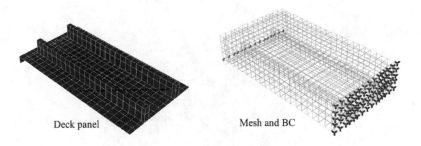

Figure 12 – Finite element modeling

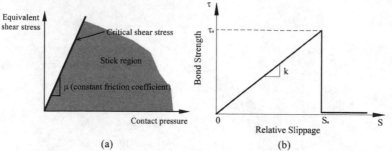

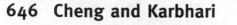

(a)   (b)

Figure 13 – (a) Friction model; (b) Proposed bond strength-slippage model

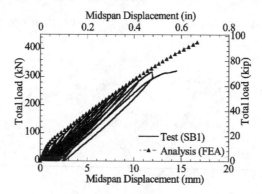

Figure 14 – Verification of FEA model for SB1 from the load-displacement response

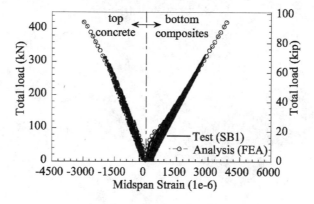

Figure 15 – Verification of FEA model for SB1 from load-strain response

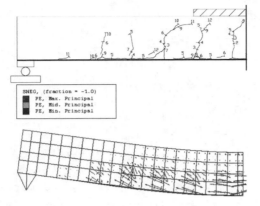

Figure 16 – Comparison of crack pattern in SB1

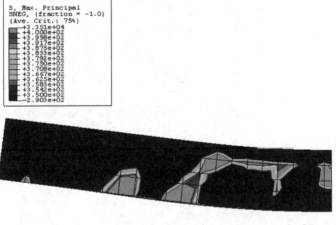

Figure 17 – Comparison of principal stress contour in SB1

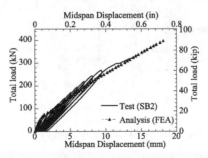

Figure 18 – Verification of FEA model for SB2 from the load-displacement response

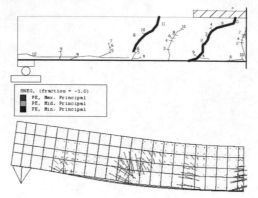

Figure 19 – Comparison of crack pattern in SB2

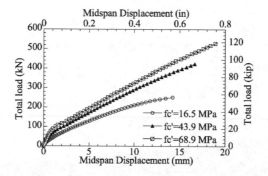

Figure 20 – Effect of concrete strength with rib spacing of 152 mm

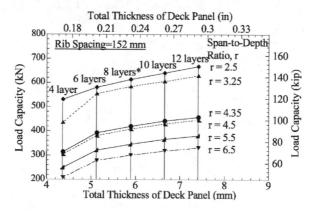

Figure 21 – Effect of carbon fiber amount for cases with rib spacing of 152 mm

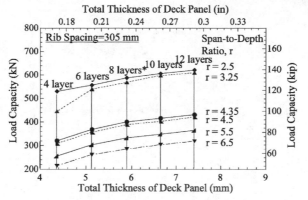

Figure 22 – Effect of carbon fiber amount for cases with rib spacing of 305 mm

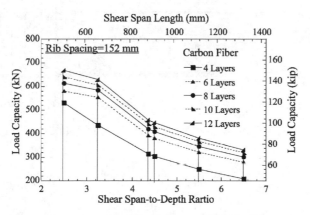

Figure 23 – Effect of span-to-depth ratio for cases with rib spacing of 152 mm

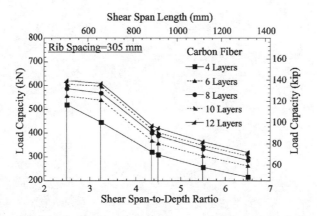

Figure 24 – Effect of span-to-depth ratio for cases with rib spacing of 305 mm

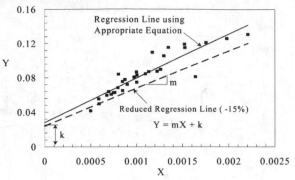

Figure 25 – Plot for regression analysis

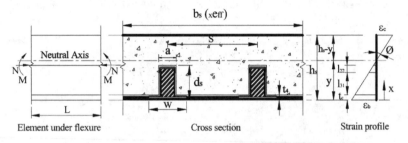

Figure 26 – Typical cross section of FRP-concrete deck system

Rationale for the ACI 440.1R-06 Indirect Deflection Control Design Provisions

by C.E. Ospina and S.P. Gross

Synopsis: Compared to ordinary steel reinforcement, Fiber-Reinforced Polymer (FRP) reinforcing bars have a lower stiffness, display a brittle-elastic response, and possess particular bond characteristics. The dependence on these distinctive features makes deflection control in FRP-reinforced concrete beams and one-way slabs a more elaborate process compared to the traditional serviceability design of steel-reinforced members. This paper reports the rationale and fundamental concepts backing the indirect deflection control procedure for concrete beams and one-way slabs reinforced with FRP bars adopted by ACI 440.1R-06. The fundamental procedure can be applied regardless of the type of reinforcement; it is independent of the member's stiffness through the cracked stage; and it is expressed as a function of the deflection-span ratio, which allows designers to fully control deflections depending on applicable serviceability limits. The paper also explains the simplifications made to the fundamental procedure that led to the development of the indirect deflection control procedure in tabular form found in ACI 440.1R-06, including the method by which tension stiffening effects are accounted for.

Keywords: deflection; FRP-reinforced members; indirect deflection control; one-way slabs; serviceability design; tension stiffening

Carlos E. Ospina, Ph.D., P.E., is a Senior Engineer with Berger/ABAM Engineers Inc., USA, where he has been involved in the structural design/assessment of industrial buildings, monorail guideways, nuclear plant cranes, underground facilities and waterfront structures. He is a Member of ACI 314, Simplified Design of Concrete Buildings, and an Associate Member of ACI 440, Fiber-Reinforced Polymers, ASCE-ACI 423, Prestressed Concrete, and ACI 318S, Spanish Translation of ACI 318.

Shawn P. Gross, Ph.D., is an Associate Professor in the Department of Civil and Environmental Engineering at Villanova University in Villanova, PA, USA. He is Secretary of Joint ASCE-ACI Committee 423, Prestressed Concrete; a Member of ACI Committee 363, High Strength Concrete, a Member of ACI Committee E803, Faculty Network Coordinating Committee, and an Associate Member of ACI Committee 440, Fiber-Reinforced Polymers.

INTRODUCTION

For serviceability design of FRP-reinforced members, ACI 440.1R-06 provides guidance on preliminary sizing of concrete beams and one-way slabs through an indirect deflection control procedure that is expressed in terms of minimum beam and one-way slab thickness requirements. The designer is then required to calculate deflections directly based on a modification of the traditional Branson's effective moment of inertia equation because, unlike in ACI 318-05, the indirect deflection control procedure in ACI 440.1R-06 is not intended to waive the direct deflection control calculation.

The procedure used here to develop the indirect deflection control provisions in ACI 440.1R-06 can be applied to any type of reinforcement. It is also independent of the member's stiffness, which is difficult to evaluate through the cracked stage, and it is directly related to a deflection-span ratio, which allows designers to control the serviceability design depending on applicable deflection limits. After presenting the general procedure, the paper **explains** the development of the minimum thickness table found in ACI 440.1R-06. The fundamental assumptions used in the simplification of the general procedure are identified, and the method by which the procedure is adjusted for tension stiffening effects is presented.

BACKGROUND

Deflections of reinforced concrete beams and one-way slabs can be controlled directly or indirectly. Direct deflection control refers to the calculation of deflections and their comparison with allowable limits. Direct methods span from traditional elastic theory to advanced finite element analyses. ACI Committee 435 (1974) and Branson (1977) report comprehensive summaries of classic direct deflection control procedures for steel-reinforced concrete beams and flat plates. Indirect deflection control procedures limit deflections by determining maximum span-depth ratios, minimum depths, or minimum tension reinforcement ratios (ACI Committee 435, 1978) that satisfy a given deflection-span ratio, Δ_m/L. The latter is defined by experience. In the context of steel-reinforced members, Branson (1977) recommends using indirect procedures for initial member proportioning and then checking deflections directly. Deflections can also be controlled by means of appropriate construction practice. Precambering and delaying removal of forms are some of the preferred options.

Deflection control provisions for steel-reinforced beams and one-way slabs in ACI 318-05 are concerned with deflections that occur at service levels due to immediate and sustained static loads. Effects from dynamic loads such as earthquakes, winds, or vibration of machinery are not considered. Two methods are given: i) the indirect method of controlling the minimum thickness of the member (ACI 318 Table 9.5a); and ii) the direct method of limiting computed deflections (ACI 318 Table 9.5b). The choice of method is left to the discretion of the designer. Table 1 shows values of minimum thickness for non-prestressed steel-reinforced beams and one-way slabs per ACI 318-05 expressed as maximum span-depth ratios. Table 2 shows the allowable deflections per ACI 318-05.

The direct deflection control design provisions for FRP-reinforced concrete beams and one-way slabs in ACI 440.1R-06 follow a format similar to that of ACI 318-05. Deflections are calculated directly using a modified version of Branson's effective moment of inertia equation, developed by Gao, Benmokrane and Masmoudi (1998). Based on recent data published by Yost, Gross, and Dinehart (2003) and other researchers, the equation used in earlier versions of ACI 440.1 was modified by the second author and adopted by ACI Committee 440 for inclusion in ACI 440.1R-06.

At its Washington D.C. 2004 Spring meeting, ACI Sub-committee 440H commissioned the authors to develop an indirect deflection control table similar to Table 9.5(a) of ACI 318 to define maximum span-depth ratios for FRP-reinforced concrete beams and one-way slabs, based on the indirect deflection control approach proposed by Ospina, Alexander, and Cheng (2001). The goal of the sub-committee was to provide designers guidance for preliminary sizing of members in the form of typical span-depth ratios required to satisfy serviceability design criteria. The indirect deflection control table developed by the authors, which may be found in Appendix A, was approved by ACI 440 in its San Francisco 2004 Fall meeting. Presentation of the rationale behind the development of this table constitutes the main subject of this paper.

PROPOSED INDIRECT DEFLECTION CONTROL PROCEDURE

As reported by Branson (1977), the instantaneous midspan deflection, Δ_m, of a reinforced concrete beam or one-way slab subjected to a uniformly distributed load can be calculated as

$$\Delta_m = K_1 \left(\frac{5}{48} \right) \left(\frac{M_m L^2}{E_c I_e} \right) \tag{1}$$

where M_m is the midspan bending moment, L is the span length, E_c is concrete's elastic modulus, and I_e is an effective moment of inertia. The constant K_1 depends only on boundary conditions, and is defined as

$$K_1 = 1.2 - 0.2 \frac{M_o}{M_m} \tag{2}$$

where M_o is the statical moment, i.e. $M_o = \frac{wL^2}{8}$, $K_1 = 1$ for simply supported spans, $K_1 = 0.8$ for fixed-hinged beams, and $K_1 = 0.6$ for fixed-fixed beams. For cantilevered spans, $K_1 = 2.4$, M_m is replaced with the moment at the support, and Δ_m is replaced with the deflection at the free end. In Eq. 2, both M_o and M_m result from the same loading.

Due to the difficulty in evaluating I_e to account for cracking effects on flexural stiffness, Eq. 1 can be rewritten independent of I_e (Ospina, Alexander and Cheng 2001) as

$$\Delta_m = K_1 \left(\frac{5}{48} \right) \psi_m L^2 \tag{3}$$

where ψ_m is the curvature at midspan. Assuming cracked-elastic behavior,

$$\psi_m = \frac{\varepsilon_{rm}}{d(1 - k_m)} \tag{4}$$

where ε_{rm} is the reinforcement strain at midspan (or support for a cantilevered span), d is the effective flexural depth, and k_m is the ratio of the neutral axis depth to the flexural depth, also at midspan (or support for a cantilevered span), calculated as

$$k_m = \sqrt{(n\rho_r)^2 + 2n\rho_r} - n\rho_r \tag{5}$$

where n is the modular ratio, $n = E_r / E_c$, and ρ_r is the reinforcement ratio. Dividing both sides of Eq. 3 by L and substituting Eq. 4 into 3 leads to

$$\frac{\Delta_m}{L} = K_1 \left(\frac{5}{48} \left(\frac{\varepsilon_{rm}}{d(1 - k_m)} \right) \right) L \tag{6}$$

Rearranging terms and setting $\eta = \frac{d}{h}$, the maximum span-depth ratio is given by

$$\frac{L}{h} \le \frac{48\eta}{5 K_1} \left(\frac{1 - k_m}{\varepsilon_{rm}} \right) \frac{\Delta_m}{L} \tag{7}$$

where η may be assumed to vary from 0.85 to 0.95.

In Eq. 7, both the reinforcement strain and the neutral axis location define a limiting curvature that is consistent with an allowable deflection-span ratio. The interdependence of the reinforcement strain, the deflection-span ratio and the span-depth ratio is further illustrated in Fig. 1, which shows the central portion (segment ABCD) of a beam with midspan deflection Δ_m and curvature ψ_m. If the deflection, and hence curvature, is to remain unchanged, an increase in the service strain in the reinforcement from ε_{rm} to ε'_{rm} must be accompanied by deepening the beam (segment ABC'D'), reducing the span-depth ratio, as implied by Eq. 7.

The merits of Eq. 7 are evident: it can be applied to members reinforced with either ordinary steel or FRP bars; it is independent of the member's flexural stiffness; and the direct dependency on the deflection-span ratio allows the designer to control the serviceability design based upon specific deflection limits. Table 3 shows a list of allowable deflection-span limits for structural, sensorial and aesthetic reasons, adapted from ACI 435.3R-68 (ACI Committee 435 1968), which can be used in conjunction with Eq. 7, depending on the specific application. A very comprehensive list of allowable deflection-span limits can be found in Branson (1977).

VALIDATION OF PROPOSED MODEL

Figure 2 shows the effect of the reinforcement strain level at midspan, ε_{rm}, and different span fixity conditions on the maximum span-depth ratio assuming $\Delta_m/L = 1/240$, $k_m = 0.195$, $\rho_r E_r = 96$ ksi (661.9 MPa), $\eta = 0.9$, and $f'_c = 5$ ksi (34.5 MPa). Values correspond to a one-way slab with a reinforcement ratio that is about 3.5 times the balanced reinforcement ratio prescribed by ACI 440.1R-06, thereby implying a compressive failure at ultimate. The variation in the span support conditions is represented by the M_o/M_m ratio. For instance, a value of M_o/M_m of about 2.0 simulates an edge span supported by a masonry wall at the edge, with the first interior support continuous. In a prototype interior span, M_o/M_m varies from about 2.8 to 3.0. In a simple span, $M_o/M_m = 1$.

Figure 2 shows how the maximum span-depth value decreases as ε_{rm} increases, as was observed conceptually in Fig. 1. Figure 2 also shows that the effect of boundary conditions on the span-depth ratio is more noticeable at lower reinforcement strain levels.

To validate the proposed indirect deflection control procedure, let us consider a one-way slab with ordinary steel reinforcement, assuming $\varepsilon_{rm} = 0.0012$. This strain value is roughly 60% of the yield strain of steel, which is often considered a target strain level in steel-reinforced concrete serviceability design. Assuming simple support conditions, Fig. 2 shows that $\varepsilon_{rm} = 0.0012$ leads to a span-depth ratio of 24.2. This value is close to the lower bound span-depth ratio of 20 prescribed by ACI 318-05 for simply-supported steel-reinforced one-way slabs. The latter is somewhat lower, because, as pointed out by Branson (1977), the span-depth ratios in ACI 318 include several modifying factors for different conditions, including long-term effects. Generally speaking, this comparison confirms that using a deflection-span ratio, Δ_m/L, equal to 1/240, where Δ_m is calculated on the basis of an instantaneous deflection under total service load, results in span-depth ratios that are relatively consistent with those given in ACI 318-05.

Assume now that the slab is reinforced with GFRP bars and that ε_{rm} is equal to 0.002. This FRP strain value is 5/3 greater than 0.0012, which is consistent with ACI 440.1R-06 rationale of allowing larger crack widths in members with FRP due to FRP's superior corrosion resistance. For $\varepsilon_{rm} = 0.002$ and assuming simple support conditions, Fig. 2 renders a span-depth ratio of 14.5 which means that the GFRP-reinforced slab needs be about 1.7 times thicker than the steel-reinforced slab, for the same span length. The increased depth is the result of allowing a higher reinforcement strain at service level. Note that if the target GFRP strain were 0.0012, the corresponding span-depth requirement would be close to 24.2. Evidently, since the elastic modulus of GFRP is lower than steel's, say $E_r = 6,000$ ksi (41,370 MPa), the GFRP-reinforced slab would have a reinforcement ratio that is 29,000/6,000 = 4.83 times that of the steel-reinforced slab. It is worth noting, however, that this comparison is rather simplistic because bond between GFRP bars and concrete differs from that between steel bars and concrete, which implies that the resulting span-depth limit may deviate from 24.2.

For comparable span lengths, the higher member depth requirement associated with FRP-reinforced beams and one-way slabs relative to their steel-reinforced counterparts has significant economic consequences. These translate directly into higher costs due to greater concrete volumes and high FRP reinforcement ratios. Thus far, however, the proposed indirect deflection control formulation has neglected the effect of tension stiffening (i.e. the tensile contribution of concrete between cracks) on deflection. It can be hypothesized that span-depth requirements in FRP-reinforced members may be relaxed if this effect is introduced in the indirect deflection control procedure. It is worth noting that the indirect deflection control table in ACI 318-05 does not explicitly account for tension stiffening effects.

TENSION STIFFENING EFFECT ON INDIRECT DEFLECTION CONTROL

To account for concrete's tension stiffening effect on the proposed indirect deflection control procedure, it is necessary to express the span-depth requirements as a function of the average curvature instead of the curvature at a crack. The procedure (Ospina, Alexander, and Cheng 2001) makes use of the tension stiffening model given by CEB/FIB MC90 (1990). Accordingly, Eq. 1 can be rewritten as

$$\frac{\Delta_m}{L} = \frac{5}{48} K_1 L \left[(1 - \xi) \psi_1 + \xi \psi_2\right] \tag{8}$$

For members with rectangular cross-section, the midspan curvatures ψ_1 (at uncracked section level) and ψ_2 (at fully cracked level) are defined as

$$\psi_1 = \frac{2 f_r}{E_c h} \tag{9}$$

$$\psi_2 = \frac{\varepsilon_{rm}}{d(1 - k_m)} \tag{10}$$

and

$$\xi = 1 - \beta_1 \beta_2 \left(\frac{M_{cr}}{M_m} \right)^2 \geq 0.4 \tag{11}$$

where M_{cr} is the cracking moment and the coefficients β_1 and β_2 characterize, respectively, the bond quality of the bars ($\beta_1 = 1.0$ for high bond bars) and the influence of load duration or repetition ($\beta_2 = 1.0$ for first loading). For FRP-reinforced concrete members subjected to short-term first loading, Hall (2000) recommends $\beta_1\beta_2 = 0.5$.

Substituting Eqs. 9 and 10 into Eq. 8, assuming $f_r = 7.5 \sqrt{f_c'}$ (f_c in psi),

$E_c = 57000 \sqrt{f_c'}$ (f_c in psi), $\eta = d/h$, and rearranging terms, leads to

$$\frac{L}{h} \leq \frac{48\eta}{5K_1} \left[\frac{1}{\left[(1-\xi)\dfrac{15\eta}{57000} + \xi \dfrac{\varepsilon_{rm}}{(1-k_m)} \right]} \right] \frac{\Delta_m}{L} \tag{12}$$

Figure 3 illustrates the tension stiffening effect on the maximum span-depth ratio prediction for the simply-supported slab of the validation example, for different service level-to-cracking moment ratios, M_m / M_{cr}, assuming $\beta_1\beta_2 = 0.5$. Equation 12 leads to a curve that approaches asymptotically a span-depth ratio of 14.5, which is that in absence of tension stiffening effects. The figure shows that concrete's tensile contribution significantly affects the maximum span-depth requirement, especially at load levels that are slightly greater than the cracking load; within these load levels, tension stiffening leads to an increased span-depth ratio, i.e. the member deepening requirement is relaxed. At higher load levels, the tension stiffening effect attenuates due to bond degradation.

INDIRECT DEFLECTION CONTROL PROVISIONS IN ACI 440.1R-06

Taking into account the inherent difficulties associated with the standardization of Eq. 12, some fundamental assumptions and simplifications were required to develop an indirect deflection control table for FRP-reinforced beams and one-way slabs. The procedure is described in this section. The table showing the recommended minimum thicknesses for the design of beams and one-way slabs with FRP bars is shown in Table 4.

The starting point was to establish that the table would be developed using Eq. 7, which not only requires less parameters to be known than does Eq. 8 but also allows for a simple tension stiffening correction: the one used in the direct deflection control provisions of ACI 440.1R-06. The authors felt that, as critical as it was to consider tension stiffening, it was just as important to consider an equation that was consistent with other provisions of the ACI 440.1R-06 document (recall that Eq. 12 is based on the CEB/FIP MC90 tension stiffening model).

The use of Eq. 7 in design is not straight forward because neither k_m nor ε_{rm} are known prior to a detailed analysis. For this reason, it was decided to conduct a parametric analysis based on typical conditions for FRP-reinforced concrete flexural members in order to determine both k_m and ε_{rm}. In the analyses, an arbitrary deflection-span limit of 1/240 (i.e. a maximum instantaneous deflection of $L/240$) was assumed under total service load. This value was not chosen to endorse a deflection limitation of $L/240$, but rather because of the relative consistency it provides with the span-depth limitations suggested in ACI 318-05, as indicated by the results presented in Fig. 2. For simplicity, explicit modifications for time-dependent behavior and other factors were not applied in deriving the ACI 440.1R-06 table; however, such modifications can implicitly be addressed by adjusting the limiting deflection-span ratio to account for assumed time-dependent deflection multipliers and ratios of dead load to live load. The assumed deflection-span ratio is clearly stated in the text of ACI 440.1R-06 to provide the designer guidance when applying the table. Since the table is only intended for preliminary member sizing, the designer is permitted to adjust a suggested minimum thickness based on less restrictive or more restrictive deflection-span limits.

Limiting span-depth ratios were first computed in the absence of tension stiffening effects, i.e. according to Eq. 7. Then, the tension stiffening effect was accounted for by multiplying the resulting span-depth ratios by the ratio of the effective moment of inertia to the cracked section moment of inertia, I_e/I_{cr}, where I_e is calculated using the modified Branson's equation for FRP, given in ACI 440.1R-06 as

$$I_e = \left(\frac{M_{cr}}{M_a}\right)^3 \beta_d I_g + \left[1 - \left(\frac{M_{cr}}{M_a}\right)^3\right] I_{cr} \leq I_g \tag{13}$$

where

$$\beta_d = \frac{1}{5}\left(\frac{\rho_f}{\rho_{bf}}\right) \leq 1 \tag{14}$$

In Eq. 14, ρ_f and ρ_{bf} are, respectively, the reinforcement ratio and the balanced reinforcement ratio for an FRP-reinforced member.

A description of the calculations performed in the analyses, including all fundamental assumptions, are provided in Table 5, which facilitates interpretation of Table 6. The latter shows the results of a typical analysis, in this case for a simply-supported one-way slab with 5 ksi (34.5 MPa) concrete and GFRP reinforcement. Numerous other tables were generated but were omitted due to space limitations.

Table 7 provides an overall summary of the analyses conducted, and forms the basis for the development of Table 4. Sixteen basic cases were considered, resulting from four different support conditions, two different reinforcement types (GFRP and CFRP), and two different member types (beam and slab). For each case, computations were performed considering four different reinforcement quantities, equivalent to 1.0, 2.0, 3.0, and 4.0 times the balanced reinforcement ratio, respectively. For one-way slabs and beams, the assumed ratio of service moment to calculated nominal moment was taken as 0.3 and 0.4, respectively. This difference is intended to reflect the fact that slabs, on a relative basis, tend to be more lightly loaded. Reinforcement properties for GFRP and CFRP reinforcement were assumed to represent typical values for commercially available bars. As can be seen in Table 7, the type of FRP reinforcement does not have a significant effect on the computed span-depth ratios. For this reason, the number of basic cases considered reduces to eight, with each case corresponding to a value in the final version of the ACI 440.1R-06 minimum thickness table.

Values from Table 7 chosen for inclusion in ACI 440.1R-06 Table 8.2 are indicated in italics. Span-depth ratios for one-way slabs are based on the analyses of sections reinforced at 2.0 times the balanced reinforcement ratio whereas those for beams correspond to analyses of sections reinforced at 3.0 times the balanced reinforcement ratio. These choices reflect the general differences in reinforcing levels for beams and slabs. While it can be argued that lower reinforcement ratios are reasonable for designs in many cases, the data shows no significant difference in the computed values between, for example, the cases of 1.0 and 2.0 times the balanced reinforcement ratio.

CONCLUSIONS AND RECOMMENDATIONS

Defining a maximum span-depth ratio as a vehicle for indirect control of deflections in concrete beams and one-way slabs reinforced with FRP bars is affected by FRP's stiffness, brittle-elastic nature, and bond properties. To overcome the limitations imposed by the influence of many intervening variables on the deflection calculations, a general indirect deflection control model for FRP-reinforced beams and one-way slabs is proposed. The model can be applied to a wide variety of support conditions regardless of the type of reinforcement. The procedure is also independent of the member's effective moment of inertia, which is difficult to quantify across the cracked stage, and it is expressed in terms of an allowable deflection-span ratio, which allows designers to have full control of the serviceability design depending on applicable deflection limits.

According to the proposed method, the maximum span-depth ratio in concrete beams or one-way slabs with FRP reinforcement is particularly affected by the level of FRP strain at a crack at service load level. For comparable span lengths, concrete beams and one-way slabs with FRP need be deepened to satisfy the same maximum deflection-span ratios for steel-reinforced members. If concrete's tension stiffening effect is accounted for, the member deepening penalty can be relaxed, especially at load levels that are roughly greater than that at first flexural cracking.

A series of indirect deflection control parametric analyses were performed on beams and one-way slabs reinforced with FRP bars based on the proposed model. Concrete's tension stiffening effect was accounted for through some simplifications. Based on these results, the maximum span-depth ratios reported in Table 4 are proposed. This table was adopted by ACI 440 for inclusion in ACI 440.1R-06. The table is intended only for use in the preliminary sizing of members, and does not supersede the requirement for designers to check deflections directly, as stipulated by ACI 440.1R-06.

660   Ospina and Gross

Experimental evidence studying deflections in FRP-reinforced concrete beams and one-way slabs under uniformly distributed gravity loads for different support conditions is needed to further examine the quality of the proposed deflection control procedures.

ACKNOWLEDGMENTS

The authors would like to thank Dr. Scott D. B. Alexander for his comments in discussions held on the subject of indirect deflection control of reinforced concrete flexural members.

REFERENCES

ACI Committee 435, "Allowable Deflections", *ACI Journal*, Proceedings V. 65, No. 6, 1968, pp. 433-444.

ACI Committee 435, "State-of-the-Art Report, Deflection of Two Way Reinforced Concrete Floor Systems," *ACI SP 43-3, Deflections of Concrete Structures*, 1974, pp. 55-81.

ACI Committee 435, "Proposed Revisions by Committee 435 to ACI Building Code and Commentary Provisions on Deflections," *ACI Journal*, Proceedings V. 75, No. 6, 1978, pp. 229-238.

ACI Committee 440, "Guide for the Design and Construction of Concrete Reinforced with FRP Bars," *ACI 440.1R-06*, American Concrete Institute, Farmington Hills.

Branson, D.E., *Deformation of Concrete Structures*, McGraw-Hill, New York, 1977, 546 pp.

Comité Euro-International du Béton (CEB) / Fédération Internationale de la Précontrainte (FIP), *Model Code for Concrete Structures, MC-90*, CEB, Thomas Telford House, 1990, London.

Gao, D., Benmokrane, B., and Masmoudi, R., "A Calculating Method of Flexural Properties of FRP-reinforced Concrete Beam: Part 1: Crack Width and Deflection," Technical Report, Department of Civil Engineering, Université de Sherbrooke, Québec, 1998, 24 pp.

Hall T., "Deflections of Concrete Members Reinforced with Fibre Reinforced Polymer (FRP) Bars," M.Sc. Thesis, Department of Civil Engineering, The University of Calgary, Calgary, Canada, 2000, 292 pp.

Ospina, C.E., Alexander, S.D.B., and Cheng, J.J.R., "Behaviour of Concrete Slabs with Fibre-reinforced Polymer Reinforcement," *Structural Engineering Report No. 242*, Department of Civil and Environmental Engineering, University of Alberta, Edmonton, Canada, 2001, 356 pp.

Yost, J.R., Gross, S.P., and Dinehart, D.W., "Effective Moment of Inertia for GFRP Reinforced Concrete Beams," *ACI Structural Journal*, Vol. 100, No. 6, Nov.-Dec. 2003, pp. 732-739.

LIST OF SYMBOLS

d	Effective flexural slab or beam depth
E_c	Elastic modulus of concrete, ksi (MPa)
E_f	Elastic modulus of FRP bars, ksi (MPa)
f_r	Concrete's modulus of rupture, ksi (MPa)
f'_c	Specified cylinder compressive strength of concrete, ksi (MPa)
h	Slab thickness or beam depth
I_{cr}	Cracked moment of inertia
I_e	Effective moment of inertia
k_m	Ratio of neutral axis-to-flexural depth at midspan, for cracked-elastic conditions
K_l	Constant depending on boundary conditions
L, ℓ	Span length
M_a	Applied moment
M_{cr}	Cracking moment
M_m	Midspan moment
M_o	Statical moment
n	Modular ratio
β_1	Bond coefficient in CEB/FIP MC90
β_2	Performance coefficient in CEB/FIP MC90
Δ_m	Midspan deflection
ε_{rm}	Reinforcement strain at midspan
ρ_{bf}	Balanced reinforcement ratio for FRP reinforced member in ACI 440.1R-06
ρ_f	Reinforcement ratio for FRP reinforced member in ACI 440.1R-06
ρ_r	Reinforcement ratio
ψ_m	Midspan curvature
ψ_1	Curvature at uncracked section level in CEB/FIP MC90
ψ_2	Curvature at fully cracked section level in CEB/FIP MC90
ξ	Tension stiffening factor in CEB/FIP MC 90

8.3.2.1 *Recommended minimum thicknesses for design*—Recommended minimum thicknesses for design of one-way slabs and beams are provided in Table 8.2. The table is only intended to provide guidance for initial design, and use of these recommended minimum thicknesses does not guarantee that all deflection considerations will be satisfied for a particular project.

Table 8.2 – Recommended minimum thickness of nonprestressed beams or one-way slabs

	Minimum Thickness, h			
	Simply-supported	One end continuous	Both ends continuous	Cantilever
Solid one-way slabs	$\ell/13$	$\ell/17$	$\ell/22$	$\ell/5.5$
Beams	$\ell/10$	$\ell/12$	$\ell/16$	$\ell/4$

Values in Table 8.2 are based on a generic maximum span-to-depth ratio limitation (Ospina, Alexander, and Cheng 2001) corresponding to the limiting curvature associated with a target deflection-span ratio (Eq. 8-10). The procedure can be applied for any type of reinforcement.

$$\frac{\ell}{h} \leq \frac{48\eta}{5K_1}\left(\frac{1-k}{\varepsilon_f}\right)\left(\frac{\Delta}{\ell}\right)_{max} \qquad (8\text{-}10)$$

In Eq. (8-10), $\eta = d/h$, k is as defined in Eq. (8-12), and $(\Delta/\ell)_{max}$ is the limiting service load deflection-span ratio. K_1 is a parameter that accounts for boundary conditions. It may be taken as 1.0, 0.8, 0.6, and 2.4 for uniformly loaded simply-supported, one-end continuous, both ends continuous, and cantilevered spans, respectively. The term ε_f is the strain in the FRP reinforcement under service loads, evaluated at midspan except for cantilevered spans. For cantilevers, ε_f shall be evaluated at the support.

Eq. (8-10) assumes no tension stiffening. To consider the effects of tension stiffening in developing Table 8.2, the values resulting from Eq (8-10) were modified by the ratio of effective and fully cracked moments of inertia, computed using Eq. (8-13a) and (8-11), respectively. Tabulated values are based on an assumed service deflection limit of $l/240$ under total service load, and assumed reinforcement ratios of $2.0\rho_{bf}$ and $3.0\rho_{bf}$ for slabs and beams, respectively.

Table 1 -- Minimum Thickness of Nonprestressed Beams or One-way Slabs Unless Deflections are Computed (ACI 318-05 Table 9.5(a))

| | Minimum Thickness, h | | | |
	Simply supported	One end continuous	Both ends continuous	Cantilever
Member	Members not supporting or attached to partitions or other construction likely to be damaged by large deflections			
Solid one-way slabs	$\ell/20$	$\ell/24$	$\ell/28$	$\ell/10$
Beams or ribbed one-way slabs	$\ell/16$	$\ell/18.5$	$\ell/21$	$\ell/8$

Note: Span length ℓ in inches

Table 2 -- Maximum Permissible Computed Deflections (ACI 318-05 Table 9.5(b))

Type of Member	Deflection to be considered	Deflection limitation
Flat roofs not supporting or attached to non-structural elements likely to be damaged by large deflections	Immediate deflection due to live load	$l/180$
Floors not supporting or attached to nonstructural elements likely to be damaged by large deflections	Immediate deflection due to live load	$l/360$
Roof or floor construction supporting or attached to nonstructural elements likely to be damaged by large deflections	That part of the total deflection occurring after attachment of nonstructural elements (sum of the long-term deflection due to all sustained loads and the immediate deflection due to any additional live load)	$l/480$
Roof or floor construction supporting or attached to nonstructural elements not likely to be damaged by large deflections		$l/240$

Table 3 -- Miscellaneous Deflection Limitations (Adapted from ACI 435 1968)

Reasons for limiting deflections	Examples	Deflection Limitation*	Portion of total deflection on which the deflection limitation is based
Sensory Acceptability			
Visual	Droopy cantilevers and sag in long span beams	By personal reference	Total deflection
Tactile	Vibrations of floors that can be felt	L/360	Full live load
	Lateral building vibrations	No recommendation	Gust portion of wind
Auditory	Vibrations producing audible noise	Not permitted	
Serviceability of Structure			
Surfaces which should drain water	Roofs, outdoor decks	L/240	Total deflection
Floors which should remain plane	Gymnasia and bowling alleys	L/360 + camber or	Total deflection
		L/600	Incremental deflections after floor is installed
Members supporting sensitive equipment	Printing presses and certain building mechanical equipment	Manufacturer's recommendations	Incremental deflections after equipment is leveled
Effect on nonstructural elements			
Walls	Masonry and plaster	L/600 or 0.3 in. (7.6 mm) max. or $\phi = 0.00167$ rad.	Incremental deflections after walls are constructed
	Metal movable partitions and other temporary partitions	L/240 or 1 in. (25.4 mm) max.	Incremental deflections after walls are constructed
	Lateral building movement	0.15 in. (3.8 mm) offset per story 0.002 x height	Five min. sustained wind load
	Vertical thermal movements	L/300 or 0.60 in. (15.2 mm) max.	Full temperature differential movement
Ceilings	Plaster Unit ceilings such as accoustic tile	L/360 L/180	Incremental deflections after ceiling is built
Adjacent building elements supported by other members	Windows, walls and folding partitions on unyielding supports below the deflecting member	Absolute deflection limited by tolerances built into the element in question	Incremental deflection after building element in question is constructed
Reasons for limiting deflections	**Examples**	**Deflection Limitation***	**Portion of total deflection on which the deflection limitation is based**
Effect on structural elements			
Deflections causing instability of primary structure	Arches and shells Long columns	Effect of deflections on the stresses and stability of the structure should be taken into account in the design	
Deflections causing different force system or change in stresses in some other element	Beam bearing rotation on masonry wall	Effect of deflections on the stresses and stability of the structure should be taken into account in the design	
Deflections causing dynamic effects	Resonant vibrations which increase static deflections and stresses such as those produced by wind, dancing, moving loads and machinery	Dynamic deflections should be added to static deflections and the total should be less than the limitations imposed for other reasons	

*   Deflection limitations are given for members supported at both ends and for cantilevers, except as noted. It is assumed that the supports do not move.

**Table 4 – Recommended Minimum Thickness of Nonprestressed Beams
or One-way Slabs Reinforced with FRP Bars**

	Minimum Thickness, h			
	Simply-supported	One end continuous	Both ends continuous	Cantilever
Solid one-way slabs	$\ell/13$	$\ell/17$	$\ell/22$	$\ell/5.5$
Beams	$\ell/10$	$\ell/12$	$\ell/16$	$\ell/4$

Table 5 – Description of Parametric Analysis Calculations

Row in Table 6	Description
1 to 5	Specify material properties. A concrete strength of 5 ksi (34.5 MPa) was used for all analyses presented here. However, it should be noted that the analyses were not very sensitive to the use of higher strength concrete. The use of 10 ksi concrete (69 MPa), as opposed to 5 ksi, generally increased the calculated minimum thicknesses by about 10 to 20%.
6	Identifies the K_1 factor per selected span support conditions. Values of 1.0, 0.8, 0.6, and 2.4 were used for simply-supported, one-end continuous, both ends continuous, and cantilevered spans, respectively.
7	Refers to the assumed d/h ratio, taken as 0.90 for all analyses.
8	Identifies the allowable immediate deflection under total service load, assumed to be $L/240$ for all analyses.
9	Identifies the assumed ratio between service and nominal moments. This ratio is used to establish the service moment after calculating the nominal moment directly. For slabs and beams, this ratio was taken as 0.30 and 0.40, respectively, to reflect the fact that slabs tend to be more lightly reinforced.
10 to 12	Identify the balanced reinforcement ratio and the reinforcement ratios assumed for the analyses (1.0, 2.0, 3.0, and 4.0 times the balanced ratio)
13	Computes the normalized neutral axis depth, k_m
14 to 16	Respectively, compute the reinforcement stress at ultimate, the reinforcement stress at service, and the reinforcement strain at service (ε_{rm} in Eq. 7)
17 to 19	Respectively, compute normalized cracking, service, and nominal moments
20 to 22	Respectively, compute normalized gross, cracked, and effective moments of inertia. The effective moment of inertia is calculated using Eq. 13.
23 and 24	Used in the computation of the effective moment of inertia (column 22)
25	Quantifies the tension stiffening effect as a multiplier, in terms of the ratio between the effective and cracked moments of inertia
26	Computes the recommended span-depth ratio (without consideration of tension stiffening) per Eq. 7.
27	Computes the recommended span-depth ratio based on Eq. 7, with a correction made to account for tension stiffening effects (Col. 27 = Col. 26 x Col. 25)

Table 6 -- Maximum Span-Depth Ratios for Simply-Supported Slab
($f'_c = 5$ ksi and GFRP Reinforcement)

Row	Description	Parameter	Unit	Values			
1	Material Properties	f_{fu}	ksi	100			
2		E_f	ksi	6000			
3		f'_c	ksi	5			
4		β_1		0.80			
5		E_c	ksi	4031			
6	Assumptions	K_1		1.00			
7		η		0.90			
8		L/Δ_m		240			
9		M_s/M_n		0.30			
10	Reinforcement Ratios	ρ_{bf}		0.0052			
11		ρ_f/ρ_{bf}		**1.00**	**2.00**	**3.00**	**4.00**
12		ρ_f		0.0052	0.0104	0.0156	0.0207
13	Neutral Axis	k_m		0.117	0.161	0.193	0.220
14	Reinforcement Stresses/Strains	$f_{f,ult}$	ksi	100.00	68.34	54.36	46.05
15		$f_{rm,serv}$	ksi	29.30	19.85	15.69	13.23
16		$\varepsilon_{rm,serv}$		0.0049	0.0033	0.0026	0.0022
17	Normalized Moments	M_{cr}/bd^2	ksi	0.109	0.109	0.109	0.109
18		M_s/bd^2	ksi	0.146	0.195	0.228	0.254
19		M_n/bd^2	ksi	0.487	0.650	0.761	0.848
20	Normalized Moments of Inertia	I_g/bd^3		0.1143	0.1143	0.1143	0.1143
21		I_{cr}/bd^3		0.0066	0.0123	0.0175	0.0223
22		I_e/bd^3		0.0134	0.0181	0.0231	0.0278
23	Tension Stiffening Effect	M_s/M_{cr}		1.34	1.79	2.09	2.33
24		β_d		0.200	0.400	0.600	0.800
25		I_e/I_{cr}		2.04	1.48	1.32	1.24
26	Recommended Span to Depth Ratios	$(L/h)_{NO\ TS}$		6.5	9.1	11.1	12.7
27		$(L/h)_{w/\ TS}$		**13.3**	**13.5**	**14.6**	**15.9**

Table 7 – Summary of Maximum Span-Depth Ratio Computations

Case			f_{fu} (ksi)	E_f (ksi)	f_c (ksi)	K_1	η	L/Δ_m	M_s/M_n	$(L/h)_{w/TS}$ at $\rho_f/\rho_{bf} =$			
										1.00	2.00	3.00	4.00
Slab	Simply-Supported	GFRP	100	6000	5	1.0	0.90	240	0.30	13.3	13.5	14.6	15.9
Slab	Simply-Supported	CFRP	300	20000	5	1.0	0.90	240	0.30	11.6	13.0	14.7	16.2
Beam	Simply-Supported	GFRP	100	6000	5	1.0	0.90	240	0.40	7.0	8.2	9.4	10.5
Beam	Simply-Supported	CFRP	300	20000	5	1.0	0.90	240	0.40	6.8	8.5	9.9	11.2
Slab	One End Continuous	GFRP	100	6000	5	0.8	0.90	240	0.30	16.6	16.9	18.3	19.8
Slab	One End Continuous	CFRP	300	20000	5	0.8	0.90	240	0.30	14.5	16.3	18.3	20.2
Beam	One End Continuous	GFRP	100	6000	5	0.8	0.90	240	0.40	8.8	10.3	11.8	13.2
Beam	One End Continuous	CFRP	300	20000	5	0.8	0.90	240	0.40	8.5	10.6	12.4	14.0
Slab	Both Ends Continuous	GFRP	100	6000	5	0.6	0.90	240	0.30	22.1	22.5	24.4	26.4
Slab	Both Ends Continuous	CFRP	300	20000	5	0.6	0.90	240	0.30	19.4	21.7	24.4	27.0
Beam	Both Ends Continuous	GFRP	100	6000	5	0.6	0.90	240	0.40	11.7	13.7	15.7	17.6
Beam	Both Ends Continuous	CFRP	300	20000	5	0.6	0.90	240	0.40	11.3	14.1	16.5	18.6
Slab	Cantilevered	GFRP	100	6000	5	2.4	0.90	240	0.30	5.5	5.6	6.1	6.6
Slab	Cantilevered	CFRP	300	20000	5	2.4	0.90	240	0.30	4.8	5.4	6.1	6.7
Beam	Cantilevered	GFRP	100	6000	5	2.4	0.90	240	0.40	2.9	3.4	3.9	4.4
Beam	Cantilevered	CFRP	300	20000	5	2.4	0.90	240	0.40	2.8	3.5	4.1	4.7

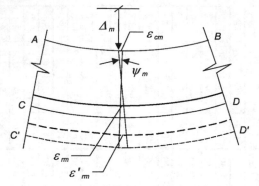

Figure 1 — Effect of Reinforcement Strain on Span-Depth Ratio

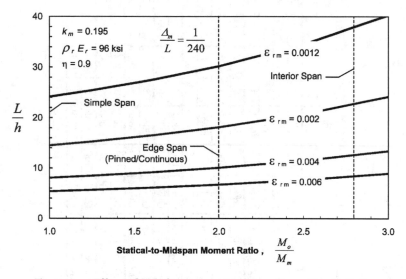

Figure 2 — Effect of Reinforcement Strain on Span-Depth Ratio

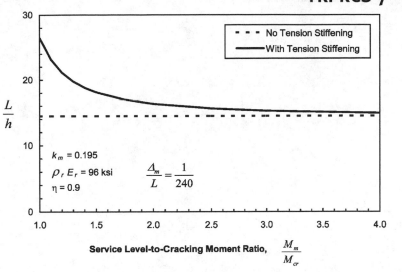

Figure 3 — Tension Stiffening Effect on Span-Depth Ratio

Time Depending Thermo Mechanical Bond Behavior of Epoxy Bonded Pre-Stressed FRP-Reinforcement

by K. Borchert and K. Zilch

Synopsis: The development of different pre-stressed epoxy bonded CFRP-reinforcement systems for retrofitting RC-structures continues. In addition to requirements concerning the ultimate limit state (ULS) pre-stressed systems could satisfy demands in regard to the serviceability state (SLS) like crack limitation or fatigue. In comparison with the well known material behavior of concrete under sustained loading epoxy resins show material properties more sensitive to thermal effects, for example loss of strength or creep deformations. In experimental tests the influence of permanent load and temperature is examined. The pure epoxy resin adhesive was evaluated in lab-shear tests and the structural behavior of the system in the form of epoxy bonded pre-stressed near surface mounted CFRP-strips. The experiments reveal a significant influence of the epoxy resin adhesive behavior on the long-term structural behavior in service. To evaluate the effects with regard to designs relevant problems as long-term development of transfer lengths of pre-stressed reinforcement or of the slip between reinforcement and concrete under varying load and thermal conditions results of a simulation model are presented.

Keywords: CFRP reinforcement; creep; epoxy resin; long-term behavior; pre-stressing; SLS; strengthening; sustained load; temperature

Dipl.-Ing. Kurt Borchert, born in 1975, received his engineering degree in 2001. Currently he is research assistant at the Departement of Concrete Structures of the Technical University Munich. His research work comprises strengthening with externally bonded and near surface mounted reinforcement. He is member of fib TG 9.3.

Prof. Dr.-Ing. Konrad Zilch, born in 1944, graduated at the Technical University Darmstadt in 1969. Following post graduate studies at the UCB and at the Boundary Layer Wind Tunnel Laboratory (University of Western Ontario) he worked at T.Y Lin International and the STRABAG AG. Since 1993 he is head of the Department of Concrete Structures at the Technical University Munich. He is member of several national and international committees, e.g. CEN TC250/SC2 – EUROCODE 2.

RESEARCH SIGNIFICANCE

Epoxy bonded pre-stressed FRP-strengthening systems could often accomplish the requirements more efficiently and reasonably than non pre-stressed FRP reinforcement. Pre-stressed near surface mounted CFRP-reinforcement holds a bond behavior that allows the abdication of mechanical anchors. The performance of pre-stressed NSM depends directly on the properties of the epoxy resin (Blaschko and Zilch 1999). Hence the time depending thermo-mechanical material behaviour of the adhesive is decisive for a long-lasting design. In order to examine the influence of the epoxy resin on the performance of pre-stressed NSM long-term tests on the pure adhesive, the system and the results of a simulation model are shown.

INTRODUCTION

The increasing demands on existing concrete-structures lead to enhanced strengthening methods. These methods have to meet requirements in ultimate limit state (ULS) and additionally in service ability state (SLS). In recent years the bonding of carbon fiber reinforced polymers CFRP by epoxy resins has shown to be an appropriate method. In many cases the existing needs can be fulfilled more effectively and economically by applying pre-stressed CFRP-reinforcement, for example fatigue problems.

Due to the brittle bond behavior of pre-stressed CFRP laminates bonded to the surface of RC structures a mechanical anchorage is required. In contrast pre-stressed near surface mounted reinforcement provides a remarkable ductile bond characteristic, that the CFRP-strips can be anchored only by adhesive. Pre-stressed reinforcement causes significantly elevated permanent bond-stresses. Hence the epoxy resin is exposed to long-term stresses. In comparison to concrete the material properties of epoxy resins are strongly depending on time and temperature. In series the time depending thermo-mechanical material behaviour of the epoxy resin has to be taken into account for a safe and durable performance of the retrofitting system. In case of strengthening existing RC members with epoxy bonded pre-stressed FRP reinforcement the long-term developments of the stress distribution along the bond length, of required transfer lengths and of the slip between reinforcement and concrete have to be known.

This paper reports on experimental long term tests on the temperature depending thermo-mechanical properties of epoxy resins and their influences on the system performance of pre-stressed near surface mounted CFRP-strips – pre-stressed near surface mounted CFRP-strips are patented by the Bilfinger Berger AG. Results of a simulation model are presented.

MATERIALS AND STRENGHTENING SYSTEM

In the following the basic material properties of epoxy resin adhesives and the principle structural behavior of near surface mounted CFRP-strips are described.

Epoxy Resin Adhesives

Epoxy resin adhesives belong to the organic binders (plastics) and are recognized as a cast-in-place adhesive featuring remarkable functionality. It is possible to obtain adhesive joints with excellent cohesion and outstanding adhesion to nearly all kinds of substrates. For strengthening cold curing, two part liquids with low shrinkage on curing are normally used. The curing step transforms the two liquids to a highly cross-linked space framework by a polyaddition reaction. Epoxy resins can be formulated to meet various specifications and use criteria. Usually aggregate fillers are added to the resins to control the mechanical properties like strength or creep. Epoxy resin adhesives show a distinctive visco-elastic material behavior.

One of the most important technological aspects of adhesive joints is their durability under long-term loading and environmental conditions. A short term test does not yield information on the maximum long-term load capacity nor is it possible to predict the performance of a joint during service exposure to elevated temperature or significant permanent loads. In practice a few compromises must be made. Not all mechanical demands can be simultaneously met with one epoxy system. Because the performance of epoxy resins is dependent on environmental conditions, it is helpful to classify primarily decisive properties (Hugenschmidt 1974). Primarily critical for the endurance is the long-term behavior in service like enduring strength, creep or influence of field temperature and a relative insensitivity to incorrect mixing ratio. Not primarily decisive for the durability is the reactivity and the short term mechanical strength. At room temperature the long term strength of epoxy resin adhesives is assumed to be between 50% and 70% of the short-term ultimate load (Habenicht 1986). For structural bonding often the lower limit value of 50% is recommended (Rehm and Franke 1982).

With rising temperatures the material behavior of epoxy resins changes noticeably at the glass transition temperature T_g (Fig. 1). At the T_g the resin leaves the glass state and reaches a state with rubber-like properties. The T_g is the critical point for all cold-curing binders, not only for epoxy resins. With reaching the glass transition zone the strength and the elastic-modulus fall significantly (Fig. 2). The T_g is the maximum service temperature for structural bonding using an epoxy resin (Habenicht 1986).

Common cold-curing epoxy adhesives show a T_g about 55°C (131°F) in the first heating. Below the T_g the short-term behavior for small loadings can be assumed as linear-elastic. For increasing or permanent loads the linear or non-linear visco-elastic behavior has to be accounted.

One of the most important properties is the creep deformation of the resin. It can be assumed that, at room temperature, the ultimate creep rate of highly cross-linked epoxy-systems is about three to four times as high as that of the concrete (Hugenschmidt 1974). Concerning creep the T_g is not the only decisive parameter. About 20°C below the glass transition the micro-brown movements of the molecules start in the epoxy bulk. Through this all time-depending effects like the loss of strength or creep accelerate (Letsch 1983). So the maximum long-term load capacity at elevated temperatures is reached after a period of about 700 h – 1000 h (Rehm and Franke 1982).

As shown, the mechanical behaviour of epoxy resins is greatly time and temperature dependent. In cases of significant long-term stresses in the adhesive layer the thermo mechanical behavior has a direct influence on the structural performance of the retrofitting system. Therefore it has to be accounted for a safe and durable design.

Near Surface Mounted CFRP-Reinforcement

The method consists in gluing CFRP-strips into grooves in the existing concrete cover (Fig. 3). The strips are between 1 to 2.5 mm in thickness and about 15 mm - 20 mm in width surrounded by an about 1 to 2 mm thick adhesive layers (Blaschko 2003).

The failure modes of NSM reinforcement are depending from various parameters. If the bonded area is long enough, the CFRP-strips fail in tension. Due to the ductile bond behaviour the complete tensile strength of the strips can be anchored at a single crack. If the slits are too close to the concrete edge the concrete corner could split off. In all other cases the bond fails inside the adhesive layer (Blaschko 2001, Blaschko and Zilch 1999, De Sena Cruz and de Barros 2004, fib 2001). In figure 4, an experimental bond stress-slip relationship is shown. Typical for NSM reinforcement are the obviously increased bond stresses in comparison with externally bonded reinforcement (EBR) and the activated friction shear stresses at large slips. The bond stress-slip relationship is affected by the deformations of the concrete perpendicular to the strip. These deformations create lateral pressure or tension in the adhesive.

Blaschko developed a conclusive bond model by combining the differential equation of the shifted bond, the differential equation of the elastic supported girder – to take the concrete deformations into account – with a failure criterium with integrated friction model for the adhesive (Blaschko 2001). On the basis of this model the measured distribution of the tensile force in the CFRP strip along the bond length and the bond capacity were ascertainable.

The ultimate load of near surface mounted CFRP-strips and is primary dependent on the properties of the epoxy resin. The method of strengthening with pre-stressed near surface mounted CFRP-strips is patented by the Bilfinger Berger AG.

EXPERIMENTAL TESTS

To examine the influence of the epoxy resin to highly-stressed adhesive bonded reinforcement two test series were conducted. In a first series, long term lap-shear tests at variable temperatures and load levels were performed to examine strength and creep processing of the epoxy resin. The second series consisted of pre-stressed NSM CFRP-strips, that were exposed between room temperature and about 5° below the glass-transition temperature T_g of the resin. For the tests the highly filled epoxy resin MC-DUR 1280 from the MC-Müller Bauchemie; Bottrop (Germany) was used. This adhesive is approved for the strengthening of RC structures with NSM and EBR in Germany (DIBt 2004).

Long-term lap shear tests

The creep properties of the highly-filled epoxy resin adhesive were tested with long-term steel lap-shear tests (Fig. 5) under varied loads and thermal conditions. The test parameters are shown in table 1. The test set up should guarantee a nearly constant stress distribution in the adhesive layer. Hence the pure epoxy performance was tested. The tests started after a curing period of 72 h at room temperature. The results affirmed the typical creep behavior of epoxy resins.

For 20°C (68°F) an enhancement of load from 15% to 80% of the long term ultimate load causes non-linear creep. After a period of about 5000 h the creep modulus ϕ at 80% was two times higher in comparison to a load level of 15%. Also at a constant load level of 15% the creep modulus raises with rising temperatures. After 5000h in 38°C (100°F) it was about 2.5 times higher and approximately 4.0 times higher after 3400h in 50°C (122°F) as the corresponding value in 20°C (68°F).

For 50°C (122°F) the elastic short-term value of the displacement was increased. This bigger deformations results from the starting reduction of the short-term stiffness of the epoxy resin in the range of the T_g.

Pre-stressed NSM Reinforcment tests

The influence of the epoxy resin on the system behavior of pre-stressed NSM CFRP reinforcement was examined with test specimens shown in figure 6 and 7. The test parameters and the material properties are displayed in table 1 and 2. The strips were subjected to pre-stressing force of 60% of the bond capacity corresponding with a bond length l_b of 100 mm – this caused a strain about 4.2‰ or stress of 714 MPa in the strip. After a curing for 24h the temporary anchorages were removed and from that time on the bond was only guaranteed by the adhesive. Another 70h to 100h later the temperature exposition started. The loss of pre-stress was detected by strain gauges in the bondless length between the two bond areas. Additional strain gauges were assembled on the strips along the bond length l_b in the case of the specimen with $l_b = 450$ mm. This set-up allows detecting redistributions of the shear-stresses inside the bond length in consequence of creeping and loss of strength of the epoxy resin. The loss of pre-stressing force can be nearly completely assigned to the adhesive layer. On account of the low pre-stressing force in comparison with the compression strength of the concrete and of the negligible creeping of CFRP-strips (for example Ando et al. 1997) the long-term behavior is primarily defined by the adhesive.

Figure 8 displays on the basis of the measured strain loss in the middle the calculated sum of the slips between strip and concrete at the inner end of both bond lengths. In the tests the exposure to higher temperatures effects a significant augmentation of the slip provoked by the described time and thermal depending material properties of the epoxy resin.

NSM REINFORCEMENT – MODELLING

For a safe and durable design of epoxy bonded pre-stressed reinforcement the development of the necessary transfer lengths (ULS) and of the slip between reinforcement and concrete (SLS) in service has to be accounted.

Simulation model

For the application of the experimental results a numerical incremental integration algorithm (Rehm 1961) was developed, that describes the structural behaviour in the SLS and the ULS along the reinforcement axis. The algorithm bases on the differential equation of the shifted bond and uses a framework model (Fig. 9).

$$\frac{d^2 s}{dx^2} - \frac{\tau_b(s)}{(E_l \cdot t_l)} = 0 \tag{1}$$

where s is the slip between CFRP-strip an concrete, E_l and t_l are the elastic modulus and the thickness of the CFRP-strip and $\tau_b(s)$ is the bond shear stress. The shear-slip relationship is taken from tests described in (Blaschko 2001). The bond characteristic includes already the effects of edge distance and strength of the epoxy resin for short term loads. This relationship was implemented idealized subdivided into four sections according to an approach for reinforcement steel bars (Eligehausen, Popov and Bertero 1983). The shear-slip relationship between the CFRP-strip and the concrete is also affected by the shear modulus and the thickness of the adhesive layer. The numerical model takes the properties of the adhesive layer in a linear way into account.

In addition to the short term material parameters the time depending thermo mechanical behaviour of the adhesive was simulated by adapting a creep law of aging concrete to the adhesive properties. The used rate-type creep law bases on a Maxwell chain model of time-variable viscosities and spring moduli (Bazant and Wu 1974). This formulation is useful for the chosen step by step time integration, because it makes the storage of the stress history unnecessary. On the part of the epoxy resin the simulation model includes also the loss of strength with time and temperature and the changing elastic modulus in range of the glass transition according to (Rehm and Franke 1982) and own accompanying tests respectively.

Comparison with experimental results

Short-term behavior – In addition to the ULS the simulation model aims at the description of the SLS, especially at the occurring deformations. Concerning the slip between CFRP-strips and concrete the geometrical boundary conditions – like thickness and width of the strips, thickness of the groove and the adhesive layer – influence the structural behavior in addition to the material properties of the join partners. In figure 10 results of the strip-force-slip-relationship with varying geometrical parameters of tests performed by *Blaschko* (Blaschko 2001) are shown. The results show a wide range of possible bond-stiffness. The mentioned deformation parameters are directly integrated in the simulation model. In addition to the experimental results the calculated outcomes are displayed.

Long-term behavior – Figure 8 also displays the calculated total slip between strip and concrete of both bond areas in the creep tests mentioned above. In addition the chronological changes of the pre-stressing force along the bond length l_b of a test with $l_b = 450$ mm is shown in figure 11. The simulation obviously is in good agreement with the measured parameters.

CONCLUSIONS AND PERSPECTIVES

The experiments revealed a significant influence of the time depending thermo mechanical material properties of the epoxy adhesive on the long-term structural behavior of pre-stressed near surface mounted CFRP strips in service. To evaluate the effects with regard to designs relevant problems under varying load and thermal conditions a simulation model was presented. These tests were part of an experimental program that was conducted to reveal the effects of elevated service temperatures on the structural behavior of the with NSM CFRP-strips strengthened bridge-decks of the Röslau valley bridge in Bavaria, Germany (Zilch et al. 2004). Among the future tasks are the further systematic analysis of supplementary parameters like other epoxy resin formulations or different load levels and the development of a simple design approach for practice use.

REFERENCES

Ando, N.; Matsakawa, H.; Hattori, A.; and Mashima, M., 1997, October, "Experimental studies on the long-term tensile properties of FRP-tendons", *Proceedings of the Third International Symposium on Non-metallic (FRP) Reinforcement for Concrete Structures*, Sapporo, pp. 120 – 146.

Blaschko, M., 2001, *On the mechanical behaviour of concrete structures with CFRP strips glued into slits*, PhD thesis, Technische Universität München, (in German).

Blaschko, M. and Zilch, K., 1999, July 5 – 9, "Rehabilitation of concrete structures with CFRP strips glued into slits", *Proceedings of the 12th International Conference on Composite Materials*, Paris.

Bazant, Z.P. and Wu, S T., 1974, "Rate Type Creep Law of Aging Concrete based on an Maxwell Chain", Materials and Structures, Vol. 7, No. 37, pp 45-60.

Blaschko, M., 2003, "Bond behaviour of CFRP strips glued into slits", *Proceedings of the Sixth International Symposium on FRP Reinforcement for Concrete Structures*, edited by Kiang Hwee Tan, Singapore, pp. 397 – 401.

De Sena Cruz, J. M. and de Barros, A. O., 2004, "Bond between Near Surface Mounted Carbon-Fiber-Reinforcemed Polymer Laminate Strips and Concrete", *Journal of composites for constructions*, Vol. 8, No. 6, pp. 519-527.

DIBt – German Institute of Construction Technology, 2004, "Strengthening of RC- and PC-members with NSM Carboplus CFRP (Z-36.12-60)", Technical approval, Berlin, (in German).

DIN – German Institute of Standardization, 1978, "Tensile shear tests for the determination of the shear stress – strain diagram of an adhesive in a bonded joint (DIN 54451)", Berlin, (in German).

Eligehausen, R; Popov, E. and Bertero, V., 1983, "Local bond stress-slip relationship of deformed bars under generalized excitations", *Report UCB/EERC-83/23*, Earthquake Engeneering Research Center, University of California at Berkley.

Fédération international du béton (fib), 2001, *Externally bonded FRP reinforcment for RC structures*, fib bulletin no 14, Lausanne.

Franke, L., 1976, "Influence of the load term on the bond behaviour of steel and concrete", DAfStb – German Committee for RC-Concrete, Vol. 268, Ernst & Sohn, Berlin, (in German).

Habenicht, G., 1986, *Bonding – Basic principles, technology and application*, Springer-Verlag, Berlin, (in German).

Hugenschmidt, F., 1974, "Epoxy adhesives in precast prestressed concrete structures", *Journal of the Prestressed Concrete Institute*, Vol. 11, No. 10, (reprint).

Letsch, R., 1983, *About the deformations of epoxy resins and epoxy mortars under static and non-static temperatures,* PhD thesis, Technische Universität München, (in German).

Rehm, G., 1961, "About the basic principles of bond between steel and concrete", DAfStb – German Committee for RC-Concrete, Vol. 138, Ernst & Sohn, Berlin, (in German).

Rehm, G. and Franke, L., 1982, "Structural bonding of concrete constructions", DAfStb – German Committee for RC-Concrete, Vol. 331, Ernst & Sohn, Berlin, (in German).

Zilch, K.; Zehetmaier, G.; Borchert, K.; Endress, B.; Fischer, O., 2004, "The bridge over the Röslau valley near Schirnding – innovative methods for the strengthening of a PC bridge", *Bauingenieur*, Vol. 79, No. 12, pp.589 - 595, (in German).

NOTATION

E_l	elastic modulus of the CFRP-strip
T_g	glass transition temperature
a_r	edge distance of a CFRP-strip
f_c	compression strength
f_t	tensile strength
l_b	bond length
s	slip between CFRP-strip and concrete
t_l	thickness of the CFRP-strip
v (t)	deformation at time t
v_e	elastic deformation
ϕ	creep modulus $\phi = \dfrac{v(t)}{v_e} - 1$
τ_b	bond shear stress

Table 1 – lap shear test parameters

Temperature		Load
[°C]	[°F]	in % of the long-term strength
20	68	15
		80
38	100	15
50	122	15

Table 2 – tests with pre-stressed NSM

Temperature		bond length l_b	bondless lenght in center
[°C]	[°F]	[mm]	[mm]
20	68	100	100
38	100	100	500
50	122	100	100
50	122	450	50

Table 3 – material parameters

Short-term material Properties	compressive strenght f_c	Age	tensile strength f_t	Elastic Modulus E	width b_1/ thickness t_1
	[MPa]	[year]	[MPa]	[GPa]	[mm]
concrete	55 (28d)	approx. 4	-	-	-
CFRP-strips	-	-	2800	170	20/2
epoxy resin	105 (7d-20°)	-	32 (7d-20°)	8	-

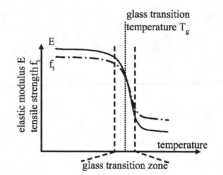

Figure 1 – schematic sketch of temperature-depending tensile strength and elastic modulus of epoxy resin adhesives

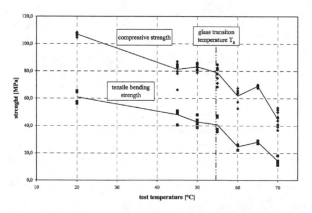

Figure 2 – compressive and tensile bending strength of epoxy resin with rising test temperature

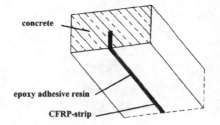

Figure 3 – near surface mounted CFRP strips glued into slits (Blaschko and Zilch 1999)

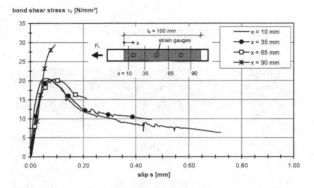

Figure 4 – shear-slip-relationships for different sections of the bond length (bond length l_b = 100 mm, distance a_r = 150 mm) (Blaschko 2003)

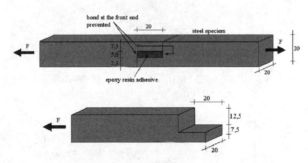
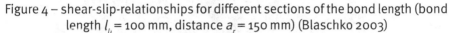

Figure 5 – lap shear test specimen according to DIN 54451-11.1978

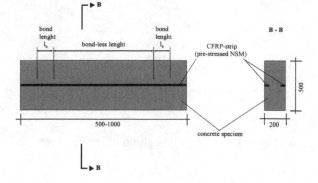

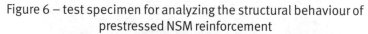

Figure 6 – test specimen for analyzing the structural behaviour of prestressed NSM reinforcement

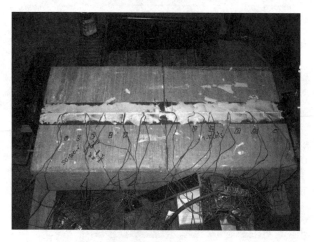

Figure 7 – test specimen with a bond length l_b of 450 mm with strain gauges within the bond length l_b – while curing

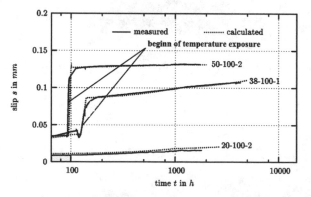

Figure 8 – total slip between strip and concrete at the inner end of the bond length – experiment and calculation

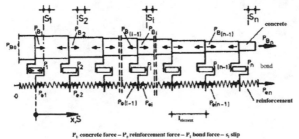

Figure 9 – framework bond model (Franke 1976)

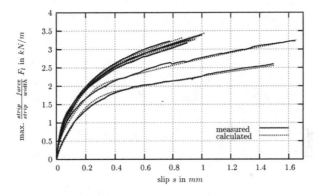

Figure 10 – strip-force-slip-relationship with varying geometrical parameters – experiment (Blaschko 2001) and calculation

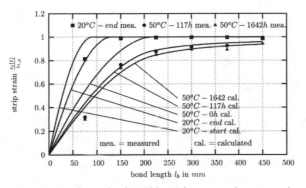

Figure 11 – strip strain along the bond length – experiment and calculation

Long-Term Performance of a CFRP Strap Shear Retrofitting System

by N. Hoult and J. Lees

<u>Synopsis:</u> A shear retrofitting method for reinforced concrete (RC) beams has been developed that uses external CFRP straps to provide additional shear capacity. Research has been undertaken to develop an installation technique that allows the CFRP strap to encompass the full depth of the beam, without requiring access to the top surface of the beam. The current testing scheme investigates the durability of the CFRP strap system using the new installation technique. A long-term load test was conducted on a RC T-beam which indicated that the straps continued to provide shear capacity after 7 months under a load equivalent to 80% of the ultimate capacity of the retrofitted beam. A cyclic test conducted on another similar T-beam specimen demonstrated that after 1,000,000 cycles, under a load that varied between 0.5 and 0.8 times the ultimate retrofitted beam capacity, the straps continued to provide effective shear enhancement.

<u>Keywords:</u> beams; CFRP; concrete; fatigue; long-term; retrofitting; shear

Neil Hoult is a PhD candidate at the University of Cambridge Department of Engineering, U.K. He did his BASc. and MASc. at the University of Toronto. His MASc. research focused on the effects of transverse reinforcement on the confinement of rectilinear columns. His current work involves the testing and modelling of shear retrofitting systems.

Dr Janet Lees is a Senior Lecturer in the Department of Engineering, University of Cambridge. Her interests include the use of advanced composite materials in infrastructure applications. Her research has investigated the use of FRPs as prestressing tendons, and for retrofitting and repairing existing structures, GFRP pipelines and recycled materials in concrete.

INTRODUCTION

It has been estimated that that cost of retrofitting reinforced concrete (RC) structures is $40 billion a year in the US alone (PCI 2003). While not all of this money is spent upgrading shear-deficient structures, even a small percentage still represents a considerable investment. Shear deficiencies are generally the result of increased load requirements or, as Collins and Mitchell (1997) note, the fact that previous design codes were less conservative than current codes. In some climates, corrosion of internal reinforcing steel due to the use of deicing salts is a further issue that can have a detrimental effect on shear capacity. Considering both the cost, and the number of structures involved, there is a need to find effective shear retrofitting systems for RC structures.

Current research efforts focus primarily on the use of Fibre Reinforced Polymers (FRPs) to enhance the shear capacity of RC beams. FRPs are not susceptible to corrosion and thus have an advantage over more traditional retrofitting techniques that use steel. FRPs are also lighter and will not contribute greatly to the dead weight of the structure. Triantafillou (1998), as well as many others, has looked at the use of FRP laminates or sheets that are bonded to the side of the specimen. The amount of shear force that can be transferred to the beam is dependent on the strength of the bond, the anchorage length, and the FRP strength and stiffness. If the anchorage length is insufficient, the FRP retrofit will delaminate before the ultimate capacity of the FRP is reached. However, anchorage length issues can be mitigated if the FRP laminates are wrapped around the full cross section. This also potentially provides better shear resistance according to Kani, Huggins, and Wittkopp (1979). They suggest that reinforced concrete beams can be considered as a series of arches, as shown in Figure 1, and that in order to prevent shear failure each of these arches must be tied together with transverse reinforcement. Thus failure of one or more of these internal arches is possible if the shear reinforcement does not encompass enough of the beam depth. In addition, with bonded systems there exists the potential for failure if the local strain in the FRP induced by crack formation exceeds the ultimate strain of the FRP or results in excessive debonding.

Another type of shear retrofitting technique uses Near Surface Mounted (NSM) FRPs. In this system, grooves are cut into the sides of concrete beams, FRP reinforcing bars are installed and then the grooves are filled with epoxy (De Lorenzis and Nanni 2001). Once again failure tends to be due to breakdown of the epoxy bond. De Lorenzis and Nanni also discovered that whilst strength enhancements of 41% were possible without penetrating the flange of their T-beam sections, increases of 106% were possible if the FRP bars were embedded into the flange. This further demonstrates the benefits of stirrups encompassing the full depth of the beam.

Previous research has highlighted the need to develop a system that is not susceptible to bond breakdown, loss of anchorage or localized strain concentrations due to crack formation. At the same time this system should achieve the beneficial effects that come from being able to encompass the complete web depth. A shear retrofitting system that meets this criteria was developed by Winistoerfer (1999) that involves wrapping thin CFRP thermoplastic tapes around a beam to form an external reinforcing element. The tapes are 12mm wide and 0.16mm thick. Because the tapes are so thin, they can be wrapped around a section with a minimum radius of 12.5mm up to 40 times with each layer providing a linear increase in tensile capacity. After 40 loops, the tensile capacity added by each additional layer is not as significant. The outermost layer of the tape is welded to the next outermost layer but the inner layers remain non-laminated. Winistoerfer discovered that although the outermost layer forms a closed loop, the inner layers act independently, each one reaching its maximum tensile strength. In contrast, if all the layers are fused together, through thickness effects reduce the capacity of each layer and the overall capacity of the strap. A further benefit is that the non-laminated straps can be prestressed. In the following, results from a series of experimental studies that have been carried out at the University of Cambridge using these straps as a shear retrofitting system for RC beams are presented. Of particular interest is the long-term durability of the CFRP strengthening system.

RESEARCH SIGNIFICANCE

Whilst previous experimental programs have shown that the prestressed CFRP strap retrofitting system enhances the shear capacity of shear-deficient concrete beams, the current work seeks to test the durability of these straps. This study takes initial steps towards establishing the long-term durability of CFRP strap systems and identifying key considerations that may need to be considered.

EXPERIMENTAL BACKGROUND

Kesse, Chan and Lees (2001) investigated the results of two RC T-beams tested in four point bending. The first served as a control beam and failed in shear at 100kN with a diagonal crack forming between the support and one of the load points. The second beam had three of the CFRP straps developed by Winistoerfer placed in each shear span. The straps were wrapped around the entire web and supported on profiled steel pads on the top and bottom of the beam as illustrated in Figure 2. The straps were prestressed with an initial force of 30kN (or approximately 60% of the ultimate strap capacity). The addition of these straps provided the beam with enough shear capacity to achieve a ductile flexural failure at a load of 150kN.

Kesse (2003) tested a series of rectangular RC cantilever sections. He looked at the effect of varying several strap parameters such as the strap spacing, strap stiffness and initial prestress. Once again, the straps were supported on metal pads on the top and bottom of the beam. Kesse concluded that in order to utilize the full capacity of the strap, they should be placed at a spacing no further than d apart, where d is the effective depth of the beam. He also concluded that stiffer straps, which are a direct function of the number of loops, were more likely to force a ductile flexural failure. Finally, he concluded that the level of initial prestress did not play a significant role in terms of the strengthened capacity of the beam. However, this result seems to be dependent on the beam geometry as work by Stenger (2003) showed that the shear capacity of deep beams wrapped with CFRP straps was influenced significantly by the initial amount of prestress.

Although the previous experimental studies have shown that CFRP straps have incredible potential as a retrofitting system, they have all been limited by the use of metal support pads. The use of these pads to support the straps, especially on the top surface of the beam, leads to serviceability issues for the structure. A freeway bridge with these pads, for example, would require an extra topping in order to ensure a smooth running surface for vehicles. The extra weight of the running surface could negate any improvement achieved by using this lightweight material. As well, any installation would require access to the top surface of the bridge, which could lead to significant traffic interruptions and additional costs.

In order to mitigate these problems, Hoult and Lees (2004) proposed a method of installing the CFRP straps that only required access to the under surface of a T-beam. The underslab retrofitting technique involved drilling holes into the flange of the beams. Although a number of hole configurations and installation methods were investigated, a system with grouted holes where the CFRP strap rested in a groove cast into the filling material showed the most promise. In practice, while the holes are being drilled and before the grout reaches full strength, temporary shoring should be provided if necessary. The study concluded that the depth of penetration into the compressive flange was important if full shear enhancement were to be achieved. Based on these experiments, the strap configuration shown in Figure 3 was developed after the paper by Hoult and Lees was published. Using this configuration, a ductile flexural failure was achieved with an ultimate specimen capacity of 135kN (an equivalent unstrengthened control beam failed at a load of 88kN).

While the static behaviour of the strengthening system was encouraging, the long-term performance is also important. Tests were therefore devised to investigate the behaviour of the strengthened beams under sustained load (long-term test) and also due to cyclic loading.

LONG-TERM TEST

Test Specimens

The beam cross section is illustrated in Figure 3. The material properties of the steel reinforcement are given in Table 1. The concrete strengths at the beginning of testing are given in Table 2 whilst Table 3 contains the CFRP strap properties. The strap layout is illustrated in Figure 4.

Strap Installation

In order to achieve the strap path illustrated in Figure 3, four holes in the concrete flange must be created. Although in the field all holes will have to be drilled, in order to protect the internal strain gauges, the two diagonal holes in the flange were pre-cast into the beam. This was achieved by wrapping round metal inserts in bicycle inner tubes and placing them in the formwork before casting. Two days after casting the metal tubes were removed and the bicycle inner tubes were pulled out, leaving behind two 30° holes in the flange. The vertical holes were drilled into the flange using a hammer drill with a 25mm diameter drill bit. The holes were drilled a minimum of two weeks after casting. Once the holes were drilled, a strip of 3mm thick and 15mm wide polytetrafluoroethylene (PTFE) was placed in the holes to create the void that the straps would later pass through. The holes were then filled with a high early strength, non-expansive concrete repair product. The product was vibrated into the holes to ensure maximum contact with the existing concrete and minimum voids. After one day the PTFE strips were removed leaving a void in the grout.

The CFRP straps were then threaded into this void and around the beam. On the underside of the web they were wrapped around a metal pad that prevented the strap from bending below its minimum critical radius of 12.5mm. Ten loops of CFRP tape were used for both the long-term and cyclic load test specimens. The CFRP thermoplastic tape was then welded together. As mentioned earlier, only the outer two layers need to be welded together to create a closed outer loop. Once the straps were welded, they were prestressed using the set-up shown in Figure 5. For convenience the experimental beam was turned over to install and prestress the straps, but this is not a necessity. The system uses an hydraulic jack to apply force through a threaded rod that lifts up the metal plate supporting the CFRP strap. The applied prestress was 15kN resulting in a strap stress of 390MPa (approximately 25% of the total strap capacity).

Test Set-Up

Two test specimens were considered. The first specimen was strengthened but unloaded and the strap strains were monitored with time. The second specimen was subjected to a shear force of 110kN. This force was chosen for several reasons. First of all, it was between the capacity of the unretrofitted specimen (Hoult and Lees 2004) of 88kN and the capacity of the retrofitted specimen, that failed in flexural, of 135kN. Work by Tilly (1979) suggests that the ratio of dead to live load applied to typical RC bridges is between 0.2 and 0.4. As a typical live load, the standard fatigue vehicle from BS 5400 (1980) was used. This vehicle has a weight of 80kN per axle, which when combined with a dead load calculated using Tilly's ratio results in two point loads of 110kN each (unfactored dead plus live load). It should be noted that the standard fatigue vehicle loading would not normally be applied to a beam of these dimensions. However, the loading provided for an extreme test of capacity as well as corresponding to the point at which significant strap strains were developed during the static tests.

The test set-up for the long-term test is illustrated in Figure 6. The beam was 3m long and had a clear span of 2.5m. The two point loads were applied 0.75m in from each support. In order to apply the 110kN of shear (and 220kN of total load) two special threaded rods were used. Each 30mm diameter rod was turned down in the middle and a full bridge strain gauge configuration was placed on the turned down area. The full bridge was then used as a load cell allowing the total force being applied to the beam to be measured. Each rod's 'load cell' was calibrated using a separate tensile testing machine. The rods were then screwed into the threaded sockets in the strong floor.

A cross beam and a spreader beam were used to transfer the load from the strong floor into the test specimen as illustrated in Figure 6. Once these beams were in place, the load was applied by tightening nuts on the threaded rods against the cross beam. In order to provide enough force, a wrench with a 2m long scaffold pole extension was used. The disadvantage of this system was that as the beam crept, the tension in the rods would reduce, which would in turn reduce the total force being applied to the beam. This meant that the nuts had to be periodically tightened to ensure that the 110kN of shear force was maintained in the test specimen. Although this situation is not ideal, the amount of shear force required made the use of dead weight for loading impractical.

The beam displacement was measured with a mechanical dial gauge. The strains on the internal longitudinal and transverse reinforcement were measured using 6mm strain gauges. The strains in the CFRP straps were also measured using 6mm strain gauges placed at the mid-height of outer layer on both sides of the beam. The applied loads as well as the strains were data logged at 6 hour intervals (see Figure 4 for the strain gauge locations).

Long-term behaviour

The strap strains plotted over time in the unloaded beam are given in Figure 7. In this case the strap strains have decreased with time by about 5% over 77 days. This reduction in strain is believed to be due to creep in the concrete under the prestressing force of the strap. At the same time, there may also be further losses due to relaxation in the CFRP strap. Whilst the strain gauge readings will not indicate relaxation, experimental work by Saadatmanesh and Tannous (1999) on CFRP prestressing rods indicate that relaxation losses can be between 5 and 10% of the prestressing force over a 50 year period. The strap strain with time results for the loaded beam can be seen in Figure 8. In this case, the CFRP strap strains have increased over time. The maximum strap strains have increased by approximately 0.001 or 23% after 220 days. As such, the true long-term strap strain increases would if anything be slightly larger than shown if the strain reductions due to concrete creep were removed. Interestingly, if the strap strains are plotted versus time on a logarithmic scale, the relationship appears to be linear as illustrated in Figure 9. Based on this relationship the maximum strap strains after 100 years would be 0.0076, which is less than the rupture strain of 0.01 suggesting that the strap may have the required long-term capacity. However, more work is still required to establish which parameters contribute to this relationship. If the strap strains were to eventually exceed the ultimate strain, the straps could fail, resulting in collapse of the structure. Thus, it is necessary to develop a method of calculating the long-term increase in strap strain so that the designer can have reasonable confidence in the durability of the straps.

The mid-span deflection vs. time relationship for the loaded beam is presented in Figure 10. After approximately 220 days, the loading has resulted in a total deflection of 24.1mm. Of greater interest is the increase in deflection of 8.7mm above the initial deflection of 15.4mm, representing an increase factor of 1.57.

A comparison of Figures 8 and 10 shows that the long-term deflection and strap strain relationships have the same general shape. The major difference is the magnitude of the increase, or factor, of the relationship. It is possible that a relationship could be developed to determine the long-term strap strains based on the deflection. However, a viable method of calculating deflections must first be determined before any such relationship could be considered. As such, the available methods for calculating deflections need to be examined.

Deflections due to flexure – According to most approaches, the deflection of reinforced concrete beams increases over time due to the creep and shrinkage of concrete (ACI Committee 435 2003). The creep in the concrete causes increasing strains in the longitudinal reinforcement, which translates to an overall increase in the beam deflection. There exist approaches (ACI Committee 435 2003) that are capable of estimating flexural deflections to a high degree of accuracy. Unfortunately they require a knowledge of the materials and environmental conditions that typically would not be available to a designer hoping to retrofit an existing 30-year-old structure. Instead it would seem more practical to use a simplified approach such as the ACI factor approach (ACI Committee 318 1995), given in equation 1, despite the lack of accuracy of such an approach (Espion and Halleux 1990). Safety factors could then be used to account for the inaccuracies in the method.

$$f_t = f_o \left(1 + \frac{\xi}{1 + 50\rho_c} \right) \tag{1}$$

f_t = the deflection in mm at time t
f_o = the initial deflection in mm at time $t = 0$
ξ = a factor to account for the duration of loading
$\rho_c = A_c/bd$ = the compressive reinforcement ratio
A_c = the area of compressive reinforcement (mm^2)
b = the width of the web (mm)
d = the effective depth (mm)

The factor ξ is set to 2 for loads applied for more than 5 years.

Deflections due to shear -- Interestingly, the long-term deflection calculations in design codes are based on flexural effects. However, if the deflections of the long-term specimen investigated in this study were due exclusively to flexural effects, one would not anticipate any increase in strap strain because all the strain redistribution due to long-term effects would be primarily longitudinal. In fact, if anything, a slight decrease in strap strain due to creep in the concrete might be expected as illustrated in Figure 7. Thus the increase in strap strain suggests that there is a shear component to the deflection. This result has been verified by Nie and Cai (2000) who calculated deflections due to shear to be between 13 and 35% of the total long term deflection. However, Nie and Cai also suggest that the long-term change in transverse strain is negligible and that the increase in shear deflection is the product of creep in the concrete. As can be seen from Figure 8, this assumption is incorrect in this case, with the maximum strap strains having increased by approximately 23%. A similar increase in strain in the middle steel stirrup of 31% was also observed. An increase in transverse strains would seem logical if shear deformations are significant. Without an increase in the applied shear force, the only other way for the shear strains to increase (and thus the corresponding shear displacements) would be for the effective shear modulus of the concrete to decrease. An approach that is employed in Finite Element Analysis (FEA) is to reduce the shear modulus, G, by a factor, β, to account for cracking. A decrease in the shear modulus would suggest a reduction in the concrete's shear capacity and a corresponding increase in the transverse reinforcement stresses and strains under the same load.

Long-term strap strains – The preceding discussion has illustrated the need for a method of calculating long-term deflections that is both straightforward and accounts for the shear component of the deflection. Unfortunately there is currently not enough test data to develop such a method for determining the long-term CFRP strap strains. Future research into long-term deflections should look at variables such as span to depth ratio and strap prestress to better understand the potential increases in strap strain and the relationship to shear deflection.

Fortunately, since most structures where this retrofitting technique would be employed will have already undergone substantial long-term creep deflection, significant increases in strap strain should not be an issue. The long-term test also considered the most severe loading case of a live load applied over a long period of time. Most structures are unlikely to see the full live load applied for significant periods of time and so increases in deflection and strap creep strains should be lower. Since this technique is not limited by the constraints of other FRP retrofitting techniques (bond, anchorage, and localized stress concentrations), the number of strap loops can be increased to reduce the strain required to obtain the same level of prestress and thus the total strain.

The long-term deflection test has further demonstrated the potential of these straps. The straps have not failed after approximately seven months under a load that exceeds the capacity of a unretrofitted RC beam of the same design by 25%. This bodes well for the future use of this CFRP strap retrofitting system.

CYCLIC LOADING TEST

Test Set-Up

The specimen used for the cyclic load test had the same design as that used in the long-term test so a comparison could be made between their performances. The shear force was cycled between 70 and 110kN in this test at a frequency of 2Hz as illustrated in

Figure 11. Ideally the shear force would have been cycled between 30 and 110kN to represent the 80kN fatigue vehicle specified in BS 5400. However, it was not possible to design a testing rig with the available equipment to accommodate such a large load differential. The current loading was deemed adequate as the lower limit of the loading range, 70kN, was below the capacity of the control specimen (88kN) whilst the upper limit, 110kN, was above this value. This loading range is perhaps also more realistic for many types of structures, where the dead load is a significant portion of the total load. Based on the strap strain readings from static tests, this load range creates a significant difference between minimum and maximum CFRP strap strains.

The testing rig was a self-reacting frame as illustrated in Figure 12. The specimen was placed on top of a steel beam, which served as the reaction beam. The reaction beam was then attached to specially made channel sections that were bolted to the columns. The columns supported the dynamic jack making the system completely self-contained preventing fatigue of the strong floor.

An Amsler testing machine capable of applying both a constant base load as well as a pulsating dynamic load was used. The applied load was measured using a 500kN load cell. Five LRDTs were used with one at each support, one at each load point, and one at the midpoint of the beam. Strain gauges were placed at the midpoint of the top and bottom longitudinal reinforcing bars. Gauges were also placed at the mid-height of each stirrup in one of the shear spans as well as on each CFRP strap. The results were monitored using a high-speed data acquisition system capable of scanning each data channel 100 times a second. The program was designed to record two seconds of data every hour so that changes with time could be measured.

Cyclic Behaviour

Deflections -- Although the beam will eventually be tested to two million cycles, only the results of the first million load cycles will be reported here. However, most of the trends are evident by this point in the testing. The maximum and minimum mid-span deflections are plotted against the number of cycles in Figure 13. The displacement increases over time, which is quite similar to the behaviour exhibited by the long-term specimen. The overall deflection increase factor of 1.15 is not nearly as significant as the 1.57 noted for the long-term test. However, the average load applied to the specimen is lower, at 90kN, than was used in the long-term test. This results in lower concrete stresses and reduced creep. The load is also applied over a much shorter time frame, further decreasing the amount of concrete creep and resulting deflections. Interestingly there is little change in the difference between maximum and minimum displacements over time. This indicates that for the given time, t, the beam stiffness over this load range remains relatively constant since a drop in stiffness would result in a larger change in the maximum displacement versus the minimum displacement. At the same time the overall beam stiffness is decreasing as the maximum and minimum displacements are increasing with time. This result suggests that the change in stiffness is a function of the constantly applied base load, and not the additional cyclic load.

There is a slight drop in the overall deflection at approximately 920000 cycles and then an increase thereafter. This corresponds to a period of 4 days when the beam had to be completely unloaded and realigned in the rig. The initial decrease in deflection is believed to be due to the cracks closing slightly while the beam was unloaded. An examination of the steel stirrup data over the same period showed that the strain in the stirrups had decreased in a similar fashion to the displacements, which also suggests that the crack widths decreased during the period of unloading.

The fatigue capacity of a composite beam is limited by the fatigue capacity of its component parts. As such, when considering whether to retrofit a structure, the designer must not only consider the fatigue capacity of the retrofitting material (the CFRP straps) but also the reinforcement and concrete in the existing beam.

CFRP Straps -- The strap strains are plotted against the number of cycles in Figure 14. Once again the shape of the plot is similar to the long-term tests, with the strains increasing with time. The strains in the middle straps have increased by a factor of 1.14, which is less than the 1.23 observed in the long-term tests. It suggests that in terms of a strain increase, a sustained long-term load is more critical. The difference between the minimum and maximum strain, and corresponding stress range, in the straps remains constant for the duration of the test. The middle straps have the largest stress range of approximately 85MPa. This constant stress range indicates that the cracks in the concrete are not growing significantly, which validates the conclusion that the stiffness over this loading range is also not changing significantly. Despite exhibiting slight increases in average strain, the straps appear to have the required fatigue capacity. However, even under the maximum load the stresses in the straps are only about 50% of the ultimate strap tensile capacity, so future work with higher average stresses and with higher stress ranges should be performed to validate this result.

Reinforcement fatigue -- The minimum stress level in the middle steel stirrups is approximately 275MPa whilst the maximum is 400MPa. This leads to a stress range of 125MPa, which is below the 280MPa that design codes (BS 5400 1996) deem to be acceptable. Tilly (1979) suggested that the fatigue properties of the reinforcement were dependant upon both the stress range and the average mean stress. For the case of the transverse reinforcement, the average mean stress is 338MPa, which is higher than the values tested by Tilly. He observed that increasing the average stress from 159MPa to 275MPa resulted in a 40MPa reduction in the stress range that caused failure at 10^6 cycles. The stress range to cause failure at an average stress of 275MPa was found to be approximately 210MPa. The data given by Tilly seems to exhibit fairly linear behaviour (although this requires confirmation) so by extrapolation another 40MPa reduction in stress range will occur between an average stress of 275MPa and 391MPa, making the stress range to cause failure 170MPa. As such, a fatigue failure in the transverse reinforcement should not be critical.

Unfortunately the strain gauges on the longitudinal reinforcement gave erroneous results during the cyclic test, which was probably due to loss of bond between the gauge and reinforcement bar. Results from previous static tests indicate that the minimum stress in the longitudinal reinforcement is approximately 310MPa with a maximum of approximately 475MPa at these load levels. This leads to a stress range of 165MPa, which is within the code limits of 220MPa for this bar diameter. However, when one considers the extrapolation of Tilly's work given above, the average stress of 393MPa results in a stress range of 170MPa to cause failure at 10^6 load cycles. This extrapolated stress range is quite close to the actual stress range, which suggests the possibility of a fatigue failure in the longitudinal reinforcement. If the full intended loading between 30 and 110kN had been used, then the stress would definitely have exceeded these limits. A method of predicting fatigue failure in the tensile reinforcement has been presented by Heffernan, Erki, and DuQuesnay (2004). Their method indicates that the stresses in the reinforcement need to be increased by a factor of 1.2 to allow for stress concentrations at the concrete cracks and a further 1.05 to account for tensile strain increases due to concrete softening. This would further reduce the potential stress range in the longitudinal reinforcement. Thus, any designer hoping to employ this shear retrofit system should give careful consideration to fatigue of the existing reinforcement as well as the straps.

Concrete fatigue -- Although not specifically investigated in this experiment, Czaderski and Motavalli (2004) also noted the importance of the concrete capacity. In their experiments on T-beams retrofitted with CFRP L-shaped plates, they were able to load the specimen between 39 and 59% of the ultimate specimen capacity for five million cycles. They noted that the concrete compression strain had increased significantly during the course of their test. As such, Czaderski and Motavalli recommend that the designer consider carefully whether there is enough remaining concrete capacity in the structure to be retrofitted.

Whilst the designer must exercise care when applying any retrofitting method, the cyclic testing has demonstrated the durability of the CFRP strap retrofitting system. The straps were used to strengthen a beam that was subjected to loads between 0.8 and 1.25 times its unretrofitted capacity for one million cycles without displaying any signs of fatigue failure.

CONCLUSIONS

The long-term test illustrated that the CFRP straps can provide significant shear enhancement over long periods of time. Nevertheless, the designer must be aware of the possibility of increases in strap strains as the RC beam creeps. The increases in strap strain can potentially be estimated and taken into account in the design of the strengthening system. Furthermore, the strain increases should be relatively small due to the level of loading and extent of creep in existing structures. The development of a database of long-term deflections due to flexure and shear is also required.

The cyclic load tests also indicated that the CFRP strap system has the required capacity to provide long-term shear enhancement. The increases in both deflection and strap strains were not as significant as for the long-term load test. As such, the long-term load would appear to be more critical in terms of strap strain and deflections. The designer also needs to consider the fatigue capacity of the existing structure, as it is possible that increased load levels could cause fatigue failure of the internal steel longitudinal reinforcement or concrete.

ACKNOWLEDGMENTS

The first author is grateful for the financial support provided by the Canada Cambridge Trust as well as Universities UK. The authors would also like to thank EMPA for their continued support of this research.

REFERENCES

ACI Committee 318, 1995, *Building code requirements for structural concrete*, American Concrete Institute, Farmington Hills, MI.

ACI Committee 435, 2003, *Control of Deflection in Concrete Structures (Appendix B)*, American Concrete Institute, Farmington Hills, MI, pp. B1-B13.

BS 5400, 1980, *Steel, concrete and composite bridges Part 10. Code of practice for fatigue*, British Standards Institution, London, England.

BS 5400, 1996, *Amendments to BS 5400: Part 4:1990*, British Standards Institution, London, England, 4 pp.

Collins, M.P., and Mitchell, D., 1997, *Prestressed Concrete Structures*, Response Publications, Toronto, 766 pp.

Czaderski, C. and Motavalli, M., 2004, "Fatigue behaviour of CFRP L-shaped plates for shear strengthening of RC T-beams," Composites Part B: Engineering, V. 35, No. 4, pp. 279-290.

De Lorenzis, L. and Nanni, A., 2001, "Shear Strengthening of Reinforced Concrete Beams with Near-Surface Mounted Fiber-Reinforced Polymer Rods," ACI Structural Journal, V. 98, No. 1, pp. 60-68.

Espion, B. and Halleux, P., 1990, "Long-Term Deflections of Reinforced Concrete Beams: Reconsideration of Their Variability," ACI Structural Journal, V. 87, No. 2, pp. 232-236.

Heffernan, P.J., Erki, M-A. and DuQuesnay, D.L, 2004, "Stress Redistribution in Cyclically Loaded Reinforced Concrete Beams," ACI Structural Journal, V. 101, No. 2, pp. 261-268.

Hoult, N.A., and Lees, J.M., 2004, "Shear Retrofitting of Reinforced Concrete Beams Using CFRP Straps," Proceedings of the 4th International Conference on Advanced Composites in Bridges and Structures, Calgary, Canada, 2004, 8 pp.

Kani, M.W., Huggins, M.W. and Wittkopp, R.R., 1979, *Kani on Shear in Reinforced Concrete*, Department of Civil Engineering, University of Toronto, 226 pp.

Kesse, G. Chan, C. and Lees, J., 2001, "Non-linear Finite Element Analysis of RC Beams Prestressed with CFRP Straps," *FRPRCS-5*, Thomas Telford, Cambridge, England, Vol. 1, pp. 281-290.

Kesse,G., 2003, *Concrete Beams with External Prestressed Carbon FRP Shear Reinforcement*, PhD Thesis, Department of Engineering, University of Cambridge, 222 pp.

Nie, J. and Cai, C.S., 2000, "Deflection of Cracked RC Beams under Sustained Loading," Journal of Structural Engineering, Vol. 126, No. 6, pp. 708-716.

PCI, 2003, "ACI releases second edition of 'Concrete Repair Manual'," Concrete Monthly, http://concretemonthly.com/monthly/art.php/105, accessed on February 15, 2005.

Saadatmanesh, H. and Tannous, F.E., 1999, "Relaxation, Creep, and Fatigue Behavior of Carbon Fiber Reinforced Plastic Tendons," ACI Materials J., V. 96, No. 2, pp. 143-153.

Stenger, F., 2003, "Tragverhalten von Stahlbetonscheiben mit vorgespannter externer Schubbewehrung aus Kohlenstofffasern," *CFK im Bauwesaen – heute Realitat!*, EMPA Akademie, Dubendorf, Switzerland, pp. 59-67.

Tilly, G.P., 1979, "Fatigue of Steel Reinforcement Bars in Concrete: A Review," Fatigue of Engineering Materials and Structures, Vol. 2, pp. 251-268.

Triantafillou, T.C., 1998, "Shear Strengthening of Reinforced Concrete Beams Using Epoxy-Bonded FRP Composites," ACI Structural Journal, Vol. 95, No. 2, pp. 107-115.

Winistoerfer, A.U., 1999, *Development of non-laminated advanced composite straps for civil engineering applications*, PhD Thesis, Department of Engineering, University of Warwick, 170 pp.

Table 1 – Steel reinforcement properties

Bar Diameter (mm)	Yield Strength (MPa)	Yield Strain	Elastic Modulus (GPa)	Ultimate Strength (MPa)
6	578* ± 5	0.00501*	187.4 ± 6.8	646 ± 0
8	467 ± 6	0.00233	200.3 ± 3.2	540 ± 5
16	505 ± 7	0.00262	192.9 ± 13.3	586 ± 5
20	523 ± 2	0.00263	198.7 ± 6.5	633 ± 1

- using the 0.2% offset method

Table 2 – Concrete strengths

Specimen	Compressive Cube Strength (MPa)	Split Cylinder Strength (MPa)
Long-term load	42.9 ± 1.5	2.7 ± 0.9
Cyclic load	42.2 ± 1.7	3.0 ± 0.1

Table 3 – CFRP strap properties

Modulus of Elasticity (GPa)	Ultimate Strain (mm/mm)*	Experimental Tensile Capacity (kN)**
121.0 ± 4.7	0.01	59.3 ± 2.3

* based on tests performed by Winistoerfer (1999)
** based on the average results of the 10 loop tests using the steel support pads

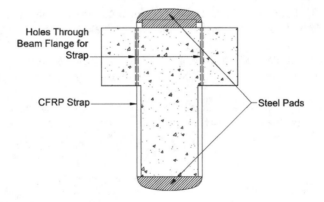

Figure 1 – Kani arch model

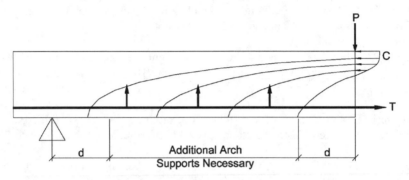

Figure 2 – CFRP strap configuration used in Kesse, Chan and Lees

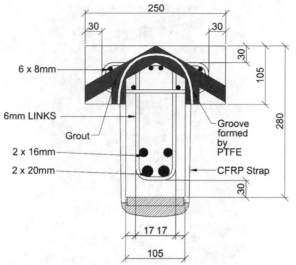

Figure 3 – Beam cross-section

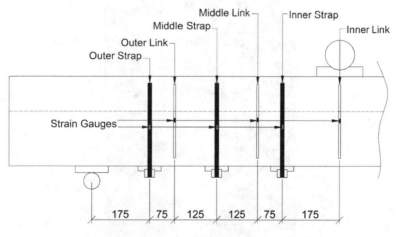

Figure 4 – Strap layout

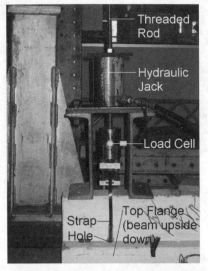

Figure 5 – Strap prestressing system

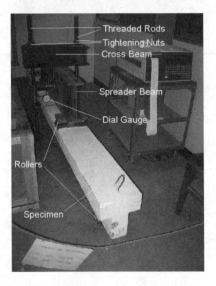

Figure 6 – Long-term beam test set-up

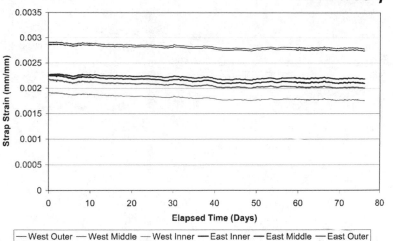

Figure 7 – Long-term CFRP strap strain vs. time under no applied load

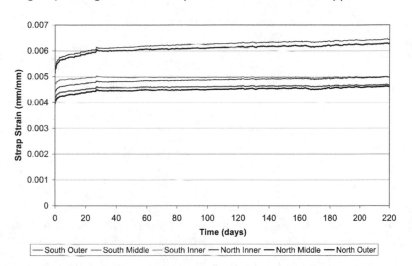

Figure 8 – Long-term CFRP strap strain vs. time under applied load

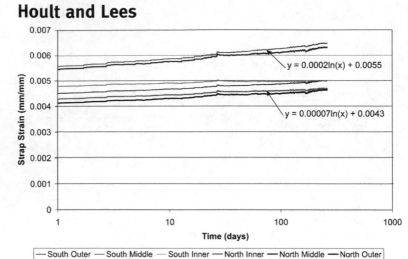

Figure 9 – Logarithmic plot of strap strain vs. time

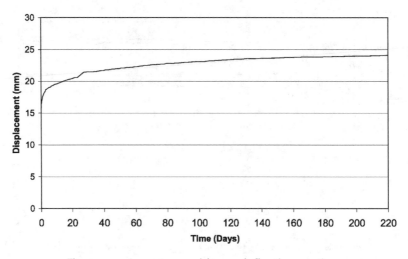

Figure 10 – Long-term midspan deflection vs. time

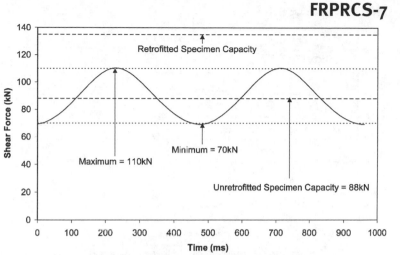

Figure 11 – Cyclic Loading

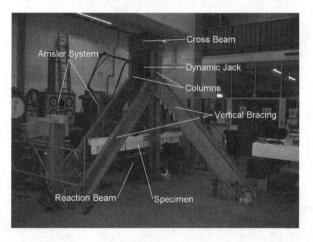

Figure 12 – Self-reacting frame for cyclic test

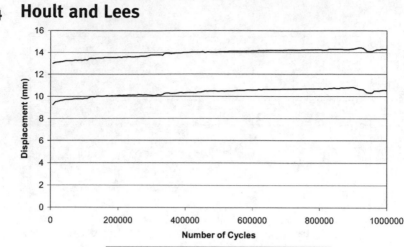

Figure 13 – Mid-span deflection vs. number of cycles

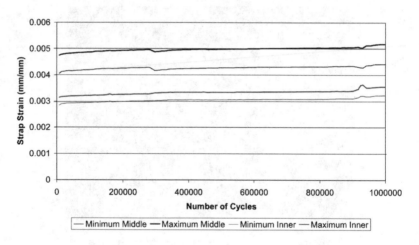

Figure 14 – Strap strains vs. number of cycles

Freeze-Thaw Behavior of FRP-Confined Concrete under Sustained Load

by A. Kong, A. Fam, and M.F. Green

Synopsis: Fiber-reinforced polymers (FRPs) are effective in strengthening concrete structures. However, little work has examined the effects of cold regions on the behavior of the strengthened members, particularly the combined effects of sustained loading and freeze-thaw exposure. This paper presents the results of an experimental study on the durability of 70 normal weight, low strength, and non-air entrained concrete cylinders (150 x 300mm). The cylinders were confined with glass-FRP (GFRP) sheets or carbon-FRP (CFRP) sheets and exposed to 300 freeze-thaw cycles while under sustained axial compression loads. FRP-wrapped cylinders showed exceptional durability performance after their extreme exposure to freeze-thaw and sustained loading with a maximum of 12% reduction in strength. Some CFRP wrapped cylinders that were exposed to freeze-thaw without longitudinal restraint, by means of sustained loads, and all the plain concrete cylinders were completely disintegrated with virtually zero residual strength.

Keywords: concrete; confinement; creep; durability; fiber reinforced polymer (FRP); freeze-thaw; sustained load; wraps

706   Kong et al.

Andrew Kong is currently a Masters Student in the Department of Civil Engineering at Queen's University at Kingston, Ontario, Canada. He received a B.Sc. from North Carolina A&T State University, U.S.A., 2002. Current research involves creep and durability of fiber reinforced polymer (FRP) used as external reinforcement for concrete cylinders.

ACI member **Amir Fam** is an Assistant Professor and Canada Research Chair in Innovative and Retrofitted Structures at Queen's University, Canada. He is a voting member of ACI Committee 440, FRP Reinforcement and a Co-Chairman of Sub-Committee 440-J, FRP Stay-in-Place Formwork. His research includes applications of FRP in new construction and retrofit of existing structures.

ACI member **Mark F. Green** is a Professor in the Department of Civil Engineering at Queen's University at Kingston, Canada. He received his Ph.D. in Engineering from Cambridge University, UK, in 1991. He is an active member of ACI committee 440 Fiber Reinforced Polymer. His research interests include strengthening concrete structures with FRPs, prestressing concrete beams with FRPs, and bridge-vehicle dynamics.

INTRODUCTION

There has been an ever-increasing need for structural repair of existing infrastructure that is now rapidly deteriorating. Over the past fifteen years, engineers and material scientists have made great strides by incorporating Fiber Reinforced Polymer (FRP) composite materials into design and construction to meet this need. These materials have been successful in many applications, but little information is available on their behavior in cold regions. In particular, the combined effects of sustained loads and freeze-thaw exposure have not been studied by researchers (Karbhari et al. 2003).

One application of FRPs is to confine concrete columns to increase their axial strength and ductility. In cold climates, these columns will be exposed to low temperature, and freeze-thaw action. FRP composites exposed to freezing temperatures and alkaline solutions may suffer matrix microcracking, degradation in the fiber-matrix bond, and even damage to the fibers themselves (Karbhari et al. 2000). Some researchers have examined confined concrete cylinders exposed to freeze-thaw or to sustained loading, independently. Karbhari and Eckel (1993) tested FRP wrapped cylinders at low temperature (-18°C), and found increased brittleness of FRP fibres at low temperature. Soudki and Green (1996) found good performance of carbon FRP strengthened columns when subjected to up to 200 freeze-thaw cycles consisting of 16 hours of freezing in air (-18°C), followed by 8 hours of thawing in a water bath (+15°C). Toutanji and Balaguru (1998) exposed FRP wrapped concrete cylinders to 300 freeze-thaw cycles and found slight deterioration due to freeze-thaw cycles, with carbon FRPs performing better than glass FRPs.

Other researchers have studied the creep behavior of FRP-wrapped cylinders under sustained axial loads. However, no research has been reported on FRP-wrapped cylinders subjected to the two conditions combined, as in actual structures, which may have a synergetic effect. Creep in polymers is commonly known as viscoelastic deformation, which is a combination of elastic deformation and viscous flow (Mallick 1988). Stress level and temperature are two factors that influence creep-strain in composites. Orientation of the fibers is another factor that needs to be considered when analyzing creep in polymers. If the main fibers are in the direction of the load, then the composite creep is dependent on creep in the fibers. Studies of creep were conducted by Naguib and Mirmiran (2002) on concrete-filled FRP tubes showed that shrinkage and creep of concrete in this case is much lower than exposed concrete due to the sealing effect.

In the research project described in this paper, FRP-wrapped concrete cylinders (152 x 305 mm) have been subjected to 300 freeze-thaw cycles, while simultaneously being subjected to sustained axial compression loads. This small size of specimen was selected for ease in repeating the results on many specimens, and so that the capacity of testing machines at Queen's University was not exceeded. Further, most tests in the literature on FRP wrapped concrete cylinders have been conducted with this size of specimen (Bisby et al. 2005). Tests on larger-scale specimens should be conducted to better understand the behavior of full-scale columns in the field. The concrete was not air entrained to simulate the worst-case scenario for concrete. Thus, better freeze-thaw resistance is expected for concrete with appropriate air entrainment. Test results of specimens subjected to the combined effects have been compared to those of other specimens subjected to sustained loading at room temperature and to those of specimens subjected to freeze-thaw without sustained loading. Two numerical models, one developed by ISIS Canada (ISIS, 2001), and the other developed by ACI Committee 440 (ACI 2001) were also used to compare the predicted confined strength of the control cylinders at room temperature with the experimental data.

RESEARCH SIGNIFICANCE

Durability of FRP-retrofitted concrete structures is a major concern. Research conducted on durability of FRP-confined concrete cylinders has concentrated on the effect of harsh environments but without the effect of sustained loads. By applying sustained loads to FRP-wrapped cylinders, the effect of freeze-thaw cycles may be more critical due to the development of internal micro cracks in the concrete under the loads. This paper discusses experimental results examining the freeze-thaw durability of FRP-wrapped cylinders while under sustained loads. The results will provide valuable data for design procedures and for engineers considering FRP repair of columns in cold climates.

EXPERIMENTAL PROGRAM

Seventy (70) concrete cylinders, (152 mm (6in) in diameter and 305mm (12in) in length) were cast. No air-entrainment was used to simulate a worst-case scenario for concrete in harsh weather conditions. The cylinders were removed from the molds after 7 days and allowed to air cure at room temperature. The cylinders were designed to have a compressive strength of 20 MPa after 28 days; the average tested strength at 28 days was 22 MPa. Low strength concrete was used to simulate strength of concrete in old structures. Two sets of 25 cylinders each were wrapped with one layer of CFRP sheet or with one layer of GFRP sheet. The remaining 20 cylinders were left unconfined and used for verification of concrete strength after the completion of freeze-thaw cycles. The manufacturer's properties of the CFRP and GFRP composites are shown in Table 1. After the wrapped cylinders were fully cured, their ends were carefully ground to achieve a smooth surface perpendicular to the main axis of the cylinder.

Before applying sustained loads and freeze-thaw exposure, all test cylinders were first subjected to one monotonic loading cycle, up to a load level of about 45 percent of their respective ultimate strength, unloaded, and then were submerged in water in room temperature for a week. This practice was adopted to promote the worst possible conditions, where microcracks were induced and the internal moisture content was maximized.

Four loading frames were built with the capability to apply sustained loads of 1200 kN each to a series of 5 concrete cylinders. Each frame consisted of 4 # 27mm B16 threaded rods, 2.7m long, and three 50 mm thick mild steel square plates that were 250 x 250 mm as shown in Figure 1. Two frames were placed in the cold room and the other two were always kept at room temperature, simultaneously. Each frame accommodated 5 cylinders placed concentrically in a series. The two frames in the environmental chamber each had 5 CFRP-wrapped and 5 GFRP-wrapped cylinders (Figure 1(a)). This was also the case for the two frames at room temperature (Figure 1(b)). Fifteen control cylinders (5 with CFRP, 5 with GFRP and 5 unconfined) were also placed in the environmental chamber but without sustained loads. Six control (unconfined) cylinders were left at room temperature to determine any increase in concrete strength due to aging by the end of testing period. Three of the six cylinders were submerged in water for a period equivalent to the total number of thawing hours to simulate the curing effect during water-thawing in the environmental chamber, while the other three were kept dry.

A few cylinders were cast with thermocouples inside the concrete core, to measure the concrete core temperature, to help determine the appropriate program to be used for controlling the environmental chamber (cold room), according to ASTM C666 (Standard Test Method for Resistance of Concrete to Rapid Freezing and Thawing) (ASTM 1997). All test specimens were subjected to 300 freeze-thaw cycles, with the temperature of the concrete core varying from +4.4°C to -18°C in 5 hours and 15 minutes (Figure 2). The cycles were very repeatable due to the excellent control of temperature available in the chamber. Since no standard exists for FRP-wrapped concrete cylinders, the standard for plain concrete cylinders was employed.

The sustained load was applied to the cylinders in the frame via a 979 kN hydraulic ram. To measure and monitor the sustained load, a 445 kN load cell was fitted in the frame to continuously monitor the load. This load was maintained constant, using the hydraulic ram throughout the duration of freeze-thaw for both the frames inside and outside the cold room. After the completion of the 300 freeze-thaw cycles, all cylinders were tested under axial compression, using a 1000 kN MTS machine. Each cylinder was instrumented with three 100 mm displacement based strain gauge transducers spaced around the perimeter to measure axial strains, and two 5 mm electric resistance strain gauges to measure hoop strains on two opposite sides of each cylinder.

TEST RESULTS AND DISCUSSION

Test results, in terms of axial compressive strength and ultimate axial and hoop strains are reported in Tables 2 and 3 for CFRP-wrapped and GFRP-wrapped cylinders, respectively. In each table, results are given for both specimens subjected to freeze-thaw cycles and those kept at room temperature, with and without sustained loading for each case. Table 4 shows the test results of plain concrete cylinders at room temperature for both the water cured cylinders and the dry cylinders. For each parameter that was varied, 3 or 4 specimens were tested. All of the results are presented in Tables 2 to 4 and the tests were very repeatable. The specimens are labeled with a 3 letter code; the first letter indicates the type of wrap (C-carbon, G-glass, U-none), the second letter indicates the environmental exposure (F-freeze-thaw, R-room temperature), and the third indicates the type of loading (S-sustained load, C-control with no load). It should be noted that all the five unconfined cylinders placed in the water bath in the environmental chamber were completely disintegrated due to freeze-thaw damage with virtually zero residual strength. Careful observation during freeze-thaw cycling revealed that, at 35 cycles, these unconfined cylinders developed major cracks, and at 100 cycles, they were completely disintegrated, as shown in Figure 3(a). All the unconfined cylinders at room temperature were tested and the ones that simulated the wet-dry condition, occurring in cold room, showed a 4% increase in strength over the ones that were kept dry (Table 4).

The average strength of the CFRP-wrapped cylinders was 40 MPa, which represented a 91% strength increase over the strength of the unconfined cylinders (22 MPa). Three of the four CFRP-wrapped control cylinders (without sustained loading) were completely damaged during freeze-thaw cycling, as shown in Figure 3(b), and were not tested in axial compression. This was attributed to expansion of the concrete core in the longitudinal direction, which was not uniform and resulted in bending of the cylinders. Because all the fibers in the CFRP sheet were oriented in the hoop direction, expansion was not restricted longitudinally. At the overlap location, however, the wrap was slightly stiffer longitudinally due to more fibers and epoxy at this location. Thus, less expansion occurred at the location of the overlap, causing bending. Only one CFRP-wrapped control cylinder could be tested after the completion of the 300 freeze-thaw cycles; this cylinder had a strength of 34.6 MPa.

Also, one CFRP-wrapped cylinder failed under sustained load while undergoing freeze-thaw cycles. Without including this one result, Table 2 shows that freeze-thaw caused only a 3% strength reduction for specimens under sustained loads (CFS vs. CRS Specimens). For the one specimen that survived the freeze-thaw without sustained load, the strength reduction was 12 % (CFC vs. CRC Specimens). Since the other 3 specimens did not maintain integrity for the full duration of the freeze-thaw, the effect of freeze-thaw without sustained loading could be larger than a 12% reduction. It also appears that sustained loading increases the strength by about 7 % at room temperature.

The average strength of the GFRP-wrapped cylinders was 38 MPa which represented an 81% increase over the unconfined cylinders. Results in Table 3 indicate that freeze-thaw caused only a 3% average reduction in strength for specimens under sustained loads

(GFS vs. GRS Specimens) and a 6 % reduction for specimens without sustained loading (GFC vs. GRC Specimens). This confirms quite well the observations on the CFRP-wrapped cylinders. It was also noted that sustained loading had almost no effect on strength in this case (GRC vs. GRS and GFC vs. GFS Specimens).

Figures 4 to 8 show the stress-strain curves for the plain concrete cylinders, the CFRP-wrapped, and the GFRP-wrapped cylinders. Each graph shows the stress versus average axial and hoop strains. The plain concrete cylinders were severely damaged due to freeze-thaw only (no sustained load) because of the lack of air-entrainment. These specimens were completely disintegrated, and could not be tested. As such, Figure 4 shows the behavior of plain concrete cylinders at room temperature only. On the other hand, the confined concrete, after 300 freeze-thaw cycles, maintained approximately the same strength as the control specimens (Tables 2 and 3). Thus, the FRP-wrapping performed exceptionally well except in the case of the CFRP-wrapping without sustained load.

The cylinders that were subjected to sustained load, at both room temperature and freeze-thaw, showed a distinctly different behavior from those not subjected to sustained loading. For example, in Figures 6 and 8, the specimens that were not subjected to sustained load (CRC and GRC) started the transition in the bilinear behavior at a lower load than for the sustained loaded specimens (CRS and GRS). This change in the transition load is attributed to creep which causes axial strains to increase under constant stress. Consequently, the radial and hoop strains will also increase due to Poisson's ratio (dilation) effect. This mechanism activates the confinement imposed by the FRP jacket. At the end of sustained loading, residual strains remain due to material non-linearity (i.e., plastic flow due to creep effects). Thus, the sustained loading appears to create a state of active confinement. Further research may be needed to confirm this hypothesis. The fact that the axial and hoop strains at failure are lower for the specimens under sustained loads gives some credence to this hypothesis. It should be noted that such a state of active confinement would not be expected in a field repair since the sustained loads in the field would be present at the time of wrapping and most of the creep would have occurred before wrapping. Thus, the state of active confinement may not occur in practice.

From the overall results, the GFRP-wrapped cylinders resisted freeze-thaw effects better than the CFRP-wrapped cylinders. Two potential mechanisms for the better performance of the GFRP were identified. The first relates to the amount of epoxy used for the two different materials. Although a similar amount of epoxy was used for each type of FRP, the GFRP appeared to absorb more epoxy. Thus, the protection for the GFRP wrapped cylinders may be due to the adherence of more epoxy. Alternatively, the CFRP sheets had all the structural fibers running in the hoop direction and no fibers in the axial direction, whereas the GFRP had most of the fibers (more than 90%) in the hoop direction but had some aramid fibers in the axial direction. These longitudinal aramid fibers may have helped the GFRP-wrapped cylinders resist the axial expansion and contraction of the concrete core during freeze-thaw cycling.

The test results were obtained for small-scale cylinders under purely axial compressive load. Thus, tests on larger specimens and/or with eccentric loading would be useful to continue this work to better represent conditions in the field.

Failures Modes

The mode of failure was governed by tensile fracture of the FRP sheets in the hoop direction. After the hoop stresses induced by the confinement pressure exceeded the FRP tensile capacity, loud popping sounds were heard. At this point, the system was unable to withstand further loading and the specimen failed, as shown in Figure 9. The majority of cylinders failed by fiber rupture at the mid-height of the cylinder. The stress-strain curves clearly demonstrated that the FRP-confined concrete cylinders exhibited a bilinear response with no descending branch. GFRP-wrapped cylinders subjected to freeze-thaw without sustained load had the quietest and least dramatic failure mode, whereas CFRP-wrapped cylinders subjected to both freeze-thaw and sustained loading had the loudest and most catastrophic failure mode.

NUMERICAL MODEL

The following models developed by ISIS Canada (ISIS 2001), and ACI Committee 440 (ACI 2001), were used to predict the FRP-confined strength, f_{cc}' of control (room temperature) specimens:

Model 1 (ACI 440)

$$f_{cc}' = f_c'[2.25\{1 + 7.9(f_l / f_c')\}^{0.5} - 2(f_l / f_c') - 1.25]$$ (1)

$$f_l = [K_a \rho_f f_{fe}]/2 = [K_a \rho_f \varepsilon_{fe} E_f]/2$$ (2)

$$\rho_f = 4nt_f / h$$ (3)

$$\varepsilon_{fu} = C_E \varepsilon_{fu}^*$$ (4)

$$f_{fu} = C_E f_{fu}^*$$ (5)

$$E_f = f_{fu} / \varepsilon_{fu}$$ (6)

Model 2 (ISIS Canada)

$$f_{cc}' = f_c'(1 + \alpha_{pc}\varpi_w)$$ (7)

$$\varpi_w = 2f_{lfrp} / \phi_c f_c'$$ (8)

$$f_{lfrp} = 2N_b \phi_{frp} f_{frpu} t_{frp} / h_c$$ (9)

Tables 5 and 6 show the results for the confinement models compared against the experimental values for the specimens tested at room temperature without sustained loading. Table 7 gives the percentage differences between the experimental results and model predictions. The ISIS equations gave a slightly better prediction than the ACI 440 equations. For CFRP-wrapped cylinders, the differences between the experimental results and the two models were: ISIS – 14 % and ACI 440 – 40 %. For GFRP-cylinders the difference between experimental and the two models were: ISIS – 10% and ACI 440 – 26%. In all cases, the confinement models overestimated the strength of the FRP-confined cylinders and thus the models were not conservative when compared against these test results. Bisby et al. (2005) recently compared the ISIS and ACI confinement models against a much wider database of experimental results and found that the ISIS model was generally conservative but that the ACI model was not.

SUMMARY AND CONCLUSIONS

Overall, FRP-wrapped cylinders displayed excellent resistance to combined freeze-thaw and sustained loading. The average reduction in strength for both GFRP-wrapped and CFRP-wrapped cylinders exposed to freeze-thaw, while under sustained loads, was approximately 3%. With only freeze-thaw exposure (no sustained loading), GFRP wrapped cylinders lost an average of 6% of their confined strength while most of the CFRP-wrapped cylinders lost their structural integrity before the end of the freeze-thaw exposure, due to the excessive expansion and contraction in the longitudinal direction. The specimens subjected to combined freeze-thaw and sustained loads performed better than those under freeze-thaw only, since the sustained loading induced active confinement and end restraints, which prevented longitudinal expansion and contraction. It was also observed that the sustained loading caused a stiffening effect, where the transition in the bilinear stress-strain curves of FRP-confined concrete occurred at a higher load level. Some of these effects of sustained loading may not be evident in the field because columns would typically be wrapped when the sustained loads are acting. The confinement model proposed by ISIS Canada design guide predicted the strength of confined concrete cylinders reasonably well, but that the ACI 440 model was not accurate. More research on larger scale specimens is recommended to better reflect field conditions. Further, this current study only considered the effects of axial compressive loads. Further tests on specimens with eccentric loads would yield an understanding of columns subject to both axial and flexural stresses.

ACKNOWLEDGMENTS

The authors are members of the Intelligent Sensing for Innovative Structures (ISIS Canada) Research Network of Centres of Excellence (NCE). This work has been supported by the NCE and the Natural Sciences and Engineering Research Council (NSERC) of Canada. The authors would also like to thank the technical staff at Queen's University and the Université de Sherbrooke.

NOTATION

C_E	= environmental reduction factor
E_f	= tensile elastic modulus of FRP
f_l	= lateral stress produced by confinement
f_{fe}	= effective stress in FRP
f_{fu}	= design ultimate tensile strength of FRP
f^*_{fu}	= tensile strength of FRP by manufacture

f_{lfrp}	= confining pressure due to FRP reinforcement
f_{frpu}	= tensile strength of FRP
f'_c	= strength of unconfined concrete
f'_{cc}	= confining strength of concrete
h	= overall thickness of member
h_c	= diameter of circular column
k_a	= efficiency factor of FRP, (based on shape of section)
N_b	= number of layers,
n	= no of plies
t_f	= nominal thickness of one ply
t_{frp}	= thickness of FRP
α_{pc}	= performance coefficient of circular column
ε^*_{fu}	= ultimate rupture strain of FRP
ρ_f	= reinforcement ratio
ϕ_c	= resistance factor of concrete
ϕ_{frp}	= resistance factor of FRP
ω_w	= volumetric ratio of FRP strength to concrete strength

REFERENCES

ACI, 2001, "Guide for the design and construction of externally bonded FRP systems for strengthening concrete structures," American Concrete Institute, Committee 440.

American Society of Testing and Material Standards (ASTM), 1997, "Standard Test Method for Resistance of Concrete to Rapid Freezing and Thawing." C666-97, Annual Book of ASTM Standards, pp. 314-319.

Bisby, L.A., Dent, A.J.S., and Green, M.F. 2005. "A Comparison of Models for FRP-Confined Concrete," *ACI Structural Journal*, Vol. 102, No. 1, pp. 62-72.

ISIS 2001, "Strengthening reinforced concrete structures with externally-bonded fiber reinforced polymers," Canadian Network of Centres of Excellence on Intelligent Sensing for Innovative Structures (ISIS), Design Manual No. 4.

Karbhari, V.M., Chin, J.W., Hunston, D., Benmokrane, B., Juska, T., Morgan, R., Lesko, J.J., Sorathia, U., and Reynaud, D. 2003. "Durability Gap Analysis for Fiber-Reinforced Polymer Composites in Civil Infrastructure," *Journal of Composites for Construction*, 7(3), pp. 238-247.

Karbhari, V.M., and Eckel, D.A. II. 1993. "Effect of a Cold-Regions-Type Climate on the Strengthening Efficiency of Composite Wraps for Columns," Technical Report, University of Delaware Center for Composite Materials, Newark, Delaware, June.

Karbhari, V. M., Rivera, J., and Dutta P. K., 2000, "Effect of short-term freeze-thaw cycling on composite confined concrete," *Journal of Composites for Construction,* ASCE, V. 4, No. 4, pp. 163-217.

Mallick, P. K., 1998, "Fiber-reinforced composites: materials, manufacturing, and design." Marcel Dekker, INC., New York, New York.

Naguib, W., and Mirmiran, A., 2002, "Time-Dependent behavior of fiber-reinforced polymer confined concrete column under axial loads," *ACI Structural Journal*, V.99, No. 2, pp. 142-148.

Soudki, K. A., and Green, M. F., 1996. "Performance of CFRP Retrofitted Concrete Columns at Low Temperatures," 2nd International Conference on Advanced Composite Materials in Bridges and Structures, Montreal, Quebec, 11-14 August, pp. 427-434.

Toutanji, H., and Balaguru, P. 1998. "Durability Characteristics of Concrete Columns Wrapped with FRP Tow Sheets," *ASCE Journal of Materials in Civil Engineering*, 10(1), pp. 52-57.

Table 1—FRP properties as provided by the manufacturers

Properties	Type of FRP sheets	
	GFRP	CFRP
Tensile strength (MPa)	575	991
Elongation at break (%)	2.2	1.26
Tensile modulus (GPa)	26.1	78.6
Laminate thickness (mm)	1.3	0.89

Table 2—Test results of CFRP-wrapped concrete cylinders

FREEZE-THAW

Specimen code	Under sustained load					No sustained load			
	CFS-1	CFS-2	CFS-3	CFS-4	**Avg.**	CFC-1	CFC-2	CFC3	**Avg.**
Strength (MPa)	40	40.2	44	F.D	**41.4**	34.6	F.D	F.D	**34.6**
Avg. axial strain	0.00927	0.00875	0.00841	0	**0.00898**	0.01329	0	0	**0.01329**
Avg. hoop strain	0.00403	0.00697	0.01127	0	**0.00757**	0.00474	0	0	**0.00474**

ROOM TEMPERATURE

Specimen Code	Under sustained load					No sustained load			
	CRS-1	CRS-1	CRS-3	CRS-4	**Avg.**	CRC-1	CRC-2	CRC-3	**Avg.**
Strength (MPa)	42.1	43.2	44.2	41.2	**42.7**	39.7	38	41.1	**39.6**
Avg. axial strain	0.01662	0.01281	0.01062	0.01449	**0.01363**	0.01446	0.01212	0.02333	**0.01664**
Avg. hoop strain	0.0063	0.00689	0.0066	0.00443	**0.00606**	0.00671	0.00672	0.0081	**0.00718**

F.D = Freeze-thaw damage

Table 3—Test results of GFRP-wrapped concrete cylinders

FREEZE-THAW

Specimen code	Under sustained load					No sustained load			
	GFS-1	GFS-2	GFS-3	GFS-4	**Avg.**	GFC-1	GFC-2	GFC-3	**Avg.**
Strength (MPa)	38.6	36.7	33.6	37.7	**36.6**	38	32.6	36.4	**35.7**
Avg. axial strain	0.01665	0.01342	0.01523	0.0128	**0.01453**	0.02183	0.01521	0.01777	**0.01827**
Avg. hoop strain	0.01201	0.01611	0.00904	0.0109	**0.01202**	0.02047	0.01145	0.01712	**0.01635**

ROOM TEMPERATURE

Specimen code	Under sustained load					No sustained load			
	GRS-1	GRS-2	GRS-3	GRS-4	**Avg.**	GRC-1	GRC-2	GRC-3	**Avg.**
Strength (MPa)	38.9	37.7	36.9	38.1	**37.9**	38.4	38.5	37.1	**38.0**
Avg. axial strain	0.02045	0.01423	0.01681	0.01378	**0.01632**	0.01545	0.01565	0.0204	**0.01717**
Avg. hoop strain	0.0142	0.01274	0.01344	0.01674	**0.01428**	0.01848	0.01611	0.01567	**0.01675**

Table 4—Test results of unconfined concrete cylinders

ROOM TEMPERATURE

Specimen code	Dry condition				Wet-dry condition		
	URC-1	URC-2	URC-3	**Avg.**	UCWD-1	UCWD-2	**Avg.**
Strength (MPa)	21.9	20.1	21.7	**21.2**	22.5	21.5	**22.0**
Avg. axial strain	0.00175	0.00174	0.0014	**0.00163**	0.00137	0.00236	**0.00186**
Avg. hoop strain	0.00098	0.00036	0.00052	**0.00062**	0.00026	0.00064	**0.00045**

Table 5—Numerical results obtained using ISIS Canada confinement model

CFRP-WRAPPED CYLINDERS

Parameters								Results			Exp.
α_{pc}	N_b	Φ_{frp}	t_{frp} mm	D_g mm	f_{frpu} MPa	Φ_c	f'_c MPa	f_{lfrp} MPa	ω_w	f'_{cc} MPa	f'_{cc} MPa
1	1	1.0	0.89	152	991	1.0	22	7.41	1.12	45.2	39.6

GFRP-WRAPPED CYLINDERS

Parameters								Results			Exp.
α_{pc}	N_b	Φ_{frp}	t_{frp} mm	D_g mm	f_{frpu} MPa	Φ_c	f'_c MPa	f_{lfrp} MPa	ω_w	f'_{cc} MPa	f'_{cc} MPa
1	1	1.0	1.3	152	575	1.0	22	4.71	0.71	41.6	38.0

Table 6—Numerical results obtained using ACI 440 confinement model

CFRP-WRAPPED CYLINDERS

Parameters										Results		Exp.
f'_c MPa	K_a	n	t_f mm	h mm	f_{fe} GPa	C_E	$\varepsilon^\star_{fu}$	E_f GPa	ρ_f	f_l MPa	f'_{cc} MPa	f'_{cc} MPa
22	1	1	0.89	153	0.8	1.0	0.01	78.6	0.0232	7.33	55.5	39.6

GFRP-WRAPPED CYLINDERS

Parameters										Results		Exp.
f'_c MPa	K_a	n	t_f mm	h mm	f_{fe} MPa	C_E	$\varepsilon^\star_{fu}$	E_f MPa	ρ_f	f_l MPa	f'_{cc} MPa	f'_{cc} MPa
22	1	1	1.3	154	0.4	1.0	0.02	26.1	0.0338	3.79	47.9	38.0

Table 7—Comparison of confinement models

CFRP-wrapped cylinders

	Strength MPa	% Difference to experimental
Exp.	39.6	
ISIS	45.2	14 %
ACI 440	55.5	40 %

GFRP-wrapped cylinders

	Strength MPa	% Difference to experimental
Exp.	38.0	
ISIS	41.6	10 %
ACI 440	47.9	26 %

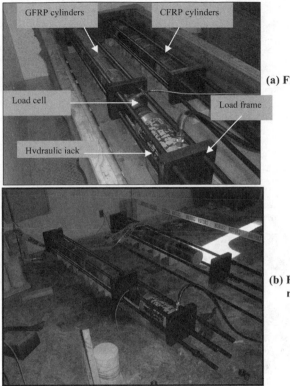

(a) Frames in cold room

(b) Frames at
    room temperature

Figure 1—Sustained loading frames

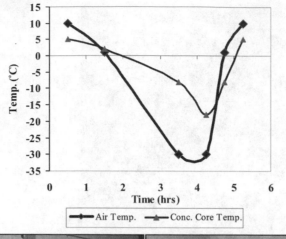

Cold Room   Inside Cold Room

Figure 2—Temperature variation during one complete freeze-thaw cycle inside the cold room, which is also shown

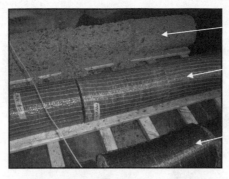

Plain concrete

GFRP cylinders

CFRP cylinders

(a)

CFRP-wrapped damage cylinders

(b)

Figure 3—(a) Unwrapped non-air entrained cylinders disintegrated after 100 freeze-thaw cycles, (b) CFRP-wrapped cylinder without end restraints after 300 freeze-thaw cycles

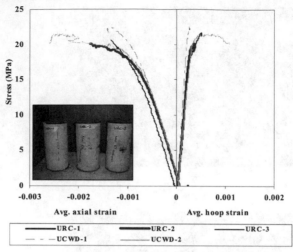

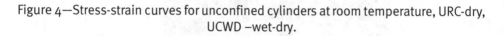

Figure 4—Stress-strain curves for unconfined cylinders at room temperature, URC-dry, UCWD –wet-dry.

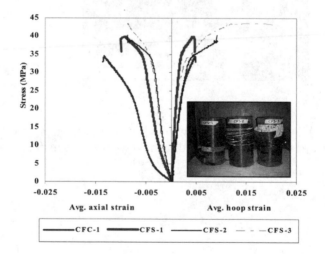

Figure 5—Stress-strain curves for CFRP-wrapped cylinders after 300 freeze-thaw cycles.

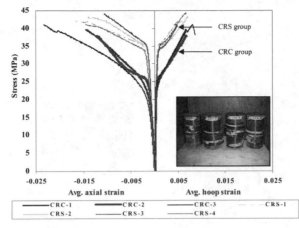

Figure 6—Stress-strain curves for CFRP-wrapped cylinders at room temperature, CRC—no load, CRS—sustained load.

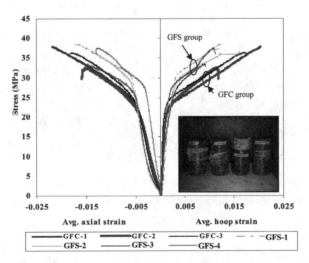

Figure 7—Stress-strain curve for GFRP-wrapped cylinders after 300 freeze-thaw cycles, GFC—no load, GFS—sustained load.

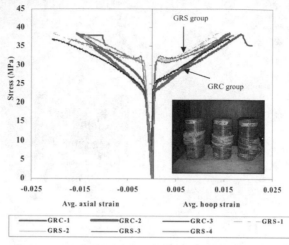

Figure 8—Stress-strain curves for GFRP-wrapped cylinders at room temperature, GRC–no load, GRS–sustained load

Figure 9—Failure modes of GFRP and CFRP-wrapped specimens tested in compression after freeze-thaw and sustained loading exposure

Chapter 7

Strengthening of Existing Concrete Structures with FRP Systems, Part II

Probabilistic Based Design for FRP Strengthening of Reinforced Concrete

by R.A. Atadero and V.M. Karbhari

Synopsis: This paper presents an approach for considering a wide variety of composite materials in reliability-based design of composite strengthening of concrete structures. The approach is based on the use of the mean value of composite properties as the design value and having a composite specific resistance factor that is a function of the COV of the composite properties. Strengthening design of a simple beam with three model composites is used to demonstrate the feasibility of this approach. For comparison, designs are created using two existing strengthening guidelines, ACI 440 and TR55. These designs are shown to have a high degree of variability in their reliability and demonstrate the need for a reliability based design procedure. Some of the critical areas of further research required for further development of this design procedure are briefly discussed.

Keywords: design guidelines; material design values; reliability-based design; strengthening

Rebecca A. Atadero is a Ph.D. student in the Department of Structural Engineering at the University of California, San Diego.

Vistasp M. Karbhari is a professor in the Department of Structural Engineering and the Material Science Program at the University of California, San Diego.

INTRODUCTION

Fiber reinforced polymer (FRP) composites are proving to be a valuable material in the retrofit and repair of reinforced concrete structures. As such, a number of guidelines for their use have been developed. [1, 2, 3, 4] These guidelines all contain provisions for topics of major concern in design of FRP strengthening, such as long-term performance and debonding. These guidelines are also similar in that though the design equations are presented in the partial factor formats that are often used in probability-based design, they are not true probabilistic codes. Instead, they typically make use of already existing design factors for loads and non-FRP materials, which in some cases are reliability based, and then provide additional factors for the FRP contribution. The factors for FRP are based on the judgment of the guideline-authoring agency based on available data. As these guidelines are not specifically calibrated from a reliability standpoint they suffer from the variation in reliability that is often associated with older design procedures. To remedy this problem it is important to get the probabilistic code development process underway.

The earliest attempt at reliability based design of composite strengthening was conducted by Plevris et. al. in 1995 [5]. Their study of flexurally strengthened beams examined the sensitivity of reliability to changes in the design variables and calibrated a set of resistance factors for use in design. Okeil et. al.[6] also examined the reliability of designs for flexural strengthening in their analysis of bridge girders. Their study used several girder designs and assumed different levels of steel degradation. Again resistance factors were calibrated based on the assumed design parameters. A common limitation of these studies is that the design factors were calibrated with one or two particular composite systems in mind. In the study by Okeil et. al. an analytically derived coefficient of variation of 2.2 percent was used for the composite failure strain. Tests of field-manufactured panels[7] suggest that this value is far too low to accurately represent wet layup composites, which are commonly used in the field due to the ease of application. Thus design factors for composite strengthening cannot be taken directly from these studies, but require further consideration of the materials involved.

The nature of composite materials provides many obstacles to the development of a reliability based design guideline. There are many different types of fiber and resin systems available, and they may be combined to produce finished composites of widely varying material properties. The variation in composite material properties is not only due to different constitutive materials; even two composites composed of the exact same constituent materials can show large differences in performance as a result of the manufacture and curing processes, particularly in the case of field manufactured wet layup.

One approach to this problem is the development of so-called "standard" composites, a set of composite materials that possess certain minimum properties. This, however, could be a shortsighted approach. Restricting strengthening design to certain established systems could hurt innovation in the materials. This approach also eliminates one of the prime features of composites, their tailorability. Finally, with so many systems currently in use how would a few standard composites be chosen? Thus the need is exposed for an approach that is broadly applicable, but which still satisfies the goals of reliability based design. This paper describes a proposed approach for creating a flexible, yet accurate, reliability based methodology for design of composite strengthening. An example of how the proposed code could be calibrated is presented, and compared to reliability results from designs based on other guidelines. Finally, some of the further research required in this field is outlined.

PROPOSED APPROACH FOR ACHIEVING BROAD APPLICABILITY OF RELIABILITY BASED DESIGN OF COMPOSITE MATERIALS

Resistance Factor as a Function of Composite Coefficient of Variation

The common approach to reliability-based design in the United States is Load and Resistance Factor Design (LRFD). In this design methodology factors are used to (generally) increase the expected load effects and decrease the predicted structural resistance. These factors are calibrated through reliability methods in an attempt to produce designs of a uniform, acceptably small, probability of failure over a range of design scenarios. For the case where the load and resistance are independent random variables, the probability of failure may be calculated using the integral shown in Equation 1, where $F_R(s)$ is the cumulative distribution function of the resistance and $f_S(s)$ is the probability distribution of the load.

$$p_f = \int_0^\infty F_R(s) f_S(s) ds \qquad (1)$$

Equation 1 clearly shows that changes in the probability distribution of either load or resistance will cause changes in the probability of structural failure. As statistical parameters of load and resistance change, it is therefore necessary to change the design factors in order to reach the same target reliability. This integral does not represent the area of overlap between the distributions of load and resistance; however, the effect of changes in these distributions on the probability of failure can be qualitatively understood by considering the area of overlap. The size and shape of the overlapping region is largely a result of the spread in the distributions, and thus when they are calibrated the design factors are largely dependent on the amount of spread in the design variables. The degree of spread is typically described by the non-dimensional coefficient of variation (COV), defined as the standard deviation divided by the mean.

In design of FRP strengthening, the probability distribution of load for a particular member is not subject to change, however the distribution of resistance will be affected by statistical properties of the FRP. For composite materials, the COV of different material properties is very changeable. It may depend on the materials used to fabricate the composite, or, more often, on the way in which the composite was manufactured. Prefabricated strips are often characterized by COVs in the low single digits, while wet layup composites with poor quality control can show variation up to approximately 20%. Clearly in order for a design code to be applicable to a broad range of composites it must be able to accommodate a wide range of COVs.

Since a change in the composite COV changes the distribution of resistance, and thus the probability of failure, in order to maintain uniform reliability different resistance factors are need for different values of the COV. The design method could assume values of COV for certain types of materials, and then provide resistance factors for the different materials. However, a more straightforward approach is to express the resistance factors directly as a function of the COV. This is the approach taken by Alqam et al.[8] for probabilistic based design of FRP compression members.

Traditionally, the resistance factor, ϕ, in LRFD is used to account for all sources of uncertainty in the resistance. This includes variation in material properties; however, it also includes for the error that is present in the models used for design and possibly geometric variations. The traditional design equation is expressed as:

$$\sum \gamma_i Q_i \leq \phi R \qquad (2)$$

In this equation, γ_i denotes the load factors specific to the type of load, Qi represents the load effect due to different types of load (dead, live, snow, wind, etc.), ϕ is the resistance factor specific to the limit state being considered, and R is the nominal member resistance with respect to the limit state in question. For use of this design equation, there would need to be a different function of composite COV for ϕ for each limit state.

In another approach, two resistance factors could be used. The traditional ϕ would be used to account for variations in the steel and concrete, as well as the design model, while a separate factor, ψ, would be specific to the COV of the composite, and would account for the variation due to the composite. This is a break from traditional LRFD implementation, but is similar in some ways to the partial factor formats used in other parts of the world such as Europe and Canada. In this case the design equation would be expressed as:

$$\sum \gamma_i Q_i \leq \phi R(..., \psi x_{FRP}) \qquad (3)$$

Here, ψ is a resistance factor specific the COV of the controlling FRP property, which is applied only to the FRP contribution to resistance. ϕ is a resistance factor specific to the limit state that considers the modeling inaccuracies, geometric variations, and variations in materials other than FRP. This approach is appealing because it isolates the effect of composite variability, making the effects of improved quality control very clear.

It would be ideal for the FRP dependent factor, ψ, to be constant for a given COV no matter what limit state is in question, however, different strengthening applications may be sensitive to variation in different composite properties. For example, for flexural strengthening, the design is usually governed by the ultimate strain of the composite, making the design quite sensitive to variation in ultimate stress. Other designs may be stiffness controlled, making the design most sensitive to modulus variation. For each type of design, the COV of the most significant composite property would be used to determine the resistance factor.

Use of Mean Value as Composite Design Allowable

The second feature of the proposed design approach for composite strengthening is to use the mean value of the composite properties as the design value. This is not a change from current practice with regard to the modulus and thickness; however, it is a very new approach with regard to strength. Currently, for most materials, the design strength is specified as a certain lower percentile of the test results, with the exact percentile varying depending on the material and the design specification. This is the approach taken by the guidelines that currently exist for FRP strengthening. By selecting the design value as a certain percentile of test results, the probability that the composite strength will fall below that value is fixed. However, the reliability is determined by the interaction of load and resistance and thus the probability of failure is not fixed by selecting a lower percentile for the design strength.

In a recent paper [9], the authors investigate the reliability implications of the current procedures used to define the material allowable for composite strength. The design procedures in ACI 440[1] are used to design strengthening for a sample beam. It is found that not only does the current design allowable not produce uniform levels of reliability as the COV of the composite strength is allowed to vary from 0.10 to 0.20, but in fact the designs for the COV equal to 0.20 are more reliable than the designs with the COV equal to 0.10. The higher reliability of the design created for the higher value of COV is explained in terms of the reliability integral in Equation 1. For this example, in order to have designs of uniform reliability, the material with a COV of 0.10 would require a smaller resistance factor than the material with a COV of 0.20. Clearly, this is an undesirable result. It implies that a composite that would generally be considered inferior due to the high variation is actually better. This effect is only an artifact of the total design procedure, which has conservatism built in with regard to the material allowable. However, if this design allowable is used for a probability-based code, designers will see that less demanding values of the resistance factor are used for materials with higher COVs. Since designers may not be familiar with the full reliability background, this could have unintended consequences such as giving no incentive for using higher quality materials, or lax quality control standards.

Another issue exposed by Atadero and Karbhari[9] and affecting the current design allowable is that designs where a higher percentage of the load is carried by FRP are more reliable than designs with smaller amounts of required FRP. This can be attributed to the small design values that often result when selecting a certain percentile of the strength distribution. These small design values result in bias factors, (ratio of mean to design value) that range up to 1.82 for a COV of 0.15 and 2.5 for a COV of 0.20 based on the material allowable definition in ACI 440[1]. For comparison, reinforcing steel has a bias factor of approximately 1.1. Based on these factors, it is obvious that when using current guidelines for designing FRP to carry load, much more conservatism is introduced than when using steel. This conservatism with regard to the FRP properties begins to assert itself as the amount of strengthening is increased. As more FRP is applied to carry a higher portion of the load, more reserve capacity is built into the structure due to these high bias factors. In their investigation of a slab strengthened to account for assumed levels of steel deficiency, Atadero and Karbhari[9] found that the reliability index increased from 3.16 for assumed 10 percent steel degradation to 4.50 for assumed 40 percent degradation.

These exposed difficulties with the current approach to material allowable specification could be handled with a complicated set of additional factors in the design procedure. However by using the mean value as the design value, the bias factor for the composite would be equal to 1, preventing the build-up of excess capacity as the percentage of load carried by the FRP increases. Furthermore, the inconsistency where higher COVs result in higher reliabilities would also be eliminated because the design value would no longer be dependent on the COV. This would eliminate the need for additional factors, resulting in a design method that is simpler to apply and one where even engineers who are not experts in reliability can understand the basis of the design approach.

Other advantages also exist to using the mean. For one, it is a value that is easy to obtain from manufacturer's data. The use of the mean also facilitates the use of additional design factors that can account for composite properties that may differ from test results. For example, a composite system for use in a rehabilitation project may be tested as a one-layer laminate, however during the design process it may be found that two layers are necessary to provide the required strength. Composites of different number of layers typically possess some differences in their properties, rather than requiring testing of the two-layer laminate a factor could be used to relate the mean of the one-layer property to the mean of the two-layer property. This idea could also be extended to account for specifics of manufacturing, such as worker experience or unusual curing conditions, that aren't generally considered when determining properties for design. It could even be used to reduce the design properties for the long-term degradation expected in service.

DESCRIPTION OF SAMPLE CALIBRATION

A simple example is presented to show how the features of the proposed approach could work. A generic beam is selected and strengthened with three different model composites, using the mean value as the design value for the composite properties. Different amounts of strengthening are applied depending on the composite resistance factor, which is allowed to range from 0.95 to 0.5 in increments of 0.05. The reliability of each design is assessed using Monte Carlo Simulation. Several different cases are evaluated, two different loading scenarios are used and for each scenario the reliability is evaluated at strength COVs of 0.05, 0.10, 0.15, and 0.20.

Sample Beam

The beam chosen for this demonstration is taken from Example 14.3 of ACI 440[1]. It is a simply supported beam with a length of 7.32 m (24 ft). The depth and width are 0.61 m (24 in) and 0.305 m (12 in), respectively, with a 0.546 m (21.5 in) depth to the reinforcing steel.   The concrete compressive strength is assumed to be 34.5 MPa (5.0 ksi). The beam is reinforced with 3 ϕ 28 bars (No.9) and the reinforcement has a yield strength of 414 MPa (60 ksi). Strengthening is designed for the beam in anticipation of an assumed increase in live load of 50 percent using standard ACI load factors for dead and live load.   The beam is sufficient in shear, and thus only flexural strengthening is required. The existing and anticipated moment effects on the beam are shown in Table 1.

Properties of Model Composites

Model carbon fiber reinforced composites are developed to represent three different levels of strength. All are assumed to have a rupture strain of 1.2 percent, thus the stronger composites are also stiffer. The thickness is assumed to be two-layers for all three cases. The assumed mean values for the properties for the three materials are shown in Table 2.   As per the proposed approach, these values are used as the design values.

Design of Strengthening

The strengthening is designed following the method outlined in ACI 440. This method involves basic sectional analysis assuming a linear strain profile through the beam. The failure strain of the concrete is taken as 0.003. A bond-dependent coefficient, κ_m, is used to define a limit on the strain developed in the FRP to prevent the beam from failing due to composite debonding. The bond-dependent coefficient is a function of the composite modulus, thickness and rupture strain. The factor, κ_m, is multiplied by the rupture strain of the FRP to determine the limiting strain. The maximum value it may take is 0.9. For most designs, the composite strain limitation is the controlling criteria. Since the concrete is not crushing, the common stress block factors do not accurately describe the force carried in the concrete.   In order to more accurately assess the force in the concrete, a parabolic stress distribution is assumed, and the stress block factors are computed using the equations found in Collins and Mitchell[10].

For this example, the format of the LRFD equation expressed in Equation 3 is used, with two factors ϕ and ψ. The value of ϕ is assumed fixed for this example at 0.9, a common resistance factor for the flexural failure mode. The value of ψ is varied from 0.95 to 0.5 in increments of 0.05. At each value of ψ the amount of composite needed to satisfy the design equation is determined. As all the composites are assumed to be two-layer laminates, the amount of material is controlled by adjusting the width of the composite strip. The needed width is computed in increments of 6.35 mm (0.25 in) so as to be as close as reasonable to the minimum required amount. The strengthening designs are summarized by the required composite width in Table 3. There is a significant decrease in the required amount of composite when switching from Material 1 to Material 2, however a similar decrease does not occur between Material 2 and Material 3. This is due to the high stiffness of Material 3, which results in a value of κ_m that is significantly smaller than the value used for Materials 1 and 2. With the strain of the composite severely limited by the low value of κ_m, nearly the same amount of Material 3 is needed as Material 2. For some smaller values of ψ it is not possible to satisfy the design equation.

Random Variables

Statistical descriptions of the design variables are necessary to analyze the reliability of the strengthening designs. For this analysis the geometric characteristics of the problem are considered deterministic, while the material properties and loads are allowed to vary.

Description of FRP Properties – The ultimate strength, modulus and thickness of the composite are considered as random variables. The strength is modeled with a Weibull distribution, a common choice for modeling composite strength. Modulus and thickness are modeled as lognormal distributions based on unpublished analysis of several FRP data sets. All distributions are fit using the method of moments, whereby the distribution parameters are determined by equating the assumed mean and standard deviation to the mean and standard deviation of the chosen distribution form. These variables are modeled as independent variables, assuming a complete lack of correlation. It is likely that there is indeed some correlation between the composite design variables, however without better data, assuming appropriate amounts of correlation is impossible. All of the statistical descriptions used here are assumptions based on limited actual data. As reliability based design of composite strengthening progresses, a key area of study will be statistical descriptions of the material properties.

The means of the composite properties are given in Table 2. Four different values are assumed for the COV of the strength; 0.05, 0.10, 0.15 and 0.20; to show how the reliability of a single design changes as the COV of strength is changed, and also to determine a relation between COV and the resistance factor, ψ. The COV of the modulus is taken as 12 percent and the COV of thickness is assumed to be 7 percent. The distribution parameters for the composite properties based on these values of mean and COV are presented in Table 4.

Description of Other Materials – The statistical description for steel is modeled on previous reliability studies. The yield strength of the steel is modeled as a lognormal variable with a bias factor of 1.1 and a coefficient of variation of 0.1 [11]. The modulus of steel is considered to be deterministic with a value of 200 GPa (29,000 ksi). The compressive strength of concrete is modeled as a normal variable with a bias factor of 1.14 [12] and a coefficient of variation of 0.15 [13]. This coefficient of variation corresponds to a normal level of quality control.

Description of Loads – Two different descriptions of loading are considered for reliability analysis. The first is appropriate for the design of new structures, with load statistics found in [14]. Dead load is modeled as a normal variable with a bias factor 1.05 and coefficient of variation 0.10. The live load is modeled as an extreme value type I distribution of maxima, also known as the Gumbel distribution. The live load has a bias factor of 1 and a coefficient of variation of 0.25.

The second loading scenario is the same as case 1, except the live load has a coefficient of variation of only 5 percent. This value is very low and is intended to represent the increased knowledge of a structure and the load demands placed on it at the evaluation and strengthening stage. The higher value of COV in the previous load case is appropriate for new design when the load demands are much more uncertain, and because conservatism at the original design stage adds very little to the cost of a structure. However, at the strengthening stage conservatism can be costly and thus it is often worth the extra effort to more accurately characterize the loads. Since the current load factors were not calibrated for such low levels of variation they are not applicable. Furthermore, it was desired to see the impact of the change in load variation on the same strengthening designs. However, since the variation is so low the designs had very high reliability. To lower the reliability into the typical range, the mean live load was increased from 173.6 kN-m (130 kip-ft) to 280.4 kN-m (210 kip-ft).

Reliability Analysis

Monte Carlo Simulation is used to determine the probability of failure. This is a procedure where random values of the design variables are selected from the appropriate distributions and are used to predict the resistance of the section. This resistance is then compared to load values randomly selected from the distributions describing loading. If the computed resistance is less than the randomly selected loads, the trial is considered a failure. This procedure is repeated a large number of times, and the probability of failure is estimated as the number of trials which failed divided by the total number of trials. For ease of comparison, this probability of failure is then related to the reliability index, β, through the approximate relation in Equation 4. In this equation Φ^{-1} represents the inverse of the standard normal distribution and p_f is the probability of failure

$$\beta = -\Phi^{-1}(p_f)$$
(4)

In the implementation of composite strengthening, failure through composite rupture is rarely witnessed, thus it was deemed unrealistic to compute the reliability of the beams against failure through composite rupture. Rather, the resistance of the beams was computed using the strain limitation provided by κ_m from ACI 440. This assumes that the ACI equation is a good predictor of the debonding strain, and implies that the reliability computed is the reliability against composite debonding (or very rarely concrete crushing), not against FRP rupture. As in the design stage, the strain limitation on the FRP resulted in concrete stress distributions that could not be accurately approximated using the common stress block factors in ACI 318. Thus the stress block factors for an assumed parabolic distribution were also used in the reliability evaluation.

RESULTS OF SAMPLE CALIBRATION

The resulting reliability indices are shown in Table 5. The results show that as the COV of the strength is increased there is very little difference in the reliability for load case 1. This is attributed to the high variation in the live load "drowning out" the effect of resistance variation. For load case 2 there is a clear decrease in reliability as the COV of strength is increased. This is one of the desired trends that the proposed method was selected to produce. For both load cases, as the resistance factor, ψ, is decreased the reliability of the beam is increased.

By selecting a target value for the reliability index, β_T, values of the resistance factor, ψ, needed to obtain this target can be found. For this example a target of 3.5 is used. Figures 1-2 show the deviation from the target reliability as a function of the resistance factor for load case 1. In Figure 1 the COV of strength is equal to 0.05, in Figure 2 it is 0.20. Here we see very little change in the required resistance factor as the strength COV is increased, implying that a single value of the resistance factor may be acceptable. Figures 3-6 show the deviation from target reliability as a function of the resistance factor for load case 2. In these figures it is clear that as the strength COV increases the resistance factor needed to reach the target reliability decreases. It is also clear that Material 3 shows significantly higher reliabilities than Materials 1 and 2. This indicates that the relative strength and stiffness of the composite can impact the reliability. The impact of higher strength variation and of higher strength and stiffness on reliability were not visible in load case 1 where the high live load variation dominated the reliability problem. However, load case 2 reveals these issues in reliability-based design. The values of ψ for load cases 1 and 2 are different from each other because case 2 is a specially selected example intended to highlight issues encountered in reliability based design.

A function of COV for determination of the composite resistance factor can be found for load case 2. For each COV the value of ψ necessary to provide a reliability of β_T is approximated from Figures 3-6. Figure 7 shows the equation fit to the relation between COV and ψ. In the range of COV from 0.05 to 0.10, ψ is constant at 0.8. For higher values of COV, ψ decreases linearly. The equation fitted to this line gives the value of ψ needed to reach the target reliability of 3.5 for COVs ranging from 0.05 to 0.20. A function such as this would be part of the design procedure and would allow designers to compute the resistance factor required to reach the target reliability specific to the COV that is anticipated in their design. This method could even allow designers to set their own target reliability index, by offering different functions for computing ψ for different values of the target reliability. Functions such as this would also allow designers to easily determine the effect on design of changing the COV through selection of higher quality materials or changes in quality control standards.

For all three materials at each level of COV the target value of reliability can be reached with use of the proper resistance factor, this shows that it is possible to use the mean value as the design value for composite material properties and still achieve safe designs as long as the proper resistance factor is used. For Materials 1 and 2, the relation between COV and reliability is very similar. However, Material 3 demonstrates different behavior. This change in behavior may be directly caused by the enhanced strength and stiffness of Material 3, or it may be a result of changes in the accuracy of the bond model at this higher stiffness. The cause for this change must be assessed. In order to make the proposed code widely applicable, it will be necessary to calibrate over a large range of possible properties. In order to select appropriate factors, it may be necessary to define ranges of properties where the behavior is similar and calibrate unique values of ψ for each of those ranges.

COMPARISON TO EXISTING DESIGN GUIDELINES

In order to highlight the advantages of the proposed procedure designs were also created with two existing approaches, ACI 440[1] and TR 55[2]. The ACI 440 design procedure requires that the characteristic value of the composite strength and ultimate strain be calculated as the mean less three standard deviations. This characteristic value is then multiplied by an environmental reduction factor, C_E, which in this case was taken to be 0.95 for a beam with interior exposure strengthened with CFRP. The ultimate strain is limited through κ_m, to prevent debonding.   The design format includes an overall resistance factor of 0.9 and a composite specific reduction factor of 0.85.   TR 55 defines the characteristic value as the mean less two standard deviations. Two safety factors are applied to the composite strength, one for the material, γ_{mF}, and one for the type of manufacture, γ_{mm}. In this example γ_{mF} was equal to 1.4 for CFRP and γ_{mm} was also set at 1.4 for wet layup manufacture. An additional factor, γ_{mE}, is applied to the modulus to account for long-term effects.   Design using TR55 also used material partial safety factors of 1.15 for steel and 1.5 for concrete. For both the ACI 440 and TR55 designs FRP rupture or debonding controlled the design, and the concrete stress at that point was modeled using the stress block factors for a parabolic stress distribution. The resulting designs are summarized with the required composite width in Table 6. In many cases, particularly for the high COVs, it was not possible to create a design using these guidelines.

The calculated values of the reliability index for these designs are shown in Table 7. It is clear that there is significant variation in the reliability index. In particular, as the COV increases, so does the reliability (with the exception of TR55 designs at a COV of 0.20). This is the same result found by Atadero and Karbhari in[9]. The ACI 440 designs for strength COV equal to 0.05 are close to the target reliability of 3.5. However, it is important to note that while the design procedure took into account long-term effects, the reliability analysis did not consider time dependent degradation. Had degradation been considered all of the reliability indices in Table 7 would be lower, and many of these designs may have fallen below the target reliability.

It is important that a design code produces safe designs for all composite systems that may be designed using it. With the ACI 440 and TR55 design approaches, some of the designs are likely to be safe, but with such variation in reliability from case to case, it is impossible to assert that all designs will be safe. In contrast, reliability based design is specifically intended to explicitly consider material variability and its impact on the reliability. The sample calibration did not consider time-dependent behavior, however, this feature could be easily added and since the same principles would be used it would still be possible to provide relatively uniform reliability.

FURTHER WORK

The example presented here falls far short of a true calibration. In a true calibration a large collection of design cases would be represented, including beams of different sizes, various live to dead ratios and various degrees of strengthening. The intent of this example was simply to show the potential for COV specific resistance factors and the use of the mean as the design value. The example cases presented clearly demonstrate the need to consider many different design scenarios when calibrating a reliability-based code. For example, the differences witnessed between load cases 1 and 2 could also occur when the live to dead load ratio is very low.

In addition to the extended range of design cases that would be required for a full calibration, there are many other areas that require further development for creation of a complete code.

1. The design checking equation used in this example has two resistance factors, ϕ and ψ. For simplicity, only ψ was varied, however for a full code calibration ϕ could also be adjusted. The flexibility of two resistance factors may assist in making the code apply to a larger range of design cases.

2. Since the value of ψ depends on the COV it will be important to give designers a basis for estimating the COV the material is likely to exhibit in the field. This will require analysis of many sets of field data.

3. The limit state equations require further development. The present example looked only at flexural strengthening, and used a fairly simple equation to deal with the possibility of composite debonding. The reliability results found here suggest that stiffer composites may need to be treated slightly different from less stiff composites. Stiffer composites are more prone to debonding; however, if the bond model used for design is perfectly accurate, changes in stiffness should not affect the reliability of designs to the extent witnessed here. ACI 440 describes its bonding model as preliminary and likely to change with further research, thus it would be wise to settle on a more accurate model of debonding before dividing composite properties into different ranges of behavior.

4. One of the most important areas for further development is characterization of composite properties and the variation in those properties. The assumptions made here are based on the data available, but remain only assumptions. Full characterization of composite properties would include consideration of inherent variation; differences between test specimens and field specimens; and time-dependent degradation.

CONCLUSION

In this paper, an approach for creating a reliability based design code considering a variety of composite materials with different mean property values and different degrees of property variation is presented. This approach is based on the ideas of using the mean property values as the design values and having a composite specific resistance factor that is a function of the COV of the composite. A simple calibration example has shown the effectiveness of this approach in achieving uniform levels of reliability and how this code could be developed. This example shows, in at least a preliminary manner, the feasibility of the reliability based design approach presented. It also suggests that it is truly possible to create a reliability based design code for composite strengthening that applies to a wide range of composite materials.

Designs created using existing composite guidelines are also presented. It is shown that the existing guidelines have large degrees of variation in the reliability of their designs. This variation makes it impossible to assert the safety level of designs created with these guidelines without conducting complete reliability evaluations. As design moves toward risk based considerations it is important that the reliability can be accurately quantified.

Continuation of the code development process requires large amounts of further information. Research priorities are more information about composites including properties at manufacture and after exposure to environmental stresses and the development of more accurate models for various strengthening limit states. Efforts in this direction will be well worth the work, as a well-developed design code will make composite strengthening technology accessible to many engineers.

ACKNOWLEDGMENTS

The authors would like to acknowledge discussions with Dr. Charles Sikorsky and Susan Hida of Caltrans.

REFERENCES

1. ACI Committee 440, 2002, "Guide for the Design and Construction of Externally Bonded FRP Systems for Strengthening Concrete Structures (ACI 440.2R-02)," American Concrete Institute, Farmington Hills, MI, 45 pp.

2. The Concrete Society, 2000, "Design guidance for strengthening concrete structures using fibre composite materials," *Technical Report No. 55*, The Concrete Society, Crowthorne, Berkshire, UK, 71 pp.

3. Täljsten, B., 2002, *FRP Strengthening of Existing Concrete Structures: Design Guidelines,* Luleå University Printing Office, Luleå, Sweden, 228 pp.

4. Neale, Kenneth, 2001, *Strengthening Reinforced Concrete Structures with Externally-Bonded Fibre Reinforced Polymers,* ISIS Canada, Winnipeg, Manitoba, Canada, 202 pp.

5 . Plevris, N., Triantafillou, T.C., and Veneziano, D., 1995, "Reliability of RC Members Strengthened with CFRP Laminates," *Journal of Structural Engineering*, V. 121, No. 7, July, pp. 1037-1044.

6. Okiel, A.M., El-Tawil, S., and Shahawy, M., 2002, "Flexural Reliability of Reinforced Concrete Bridge Girders Strengthened with Carbon Fiber-Reinforced Polymer Laminates," *Journal of Bridge Engineering,* V. 7, No. 5, September 1, pp. 290-299.

7. Atadero, R.A., Lee, L.S., and Karbhari, V.K., 2004, "Reliability Based Assessment of FRP Strengthened Slabs," *International SAMPE Technical Conference (Proceedings),*

8. Alqam, M., Bennett, R.M., and Zureick, A.H., 2004, "Probabilistic Based Design of Concentrically Loaded Fiber-Reinforced Polymeric Compression Members," *Journal of Structural Engineering,* V. 130, No. 12, December 1, pp 1914-1920.

9. Atadero, R.A., and Karbhari, V.M., 2005, "Determination of Design Values for FRP Used for Strengthening," *International SAMPE Technical Conference (Proceedings),* May 1-5.

10. Collins, M.P., and Mitchell, D., 1991, *Prestressed Concrete Structures,* Prentice Hall, Englewood Cliffs, NJ, pp. 175-177.

11. Val, D. V., 2003, "Reliability of Fiber-Reinforced Polymer Confined Reinforced Concrete Columns," *Journal of Structural Engineering,* v. 129 No. 8, pp. 1122-1130.

12. Barker, R.M. and Puckett, J.A., 1997, *Design of Highway Bridges, Based on AASHTO LRFD Bridge Design Specifications,* John Wiley & Sons, Inc., New York, p. 114.

13. MacGregror, J.G., 1997, *Reinforced Concrete, Mechanics and Design*, Third Edition. Prentice Hall, Englewood Cliffs, NJ, pp. 39-40.

14. Ellingwood, B., T.V. Galambos, J.G. MacGregor and C.A. Cornell. 1980. *Development of a Probability Based Load Criterion for American National Standard A58.* National Bureau of Standards, pp. 37-39.

Table 1 -- Load Effects on the Sample Beam

Load Effect	Existing Load	Anticipated Load
Unfactored Dead Load Effect	96.2 kN-m (72 k-ft)	96.2 kN-m (72 k-ft)
Unfactored Live Load Effect	114.9 kN-m (86 k-ft)	173.6 kN-m (130 k-ft)
Total Factored Moment	331.3 kN-m (248 k-ft)	428.8 kN-m (321 k-ft)

Table 2 -- Properties of Model Composites

Material	Ultimate Strength MPa (ksi)	Modulus GPa (ksi)	Thickness mm (in)
1	620.5 (90)	51.71 (7500)	2.03 (0.08)
2	827.4 (120)	68.95 (10000)	2.03 (0.08)
3	1034.2 (150)	86.18 (12500)	2.03 (0.08)

Table 3 -- Required Composite Width as a Function of ψ in mm (in)

ψ	Material 1	Material 2	Material 3
0.95	146 (5.75)	114 (4.5)	114 (4.5)
0.90	159 (6.25)	121 (4.75)	121 (4.75)
0.85	171 (6.75)	133 (5.0)	127 (5.0)
0.80	184 (7.25)	140 (5.5)	140 (5.5)
0.75	203 (8.0)	152 (5.75)	146 (5.75)
0.70	222 (8.75)	165 (6.25)	159 (6.25)
0.65	222 (9.75)	184 (6.75)	171 (6.75)
0.60	248 (11.25)	210 (7.5)	191 (7.5)
0.55	NA	241 (8.25)	210 (8.25)
0.50	NA	286 (9.5)	241 (9.5)

Table 4 -- Parameters of Statistical Distributions for FRP Properties

		Strength MPa (ksi)		Modulus GPa (ksi)		Thickness mm (in)	
	COV	α	β	λ	ζ	λ	ζ
MAT 1	0.05	24.950	634.23 (91.99)	3.9456 (8.9155)	0.1196	0.70658 (-2.5282)	0.0699
	0.1	12.153	647.23 (93.87)				
	0.15	7.907	659.29 (95.62)				
	0.2	5.797	670.15 (97.20)				
MAT 2	0.05	24.950	845.64 (122.65)	4.2333 (9.2032)	0.1196	0.70658 (-2.5282)	0.0699
	0.1	12.153	862.98 (125.16)				
	0.15	7.907	879.05 (127.50)				
	0.2	5.797	893.54 (129.60)				
MAT 3	0.05	24.950	1057.05 (153.31)	4.4565 (9.4263)	0.1196	0.70658 (-2.5282)	0.0699
	0.1	12.153	1078.72 (156.46)				
	0.15	7.907	1098.82 (159.37)				
	0.2	5.797	1116.93 (162.00)				

Table 5 — Reliability Indices for Different Values of ψ and COV

		COV	ψ									
			0.95	0.90	0.85	0.80	0.75	0.70	0.65	0.60	0.55	0.50
Load Case 1	Material 1	0.05	3.13	3.17	3.24	3.27	3.35	3.42	3.49	3.61	--	--
		0.10	3.12	3.17	3.23	3.27	3.34	3.41	3.49	3.58	--	--
		0.15	3.11	3.16	3.21	3.26	3.32	3.40	3.46	3.57	--	--
		0.20	3.09	3.14	3.19	3.22	3.29	3.36	3.43	3.54	--	--
	Material 2	0.05	3.12	3.16	3.23	3.26	3.33	3.38	3.46	3.58	3.69	3.84
		0.10	3.12	3.15	3.22	3.26	3.31	3.39	3.46	3.54	3.65	3.82
		0.15	3.10	3.14	3.19	3.22	3.29	3.36	3.44	3.54	3.64	3.80
		0.20	3.09	3.11	3.18	3.21	3.27	3.33	3.41	3.50	3.64	3.73
	Material 3	0.05	3.18	3.21	3.25	3.33	3.36	3.43	3.50	3.59	3.65	3.81
		0.10	3.17	3.21	3.25	3.31	3.35	3.43	3.51	3.58	3.65	3.79
		0.15	3.16	3.21	3.23	3.31	3.35	3.42	3.49	3.55	3.67	3.76
		0.20	3.15	3.18	3.23	3.29	3.33	3.40	3.44	3.56	3.64	3.75
Load Case 2	Material 1	0.05	3.19	3.31	3.41	3.50	3.62	3.72	3.83	4.00	--	--
		0.10	3.15	3.27	3.36	3.48	3.59	3.70	3.81	3.93	--	--
		0.15	3.09	3.20	3.30	3.39	3.52	3.66	3.76	3.92	--	--
		0.20	2.97	3.08	3.18	3.25	3.38	3.48	3.61	3.72	--	--
	Material 2	0.05	3.18	3.26	3.42	3.49	3.60	3.67	3.82	3.93	4.08	4.18
		0.10	3.15	3.24	3.37	3.46	3.59	3.67	3.81	3.93	4.02	4.16
		0.15	3.09	3.16	3.32	3.40	3.50	3.63	3.76	3.87	4.04	4.17
		0.20	3.01	3.07	3.20	3.26	3.38	3.47	3.58	3.73	3.81	3.94
	Material 3	0.05	3.32	3.41	3.49	3.66	3.77	3.85	3.98	4.05	4.14	4.19
		0.10	3.31	3.41	3.50	3.64	3.72	3.87	3.91	4.00	4.07	4.18
		0.15	3.29	3.38	3.46	3.59	3.67	3.78	3.85	3.99	4.09	4.23
		0.20	3.20	3.27	3.34	3.47	3.53	3.64	3.73	3.85	3.96	3.96

Table 6 -- Strengthening Designs Created with ACI 440 and TR55 mm (in)

COV	ACI 440			TR 55		
	Material 1	Material 2	Material 3	Material 1	Material 2	Material 3
0.05	203 (8.0)	165 (6.5)	159 (6.25)	248 (9.75)	203 (8.0)	165 (6.5)
0.1	248 (9.75)	203 (8.0)	197 (7.75)	--	235 (9.25)	191 (7.5)
0.15	--	260 (10.25)	260 (10.25)	--	273 (10.75)	216 (8.5)
0.2	--	--	--	--	305 (12.0)	260 (10.25)

Table 7 — Reliability Indices for ACI 440 and TR55 Designs

COV	ACI 440			TR 55		
	Material 1	Material 2	Material 3	Material 1	Material 2	Material 3
0.05	3.63	3.69	3.83	3.84	3.89	3.91
0.1	3.80	3.92	4.07	--	4.04	4.02
0.15	--	4.04	4.18	--	4.13	4.17
0.2	--	--	--	--	3.97	4.03

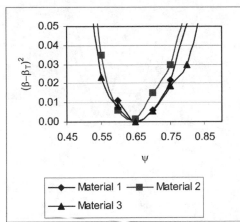

Figure 1 — Deviation from β_T as a Function of ψ for Load Case 1, COV=5.0%

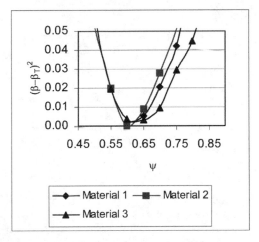

Figure 2 — Deviation from β_T as a Function of ψ for Load Case 1, COV=20.0%

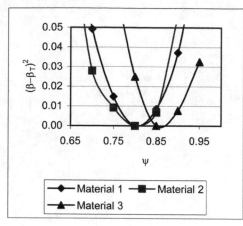

Figure 3 — Deviation from β_T as a Function of ψ for Load Case 2 COV=5.0%

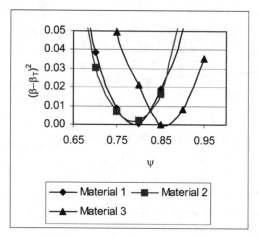

Figure 4 — Deviation from β_T as a Function of ψ for Load Case 2 COV=10.0%

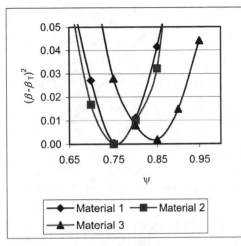

Figure 5 — Deviation from β_T as a Function of ψ for Load Case 2 COV=15.0%

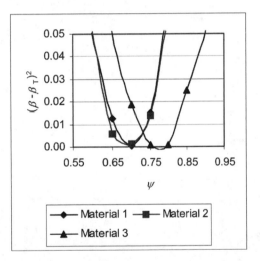

Figure 6 — Deviation from β_T as a Function of ψ for Load Case 2 COV=20.0%

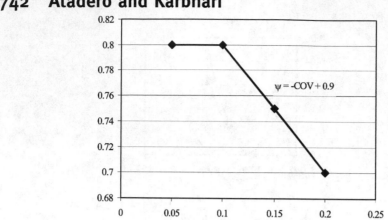

$\psi = -COV + 0.9$

Figure 7 — ψ as a Function of the COV of Ultimate Strength for Materials 1 and 2, Load Case 2

Strengthening of Concrete Structures using Steel Wire Reinforced Polymer

by W. Figeys, L. Schueremans, K. Brosens, and D. Van Gemert

Synopsis: This paper deals with a new material for external reinforcement: Steel Wire Reinforced Polymer (SWRP). It consists of thin high-strength steel fibres embedded in a polymer laminate. This innovative material combines the advantages of steel plates and CFRP, which are already used today. The material cost of SWRP is relatively low, and the laminate is quite flexible. In the feasibility part, the practical use of SWRP is studied. Further, the available design model for externally bonded reinforcement for concrete elements is confronted with the results of an experimental program, carried out at the Reyntjens Laboratory of KULeuven. The model is adapted accordingly.

Keywords: design model; external reinforcement; steel wire reinforced polymer (SWRP)

Wine Figeys, ir., obtained Master of Science in Architectural Engineering in 2004. She is PhD researcher at the Building Materials Division at the KULeuven. Her research topic concerns structural restoration techniques, focusing on epoxy bonded external reinforcement. She has a research grant of the Institute for the Promotion of Innovation through Science and Technology in Flanders (IWT-Vlaanderen).

Luc Schueremans, dr. ir., graduated as a civil engineer at the KULeuven in 1995. He defended his Ph. D. thesis at the end of 2001. Since 2002, he is a post-doctoral researcher at the Building Materials Division of KULeuven, Belgium. His research focuses on structural behavior of masonry, material modeling, and reliability based design and safety assessment of existing structures.

Kris Brosens, dr. ir., obtained his PhD at the Department of Civil Engineering, K.U.Leuven, Belgium in 2001. His research concerned the strengthening of concrete structures with externally bonded steel plates and fibre reinforced materials. Nowadays he works as a project engineer for Triconsult N.V., a spin-off company of the K.U.Leuven, specialized in the structural restoration of historical monuments.

Dionys Van Gemert, dr. ir., is professor of building materials science and renovation of constructions at the Department of Civil Engineering of KULeuven, Belgium. He is head of the Reyntjens Laboratory for Materials Testing. His research concerns repairing and strengthening of constructions, deterioration and protection of building materials, and concrete polymer composites.

1. INTRODUCTION

The capacity of a structure can be enhanced using the technique of externally bonded reinforcement. By adding extra reinforcement the flexural capacity, the flexural stiffness and the shear capacity are influenced. Today, mostly steel plates and carbon fibre reinforced polymer (CFRP) sheets and laminates are used. Since both materials have their own properties, they also have their preferable applications [1; 2]. Steel plates have a low material cost and can easily be applied in larger sections. Therefore, deformation problems are often tackled with steel plates. As steel plates have no fibrous structure, bolts can be used to reduce the anchorage length. Disadvantages are the high density of steel which hampers the application, and steel plates need a special treatment against corrosion. CFRP is a lightweight flexible composite. This makes it easier to apply. It has an E-modulus comparable to steel. The tensile strength is 5 to 10 times higher than standard steel. CFRP is used for strengthening of concrete plates, because their strength can be better exploited. As the flexible sheets can easily be wrapped, CFRP is also often applied as external shear reinforcement. But, as carbon can not take shear stresses, beams have to be rounded with a radius of 3 cm. Disadvantages are the high material cost and its brittleness. Therefore, large safety factors are required.

Steel wire reinforced polymer (SWRP) is a new material that can be used as external reinforcement. It consists of thin high-strength steel fibres which are bundled into cords (see Fig. 1). These cords are woven into unidirectional sheets with a synthetic textile. SWRP combines the advantages of steel and CFRP: the composite has the same strength as CFRP but is ductile. The material cost is low and the SWRP-laminate remains quite flexible. The new composite enables the same applications as steel plates and CFRP sheets and laminates, but also new application challenges can be tackled: shear strengthening of complex shapes, wrapping of rectangular beams, and improved uses of pre-stressing.

This paper deals with the use of SWRP as external reinforcement. It consists of three parts. In the first part a feasibility study is made. Problems for the practical use of SWRP as external reinforcement are recognized. In this part, the impregnation and the flexural stiffness of SWRP are studied. In the second part, the behaviour of the bonded connection with SWRP is examined. If this behaviour can be predicted, anchorage length and transferable load can be calculated. Therefore, non-linear fracture mechanics is applied to model and describe the shear-slip behaviour of the bonded connection. Some model parameters, specific for the new material, are determined by means of direct shear tests. In the last section, the combination of bending and shear is studied. When a beam is strengthened, some extra failure modes are introduced. One of them, so called delamination, is studied in this section. The used model is tested by means of four point bending tests on beams strengthened with SWRP.

2.   MATERIAL PROPERTIES

The SWRP used in the experiments of this paper is a prototype, produced by Bekaert Inc., Fig. 2. One sheet (width: 95 mm) consists of 65 steel cords. 19 filaments are twisted in a cord. The filament in the middle has a diameter of 0.25 mm, the other 18 have a diameter of 0.22 mm. By means of five tensile tests, the tensile strength and E-modulus of the SWRP are determined. The average tensile strength is 2775 N/mm², the average E-modulus is 177 600 N/mm².

To ensure a good impregnation of the cords, SWRP can be pre-impregnated. This kind of SWRP can be compared with the pre-cured CFRP laminates where impregnation is also more complete. In this research, some SWRP is pre-impregnated with an epoxy resin through vacuum impregnation followed by autoclaving at 125°C and 3 bar. Experiments are done on double layer SWRP with the epoxy resin F533 from Hexel Composites [3] with E-modulus 2900 N/mm².

3.   FEASIBILITY

In practical applications, it is desirable that SWRP is glued as easily as CFRP. Such sheets can be easily wrapped because of the low flexural stiffness of the fibres. The stiffness of SWRP is investigated first, as well as the impregnation of SWRP with the adhesive.

3.1   Flexural stiffness of SWRP

The new composite enables the same applications as steel plates and CFRP but also new application domains are targeted: shear strengthening of complex forms, wrapping of rectangular beams. Since the sheets are flexible, CFRP can be easily applied, which is an important advantage in practice. When wrapping CFRP, the reinforcement sticks to the beam or column without extra auxiliary actions. Also SWRP seeks an easy application. For the proposed applications, it is important that SWRP can be wrapped and kept in place without special arrangements.

The stiffness of SWRP can be checked by calculating the cross sectional moment of inertia. Without taking into account the torsion of the filaments, the moment of inertia equals 4.05 mm^4, Equation (1), for the type of SWRP studied in this paper.

$$I_{cord} = \sum_i (I_{filament} + A_i y_i^2) \tag{1}$$

with A_i section of filament i   [mm²]
         y_i distance of filament i to the neutral axis   [mm]

When the shape of a CFRP sheet is assumed to be a rectangular plate (0.0167mm x 95 mm), the moment of inertia is 0.037 mm^4. This means that the studied type of SWRP is more than 100 times stiffer than CFRP. A decrease in stiffness can be reached if less filaments belong to a cord, or if the filaments are used single, or if other types of SWRP are used.

3.2   Impregnation

It is important that the external reinforcement can be glued, without causing a weak link in the connection. As the reinforcement is often applied above the head, viscous adhesive is used. When using this kind of adhesive, it has to be checked that all steel fibres are surrounded by the adhesive. If not, the glue is the weakest link in the connection and can cause premature failure.

The impregnation of SWRP is tested by means of six pull-off tests. Several pieces of SWRP are glued on concrete with Epicol U [4], a viscous adhesive. It is necessary that the external reinforcement is pushed into the glue, which is not convenient in practice. Therefore, new adhesives have to be developed. Pull-off test were carried out on the SWRP-laminates. On the SWRP, a cylindrical element is glued. After hardening of the adhesive, a cylindrical saw cut is made to define the failure area. Afterwards, the cylindrical element is pulled off. In Fig. 3, test samples are presented after the pull-off test. The impregnation of the cords is sufficient. All test samples have failed in the concrete, not in the adhesive or in the connection between cylinder and SWRP or between SWRP and concrete. But in the laboratory, the work is more accurately done than in practice. Therefore, it seems that enhanced impregnation is necessary. This is possible by developing new adhesives or with pre-impregnated laminates. These laminates have a very good impregnation which does not depend on the circumstances of the site.

4. SHEAR BEHAVIOUR

When using SWRP as external reinforcement, the required anchorage length and transferable force must be known. Therefore, non linear fracture mechanics are applied to a model which describes the shear-slip behaviour of the bonded connection. At the Reyntjens laboratory of the KULeuven, model parameters are determined in a test program by means of direct shear tests.

4.1 Pure shear model

Using the equilibrium of forces, Volkersen [5] derived the following differential equation which describes the shear stresses as function of the slip of the external reinforcement:

$$\frac{d^2 s_l(x)}{dx^2} - \frac{1 + m_l \gamma_l}{E_l h_l} \tau_l(x) = 0 \tag{2}$$

in which $m_l = \dfrac{E_l}{E_c}$ $\qquad$ (3)

$$\gamma_l = \frac{A_l}{A_c} \tag{4}$$

and $s_l (x)$ slip of the external reinforcement at x [mm]
$\qquad$ $\tau_l (x)$ shear stresses in the adhesive at x [N/mm²]
$\qquad$ E_l E-modulus of the external reinforcement [N/mm²]
$\qquad$ E_c E-modulus of the concrete [N/mm²]
$\qquad$ h_l thickness of the external reinforcement [mm]
$\qquad$ A_l section of the external reinforcement [mm²]
$\qquad$ A_c concrete section [mm²]

The best results are given by the assumption of a bilineair shear-slip relationship, Fig. 4, as shown in [6; 7]. Initially, the shear stress increases, until the maximum shear stress τ_{lm} is reached. Afterwards, concrete cracks appear in the concrete and consequently the shear stress decreases. When the ultimate slip s_{l0} is reached, no forces are transferred anymore and the connection fails. With this assumption, the solution of Equation (3) becomes:

$$s_l(x) = A \sinh(\omega x) + B \cosh(\omega x) \tag{5}$$

with: $\qquad$ $\omega^2 = \dfrac{\tau_{lm}}{s_{lm}} \dfrac{1 + m_l \gamma_l}{E_l h_l}$ $\qquad$ (6)

Three model parameters τ_{lm}, s_{lm}, s_{l0} are introduced in the bilinear shear-slip relationship. The shear peak stress τ_{lm} depends only on the strength properties of the concrete because failure will occur in the concrete. Applying a linear Mohr-Coulomb failure criterion [6],

Equation (7) can be derived from the Mohr's circle for pure shear and a tangential intrinsic curve [1].

$$\tau_{lm} = \frac{f_{ctm} f_{cm}}{f_{ctm} + f_{cm}} \tag{7}$$

with:

f_{cm}	compressive strength of concrete	[N/mm²]
f_{ctm}	tensile strength of concrete	[N/mm²]

However, additional parameters must be introduced to account for the relationship between the lab test and the reality, Equation (8).

$$\tau_{lm} = k_{b1} k_{b2} k_c \frac{f_{ctm} f_{cm}}{f_{ctm} + f_{cm}} \tag{8}$$

k_c is the concrete influence factor, varying between 0,65 and 1. k_c is 1 in case of good workmanship, reducing to 0.65 [1] in case of bad workmanship. k_{b1} describes the size effect for brittle materials and is given by Equation (9) [7]. The mechanical strength increases when the test sample becomes smaller.

$$k_{b1} = \sqrt{\frac{k}{1 + \frac{b_l(k-1)}{h_{ref}}}} \tag{9}$$

where b_l is the width of the external reinforcement and h_{ref} is an empirical factor. h_{ref} is the depth of the concrete that is influenced by the shear stresses in the adhesive. Holzenkämpfer [7] proposed a value of h_{ref} of 2.5 – 3 times the size of the biggest gravel stones. k is an empirical factor which takes into consideration the multi-axial stress situation. This parameter has to be determined from experiments. k_{b2} introduces a second width effect: it accounts the spreading out of the forces in the concrete, Fig. 5.

$$k_{b2} = \sqrt{2 - \frac{b_l}{b_c}} \tag{10}$$

in which b_c is the width of the concrete.

The second model parameter s_{lm} is the value of the slip at peak shear stress. The slip is determined as the sum of the slip in the different layers: the concrete (height h_{ref}), the adhesive and the external reinforcement:

$$s_{lm} = \sum_i \frac{h_i}{G_i} \tau_i = \tau_{lm} \left(2,4 \frac{h_{ref}}{E_c} + 2,5 \frac{h_g}{E_g} + 2 \cdot (1 + v_l) \frac{h_l}{E_l} \right) \tag{11}$$

with:	h_l	thickness of the external reinforcement	[mm]
	h_g	thickness of the adhesive layer	[mm]

$$G_i = \frac{E_i}{2(1+\nu)} \qquad \text{shear modulus of the i}^{\text{th}} \text{ layer} \qquad [\text{N/mm}^2]$$

ν Poisson's ratio;   [-]

concrete: $\nu_c = 0.2$; adhesive : $\nu_g = 0.25$; steel wire : $\nu = 0.3$

As s_{lm} depends on τ_{lm}, it also depends on the parameters k_{b1}, k_{b2}, k_c.

The last model parameter is the slip s_{l0} at which no more forces can be transferred. Therefore, a new parameter is introduced: the fracture energy G_f. This fracture energy is the energy per unit area, needed to bring a connection into complete failure. It is given by the area under the shear-slip curve, Fig. 4. In Equation (12) gives the fracture energy for a bilinear shear-slip relationship.

$$G_f = \int_0^\infty \tau(s_l)ds_l = \frac{\tau_{lm}s_{l0}}{2} \tag{12}$$

If the fracture energy is known, the slip s_{l0} can easily be calculated. To determine the fracture energy, following expression is prosposed by Holzenkämpfer [6]:

$$G_f = k_{b1}^2 k_{b2}^2 k_c^2 C_f f_{ctm} \tag{13}$$

G_f depends on the tensile strength of the concrete. Again, the parameters k_{b1}, k_{b2}, k_c are included. A new empirical factor is introduced. C_f is a parameter fits the experiments to the model.

When the three model parameters are known, the differential equation of Volkersen can be solved. The shear stresses can be calculated at every location as a function of the external loading. Also the anchorage length and the corresponding maximum transferable load can be derived [1].

4.2 Experiments

Eight direct shear tests were executed at the Reyntjens laboratory. On four of them single SWRP is tested. Pre-impregnated double layer SWRP (marked with I) are used in the four others. The test set-up is shown in Fig. 6.

Two concrete prisms are bonded together with SWRP on two opposite sides which are grit blasted. Between the two prisms, there is a gap of 18 mm. Bonding length is 150 mm or 200 mm. On the other sides, steel plates are glued. They transfer the load from the testing machine to the test samples. Commercially available adhesive Epicol U [4] is used. This adhesive is a glue paste, its E-modulus equals 7000 MPa. After hardening, a saw-cut is made alongside the SWRP, to prohibit spreading out of forces. Therefore, k_{b2} equals 1. Since the test specimens are carefully prepared under laboratory conditions, k_c is also 1.

The test is deformation controlled. The change of the gap between the two concrete blocks is monitored by means of two Linear Voltage Differential Transformers (LVDT) at two opposite sides of the block, Fig. 6b. The gap increases at a constant mean rate of 0,001 mm/s. By comparing the two individual signals, one can check whether or not the tensile force acts centrically. Test 150a was oblique and is not taken in consideration.

In all test specimens, failure was due to failure of concrete. This means that the concrete is the weakest link in the connection. Test specimen I150a after failure is shown in Fig. 7.

Brosens [1] determined empirically the values of the parameters for CFRP: $k = 1,47$ mm and $C_f = 0,40$. With these parameters, a first estimation of the transferable load can be made, table 1. The model parameters are calculated with the data from table 1.

In Fig. 8, the measured tensile force and slip are compared to the model using the CFRP model parameters. In the test with single SWRP a small deviation is observed. At the end of the test, the measured curve diverges more from the model and failure took place later than predicted. The deviation on the force is about 12% and the test results are always underestimated. The SWRP seems to behave stronger and stiffer compared to elements strengthened with CFRP. The model parameters for CFRP predict the results relatively good, but are a rather conservative approach. The test results on pre-impregnated double layer SWRP are presented in Fig. 8 b. Smaller slip is measured, compared to the elements strengthened with single SWRP because of the higher amount of reinforcement. The measured curves diverge more from the CFRP-model (average deviation for the transferred forces is +20%).

In [8] a sensitivity analysis on the transferable forces is performed. Based on this analysis, it is concluded that a deviation on the material characteristics can not explain a deviation of 12% and 20% from the model. Also the sensitivity of the modelparameters h_{ref}, k, C_f is examined. Changes on h_{ref} cause only small changes in the transferable forces. Parameter C_f is a factor, needed to calculate the fracture energy. It has almost no effect at small bonding lengths. The influence of C_f is only observed for long bonding length, especially longer than the theoretical anchorage length. Anchorage length can be calculated as explained in [1] : with the CFRP parameters it becomes 285 mm for the single SWRP and 399 mm for pre-impregnated SWRP with 2 layers. Finally, the parameter k takes into consideration the multi-axial stress situation. When changing this parameter, it has an influence on all bonding lengths.

All test results were systematically underestimated. The predictions of the test results on single SWRP are rather well, but in a conservative manner. The pre-impregnated SWRP laminates have a bigger deviation, so that adapted model parameters had to be proposed. The tests were made on test specimens with a small bonding length. As concluded in the sensitivity analysis, only the parameter k can be varied. Figeys proposed to decrease parameter k [8] : for single SWRP, k can be taken 1.2 mm. For pre-impregnated double SWRP, calculations with $k = 1.0$ mm give improved results. In table 2, the transferable force, calculated with the CFRP-model (F_{CFRP}) and with the adapted model parameters for SWRP (F_{SWRP}), are compared. To evaluate the parameter C_f, additional experiments are needed for confirmation, especially tests with a longer bonding length.

5. BENDING

When a beam is strengthened, the required section is calculated from the equilibrium of internal forces. A sufficient bonding length can be provided to ensure that the forces can be introduced. However, the beam can fail in a different failure mode [9]:

- debonding of the external reinforcement,
- delamination,
- plate end shear failure,
- peeling-off of laminates at intermediate locations…

In this section the delamination is investigated for applications with SWRP.

5.1 Delamination

This failure occurs when the beam end remains unstrengthened, e. g. when the support rests on a column. At the beginning of the reinforcement, stress concentrations cause the external reinforcement to peel off. Three stresses act at the plate end: τ_l, s_l, s_n, see Fig. 9. Additional interfacial shear stresses are developed on top of the Jourawski shear stresses. Also, the plate experiences a normal stress, s_n, which peels the laminate away from the beam. Malek, Sadaatmanesh and Ehsani [10] derived the Equations (14), (15) and (16) to describe these stresses. These are derived from the equilibrium of forces, Fig. 9, assuming linear elastic and isotropic materials, perfect bond between plate and concrete, a linear strain distribution through the full depth of the section, and no interaction between shear strains and normal strains.

$$\tau_l(x) = \frac{d\sigma_l(x)}{dx} h_l \tag{14}$$

$$\frac{d^2\sigma_l(x)}{dx^2} - \frac{G_g}{h_g} \frac{1}{h_l E_l} \sigma_l(x) = -\frac{G_g}{h_g} \frac{1}{h_l E_l} \frac{E_l}{E_c} \sigma_c(x) \tag{15}$$

$$\frac{d^4\sigma_n(x)}{dx^4} + 4\beta^4 \sigma_n(x) = \frac{K_n}{E_c I_c} q(x) \tag{16}$$

in which

$\sigma_l(x)$	normal stress in the external reinforcement	[N/mm²]
$\sigma_c(x)$	normal stress at the bottom of the concrete section	[N/mm²]
$\tau_l(x)$	shear stresses in the external reinforcement	[N/mm²]
$\sigma_n(x)$	normal stresses acting between the concrete and plate	[N/mm²]
E_l	E-modulus of the external reinforcement	[N/mm²]
E_c	E-modulus of the concrete	[N/mm²]
E_g	E-modulus of the adhesive	[N/mm²]
I_l	inertial moment of the external reinforcement	[mm⁴]

I_c	inertial moment of the concrete	[mm^4]
h_l	thickness of the external reinforcement	[mm]
h_g	thickness of the adhesive	[mm]
G_g	shear modulus of the adhesive	[N/mm^2]
$q(x)$	distributed load on the concrete beam	[N/mm]

$$\beta = \sqrt[4]{\frac{K_n b_l}{4E_l I_l}} \qquad\qquad\qquad\qquad \text{[1/mm] (17)}$$

$$K_n = \frac{E_g}{h_g} \qquad\qquad\qquad\qquad\qquad \text{[N/mm}^3\text{] (18)}$$

These equations are solved at the plate end [10]. The stresses are given by eq. (21), (22) and (23), given an internal bending moment M (Equation (19)) and a shear force V (Equation (20)), Fig. 10:

$$M(x_0) = a_1 x_0^2 + a_2 x_0 + a_3 \qquad\qquad\qquad (19)$$

$$V(x_0) = \frac{dM(x_0)}{dx_0} = 2a_1 x_0 + a_2 \qquad\qquad\qquad (20)$$

The stresses are:

$$\sigma_l(x) = b_3 \left[\sinh(\omega x) - \cosh(\omega x)\right] + b_1 x^2 + b_2 x + b_3 \qquad (21)$$

$$\tau_l(x) = b_3 h_l \omega \left[\sinh(\omega x) - \cosh(\omega x)\right] + h_l \left[2b_1 x + b_2\right] \qquad (22)$$

$$\sigma_n(x) = \frac{K_n M_0}{2\beta^2 E_c I_c} e^{-\beta x} \left[\cos(\beta x) - \sin(\beta x)\right] + \frac{q(x) E_l I_l}{b_l E_c I_c} \qquad (23)$$

in which

$$\omega^2 = \frac{G_g}{h_g} \frac{1}{E_l h_l} \qquad\qquad\qquad\qquad (24)$$

$$b_1 = m_l \frac{\overline{y_c}}{I_{tr}} a_1 \qquad\qquad\qquad\qquad (25)$$

$$b_2 = m_l \frac{\overline{y_c}}{I_{tr}} (2a_1 l_0 + a_2) \qquad\qquad\qquad (26)$$

$$b_3 = m_l \frac{\overline{y_c}}{I_{tr}} (a_1 l_0^2 + a_2 l_0 + a_3) + \frac{2b_1}{\omega^2} \qquad\qquad (27)$$

$$M_0 = a_1 l_0^2 + a_2 l_0 + a_3$$

l_0	unplated length, Fig. 10	
$\overline{y_c}$	distance between the centre of the transformed section and the bottom of the beam	[mm]
I_{tr}	inertial moment of the beam	[mm^4]

One could derive that a peak shear stress (τ_{lmax}) and a peak normal peeling stress (σ_{nmax}) occur at the laminate end. Stresses are higher when the unplated length l_0 increases. If these two stresses make a critical combination, delamination will take place. A possible failure criterion is the Mohr-Coulomb criterion. The Mohr Coulomb line is given as Equation (28)[1]. From this value, the maximum load of the beam at delamination can be derived.

$$\tau = \frac{1}{2}\sqrt{f_{cm}f_{ctm}}\left[1+(\frac{f_{cm}}{f_{ctm}}-1)\frac{\sigma}{f_{cm}}\right] \qquad (28)$$

σ	normal stress in an infinitesimal part of the beam	[N/mm²]
τ	shear stress in an infinitesimal part of the beam	[N/mm²]
f_{cm}	mean compressive strength of concrete	[N/mm²]
f_{ctm}	mean tensile strength of concrete	[N/mm²]

5.2   Experiments

Five beams are examined in a four point bending test. The test setup is given in Fig. 11. The span of the beam is 1500mm. The cross section of the beam is 225mm by 125mm. The beam has a length of 1700mm. Internal reinforcement (BE 500) is provided: 3 ø8mm as tensile reinforcement, 2 ø6mm as compressive reinforcement and stirrups of ø6mm every 100mm. Concrete cover is 20mm. One beam remains unstrengthened and acts as a reference. Three beams are strengthened with one layer of SWRP (A_l = 47,67mm²), the last beam is strengthened with two layers SWRP. The beams have an unplated length l_0 of 100mm or 250mm. The test specimens are listed in table 3.

The beams were tested in a load controlled testing device. The load increases with steps of 5 kN (rate: 5 kN/minute). After each step the mid span deflection is measured. At both sides of the beam demec strain gauges are applied, Fig. 12, so that strain can be followed during the test. Also, the crack pattern is recorded. The reference beam failed by yielding of the internal steel. This occurs at a load of 53 kN. All strengthened beams fail through delamination, e.g. beam 1L100b in Fig. 13. All test results are given in table 3.

5.3   Model versus experiments

A beam can fail through concrete crushing or yielding of the internal reinforcement. The load at which this kind of failure occurs can be easily calculated from the equilibrium of forces. In all cases, the internal tensile reinforcement yields first. This happens at a load of 57 kN for the unstrengtened beam, 87 kN for beams strengthened with one layer SWRP and 119 kN for the beam strengthened with two layers, table 4. These loads have to be reduced because of delamination. E.g. a beam strengthened with one layer SWRP and an unstrengtened length of 100mm will fail at a load of 77 kN. The delamination loads are listed in table 4. Remarkable is the small load, 34 kN, when the unplated length becomes large, 250 mm. This means that the beam will act as an unstrengthened beam for loads higher than 34 kN. The beam should fail through yielding of the internal reinforcement at a load of 57 kN.

The maximum loads of the beams 1L100a, 2L100 and the reference beam, are predicted well, with an error of respectively 4%, 4% and 7%. Beam 1L100b has a larger deviation (17%) but this remains within acceptable limits.

Beam 1L250 fails in a different way than predicted. Theoretically, the beam should fail as an unstrengtened beam because of yielding of the internal reinforcement at a load of 57kN. However, the experimental maximum load is 70kN at delamination. An unplated length of 250 mm is rather large with a shear span of 500 mm. This means that the external reinforcement starts at the middle of the span between the support and the point of application of the load. In practice, this is unusual. Possibly, this is caused by the interaction between shear, bending and spreading out of the load under the application point.

It can be concluded that the described model predicts the maximum load quite well, for reduced unplated lengths.

6.   CONCLUSION

SWRP is a new material that can combine the advantages of steel plates and CFRP. It combines a relatively low material cost with a high strength and a flexible shape. For some applications (wrapping), it is necessary that SWRP has a low flexural stiffness. Therefore, it is necessary to reduce the stiffness of SWRP. Hence, new types of SWRP are required. Using SWRP as external reinforcement, the application with viscous adhesive is difficult. The impregnation is sufficient if accurately applied. The development of a new adhesive is necessary to improve impregnation. Also, pre-impregnated laminates are an interesting option.

In this paper non-linear fracture mechanics is applied to model and describe the shear-slip behaviour of the bonded connection. Experiments are presented which demonstrate that the new material behaves stronger and stiffer than elements strengthened with CFRP. Adapted material dependent parameters for SWRP for design purposes are proposed. Further experiments will be performed to confirm the general applicability of these model parameters.

When a beam is strengthened, additional failure modes have to be considered. One of them is delamination. By means of four point bending tests, it is verified that the model needs no adaptation when using SWRP. Experiments show that the model well predicts the maximum load at reduced unplated lengths.

7.   ACKNOWLEDGMENT

The authors would like to thank the Flemish Institute for Promotion of Scientific and Technological Research in the Industry (IWT − Vlaams Instituut voor de Bevordering van Wetenschappelijk-Technologisch Onderzoek in de Industrie) for their financial support and to Bekaert N.V. for the materials and the support.

8. REFERENCES

[1] Brosens, K., *Anchorage of externally bonded steel plates and CFRP laminates for strengthening of concrete elements*, doctoral thesis, Katholieke Universiteit Leuven, 2001.

[2] Matthijs, S., *Structural Behaviour and Design of Concrete Members Strengthened with Externally bonded FRP Reinforcement,* doctoral thesis, Universiteit Gent, 2000.

[3] Hexcel Composites, *technical data sheet F185*, 2000.

[4] Resiplast, *Product catalogus betonherstelling*, 2003, Wommelgem.

[5] Volkersen, O., 1938, *Die Nietkraftverteilung in zugbeanspruchten Nietverbindungen mit konstanten Laschenquerschnitten*, Luftfahrtforschung, 15, pg. 41-47.

[6] Ranisch, E.H., Zur Trägfähigkeit von Verklebungen zwischen Baustahl und Beton-Geklebte Bewehrung, Heft 54, T.U. Braunschweig, 1982.

[7] Holzenkämpfer, P., *Ingenieurmodelle des Verbunds geklebter Bewehrung für Betonbauteile*, doctoraatsverhandeling, Heft 108, IBMB, Braunschweig, pg 5-86, 1994.

[8] Figeys, W., *Strengthening of reinforced concrete structures with bandweave (in Dutch: Versterking van gewapend beton met bandweefsel)*, Master of Science thesis, Katholieke Universiteit Leuven, 2004.

[9] Teng, J.G.; Chen, J.F., FRP strengthened RC structures, John Wiley & Sons, Weinheim, 2001.

[10] Malek, A., Sadaatmanesh; H., Ehsani, R., *Prediction of Failure Load of R/C Beams strengthened with FRP Plate Due to Stress Concentration at the Plate End*, ACI structural Journal, vol. 95, nr 2, 1998, pg 142-152.

Table 1– Prediction of transferable load with the CFRP model parameters, (F_{CFRP})

Type*	f_{ctm} (N/mm²)	f_{cm} (N/mm²)	h_g(mm) (epoxy adhesive)	E_g (N/mm²)	h_{g2} (mm) (epoxy resin)	E_{g2} (N/mm²)
1	2,65	35,0	0,98	7000	-	-
2	2,65	35,0	1,39	7000	0,65	2900

Type*	Bonding length (mm)	G_f (N/mm)	τ_{lm} (N/mm²)	s_{lm} (mm)	s_{l0} (mm)	F_{CFRP} (kN)
1	150	0,736	2,06	0,0067	0,7134	25,9
	200	0,736	2,06	0,0067	0,7134	31,0
2	150	0,736	2,06	0,0084	0,7134	27,7
	200	0,736	2,06	0,0084	0,7134	35,0

* 1: 1 layer, 2: Pre-impregnated + 2 layers

Table 2 – Comparison between CFRP-model and adapted model

Test	Bonding length (mm)	Type *	F_{exp} (kN)	F_{CFRP} (kN)	F_{exp}/F_{CFRP} (-)	F_{SWRP} (kN)	F_{exp}/F_{SWRP} (-)
150 b	150	1	29,5	25,9	1,14	28,1	1,05
200a	200	1	34,4	31,0	1,10	33,6	1,02
200b	200	1	35,1	31,0	1,13	33,6	1,04
I 150 a	150	2	33,4	27,7	1,21	33,2	1,01
I 150 b	150	2	35,1	27,7	1,27	33,2	1,06
I 200 a	200	2	39,5	35,0	1,13	41,9	0,94
I 200 b	200	2	41,7	35,0	1,19	41,9	1,00

* 1: 1 layer, 2: Pre-impregnated + 2 layers

Table 3 – Results of the four point bending test.

Beam	Layers; l_0	$P_{exp.}$ [kN]	Max. midspan deflection [mm]	failure
reference	-	53	5,02	Yielding internal reinforcement
1L250	One layer; 250 mm	70	4,77	Delamination
1L100a	One layer; 100 mm	80	6,51	Delamination
1L100b	One layer; 100 mm	90	6,96	Delamination
2L100	Two layers; 100 mm	75	3,44	Delamination

Table 4 – Calculated loads.

	Reference	One layer	One layer	Two layers
l_0 [mm^2]	-	250	100	100
τ_{lmax} [P/mm^2]	-	$8.80 \cdot 10^{-5}$	$3.85 \cdot 10^{-5}$	$4.05 \cdot 10^{-5}$
σ_{nmax} [P/mm^2]	-	$1.09 \cdot 10^{-6}$	$0.44 \cdot 10^{-6}$	$1.20 \cdot 10^{-6}$
$P_{delamination}$ [kN]	-	34	77	72
$P_{bending}$ [kN]	57	87	87	119

Table 5 – Model versus experiments.

Beam	P_{theo} [kN]	Theoretical failure mode	P_{exp} [kN]	Experimental failure mode	$P_{exp}/$ P_{theo} [-]
reference	57	Yielding internal reinforcement	53	Yielding internal reinforcement	0,93
1L250	57 (34)	Yielding internal reinforcement	70	Delamination	1,23 (2,06)
1L100a	77	Delamination	80	Delamination	1,04
1L100b	77	Delamination	90	Delamination	1,17
2L100	72	Delamination	75	Delamination	1,04

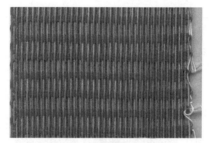

Figure 1 – SWRP, top view.

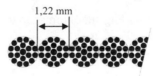

Figure 2 – SWRP, type studied in this paper, 65x [1x 0.25mm + 18x 0.22mm].

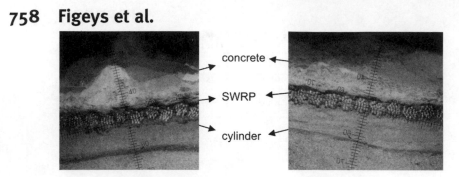

Figure 3 – SWRP, section after pull-off test.

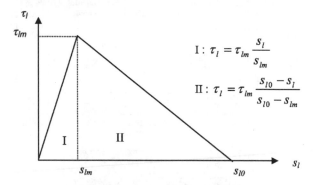

Figure 4 – Bilinear t_l - s_l relationship.

$$I: \tau_l = \tau_{lm} \frac{s_l}{s_{lm}}$$

$$II: \tau_l = \tau_{lm} \frac{s_{l0} - s_l}{s_{l0} - s_{lm}}$$

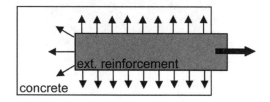

Figure 5 – Spreading out of forces in the concrete.

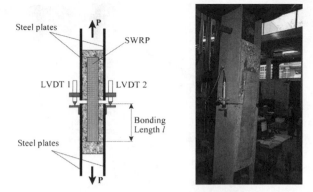

Figure 6 – a. Scheme of shear test [1]; b. photograph of test set sample.

Figure 7 – Test l150a after failure.

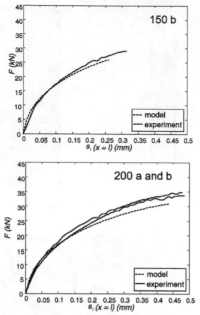

Figure 8 a – Comparison between experiment and model. Parameters single SWRP: h_{ref} = 40 mm, k = 1.47 mm, C_f = 0.40, τ_{lm} = 2.06 N/mm², s_{lm} = 0.0067 mm, s_{lo} = 0.7134 mm.

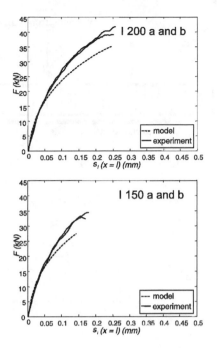

Figure 8 b – Comparison between experiment and model. Parameters pre-impregnated double SWRP: h_{ref} = 40 mm, k = 1.47 mm, C_f = 0.40; τ_{lm} = 2.06 N/mm², s_{lm} = 0.0087 mm, S_{lo} = 0.7134 mm.

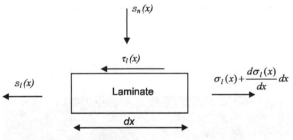

Figure 9 – Stress acting on an external laminate.

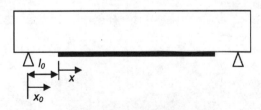

Figure 10 – Strengthened concrete beam.

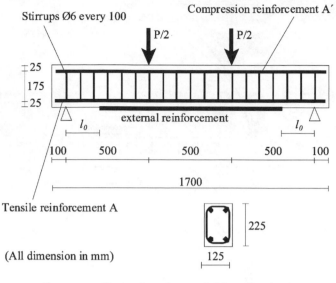

Figure 11 – Test set up four point bending test.

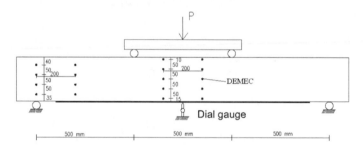

Figure 12 – Position of Demec strain gauges.

Figure 13 – Beam 1L100b, after failure.

Performance of Double-T Prestressed Concrete Beams Strengthened with Steel Reinforced Polymer

by P. Casadei, A. Nanni, T. Alkhrdaji, and J. Thomas

Synopsis: In the fall of 2002, a two story parking garage in Bloomington, Indiana, built with precast prestrestressed concrete (PC) double-T beams, was decommissioned due to a need for increased parking-space. This led to the opportunity of investigating the flexural performance of the PC double-T beams, upgraded in the positive moment region with steel reinforced polymer (SRP) composite materials, representing the first case study where this material has been applied in the field. SRP makes use of high-strength steel cords embedded in an epoxy resin. This paper reports on the test results to failure of three beams: a control specimen, a beam strengthened with one ply of SRP and a third beam strengthened with two plies of SRP anchored at both ends with SRP U-wraps. Results showed that SRP can significantly improve both flexural capacity and enhance pseudo-ductility.

Keywords: double-T beams; ductility; flexure; in-situ load test; prestressed concrete; steel reinforced polymer; strengthening

Paolo Casadei, is a Lecturer in Structural Engineering at the University of Bath, UK. His research interests include repair and assessment of reinforced concrete and masonry structures using conventional methods and FRP strengthening as well as full-scale in-situ load testing. He is a PE in Italy and an EIT in the US. He is a member of ACI, and serves in ACI Committee 437 - Strength Evaluation of Existing Concrete Structures.

Antonio Nanni, is the V & M Jones Professor of Civil Engineering at the University of Missouri-Rolla, USA. He is an active member in the technical committees of ACI (Fellow), ASCE (Fellow), ASTM and TMS. He was the founding Chairman of ACI Committee 440 - FRP Reinforcement and is the current Chairman of ACI Committee 437 - Strength Evaluation of Existing Concrete Structures.

Tarek Alkhrdaji, Ph.D., is a design engineer at the Strengthening Division of the Structural Group. He has been involved in numerous projects involving structural upgrade and rehabilitation of concrete structures using conventional methods and FRP strengthening as well as full-scale in-situ load testing. He is a member of ACI Committee 437, currently Co-chair of ACI Committee 440, Subcommittees F (FRP Strengthening), and also a member of ASCE, and ICRI.

Jay Thomas, is vice president Structural Preservation Systems, Inc. He has more than 20 years experience in the concrete repair and strengthening industry. He is currently a member of ACI (American Concrete Institute) committees 440 (FRP) and ACI 437 (Strengthening Evaluation of Existing Concrete Structures), and a member of ICRI (International Concrete Repair Institute).

INTRODUCTION

The use of advanced composite materials in the construction industry is nowadays a mainstream technology[1], supported by design guidelines such as the ACI 440.2R-02 (ACI 440)[2] in the United States, the Fib-Bulletin 14 (2001)[3] in Europe and the recently published TR55 (2004)[4] in the United Kingdom. Fiber reinforced polymer (FRP) composite materials, even though very attractive, may be hindered by lack of ductility[5] and fire resistance[6]. Both issues are currently under study by the research community, in order to provide on one hand, better knowledge in terms of overall structural performance and, on the other, remedies such as coatings that could prolong fire resistance.

A new family of composite materials based on unidirectional high strength twisted steel wires (about 7 times stronger than typical common reinforcing bars) of fine diameter *(0.20~0.35 mm (0.0079~0.0138 in)* see Figure 1), that can be impregnated with thermo-set (referred to as steel reinforced polymer, SRP) or cementitious (referred to as steel reinforced grout, SRG) resin systems is presented in this work (Hardwire 2002)[7]. SRP/G has the potential to address the two shortcomings mentioned for FRP, indeed: a) steel cords have some inherent ductility; and b) impregnation with cementitious paste may overcome the problems of fire endurance and lowering down the application cost considerably.

The steel cords used in SRP are identical to those used for making the reinforcement of automotive tires, and manufactured to obtain the shape of the fabric tape prior to impregnation (Hardwire, 2002). The twisting of the wires allows some mechanical interlock between the cords and the matrix, and may also induce an overall ductile behavior upon stretching. The cords are also coated with either brass or zinc making the material potentially free of any corrosion and suitable for different kind of environmental exposure. Characterization work, including durability as well as bond related issues, is currently in progress as necessary for implementation in future design guidelines. Recent test results[8] showed that the material does not experience a substantial yielding, but rather a similar behavior to the one experienced by high-strength steel used in prestressed concrete (PC) construction, with a slight non-linear range prior to rupture of the cords (see Figure 2).

The opportunity for experimenting this new material in the field, became available in the winter of 2003 when the City of Bloomington, Indiana, decommissioned an existing parking garage near the downtown area, built with double-T PC beams. The concrete repair contractor, Structural Preservation Systems, Hanover, MD, strengthened in flexure the bottom stem of several double-T beams with epoxy-based SRP. This paper reports on the experimental results of tests to failure conducted on three beams: a control specimen, a beam strengthened with one ply of SRP and a third beam strengthened with two plies of SRP anchored at both ends with U-wraps.

EXPERIMENTAL PROGRAM

Building Characteristics

The parking garage used for the tests was a two story structure constructed in the 1980s (see Figure 3a and b). It consisted of a reinforced concrete (RC) frame, cast in place columns and precast reversed-T PC beams, supporting double-T PC beams, of span length varying from *4.66 m (15.3 ft)* to *13.41 m (44 ft)*.

Since no maintenance or construction records were available for the materials and the layout of the prestressing tendons, a field investigation was carried out. Based on the survey, it was determined that the double-T PC beams were of type *8DT32* (see Figure 3c) according to the Prestressed Concrete Institute (1999)[9] specifications with concrete topping of *76 mm (3in)*, and with an arrangement of the tendons different from current specifications. For the span of *4.66 m (15.3 ft)*, two straight 7-wire strands were found in each stem, each with a diameter of *12.7 mm (0.5 in)*, corresponding to an area of *112 mm²* *(0.174 in²)*, the first at *248 mm (9.75 in)* from the bottom of the stem and the second spaced *305 mm (1 ft)* from the first one. No mild reinforcement was found at any location. Welded pockets, connecting two adjacent beams, were positioned every *910 mm (3ft)* at a depth of *76 mm (3 in)* from top surface. Concrete properties were evaluated using three cores taken from three different beams at the location of the stem and an avarage concrete cylinder strength of $f_c'=34$ *N/mm2 (f_c'=5000 psi)* was found and its modulus of elasticity was determined according to ACI 318-02 Section 8.5.1[10] provisions.

The strands properties were assumed to be conventional *1861 MPa (270 ksi)* strength. Table 1 summarizes construction material properties.

Specimens and Installation of Steel Reinforced Polymer

A total of three double-T PC beams were tested (see Figure 4*a* and *b*): beam DT-C is the control beam, beam DT-1 represents the beam strengthened with one ply of SRP and DT-2U the one strengthened with 2 plies of SRP anchored with SRP U-wraps. Figure 4*c* illustrates the strengthening geometry details for beam DT-1 and DT-2U.

The epoxy resin for both strengthened beams was SikaDur Resin 330[11]. The choice of the resin was based on constructability so that it could be rolled onto the surface for overhead applications, while having enough consistency, even before curing, to be able to hold the weight of the steel tape during cure.

Table 2 reports the resin properties supplied by the manufacturer and verified by testing according to ASTM standards by Huang et al. 2004[8]. Figure 5*a* shows the mixing of the resin prior to installation.

The tape was medium density consisting of *6.3 cords per cm (12 WPI)*, with material properties defined in Table 3. The typical stress-strain diagram for an impregnated medium density tape, tested following the ASTM D 3039[12] recommendations, is reported in Figure 2 (properties based on steel net area).

SRP was installed following the recommendations of ACI 440[2] provisions for FRP materials. The sequence of installation steps is reported in Figure 5. The bottom stem of the double-T beams was first abrasive-blasted to ensure proper bond of the SRP system. With the surface roughened and cleaned, the first layer of epoxy was directly applied (see Figure 5*b*), without primer coating. The steel tape was cut to length of *4.57 m (15 ft)* and width of *102 mm (4 in)*, covering the bottom of the stem length and width entirely. A rib-roller was then utilized to press onto the tape to ensure epoxy impregnation and encapsulation of each cord and allow excess resin to squeeze out. The excess resin was spread with a putty-knife to create an even surface (see Figure 5*c*) and a synthetic scrim was applied to avoid any dripping of the resin (see Figure 5*d*). For the two ply application, once the first ply was in place and the excess resin leveled, the second ply was installed, following an identical procedure. This time the ply started *152 mm (6 in)* away from the terminations of the first ply, making it *4.27 m (14 ft)* long. To provide a mechanical anchorage for the two longitudinal plies, an SRP U-wrap *914 mm (3 ft)* wide was installed at both ends of the stems (see Figure 5*e*). Due to the stiffness of the steel tape, pre-forming is done with a standard sheet metal bender before installation. For this reason, the U-wrap was obtained by overlapping two L-shaped wraps. A final coat of epoxy resin was then applied on top of the U-wrap as final layer to impregnate the tape (see Figure 5*f*).

Test Setup and Instrumentation

The experimental setup is shown in Figure 6*a* and Figure 6*b*. The beams were tested under simply supported conditions and subject to a single concentrated load spread over both stems at mid-span, that is, 3-point bending at mid-span (see Figure 6*c*).

All three tests were conducted using a close-loop load configuration, where no external reaction is required. The load was applied in cycles by one hydraulic jack of *890 kN (200 kip)* capacity connected to a hand-pump. The load was transferred to the PC beam in two points through one spreader steel beam (see Figure 6*b*). The reverse-T PC-Ledger beams, on which the double-T beam rests, supplied the reaction. As the hydraulic jack extended, it pulled on the high-strength steel bars, which lifted the reaction bailey-truss below. The reaction truss was built with three bailey-truss frames *6.09 m (20 ft)* long assembled as per manufacturer's specifications (Mabey Bridge and Shore, Baltimore, MD) [13], and properly designed to carry the test load (see Figure 6*a*). Plywood was placed at each contact point to protect the concrete. The load was measured using a *890 kN (200 kip)* load cell placed on top of the jack (see Figure 6*c*). The preparation work consisted of drilling one hole of small diameter (~*50 mm (2 in)*) necessary for passing the high-strength steel bar through the flange of the double-T PC beam and isolating each test specimen from the adjacent beams originally joined by the welded-pockets.

An electronic data acquisition system recorded data from four linear variable differential transducers (LVDTs) and two electrical strain-gages applied to the SRP in beams DT-1 and DT-2U. Two LVDTs were placed at mid-span, and the remaining two LVDTs, were placed under the reverse-T ledger beams to verify potential support settlements. Strain gages were installed at mid-span on the bottom flange of the two strengthened double-T beams, directly onto the SRP material.

RESULTS AND DISCUSSION

All beams failed in flexure and had a similar behavior up to the cracking load. Beam DT-C failed due to fracture of the lowest tendon. In beam DT-1, since the SRP ply was not mechanically anchored, failure was dictated by peel off of the ply from each stem almost simultanuously. Beam DT-2U, strengthened with two anchored plies per stem, failed due to rupture of the lower tendon. Table 4 reports the test results.

In beam DT-C flexural cracks were concentrated in the mid-span region where the point load was applied (see Figure 7*a*). As soon as cracking occurred, since no mild reinforcement was present and tendons were placed far away from the bottom of the stem, cracks developed throughout the entire stem and deck (see Figure 7*b*). In beams DT-1 and DT-2U a similar behavior occurred with the difference that the presence of the SRP allowed the formation of additional flexural cracks. In beam DT-1 the SRP laminate started debonding at mid-span initiated by the widening of mid-span cracks (see Figure 7*c*) and then progressed towards the supports (see Figure 7*d*). Complete detachment of the laminate occurred at one end of the beam with part of the concrete substrate attached to the laminate, denoting a good interface bond between the concrete and the SRP (see Figure 8*a* and *b*). In beam DT-2, SRP could not completely peel off due to the presence of U-wraps. Delamination propagated from mid-span towards the supports similarly to Beam DT-1, until rupture of the lower tendon occurred (see Figure 7*e*) and immediately followed by SRP rupture exactly at the location where the SRP U-wrap started (see Figure 7*f*). No shear cracks were noted on any of the three beams.

Figure 9 shows the Load-vs-mid-pan Deflection curves for all three beams. The capacities of beams DT-1 and DT-2U increased approximately 12% and 26% with respect to the control specimen DT-C.

Figure 10 report the Load-vs-Mid-Span Strain responses for beams DT-1 and DT-2U. Two distinct phases, pre- and post-cracking, characterize the behavior of each specimen. Up to cracking there was practically no strain in the SRP. Past the cracking load, the presence of the SRP significantly affected performance.

Beam DT-C cracked at a considerably lower load (250.8 kN (56.4 kip)), with respect to the other two strengthened specimens. The occurrence of the first crack, at mid-span only, corresponds to the load drop in the Load-vs-Displacement plot. Upon unloading, the beam remained almost perfectly elastic, recovering almost all deflection. At the third loading cycle the lower strand suddenly fractured at a load of 344.3 kN (77.4 kip). For beams DT-1 and DT-2U the cracking load increased of approximately 23% and 17% with respect to DT-C. The lower cracking load for DT-2U may be explained by the fact that the beam had been previously repaired by means of epoxy injection.

Beam DT-1 reached the peak load of 387 kN (87 kip) and held it constant with increasing deflection, while SRP progressively delaminated from mid-span towards the support. The strain profile reported in Figure 10 shows how the SRP was not engaged until cracking occurred and as soon as the first crack opened at mid-span, the SRP bridged the crack and strain suddenly increased to approximately 5500 με (strain-gauge was placed at mid span where the first crack occurred). The maximum strain recorded in the steel tape (12300 με), prior to complete peel-off, shows how the material was well bonded to the concrete substrate. The ductility reported in the load-deflection curve, is the result of the slow peeling propagation rather than to the yielding of the reinforcing steel tape itself. Figure 2 shows in fact an almost elastic behavior till rupture of the SRP laminate.

Past the cracking load (see Figure 10), beam DT-2U behaved almost linearly, although with a lower stiffness, until it reached the load of 400 kN (90 kip) then, stiffness decreased significantly till the peak load was reached. When the load of 434 kN (97.6 kip) was reached, the lower tendon ruptured and a sudden drop in the load-deflection curve was recorded. The strain in the SRP material at time which the tendon ruptured was 6400 με. At this stage, once the lower tendon ruptured, the SRP laminate was completely debonded except for the region where anhoring was provided by the U-wraps. The test was continued untill suddenly the SRP laminate ruptured at 388 kN (87.2 kip). The strain recorded in the SRP laminate at failure was 12000 με, similarly the values attained in beam DT-1.

CONCLUSIONS

The following conclusions may be drawn from this experimental program:

- SRP composite materials have shown to be effective in increasing the flexural capacity of the double-T PC beams.

- End anchors in the form of SRP U-wraps have shown to be effective by preventing a complete detachment, once debonding has occurred throughout the concrete-SRP interface.

- SRP is similar to FRP in terms of ease of installation, although self weight should not be ignored when selecting the resin system in overhead applications.

- A good bond between the steel tape and the concrete substrate was achieved using epoxy resin.

FUTURE WORK

Since the completion of this test campaign, a bridge in Missouri has been strengthened with SRP materials as part of a joint MODOT – University of Missouri-Rolla (UMR) initiative[14] and studies[15] at UMR are underway to characterize the material and to properly calibrate the design factors.

ACKNOWLEDGMENTS

This research study was sponsored by the National Science Foundation Industry/University Cooperative Research Center on Repair of Buildings and Bridges (RB2C) at the University of Missouri – Rolla. Hardwire LLC., Pocomoke City, MD, provided the steel tapes and Sika Corporation, Lyndhurst, NJ, the resins for the installation. The City of Bloomington, IN, provided the opportunity for testing the structure.

REFERENCES

[1] Rizkalla, S. and Nanni, A. (2003) "Field Applications of FRP Reinforcement: Case Studies" ACI Special Publication 215, Published by the American Concrete Institute, Farmington Hills, MI.

[2] ACI 440.2R-02, 2002: "Guide for the Design and Construction of Externally Bonded FRP Systems for Strengthening Concrete Structures," Published by the American Concrete Institute, Farmington Hills, MI, pp. 45.

[3] FIB Bullettin 14 (2001). "Design and use of externally bonded fibre reinforced polymer reinforcement (FRP EBR) for reinforced concrete structures, by 'EBR' working party of FIB TG 9.3, July 2001, 138 pp.

770 Casadei et al.

[4] The Concrete Society, Technical Report No. 55, 2004: "Design Guidance for strengthening concrete structures using fibre composite materials (Second Edition)" The Concrete Society, 102 pp.

[5] Seible, F.; Priestley, M. J. N.; Hegemier, G. A.; and Innamorato, D., 1997, "Seismic Retrofit of RC Columns with Continuous Carbon Fiber Jackets," Journal of Composites for Construction, No. 1, pp. 52-62.

[6] Williams, B.K., Kodur, V.K.R., Bisby, L.A., and Green, M.F. "The Performance of FRP-Strengthened Concrete Slabs in Fire," Fourth International Conference on Advanced Composite Materials in Bridges and Structures - ACMBS-IV July 20-23, 2004 The Westin Hotel, Calgary, Alberta, Canada.

[7] Hardwire LLC, 2002, "What is Hardwire," www.hardwirellc.com, Pocomoke City, MD.

[8] Huang, X., Birman, V., Nanni, A., and Tunis, G., "Properties and potential for application of steel reinforced polymer and steel reinforced grout composites," Composites, Part B: Engineering, Volume 36, Issue 1, January 2004, Pages 73-82.

[9] PCI (1999): "PCI Design Handbook: Precast and Prestressed Concrete", Published by the Precast/ Prestressed Concrete Institute, Chicago, IL.

[10] ACI 318-02, 2002: "Building Code Requirements for Structural Concrete and Commentary (318R-02)," Published by the American Concrete Institute, Farmington Hills, MI, pp. 443.

[11] Sika, 2004, "Sikadur 330", www.sikausa.com, Lyndhurst, NJ.

[12] ASTM D 3039, 2002: "Test Method for Tensile Properties of Fiber Resin Composites" Published by the American Society for Testing and Materials, West Conshohocken, PA, pp. 13.

[13] Mabey Bridge & Shore, Inc., www.mabey.com, Baltimore, MD.

[14] Lopez, A., and Nanni, A., "Validation of FRP Composite Technology Through Field Testing" 16th World Conference on Nondestructive Testing. Montreal, Canada. August 30- September 3, 2004.

[15] Wobbe, E., Silva, P.F., Barton, B.L., Dharani, L.R., Birman, V., Nanni, A., Alkhrdaji, T., Thomas, J., and Tunis, T., "Flexural Capacity of RC Beams Externally Bonded with SRP and SRG" Proceedings of Society for the Advancement of Material and Process Engineering 2004 Symposium, 16-20 May 2004, Long Beach, Ca., 20pp.

Table 1 - Properties of Construction Materials

Material	Cylinder Compressive Strength, MPa (psi)	Yield Strength MPa (ksi)	Rupture Strength MPa (ksi)	Elastic modulus[2] MPa (ksi)	7 wire Tendon Cross Section, Ap mm² (in²)
Concrete [1]	34.4 (5,000)	-	-	27,600 (4,000)	-
Steel	-	1585 (230)	1862 (270)	200,000 (29,000)	112 (0.174)

[1] Average of 3 specimens [76.2 mm×152.4 mm (3 in×6 in) cylinders].

[2] $E_c = 4700\sqrt{f_c'}$ ACI 318-02 Section 8.5.1

Table 2 - Mechanical Properties of Epoxy Resin

Matrix	Tensile Strength, MPa (psi)	Ultimate Rupture Strain ε_{fu}^{*} (%)	Tensile Modulus of Elasticity, MPa (ksi)
SikaDur 330[1]	30 (4350)	1.5	3800 (551)

[1] Values provided by the manufacturer (Sika, 2002)

Table 3 - Material Properties of Steel Tape[8]

Cord Coating	Cord Area per 12 Wires, mm² (in²)	Cords per cm (in)	Nominal Thickness[1], t_{SRP} mm (in)	Tensile Strength f_{fu_SRP}, MPa (ksi)	Ultimate Rupture Strain ε_{fu_SRP} (mm/mm)	Tensile Modulus of Elasticity, GPa (ksi)
Brass	0.396 (0.000615)	3.7 (9.5)	0.148 (0.0058)	3070 (447)	0.0167	184 (26700)

[2] The nominal thickness has been computed assuming the area of each cord and counting the number of cords in each ply, reported in *cords per cm*

Table 4 – Beam Test Results

Beam	Failure load kN (kip)	Load Capacity Increase	SRP Strain at Failure ε_{SRP}	Failure Mode
DT-C	344 (77.4)	1	-	Rupture of Lower Tendon
DT-1	387 (87)	1.12	0.0123	SRP Delamination
DT-2U	434 (97.6)	1.26	0.0064	Rupture of Lower Tendon

a) Front view of Tape with Cords Held Together by a Polyester Scrim

b) Back view of Tape with detail of Polyester Scrim

c) Steel Cord with Wires Wrapped by One Wire

Figure 1 – Example of 12X Steel Cord and Medium Density Tape

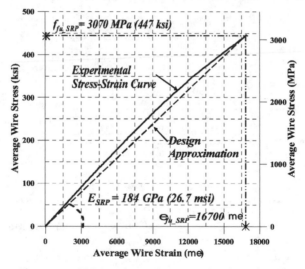

Figure 2 – SRP Laminate Stress vs Strain Behavior

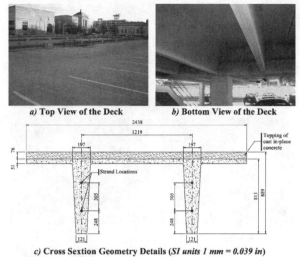

a) **Top View of the Deck**   *b)* **Bottom View of the Deck**

c) **Cross Sextion Geometry Details** *(SI units 1 mm = 0.039 in)*

Figure 3 – Bloomington Parking Garage

a) **Saw-Cut Marks on Top of Deck**

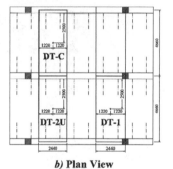

b) **Plan View**

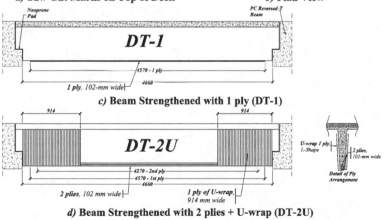

c) **Beam Strengthened with 1 ply (DT-1)**

d) **Beam Strengthened with 2 plies + U-wrap (DT-2U)**

Figure 4 – Test Beams *(SI units 1 mm = 0.039 in.)*

Casadei et al.

a) Mixing of the Epoxy Resin   *b)* Application of Longitudinal Ply

c) Squeezing Out the Resin Excess   *d)* Application of Scrim on Longitudinal Ply

e) Application of U-Wraps   *f)* Application of Epoxy on U-Wrap

Figure 5 – SRP Installation Procedure

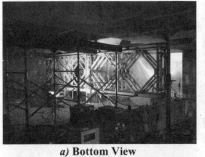

a) **Bottom View**

b) **Top View**

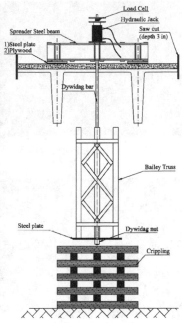

c) **Cross Section at Mid-Span**

Figure 6 – Test Set Up

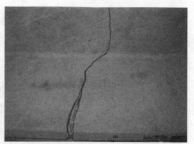

a) Initial Crack Propagation

b) Crack Propagation Through the Deck

Beam DT-C

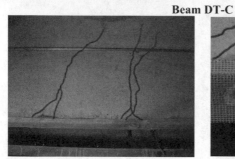

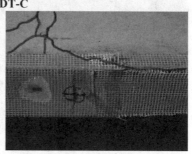

c) Crack Propagation Prior to Complete Peeling

d) Debonding Propagation from Mid-Span

Beam DT-1

e) SRP Rupture

f) Rupture of the Lower Tendon

Beam DT-2U

Figure 7 – Failure Mechanisms in the Three Beams

| *a)* Close View of Concrete Substrate | *b)* Close View of SRP Laminate |

Figure 8 – Concrete and SRP Laminate Condition After Debonding Has Occurred

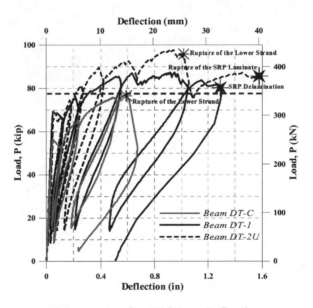

Figure 9 – Load vs Mid-Span Deflection

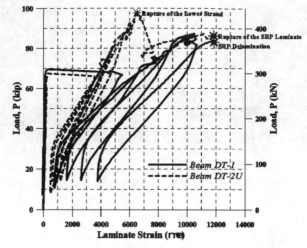

Figure 10 – Load vs Mid-Span Strain

Analytical Evaluation of RC Beams Strengthened with Near Surface Mounted CFRP Laminates

by J.-Y. Kang, Y.-H. Park, J.-S. Park, Y.-J. You, and W.-T. Jung

Synopsis: To assess the strengthening efficiency of near-surface mounted (NSM) carbon fiber reinforced polymer (CFRP) laminates according to their groove depth and disposition, 4-point bending tests were performed on 4 specimens strengthened with NSM CFRP. A structural model for the finite element method (FEM) able to simulate accurately the experimental results was determined to analyze the strengthening efficiency of the NSM technique analytically. Applying the model, parametric analysis was performed considering the groove depth and spacing of CFRP laminates. Analytical study on the groove depth revealed the existence of a critical depth beyond which the increase of the ultimate load becomes imperceptible. In other words, this means that there exists a limit of strengthening efficiency where it remains in a definite level even if the groove depth is increased. Analytical results regard to the spacing of the CFRP laminates showed that comparatively smooth fluctuations of the ultimate load were produced by the variation of the spacing and the presence of an optimal spacing range for which relatively better strengthening efficiency can be obtained. Particularly, a spacing preventing the interference between adjacent CFRP laminates and the influence of the concrete cover at the edges as well as allowing the CFRP laminates to behave independently was derived. Using the analytical results, various strengthening schemes could be established with different numbers of CFRP laminates, groove depths and dispositions of the reinforcements for a determinate quantity of reinforcements.

Keywords: carbon fiber reinforced polymer (CFRP); groove depth; near surface mounted (NSM); spacing; strengthening

780   Kang et al.

Jae-Yoon Kang received his MS degree in Civil Engineering from Dongguk University. He is a Senior Researcher at the Structure Research Department of the Korea Institute of Construction Technology.

Young-Hwan Park received his MS degree and Ph.D. in Civil Engineering from Seoul National University. He is a Research Fellow at the Structure Research Department of the Korea Institute of Construction Technology.

Jong-Sup Park received his MS degree in Civil Engineering from Myongji University. He is a Senior Researcher at the Structure Research Department of the Korea Institute of Construction Technology.

Young-Jun You received his MS degree in Civil Engineering from Yonsei University. He is a Researcher at the Structure Research Department of the Korea Institute of Construction Technology.

Woo-Tai Jung received his MS degree in Civil Engineering from Myongji University. He is a Researcher at the Structure Research Department of the Korea Institute of Construction Technology.

INTRODUCTION

Near surface mounted (NSM) carbon fiber reinforced polymer (CFRP) technique becomes attractive for flexural strengthening of slabs and girders, as it has a greater bonding capacity compared to externally bonded CFRP composites (El-Hacha et al. 2004; Hassan and Rizkalla 2003; Lorenzis and Nanni 2002). In the NSM CFRP technique, CFRP composites are embedded into previously epoxy filled grooves pre-cut on the concrete cover. It is a strengthening method utilizing the bonding force developed at the interface between the epoxy and concrete as a stress transferring mechanism (Täljsten et al. 2003; Lorenzis and Nanni 2002). Therefore, the groove depth and disposition of CFRP laminates have a significant effect on the strengthening efficiency of NSM as well as on the workability and the cost (Hassan and Rizkalla 2003). In the case of confined construction areas like concrete beams which present narrow width and restrained concrete cover, one should be able to determine the optimal amount of reinforcement and optimal location of embedment that produce sufficient strengthening efficiency.

This study examines experimentally the flexural behavior of reinforced concrete beams strengthened by NSM CFRP technique regard to the reinforcement groove depth and disposition. An analytical model which simulates accurately such behavior has also been determined. Applying the analytical model, parametric evaluation of the strengthening efficiency is achieved considering the variations of the groove depth and spacing of the CFRP laminates. The range of groove depth and spacing producing the maximum strengthening efficiency has been proposed.

EXPERIMENTS AND ANANLYSIS

Preliminary experiments

To assess the efficiency of NSM technique and to determine the finite element model that can simulate the flexural behavior of strengthened beam, preliminary 4-point bending tests were carried out according to the groove depth and disposition of CFRP laminates. The specimens illustrated in Figure 1 consisted of 1 unstrengthened control specimen, 2 specimens with different groove depths (TYPE 1-1 and TYPE 1-2) and 2 specimens with CFRP strips disposed at different spacing (TYPE 2-1 and TYPE 2-2).

The experimental results, summarized in Table 1, showed that the ultimate loads of the NSM strengthened specimens were increased by 40 % to 95 % compared to the unstrengthened specimen. Especially, the ultimate load of TYPE 1-2 was 1.1 times larger than that of TYPE 1-1 while the sectional area of the reinforcements was 1.65 times larger. Areas of CFRP reinforcements for TYPE 2-1 and TYPE 2-2 were 2 times larger than that of TYPE 1-2; the measured ultimate loads, however, were only 1.27 and 1.24 larger, respectively. In view of the results for TYPE 2-1 and TYPE 2-2, it appears that strengthening efficiency also varies with the disposition of the CFRP reinforcements for the same area of reinforcements. It is seen that, consequently, strengthening efficiency is not directly proportional to the area of CFRP reinforcement and also varies according to the disposition of the reinforcements.

Analysis scheme

According to the experimental results, the differences in strengthening efficiency with respect to the disposition of the reinforcements for the same groove depth and the non-direct proportionality of increase in load carrying capacity with the variation of the groove depth have been observed. To verify the above results, using the analysis software ABAQUS v.6.4, nonlinear finite element analysis has been performed for RC beams strengthened by NSM technique. Three-dimensional finite element model has been implemented as structural model in order to consider the disposition of the reinforcements within the section of the beams. Prior to the parametric analysis, structural analysis has been carried out on the 5 specimens used in the test and the feasibility of the analysis model has been verified through the comparison of the analytical and experimental results.

The conditions considered in the parametric analysis can be roughly classified into two cases: the arrangement of a single CFRP laminate and the arrangement of two CFRP laminates. In the former case, analysis has been performed with respect to the variation of groove depth from 5 mm to 35 mm by 5 mm. In the latter case, the groove depth and disposition of the reinforcements were simultaneously considered; the embedment depth was fixed to 15 mm or 25 mm and, for each case, the spacing between the two reinforcements was varied from 20 mm to 180 mm.

Description of finite element model

For the finite element analysis of RC beams, 8-node continuum element was used for the concrete and 'hourglass control' has been applied to prevent excessive deformation of the elements caused by reduced integration. The reinforcing steel bar was modeled by 2-node 3D truss element and applied as embedded element inside the concrete element. The material properties of concrete and steel rebar have been assumed by the stress-strain relationships plotted in Figures 2 and 3. Similarly to the concrete member, 8-node continuum element was applied for the epoxy block and perfect bonding has been assumed at the concrete-epoxy interface. The CFRP laminates were also modeled by 8-node continuum element assumed as embedded element in the epoxy element. This assumption has been done so as to reflect the failure mode exhibited by the epoxy and CFRP laminate reinforcement, which behaved as a block in the experiments. The stress-strain relationship of CFRP laminate was shown in Figure 4. The boundary conditions of the concrete, epoxy and carbon laminate interface are illustrated in Figure 5.

Three-dimensional nonlinear analysis has been carried out using the modified Riks solver, a nonlinear static equilibrium solver supplied by ABAQUS, in order to consider the instability of the boundary conditions provoked by the failure of the reinforcement element or the concrete element in the RC beam strengthened by NSM technique and to trace exactly the path of the load-displacement curve.

Verification of the analysis model

Comparison of the experimental and analytical results has been performed to verify the feasibility of the analysis model. As shown in Figure 6, the relative errors of the analytical results for CONTROL, were about 2 % for the yield load and 8 % for the ultimate load, which approached fairly the experimental results.

Figure 7 plots the analytical results for TYPE 1-1 and TYPE 1-2, the specimens with groove depth of 15 mm and 25 mm, respectively. An error of 10 % occurred for the maximum displacement. The ultimate load approximated the experimental results by a relative error of less than 8 %.

Figure 8 plots the analytical results for TYPE 2-1 and TYPE 2-2, the specimens with two CFRP laminates disposed at spacing of 60 mm and 120 mm, respectively. As seen in the figure, the load-displacement curve obtained through the analysis approximates the experimental results with an error of less than 10 % for the displacement. Especially, the error for the load being lower than 4 %, the model can be said to simulate the actual behavior with accuracy.

Analysis on groove depth

For NSM technique, the maximum groove depth is limited to about 50 mm corresponding to the thickness of the concrete cover. It can be expected that the strengthening efficiency and load carrying capacity of the strengthened member increase in proportion to the groove depth or the quantity of reinforcements. However, experimental results revealed that the strengthening efficiency is not directly proportional to the groove depth or the quantity of reinforcements. It means that there exists an optimal groove depth which maximizes strengthening efficiency. This study investigated analytically the fluctuations of the increase of load carrying capacity according to the variation of the groove depth for the case of strengthening with a single CFRP laminate;

and a critical value for the groove depth has been derived to produce efficient strengthening.

Table 2 summarizes the results obtained from the parametric analysis on the groove depth varying from 5 mm to 35 mm. It showed that increase in yield load varies from 5 % to 29 % compared to the unstrengthened beam; and increase in ultimate load varies from 5 % to 60 %. Analytical results showed the poor influence of the variation of the groove depth or the variation of the amount of reinforcement on the increase of the yield load; however, it appeared also that these variations have large effect on the increase of the ultimate load after yielding. Table 3 compares the variations of load carrying capacities referring to the area of reinforcement with respect to the groove depth.

Figure 9 plots the increase rate of the ultimate load according to the variation of groove depth. The increase rate of the ultimate load is shown to converge toward a determinate value beyond a certain groove depth; hence, the ultimate load-groove depth relationship can be represented as a second-order function shown in Figure 9. For the dimension of the beam adopted in this study, the ultimate load tended to increase until the critical groove depth of 35 mm, and beyond that depth, no additional strengthening effect was observed; although this tendency and critical groove depth can be altered for different dimensions of beams.

Analysis on disposition of CFRP reinforcements

In cases where insufficient load carrying capacity is expected for a beam to be strengthened with a single CFRP laminate, two or more CFRP laminates disposed at regular spacing should be installed. In such cases, attention should be paid on securing a certain spacing between the CFRP laminates so as to prevent mutual interference which causes reduction of strengthening efficiency. On the other hand, if this spacing becomes excessive, adverse effects may drop off the strengthening efficiency because the CFRP laminate near the concrete edge can not acquire bonding capacity; furthermore, it causes surface spalling of the edge of concrete cover. Consequently, deciding an optimal spacing that is able to maximize the strengthening efficiency, is necessary; in other words, by deciding an optimal disposition, each of the CFRP laminates behaves independently and, at the same time, the influence of the distance to the concrete edge can be avoided.

Table 4 summarizes the analytical results on spacing of the CFRP laminates when two rows of laminate are disposed. The increase rate of the ultimate load is seen to range between 65 % and 98 % according to the spacing for the groove depth of 25 mm; and between 52 % and 75 % for the groove depth of 15 mm. The spacing exhibiting the maximum strengthening efficiency was 80 mm around which the strengthening efficiency was seen to decline.

Figure 10 plots two cases of analytical results together to derive the optimal spacing range that enables the maximum strengthening efficiency. It can be seen that the highest strengthening efficiency is developed for spacing between the reinforcements ranging between 80 mm and 120 mm. The increase rate of ultimate load reaches up to 65 % for the groove depth of 15 mm; and runs up to 95 % for the groove depth of 25 mm. In addition, parametric analysis showed that the strengthening efficiency appears to decline relatively if either the disposed spacing or the distance to concrete edge is less than 40 mm. Consequently, the following preconditions should be satisfied for NSM technique; (1) to ensure a minimal spacing of 40 mm between adjacent CFRP laminates for independent behavior; (2) to dispose CFRP laminate at a distance of more than 40 mm from concrete edge.

DISCUSSION

This paper deals with the strengthening efficiency of NSM for particular RC beams used in experiments and parametric analysis. Since the behavior of RC beams with arbitrary dimensions may be different from the results of this study, it is highly recommended to derive the general relationships between the increase of load carrying capacity and the amount of reinforcements. To that goal, wider range of experimental and analytical data considering the following should be secured: (1) the behavior considering the spacing of reinforcements with respect to the groove depth; (2) the behavior according to varying number of reinforcements; (3) the behavior for different dimensions and steel reinforcement ratios; and (4) the behavior regard to the strength of material. Once generalized relationships of reinforcement ratio versus increase rate of load carrying capacity will be prepared through future tests or parametric studies, it will be possible to establish easily the strengthening scheme for NSM, as shown in Figure 12.

CONCLUSIONS

Analytical investigation on the ultimate load versus groove depth relationship revealed the existence of a critical groove depth beyond which the increase of the ultimate load of strengthened member becomes very slight. In other words, the existence of a critical load carrying capacity, beyond which additional strengthening effect could not be obtained even if the groove depth was continuously increased, could be presumed. In terms of the dimensions of the specimens applied in this parametric study, the maximum groove depth being 35 mm, the critical load carrying capacity that could be secured by NSM technique reached 1.6 times that of the unstrengthened beam.

Analytic results on spacing of the CFRP laminates showed that, for a definite amount of reinforcement, the yield load increased uniformly indifferently from the variation of spacing; and it showed relatively smooth fluctuations of the ultimate load regard to the variation of spacing. The presence of an optimal range of spacing for which relatively high strengthening efficiency was obtained has also been found. Especially, by the parametric study, the particular conclusions were also ascertained as follows: (1) there exists a minimum spacing between adjacent CFRP laminates to be secured to prevent mutual interference; and (2) there exists a minimum distance to the concrete edge that avoids the influence of the concrete cover in the vicinity of the edges of the beam. It has been verified that this minimum spacing and minimum distance to edge should exceed 40 mm to ensure that each of CFRP laminates behaves independently. As these features are supposed to occur regardless of the dimensions of the beam to be strengthened, these requisites should be preceded when deciding the disposition of more than 2 rows of

CFRP laminates.

Results of this parametric analysis made it possible to know that the strengthening efficiency depends on the groove depth, the amount of reinforcements according to the groove depth, and the disposition of the reinforcements. The possibility to derive several alternatives of strengthening scheme with respect to the required increase of load carrying capacity of a member to be strengthened by NSM technique has also been highlighted. It has been seen that, for a prescribed amount of reinforcement, larger strengthening efficiency could be realized by disposing several reinforcements at regular spacing near the surface instead of a single reinforcement. However, because it is rational to reduce the processes such as the cutting of the grooves or the epoxy filling, it is advisable to decide the disposition of the reinforcements after consideration of the workability and economical efficiency.

REFERENCES

El-Hacha, R., Silva Filho, J. N., Melo, G. S. and Rizkalla, S. H., 2004, "Effectiveness of Near-Surface Mounted FRP Reinforcement for Flexural Strengthening of Reinforced Concrete Beams," *Proceedings of the 4th International Conference on Advanced Composite Materials in Bridges and Structures (ACMBS IV)*, Calgary, Alberta, Canada, July 20-23

Hassan, T. and Rizkalla, S., 2003, "Investigation of Bond in Concrete Structures Strengthened with Near Surface Mounted Carbon Fiber Reinforced Polymer Strips," *Journal of Composites for Construction*, Vol. 7, No. 3, pp. 248-257.

Lorenzis, L.D. and Nanni, A., 2002, "Bond Between Near Surface Mounted FRP Rods and Concrete in Structural Strengthening," *ACI Structures Journal*, Vol. 99, No. 2, pp. 123-133.

Täljsten, B., Carolin, A. and Nordin, H., 2003, "Concrete Structures Strengthened with Near Surface Mounted Reinforcement of CFRP," *Advances in Structural Engineering*, Vol. 6, No. 3, pp. 201-213

Table 1 – Summary of test results

Specimen	P_y (kN)	d_y (mm)	P_u (kN)	d_u (mm)	% increase in P_y	% increase in P_u
CONTROL	46.7	12.78	56.2	71.68	–	–
TYPE 1-1	57.5	15.5	78.4	58.9	23.1	39.5
TYPE 1-2	62.0	16.1	86.2	54.0	32.8	53.4
TYPE 2-1	72.0	16.5	109.7	46.9	54.2	95.2
TYPE 2-2	70.5	14.2	107.0	44.4	51.0	90.4

P_y : yield load
d_y : midspan deflection at yielding
P_u : ultimate failure load
d_u : midspan deflection at failure

Table 2 – Comparison of the increase of P_u according to groove depth

d (mm)	P_y (kN)	P_u (kN)	$P_y/P_{y(control)}$	$P_u/P_{u(control)}$	Specimen
CONTROL	46.69	56.19	1.00	1.00	CONTROL
5	48.87	59.04	1.05	1.05	–
10	51.53	69.49	1.10	1.24	–
15	53.98	76.33	1.16	1.36	TYPE 1-1
17.5	54.76	78.27	1.17	1.39	–
20	55.34	80.58	1.19	1.43	–
25	56.84	84.37	1.22	1.50	TYPE 1-2
27.5	58.15	86.62	1.25	1.54	–
30	59.05	84.7	1.26	1.51	–
35	60.37	89.76	1.29	1.60	–

d : groove depth
P_y : yield load
d_y : midspan deflection at yielding
P_u : ultimate failure load
d_u : midspan deflection at failure

Table 3 – Comparison of the increase rate of P_u according to amount of reinforcements

d (mm)	Dimension of CFRP laminate	A_f (mm²)	d_f (mm)	ρ_f (×10⁻³)	ρ_f/ρ_s	$P_y/P_{y(control)}$	$P_u/P_{u(control)}$
5	1.2 × 5	6	297.5	0.101	0.0245	1.047	1.051
10	1.2 × 10	12	295.0	0.203	0.0494	1.104	1.237
15	1.2 × 15	18	292.5	0.308	0.0748	1.156	1.358
17.5	1.2 × 17.5	21	291.3	0.361	0.0876	1.173	1.393
20	1.2 × 20	24	290.0	0.414	0.1006	1.185	1.434
25	1.2 × 25	30	287.5	0.522	0.1268	1.217	1.501
27.5	1.2 × 27.5	33	286.3	0.576	0.1401	1.245	1.541
30	1.2 × 30	36	285.0	0.632	0.1535	1.265	1.507
35	1.2 × 35	42	282.5	0.743	0.1806	1.293	1.597

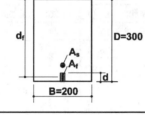

d : groove depth
A_f (=1.2d) : section area of CFRP laminate
ρ_f (=A_f/Bd_f) : ratio of CFRP laminate
ρ_s (=A_s/BD) : ratio of reinforcing steel bar
P_y : yield load
P_u : ultimate load

Table 4 – Analytic results according to groove depth and spacing of reinforcements

d (mm)	A_f (mm^2)	b_g (mm)	b_e (mm)	b_g/B	P_y (kN)	P_u (kN)	$P_y/P_{y(control)}$	$P_u/P_{u(control)}$
0 (CONTROL)	0	0	0	0	46.69	56.19	-	-
25	30	20	90	0.1	72.25	97.89	1.55	1.74
		40	80	0.2	73.24	98.06	1.57	1.75
		50	75	0.25	73.91	103.31	1.58	1.84
		60	70	0.3	72.61	108.43	1.56	1.93
		80	60	0.4	72.93	111.19	1.56	1.98
		100	50	0.5	71.62	106.51	1.53	1.90
		120	40	0.6	72.64	107.89	1.56	1.92
		140	30	0.7	71.27	103.77	1.53	1.85
		160	20	0.8	71.63	105.05	1.53	1.87
		180	10	0.9	72.61	92.44	1.56	1.65
15	18	20	90	0.1	59.12	88.59	1.27	1.58
		40	80	0.2	59.02	86.17	1.26	1.53
		60	70	0.3	59.43	87.51	1.27	1.56
		80	60	0.4	58.14	98.41	1.25	1.75
		100	50	0.5	58.14	89.06	1.25	1.58
		120	40	0.6	57.12	95.25	1.22	1.70
		140	30	0.7	56.8	89.79	1.22	1.60
		160	20	0.8	57.47	90.01	1.23	1.60
		180	10	0.9	57.47	85.56	1.23	1.52

d : groove depth
A_f (=1.2d) : section area of CFRP laminate
b_g : spacing of CFRP laminates
b_e : distance to concrete edge
P_y : yield load
P_u : ultimate load

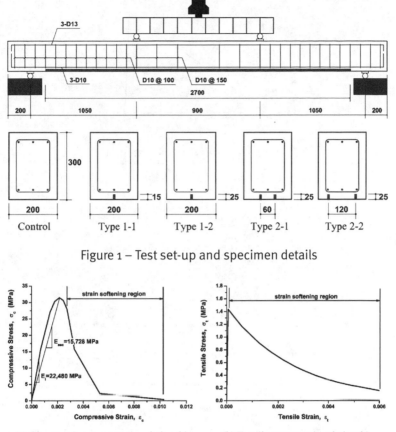

Figure 1 – Test set-up and specimen details

(a) Compressive stress-strain relationship   (b) Tensile stress-strain relationship

Figure 2 – Property of concrete

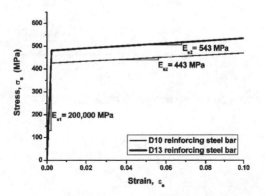

Figure 3 – Property of reinforcing bar

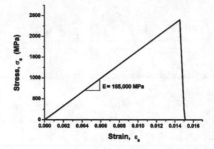

Figure 4 – Property of CFRP laminate

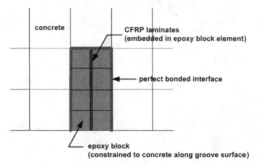

Figure 5 – Boundary conditions at interface

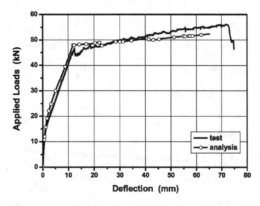

Figure 6 – Comparison of analytic and experimental results for CONTROL

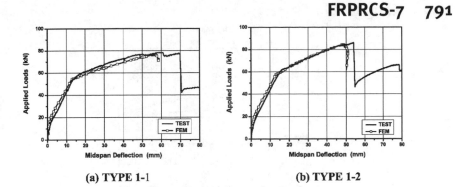

(a) TYPE 1-1   **(b) TYPE 1-2**

Figure 7 – Comparison of analytical and experimental results according to groove depth

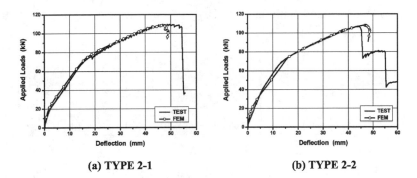

(a) TYPE 2-1   **(b) TYPE 2-2**

Figure 8 – Comparison of analytical and experimental results according to spacing

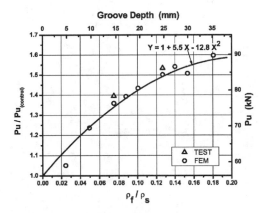

Figure 9 – Increase rate of ultimate load vs. reinforcement area ratio

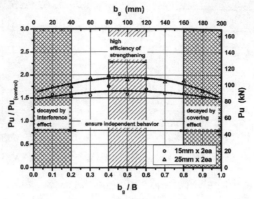

Figure 10 – Comparison of ultimate load by groove depth relative to spacing ratio

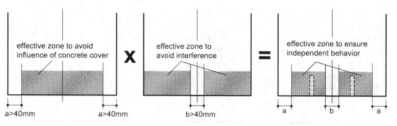

Figure 11 – Disposition to ensure independent behavior of the CFRP reinforcements

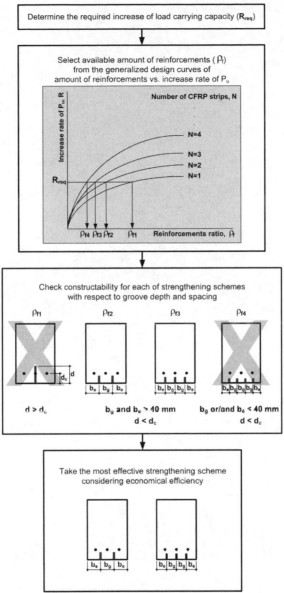

Figure 12 – Conceptual design flow of NSM reinforcement

Experimental Investigation on Flexural Behavior of RC Beams Strengthened by NSM CFRP Reinforcements

by W.-T. Jung, Y.-H. Park, J.-S. Park, J.-Y. Kang, and Y.-J. You

Synopsis: This study presents the results of experiments performed on RC (Reinforced Concrete) beams strengthened with NSM(Near Surface Mounted) reinforcement. A total of 8 specimens have been tested. The specimens can be classified into EBR(Externally Bonded Reinforcement) specimens and NSM(Near Surface Mounted) reinforcements specimens. Two of NSM specimens were strengthened with 12 mechanical interlocking grooves with a width of 20 mm and spacing of 200 mm in order to prevent debonding failure of the CFRP(Carbon Fiber Reinforced Polymer) reinforcement. Experimental results revealed that NSM specimens used CFRP reinforcements more efficiently than the EBR specimens, but debonding failure between adhesive and concrete occurred. This showed that, similarly to EBR specimens, NSM specimens also required countermeasures against debonding failure. Failure mode of NSM specimens added with mechanical interlocking grooves failed by rupture of CFRP rod and strip. The measured ultimate load showed an increase of 15% compared with the common NSM specimens. The application of mechanical interlocking grooves made it possible to avoid debonding failure and to enhance strengthening performance.

Keywords: carbon fiber reinforced polymer; externally bonded CFRP reinforcements; near surface mounted CFRP reinforcements; strengthening

Woo-Tai Jung received his MS degree in Civil Engineering from Myongji University. He is a Researcher at the Structure Research Department of the Korea Institute of Construction Technology. His research interests are in the area of strengthening with FRP reinforcements of deteriorated concrete structures.

Young-Hwan Park received his MS degree and Ph.D. in Civil Engineering from Seoul National University. He is a Research Fellow at the Structure Research Department of the Korea Institute of Construction Technology.

Jong-Sup Park received his MS degree in Civil Engineering from Myongji University. He is a Senior Researcher at the Structure Research Department of the Korea Institute of Construction Technology.

Jae-Yoon Kang received his MS degree in Civil Engineering from Dongguk University. He is a Senior Researcher at the Structure Research Department of the Korea Institute of Construction Technology.

Young-Jun You received his MS degree in Civil Engineering from Yonsei University. He is a Researcher at the Structure Research Department of the Korea Institute of Construction Technology.

INTRODUCTION

Among the various strengthening techniques that have been developed and applied to strengthen deteriorated RC structures, a number of applications using FRP reinforcements have significantly increased recently. FRP reinforcements are bonded to concrete surfaces by adhesives but frequently experience debonding failure at the interface between FRP reinforcements and concrete. Most research, to date, has focused on investigating the strengthening effects and failure modes of EBR system.

The problem of premature failure of EBR system may be solved by increasing the interface between FRP and concrete. Using this principle, the NSM system has been introduced recently. The NSM system for concrete structure using steel reinforcement already began in 1940s. However, the corrosion of the steel reinforcement and the poor bonding performance of the grouting material largely impaired its application. The development of improved epoxy and the adoption of FRP reinforcement offered the opportunity to implement NSM system (Hassan and Rizkalla 2003; Täljsten and Carolin 2001). Because of their light weight, ease of installation, minimal labor costs and site constraints, high strength-to-weight ratios, and durability, FRP repair systems can provide an economically viable alternative to traditional repair systems and materials (Mirmiran et al. 2004). Rizkalla and Hassan (2002) have compared EBR and NSM system in terms of cost, including costs of materials and labor, and strengthening effect. They concluded that the NSM system was more cost-effective than the EBR system using CFRP strips.

One of failure modes in the NSM system is the debonding failure as the EBR system. Two different types of debonding failures can occur for NSM system. One is due to spiltting of the adhesive cover as a result of high tensile stresses at the FRP-adhesive interface, and is termed "adhesive split failure". The other is due to cracking of the concrete surrounding the adhesive, and is termed "concrete split failure"(Hassan and Rizkalla 2004). Concrete split failure is the governing debonding failure mode for NSM system because the tensile strength of the adhesive is generally greater than that of the concrete. If concrete spilt failure is delayed or prevented, the performance of the NSM system will be greatly enhanced. Epoxy keys or deformations are proposed in this study to provide the mechanical interlocking between the adhesive and the concrete.

This experimental study investigates the applicability and strengthening performances of NSM using CFRP rods and strips. For comparison, flexural tests on RC beams strengthened by EBR and by NSM have been performed. In addition, a mechanical interlocking technique has been studied to determine its effectiveness in preventing the debonding failure.

EXPERIMENTAL PROGRAM

Manufacture of specimens

A total of 8 specimens of simply supported RC beams with span of 3m have been cast. The details and cross-section of the specimens are illustrated in Fig. 1. A concrete with compressive strength of 31.3 MPa at 28 days has been used. Steel reinforcements D10(ϕ9.53mm) of SD40 have been arranged with steel ratio of 0.0041 and a layer of three D13(ϕ12.7mm) has been arranged as compression reinforcements. Shear reinforcements of D10 have been located every 10 cm in the shear zone to avoid shear failure. Table 1 summarizes the material properties used for the test beams.

Experimental parameters

Table 2 lists the experimental parameters. The control specimen, an unstrengthened specimen, has been cast to compare the strengthening performances of the various systems. SH-BOND and CPL-50-BOND, EBR specimens, have been strengthened with CFRP sheet and CFRP strip, respectively. The remaining 5 specimens were strengthened with NSM CFRP rods and CFRP strips. Among the specimens strengthened by NSM, ROD-MI-20 and PL-MI-20 included the addition of 12 mechanical interlocking grooves with a width of 20 mm and spacing of 200 mm, as shown in Fig. 2, to prevent debonding failure of the FRP reinforcement. The strengthened length of all the specimens has been fixed to 2,700 mm.

Installation of the FRP reinforcements

Fig. 3 shows the details of the strengthened cross-sections. The strengthening process of EBR specimens (SH-BOND, CPL-50-BOND) was proceeded by coating the beam with a primer after surface treatment, followed by the bonding of the CFRP sheet or strip. The strengthened beams were cured at ambient temperature for 7 days for the curing of epoxy adhesive. The process for NSM strengthening progressed by cutting the grooves at the bottom of the beams using a grinder, cleaning the debris, and embedding the CFRP rod or strip after application of the adhesive. The strengthened beams were cured for 3 days so that the epoxy adhesive achieves its design strength.

<u>**Loading and measurement methods**</u>

All specimens were subjected to 4-point bending tests to failure by means of UTM (Universal Testing Machine) with capacity of 980 kN. The loading was applied under displacement control at a speed of 0.02 mm/sec until the first 15 mm and 0.05 mm/sec from 15 mm until failure. The measurement of all test data was recorded by a static data logger and a computer at intervals of 1 second. Electrical resistance strain gauges were fixed at mid-span and L/4 to measure the strain of steel reinforcements. Strain gauges to measure the strain of concrete were located at the top, 5 cm and 10 cm away from the top on one side at mid-span. Strain gauges were also placed on the FRP reinforcement located at the bottom of the mid-span and loaded points to measure the strain according to the loading process.

EXPERIMENTAL RESULTS

<u>**Failure modes**</u>

Before cracking, all the strengthened specimens exhibited bending behavior similar to the unstrengthened specimen. This shows that the FRP reinforcement is unable to contribute to the increase of the stiffness and strength in the elastic domain. However, after cracking, the bending stiffness and strength of the strengthened specimens were seen to increase significantly until failure compared to the unstrengthened specimens.

Examining the final failure, the unstrengthened control specimen presented typical bending failure mode which proceeds by the yielding of steel reinforcement followed by compression failure of concrete. The failure of SH-BOND and CPL-50-BOND, EBR specimens, began with the separation of CFRP reinforcement and concrete at mid-span to exhibit finally brittle debonding failure (Figs. 4 and 5). Failure of CRD-NSM and NSM-PL-25, pertaining to NSM specimens, occurred through the simultaneous separation of the CFRP reinforcement and epoxy from concrete (Figs. 6 and 7). Failure of the remaining NSM specimens (NSM-PL-15, ROD-MI-20, PL-MI-20) occurred with the rupture of the FRP reinforcement(Figs. 8 and 9). Table 3 summarizes the failure modes.

<u>**Strengthening performances of EBR and NSM**</u>

Figs. 10 and 11 plot the load-deflection curves of EBR and NSM specimens, respectively. The specimens with EBR, SH-BOND and CPL-50-BOND, presented ultimate load increased by 30 to 47% compared to the unstrengthened specimen, while NSM specimens (CRD-NSM, NSM-BAR, NSM-PL-25) increased the ultimate load by 39 to 65%.

Observation of Fig. 10 reveals that even if CPL-50-BOND with relatively large cross-sectional area of CFRP reinforcement developed larger initial stiffness, premature debonding failure occurred because its bonding area is much smaller than SH-BOND. EBR specimens behaved similarly to the unstrengthened control specimen after debonding failure.

In Fig. 11, the stiffness of NSM specimens before yielding of steel reinforcement was smaller than the stiffness developed by EBR specimens. The ultimate load and yield load are seen to increase with the cross-sectional area of NSM reinforcement.

Examining the ultimate strain of FRP summarized in Table 3, the maximum strain for EBR specimens appears to attain 30 to 50% of the ultimate strain, and 84 to 100% for NSM specimens. This proves that the NSM system is utilizing FRP reinforcement efficiently.

Mechanical interlocking

Experimental results showed that NSM specimens are exploiting the FRP reinforcement more efficiently than EBR specimens. However, they still showed debonding failure (Figs. 6 and 7). Therefore, so-called mechanical interlocking grooves (epoxy grooves) were supplemented in order to maximize the utilization of the CFRP reinforcements and prevent debonding failure (ROD-MI-20, PL-MI-20).

Figs. 12 and 13 compare the load-deflection curves developed by NSM specimens with added mechanical interlocking grooves and specimens strengthened with common NSM. As seen in Table 3, the tests performed on ROD-MI-20 and PL-MI-20 ended following the rupture of NSM CFRP rod and strip. The measured ultimate load showed an increase of 15% compared with the common NSM specimens CRD-NSM and NSM-PL-25. Consequently, it has been seen that prevention of debonding failure and improvement of strengthening performance can be achieved by adding mechanical interlocking effects to the NSM system.

CONCLUSIONS

Performance tests have been carried out on RC beams strengthened with NSM systems. The following conclusions were derived from the experimental results.

It has been seen that NSM specimens utilized the CFRP reinforcement more efficiently than the externally bonded strengthening specimens.

According to the static loading test results, the strengthening performances were improved in NSM specimens compared with EBR specimens. However, the specimens CRD-NSM and NSM-PL-25 failed by the separation of the CFRP reinforcements and epoxy adhesive from the concrete. Consequently, it is necessary to take some countermeasures to prevent debonding failure for NSM specimens.

The use of mechanical interlocking grooves made it possible to prevent debonding failure of the CFRP reinforcements and to increase the strengthening performance. Further research will investigate the influence of the width and spacing of such mechanical interlocking grooves on the flexural behavior of strengthened RC beams so as to derive their optimal dimensions.

800 Jung et al.

REFERENCES

Hassan, T. and Rizkalla, S. (2003), "Investigation of Bond in Concrete Structures Strengthened with Near Surface Mounted Carbon Fiber Reinforced Polymer Strips", *Journal of Composites for Construction*, Vol 7, No. 3, pp. 248-257

Hassan, T., and Rizkalla, S. (2004), "Bond Mechanism of Near-Surface-Mounted Fiber-Reinforced Polymer Bars for Flexural Strengthening of Concrete Structures", *ACI Structural Journal*, Vol. 101, No. 6, pp. 830-839

Mirmiran, A., Shahawy, M., Nanni, A., and Karbhari, V. (2004), "Bonded Repair and Retrofit of Concrete Structures Using FRP Composites", *Recommended Construction Specifications and Process Control Manual, NCHRP Report 514*, Transportation Research Board

Rizkalla, S., and Hassan, T. (2002), "Effectiveness of FRP for Strengthening Concrete Bridges", *Structural Engineering International*, Vol. 12, No. 2, pp. 89-95

Täljsten, B. and Carolin, A. (2001), "Concrete Beams Strengthened with Near Surface Mounted CFRP Laminates", *Proceeding of the fifth international conference on fibre-reinforced plastics for reinforced concrete structures* (FRPRCS-5), Cambridge, UK, 16-18 July 2001, pp. 107-116

Table 1 — Summary of material properties

Material	Property	
Concrete[1]	Compressive strength(MPa)	31.3
Tension steel reinforcement (D10) [1]	Yield strength (MPa)	426
	Tensile strength (MPa)	562
	Diameter(mm)	9.53
	Area(cm^2)	0.7133
Compression steel reinforcement(D13) [1]	Yield strength (MPa)	481
	Tensile strength (MPa)	608
	Diameter(mm)	12.7
	Area(cm^2)	1.267
CFRP strip[1] (Smooth surface)	Thickness (mm)	1.4
	Tensile strength (MPa)	2452.59
	Elastic modulus (GPa)	165.49
	Ultimate strain (%)	1.48
CFRP sheet[2] (Smooth surface)	Design thickness (mm)	0.11
	Tensile strength (MPa)	3479
	Elastic modulus (GPa)	230.3
CFRP rod [2] (Deformed surface)	Diameter (mm)	9
	Tensile strength (MPa)	1878
	Elastic modulus (GPa)	121.42
	Ultimate strain (%)	1.55

1) from tests carried out by the authors
2) from the supplier

Table 2 — Experimental parameters

Specimen	CFRP area (mm²)	CFRP type	ρ_{CFRP}(%)	Strengthening method
CONTROL	–	–	–	unstrengthened
SH-BOND	26.4	Sheet	0.0489	EBR[1]
CPL-50-BOND	70	Strip	0.1296	EBR[1]
CRD-NSM	63.6	Rod	0.1178	NSM[2]
NSM-PL-25	35	Strip	0.0648	NSM[2]
NSM-PL-15	21	Strip	0.0389	NSM[2]
ROD-MI-20	63.6	Rod	0.1178	NSM[2]+MI[3]
PL-MI-20	35	Strip	0.0648	NSM[2]+MI[3]

1) EBR: Externally Bonded Reinforcement
2) NSM : Near Surfaced Mounted Reinforcement
3) MI : Mechanical Interlocking grooves; 12ea, width : 20mm, space : 200mm

Table 3 — Summary of experimental results

Specimen	P_y (kN)	d_y (mm)	P_u (kN)	d_u (mm)	Increase in P_u(%)	Failure mode	ε_{uFRP}
CONTROL	46.69	12.78	56.19	71.68	-	(a)	-
SH-BOND	60.82	13.34	82.38	34.98	47	(c)	8473
CPL-50-BOND	61.04	10.52	73.24	16.00	30	(c)	4449
CRD-NSM	62.58	15.36	92.63	43.88	65	(b)	13071
NSM-PL-25	61.99	16.06	86.18	53.98	53	(b)	12350
NSM-PL-15	57.47	15.50	78.49	58.94	40	(d)	15417
ROD-MI-20	65.10	15.10	106.20	53.84	89	(d)	15710
PL-MI-20	61.83	15.68	98.72	59.8	76	(d)	15614

(a) : Steel yielding followed by crushing of concrete
(b) : Debonding of the NSM FRP reinforcement and the epoxy
(c) : Debonding of the externally bonded FRP reinforcement
(d) : Rupture of the NSM FRP reinforcement

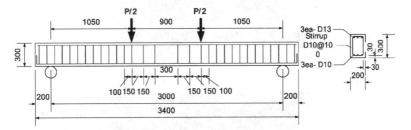

Figure 1 —Details and cross section of the specimen (mm)

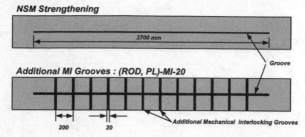

Figure 2 —Bottom schemes of RC beams strengthened with NSM reinforcement (mm)

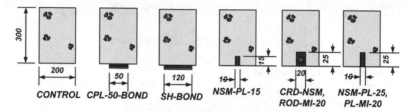

Figure 3 — FRP strengthening schemes (mm)- Cross section

Figure 4 — Failure mode (SH-BOND)

Figure 5 —Failure mode (CPL-50-BOND)

Figure 6 — Failure mode (CRD-NSM)

Figure 7 — Failure mode (NSM-PL-25)

Figure 8 — Failure mode (PL-MI-20)

Figure 9 — Failure mode (ROD-MI-20)

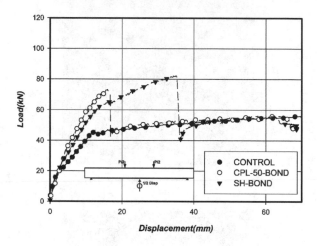

Figure 10 — Load-deflection curve: beam with EBR specimens

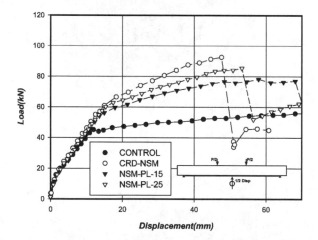

Figure 11 — Load-deflection curve: beam with NSM specimens

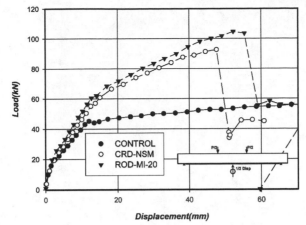

Figure 12 — Load-deflection curve: mechanical interlocking effect (CFRP rod)

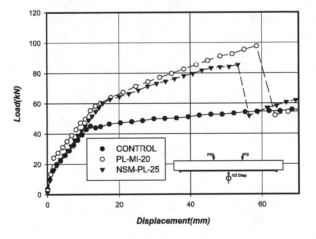

Figure 13 — Load-deflection curve: mechanical interlocking effect (CFRP strip)

Shear Strengthening of RC Beams with Near-Surface-Mounted CFRP Laminates

by S. Dias and J. Barros

Synopsis: The efficacies of the Near Surface Mounted (NSM) and Externally Bonded Reinforcing (EBR) techniques for the shear strengthening of rectangular cross section RC beams are compared. Both techniques are based on the use of carbon fiber reinforced polymer (CFRP) materials. The NSM was the most effective technique, and was also the easiest and fastest to apply, and assured the lowest fragile failure modes. The performance of the *ACI* and *fib* analytical formulations for the EBR shear strengthening was appraised. In general, the contribution of the CFRP systems predicted by the analytical formulations was larger than the values registered experimentally. The capability of the De Lorenzis formulation of predicting the contribution of the NSM technique for the shear strengthening of RC beams was appraised using bond stress and CFRP effective strain values obtained in pullout bending tests. This formulation provided values 61% lower than the values obtained experimentally.

Keywords: carbon fiber reinforced polymers; externally bonded reinforcing; near surface mounted; shear strengthening

808 Dias and Barros

Salvador Dias is a PhD Student at the Department of Civil Engineering of Minho University, Portugal. In 2001 he concluded his MSc from the Faculty of Engineering of Porto University, Portugal. His research interests include the application of fiber-reinforced polymer materials for the structural strengthening and rehabilitation.

ACI member **Joaquim Barros** is Professor of the Structural Division of Minho University, Portugal. He received his BS, MSc and PhD from Faculty of Civil Engineering of Porto University, Portugal. He is a member of ACI Committees 440 (Fiber reinforced polymers), 506 (Shotcrete) and 544 (Fiber reinforced concrete). His research interests include structural strengthening, composite materials, fiber reinforced concrete and finite element method.

INTRODUCTION

The use of fiber reinforced polymer (FRP) materials for structural repair and strengthening has continuously increased in the last years, due to several advantages resulting from opting for these composites in detriment of traditional construction materials such as steel, wood and concrete. These benefits include low weight, easy installation, high durability (non corrosive) and tensile strength, electromagnetic neutrality and practically unlimited availability in size, geometry and dimension [1, 2, 3].

Externally bonded reinforcing (EBR) technique using FRP laminates and wet lay-up sheets has been used to increase the shear resistance of RC beams [4]. The analysis of research studies confirmed that the shear resistance of RC beams can be significantly increased from applying the EBR technique. The carried out research has, however, revealed that this technique cannot mobilize the full tensile strength of FRP materials, due to their premature debonding. Furthermore, EBR reinforcements could be highly susceptible to damage from collision, fire and temperature variation, ultraviolet rays, and moisture absorption [5]. In an attempt at overcoming these drawbacks, a strengthening technique designated by near surface mounted (NSM) was proposed, where FRP rods are fixed into pre-cut grooves opened on the concrete cover of the elements to be strengthened [6]. Barros and Dias [7] proposed a similar strengthening technique based on installing CFRP laminate strips into pre-cut slits opened on the concrete cover. The CFRP was bonded to concrete by epoxy adhesive. This strengthening technique has already been used to increase the load carrying capacity of concrete structures failing in bending [8, 9, 10]. The obtained results showed that this technique is more efficient and easy to apply than EBR technique. This higher effectiveness is derived from the larger CFRP laminate-concrete bond stress values that can be mobilized in the NSM technique [11].

To assess the efficacy of the NSM technique for increasing the shear resistance of RC beams, an experimental program of four-point bending tests was carried out. Influences of the longitudinal tensile steel reinforcement, ρ_{sl}, laminate strip inclination and beam depth on the efficacy of the NSM technique were analyzed. This efficacy was assessed not only in terms of the increase of maximum load and deflection at beam rupture, but also in terms of the beam strengthening performance per unit length of the applied material. The performance of the analytical formulations proposed by ACI[1], fib[2] and De Lorenzis[6] for the shear strengthening was appraised.

RESEARCH SIGNIFICANCE

The efficacies of the NSM and EBR techniques for the shear strengthening of rectangular cross section RC beams are compared in terms of the increase of maximum load, the deflection at beam rupture and the failure mode. For the NSM technique, the beam strengthening performance per unit length of the applied material is assessed. The performance of the ACI and fib analytical formulations for the shear strengthening with externally bonded wet lay-up FRP systems is appraised. The capability of the De Lorenzis formulation of predicting the contribution of the NSM technique for the shear strengthening of RC beams is checked.

EXPERIMENTAL PROGRAM

The experimental program is composed of the four test series represented in Fig. 1. Each series is made up of a beam without any shear reinforcement (R) and a beam for each of the following shear reinforcing systems: steel stirrups of ϕ6 mm (S), U shaped strips of wet lay-up CFRP sheet (M) and CFRP laminate strips at 45° (IL) or at 90° (VL) in relation to the beam axis. The M beams were strengthened by EBR technique, while in IL and VL beams laminate strips of CFRP were installed into pre-cut slits opened on the concrete cover of the beam's lateral surfaces (NSM technique), see Fig. 2. Series A10 and A12 are composed of beams with a cross section of 0.15x0.30 m^2 and a span length of 1.5 m. Series B10 and B12 are constituted of beams with a cross section of 0.15x0.15 m^2 and a span length of 0.9 m. To evaluate the influence of ρ_{sl}, series A10 and B10 had 4ϕ10 steel bars at bottom surface, while A12 and B12 series had 4ϕ12. The shear span, a, (Fig. 1) in both series of beams was two times the depth of the corresponding beams. At top surface, the beams of all series were reinforced with 2ϕ6 steel bars. The concrete clear cover for the top, bottom and lateral faces of the beams was 15 mm. Table 1 includes general information of the beams composing the four series. Further information can be found elsewhere[7].

The amount of shear reinforcement applied on the four reinforcing systems was evaluated in order to assure that all beams would fail in shear, at a similar load carrying capacity. The percentage of the CFRP shear reinforcing systems was evaluated to provide a contribution for the beam shear resistance similar to the one of the steel stirrups. For the strips of wet lay-up CFRP sheets of U shape, the recommendations of the ACI Committee 440 were followed[1]. For the NSM CFRP laminate strips, the formulation used for the steel stirrups was adopted, but the yield stress was replaced by an effective stress that was determined assuming a CFRP strain value of 4‰, that is the maximum effective strain value recommended by ACI Committee 440 for the EBR shear reinforcing systems. Steel stirrups were not applied in the series reinforced with CFRP systems. The authors are aware that this scenario would probably never be encountered in practical situations since a certain percentage of steel stirrups always exist in reinforced concrete elements, even if it is inadequate. However, since the main purpose of the present research is to assess the effectiveness of NSM shear strengthening technique, the interaction between the CFRP shear reinforcement and the steel stirrups will be only investigated in future experimental programs.

Tables 2 and 3 include the main properties of the concrete and steel bars used in the experimental program, respectively. The average values of the concrete compression strength at 28 days and at the date of testing the beams were evaluated from uniaxial compression tests with cylinders of 150 mm diameter and 300 mm height. The properties of the steel bars were obtained from uniaxial tensile tests. Two CFRP systems were used on the present work: unidirectional wet lay-up sheets of 25 mm width and precured laminates of 1.4×10 mm^2 cross-section. These CFRP systems have the properties indicated in Table 4.

The relationship between the force and the deflection at mid span of the tested beams is represented in Figure 3. Table 5 includes the main results obtained in the four tested beam series. Adopting the designation of F_{max,K_R} and F_{max,K_S} for referring the maximum load of a beam without shear reinforcement and a beam reinforced with steel stirrups, respectively, (K represents the beam series) the ratios $F_{max}/F_{max,K\_R}$ and $F_{max}/F_{max,K\_S}$ were determined for assessing the efficacy of the shear strengthening techniques, in terms of increasing the beam load carrying capacity.

From the results obtained in the experimental program, the following main conclusions can be pointed out:
- The CFRP shear strengthening systems applied in the present work increased significantly the shear resistance of concrete beams;
- The NSM shear strengthening technique was the most effective of the CFRP systems. This effectiveness was not only in terms of the beam load carrying capacity, but also in terms of the deformation capacity at beam's failure. Using the load carrying capacity of the unreinforced beams for comparison purposes, the beams strengthened by EBR and NSM techniques showed an average increase of 54% and 83%, respectively;
- Increasing the beam depth, laminates at 45° became more effective than vertical laminates;
- F_{max} of the beams reinforced with steel stirrups and F_{max} of the beams strengthened by NSM technique were almost similar;
- Failure modes of the beams strengthened by the NSM technique were not so fragile as the ones observed in the beams strengthened by the EBR technique.

APPRAISAL THE PERFORMANCE OF ANALYTICAL FORMULATIONS

Taking the results obtained in the tested beams strengthened with EBR technique, the performance of the analytical formulations proposed by ACI[1] and fib[2] was appraised. The documents published by these institutions are not yet dealing with the NSM technique. Thereby, the applicability of the analytical formulation proposed by De Lorenzis[6] was checked, using for this purpose the experimental results obtained in the beams strengthened with NSM laminate strips. Since De Lorenzis's formulation was developed for FRP reinforcing rod elements, the necessary adjustments were introduced to take into account that FRP elements are now laminate strips. New estimates for the parameters of this model are proposed in order to take into account the bond stress and the CFRP effective strain values recorded in the pullout bending tests [11] and to obtain an appropriate safety factor for the CFRP contribution towards shear resistance of concrete beams.

ACI recommendations for EBR technique

According to ACI[1], the design value of the contribution of the FRP shear reinforcement is given by,

$$V_{fd} = \phi \, \psi_f \, \frac{A_{fv} \, f_{fe} \, d_f}{s_f} \tag{1}$$

where ϕ is the strength-reduction factor required by ACI[12] that, for shear strengthening of concrete elements, has a value of 0.85, ψ_f is an additional reduction factor of 0.85 for the case of three-sided U-wraps (see Fig. 4), s_f is the spacing of the wet lay-up strips of FRP sheets, A_{fv} is the area of FRP shear reinforcement within spacing s_f,

$$A_{fv} = 2 n t_f w_f \tag{2}$$

with n, t_f and w_f being the number of layers per strip, the thickness of a layer and the width of the strips. The effective stress in the FRP, f_{fe}, is obtained multiplying the elasticity modulus of the FRP, E_f, by the effective strain,

$$\varepsilon_{fe} = k_v \varepsilon_{fu} \leq 0.004 \quad \text{(for U-wraps)} \tag{3}$$

where k_v is a bond-reduction coefficient that is a function of the concrete strength, the type of wrapping scheme used, and the stiffness of the FRP,

$$k_v = \frac{k_1 k_2 L_e}{11900 \varepsilon_{fu}} \leq 0.75 \tag{4}$$

with,

$$L_e = \frac{23300}{\left(n t_f E_f \right)^{0.58}} \tag{5}$$

$$k_1 = \left(\frac{f_c'}{27} \right)^{2/3} \tag{6}$$

$$k_2 = \frac{d_f - L_e}{d_f} \quad \text{(for U-wraps)} \tag{7}$$

In (1) and (7) d_f is the depth of FRP shear reinforcement (see Fig. 4), and f_c' is the characteristic value of the concrete compression strength [12]. The length and the force unities of the variables in (4) to (7) are millimeter and Newton, respectively.

812   Dias and Barros

In Table 6, the values obtained with this formulation are compared to those registered experimentally. Apart beam B10_M, the *ACI* formulation has estimated a FRP contribution for the shear strengthening that was larger than the contribution recorded experimentally. A deficient bond of the strip crossed by the shear failure crack might have caused the high abnormal value of $V_{fd}^{ana.}/V_f^{exp.}$ of A10_M beam, since this strip has debonded prematurely (see Fig. 5).

Fib recommendations for EBR technique

According to *fib* recommendations [2], the contribution of wet lay-up strips of FRP sheets for shear strengthening is evaluated by the following expression,

$$V_{fd} = 0.9\,\varepsilon_{fe,d}\,E_f\,\rho_f\,b_w\,d \qquad (8)$$

where b_w and d are the width of the beam cross section and the distance from extreme compression fiber to the centroid of the nonprestressed steel tension reinforcement. In (8) ρ_f is the FRP shear reinforcement ratio,

$$\rho_f = \frac{A_{fv}}{b_w s_f} \qquad (9)$$

and $\varepsilon_{fe,d}$ is the design effective strain in the FRP, that can be obtained from ε_{fe},

$$\varepsilon_{fe} = \min\left[0.65\left(\frac{f_{cm}^{2/3}}{E_f\rho_f}\right)^{0.56}\times 10^{-3};\ 0.17\left(\frac{f_{cm}^{2/3}}{E_f\rho_f}\right)^{0.30}\varepsilon_{fu}\right] \quad (f_{cm}\ \text{in MPa and}\ E_f\ \text{in} \qquad (10)$$

GPa)

applying two safety factors, $\varepsilon_{fe,d} = 0.8\varepsilon_{fe}/1.3$, the first one, 0.8, to convert ε_{fe} in a characteristic value and the second one, 1.3, that depends on the FRP failure mode (debonding in the present case). In (10) f_{cm} is the cylinder average concrete compression strength and ε_{fu} is the ultimate FRP strain. The analytical and the experimental results are compared in Table 7. Apart beam B12_M, *fib* formulation has also predicted an FRP contribution larger than the experimentally registered values. Like in the *ACI* formulation, an abnormal high $V_{fd}^{ana.}/V_f^{exp.}$ value was also obtained in A10_M beam, which stresses the suspicious that the strip crossing the shear failure crack was deficiently bonded.

Fig. 6 compares the values of the CFRP contribution for the shear strengthening according to *ACI* and *fib* formulations. In general, all the formulations have estimated large values than the ones registered experimentally. Apart B12_M beam, in the remaining beams the *ACI* formulation estimated lower values than *fib*. The differences between the values from *ACI* and *fib* are, however, not too significant.

De Lorenzis analytical formulation for NSM technique

According to De Lorenzis [6], the contribution of the NSM FRP elements for shear strengthening is the minimum value of V_{1f} and V_{2f},

$$V_f = \min\,(V_{1f},\ V_{2f}) \qquad (11)$$

where V_{1f} is the term associated with the FRP-concrete bond strength, while V_{2f} derives from a strain limit of ε_{fe} imposed on the FRP. The De Lorenzis formulation was developed for NSM FRP rod systems. To adjust this formulation for the case of laminate strips, the diameter of the rod cross section was conveniently replaced by the dimensions of the laminate cross section, a_l and b_l, resulting in the following expression for the V_{1f} term,

$$V_{1f} = 4 \cdot (a_l + b_l) \cdot \tau_b \cdot L_{tot\ min} \tag{12}$$

In this expression τ_b represents the average bond stress of the FRP elements intercepted by the shear failure crack and, for vertical laminates, $L_{tot\ min}$ is obtained from,

$$L_{tot\ min} = d_{net} - s_f \qquad \text{if} \qquad \frac{d_{net}}{3} \le s_f < d_{net}$$

$$L_{tot\ min} = 2d_{net} - 4s_f \qquad \text{if} \qquad \frac{d_{net}}{4} < s_f < \frac{d_{net}}{3} \tag{13}$$

while for laminates at 45°,

$$L_{tot\ min} = \left(2d_{net} - s_f\right)\frac{\sqrt{2}}{2} \qquad \text{if} \qquad \frac{2d_{net}}{3} \le s_f < 2d_{net}$$

$$L_{tot\ min} = \left(d_{net} - s_f\right)2\sqrt{2} \qquad \text{if} \qquad \frac{d_{net}}{2} < s_f < \frac{2d_{net}}{3} \tag{14}$$

where s_f is the FRP spacing and d_{net} is a reduced value for the effective length of the laminate (see Fig. 7),

$$d_{net} = d_r - 2c \tag{15}$$

with d_r being the actual length of the laminate and c the concrete clear cover.
The term V_{2f} is evaluated from,

$$V_{2f} = 4 \cdot (a_l + b_l) \cdot \tau_b \cdot \overline{L_i} \qquad \text{if} \qquad \frac{d_{net}}{2} \le s_f < d_{net}$$

$$V_{2f} = 4 \cdot (a_l + b_l) \cdot \tau_b \cdot \overline{L_i} \cdot \frac{3d_{net} - 4s_f}{d_{net}} \qquad \text{if} \qquad \frac{d_{net}}{4} < s_f < \frac{d_{net}}{2} \tag{16}$$

for the vertical laminates, and

$$V_{2f} = 4 \cdot (a_l + b_l) \cdot \tau_b \cdot \overline{L_i} \qquad \text{if} \qquad d_{net} \le s_f < 2d_{net}$$

$$V_{2f} = 4 \cdot (a_l + b_l) \cdot \tau_b \cdot \overline{L_i} \cdot \frac{3d_{net} - 2s_f}{d_{net}} \qquad \text{if} \qquad \frac{d_{net}}{2} < s_f < d_{net} \tag{17}$$

for the laminates at 45°, where

$$\overline{L_i} = \frac{\varepsilon_{fe}}{2} \cdot \frac{a_l \cdot b_l}{a_l + b_l} \cdot \frac{E_f}{\tau_b} \tag{18}$$

According to De Lorenzis [6], if

$$d_{net} < 2\overline{L}_i \tag{19}$$

in the case of vertical laminates, or if

$$d_{net} < \sqrt{2}\,\overline{L}_i \tag{20}$$

in the case of laminates at 45°, it is not necessary to calculate V_{2f}, since V_{1f} is the conditioning term, giving the lowest value. The design shear contribution of FRP to the RC beam shear capacity is evaluated from,

$$V_{fd} = 0.7 \times V_f \tag{21}$$

The average bond stress, τ_b, was obtained from the results registered in pullout-bending tests [11]. From the obtained peak pullout forces a τ_b of 16.1 MPa was determined, which is much larger than the value recommended by De Lorenzis for the NSM FRP rod strengthening system ($\tau_b = 6.9$ MPa). The CFRP average strain (ε_{fe}) in the bond length at peak pullout force was 5.9‰, which is larger than the value recommended by De Lorenzis for the NSM FRP rod strengthening system ($\varepsilon_{fe} = 4.0$‰).

Assuming that τ_b, ε_{fe} and E_f are equal to 16.1 MPa, 5.9‰ and 166 GPa, respectively, the analytical results indicated in Table 8 were obtained. This table does not include the data of the B10_VL beam since, according to the De Lorenzis formulation, the FRP contribution is null in beams with s_f larger than d_{net}. If the experimental results ($V_f^{exp.}$) are compared to the analytical ones ($V_{fd}^{ana.}$), an average $V_f^{exp.}/V_{fd}^{ana.}$ ratio of about 1.65 was obtained. Since a safety factor of 1.79 ($V_c^{exp.}/V_{cd}^{ana.} = 1.79$) was obtained in the beams without any shear reinforcement, and a safety factor of 1.24 ($V_{sw}^{exp.}/V_{swd}^{ana.} = 1.24$) was determined for the contribution of the steel stirrups for the shear resistance, the safety factor of 1.65 seems to be an appropriate value for the contribution of the NSM CFRP systems.

PROFITABILITY OF THE NSM TECHNIQUE

To assess the influence of the CFRP laminate orientation, not only in terms of increasing the beam load carrying capacity (F_{max}), but also in terms of the amount of consumed CFRP, the ratio $\Delta F/l_{CFRP}$ of the beams strengthened by the NSM technique was evaluated (designated by profitability index), where ΔF is the increase in the F_{max} and l_{CFRP} is the total length of the laminates applied in the beam. The values included in Table 9 show that (see also Fig. 8), independent of the beam height and the longitudinal steel reinforcement ratio (ρ_{sl}), the profitability index was larger in the beams with laminates at 45°. For both the A series, the profitability index increased with the increase of ρ_{sl}. This tendency was not observed in both B series since the reduced bonded lengths of the CFRP laminates in these shallow beams limited the increase on the ΔF.

CONCLUSIONS

The main purpose of the present research is to assess the effectiveness of the near surface mounted (NSM) technique for the shear strengthening of RC beams. In comparison to the performance of the experimentally bonded reinforcing (EBR) technique, the NSM was the more effective technique, and was also easier and faster to apply, and assured lower fragile failure modes.

Using the *ACI* and *fib* formulations for the evaluation of the contribution of the CFRP EBR strengthening systems for the beam shear resistance, it was verified that these formulations have given design values 2% and 8% higher than the values registered experimentally, respectively, (a beam with a deficient bonding was not considered in this analysis). Using similar EBR shear strengthening configuration, other researchers have obtained larger safety factors. However, these researchers have used wet lay-up CFRP sheets of Young's modulus (E_f) of about 220 GPa, and, in the major cases, the shear CFRP strips were formed of one layer. In the present research a CFRP sheet of E_f=390 GPa and strips of two layers were used. This indicates that the expressions of ACI and *fib* formulations defining the FRP effective strain were not well calibrated for this situation, since they are providing too high effective strain values when using stiffer shear CFRP systems. Therefore, more research is needed in this field.

Assuming a bond stress of 16.1 MPa and an effective strain of 5.9‰ (average values of the data recorded in pullout bending tests), the De Lorenzis formulation predicted a CFRP contribution around 61% of the experimentally registered values, which seems to provide an appropriate safety factor (1.65).

ACKNOWLEDGMENTS

The authors of the present work wish to acknowledge the materials provided by the degussa® Portugal, S&P® and Unibetão (Braga). The study reported in this paper forms a part of the research program "CUTINSHEAR - Performance assessment of an innovative structural FRP strengthening technique using an integrated system based on optical fiber sensors" supported by FCT, POCTI/ECM/59033/2004.

REFERENCES

1. ACI Committee 440, 2002, *"Guide for the design and construction of externally bonded FRP systems for strengthening concrete structures"*, American Concrete Institute, 118 pp.

2. CEB-FIP Model Code, 1993, Comite Euro-International du Beton, Bulletin d'Information nº 213/214.

3. Bakis, C.E., Bank, L.C., Brown, V.L., Cosenza, E., Davalos, J.F., Lesko, J.J., Machida, A., Riskalla, S.H. and Triantafillou, T.C., 2002, *"Fiber-Reinforced Polymer Composites for Construction – State-of-the-art Review"*, Journal of Composites for Construction, Vol. 6, Nº2, May, pp. 73-87.

4.   Bousselham A. and Chaallal, O., 2004, *"Shear Strengthening Reinforced Concrete Beams with Fiber-Reinforced Polymer: Assessment of Influencing Parameters and Required Research"*, ACI Structural Journal, Vol. 101, N° 2, March-April, pp. 219-227.

5.   ACI Committee 440, 1996, *"State of the art report on fiber reinforced plastic reinforcement for concrete structures"* (Reapproved 2002), ACI Committee 440, 68 pp.

6.   De Lorenzis, Laura, 2002, *"Strengthening of RC Structures with Near-Surface Mounted FRP rods"*, PhD Thesis in Civil Engineering, Universita' Degli Studi di Lecce, Italy, May, 289 pp.

7.   Barros, J.A.O. and Dias, S.J.E., 2003, *"Shear strengthening of reinforced concrete beams with laminate strips of CFRP"*, Proceedings of the International Conference Composites in Constructions - CCC2003, Italia, September, pp. 289-294.

8.   Blaschko, M. and Zilch, K., 1999, *"Rehabilitation of concrete structures with CFRP strips glued into slits"*, Proceedings of the Twelfth International Conference of Composite Materials, ICCM 12, Paris, France (CD-ROM).

9.   El-Hacha, R. and Riskalla S.H., 2004, *"Near-Surface-Mounted Fiber-Reinforced Polymer Reinforcements for Flexural Strengthening of Concrete Structures"*, ACI Structural Journal, Vol. 101, N°5, September-October, pp. 717-726.

10.   Barros, J.A.O., Sena-Cruz, J.M., Dias, S.J.E., Ferreira, D.R.S.M. and Fortes, A. S., 2004, *"Near surface mounted CFRP-based technique for the strengthening of concrete structures"*, Workshop on R+D+I in Technology of Concrete Structures - tribute to Dr. Ravindra Gettu, Barcelona, Spain, October, pp. 205-217 (CD-ROM).

11.   Sena-Cruz, J.M. and Barros, J.A.O., 2004, *"Bond between near-surface mounted CFRP laminate strips and concrete in structural strengthening"*, Journal of Composites for Construction, Vol. 8, N° 6, pp. 519-527.

12.   ACI Committee 318, 2002, *"Building code requirements for structural concrete and commentary"*, American Concrete Institute, Reported by ACI Committee 118.

Table 1 — Shear strengthening arrangements of the tested series

<table>
<tr><th colspan="2" rowspan="2">Beam's
designation</th><th colspan="4">Shear strengthening systems</th></tr>
<tr><th>Material</th><th>Quantity</th><th>Spacing
(mm)</th><th>Angle
(°)</th></tr>
<tr><td rowspan="10">A series</td><td rowspan="5">A10</td></tr>
</table>

Beam's designation		Shear strengthening systems			
		Material	Quantity	Spacing (mm)	Angle (°)
A series / A10	A10_R	-	-	-	-
	A10_S	Steel stirrups	6φ6 of two branches	300	90
	A10_M	Strips of S&P C-Sheet 530	8×2 layers of 25 mm (U shape)	190	90
	A10_VL	S&P laminate strips of CFK 150/2000	16 CFRP laminates	200	90
	A10_IL	S&P laminate strips of CFK 150/2000	12 CFRP laminates	300	45
A series / A12	A12_R	-	-	-	-
	A12_S	Steel stirrups	10φ6 of two branches	150	90
	A12_M	Strips of S&P C-Sheet 530	14×2 layers of 25 mm (U shape)	95	90
	A12_VL	S&P laminate strips of CFK 150/2000	28 CFRP laminates	100	90
	A12_IL	S&P laminate strips of CFK 150/2000	24 CFRP laminates	150	45
B series / B10	B10_R	-	-	-	-
	B10_S	Steel stirrups	6φ6 of two branches	150	90
	B10_M	Strips of S&P C-Sheet 530	10×2 layers of 25 mm (U shape)	80	90
	B10_VL	S&P laminate strips of CFK 150/2000	16 CFRP laminates	100	90
	B10_IL	S&P laminate strips of CFK 150/2000	12 CFRP laminates	150	45
B series / B12	B12_R	-	-	-	-
	B12_S	Steel stirrups	10φ6 of two branches	75	90
	B12_M	Strips of S&P C-Sheet 530	16×2 layers of 25 mm (U shape)	40	90
	B12_VL	S&P laminate strips of CFK 150/2000	28 CFRP laminates	50	90
	B12_IL	S&P laminate strips of CFK 150/2000	24 CFRP laminates	75	45

Table 2 — Concrete properties

Beam's series	f_{cm} (MPa)	
	28 days	At beam testing
A	37.6	49.2 (227 days)
B	49.5	56.2 (105 days)

Table 3 — Properties of the conventional steel bars

Beam's series	φ6 (longitudinal)	φ6 (stirrups)	φ10	φ12
A	$f_{sym} = 622$ MPa	$f_{sym} = 540$ MPa	$f_{sym} = 464$ MPa	$f_{sym} = 574$ MPa
	$f_{sum} = 702$ MPa	$f_{sum} = 694$ MPa	$f_{sum} = 581$ MPa	$f_{sum} = 672$ MPa
B	$f_{sym} = 618$ MPa	$f_{sym} = 540$ MPa	$f_{sym} = 464$ MPa	$f_{sym} = 571$ MPa
	$f_{sum} = 691$ MPa	$f_{sum} = 694$ MPa	$f_{sum} = 581$ MPa	$f_{sum} = 673$ MPa

Table 4 — Properties of the CFRP systems

CFRP system		Main properties			
Type	Materials	Tensile strength (MPa)	Young's modulus (GPa)	Ultimate strain (‰)	Thickness (mm)
Wet lay-up sheet	Primer	12	0.7	30	-
	Epoxy	54	3	25	-
	Sheet (S&P C-Sheet 530)	3000	390	8	0.167
Precured laminate	Adhesive	-	7	-	-
	Laminate (S&P laminate CFK 150/2000)	2200	150	14	1.4
		2286^1	166^1	13^1	1.4^1

^{1} Evaluated from experimental tests carried out in the present research program

Table 5 — Main results of the four tested beam series

Beam's A series (4ϕ10)	F_{max}^* (kN)	$\dfrac{F_{max}}{F_{max,A10\_R}}$	$\dfrac{F_{max}}{F_{max,A10\_S}}$	Beam's A series (4ϕ12)	F_{max}^* (kN)	$\dfrac{F_{max}}{F_{max,A12\_R}}$	$\dfrac{F_{max}}{F_{max,A12\_S}}$
A10_R	100.40	1.00	0.59	A12_R	116.50	1.00	0.54
A10_S	169.35	1.69	1.00	A12_S	215.04	1.85	1.00
A10_M	122.06	1.22	0.72	A12_M	179.54	1.54	0.83
A10_VL	158.64	1.58	0.94	A12_VL	235.11	2.02	1.09
A10_IL	157.90	1.57	0.93	A12_IL	262.38	2.25	1.22

Beam's B series (4ϕ10)	F_{max}^* (kN)	$\dfrac{F_{max}}{F_{max,B10\_R}}$	$\dfrac{F_{max}}{F_{max,B10\_S}}$	Beam's B series (4ϕ12)	F_{max}^* (kN)	$\dfrac{F_{max}}{F_{max,B12\_R}}$	$\dfrac{F_{max}}{F_{max,B12\_S}}$
B10_R	74.02	1.00	0.61	B12_R	75.7	1.00	0.48
B10_S	120.64	1.63	1.00	B12_S	159.1	2.10	1.00
B10_M	111.14	1.50	0.92	B12_M	143.0	1.89	0.90
B10_VL	131.22	1.77	1.09	B12_VL	139.2	1.84	0.87
B10_IL	120.44	1.63	1.00	B12_IL	148.5	1.96	0.93

$^* F_{max} = 2P_{max}$ (see Fig. 1)

Table 6 — Analytical *vs* experimental results (*ACI* analytical formulation)

Beam's designation	Experimental $V_f^{exp.}$ (kN)	Analytical $V_{fd}^{ana.*}$ (kN)	$V_{fd}^{ana.} / V_f^{exp.}$
A10_M	10.8	17.0	1.57
A12_M	31.5	33.8	1.07
B10_M	18.6	17.7	0.95
B12_M	33.7	35.0	1.04

$^* f_c'$ values were obtained at the age of the beam tests ($f_c' = 40.2$ MPa for A series and $f_c' = 46.5$ MPa for B series).

Table 7 — Analytical *vs* experimental results (*fib* analytical formulation)

Beam's designation	Experimental $V_f^{exp.}$ (kN)	Analytical $V_{fd}^{ana.}$ * (kN)	$V_{fd}^{ana.} / V_f^{exp.}$
A10_M	10.8	24.0	2.22
A12_M	31.5	38.9	1.23
B10_M	18.6	20.5	1.10
B12_M	33.7	30.9	0.92

* f_{cm} values were obtained at the age of the tested beams.

Table 8 — Analytical *vs* experimental results

(De Lorenzis analytical formulation with $\varepsilon_f = 5.9‰$, $\tau_b = 16.1$ MPa and $E_f = 166$ GPa)

Beam's designation	Series	Experimental $V_f^{exp.}$ (kN)	Analytical $V_{fd}^{ana.}$ (kN)	$V_f^{exp.} / V_{fd}^{ana.}$
A10_VL	A (4φ10)	29.1	19.2	1.52
A10_IL	A (4φ10)	28.8	19.2	1.50
A12_VL	A (4φ12)	59.3	26.4	2.25
A12_IL	A (4φ12)	72.9	34.2	2.13
B10_IL	B (4φ10)	23.2	19.2	1.21
B12_VL	B (4φ12)	31.8	19.2	1.66
B12_IL	B (4φ12)	36.4	27.6	1.32

Table 9 — Profitability of the NSM technique

Series		Beam's designation	F_{max} (kN)	ΔF (kN)	l_{CFRP} (m)	$\Delta F / l_{CFRP}$ (kN/m)
A (h = 0.30m)	(4φ10)	A10_R	100.4	-	-	-
		A10_VL	158.64	58.24	4.8	12.13
		A10_IL	157.9	57.5	3.68	15.63
	(4φ12)	A12_R	116.5	-	-	-
		A12_VL	235.11	118.61	8.4	14.12
		A12_IL	262.38	145.88	7.35	19.85
B (h = 0.15m)	(4φ10)	B10_R	74.02	-	-	-
		B10_VL	131.22	57.2	2.4	23.83
		B10_IL	120.44	46.42	1.97	23.56
	(4φ12)	B12_R	75.7	-	-	-
		B12_VL	139.2	63.5	4.2	15.12
		B12_IL	148.5	72.8	3.91	18.62

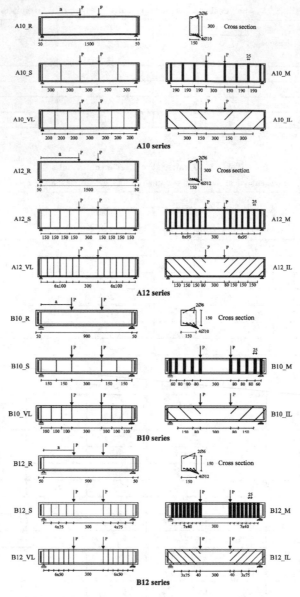

Figure 1 — Tested series

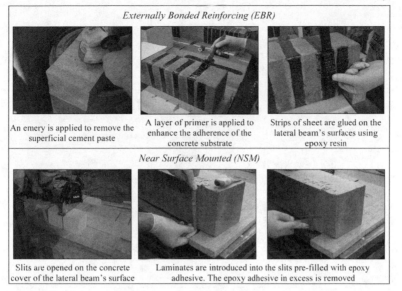

Figure 2 — Techniques for the shear strengthening of reinforced concrete beams

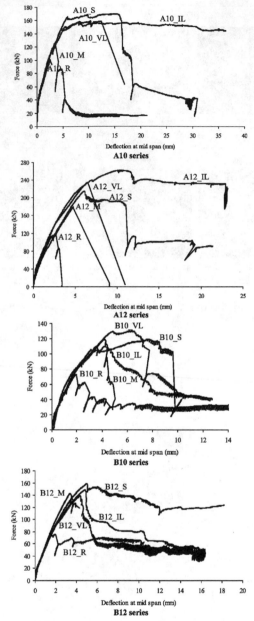

Figure 3 — Force-deflection relationship of the four tested beams series

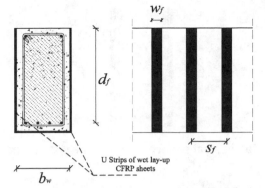

Figure 4 — Data for the externally bonded shear strengthening technique

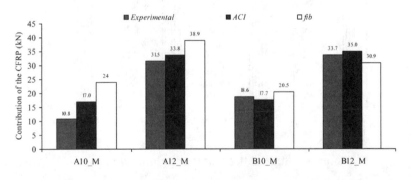

Figure 5 — Failure of A10_M beam

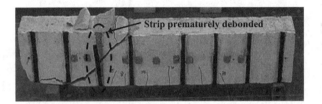

Figure 6 — Analytical *vs* experimental results (*ACI* and *fib* analytical formulation)

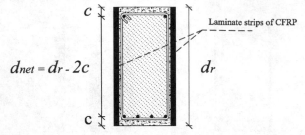

$$d_{net} = d_r - 2c$$

Figure 7 — Data for the near surface mounted shear strengthening technique

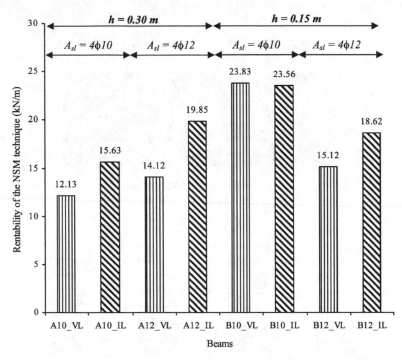

Figure 8 — Representation of the profitability index for the NSM technique

Shear Assessment and Strengthening of Contiguous-Beam Concrete Bridges Using FRP Bars

by P. Valerio, T.J. Ibell and A.P. Darby

Synopsis: Many concrete bridges related to railways in the U.K. consist of prestressed rectangular concrete beams, post-tensioned together transversely to aid lateral distribution of load; this bridge type has been repeatedly flagged as having insufficient shear capacity. Sixteen tests on small-scale beams, which are scaled-down replica models of the actual bridge beams, are presented. The specimens are tested under a four-point loading system and are both prestressed (PRC) with and without stirrups and non-prestressed (RC) with and without stirrups, to provide full understanding of their shear behaviour. Four further tests are then presented on RC beams strengthened in shear with FRP bars inserted from the soffit into pre-drilled holes and fixed in place using epoxy resin; this method allows strengthening in cases where the webs are inaccessible. Comparisons are made with current code predictions for the strength of all specimens. The results show that unstrengthened RC beams behave mostly as expected and as predicted by codes, while for PRC beams a great variation in shear-carrying capacity following shear cracking is observed for different span-to-depth loading ratios. The proposed FRP strengthening scheme is effective and provides significant improvement to the shear-carrying load capacity.

Keywords: bridges; fiber-reinforced polymers; prestressed concrete; reinforced concrete; shear; strengthening

826   Valerio et al.

Pierfrancesco Valerio is a Research Officer in the Department of Architecture and Civil Engineering at the University of Bath, UK, where he started a PhD in Structural Engineering in October 2003. He received his MEng in Italy in 2001 and worked as a consultant in an Italian bridge design company from 2001 to 2003. He is a Professional Engineer in Italy since 2002. His interests and fields of expertise include the design of steel and concrete structures and bridge engineering.

Tim Ibell is an ACI Member. He obtained his PhD from the University of Cambridge, UK, in 1992. Tim is now Chair of the Department of Architecture and Civil Engineering at the University of Bath, UK. His research interests include the internal prestressing and strengthening of concrete structures using fiber-reinforced polymer materials.

Antony Darby is a Lecturer in Structural Engineering at the Department of Architecture and Civil Engineering, University of Bath, UK. He obtained his PhD from Cambridge University and spent two years at Oxford University gaining post-doctorial research experience. He has also worked in Industry for a number of years as a Design Engineer. His research interests are in FRP strengthening and structural dynamics.

INTRODUCTION

Due to increased traffic, higher allowable truck weights and deterioration of materials, most bridges in the U.K. have been assessed recently for both bending and shear capacity. A steady-state assessment program of these bridges is now moving into place. During this process, various bridge owners have identified two areas of concern which need to be addressed: the first is the realistic shear strength assessment of prestressed concrete bridge beams when made contiguous within a deck, and the second is the necessity for quick, cheap and practical shear strengthening measures for such bridges when the webs of the beams are inaccessible.

As plasticity-based yield line methods for flexural strength assessment of concrete bridges become ever more popular, so the demand on shear strength in these bridges increases, leading to many shear assessment failures. The priority is therefore to ensure that a realistic shear assessment tool is in place, which is able to predict the real strength of such bridges adequately. Thereafter, if it turns out that the bridge has inadequate shear strength, a viable shear strengthening scheme is sought, which ensures minimal disruption to the bridge users. By combining these two aspects in one project, a powerful management tool for owners of concrete bridges with shear concerns will be produced.

The shear assessment tool will be achieved through large- and small-scale tests combined analytically with a plasticity-based shear assessment method for beams, slabs and beam-and-slab bridges developed previously[1], and through laboratory testing of the same specimens, but in the strengthened situation.

There are several approaches to retrofitting existing concrete bridges in shear[2], often involving the use of FRP laminates fixed to the webs of the bridge beams or FRP bars mounted near the surface of the webs[3]. However, in the particular case where the webs of the beams are inaccessible, an altogether new approach must be adopted. Vertical holes are drilled into the bridge beams from the soffit level. FRP bars are then inserted and embedded in place using high-viscosity epoxy resin injected into the drilled holes. In this way, shear enhancement is shown to be possible.

In this paper, the tests on unstrengthened small-scale beams (which are replica scaled-down models of actual bridge beams) are presented alongside four further tests on RC beams strengthened in shear with FRP bars using the technique described above. Comparisons are made against current code predictions for the strength of all specimens in both the unstrengthened and strengthened situations.

The next step in this research project will be the completion of tests on complete small-scale bridge specimens as well as large-scale beams, along with the validation of the proposed FRP-strengthening technique for PRC beams, small-scale bridge specimens and large-scale beams.

EXPERIMENTAL PROGRAM

Design of the test specimens

The existing bridge taken as a reference for the construction of the test specimens for assessing the unstrengthened situation is a simply-supported railway underbridge consisting of ten pre-tensioned rectangular concrete beams transversely post-tensioned together to form a deck (see Figure 1). Each beam, whose minimum concrete compressive cube strength at 28 days is 60MPa, is 900mm wide and 762mm deep and is pre-tensioned longitudinally with a total of 38 tendons at mid-span (21 of which are debonded towards the support): each tendon is stressed to an initial force of 159kN (see Figure 2). The percentage of transverse reinforcement equals 0.35% at support, 0.175% in the shear zone and 0.098% in the middle zone of the beam.

Sixteen small-scale specimens were designed in such a way as to match, as much as possible, the average values in the shear zone of the real beams in terms of geometric and sectional characteristics (effective depth to total depth ratio, percentage of longitudinal and transverse reinforcement) and in terms of tension state due to the prestressing forces and permanent loads. The beams were 110mm wide, 190mm deep and 3000mm long, with four 7mm wires (of which only the upper two were pre-tensioned to 45kN each, in the case of the PRC beams) as longitudinal reinforcement and 3mm mild steel bars as vertical stirrups, whose spacing was 175mm in the central zone, 100mm in the shear zone and 50mm at the support (see Figures 2 and 3).

Four RC specimens were then designed in order to verify the feasibility of the proposed shear strengthening scheme. The FRP reinforcement used was 10mm diameter aramid ARAPREE[4] bars with a tensile strength of 1.5GPa, Young's modulus of 60GPa, ultimate strain of 2.4% and density of 12.5kN/m^3. These beams were 110mm wide,

220mm deep and 3000mm long, with no stirrups and two 12mm high yield steel bars as longitudinal reinforcement (see Figure 4).

Test program

All beams were tested under a four-point loading system with a hinge and roller respectively at each end, to provide constant shear within the shear spans. For the sixteen unstrengthened small-scale specimens the load position was varied along the length of the beam, at 3, 4, 5 and 6 times the effective depth d of the beam, to ensure that a range of shear span lengths was examined. Furthermore, to understand the numerous parameters involved in the shear-carrying capacity of concrete beams (especially if prestressed), they were tested in all their four different configurations, i.e. prestressed with stirrups (PRCst), prestressed without stirrups (PRC), non-prestressed with stirrups (RCst) and non-prestressed without stirrups (RC). Figure 5 shows the test layout of the sixteen unstrengthened small-scale beams (USB).

All four beams to be FRP-strengthened were non-prestressed without stirrups (RC) and were tested at 3 times their effective depth. The first specimen contained no FRP bars, while the second, third and fourth specimens contained, respectively, three, two and one ARAPREE bars in each shear span, as shown in Figure 6 (where SB stands for strengthened beams).

Casting and testing procedure

The beams were cast in a steel formwork, which forms the prestressing rig for the PRC beams (see Figure 7). The prestressing force load was applied by a 200kN hydraulic jack bearing against the end plate and each jack was tensioned to 45kN. The characteristic compressive cube strength required for the concrete was 60MPa. For the beams due to be strengthened, two weeks after casting the required holes were drilled in the shear spans and the FRP bars were inserted and epoxy-resined into place.

The loading arrangement for the tests is shown in Figure 8. The loads were applied on the beams via two steel bearings loaded by 100kN hydraulic jacks, whose reactions were provided by a rigid steel beam, itself supported over two rigid transversal steel frames. Prior to testing, six displacement transducers were set on each specimen to obtain the load-displacement plots. All beams were loaded monotonically until failure, with crack patterns marked throughout the testing procedure. The readings from the transducers were taken every second with a measurement group System 500 data logger.

TEST RESULTS AND DISCUSSION

Samples of the 7mm prestressing wires, the 3mm mild steel bars used for the stirrups and of the 12mm high yield steel bars for the FRP-strengthened specimens were tested in tension and the corresponding stress-strain plots were obtained. Average values of 1610MPa, 700MPa and 635MPa were found for the tensile strength of the wires and the yield stress of the 3mm and 12mm bars respectively.

Tests on the unstrengthened small-scale beams

Table 1 summarizes the results of the sixteen tests on the unstrengthened small-scale beams; f_{cu} is the average of the characteristic value of the compressive cube strength of the four cubes tested for each beam. Figures 9 to 28 show pictures at failure and the load-displacement plots for each beam.

The four RC beams, which all failed in shear, behaved similarly in terms of peak load and maximum deflection, and the failure load in all cases was just slightly higher (4 to 10%) than the first shear cracking load, demonstrating that for a non-prestressed beams without stirrups the first shear cracking initiates beam collapse. A reduction in the ultimate failure load of about 15% was noted for the beams loaded at $5d$ and $6d$ with respect to the ones loaded at $3d$ and $4d$, showing a tendency for the ultimate shear failure load to decrease with the shear span length. The four RC beams with stirrups, which again all failed in shear, behaved similarly in terms of peak load and maximum deflection, and their failure load was 6.7 to 8.7kN higher than the failure load of the equivalent beams without stirrups, showing an average increase of 7.6kN in the ultimate carrying capacity due to the presence of the stirrups.

The four PRC beams without stirrups all failed in shear, but this time with notable differences in terms of peak load, maximum displacement and stiffness. A great enhancement in stiffness, shear carrying capacity and maximum displacement was observed as the shear span length reduced; the peak load ranged from 28.5kN (at $6d$) to 54.1kN (at $3d$), (a remarkable 90% increase), while at $4d$ and $5d$ the peak load was 37.5 and 31.3kN respectively, values in between the two extremes. When compared with the equivalent non-prestressed beams, the enhancement in shear capacity due to the prestressing effect reduced greatly with the shear span length, ranging from 230% at $3d$ to 31% at $6d$. In these four prestressed beams, the shear failure load was substantially higher than the first shear cracking load, proving the existence of an arching action of the prestressing force which helps the shear carrying mechanism. Furthermore, the first shear cracks for the beam loaded at $6d$ appeared at a load of 21kN, just 1kN more than the non-prestressed case, showing that for such high shear slenderness the enhanced shear capacity of prestressed beams can only be related to the previously mentioned arch action.

The four PRC beams with stirrups all failed in flexure, apart from the one loaded at $3d$, which failed in shear at 54.5kN, a value practically coincident with the equivalent PRC beam without stirrups. A strong reduction in stiffness with increasing shear span length was noted for these four beams. The stirrups in the beams loaded at 4, 5 and 6 times d were able to change the failure mode from brittle shear to ductile flexure and, therefore, were very effective. The lack of enhancement of capacity in the beam loaded at $3d$ was probably due to the fact that, for such high loads, amongst the three shear carrying mechanisms that interact in PRC beams with stirrups (namely the concrete, the stirrups and the arch action contributions), the arch action is dominant. Once this action expires the load level is so high that the stirrups cannot be effective anymore. In the beam loaded at $6d$ the flexural failure was not as ductile as the other two: signs of shear failure were observed in the test specimen, again confirming the tendency for the ultimate shear failure load to decrease with the shear span length.

Tests on the FRP-strengthened beams

Table 2 summarizes the results of the four tests on the FRP-strengthened beams, while Figures 29 to 33 show their the load-displacement plots and pictures at failure.

The plain control specimens (SB 1) failed in brittle shear at 23kN. The beam strengthened with 3 bars (SB 2) failed in flexure at 41.5kN, while the ones with two and one FRP bar (SB 3 and SB 4) failed in shear at 32kN and 30kN respectively. The strengthening was clearly effective. With respect to the plain specimen, the three FRP bars contained in beam SB 2 altered the failure mechanism leading from a brittle shear to a ductile flexural failure, increasing considerably the ultimate load capacity and ductility of the beam. Again, the increase in ultimate capacity was remarkable even in specimens SB 3 and SB 4, which failed in shear but with a 40% and 30% increase respectively when compared to beam SB 1.

Looking at the shape of the shear cracks that led to failure in SB 1 and SB 3, it is clear that the slope of the shear discontinuity in beam SB 3 was steeper than that in SB 1: the presence of the bars forced the shear discontinuity to be steeper and form within the constricted space between the vertical bars. This permitted greater internal energy dissipation in the failure mechanism, leading to a higher ultimate load capacity.

ANALYSIS OF TEST RESULTS

Flexural analysis

Following the symbols of EC2[5], in the uncracked phase the theoretical first cracking load for all beams is given by:

$$P_{cr} = \frac{M_{cr}}{a} = \frac{I}{a\,y_{bot}}\left(f_{ctm} + \sigma_{cp,bot}\right) \tag{1}$$

where $f_{ctm} = 0.3 f_{ck}^{2/3}$ is the mean value of the axial tensile strength of concrete, f_{ck} is the characteristic compressive cylinder strength of concrete, $\sigma_{cp,bot}$ is the stress due to prestressing force at the bottom fibre of the section after all the prestressing losses, I is the transformed second moment of area of the cross-section, y_{bot} is the distance from the centroid to the bottom fibre and a is the shear span length.

At the ultimate limit state, assuming all partial safety factors equal to unity, the theoretical ultimate load is given by:

$$P_{ult} = \frac{M_{ult}}{a} = \frac{\left(C_u\,0.6x\right) + \sum T_{s,i}\left(d_i - x\right)}{a} \tag{2}$$

where C_u is the ultimate compressive force in concrete, x is the depth to the neutral axis, d_i is the effective depth of each steel layer and T_i is the ultimate force in each steel layer.

Table 3 shows the correlation between the flexural analysis and test results; it is evident that both the theoretical first flexural cracking load and the ultimate load (when applicable, i.e. for the specimens which failed in flexure) produce very accurate predictions, with a maximum discrepancy of 3% for the ultimate load and 10% for the cracking load.

Shear predictions from codes of practice

Comparisons between the ultimate load observed and code predictions are now made for all beams. Codes-of-practice EC2[5], BS 8110-1[6], BD 44/95[7] and DIN 1045-1[8] are used for comparison purposes.

In determining the ultimate shear capacity of the beams strengthened with FRP bars, the term V_f (the shear strength contribution from the FRP) is added to the concrete contribution V_c. It is assumed that the vertically-embedded FRP bars will strain to 0.004 at the ultimate shear capacity of the beams[3]. With a Young's modulus of the FRP bars of 60GPa, the limit stress in the FRP bars at shear collapse is 240MPa. Thus, the term for V_f, based on a 45° truss analogy, is:

$$V_f = \frac{240\,A_f\,z}{s_v} \tag{3}$$

where A_f is the cross-sectional area of each bar, z is the effective lever arm of the truss and s_v is the spacing between vertical bars.

Although the codes-of-practice adopted here limit the spacing between stirrups to various fractions of the effective depth, d, the calculations conducted here ignore this limitation. Naturally, the design of an adequate shear-strengthening scheme would require closely-spaced vertical bars in reality. The value of z is taken to be the fully-anchored length of cach embedded bar, which is the overall length of each bar minus the anchorage length at each end. It is assumed that the average bond strength between epoxy-resined bar and concrete (where $\tau_{av}=0.5\,\tau_{max}$ with a triangular stress distribution) is of the order of 12MPa[9]. This translates to an anchorage length, l_b, of 50mm for the 10mm diameter bars, so $z=h-2l_b=120$mm.

Table 4 shows details of comparison between the various code predictions and the actual results. For the unstrengthened specimens, all codes give good correlation for the RC beams; for RC beams with stirrups the predictions of BS 8810-1 and BD 44/95 are good, while EC2 and DIN 1045-1, which consider a variable angle truss model, are close to actual only if considering the minimum allowed inclination θ of the compression struts (for DIN 1045-1 the suggested value of $\theta=40°$ is considered in the Table, resulting in very conservative predictions). For PRC beams, EC2 and DIN 1045-1 give good predictions for long shear spans, but are too conservative for short shear spans. BS 8110-1 and BD 44/95 are very conservative for PRC beams. For PRC beams with stirrups all predictions are extremely conservative and poor, because no code is able to catch the flexural failure. Only DIN 1045-1 predictions are reasonable for long shear spans.

All codes-of-practice predict the shear capacity of the FRP-strengthened RC beams reasonably accurately, especially DIN 1045-1, whose predictions are always within a 5% error margin from the actual failure load. However, for specimens SB 3 and SB 3, in particular, there is an overestimation of the effectiveness of the vertical bars. This is almost certainly due to their wide spacing, which is close to one effective depth for specimen SB 3 and substantially more than one effective depth for specimen SB 4. Therefore, clearly it is essential that in order for this strengthening scheme to be efficiently used in reality, the vertically-embedded bars should be spaced sufficiently closely in order for shear predictions to be valid. It seems sensible that this minimum spacing should be in the region of 0.5 to 0.75 times the effective depth (as in SB 2), just as recommended by present codes of practice. However, in specimens SB 3 and SB 4, although the bars were spaced too widely to be fully effective, they enhanced shear capacity by altering the shear discontinuity geometry.

It is then observed that, if considering the average value between all 20 tests, EC2 is the code with the better coefficient of variation, while the poor performance of DIN 1045-1 is caused mostly by the excessively conservative predictions for the RC beams with stirrups.

CONCLUSIONS

With regard to the unstrengthened small-scale beams, it has been shown that the RC beams behaved mostly as expected and as predicted by codes of practice on condition that a minimum value for the inclination of the compression struts is chosen for the beams containing stirrups. For PRC beams, a great variation in shear-carrying capacity after shear cracking is observed when loading with different span-to-depth ratios. In particular the shear capacity decreased greatly with increasing shear span lengths, pointing to the fact that in PRC beams, along with a great shear enhancement near the support, the critical zone for shear failure is located at 5 to 6 times the effective depth. This fact should be considered when providing shear strengthening for PRC beams. For PRC beams, the codes of practice proved to be excessively conservative, especially for the PRC beams containing stirrups.

The proposed shear-strengthening approach has been shown to be feasible and successful for RC beams. The spacing between embedded bars should be close enough so that the shear discontinuity cannot form between bars. It is suggested that existing requirements for maximum spacing of vertical reinforcement, which vary between 0.5 and 0.75 times the effective depth, should be adequate for such strengthening. It can therefore be concluded that it is possible to design a shear-strengthening scheme using embedded FRP bars by assuming the reinforcement contribution according to Eq. (3). Validation of the technique for the case of PRC beams and large-scale beams will be the next task of this research project.

ACKNOWLEDGEMENTS

The authors gratefully acknowledge the help of the laboratory staff and the financial support of the Engineering and Physical Sciences Research Council (EPSRC) and Network Rail.

REFERENCES

1. Ibell, T. J., Morley, C. T. and Middleton, C. R., "A plasticity approach to the assessment of shear in concrete beam-and-slab bridges". The Structural Engineer, Vol. 75, No. 19, pp. 331-338, 1997.

2. Concrete Society TR 55, "Design guidance for strengthening concrete structures using fibre composite materials". Second Edition. The Concrete Society, Camberley, UK, 2004.

3. De Lorenzis, L. and Nanni, A., "Shear strengthening of reinforced concrete beams with near-surface mounted fiber-reinforced polymers rods". ACI Structural Journal, Vol. 98, No. 1, pp. 60-68, 2001.

4. Sireg SpA, "ARAPREE-CARBOPREE bars". Sireg Geotechnical Division Catalogue, Arcore, Italy, 2001.

5. Eurocode 2: EN 1992-1-1 Final draft, "Design of concrete structures. Part 1-1: General rules and rules for buildings". European Committee for Standardization, 2003.

6. BS 8110-1, "Structural use of concrete. Part 1: Code of practice for design and construction". British Standard Institution, London, 1997.

7. BD 44/95, "The assessment of concrete highway bridges and structures". Department of Transport, London, 1995.

8. DIN 1045-1, "Reinforced an prestressed concrete structures – Part 1: Design". Deutsches Institut für Normung (DIN), Berlin, 2001.

9. De Lorenzis, L. and Nanni, A., "Characterization of FRP rods as near surface mounted reinforcement". ASCE Journal of Composites for Construction, Vol. 5, No. 2, pp. 114-121, 2001.

Table 1 – Tests on the unstrengthened small-scale beams

Specimen	Test Layout	f_{cu} (MPa)	P at flex. cracking (kN)	P at shear cracking (kN)	Peak failure load P (kN)	Max span deflection (mm)
USB 1.1	RC@3d	62	6	22.5	23.5 (shear)	25.9
USB 1.2	RCst@3d	54	6	22.5	31.5 (shear)	32.3
USB 1.3	PRC@3d	61	18	34.5	54.1 (shear)	66.5
USB 1.4	PRCst@3d	59	18	34.5	54.5 (shear)	64.3
USB 2.1	RC@4d	59	5.5	23	25.1 (shear)	30.9
USB 2.2	RCst@4d	59	5.5	23	33.8 (shear)	44.4
USB 2.3	PRC@4d	59	15	30	37.5 (shear)	45.6
USB 2.4	PRCst@4d	60	15	30	46.4 (flexure)	91.7
USB 3.1	RC@5d	64	4	20	21.3 (shear)	27.8
USB 3.2	RCst@5d	64	4	20	28.3 (shear)	38.6
USB 3.3	PRC@5d	60	12	26	31.3 (shear)	45.2
USB 3.4	PRCst@5d	64	12	26	37.9 (flexure)	86.1
USB 4.1	RC@6d	60	3	20	21.7 (shear)	34.7
USB 4.2	RCst@6d	54	3	20	28.4 (shear)	44.3
USB 4.3	PRC@6d	63	10	21	28.5 (shear)	42.0
USB 4.4	PRCst@6d	57	10	21	31.2 (flexure)	70.3

Table 2 – Tests on the FRP-strengthened beams

Specimen	Shear zone layout	f_{cu} (MPa)	P at flex. cracking (kN)	P at shear cracking (kN)	Peak failure load P (kN)	Max span deflection (mm)
SB 1	Plain	52	7.5	22.5	22.5 (shear)	12
SB 2	3+3 bars	52	7.5	23	41.5 (flexure)	>40
SB 3	2+2 bars	50	7.5	23	32 (shear)	13
SB 4	1+1 bar	51	7.5	22	30 (shear)	12

Table 3 – Correlation between flexural analysis and test results for all beams

Specimen	Actual P_{cr} (kN)	Predicted P_{cr} (kN)	Actual P_{ult} (kN)	Predicted flexural P_{ult} (kN)
USB 1.1	6.0	5.6	23.5 (shear)	63.0
USB 1.2	6.0	5.6	31.5 (shear)	63.0
USB 1.3	18.0	18.0	54.1 (shear)	64.1
USB 1.4	18.0	18.0	54.5 (shear)	64.1
USB 2.1	5.5	4.1	25.1 (shear)	47.1
USB 2.2	5.5	4.1	33.8 (shear)	47.1
USB 2.3	15.0	13.5	37.5 (shear)	48.0
USB 2.4	15.0	13.5	46.4 (flexure)	48.0
USB 3.1	4.0	3.2	21.3 (shear)	37.7
USB 3.2	4.0	3.2	28.3 (shear)	37.7
USB 3.3	12.0	10.8	31.3 (shear)	38.3
USB 3.4	12.0	10.8	37.9 (flexure)	38.3
USB 4.1	3.0	2.6	21.7 (shear)	31.3
USB 4.2	3.0	2.6	28.4 (shear)	31.3
USB 4.3	10.0	9.0	28.5 (shear)	31.9
USB 4.4	10.0	9.0	31.2 (flexure)	31.9
SB 1	7.5	7.6	22.5 (shear)	40.5
SB 2	7.5	7.6	41.5 (flexure)	40.5
SB 3	7.5	7.6	32 (shear)	40.5
SB 4	7.5	7.6	30 (shear)	40.5

Table 4 – Correlation between codes predictions and test results for all beams

Specimen	Actual P_{ult} (kN)	EC 2 P_{ult} (kN)	BS 8110 P_{ult} (kN)	BD 44 P_{ult} (kN)	DIN 1045 P_{ult} (kN)
USB 1.1	23.5 (shear)	21.0	18.6	20.5	17.4
USB 1.2	31.5 (shear)	28.4	27.8	30.6	15.8
USB 1.3	54.1 (shear)	29.5	22.8	26.3	24.2
USB 1.4	54.5 (shear)	28.4	32.6	37.1	39.7
USB 2.1	25.1 (shear)	21.0	18.6	20.5	17.4
USB 2.2	33.8 (shear)	28.4	27.8	30.6	15.8
USB 2.3	37.5 (shear)	29.5	19.8	20.3	24.2
USB 2.4	46.4 (flexure)	28.4	26.4	31.1	39.7
USB 3.1	21.3 (shear)	21.0	18.6	20.5	17.4
USB 3.2	28.3 (shear)	28.4	27.8	30.6	15.8
USB 3.3	31.3 (shear)	29.5	18.1	17.0	24.2
USB 3.4	37.9 (flexure)	28.4	27.9	27.8	39.7
USB 4.1	21.7 (shear)	21.0	18.6	20.5	17.4
USB 4.2	28.4 (shear)	28.4	27.8	30.6	15.8
USB 4.3	28.5 (shear)	29.5	16.9	14.9	24.2
USB 4.4	31.2 (flexure)	28.4	26.7	25.6	39.7
SB 1	23 (shear)	26	24	24.5	22
SB 2	41.5 (flexure)	40.5 (flex)	40.5 (flex)	40.5 (flex)	40.5 (flex)
SB 3	32 (shear)	37	35	36	33
SB 4	30 (shear)	33.5	31.5	32	29.5
Mean ratio of predicted to actual capacity	--	0.895	0.798	0.843	0.785
Average coefficient of variation	--	19.7%	24.1%	24.2%	27.3%

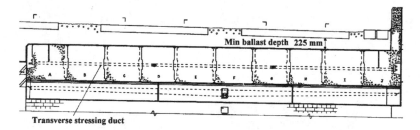

Figure 1 – Cross section of the existing bridge

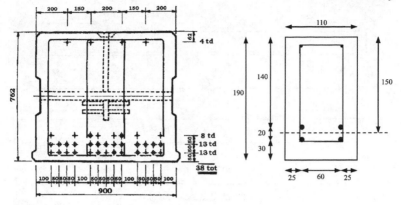

Figure 2 – Cross-section of the existing bridge beams and of the small-scale beams (mm)

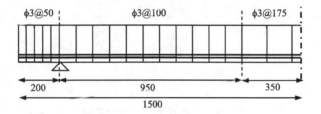

Figure 3 – Half length elevation of the small-scale beams (mm)

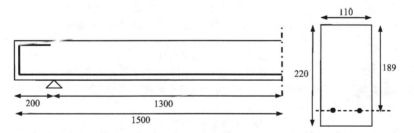

Figure 4 – Half length elevation and cross section of the FRP-strengthened beams (mm)

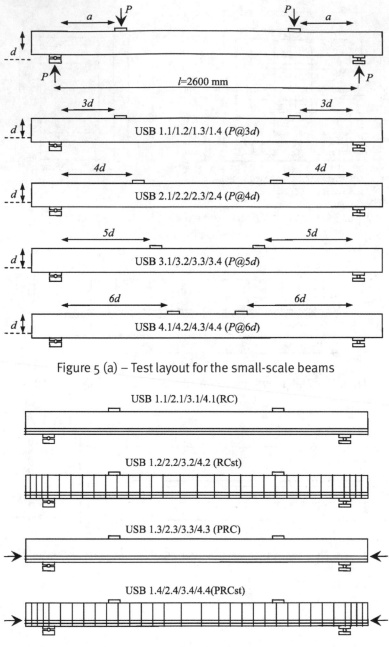

Figure 5 (a) – Test layout for the small-scale beams

Figure 5 (b) – Test layout for the small-scale beams

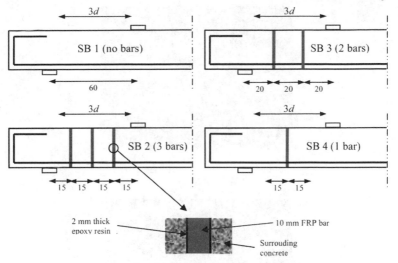

Figure 6 – Test layout for the FRP-strengthened beams

Figure 7 – Casting/prestressing rig

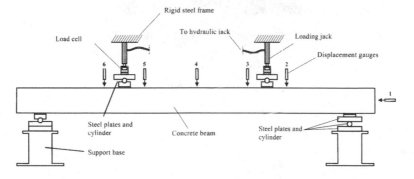

Figure 8 – Test set-up

Figure 9 – USB 1.1 at failure

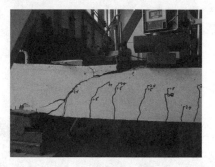

Figure 10 – USB 1.2 at failure

Figure 11 – USB 1.3 at failure

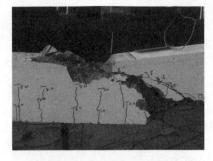

Figure 12 – USB 1.4 at failure

Figure 13 – USB 2.1 at failure

Figure 14 – USB 2.2 at failure

Figure 15 – USB 2.3 at failure

Figure 16 – USB 2.4 at failure

Figure 17 – USB 3.1 at failure

Figure 18 – USB 3.2 at failure

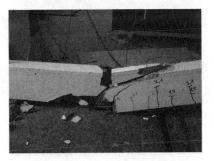

Figure 19 – USB 3.3 at failure

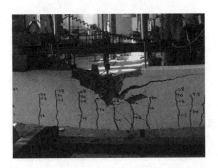

Figure 20 – USB 3.4 at failure

Figure 21 – USB 4.1 at failure

Figure 22 – USB 4.2 at failure

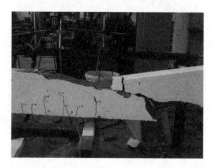

Figure 23 – USB 4.3 at failure

Figure 24 – USB 4.4 at failure

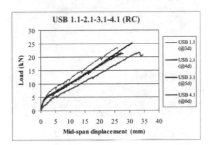

Figure 25 – Load displacement plot for the unstrengthened RC beams

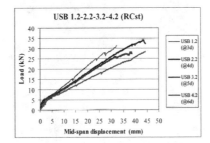

Figure 26 – Load displacement plot for the unstrengthened RC beams with stirrups

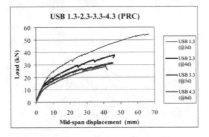

Figure 27 – Load displacement plot for the unstrengthened PRC beams

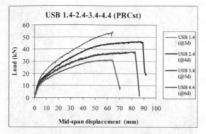

Figure 28 – Load displacement plot for the unstrengthened PRC beams with stirrups

Figure 29 – SB 1 at failure

Figure 30 – SB 2 at failure

Figure 31 – SB 3 at failure

Figure 32 – SB 4 at failure

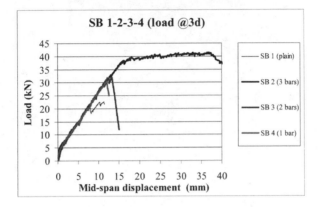

Figure 33 – Load-displacement plot for the FRP-strengthened beams

CONVERSION FACTORS—INCH-POUND TO SI (METRIC)*

To convert from	to	multiply by
Length		
inch	millimeter (mm)	25.4E†
foot	meter (m)	0.3048E
yard	meter (m)	0.9144E
mile (statute)	kilometer (km)	1.609
Area		
square inch	square centimeter (cm^2)	6.451
square foot	square meter (m^2)	0.0929
square yard	square meter (m^2)	0.8361
Volume (capacity)		
ounce	cubic centimeter (cm^3)	29.57
gallon	cubic meter (m^3)‡	0.003785
cubic inch	cubic centimeter (cm^3)	16.4
cubic foot	cubic meter (m^3)	0.02832
cubic yard	cubic meter (m^3)‡	0.7646
Force		
kilogram-force	newton (N)	9.807
kip-force	newton (N)	4448
pound-force	newton (N)	4.448
Pressure or stress (force per area)		
kilogram-force/square meter	pascal (Pa)	9.807
kip-force/square inch (ksi)	megapascal (MPa)	6.895
newton/square meter (N/m^2)	pascal (Pa)	1.000E
pound-force/square foot	pascal (Pa)	47.88
pound-force/square inch (psi)	kilopascal (kPa)	6.895
Bending moment or torque		
inch-pound-force	newton-meter (Nm)	0.1130
foot-pound-force	newton-meter (Nm)	1.356
meter-kilogram-force	newton-meter (Nm)	9.807

To convert from	to	multiply by
Mass		
ounce-mass (avoirdupois)	gram (g)	28.34
pound-mass (avoirdupois)	kilogram (kg)	0.4536
ton (metric)	megagram (Mg)	1.000E
ton (short, 2000 lbm)	megagram (Mg)	0.9072
Mass per volume		
pound-mass/cubic foot	kilogram/cubic meter (kg/m^3)	16.02
pound-mass/cubic yard	kilogram/cubic meter (kg/m^3)	0.5933
pound-mass/gallon	kilogram/cubic meter (kg/m^3)	119.8
Temperature§		
deg Fahrenheit (F)	deg Celsius (C)	$t_C = (t_F - 32)/1.8$
deg Celsius (C)	deg Fahrenheit (F)	$t_F = 1.8t_C + 32$

* This selected list gives practical conversion factors of units found in concrete technology. The reference source for information on SI units and more exact conversion factors is "Standard for Metric Practice" ASTM E 380. Symbols of metric units are given in parentheses.

† E indicates that the factor given is exact.

‡ One liter (cubic decimeter) equals 0.001 m^3 or 1000 cm^3.

§ These equations convert one temperature reading to another and include the necessary scale corrections. To convert a difference in temperature from Fahrenheit to Celsius degrees, divide by 1.8 only, i.e., a change from 70 to 88 F represents a change of 18 F or 18/1.8 = 10 C.

Index